FUZZY LOGIC WITH ENGINEERING APPLICATIONS

FUZZY LOGIC WITH ENGINEERING APPLICATIONS

Timothy J. Ross
Professor and Regents' Lecturer
University of New Mexico

McGraw-Hill, Inc.
New York St. Louis San Francisco Auckland Bogotá Caracas Lisbon
London Madrid Mexico City Milan Montreal New Delhi
San Juan Singapore Sydney Tokyo Toronto

This book was set in Times Roman by Publication Services, Inc.
The editors were Lynn Cox and John M. Morriss;
the production supervisor was Richard A. Ausburn.
The cover was designed by Carla Bauer;
cover photo by Chuck Levey.
Project supervision was done by Publication Services, Inc.
R. R. Donnelley & Sons Company was printer and binder.

FUZZY LOGIC WITH ENGINEERING APPLICATIONS

This book is printed on acid-free paper.

1 2 3 4 5 6 7 8 9 0 DOC DOC 9 0 9 8 7 6 5

ISBN 0-07-053917-0

Library of Congress Cataloging-in-Publication Data

Ross, Timothy J.
Fuzzy logic with engineering applications / Timothy J. Ross.
p. cm.
Includes bibliographical references and index.
ISBN 0-07-053917-0
1. Engineering mathematics. 2. Fuzzy logic. I. Title.
TA331.R74 1995
620'.001'5113—dc20 95-2795

INTERNATIONAL EDITION

When ordering this title, use ISBN 0-07-113637-1.

ABOUT THE AUTHOR

Timothy J. Ross is Professor of Civil Engineering and Director of the Environmental Scanning Electron Microscope Laboratory at the University of New Mexico. He received his Ph.D. degree in Civil Engineering from Stanford University, his M.S. from Rice University, and his B.S. from Washington State University. Professor Ross has held previous positions as President, IntelliSys Corporation from 1986 to 1988; Senior Research Structural Engineer, Air Force Weapons Laboratory from 1978 to 1986; and Vulnerability Engineer, Defense Intelligence Agency from 1973 to 1978. Professor Ross has more than 100 publications and has been active in the research and teaching of fuzzy logic since 1983. He is the founding Co-Editor-in-Chief of the *Intelligent and Fuzzy Systems* international journal and the co-editor of *Fuzzy Logic and Control: Software and Hardware Applications.* Professor Ross is a Fellow in the American Society of Civil Engineers as well as a member of the National Fuzzy Information Processing Society and the International Fuzzy Systems Association. He was recently named a University of New Mexico Regents' Lecturer for the period 1994–1997.

To
Carol, Bradley, Amethyst, Allyson, and Warren
for being so patient

CONTENTS

PREFACE

This textbook has been in the making for several years. I have been most fortunate in finding an understanding and excellent publisher in McGraw-Hill. We are pleased to be able to provide this text to the legion of professionals and academics who seek a simple, straightforward teaching tool for the expanding and important technology we summarize with the phrase *fuzzy logic.* I hope that this new text will be valuable in the training of engineers and technologists at the undergraduate, graduate, and/or professional levels to the extent that the technology becomes not just a funny phrase but rather a practical instrument in the solution of today's complex problems.

Fuzzy logic has come a long way since it was first subjected to technical scrutiny in 1965, when Dr. Lotfi Zadeh published his seminal work "Fuzzy sets" in the journal *Information and Control.* Since that time, the subject has been the focus of many independent research investigations by mathematicians, scientists, and engineers from around the world. Unfortunately, perhaps because of the term's connotations, fuzzy logic did not receive serious notice in this country until the last decade. The attention currently being paid to fuzzy logic is most likely the result of present popular consumer products employing fuzzy logic. Over the last several years, the Japanese alone have filed for well over 1000 patents in fuzzy logic technology, and they have already grossed billions of U.S. dollars in the sales of fuzzy logic-based products to consumers the world over.

The integration of fuzzy logic with neural networks and genetic algorithms is now making automated cognitive systems a reality in many disciplines. In fact, the reasoning power of fuzzy systems, when integrated with the learning capabilities of artificial neural networks and genetic algorithms, is responsible for new commercial products and processes that are reasonably effective cognitive systems (i.e., systems that can learn *and* reason). Fuzzy technology is so important in Japan that the word *fuzzy* was proclaimed by the Japanese as the "Keyword" for the 1990s. In a 1989 study focused exclusively on fuzzy logic, the international marketing research firm of Frost & Sullivan projected that fuzzy logic, with an annual growth rate of 20 percent, would be one of the world's 10 hottest technologies going into the twenty-first century. The National Technical Information Service (NTIS), in its 1990 and

1991 studies on foreign technology of interest to the United States, found that fuzzy logic will have a significant future impact.

Although it is relatively new, I believe that the technology involved in intelligent and fuzzy systems is of such a fundamental nature that by the turn of the century it will be standard knowledge for all engineers and scientists. It is this belief that has sustained me while preparing this text over the last four years. Interest in fuzzy systems is growing most rapidly among undergraduate students, who are seeking a new field for their graduate and/or professional work. And because American campuses are responsible for replenishing a significant percentage of the world's supply of technical talent, I see the young professional as the fastest-growing group of potential users of this text.

Many of the contributions in fuzzy logic and fuzzy set theory are dispersed over a broad range of scientific journals, providing limited, scattered dissemination and utility of knowledge. Most of these journals and edited texts, largely peer-reviewed and archival in nature, are written for other researchers in the field; as such, they are typically too difficult for the uninitiated reader. More simply, the bulk of publications dealing with the theory and application of fuzzy logic presents material that is too complicated and advanced to be quickly assimilated and put into practice.

The pedagogy of this textbook is designed for the professional and academic audience interested primarily in applications of fuzzy logic in engineering and technology. In the last three years of teaching courses in fuzzy logic and intelligent systems at the University of New Mexico and in delivering short courses to industry and national laboratories, I have found that the majority of students and practicing professionals are interested in the applications of fuzzy logic to their particular fields. Many of these individuals have expressed frustration with the difficulty in understanding the abstract mathematical terms presented in much of the currently available fuzzy logic literature. Hence, the book is written for an audience primarily at the senior undergraduate and first-year graduate levels. With numerous examples throughout the text, this book is designed to assist the learning process of a broad cross section of technical disciplines. The book is primarily focused on applications, but each of the book's chapters begins with the rudimentary structure of the underlying mathematics required for a fundamental understanding of the methods illustrated. Most of the text can be covered in a one-semester course at the senior undergraduate level. In fact, most science disciplines and virtually all math and engineering disciplines contain the basic ideas of set theory, mathematics, and predicate logic, which form the only knowledge necessary for a complete understanding of the text. Instructors may want to exclude some or all of the material covered in the last three sections of Chapter 4 (neural networks, genetic algorithms, and inductive reasoning), Chapter 9 (fuzzy nonlinear simulation) and the last three chapters of the text, Chapters 13 (fuzzy control), 14 (miscellaneous topics), and 15 (fuzzy measures) and reserve these topics either as introductory material for a graduate-level course or for additional coverage for graduate students taking the undergraduate course for graduate credit.

The book is organized into two broad categories of chapters. The first category, comprising Chapters 1 through 8, introduces basic concepts of fuzzy logic and operations. The second category, Chapters 9 through 15, illustrates the utility of the fundamental properties of fuzzy sets and fuzzy logic in a host of engineering paradigms, such as classification, pattern recognition, optimization, nonlinear simulation, knowledge-based systems, regression, decision making, and possibility theory.

Chapter 1 introduces the basic concept of fuzziness and distinguishes fuzzy uncertainty from other forms of uncertainty. It also introduces the fundamental idea of set membership, thereby laying the foundation for all material that follows, and presents membership functions as the format used for expressing set membership. Chapter 1 reviews Bart Kosko's "sets as points" idea as a graphical analog in understanding the relationship between classical (crisp) and fuzzy sets.

Chapter 2 reviews classical set theory and develops the basic ideas of fuzzy sets. Operations, laws, and properties of fuzzy sets are introduced by way of comparisons with the same entities for classical sets.

Chapter 3 develops the ideas of fuzzy relations as a means of both mapping fuzziness from one universe to another and developing fuzzy functions. Various forms of the composition operation for relations are presented. Again, the epistemological approach in Chapter 3 uses comparisons with classical relations in developing and illustrating fuzzy relations. This chapter also illustrates methods to determine numerical elements of fuzzy relations, and fuzzy relational equations are developed.

Chapter 4 discusses membership functions in more detail in terms of their properties and geometric form and the idea of fuzzification. The chapter provides seven methods of developing membership functions, including methods that make use of the technologies of neural networks, genetic algorithms, and inductive reasoning.

Chapter 5 deals with the routines to convert from fuzzy sets and fuzzy relations to classical sets and classical relations, respectively. Such translation, or conversion, is found to be most useful in dealing with the ubiquitous crisp (binary) world around us. In addition, the most common methods of defuzzification are developed and illustrated with examples.

Chapter 6 summarizes some typical operations in fuzzy arithmetic, fuzzy numbers, and fuzzy vectors. The extension of fuzziness to nonfuzzy mathematical forms using Zadeh's extension principle and several approximate methods to implement this principle are illustrated. The algebra of fuzzy vectors is introduced here to be used later in Chapter 12 in the area of pattern recognition.

Chapter 7 introduces the precepts of fuzzy logic, again through a review of the relevant features of classical, or first-order predicate, logic. Various logical connectives and operations are illustrated. There is a thorough discussion of the various forms of the implication operation and the composition operation provided in this chapter. Approximate reasoning, or reasoning under imprecise (fuzzy) information, is also introduced in Chapter 7.

Chapter 8 introduces natural language and fuzzy expert (rule-based) systems. Important ideas include set descriptions of linguistic data, rule construction, and logic. Graphical methods for inferencing are presented. The fuzzy rule-based systems are seen as generalized fuzzy relational equations.

Beginning the second category of chapters in the book highlighting applications, Chapter 9 continues with the rule-based format to introduce fuzzy nonlinear simulation. In this context, nonlinear functions are seen as mappings of information "patches" from the input space to information "patches" of the output space, instead of the "point-to-point" idea taught in classical engineering courses. Fidelity of the simulation is ensured with standard functions, but the power of the idea can be seen in systems too complex for an algorithmic description. This chapter formalizes fuzzy associative memories (FAMs) as generalized mappings.

Chapter 10 develops fuzzy decision making by introducing some simple concepts in synthetic evaluation, ordering, preference and consensus, and multiobjective decisions. It introduces the powerful concept of Bayesian decision methods by fuzzifying this classic probabilistic approach. This chapter illustrates the power of combining fuzzy set theory with probability to handle random and nonrandom uncertainty in the decision-making process.

Chapter 11 discusses a few fuzzy classification methods by contrasting them with classical methods of classification, and develops a simple metric to assess the goodness of the classification, or misclassification. This chapter also summarizes classification using equivalence relations.

Chapter 12 uses the information presented in Chapter 11 as a springboard to introduce fuzzy pattern recognition. A single-feature and a multiple-feature procedure are summarized. Some simple ideas in image processing and syntactic pattern recognition are also illustrated.

Chapter 13 introduces the field of fuzzy control systems. A brief review of control system design and control surfaces is provided. Three example problems in control are provided. Because of the large number of commercial applications using fuzzy control and the current interest in the subject, a few current industrial systems are illustrated from a fuzzy control rule perspective.

Chapter 14 briefly addresses some important ideas embodied in fuzzy optimization, fuzzy regression, and an approximation method for finding the inverse of fuzzy equations; the latter is critical for solving system identification problems.

Finally, Chapter 15 enlarges the reader's understanding of the relationship between fuzzy uncertainty and random uncertainty (and other general forms of uncertainty, for that matter) by illustrating the foundations of fuzzy measures. The chapter discusses fuzzy measures in the context of evidence theory and probability theory. Because this chapter is an expansion of ideas relating to other disciplines (Dempster-Shafer evidence theory and probability theory), it can be omitted without impact on the material preceding it. The chapter, and the book, concludes with an epilogue showing the axiomatic similarity of fuzzy set theory and probability theory.

The problems in this text are typically based on current and potential applications, case studies, and education in intelligent and fuzzy systems in engineering and related technical fields. The problems address the disciplines of computer science, electrical engineering, manufacturing engineering, industrial engineering, chemical engineering, mechanical engineering, civil engineering, engineering management, and a few related fields such as mathematics, medicine, operations research, technology management, the hard and soft sciences, and technical business issues. The references cited in the chapters are listed at the end of each chapter. These references will provide sufficient detail for those readers interested in learning more about particular issues in the theory of fuzzy sets and fuzzy logic, as well as in applications in control theory and signal processing, robotics and manufacturing, intelligent process control, expert systems, image processing, decision making, pattern recognition, cluster analysis, and a variety of learning methods, such as neural networks, genetic algorithms, and inductive reasoning.

The wealth of problems provided in the text at the end of each chapter allows instructors a large-enough problem base to afford instruction from this text for a multisemester or multiyear basis, without having to assign the same problems year after year. A solutions manual for all the problems is available to classroom instructors.

The preparation of a book such as this is an arduous task, and would have been exceedingly difficult without the assistance and encouragement of many colleagues and students, past and present. Space in this preface limits me to mentioning only a few, but no such limitation exists in my heart for all those who have had a positive influence on my quest to educate those with an interest in learning this subject. In particular, I am grateful for the introduction to this subject provided by Auguste Boissonnade and Haresh Shah in the late 1970s. Early teachings and practical illustrations by Weimin Dong have had an indelible and lasting effect on my efforts to organize much of the material, and early collaborations with Felix Wong had a stimulating influence on my own interest in this technology. I am most grateful to James Bezdek for his humor and insight in helping me keep perspective about the many antagonists encountered in my research and teaching over the years, and for his profound effect on the field of fuzzy pattern recognition in general. The early writings of Colin Brown and James Yao, and later collaborations with Fabian Hadipriono, convinced me of the tremendous potential of this technology in my field, civil engineering. This book would not have been possible without the strong support of my own college faculty at the University of New Mexico, especially that of my close colleagues Mo Jamshidi and Nader Vadiee who, with me, have provided fuzzy logic education and training to over 400 students and professionals during the last three years; and George Luger, whose experience as a textbook author provided me with calming advice. I am thankful to Nader Vadiee for his assistance to me in writing Chapter 13. I am thankful to the Jayanthi brothers, Subbarao and Ramasastry, for their patience in preparing many of the equations and figures; to Sunil Donald and Fan Wu for their assiduous pursuit of accuracy in the solutions of many of the problems in this text; to Ward Deng, Lewis Wagner, Tran Lai, Ken Downs, Angelos Klonis, and others for their software implementations of many of the algorithms discussed in and included with the text; and to Ashish Vasil for his steadfast support in helping me with several technical holes in the manuscript and in resurrecting many obscure references. I am grateful to Anne Brown, my initial editor, whose infectious enthusiasm for this textbook resulted in its publication a year or two sooner than I had originally planned, and to my second editor, George Hoffman, for his patience and his understanding that a quality text would take longer than expected to write, rewrite, and produce. Finally, I thank my final editor, Lynn Cox, for helping with final production.

Reviewers are most important to ensuring the quality of a book. Two people, in particular, used my text in their own classrooms and provided very useful suggestions. I am grateful to Tony Zygmont, Villanova University, and to George

Cunningham, New Mexico Institute of Technology, for their comments. I would also like to express my thanks for the many useful comments and suggestions provided by the following reviewers: Tamer Basar, University of Illinois at Urbana-Champaign; David Daut, Rutgers University; George Klir, State University of New York at Binghamton; John Painter, Texas A&M University; Kevin M. Passino, The Ohio State University; and P. Y. Ramamourthy, University of Cincinnati.

It is certainly not possible to summarize all the important facets of fuzzy set theory and fuzzy logic in a single textbook. Hundreds of edited works and thousands of archival papers have not yet, in my opinion, fully captured the potential of this rapidly growing technology, where new discoveries are being published every month. Nonetheless, it is my fervent hope that this introductory textbook will assist students and practicing professionals to learn, to apply, and to be comfortable with fuzzy set theory and fuzzy logic. I welcome comments from all readers in my quest to make this textbook a truly useful and practical instrument to the community of engineers and technologists who will become knowledgeable about the potential of fuzzy logic and fuzzy set theory. It is just this knowledge that I believe will assist those involved in improving technology that relates to human lives and human comfort.

Timothy J. Ross

FUZZY LOGIC WITH ENGINEERING APPLICATIONS

CHAPTER 1

INTRODUCTION

As the complexity of a system increases, our ability to make precise and yet significant statements about its behavior diminishes until a threshold is reached beyond which precision and significance (or relevance) become almost mutually exclusive characteristics.

Lotfi Zadeh
Professor, Systems Engineering, 1973

The real world is complex; complexity in the world generally arises from uncertainty in the form of ambiguity. Problems featuring complexity and ambiguity have been addressed subconsciously by humans since they could think; these ubiquitous features pervade most social, technical, and economic problems faced by the human race. Why then are computers, which have been designed by humans after all, not capable of addressing complex and ambiguous issues? How can humans reason about real systems, when the complete description of a real system often requires more detailed data than a human could ever hope to recognize simultaneously and assimilate with understanding? The answer is that humans have the capacity to reason approximately, a capability that computers currently do not have. In reasoning about a complex system, humans reason approximately about its behavior, thereby maintaining only a generic understanding about the problem. Fortunately, this generality and ambiguity are sufficient for human comprehension of complex systems. As the quote above from Dr. Zadeh's *principle of incompatibility* suggests, complexity and ambiguity (imprecision) are correlated: "The closer one looks at a real-world problem, the fuzzier becomes its solution" [Zadeh, 1973].

As we learn more and more about a system, its complexity decreases and our understanding increases. As complexity decreases, the precision afforded by computational methods becomes more useful in modeling the system. Figure 1.1, which

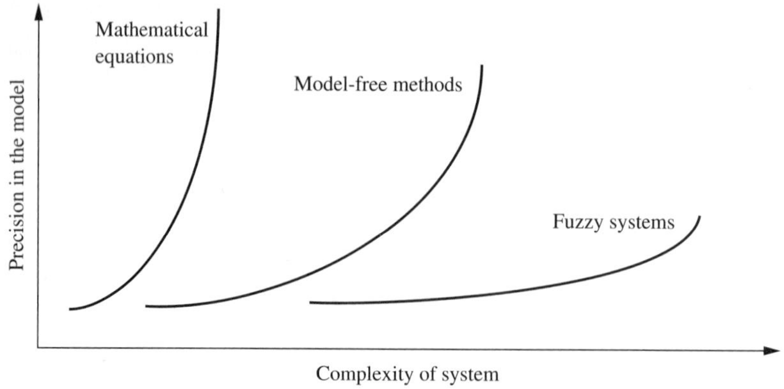

FIGURE 1.1
Complexity of a system versus precision in the model of the system.

provides a useful insight into these ideas, relates the degree of complexity of a system to the precision inherent in our models of the system. For systems with little complexity, hence little uncertainty, closed-form mathematical expressions provide precise descriptions of the systems. For systems that are a little more complex, but for which significant data exist, model-free methods, such as artificial neural networks, provide a powerful and robust means to reduce some uncertainty through learning, based on patterns in the available data. Finally, for the most complex systems where few numerical data exist and where only ambiguous or imprecise information may be available, fuzzy reasoning provides a way to understand system behavior by allowing us to interpolate approximately between observed input and output situations. The imprecision in fuzzy models is therefore generally quite high. Fuzzy systems *can* implement crisp inputs and outputs, and in this case produce a nonlinear functional mapping just as do algorithms. All of the models shown in Fig. 1.1 can be assessed with nonlinear equations or with fuzzy models or neural networks. All are mathematical abstractions of the real physical world. The point is, however, to match the model type with the character of the uncertainty exhibited in the problem. In situations where precision is apparent, for example, fuzzy systems are less efficient than the more precise algorithms in providing us with the best understanding of the problem (see Chapter 9 on nonlinear simulation). On the other hand, fuzzy systems can focus on modeling problems characterized by imprecise or ambiguous information. Skill in using the tools illustrated in the rest of this book will allow readers the capability to address the vast majority of problems that are characterized either by their complexity or by their lack of a requirement for precision.

BACKGROUND

In the decade after Dr. Zadeh's seminal paper on fuzzy sets [Zadeh, 1965], many theoretical developments in fuzzy logic took place in the United States, Europe, and Japan. From the mid-1970s to the present, however, Japanese researchers have been a primary force in advancing the practical implementation of the theory; they have done an excellent job of commercializing this technology and now have over 2000 patents in the area. Much of the success of the new products associated with the fuzzy technology is due to fuzzy logic, and some is also due to the advanced sensors used in these products.

Fuzzy logic affects many disciplines. In videography, for instance, Fisher, Sanyo, and others make fuzzy logic camcorders, which offer fuzzy focusing and image stabilization. Mitsubishi manufactures a fuzzy air conditioner that controls temperature changes according to human comfort indexes. Matsushita builds a fuzzy washing machine that combines smart sensors with fuzzy logic. The sensors detect the color and kind of clothes present and the quantity of grit, and a fuzzy microprocessor selects the most appropriate combination from 600 available combinations of water temperature, detergent amount, and wash and spin cycle times. The Japanese city of Sendai has a 16-station subway system that is controlled by a fuzzy computer. The ride is so smooth riders do not need to hold straps, and the controller makes 70 percent fewer judgment errors in acceleration and braking than human operators. Nissan introduced a fuzzy automatic transmission and a fuzzy anti-skid braking system in one of their recent luxury cars. Tokyo's stock market has had at least one stock-trading portfolio based on fuzzy logic that outperformed the Nikkei Exchange average. In Japan there are fuzzy golf diagnostic systems, fuzzy toasters, fuzzy rice cookers, fuzzy vacuum cleaners, and many other industrial fuzzy control processes. In fact, the number of fuzzy consumer products and fuzzy applications involving new patents is increasing so rapidly that, in order to stay competitive, many U.S. companies are launching their own internal fuzzy projects. Publicity on these U.S.-based products is scant at this point. The U.S. Department of Defense has been exploring the utility of fuzzy logic for some years in the area of pattern recognition and classification, but few definitive results have been published. Likewise, the National Aeronautics and Space Administration has been involved for a number of years in the use of fuzzy logic in space docking control.

So why is a paradigm invented to model imprecision in large complex systems in the United States finding numerous applications in simple consumer products in Asian countries, such as Japan? One explanation is that the Japanese spirit is entrepreneurial, and the Japanese are trained to bring basic research ideas to fruition in the commercial markets quicker than anyone else. Western cultures are mired in the black-and-white, yes-or-no, guilty-or-not-guilty, binary world of Aristotelian logic, which runs counter to the acceptance of "gray" or fuzzy ideals that Eastern cultures so naturally accommodate. The term *fuzzy* carries negative connotations in the English-speaking world that it does not in other, specifically Asian, cultures. Consider, for example, the oxymoronic descriptor *fuzzy focusing* for the new image-stabilized cameras; would the average American consumer buy such a camera? Another explanation of why *fuzzy* has been more readily accepted in Asia is related to language. In Japan, the kanji symbols adopted from Chinese use progressively narrower classifications in describing objects. For example, a fox is red, four-legged, an animal, etc. In Japanese, the word *fuzzy* does not imply disorganization and imprecision, as it does in English.

It is convenient to rationalize that U.S. industry and government have not seriously considered fuzzy logic because of its less-than-precise connotation. This explanation, however, does not convince linguistic experts. Few linguists accept a theory of strong linguistic determinism, which argues that fluency in different languages gives the speaker special cognitive advantages or disadvantages. Japanese kanji are writing symbols adapted from Chinese; they do not themselves constitute a language. Although one could argue that their origin is compositional, e.g., that one can identify symbols for "red," "animal," "four-legged," etc. in the symbol "fox," it is doubtful that this has much psychological reality for the average Japanese speaker. Perhaps it is easier to explain the popularity of fuzzy logic in Japan, where it solves a plethora of technical problems, because its use has resulted in innumerable commercial products that work [Rogers and Hoshuai, 1990]!

UNCERTAINTY AND IMPRECISION

Fuzzy set theory provides a means for representing uncertainties. Historically, probability theory has been the primary tool for representing uncertainty in mathematical models. Because of this, all uncertainty was assumed to follow the characteristics of random uncertainty. A random process is one where the outcomes of any particular realization of the process are strictly a matter of chance; a prediction of a sequence of events is not possible. What is possible for random processes is a precise description of the *statistics* of the long-run averages of the process. However, not all uncertainty is random. Some forms of uncertainty are nonrandom and hence not suited to treatment or modeling by probability theory. In fact, it could be argued that the overwhelming amount of uncertainty associated with complex systems and issues, which humans address on a daily basis, is nonrandom in nature. Fuzzy set theory is a marvelous tool for modeling the kind of uncertainty associated with vagueness, with imprecision, and/or with a lack of information regarding a particular element of the problem at hand.

Many of us, especially those in positions to develop models of physical processes, understand that we lack complete information in solving problems. Some of the information we do have about a particular problem might be judgmental, perhaps a visceral reaction on the part of the modeler, rather than hard quantitative information. How then do we incorporate intuition into a problem? This text will examine this question and others in addressing applications of fuzzy set theory.

One prevalent way to convey information is our own means of communication: natural language. By its very nature, natural language is vague and imprecise; yet it is the most powerful form of communication and information exchange among humans. Despite the vagueness in natural language, humans have little trouble understanding one another's concepts and ideas; this understanding is not possible in communications with a computer, which requires extreme precision in its instructions. For instance, what is the meaning of a "tall person?" To individual A a tall person might be anybody over 5′11″. To individual B a tall person is someone who is 6′2″ or taller. What sort of meaning does the linguistic descriptor "tall" convey to either of these individuals? It is surprising that, despite the potential for misunderstanding, the term "tall" conveys sufficiently similar information to the two individuals, even if they are significantly different heights themselves, and that understanding and correct communication are possible between them. Individuals A and B, regardless of their own heights, do not require identical definitions of the term "tall" to communicate effectively; again, a computer would require a specific height to compare with a preassigned value for "tall." The underlying power of fuzzy set theory is that it uses *linguistic* variables, rather than *quantitative* variables, to represent imprecise concepts.

The incorporation of fuzzy set theory and fuzzy logic into computer models has shown tremendous payoff in areas where intuition and judgment still play major roles in the model. Control applications, such as temperature control, traffic control, or process control, are the most prevalent of current fuzzy logic applications. Fuzzy logic seems to be most successful in two kinds of situations: (*i*) very complex models where understanding is strictly limited or, in fact, quite judgmental, and (*ii*) processes where human reasoning, human perception, or human decision making are inextricably involved. Generally, simple linear systems or naturally automated processes have not been improved by the implementation of fuzzy logic. This text will continue to reiterate this fact: Fuzzy logic is not a panacea for all problems. However, its value has been demonstrated numerous times in the two kinds of situations just mentioned, and these comprise many of the processes or products of interest to a large number of consumers.

Our understanding of physical processes is based largely on imprecise human reasoning. This imprecision (when compared to the precise quantities required by computers) is nonetheless a form of information that can be quite useful to humans. The ability to embed such reasoning in hitherto intractable and complex problems is the criterion by which the efficacy of fuzzy logic is judged. Undoubtedly this ability cannot solve problems that require precision—problems such as shooting precision laser beams over tens of kilometers in space; milling machine components to accuracies of parts per billion; or focusing a microscopic electron beam on a specimen the size of a nanometer. The impact of fuzzy logic in these areas might be years away, if ever. But not many human problems require such precision—problems such as parking a car, backing up a trailer, navigating a car among others on a freeway, washing clothes, controlling traffic at intersections, judging beauty contestants, and so on.

Requiring precision in engineering models and products translates to requiring high cost and long lead times in production and development. For other than simple systems, expense is proportional to precision: More precision entails higher cost. When considering the use of fuzzy logic for a given problem, an engineer or scientist should ponder the need for *exploiting the tolerance for imprecision.* Not only does high precision dictate high costs but it also entails low tractability in a problem. Recent articles in the popular media illustrate the need to exploit imprecision. Take the "traveling salesrep" problem, for example. In this classic optimization problem a sales representative wants to minimize total distance traveled by considering various itineraries and schedules between a series of cities on a particular trip. For a small number of cities, the problem is a trivial exercise in enumerating all the possibilities and choosing the shortest route. As the number of cities continues to grow, the problem quickly approaches a combinatorial explosion impossible to solve through an exhaustive search, even with a computer. For example, for 100 cities there are $100 \times 99 \times 98 \times 97 \times \cdots \times 2 \times 1$, or about 10^{200} possible routes to consider! No computers exist today that can solve this problem through a brute-force enumeration of all the possible routes. There are real, practical problems analogous to the traveling salesrep problem. For example, such problems arise in the fabrication of circuit boards, where precise lasers drill hundreds of thousands of holes in the board. Deciding in which order to drill the holes (where the board moves under a stationary laser) so as to minimize drilling time is a traveling salesrep problem [Kolata, 1991].

Thus, algorithms have been developed to solve the traveling salesrep problem in an optimal sense; that is, the exact answer is not guaranteed but an optimum answer is achievable—the optimality is measured as a percent accuracy, with zero percent representing the exact answer and accuracies larger than zero representing answers of lesser accuracy. Suppose we consider a transportation problem analogous to the traveling salesrep problem where we want to find the optimum path (i.e., minimum travel time) between 100,000 nodes in a travel network to an accuracy within 1 percent of the exact solution; this requires two days of CPU time on a supercomputer that runs about 40 times faster than a personal computer did in 1991. If we take the same problem and increase the precision requirement a modest amount to an accuracy of 0.75 percent, the computing time approaches seven months! Now suppose we can live with an accuracy of 3.5 percent (quite a bit more accurate than most problems we deal with), and we want to consider an order-of-magnitude more nodes in the network, say 1,000,000; the computing time for this problem is only slightly more than 3 hours [Kolata, 1991]. This remarkable reduction in cost (translating time to dollars) is due solely to the acceptance of a lesser degree of precision in the optimum solution. Can humans live with a little less precision? The answer to this question depends on the situation, but for the vast majority of problems we deal with every day the answer is a resounding yes.

The quest for a method to quantify nonrandom uncertainty (imprecision, vagueness, fuzziness) in physical processes is the basic premise of this textbook, for to understand uncertainty in a system is to understand the system itself. As understanding improves, the fidelity in modeling improves. This text is focused on the utility of fuzzy set theory and fuzzy logic as tools to improve our engineering and physical models.

STATISTICS AND RANDOM PROCESSES

The uninitiated often claim that fuzzy set theory is just another form of probability theory in disguise. This statement, of course, is simply not true (the Epilogue in Chapter 15 formally rejects this claim with an axiomatic discussion of both probability theory and fuzzy logic). Basic statistical analysis is founded on probability theory or stationary random processes, whereas most experimental results contain both random (typically noise) and nonrandom processes. One class of random processes, stationary random processes, exhibits the following three characteristics: (1) The sample space on which the processes are defined cannot change from one experiment to another; i.e., the outcome space cannot change. (2) The frequency of occurrence, or probability, of an event within that sample space is constant and cannot change from trial to trial or experiment to experiment. (3) The outcomes must be repeatable from experiment to experiment. The outcome of one trial does not influence the outcome of a previous or future trial. There are more general classes of random processes than the class mentioned here. However, fuzzy sets are not governed by these characteristics.

Stationary random processes are those that arise out of chance, where the chances represent frequencies of occurrence that can be measured. Problems like picking colored balls out of an urn, coin and dice tossing, and many card games are good examples of stationary random processes. How many of the decisions that humans must make every day could be categorized as random? How about the uncertainty in the weather—is this random? How about your uncertainty in choosing clothes for the next day, or which car to buy, or your preference in colors—-are these random uncertainties? How about your ability to park a car; is this a random process? How about the risk in whether a substance consumed by an individual now will cause cancer in that individual 15 years from now; is this a form of random uncertainty? Although it is possible to model all of these forms of uncertainty with various classes of random processes, the solutions may not be reliable. Treatment of these forms of uncertainty using fuzzy logic should also be done with caution. One needs to study the character of the uncertainty, then choose an appropriate approach to develop a model of the process. Many features of the same problem that vary in time and space should be considered. For example, when the weather report suggests that there is a 60 percent chance of rain tomorrow, does this mean that there has been rain on tomorrow's date for 60 of the last 100 years? Does it mean that somewhere in your community 60 percent of the land area will receive rain? Does it mean that 60 percent of the time it will be raining and 40 percent of the time it will not be raining? Humans often deal with these forms of uncertainty linguistically—such as, "It will likely rain tomorrow." And with this crude an assessment of the possibility of rain, humans can still make appropriately accurate decisions about the weather.

Random errors will generally average out over time, or space. Nonrandom errors, such as some unknown form of bias (often called a systematic error) in an experiment, will not generally average out and will likely grow larger with time. The systematic errors generally arise from causes about which we are ignorant, for which we lack information, or that we cannot control. Distinguishing between random and nonrandom errors is a difficult problem in many situations, and to quantify this distinction often results in the illusion that the analyst knows the extent and character of each type of error. In all likelihood nonrandom errors can increase without bounds.

It is historically interesting that the word *statistics* is derived from the now obsolete term *statist,* which means *an expert in statemanship.* Statistics were the numerical facts that statists used to describe the operations of states. To many people, statistics, and other recent methods to represent uncertainty like evidence theory and fuzzy set theory are still the facts by which politicians, newspapers, insurance sellers, and other broker occupations approach us as potential customers for their services or products! The air of sophistication that these methods provide to an issue should not be the basis for making a decision; it should be made only after a good balance has been achieved between the information content in a problem and the proper representation tool to assess it. The power of the oldest of the uncertainty methods, statistics, to deceive the unsuspecting is perhaps why W. H. Auden, in his 1946 Harvard Phi Beta Kappa Poem "Under Which Lyre," was led to include as one of ten commandments for professors, "Thou shalt not sit with statisticians, nor commit a social science" [Amis, 1978]. This verse could now be updated with "not sitting with statisticians and other logisticians!"

Popular lore suggests that the various uncertainty theories allow engineers to fool themselves in a highly sophisticated way when looking at relatively incoherent heaps of data (computational or experimental), as if this form of deception is any more palatable than just plain ignorance. All too often, scientists and engineers are led to use these theories as a crutch to explain vagaries in their models or in their data. For example, in probability applications the assumption of independent random variables is often assumed to provide a simpler method to prescribe joint probability distribution functions. An analogous assumption, called noninteractive sets (see Chapter 3), is used in fuzzy applications to develop joint membership functions from individual membership functions for sets from different universes of discourse. Should one ignore apparently aberrant information, or consider all information in the model whether or not it conforms to the engineers' preconceptions? Additional experiments to increase understanding cost money, and yet, unfortunately, they almost always increase uncertainty. A humorous corollary to one of Murphy's laws would suggest that if you have two conflicting pieces of information, the odds are 10 to 1 that whichever one you throw out will be the one you should have kept! The moral of this story is that statistics alone, or fuzzy sets alone, or evidence theory alone, is individually insufficient to explain many of the imponderables that people face every day. Collectively they could be very powerful. A poem by J. V. Cunningham [1971] titled "Meditation on Statistical Method" provides a good lesson in caution for any technologist pondering the thought that representing uncertainty (again, using statistics because of the era of the poem) in a problem will somehow make its solution seem less uncertain. It is repeated here as a point of departure in the description of fuzzy sets.

Plato despair!
We prove by norms
How numbers bear
Empiric forms,

How random wrongs
Will average right
If time be long
And error slight;

But in our hearts
Hyperbole
Curves and departs
To infinity.

Error is boundless.
Nor hope nor doubt,
Though both be groundless,
Will average out.

UNCERTAINTY IN INFORMATION

Figure 1.2 is a diagram that illustrates the information contained in a typical problem (throughout this text graphics are used to the maximum extent possible to illustrate mathematics or concepts). The collection of all information in this context shall be termed the "information world." Now consider the uncertainty in the content of this information. As Fig. 1.2 reveals, only a small portion of a typical problem might be regarded as certain, or deterministic. Unfortunately, the vast majority of the material taught in engineering classes is based on the presumption that the knowledge involved is deterministic. Most processes are neatly and surreptitiously reduced to closed-form algorithms—equations and formulas. When students graduate, it seems that their biggest fear upon entering the real world is "forgetting the correct formula." These formulas typically describe a deterministic process—one where there is no uncertainty in the physics of the process (i.e., the right formula) and there is no uncertainty in the parameters of the process (i.e., the coefficients are known with precision). It is only after we leave college, it seems, that we realize we were duped in academe, and that the information we have for a particular problem virtually always contains uncertainty. For how many of our problems can we say that the information content is known absolutely, i.e., with no ignorance, no vagueness, no imprecision, no element of chance? Uncertain information can take on many different forms. There is uncertainty that arises because of complexity; for example, the complexity in the reliability network of a nuclear reactor. There is uncertainty that arises from ignorance, from chance, from various classes of randomness, from imprecision, from the inability to perform adequate measurements, from lack of knowledge, or from vagueness, like the fuzziness inherent in our natural language.

The nature of uncertainty in a problem is a very important point that engineers should ponder prior to their selection of an appropriate method to express the uncertainty. Fuzzy sets provide a mathematical way to represent vagueness in humanistic systems. For example, suppose you are teaching your child to bake cookies and you want to give instructions about when to take the cookies out of the oven. You could say to take them out when the temperature inside the cookie dough reaches *375°F*, or you could advise your child to take them out when the tops of the cookies turn *light brown.* Which instruction would you give? Most likely, you would use the second of

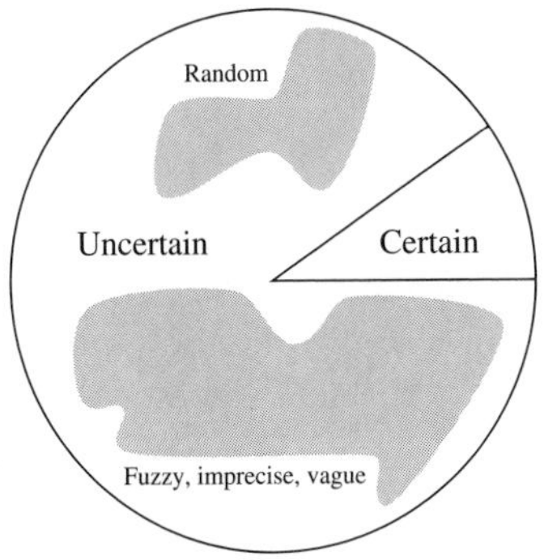

FIGURE 1.2
Forms of uncertainty in the information world.

the two instructions. The first instruction is too precise to implement practically; in this case precision is not useful. The vague term *light brown* is useful in this context and can be acted upon even by a child. We all use vague terms, imprecise information, and other fuzzy data just as easily as we deal with situations governed by chance, where probability techniques are warranted and very useful. Hence, our sophisticated computational methods should be able to represent and manipulate both fuzzy and statistical uncertainties.

FUZZY SETS AND MEMBERSHIP

The foregoing sections discuss the various elements of uncertainty. Making decisions about processes that contain nonrandom uncertainty, such as the uncertainty in natural language, has been shown to be less than perfect. The idea proposed by Lotfi Zadeh suggested that *set membership* is the key to decision making when faced with uncertainty. In fact, Dr. Zadeh made the following statement in his seminal paper of 1965:

> The notion of a fuzzy set provides a convenient point of departure for the construction of a conceptual framework which parallels in many respects the framework used in the case of ordinary sets, but is more general than the latter and, potentially, may prove to have a much wider scope of applicability, particularly in the fields of pattern classification and information processing. Essentially, such a framework provides a natural way of dealing with problems in which the source of imprecision is the absence of sharply defined criteria of class membership rather than the presence of random variables.

Suppose the height example from the previous section is carried further. We can easily assess whether someone is over 6 feet tall. In a binary sense, the person either is or is not, based on the accuracy, or imprecision, of our measuring device. For example, if "tall" is a set defined as heights equal to or greater than 6 feet, a computer would not recognize an individual of height 5′11.999″ as being a member of the set "tall." But how do we assess the uncertainty in the following question: Is the person *nearly* 6 feet tall? The uncertainty in this case is due to the vagueness or ambiguity of the adjective *nearly.* A 5′11″ person could clearly be a member of the set of "nearly 6-feet tall" people. In the first situation, the uncertainty of whether a person, whose height is unknown, is 6 feet or not is binary; the person either is or is not, and we can produce a probability assessment of that prospect based on height data from many people. But the uncertainty of whether a person is nearly 6 feet is nonrandom. The degree to which the person approaches a height of 6 feet is fuzzy. In reality, "tallness" is a matter of degree and is relative. Among peoples of the Tutsi tribe in Rwanda and Burundi a height for a male of 6 feet is considered short. So, 6 feet can be tall in one context and short in another. In the real (fuzzy) world, the set of tall people can overlap with the set of not-tall people, an impossibility in the world of binary logic (this is discussed in Chapter 7).

This notion of set membership, then, is central to the representation of objects within a universe by sets defined on the universe. Classical sets contain objects that satisfy precise properties of membership; fuzzy sets contain objects that satisfy imprecise properties of membership, i.e., membership of an object in a fuzzy set can be approximate. For example, the set of heights *from 5 to 7 feet* is crisp; the set of heights in the region *around 6 feet* is fuzzy. To elaborate, suppose we have an exhaustive collection of individual elements (singletons) x, which make up a universe of information (discourse), X. Further, various combinations of these individual elements make up sets, say A, on the universe. For crisp sets an element x in the universe X is either a member of some crisp set A or it is not. This binary issue of membership can be represented mathematically with the indicator function,

$$\chi_{\mathrm{A}}(x) = \begin{cases} 1, & x \in \mathrm{A} \\ 0, & x \notin \mathrm{A} \end{cases} \tag{1.1}$$

where the symbol $\chi_{\mathrm{A}}(x)$ gives the indication of an unambiguous membership of element x in set A, and the symbols $\in$ and $\notin$ denote contained-in and not contained-in, respectively. For our example of the universe of heights of people, suppose set A is the crisp set of all people with $5.0 \le x \le 7.0$ feet, shown in Fig. 1.3*a*. A particular individual, x_1, has a height of 6.0 feet. The membership of this individual in crisp set A is equal to 1, or full membership, given symbolically as $\chi_{\mathrm{A}}(x_1) = 1$. Another individual, say, x_2, has a height of 4.99 feet. The membership of this individual in set A is equal to 0, or no membership, hence $\chi_{\mathrm{A}}(x_2) = 0$, also seen in Fig. 1.3*a*. In these cases the membership in a set is binary, either an element is a member of a set or it is not.

Zadeh extended the notion of binary membership to accommodate various "degrees of membership" on the real continuous interval [0, 1], where the endpoints of 0 and 1 conform to no membership and full membership, respectively, just as the indicator function does for crisp sets, but where the infinite number of values in between the endpoints can represent various degrees of membership for an element x in some set on the universe. The sets on the universe X that can accommodate "degrees of membership" were termed by Zadeh as "fuzzy sets." Continuing further on the example on heights, consider a set H consisting of heights *near 6 feet.* Since the property *near 6 feet* is fuzzy, there is not a unique membership function for H.

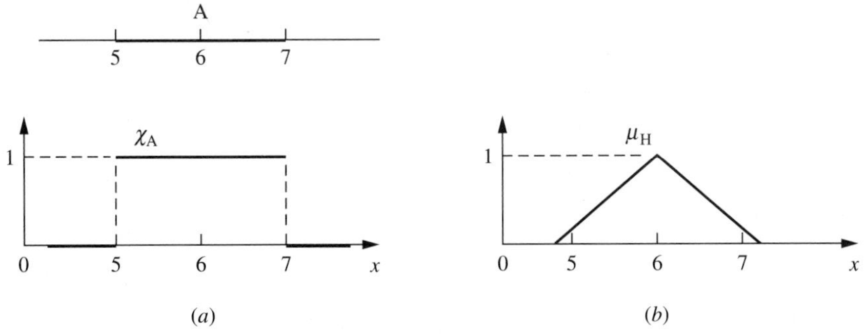

FIGURE 1.3
Height membership functions for (*a*) a crisp set A and (*b*) a fuzzy set H.

Rather, the analyst must decide what the membership function, denoted μ_H, should look like. Plausible properties of this function might be (*i*) normality [$\mu_H(6) = 1$], (*ii*) monotonicity (the closer H is to 6, the closer μ_H is to 1), and (*iii*) symmetry (numbers equidistant from 6 should have the same value of μ_H) [Bezdek, 1993]. Such a membership function is illustrated in Fig. 1.3*b*. A key difference between crisp and fuzzy sets is their membership function; a crisp set has a unique membership function, whereas a fuzzy set can have an infinite number of membership functions to represent it. For fuzzy sets, the uniqueness is sacrificed, but flexibility is gained because the membership function can be adjusted to maximize the utility for a particular application.

James Bezdek provided one of the most lucid comparisons between crisp and fuzzy sets [Bezdek, 1993]. It bears repeating here. Crisp sets of real objects are equivalent to, and isomorphically described by, a unique membership function, such as χ_A in Fig. 1.3*a*. But there is no set-theoretic equivalent of "real objects" corresponding to χ_A. Fuzzy sets are always *functions,* which map a universe of objects, say, X, onto the unit interval [0, 1]; that is, the fuzzy set H is the *function* μ_H that carries X into [0, 1]. Hence, *every* function that maps X onto [0, 1] is a fuzzy set. Although this statement is true in a formal mathematical sense, many functions that qualify on the basis of this definition cannot be suitable fuzzy sets. But they *become* fuzzy sets when, and only when, they match some intuitively plausible semantic description of imprecise properties of the objects in X.

The membership function embodies the mathematical representation of membership in a set, and the notation used throughout this text for a fuzzy set is a set symbol with a tilde underscore, say $\underset{\sim}{A}$, where the functional mapping is given by

$$\mu_{\underset{\sim}{A}}(x) \in [0, 1] \tag{1.2}$$

and the symbol $\mu_{\underset{\sim}{A}}(x)$ is the degree of membership of element x in fuzzy set $\underset{\sim}{A}$. Therefore, $\mu_{\underset{\sim}{A}}(x)$ is a value on the unit interval that measures the degree to which element x belongs to fuzzy set $\underset{\sim}{A}$; equivalently, $\mu_{\underset{\sim}{A}}(x) =$ degree to which $x \in \underset{\sim}{A}$.

CHANCE VERSUS AMBIGUITY

Suppose you are a basketball recruiter and are looking for a "very tall" player for the center position on a men's team. One of your information sources tells you that a hot prospect in Oregon has a 95 percent chance of being over 7 feet tall. Another of your sources tells you that a good player in Louisiana has a high membership in the set of "very tall" people. The problem with the information from the first source is that it is a random quantity. There is a 5 percent chance that the Oregon player is not over 7 feet tall and could, conceivably, be someone of extremely short stature. The second source of information would, in this case, contain less uncertainty for the recruiter because if the player turned out to be less than 7 feet tall there is still a high likelihood that he would be quite tall.

Another example involves a personal choice. Suppose you are seated at a table on which rest two glasses of liquid. The liquid in the first glass is described to you as having a 95 percent chance of being healthful and good. The liquid in the second glass is described as having a .95 membership in the class of "healthful and good" liquids. Which glass would you select, keeping in mind that the first glass has a 5 percent chance of being filled with nonhealthful liquids, including poisons [Bezdek, 1993]?

What philosophical distinction can be made regarding these two forms of information? Suppose we are allowed to measure the basketball players' heights and test the liquids in the glasses. The prior probability of .95 in each case becomes a posterior probability of 1.0 or .0; i.e., either the player is or is not over 7 feet tall and the liquid is either benign or not. However, the membership value of .95, which measures the extent to which the player's height is over 7 feet, or the drinkability of the liquid is "healthful and good," remains .95 after measuring or testing. These two examples illustrate very clearly the difference in the information content between chance and ambiguous events.

This brings us to the clearest distinction between fuzziness and randomness. *Fuzziness describes the ambiguity of an event, whereas randomness describes the uncertainty in the occurrence of the event.* The event will occur or not occur; but is the description of the event unambiguous enough to measure its occurrence or nonoccurrence? Consider the following geometric questions, which serve to illustrate our ability to address ambiguity with certain mathematical relations. The geometric shape in Fig. 1.4*a* can resemble a disk, a cylinder, or a rod, depending on the aspect ratio of d/h. For $d/h \ll 1$ the shape of the object approaches a long rod; in fact, as $d/h \rightarrow 0$ the shape approaches a line. For $d/h \gg 1$ the object approaches the shape of a flat disk; as $d/h \rightarrow \infty$ the object approaches a circular area. For other values of this aspect ratio, e.g., for $d/h \approx 1$, the shape is typical of what we would call a "right circular cylinder." See Fig. 1.4*b*.

The geometric shape in Fig. 1.5*a* is an ellipse, with parameters a and b. Under what conditions of these two parameters will a general elliptic shape become a circle? Mathematically, we know that a circle results when $a/b = 1$. We know that when $a/b \ll 1$ or $a/b \gg 1$ we clearly have an elliptic shape; and as $a/b \rightarrow \infty$, a line segment results. Using this knowledge, we can develop a description of the membership function to describe the geometric set we call "circle." Without a theoretical

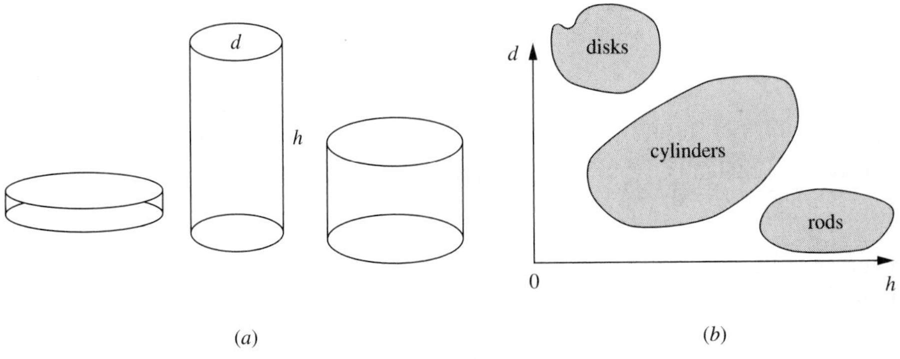

FIGURE 1.4
Relationship between (*a*) mathematical terms and (*b*) ambiguous terms.

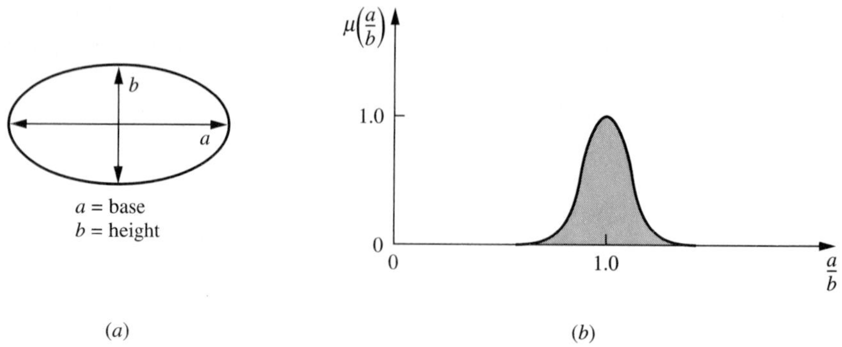

FIGURE 1.5
The (*a*) geometric shape and (*b*) membership function for a circle.

development, the following expression describing a Gaussian curve (for this membership function all points on the real line have nonzero membership; this can be an advantage or disadvantage depending on the nature of the problem) offers a good approximation for the membership function of the fuzzy set "circle," denoted $\underset{\sim}{C}$.

$$\mu_{\underset{\sim}{C}}\left(\frac{a}{b}\right) = \exp\left\{-3\left(\frac{a}{b} - 1\right)^2\right\} \tag{1.3}$$

Figure 1.5*b* is a plot of the membership function given in Eq. (1.3). As the elliptic ratio a/b approaches a value of unity, the membership value approaches unity; for $a/b = 1$ we have an unambiguous circle. As $a/b \to \infty$ or $a/b \to 0$, we get a line segment; hence, the membership of the shape in the fuzzy set $\underset{\sim}{C}$ approaches zero, because a line segment is not very similar in shape to a circle. In Fig. 1.5*b* we see that as we get farther from $a/b = 1$ our membership in the set "circle" gets smaller and smaller. All values of a/b that have a membership value of unity are called the prototypes; in this case $a/b = 1$ is the only prototype for the set "circle."

Suppose we were to place in a bag a large number of generally elliptical two-dimensional shapes and ask the question: "What is the probability of randomly selecting a circle from the bag?" We could not answer this question without first assessing the two different kinds of uncertainty. First, we would have to address the issue of fuzziness in the meaning of the term "circle" by selecting a value of membership, above which we would be willing to call the shape a circle; for example, any shape with a membership value above 0.9 in the fuzzy set "circle" would be considered a circle. Second, we would have to know the proportion of the shapes in the bag that have membership values above 0.9. The first issue is one of assessing ambiguity and the second relates to the frequencies required to address probability questions.

SUMMARY

This chapter has discussed models with essentially two different kinds of information: fuzzy membership functions, which represent similarities of objects to ambiguous properties, and probabilities, which provide knowledge about relative frequencies. The value of either of these kinds of information in making decisions is a matter of preference; popular, but controversial, contrary views have been offered [Cheeseman, 1988; Elkan, 1994]. Fuzzy models are *not* replacements for probability models. As seen in Fig. 1.3, every crisp set is fuzzy, but the converse does not hold. Fuzzy models are not that different from more familiar models. Sometimes they work better, and sometimes they do not. After all, the efficacy of a model in solving a problem should be the only criterion used to judge that model. Lately, a growing body of evidence suggests that fuzzy approaches to real problems are an effective alternative to previous, traditional methods.

REFERENCES

Amis, K. (1978). *The new Oxford book of English light verse,* Oxford University Press, New York.

Bezdek, J. (1993). "Editorial: Fuzzy models—What are they, and why?" *IEEE Trans. Fuzzy Syst.,* vol. 1, pp. 1–5.

Cheeseman, P. (1988). "An inquiry into computer understanding," *Comput. Intell.,* vol. 4, pp. 57–142 (with commentaries and replies).

Cunningham, J. (1971). *The collected poems and epigrams of J.V. Cunningham,* Swallow Press, Chicago.

Elkan, C. (1994). "The paradoxical success of fuzzy logic," *IEEE Expert,* August, pp. 3–8.

Kolata, G. (1991). "Math problem, long baffling, slowly yields," *New York Times,* March 12, p. C1.

Rogers, M. and Hoshuai, Y. (1990). "The future looks 'fuzzy,' " *Newsweek,* May 28, p. 46.

Zadeh, L. (1965). "Fuzzy sets," Inf. Control, vol. 8, pp. 338–353.

Zadeh, L. (1973). "Outline of a new approach to the analysis of complex systems and decision processes," *IEEE Trans. Syst., Man, Cybern.,* vol. SMC-3, pp. 28–44.

PROBLEMS

1.1. Develop a reasonable membership function for the following fuzzy sets based on height measured in centimeters:

(*a*) "Tall"

(*b*) "Short"

(*c*) "Not short"

1.2. Develop a reasonable membership function for the fuzzy color set "red" based on the frequencies of the color spectrum.

1.3. Develop a reasonable membership function for a square, based on the geometric properties of a rectangle. For this problem use L as the length of the longer side and l as the length of the smaller side.

1.4. For the cylindrical shapes shown in Fig. 1.4, develop a membership function for each of the following shapes using the ratio d/h, and discuss the reason for any overlapping among the three membership functions:

(*a*) Rod

(*b*) Cylinder

(*c*) Disk

1.5. The question of whether a glass of water is half-full or half-empty is an age-old philosophical issue. Such descriptions of the volume of liquid in a glass depend on the state of mind of the person asked the question. Develop membership functions for the fuzzy sets "half-full," "full," "empty," and "half-empty" using percent volume as the element of information. Assume the maximum volume of water in the glass is V_0. Discuss whether the terms "half-full" and "half-empty" should have identical membership functions. Does your answer solve this ageless riddle?

1.6. In census data, there are many reasons why every community has to be categorized as a village, a town, or a city according to its population. One of the uses of this information is in forecasting and planning new traffic routes for a region. Define membership functions on a universe of populations for each of the mentioned communities.

1.7. One of the most important features of new passenger airplanes is their characterization as short-haul, medium-haul, or long-haul aircraft. This designation is a function of their capabilities in units of passenger-miles. Define membership functions for the three kinds of commercial airplanes.

1.8. Differentiate *infrasonic* and *ultrasonic* sounds according to their frequencies using membership functions.

1.9. The levels of service of highways are of five types, namely, A, B, C, D, and E according to the capacity of the highway (capacity depends on the type of artery, number of lanes, and traffic volumes). Classify them using membership functions and traffic volumes.

1.10. Using the ratios of internal angles or sides of a hexagon, draw the membership diagrams for "regular" and "irregular" hexagons.

1.11. Compare "medium wave" and "shortwave" receivers according to their frequency range. Plot the membership functions.

1.12. Figure 1.5 shows the membership function for a circle. Determine the membership functions for an ellipse by comparing the ratio of two diametrical chords in the circle and comment on this shape.

CHAPTER 2

CLASSICAL SETS AND FUZZY SETS

Philosophical objections may be raised by the logical implications of building a mathematical structure on the premise of fuzziness, since it seems (at least superficially) necessary to require that an object be or not be an element of a given set. From an aesthetic viewpoint, this may be the most satisfactory state of affairs, but to the extent that mathematical structures are used to model physical actualities, it is often an unrealistic requirement. . . . Fuzzy sets have an intuitively plausible philosophical basis. Once this is accepted, analytical and practical considerations concerning fuzzy sets are in most respects quite orthodox.

James Bezdek
Professor, Computer Science, 1981

As alluded to in Chapter 1, the *universe of discourse* is the universe of all available information on a given problem. Once this universe is defined we are able to define certain events on this information space. We will describe sets as mathematical abstractions of these events and of the universe itself. Figure 2.1*a* shows an abstraction of a universe of discourse, say X, and a crisp (classical) set A somewhere in this universe. A classical set is defined by *crisp* boundaries; i.e., there is no uncertainty in the prescription or location of the boundaries of the set, as shown in Fig. 2.1*a* where the boundary of crisp set A is an unambiguous line. A fuzzy set, on the other hand, is prescribed by vague or ambiguous properties; hence its boundaries are ambiguously specified, as shown by the fuzzy boundary for set $\underset{\sim}{A}$ in Fig. 2.1*b*.

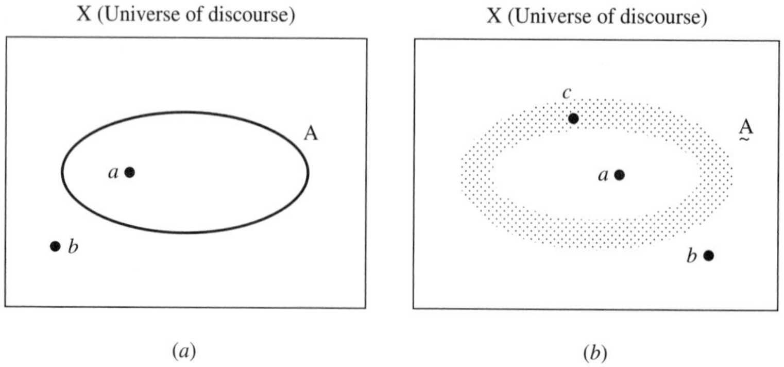

FIGURE 2.1
Diagrams for (*a*) crisp set boundary and (*b*) fuzzy set boundary.

In Chapter 1 we introduced the notion of set membership, from a one-dimensional viewpoint. Figure 2.1 again helps to explain this idea, but from a two-dimensional perspective. Point *a* in Fig. 2.1*a* is clearly a member of crisp set A; point *b* is unambiguously *not* a member of set A. Figure 2.1*b* shows the vague, ambiguous boundary of a fuzzy set $\underset{\sim}{A}$ on the same universe X: the shaded boundary represents the boundary region of $\underset{\sim}{A}$. In the central (unshaded) region of the fuzzy set, point *a* is clearly a full member of the set. Outside the boundary region of the fuzzy set, point *b* is clearly not a member of the fuzzy set. However, the membership of point *c*, which is on the boundary region, is ambiguous. If complete membership in a set (such as point *a* in Fig. 2.1*b*) is represented by the number 1, and no-membership in a set (such as point *b* in Fig. 2.1*b*) is represented by 0, then point *c* in Figure 2.1*b* must have some intermediate value of membership (partial membership in fuzzy set $\underset{\sim}{A}$) on the interval [0,1]. Presumably the membership of point *c* in $\underset{\sim}{A}$ approaches a value of 1 as it moves closer to the central (unshaded) region in Fig. 2.1*b* of $\underset{\sim}{A}$, and the membership of point *c* in $\underset{\sim}{A}$ approaches a value of 0 as it moves closer to leaving the boundary region of $\underset{\sim}{A}$.

In this chapter, the precepts and operations of fuzzy sets are compared with those of classical sets. Several good books are available for reviewing this basic material [see for example, Dubois and Prade, 1980; Klir and Folger, 1988; Zimmerman, 1991]. Fuzzy sets embrace virtually all (with one exception, as will be seen) of the definitions, precepts, and axioms that define classical sets. As indicated in Chapter 1, crisp sets are a special form of fuzzy sets; they are sets without ambiguity in their membership (i.e., they are sets with unambiguous boundaries). It will be shown that fuzzy set theory is a mathematically rigorous and comprehensive set theory useful in characterizing concepts (sets) with natural ambiguity. It is instructive to introduce fuzzy sets by first reviewing the elements of classical (crisp) set theory.

CLASSICAL SETS

Define a universe of discourse, X, as a collection of objects all having the same characteristics. The individual elements in the universe X will be denoted as *x*. The features of the elements in X can be discrete, countable integers or continuous valued quantities on the real line. Examples of elements of various universes might be as follows:

The clock speeds of computer CPUs
The operating currents of an electronic motor
The operating temperature of a heat pump (in degrees Celsius)
The Richter magnitudes of an earthquake
The integers 1 to 10

Most real-world engineering processes contain elements that are real and non-negative. The first four items just named are examples of such elements. However, for purposes of modeling, most engineering problems are simplified to consider only integer values of the elements in a universe of discourse. So, for example, computer clock speeds might be measured in integer values of megaHertz and heat pump temperatures might be measured in integer values of degrees Celsius. Further, most engineering processes are simplified to consider only finite-sized universes. Although Richter magnitudes may not have a theoretical limit, we have not historically measured earthquake magnitudes much above 9; this value might be the upper bound in a structural engineering design problem. As another example, suppose you are interested in the stress under one leg of the chair in which you are sitting. You might argue that it is possible to get an infinite stress on one leg of the chair by sitting in the chair in such a manner that only one leg is supporting you and by letting the area of the tip of that leg approach zero. Although this is theoretically possible, in reality the chair leg will either buckle elastically as the tip area becomes very small or yield plastically and fail because materials that have infinite strength have not yet been developed. Hence, choosing a universe that is discrete and finite or one that is continuous and infinite is a modeling choice; the choice does not alter the characterization of sets defined on the universe. If elements of a universe are continuous, then sets defined on the universe will be composed of continuous elements. For example, if the universe of discourse is defined as all Richter magnitudes up to a value of 9, then we can define a set of "destructive magnitudes," which might be composed (*i*) of all magnitudes greater than or equal to a value of 6 in the crisp case or (*ii*) of all magnitudes "approximately 6 and higher" in the fuzzy case.

A useful attribute of sets and the universes on which they are defined is a metric known as the cardinality, or the cardinal number. The total number of elements in a universe X is called its cardinal number, denoted n_x, where x again is a label for individual elements in the universe. Discrete universes that are composed of a countably finite collection of elements will have a finite cardinal number; continuous universes comprised of an infinite collection of elements will have an infinite cardinality. Collections of elements within a universe are called sets, and collections of elements within sets are called subsets. Sets and subsets are terms that are often used synonymously, since any set is also a subset of the universal set X. The collection of all elements in the universe is also called the *whole set*.

For crisp sets A and B consisting of collections of some elements in X, the following notation is defined:

$x \in X \Rightarrow x$ belongs to X
$x \in A \Rightarrow x$ belongs to A
$x \notin A \Rightarrow x$ does not belong to A

For sets A and B on X, we also have

$A \subset B \Rightarrow$ A is fully contained in B (if $x \in A$, then $x \in B$)
$A \subseteq B \Rightarrow$ A is contained in or is equivalent to B
$A = B \Rightarrow A \subseteq B$ and $B \subseteq A$

We define the null set, $\varnothing$, as the set containing no elements, and the whole set, X, as the set of all elements in the universe. The null set is analogous to an impossible event, and the whole set is analogous to a certain event. All possible sets of X constitute a special set called the power set, P(X). For a specific universe X, the power set P(X) is enumerated in the following example.

Example 2.1. We have a universe comprised of three elements, $X = \{a, b, c\}$, so the cardinal number is $n_x = 3$. The power set is

$$P(X) = \{\varnothing, \{a\}, \{b\}, \{c\}, \{a, b\}, \{a, c\}, \{b, c\}, \{a, b, c\}\}$$

The cardinality of the power set, denoted $n_{P(X)}$, is found as

$$n_{P(X)} = 2^{n_X} = 2^3 = 8$$

Note that if the cardinality of the universe is infinite, then the cardinality of the power set is also infinity; i.e., $n_X = \infty \Rightarrow n_{P(X)} = \infty$.

Operations on Classical Sets

Let A and B be two sets on the universe X. The union between the two sets, denoted $A \cup B$, represents all those elements in the universe that reside in (or belong to) either the set A, the set B, or both sets A and B. (This operation is also called the *logical or*; another form of the union is the *exclusive or* operation. The *exclusive or* will be described in Chapter 7.) The intersection of the two sets, denoted $A \cap B$, represents all those elements in the universe X that simultaneously reside in (or belong to) both sets A and B. The complement of a set A, denoted $\overline{A}$, is defined as the collection of all elements in the universe that do not reside in the set A. The difference of a set A with respect to B, denoted $A \mid B$, is defined as the collection of all elements in the universe that reside in A and that do not reside in B simultaneously. These operations are shown below in set-theoretic terms.

Union $$A \cup B = \{x \mid x \in A \text{ or } x \in B\} \quad (2.1)$$

Intersection $$A \cap B = \{x \mid x \in A \text{ and } x \in B\} \quad (2.2)$$

Complement $$\overline{A} = \{x \mid x \notin A, x \in X\} \quad (2.3)$$

Difference $$A \mid B = \{x \mid x \in A \text{ and } x \notin B\} \quad (2.4)$$

These four operations are shown in terms of Venn diagrams in Figs. 2.2 through 2.5.

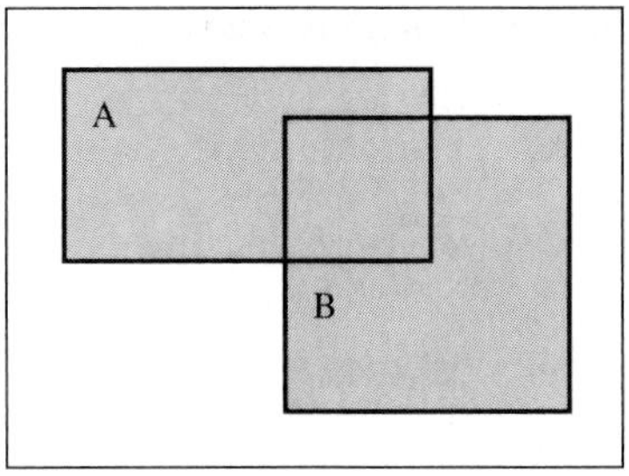

FIGURE 2.2
Union of sets A and B (logical or).

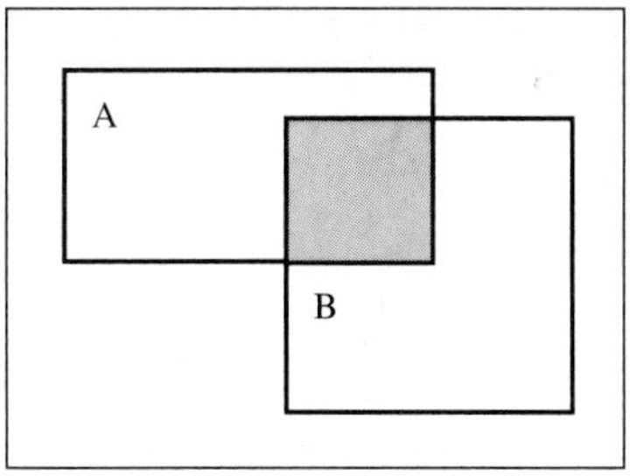

FIGURE 2.3
Intersection of sets A and B.

Properties of Classical (Crisp) Sets

Certain properties of sets are important because of their influence on the mathematical manipulation of sets. The most appropriate properties for defining classical sets and showing their similarity to fuzzy sets are as follows:

Commutativity
$$\begin{aligned} A \cup B &= B \cup A \\ A \cap B &= B \cap A \end{aligned} \tag{2.5}$$

Associativity
$$\begin{aligned} A \cup (B \cup C) &= (A \cup B) \cup C \\ A \cap (B \cap C) &= (A \cap B) \cap C \end{aligned} \tag{2.6}$$

Distributivity
$$\begin{aligned} A \cup (B \cap C) &= (A \cup B) \cap (A \cup C) \\ A \cap (B \cup C) &= (A \cap B) \cup (A \cap C) \end{aligned} \tag{2.7}$$

Idempotency
$$\begin{aligned} A \cup A &= A \\ A \cap A &= A \end{aligned} \tag{2.8}$$

Identity
$$\begin{aligned} A \cup \varnothing &= A \\ A \cap X &= A \\ A \cap \varnothing &= \varnothing \\ A \cup X &= X \end{aligned} \tag{2.9}$$

Transitivity
$$\text{If } A \subseteq B \subseteq C, \text{ then } A \subseteq C \tag{2.10}$$

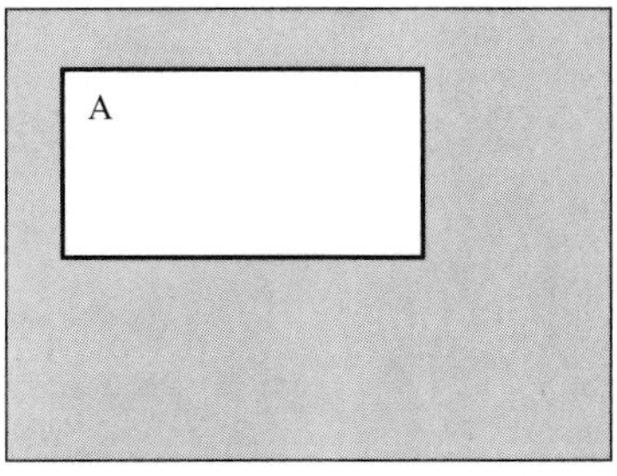

FIGURE 2.4
Complement of set A.

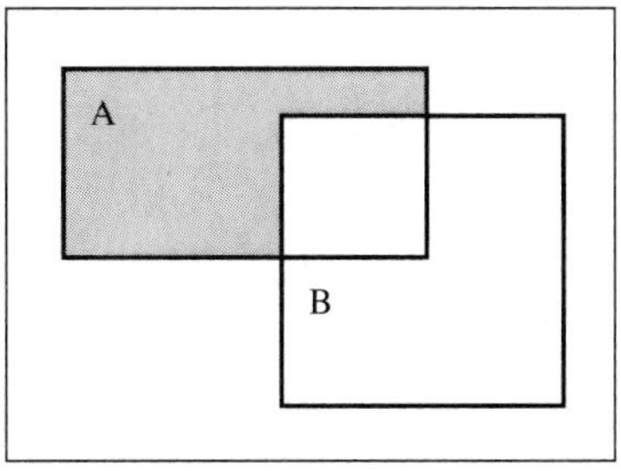

FIGURE 2.5
Difference operation A | B.

Note here that the symbol $\subseteq$ means *contained in* or *equivalent to;* a similar-looking symbol $\subset$ simply means *contained in.*

Involution $$\overline{\overline{A}} = A \tag{2.11}$$

The double-cross-hatched area in Figure 2.6 is a Venn diagram example of the associativity property for intersection, and the double-cross-hatched areas in Figs. 2.7 and 2.8 are Venn diagram examples of the distributivity property for various combinations of the intersection and union properties.

Two special properties of set operations are known as the *excluded middle laws* and *De Morgan's laws.* These laws are enumerated here for two sets A and B. The *excluded middle laws* are very important because these are the only set operations described here that are *not* valid for both classical sets and fuzzy sets. There are two excluded middle laws (given in Eqs. [2.12]). The first, called the *law of the excluded middle,* deals with the union of a set A and its complement; the second, called the *law of contradiction,* represents the intersection of a set A and its complement.

Law of the excluded middle $$A \cup \overline{A} = X \tag{2.12a}$$

Law of the contradiction $$A \cap \overline{A} = \varnothing \tag{2.12b}$$

De Morgan's laws are important because of their usefulness in proving tautologies and contradictions in logic, as well as in a host of other set operations and proofs. De Morgan's laws are displayed in the shaded areas of the Venn diagrams in Figs. 2.9 and 2.10 and described mathematically in Eq. (2.13).

$$\overline{A \cap B} = \overline{A} \cup \overline{B} \tag{2.13a}$$

$$\overline{A \cup B} = \overline{A} \cap \overline{B} \tag{2.13b}$$

In general, De Morgan's laws can be stated for n sets, as provided here for events, E_i:

$$\overline{E_1 \cup E_2 \cup \cdots \cup E_n} = \overline{E_1} \cap \overline{E_2} \cap \cdots \cap \overline{E_n} \tag{2.14a}$$

$$\overline{E_1 \cap E_2 \cap \cdots \cap E_n} = \overline{E_1} \cup \overline{E_2} \cup \cdots \cup \overline{E_n} \tag{2.14b}$$

From the general equations, Eqs. (2.14), for De Morgan's laws we get a duality relation: The complement of a union or an intersection is equal to the intersection or

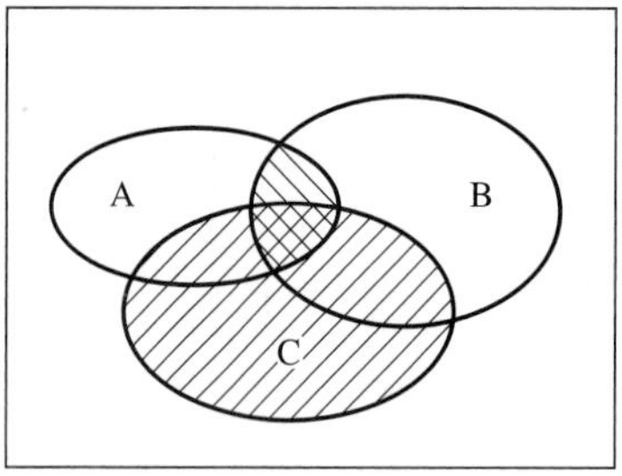

(*a*)

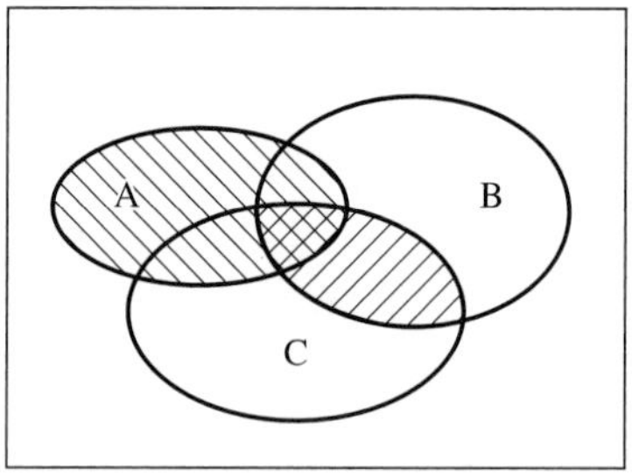

(*b*)

FIGURE 2.6
Venn diagrams for (*a*) (A ∩ B) ∩ C and (*b*) A ∩ (B ∩ C).

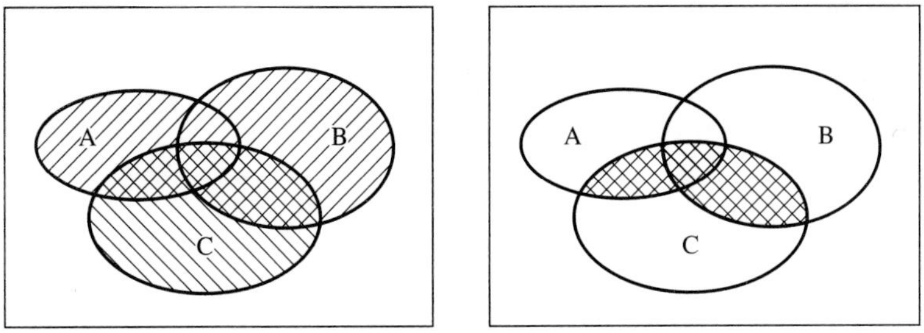

(*a*) (*b*)

FIGURE 2.7
Venn diagrams for (*a*) $(A \cup B) \cap C$ and (*b*) $(A \cap C) \cup (B \cap C)$.

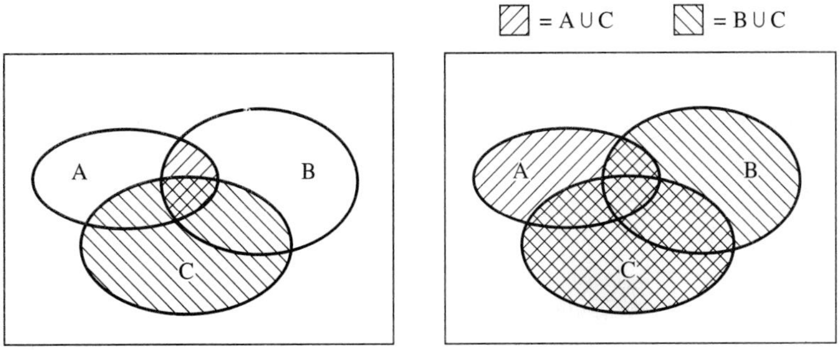

(*a*) (*b*)

FIGURE 2.8
Venn diagrams for (*a*) $(A \cap B) \cup C$ and (*b*) $(A \cup C) \cap (B \cup C)$.

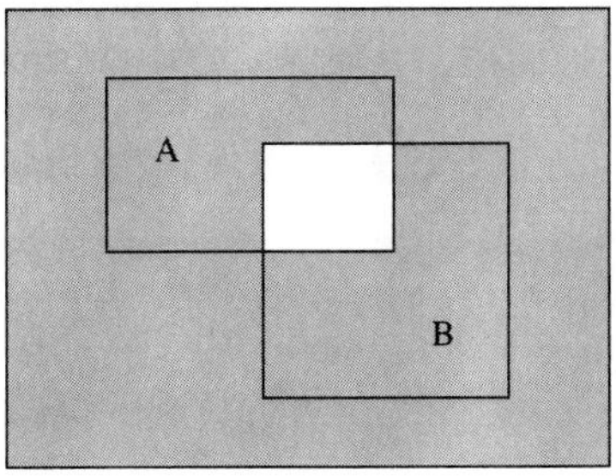

FIGURE 2.9
De Morgan's law $(\overline{A \cap B})$.

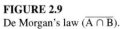

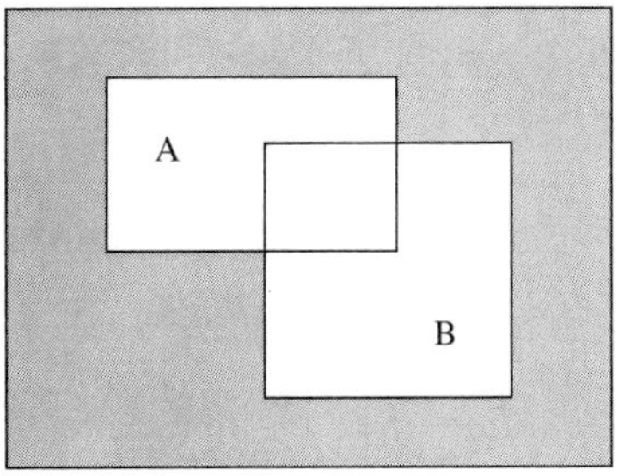

FIGURE 2.10
De Morgan's law $(\overline{A \cup B})$.

union, respectively, of the respective complements. This result is very powerful in dealing with set structures since we often have information about the complement of a set (or event), or the complement of combinations of sets (or events), rather than information about the sets themselves.

Example 2.2. A shallow arch consists of two slender members as shown in Fig. 2.11. If either member fails, then the arch will collapse. If E_1 = survival of member 1 and E_2 = survival of member 2, then survival of the arch = $E_1 \cap E_2$, and, conversely, collapse of the arch = $\overline{E_1 \cap E_2}$. Logically, collapse of the arch will occur if either of the members fails, i.e. when $\overline{E_1} \cup \overline{E_2}$. Therefore,

$$\overline{E_1 \cap E_2} = \overline{E_1} \cup \overline{E_2}$$

which is an illustration of De Morgan's law.

As Eq. (2.14) suggests, De Morgan's laws are very useful for compound events, as illustrated in the following example.

Example 2.3. For purposes of safety, the fluid supply for a hydraulic pump C in an airplane comes from two redundant source lines, A and B. The fluid is transported by high-pressure hoses consisting of branches 1, 2, and 3, as shown in Fig. 2.12. Operating specifications for the pump indicate that either source line alone is capable of supplying the necessary fluid pressure to the pump. Denote E_1 = failure of branch 1, E_2 = failure of branch 2, and E_3 = failure of branch 3. Then insuffícent pressure to operate the pump would be caused by $(E_1 \cap E_2) \cup E_3$, and sufficient pressure would be the complement of this event. Using De Morgan's laws, we can calculate the condition of sufficient pressure to be

$$\overline{(E_1 \cap E_2) \cup E_3} = (\overline{E_1} \cup \overline{E_2}) \cap \overline{E_3}$$

in which $(\overline{E_1} \cup \overline{E_2})$ means the availability of pressure at the junction, and $\overline{E_3}$ means the absence of failure in branch 3.

Mapping of Classical Sets to Functions

Mapping is an important concept in relating set-theoretic forms to function-theoretic representations of information. In its most general form it can be used to map elements or subsets on one universe of discourse to elements or sets in another universe. Suppose X and Y are two different universes of discourse (information). If an ele-

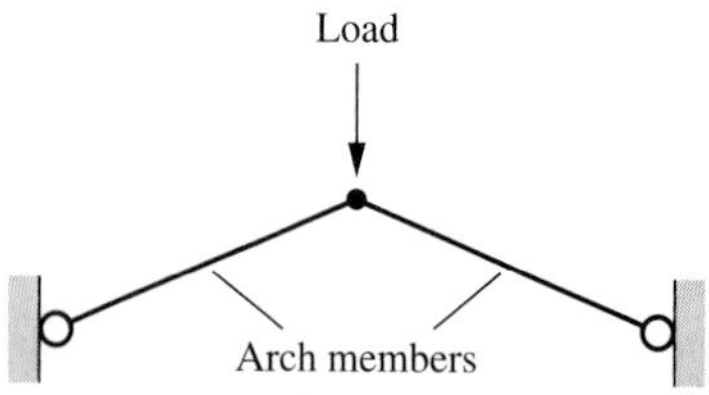

FIGURE 2.11
A two-member arch.

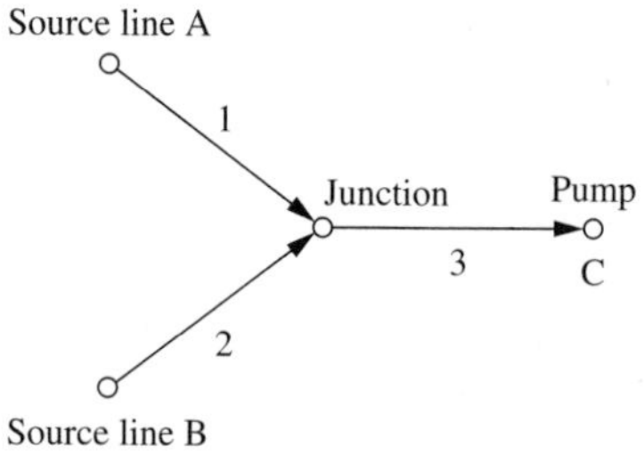

FIGURE 2.12
Hydraulic hose system.

ment x is contained in X and corresponds to an element y contained in Y, it is generally termed a mapping from X to Y, or $f : \mathrm{X} \rightarrow \mathrm{Y}$. As a mapping, the characteristic (indicator) function χ_{A} is defined by

$$\chi_{\mathrm{A}}(x) = \begin{cases} 1, & x \in \mathrm{A} \\ 0, & x \notin \mathrm{A} \end{cases} \tag{2.15}$$

where χ_{A} expresses "membership" in set A for the element x in the universe. This membership idea is a mapping from an element x in universe X to one of the two elements in universe Y; i.e., to the elements 0 or 1, as shown in Fig. 2.13.

For any set A defined on the universe X, there exists a function-theoretic set, called a value set, denoted V(A), under the mapping of the characteristic function, χ. By convention, the null set $\varnothing$ is assigned the membership value 0 and the whole set X is assigned the membership value 1.

Example 2.4. Continuing with the example (Example 2.1) of a universe with three elements, $\mathrm{X} = \{a, b, c\}$, we desire to map the elements of the power set of X, i.e., P(X), to a universe, Y, consisting of only two elements (the characteristic function),

$$\mathrm{Y} = \{0, 1\}$$

As before, the elements of the power set are enumerated.

$$\mathrm{P(X)} = \left\{\varnothing, \{a\}, \{b\}, \{c\}, \{a, b\}, \{b, c\}, \{a, c\}, \{a, b, c\}\right\}$$

Thus, the elements in the value set V(A) as determined from the mapping are

$$\mathrm{V\{P(X)\}} = \left\{\{0, 0, 0\}, \{1, 0, 0\}, \{0, 1, 0\}, \{0, 0, 1\}, \{1, 1, 0\}, \{0, 1, 1\}, \{1, 0, 1\}, \{1, 1, 1\}\right\}$$

For example, the third subset in the power set P(X) is the element b. For this subset there is no a, so a value of 0 goes in the first postion of the data triplet; there is a b, so a value of 1 goes in the second position of the data triplet; and there is no c, so a value of 0 goes in the third position of the data triplet. Hence, the third subset of the value set is the data triplet, {0, 1, 0}, as already seen. The value set has a graphical analog that is described at the end of this chapter in the section "Sets as Points in Hypercubes."

Now, define two sets, A and B, on the universe X. The union of these two sets in terms of function-theoretic terms is given as follows (the symbol $\bigvee$ is the maximum operator and $\bigwedge$ is the minimum operator):

Union $\quad \mathrm{A} \cup \mathrm{B} \rightarrow \chi_{\mathrm{A}\cup\mathrm{B}}(x) = \chi_{\mathrm{A}}(x) \bigvee \chi_{\mathrm{B}}(x) = \max(\chi_{\mathrm{A}}(x), \chi_{\mathrm{B}}(x))$ (2.16)

The intersection of these two sets in function-theoretic terms is given by

Intersection $\quad \mathrm{A} \cap \mathrm{B} \rightarrow \chi_{\mathrm{A}\cap\mathrm{B}}(x) = \chi_{\mathrm{A}}(x) \bigwedge \chi_{\mathrm{B}}(x) = \min(\chi_{\mathrm{A}}(x), \chi_{\mathrm{B}}(x))$ (2.17)

The complement of a single set on universe X, say A, is given by

Complement $\quad \overline{\mathrm{A}} \rightarrow \chi_{\overline{\mathrm{A}}}(x) = 1 - \chi_{\mathrm{A}}(x)$ (2.18)

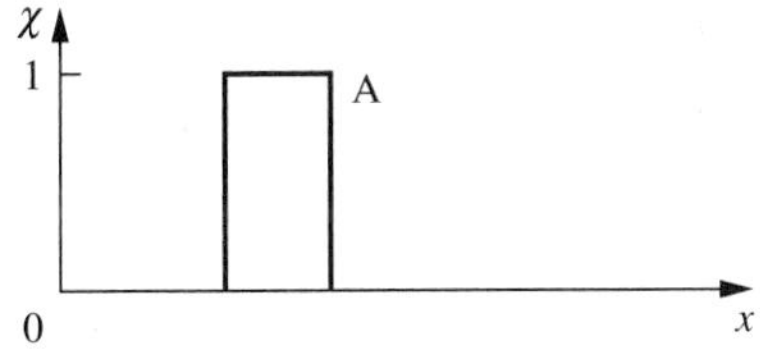

FIGURE 2.13
Membership function is a mapping for crisp set A.

For two sets on the same universe, say A and B, if one set (A) is contained in another set (B), then

Containment $$\mathrm{A} \subseteq \mathrm{B} \rightarrow \chi_{\mathrm{A}}(x) \leq \chi_{\mathrm{B}}(x) \tag{2.19}$$

Function-theoretic operators for union and intersection (other than maximum and minimum, respectively) are discussed in the literature [Gupta and Qi, 1991].

FUZZY SETS

In classical, or crisp, sets the transition for an element in the universe between membership and nonmembership in a given set is abrupt and well-defined (said to be "crisp"). For an element in a universe that contains fuzzy sets, this transition can be gradual. This transition among various degrees of membership can be thought of as conforming to the fact that the boundaries of the fuzzy sets are vague and ambiguous. Hence, membership of an element from the universe in this set is measured by a function that attempts to describe vagueness and ambiguity.

A fuzzy set, then, is a set containing elements that have varying degrees of membership in the set. This idea is in contrast with classical, or crisp, sets because members of a crisp set would not be members unless their membership was full, or complete, in that set (i.e., their membership is assigned a value of 1). Elements in a fuzzy set, because their membership need not be complete, can also be members of other fuzzy sets on the same universe.

Elements of a fuzzy set are mapped to a universe of *membership values* using a function-theoretic form. As mentioned in Chapter 1 (Eq. [1.2]), fuzzy sets are denoted in this text by a set symbol with a tilde understrike; so, for example, $\underset{\sim}{\mathrm{A}}$ would be the *fuzzy set A*. This function maps elements of a fuzzy set $\underset{\sim}{\mathrm{A}}$ to a real numbered value on the interval 0 to 1. If an element in the universe, say x, is a member of fuzzy set $\underset{\sim}{\mathrm{A}}$, then this mapping is given by Eq. (1.2), or $\mu_{\underset{\sim}{\mathrm{A}}}(x) \in [0, 1]$. This mapping is shown in Fig. 2.14 for a typical fuzzy set.

A notation convention for fuzzy sets when the universe of discourse, X, is discrete and finite, is as follows for a fuzzy set $\underset{\sim}{\mathrm{A}}$:

$$\underset{\sim}{\mathrm{A}} = \left\{ \frac{\mu_{\underset{\sim}{\mathrm{A}}}(x_1)}{x_1} + \frac{\mu_{\underset{\sim}{\mathrm{A}}}(x_2)}{x_2} + \cdots \right\} = \left\{ \sum_i \frac{\mu_{\underset{\sim}{\mathrm{A}}}(x_i)}{x_i} \right\} \tag{2.20}$$

When the universe, X, is continuous and infinite, the fuzzy set $\underset{\sim}{\mathrm{A}}$ is denoted by

$$\underset{\sim}{\mathrm{A}} = \left\{ \int \frac{\mu_{\underset{\sim}{\mathrm{A}}}(x)}{x} \right\} \tag{2.21}$$

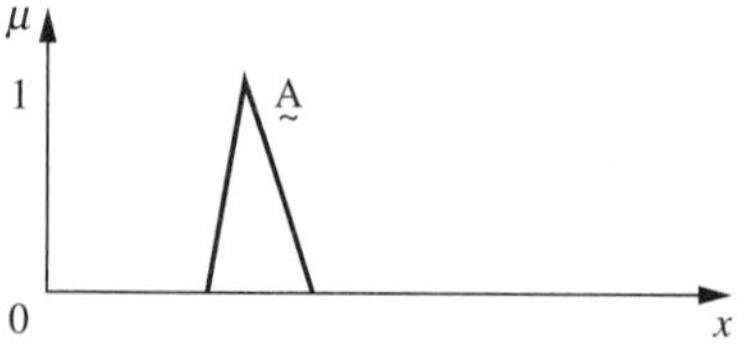

FIGURE 2.14
Membership function for fuzzy set $\underset{\sim}{\mathrm{A}}$.

In both notations, the horizontal bar is not a quotient but rather a delimiter. The numerator in each term is the membership value in set $\underset{\sim}{A}$ associated with the element of the universe indicated in the denominator. In the first notation, the summation symbol is not for algebraic summation but rather denotes the collection or aggregation of each element; hence the "+" signs in the first notation are not the algebraic "add" but are a function-theoretic union. In the second notation the integral sign is not an algebraic integral but a continuous function-theoretic union notation for continuous variables.

Fuzzy Set Operations

Define three fuzzy sets $\underset{\sim}{A}$, $\underset{\sim}{B}$, and $\underset{\sim}{C}$ on the universe X. For a given element x of the universe, the following function theoretic operations for the set-theoretic operations of union, intersection, and complement are defined for $\underset{\sim}{A}$, $\underset{\sim}{B}$, and $\underset{\sim}{C}$ on X:

Union $$\mu_{\underset{\sim}{A}\cup\underset{\sim}{B}}(x) = \mu_{\underset{\sim}{A}}(x) \vee \mu_{\underset{\sim}{B}}(x) \tag{2.22}$$

Intersection $$\mu_{\underset{\sim}{A}\cap\underset{\sim}{B}}(x) = \mu_{\underset{\sim}{A}}(x) \wedge \mu_{\underset{\sim}{B}}(x) \tag{2.23}$$

Complement $$\mu_{\overline{\underset{\sim}{A}}}(x) = 1 - \mu_{\underset{\sim}{A}}(x) \tag{2.24}$$

Venn diagrams for these operations, extended to consider fuzzy sets, are shown in Figs. 2.15–2.17.

Any fuzzy set $\underset{\sim}{A}$ defined on a universe X is a subset of that universe. Also by definition, just as with classical sets, the membership value of any element x in the null set $\varnothing$ is 0, and the membership value of any element x in the whole set X is 1. Note that the null set and the whole set are not fuzzy sets in this context (no tilde understrike). The appropriate notation for these ideas is as follows:

$$\underset{\sim}{A} \subseteq X \Rightarrow \mu_{\underset{\sim}{A}}(x) \le \mu_X(x) \tag{2.25a}$$

$$\text{For all } x \in X,\ \mu_{\varnothing}(x) = 0 \tag{2.25b}$$

$$\text{For all } x \in X,\ \mu_X(x) = 1 \tag{2.25c}$$

The collection of all fuzzy sets and fuzzy subsets on X is denoted as the fuzzy power set $\underset{\sim}{P}(X)$. It should be obvious, based on the fact that all fuzzy sets can overlap, that the cardinality, $n_{P(X)}$, of the fuzzy power set is infinite; that is, $n_{P(X)} = \infty$.

De Morgan's laws for classical sets also hold for fuzzy sets, as denoted by these expressions:

$$\overline{\underset{\sim}{A} \cap \underset{\sim}{B}} = \overline{\underset{\sim}{A}} \cup \overline{\underset{\sim}{B}} \tag{2.26a}$$

$$\overline{\underset{\sim}{A} \cup \underset{\sim}{B}} = \overline{\underset{\sim}{A}} \cap \overline{\underset{\sim}{B}} \tag{2.26b}$$

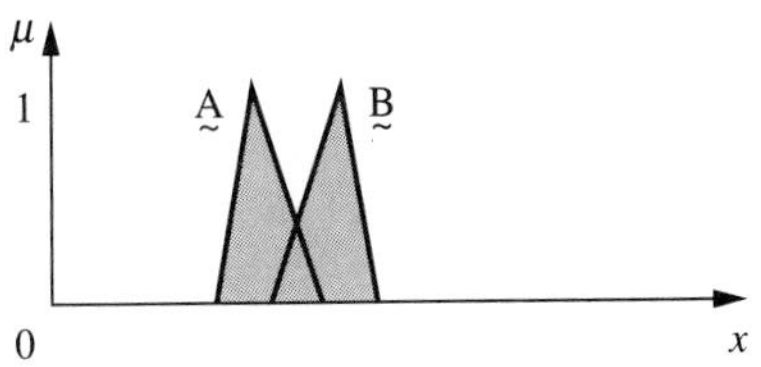

FIGURE 2.15
Union of fuzzy sets $\underset{\sim}{A}$ and $\underset{\sim}{B}$.

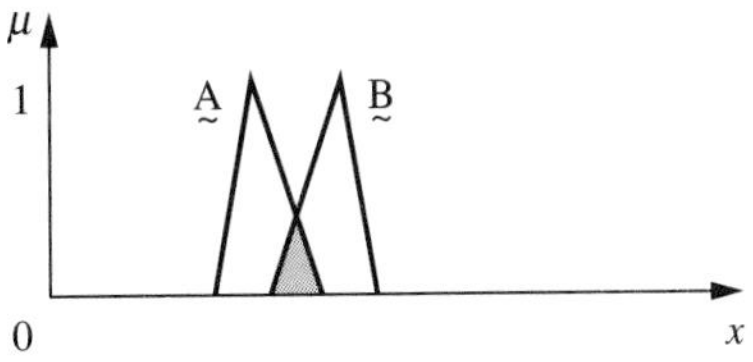

FIGURE 2.16
Intersection of fuzzy sets $\underset{\sim}{A}$ and $\underset{\sim}{B}$.

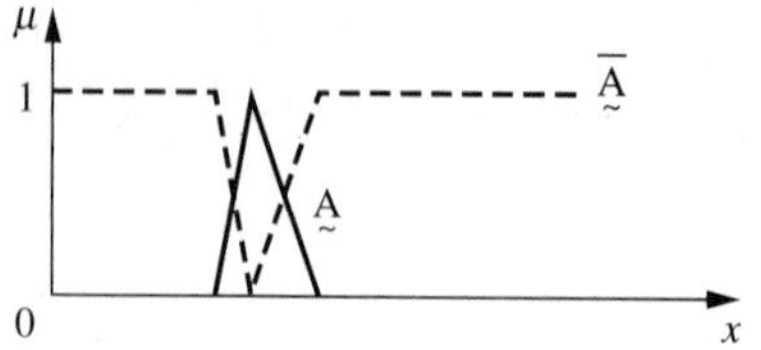

FIGURE 2.17
Complement of fuzzy set $\underset{\sim}{A}$.

As enumerated before, all other operations on classical sets also hold for fuzzy sets, except for the excluded middle laws. These two laws do not hold for fuzzy sets; since fuzzy sets can overlap, a set and its complement can also overlap. The *excluded middle laws,* extended for fuzzy sets, are expressed by

$$\underset{\sim}{A} \cup \overline{\underset{\sim}{A}} \neq X \tag{2.27a}$$

$$\underset{\sim}{A} \cap \overline{\underset{\sim}{A}} \neq \varnothing \tag{2.27b}$$

Extended Venn diagrams comparing the *excluded middle laws* for classical (crisp) sets and fuzzy sets are shown in Figs. 2.18 and 2.19, respectively.

Properties of Fuzzy Sets

Fuzzy sets follow the same properties as crisp sets. Because of this fact and because the membership values of a crisp set are a subset of the interval [0,1], classical sets can be thought of as a special case of fuzzy sets. Frequently used properties of fuzzy sets are listed below.

Commutativity

$$\begin{aligned} \underset{\sim}{A} \cup \underset{\sim}{B} &= \underset{\sim}{B} \cup \underset{\sim}{A} \\ \underset{\sim}{A} \cap \underset{\sim}{B} &= \underset{\sim}{B} \cap \underset{\sim}{A} \end{aligned} \tag{2.28}$$

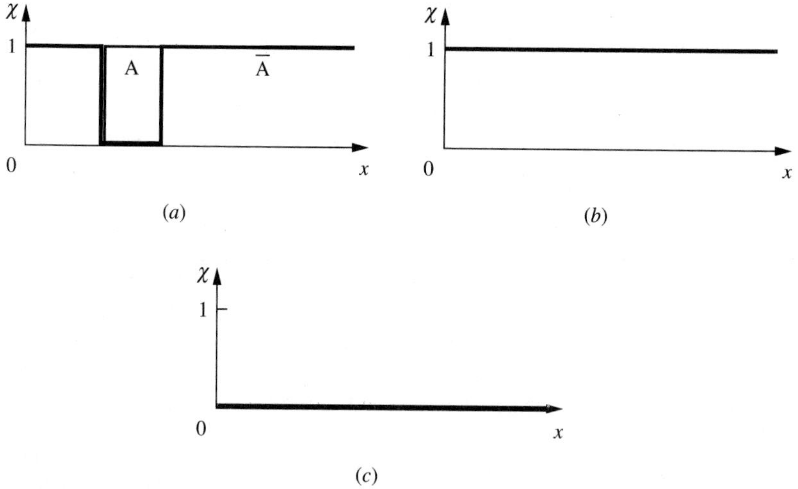

FIGURE 2.18
Excluded middle laws for crisp sets. (*a*) Crisp set A and its complement; (*b*) crisp $A \cup \overline{A} = X$ (law of excluded middle); (*c*) crisp $A \cap \overline{A} = \varnothing$ (law of contradiction).

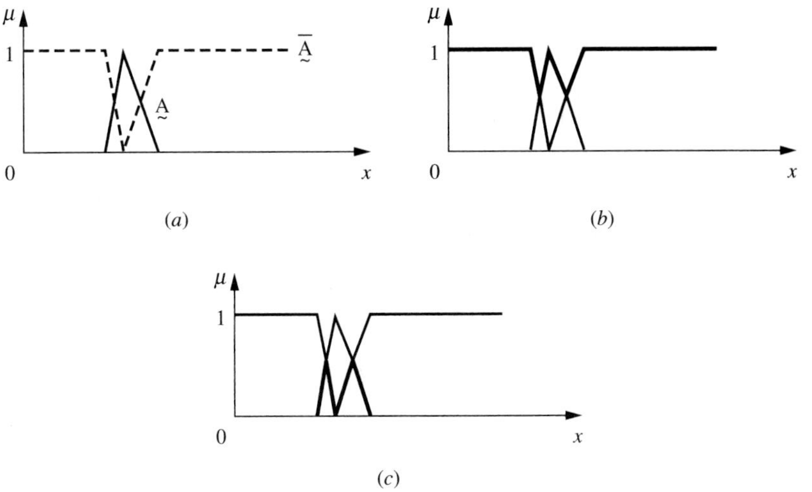

FIGURE 2.19
Excluded middle laws for fuzzy sets. (*a*) Fuzzy set $\underset{\sim}{A}$ and its complement; (*b*) fuzzy $\underset{\sim}{A} \cup \overline{\underset{\sim}{A}} \neq X$ (law of excluded middle); (*c*) fuzzy $A \cap \overline{A} \neq \varnothing$ (law of contradiction).

Associativity
$$\underset{\sim}{A} \cup (\underset{\sim}{B} \cup \underset{\sim}{C}) = (\underset{\sim}{A} \cup \underset{\sim}{B}) \cup \underset{\sim}{C}$$
$$\underset{\sim}{A} \cap (\underset{\sim}{B} \cap \underset{\sim}{C}) = (\underset{\sim}{A} \cap \underset{\sim}{B}) \cap \underset{\sim}{C} \tag{2.29}$$

Distributivity
$$\underset{\sim}{A} \cup (\underset{\sim}{B} \cap \underset{\sim}{C}) = (\underset{\sim}{A} \cup \underset{\sim}{B}) \cap (\underset{\sim}{A} \cup \underset{\sim}{C})$$
$$\underset{\sim}{A} \cap (\underset{\sim}{B} \cup \underset{\sim}{C}) = (\underset{\sim}{A} \cap \underset{\sim}{B}) \cup (\underset{\sim}{A} \cap \underset{\sim}{C}) \tag{2.30}$$

Idempotency
$$\underset{\sim}{A} \cup \underset{\sim}{A} = \underset{\sim}{A} \quad \text{and} \quad \underset{\sim}{A} \cap \underset{\sim}{A} = \underset{\sim}{A} \tag{2.31}$$

Identity
$$\underset{\sim}{A} \cup \varnothing = \underset{\sim}{A} \quad \text{and} \quad \underset{\sim}{A} \cap X = \underset{\sim}{A}$$
$$\underset{\sim}{A} \cap \varnothing = \varnothing \quad \text{and} \quad \underset{\sim}{A} \cup X = X \tag{2.32}$$

Transitivity
$$\text{If } \underset{\sim}{A} \subseteq \underset{\sim}{B} \subseteq \underset{\sim}{C}, \text{ then } \underset{\sim}{A} \subseteq \underset{\sim}{C} \tag{2.33}$$

Involution
$$\overline{\overline{\underset{\sim}{A}}} = \underset{\sim}{A} \tag{2.34}$$

Example 2.5. Consider a simple hollow shaft of approximately 1-m radius and wall thickness $1/(2\pi)$ m. The shaft is built by stacking a ductile section, D, of the appropriate cross section over a brittle section, B, as shown in Fig. 2.20. A downward force P and a torque T are simultaneously applied to the shaft. Because of the dimensions chosen, the nominal shear stress on any element in the shaft is T (pascals) and the nominal vertical component of stress in the shaft is P (pascals). We also assume that the failure properties of both B and D are not known with any certainty.

We define the fuzzy set $\underset{\sim}{A}$ to be the region in (P, T) space for which material D is "safe" using as a metric the failure function $\mu_A = f([P^2 + 4T^2]^{1/2})$. Similarly, we define the set $\underset{\sim}{B}$ to be the region in (P, T) space for which material B is "safe," using as a metric the failure function $\mu_B = g(P - \beta|T|)$, where β is an assumed material parameter. The functions f and g will, of course, be membership functions on the interval [0, 1]. Their exact specification is not important at this point. What is useful, however,

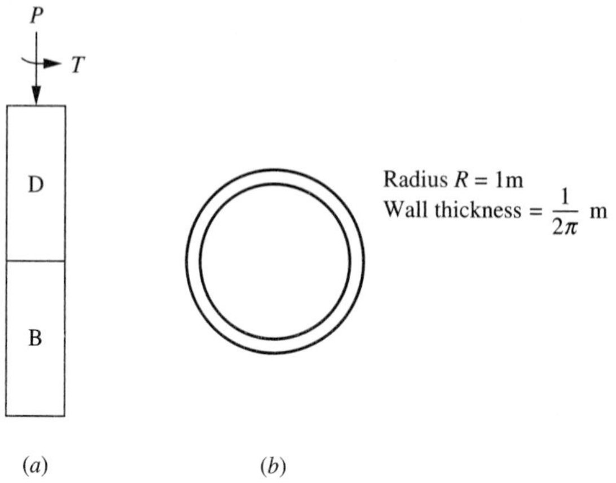

FIGURE 2.20
(*a*) Axial view and (*b*) cross-sectional view of example hollow shaft.

prior to specifying f and g, is to discuss the basic set operations in the context of this problem. This discussion is summarized below:

1. $\underset{\sim}{\mathrm{A}} \cup \underset{\sim}{\mathrm{B}}$ is the set of loadings for which one expects that either material B or material D will be "safe."
2. $\underset{\sim}{\mathrm{A}} \cap \underset{\sim}{\mathrm{B}}$ is the set of loadings for which one expects that both material B and material D are "safe."
3. $\overline{\underset{\sim}{\mathrm{A}}}$ and $\overline{\underset{\sim}{\mathrm{B}}}$ are the sets of loadings for which material D and material B are unsafe, respectively.
4. $\underset{\sim}{\mathrm{A}} \mid \underset{\sim}{\mathrm{B}}$ is the set of loadings for which the ductile material is safe but the brittle material is in jeopardy.
5. $\underset{\sim}{\mathrm{B}} \mid \underset{\sim}{\mathrm{A}}$ is the set of loadings for which the brittle material is safe but the ductile material is in jeopardy.
6. De Morgan's law $\overline{\underset{\sim}{\mathrm{A}} \cap \underset{\sim}{\mathrm{B}}} = \overline{\underset{\sim}{\mathrm{A}}} \cup \overline{\underset{\sim}{\mathrm{B}}}$ asserts that the loadings that are not safe with respect to both materials are the union of those that are unsafe with respect to the brittle material with those that are unsafe with respect to the ductile material.
7. De Morgan's law $\overline{\underset{\sim}{\mathrm{A}} \cup \underset{\sim}{\mathrm{B}}} = \overline{\underset{\sim}{\mathrm{A}}} \cap \overline{\underset{\sim}{\mathrm{B}}}$ asserts that the loads that are safe for neither material D nor material B are the intersection of those that are unsafe for material D with those that are unsafe for material B.
8. The law of the excluded middle for crisp sets asserts that loadings that are not clearly "unsafe" are therefore "safe," but when there is fuzziness in the uncertainty of $\underset{\sim}{\mathrm{A}}$ or $\underset{\sim}{\mathrm{B}}$ there are loadings that can be both safe and unsafe. This situation is realistic given the fact that structural failure is not binary in the real world.

To illustrate these ideas numerically, let's say we have two fuzzy sets, namely,

$$\underset{\sim}{\mathrm{A}} = \left\{ \frac{1}{2} + \frac{.5}{3} + \frac{.3}{4} + \frac{.2}{5} \right\} \quad \text{and} \quad \underset{\sim}{\mathrm{B}} = \left\{ \frac{.5}{2} + \frac{.7}{3} + \frac{.2}{4} + \frac{.4}{5} \right\}$$

We can now calculate several of the operations just discussed (membership for element 1 in both $\underset{\sim}{\mathrm{A}}$ and $\underset{\sim}{\mathrm{B}}$ is implicitly 0):

Complement
$$\overline{\underset{\sim}{\mathrm{A}}} = \left\{ \frac{1}{1} + \frac{0}{2} + \frac{.5}{3} + \frac{.7}{4} + \frac{.8}{5} \right\}$$

$$\overline{\underset{\sim}{\mathrm{B}}} = \left\{ \frac{1}{1} + \frac{.5}{2} + \frac{.3}{3} + \frac{.8}{4} + \frac{.6}{5} \right\}$$

Union

$$\underset{\sim}{\mathrm{A}} \cup \underset{\sim}{\mathrm{B}} = \left\{ \frac{1}{2} + \frac{.7}{3} + \frac{.3}{4} + \frac{.4}{5} \right\}$$

Intersection

$$\underset{\sim}{\mathrm{A}} \cap \underset{\sim}{\mathrm{B}} = \left\{ \frac{.5}{2} + \frac{.5}{3} + \frac{.2}{4} + \frac{.2}{5} \right\}$$

Difference

$$\underset{\sim}{\mathrm{A}} \mid \underset{\sim}{\mathrm{B}} = \underset{\sim}{\mathrm{A}} \cap \overline{\underset{\sim}{\mathrm{B}}} = \left\{ \frac{.5}{2} + \frac{.3}{3} + \frac{.3}{4} + \frac{.2}{5} \right\}$$

$$\underset{\sim}{\mathrm{B}} \mid \underset{\sim}{\mathrm{A}} = \underset{\sim}{\mathrm{B}} \cap \overline{\underset{\sim}{\mathrm{A}}} = \left\{ \frac{0}{2} + \frac{.5}{3} + \frac{.2}{4} + \frac{.4}{5} \right\}$$

De Morgan's laws

$$\overline{\underset{\sim}{\mathrm{A}} \cup \underset{\sim}{\mathrm{B}}} = \overline{\underset{\sim}{\mathrm{A}}} \cap \overline{\underset{\sim}{\mathrm{B}}} = \left\{ \frac{1}{1} + \frac{0}{2} + \frac{.3}{3} + \frac{.7}{4} + \frac{.6}{5} \right\}$$

$$\overline{\underset{\sim}{\mathrm{A}} \cap \underset{\sim}{\mathrm{B}}} = \overline{\underset{\sim}{\mathrm{A}}} \cup \overline{\underset{\sim}{\mathrm{B}}} = \left\{ \frac{1}{1} + \frac{.5}{2} + \frac{.5}{3} + \frac{.8}{4} + \frac{.8}{5} \right\}$$

Excluded middle laws

$$\overline{\underset{\sim}{\mathrm{A}}} \cup \underset{\sim}{\mathrm{A}} = \left\{ \frac{1}{1} + \frac{1}{2} + \frac{.5}{3} + \frac{.7}{4} + \frac{.8}{5} \right\}$$

$$\overline{\underset{\sim}{\mathrm{B}}} \cap \underset{\sim}{\mathrm{B}} = \left\{ \frac{.5}{2} + \frac{.3}{3} + \frac{.2}{4} + \frac{.4}{5} \right\}$$

Example 2.6. One of the crucial manufacturing operations associated with building the external fuel tank for the space shuttle involves the spray-on foam insulation (SOFI) process, which combines two critical component chemicals in a spray gun under high pressure and a precise temperature and flow rate. Control of these parameters to *near* setpoint values is crucial for satisfying a number of important specification requirements. Specification requirements consist of aerodynamic, mechanical, chemical, and thermodynamic properties.

Optimization of the SOFI process to ensure a robust product in spite of the variabilities of the process parameters is a task employing design of experiment (DOE) methods. These methods consist of structuring the key process variables in a set of experiments conducted in the laboratory. Each key factor is assigned two or three prescribed levels, and various combinations of factors and levels are arranged usually in accordance with a defined matrix array. For example, each row of the matrix defines a practical combination of factors and levels. Key process end results (called the response function) are statistically analyzed to determine the relative sensitivities of the respective process factors.

A merging of classical DOE with the fuzzy characterization techniques could be employed to enhance the initial screening experiments. For example, in the SOFI DOE, both flow and temperature are critical. The levels defined by the matrix can only be approximated in the real world. If we target a low flow rate for 48 lb/min, it may be 38 to 58 lb/min. Also, if we target a high temperature for 135°F, it may be 133° to 137°F.

How the imprecision of the experimental setup influences the variabilities of the response function could be modeled using fuzzy set methods; e.g., high flow with

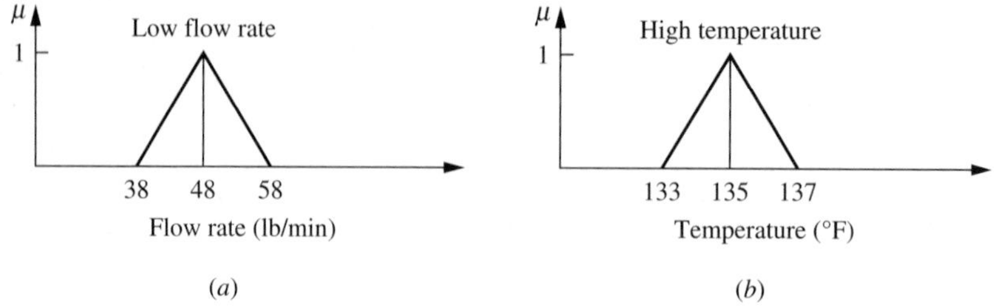

FIGURE 2.21
Foam insulation membership functions for (*a*) low flow rate and (*b*) high temperature.

high temperature, low flow with low temperature, etc. Examples are shown in Fig. 2.21, for low flow rate and high temperature.

The response function sensitivity to the fuzzy sets of flow and temperature could be evaluated using union, intersection, complement, difference, De Morgan's laws, and the excluded middle laws as they relate to the mathematical operations associated with the defined matrix of the DOE. Let us develop graphical representations of union and intersection.

Say we have two sets for flow and temperature, respectively:

$$\underset{\sim}{F} = \left\{ \frac{0}{1} + \frac{.5}{2} + \frac{1}{3} + \frac{.5}{4} + \frac{0}{5} \right\} \quad \text{and} \quad \underset{\sim}{D} = \left\{ \frac{0}{2} + \frac{1}{3} + \frac{0}{4} \right\}$$

The complements of the preceding two sets are

$$\overline{\underset{\sim}{F}} = \left\{ \frac{1}{1} + \frac{.5}{2} + \frac{0}{3} + \frac{.5}{4} + \frac{1}{5} \right\} \quad \text{and} \quad \overline{\underset{\sim}{D}} = \left\{ \frac{1}{2} + \frac{0}{3} + \frac{1}{4} \right\}$$

respectively. The union of $\underset{\sim}{F}$ and $\overline{\underset{\sim}{F}}$ is shown in Fig. 2.22 with bold lines, where the membership function of $\underset{\sim}{F}$ is represented by lines AB and BC, and the membership function of $\overline{\underset{\sim}{F}}$ is represented by lines PQ and QR. The intersection of $\underset{\sim}{D}$ and $\overline{\underset{\sim}{D}}$ is shown in Fig. 2.23 with bold lines, where the membership function of $\underset{\sim}{D}$ is represented by lines *AB* and *BC*, and the membership function of $\overline{\underset{\sim}{D}}$ is represented by lines *PQ* and *QR*. A three-dimensional image should be constructed when we take the union or intersection of sets from two different universes. For example, the intersection of $\underset{\sim}{F}$ and $\underset{\sim}{D}$ is given in Fig. 2.24. The idea of combining membership functions from two different universes

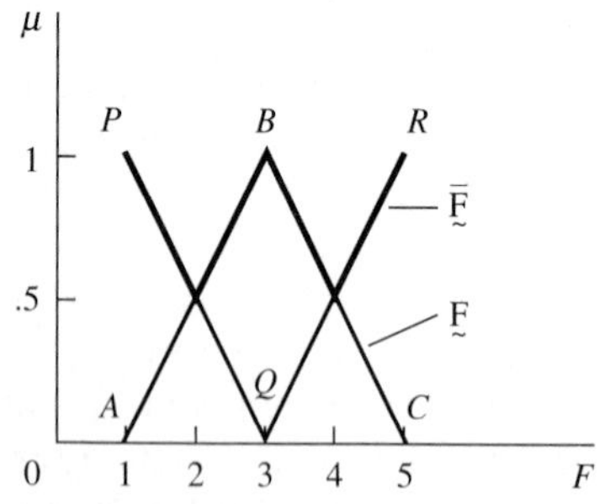

FIGURE 2.22
Membership functions for union of flow rates.

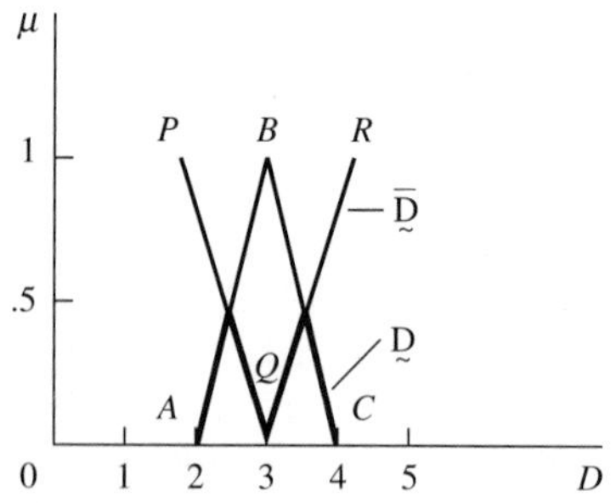

FIGURE 2.23
Membership functions for intersection of temperatures.

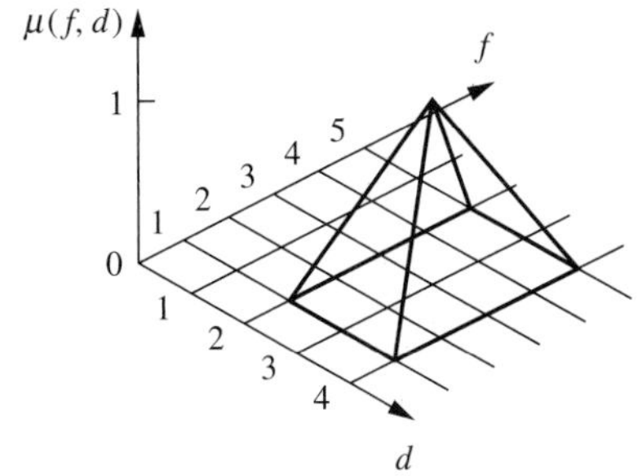

FIGURE 2.24
Three-dimensional image of the intersection of two fuzzy sets, i.e., $\underset{\sim}{F} \cap \underset{\sim}{D}$.

in an orthogonal form, as indicated in Fig. 2.24, will be shown in Chapter 3 to be associated with what is termed noninteractive fuzzy sets.

SETS AS POINTS IN HYPERCUBES

There is an interesting geometric analog for illustrating the idea of set membership [Kosko, 1992]. Heretofore we have described a fuzzy set $\underset{\sim}{A}$ defined on a universe X. For a universe with only one element, the membership function is defined on the unit interval [0, 1]; for a two-element universe, the membership function is defined on the unit square; and for a three-element universe, the membership function is defined on the unit cube. All of these situations are shown in Fig. 2.25. For a universe of n elements we define the membership on the unit hypercube, $\mathrm{I}^n = [0, 1]^n$.

The endpoints on the unit interval in Fig. 2.25*a*, and the vertices of the unit square and the unit cube in Figs. 2.25*b* and 2.25*c*, respectively, represent the possible crisp subsets, or collections, of the elements of the universe in each figure. This collection of possible crisp (nonfuzzy) subsets of elements in a universe constitutes the power set of the universe. For example, in Fig. 2.25*c* the universe comprises three elements, $\mathrm{X} = \{x_1, x_2, x_3\}$. The point (0, 0, 1) represents the crisp subset in 3-space, where x_1 and x_2 have no membership and element x_3 has full membership, i.e., the subset $\{x_3\}$; the point (1, 1, 0) is the crisp subset where x_1 and x_2 have full membership and element x_3 has no membership, i.e., the subset $\{x_1, x_2\}$; and so on for the other six vertices in Fig. 2.25*c*. In general, there are 2^n subsets in the power set of a universe with n elements; geometrically, this universe is represented by a hypercube in n-space, where the 2^n vertices represent the collection of sets constituting the power set. Two points in the diagrams bear special note, as illustrated in Fig. 2.25*c*. In this figure the point (1, 1, 1), where all elements in the universe have full membership, is called the whole set, X, and the point (0, 0, 0), where all elements in the universe have no membership, is called the null set, $\varnothing$.

The centroids of each of the diagrams in Fig. 2.25 represent single points where the membership value for each element in the universe equals $\frac{1}{2}$. For example, the point $(\frac{1}{2}, \frac{1}{2})$ in Fig. 2.25*b* is in the midpoint of the square. This midpoint in each of the three figures is a special point—it is the set of maximum "fuzziness." A membership value of $\frac{1}{2}$ indicates that the element belongs to the fuzzy set as much as it does not—that is, it holds equal membership in both the fuzzy set and its complement. In a geometric sense, this point is the location in the space that is farthest from any

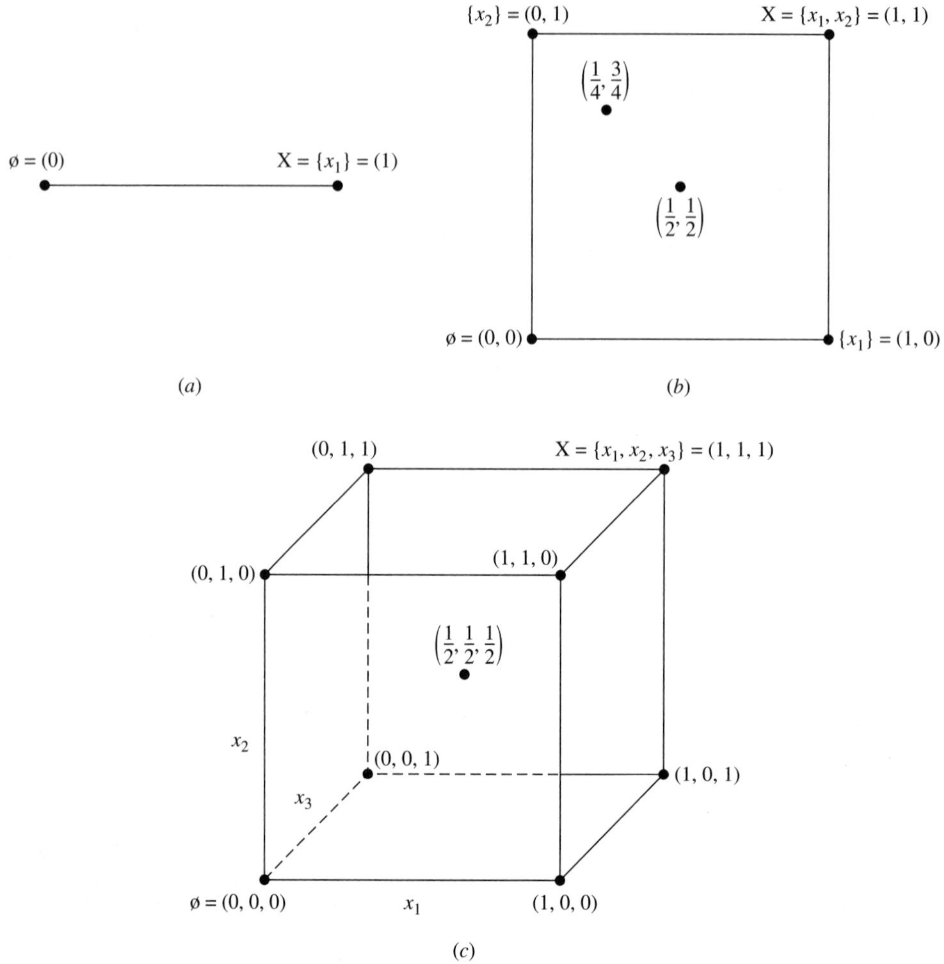

FIGURE 2.25
"Sets as points" [Kosko, 1992]: (*a*) one-element universe; (*b*) two-element universe; (*c*) three-element universe.

of the vertices and yet equidistant from all of them. In fact, all points interior to the vertices of the spaces represented in Fig. 2.25 represent fuzzy sets, where the membership value of each variable is a number between 0 and 1. For example, in Fig. 2.25*b*, the point $(\frac{1}{4}, \frac{3}{4})$ represents a fuzzy set where variable x_1 has a 0.25 degree of membership in the set and variable x_2 has a 0.75 degree of membership in the set. It is obvious by inspection of the diagrams in Fig. 2.25 that, although the number of subsets in the power set is enumerated by the 2^n vertices, the number of fuzzy sets on the universe is infinite, as represented by the infinite number of points on the interior of each space.

Finally, it can be seen that the vertices of the cube in Fig. 2.25*c* are the identical coordinates found in the value set, V{P(X)}, developed in Example 2.4 of this chapter.

SUMMARY

In this chapter we have developed the basic definitions for, properties of, and operations on crisp sets and fuzzy sets. It has been shown that the only basic axioms not common to both crisp and fuzzy sets are the two excluded middle laws. All other operations are common to both crisp and fuzzy sets. For many situations in reasoning, the excluded middle laws do present something of a constraint (see Chapter 15, Epilogue). Aside from the difference of set membership being an infinite-valued idea as opposed to a binary-valued quantity, fuzzy sets are handled and treated in the same mathematical form as are crisp sets. Again, the idea that crisp sets are special forms of fuzzy sets was illustrated graphically in the section on sets as points, where crisp sets are represented by the vertices of a unit hypercube. All other points within the unit hypercube, or along its edges, are graphically analogous to a fuzzy set.

REFERENCES

Bezdek, J. (1981). *Pattern recognition with fuzzy objective function algorithms,* Plenum Press, New York.

Dubois, D. and H. Prade. (1980). *Fuzzy sets and systems, theory and applications,* Academic, New York.

Gupta, M. and J. Qi. (1991). "Theory of T-norms and fuzzy inference methods," *Fuzzy Sets and Systems,* vol. 40, pp. 431–450.

Klir, G. and T. Folger. (1988). *Fuzzy sets, uncertainty, and information,* Prentice Hall, Englewood Cliffs, N.J.

Kosko, B. (1992). *Neural networks and fuzzy systems,* Prentice Hall, Englewood Cliffs, N.J.

Zimmermann, H. (1991). *Fuzzy set theory and its applications,* 2nd ed., Kluwer Academic Publishers, Dordrecht, Germany.

PROBLEMS

2.1. In the field of photography, the exposure time and development time of a negative are the two factors that determine how the negative will come out after processing. Define two fuzzy sets, $\underset{\sim}{A}$ = {exposure times of holographic plates}, and $\underset{\sim}{B}$ = {development times for the exposed plates}. The relative density, or relative darkness, of the processed plate, which varies between 0 and 1, is shown in Fig. P2.1. As shown by the fuzzy sets $\underset{\sim}{A}$ and $\underset{\sim}{B}$ in the figure, the two variables exposure time and development time complement each other in determining the density of the negative.

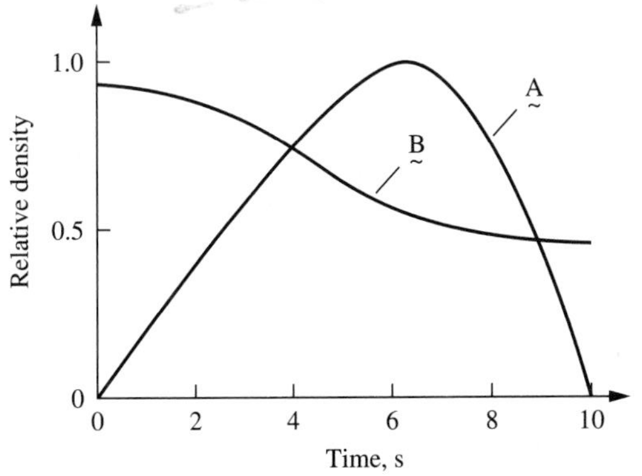

FIGURE P2.1

That is, if a negative is exposed for a shorter period of time, the plate can be developed for a longer time. Find the following graphically: (*a*) $\underset{\sim}{A} \cup \underset{\sim}{B}$, (*b*) $\underset{\sim}{A} \cap \underset{\sim}{B}$, (*c*) $\overline{\underset{\sim}{A}} \cup \underset{\sim}{B}$, (*d*) $\underset{\sim}{A} \cap \overline{\underset{\sim}{B}}$.

2.2. The speed of a hydraulic motor is a highly critical parameter, as it ultimately controls a volume of fluid to be displaced. A typical problem in the control of the hydraulic motor is that the loads placed on the motor can vary due to different circumstances. Define the load and speed as two fuzzy sets in the control of the hydraulic motor, with the membership functions shown in Fig. P2.2. Also assume that the load on the motor has an influence on the speed of the motor (finite torque output of the motor); e.g., when the motor load increases, the motor speed decreases, and vice versa. Graphically determine the following: (*a*) $\underset{\sim}{L} \cup \underset{\sim}{S}$, (*b*) $\underset{\sim}{L} \cap \underset{\sim}{S}$, (*c*) $\overline{\underset{\sim}{L}} \cup \underset{\sim}{S}$, (*d*) $\underset{\sim}{L} \cap \overline{\underset{\sim}{S}}$.

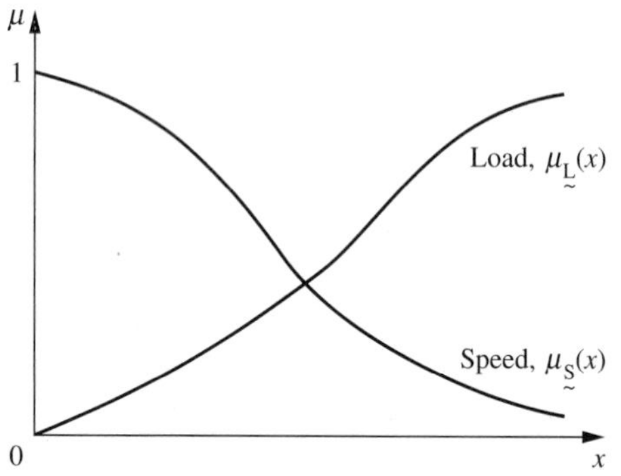

FIGURE P2.2

2.3. A common problem in statistics is to compare two sets of numbers to determine how "similar" the sets are. For example, to measure the likelihood that two sets of numbers were both generated by the same random process, one could compare the cumulative density functions (CDFs) for both processes as determined from the statistics of the two sets of numbers. Suppose the following two sets of numbers have been realized:

$$X = \{18, 3, 8, 2\} \qquad \text{and} \qquad Y = \{15, 8, 5\}$$

Define the CDFs of the two processes as membership functions as follows:

$$\mu_X(u) = \text{The proportion of the elements } x \text{ of set X such that } x \leq u$$
$$\mu_Y(u) = \text{The proportion of the elements } y \text{ of set Y such that } y \leq u$$

Since CDFs range on the same interval as membership functions, i.e., on the interval [0, 1], discuss a procedure by which you could employ basic fuzzy set operations to determine if two sets of numbers such as those just listed came from the same random process.

2.4. Given a set of measurements of the magnetic field near the surface of a person's head, we want to locate the electrical activity in the person's brain that would give rise to the measured magnetic field. This is called the inverse problem, and it has no unique solution. One approach is to model the electrical activity as dipoles and attempt to find one to four dipoles that would produce a magnetic field closely resembling the measured field. For this problem we will model the procedure a neuroscientist would use in attempting to fit a measured magnetic field using either one or two dipoles. The scientist uses a reduced chi-square statistic to determine how good the fit is. If R = 1.0, the fit is exact. If $R \geq 3$, the fit is bad. Also a two-dipole model must have a lower R than a one-dipole model to give the same amount of confidence in the model. The range of R will be taken as $R = \{1.0, 1.5, 2.0, 2.5, 3.0\}$ and we define the following fuzzy sets for D_1 = the one-dipole model and D_2 = the two-dipole model:

$$\underset{\sim}{D}_1 = \left\{ \frac{1}{1.0} + \frac{.75}{1.5} + \frac{.3}{2.0} + \frac{.15}{2.5} + \frac{0}{3.0} \right\}$$

$$\underset{\sim}{D}_2 = \left\{ \frac{1}{1.0} + \frac{.6}{1.5} + \frac{.2}{2.0} + \frac{.1}{2.5} + \frac{0}{3.0} \right\}$$

For these two fuzzy sets, find the following:

(*a*) $\underset{\sim}{D}_1 \cup \underset{\sim}{D}_2$ (*b*) $\underset{\sim}{D}_1 \cap \underset{\sim}{D}_2$
(*c*) $\overline{\underset{\sim}{D}_1}$ (*d*) $\overline{\underset{\sim}{D}_2}$
(*e*) $\underset{\sim}{D}_1 \mid \underset{\sim}{D}_2$ (*f*) $\overline{\underset{\sim}{D}_1 \cup \underset{\sim}{D}_2}$
(*g*) $\overline{\underset{\sim}{D}_1 \cap \underset{\sim}{D}_2}$ (*h*) $\underset{\sim}{D}_1 \cap \overline{\underset{\sim}{D}_1}$
(*i*) $\underset{\sim}{D}_1 \cup \overline{\underset{\sim}{D}_1}$ (*j*) $\underset{\sim}{D}_2 \cap \overline{\underset{\sim}{D}_2}$
(*k*) $\underset{\sim}{D}_2 \cup \overline{\underset{\sim}{D}_2}$

2.5. In determining credit card profitability, many lending institutions must make decisions based upon the particular consumer's spending habits, such as the amount the consumer spends and his or her capacity for spending. Many of these attributes are fuzzy. A person who charges a "large amount" is considered to be "profitable" to the credit card company. A "large" amount of charges is a fuzzy variable, as is a "profitable" return. These two fuzzy sets should have some overlap, but they should not be defined on an identical range.

$$\underset{\sim}{A} = \{\text{"large" spenders}\}$$
$$\underset{\sim}{B} = \{\text{"profitable" customers}\}$$

For the two fuzzy sets shown in Fig. P2.5, find the following properties graphically:

(*a*) $\underset{\sim}{A} \cup \underset{\sim}{B}$: all customers deemed profitable or who are large spenders.
(*b*) $\underset{\sim}{A} \cap \underset{\sim}{B}$: all customers deemed profitable and large spenders.
(*c*) $\overline{\underset{\sim}{A}}$ and $\overline{\underset{\sim}{B}}$: those persons (*i*) deemed not profitable, and (*ii*) deemed not large spenders (separately).
(*d*) $\overline{\underset{\sim}{A}} \mid \underset{\sim}{B}$: persons deemed profitable customers, but not large spenders.
(*e*) $\overline{\underset{\sim}{A} \cup \underset{\sim}{B}} = \overline{\underset{\sim}{A}} \cap \overline{\underset{\sim}{B}}$: (De Morgan's law).

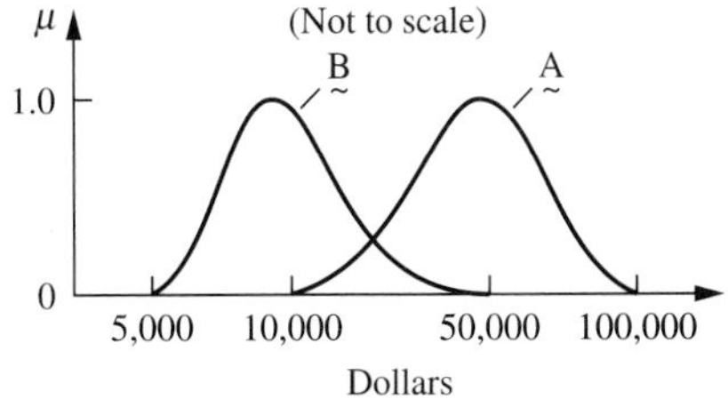

FIGURE P2.5

2.6. Suppose you are a soils engineer. You wish to track the movement of soil particles under strain in an experimental apparatus that allows viewing of the soil motion. You are building pattern recognition software to allow a computer to monitor and detect the motions. However, there are two difficulties in "teaching" your software to view the motion: (1) The tracked particle can be occluded by another particle; (2) your segmentation algorithm can be inadequate. One way to handle the occlusion is to assume that the area of the occluded particle is smaller than the area of the unoccluded particle. Therefore, when the area is changing you know that the particle is occluded. However, the segmentation algorithm also makes the area of the particle shrink if the edge detection scheme in the algorithm cannot do a good job because of poor illumination in the experimental apparatus. In other words, the area of the particle becomes small as a result of either occlusion or bad segmentation. You define two fuzzy sets: $\underset{\sim}{A}$ is a fuzzy set whose elements belong to the occlusion, and $\underset{\sim}{B}$ is a fuzzy set whose elements belong to inadequate segmentation. Let

$$\underset{\sim}{A} = \left\{ \frac{.1}{0} + \frac{.4}{1} + \frac{1}{2} + \frac{.3}{3} + \frac{.2}{4} \right\}$$

$$\underset{\sim}{B} = \left\{ \frac{.2}{0} + \frac{.5}{1} + \frac{1}{2} + \frac{.4}{3} + \frac{.1}{4} \right\}$$

Find the following:

(*a*) $\underset{\sim}{A} \cup \underset{\sim}{B}$ (*b*) $\underset{\sim}{A} \cap \underset{\sim}{B}$

(*c*) $\overline{\underset{\sim}{A}}$ (*d*) $\overline{\underset{\sim}{B}}$

(*e*) $\overline{\underset{\sim}{A} \cap \underset{\sim}{B}}$ (*f*) $\overline{\underset{\sim}{A} \cup \underset{\sim}{B}}$

2.7. You are asked to select an implementation technology for a numerical processor. Computation throughput is directly related to clock speed. Assume that all implementations will be in the same family (e.g., CMOS). You are considering whether the design should be implemented using medium-scale integration (MSI) with discrete parts, field-programmable array parts (FPGA), or multichip modules (MCM). Define the universe of potential clock speeds as $X = \{1, 10, 20, 40, 80, 100\}$ MHz; and define MSI, FPGA, and MCM as fuzzy sets of clock frequencies that should be implemented in each of these technologies. The following table defines the membership values for each of the three fuzzy sets. *Note:* The assignments made in this table reflect only the operational speed capability per cost ratio and do not include other factors.

Clock frequency, MHz	MSI	FPGA	MCM
1	1	0.3	0
10	0.7	1	0
20	0.4	1	0.5
40	0	0.5	0.7
80	0	0.2	1
100	0	0	1

Representing the three sets as MSI $= \underset{\sim}{M}$, FPGA $= \underset{\sim}{F}$, and MCM $= \underset{\sim}{C}$, find the following:

(*a*) $\underset{\sim}{M} \cup \underset{\sim}{F}$ (*b*) $\underset{\sim}{M} \cap \underset{\sim}{F}$

(*c*) $\overline{\underset{\sim}{M}}$ (*d*) $\overline{\underset{\sim}{F}}$

(*e*) $\underset{\sim}{C} \cap \overline{\underset{\sim}{F}}$ (*f*) $\overline{\underset{\sim}{M} \cap \underset{\sim}{C}}$

(*g*) $\underset{\sim}{F} \cup \overline{\underset{\sim}{F}}$ (*h*) $\overline{\underset{\sim}{C}} \cap \underset{\sim}{F}$

(*i*) $\underset{\sim}{M} \cap \overline{\underset{\sim}{C}}$ (*j*) $\underset{\sim}{M} \cup \overline{\underset{\sim}{M}}$

2.8. We want to compare two sensors based upon their detection levels and gain settings. The following table of gain settings and sensor detection levels with a standard item being monitored provides typical membership values to represent the detection levels for each of the sensors.

Gain setting	Sensor 1 detection level	Sensor 2 detection level
0	0	0
20	0.5	0.45
40	0.65	0.6
60	0.85	0.8
80	1	0.95
100	1	1

The universe of discourse is $X = \{0, 20, 40, 60, 80, 100\}$, and the membership functions for the two sensors in standard discrete form are

$$\underset{\sim}{\mathrm{S1}} = \left\{\frac{0}{0} + \frac{.5}{20} + \frac{.65}{40} + \frac{0.85}{60} + \frac{1.0}{80} + \frac{1.0}{100}\right\}$$

$$\underset{\sim}{\mathrm{S2}} = \left\{\frac{0}{0} + \frac{.45}{20} + \frac{.6}{40} + \frac{0.8}{60} + \frac{.95}{80} + \frac{1.0}{100}\right\}$$

Find the following membership functions using standard set operations:

(*a*) $\mu_{\mathrm{S1}\cup\mathrm{S2}}(x)$ (*b*) $\mu_{\mathrm{S1}\cap\mathrm{S2}}(x)$

(*c*) $\mu_{\overline{\mathrm{S1}}}(x)$ (*d*) $\mu_{\overline{\mathrm{S2}}}(x)$

(*e*) $\mu_{\overline{\mathrm{S1}\cup\mathrm{S1}}}(x)$ (*f*) $\mu_{\overline{\mathrm{S1}\cap\mathrm{S1}}}(x)$

(*g*) $\mu_{\mathrm{S1}\cup\overline{\mathrm{S1}}}(x)$ (*h*) $\mu_{\mathrm{S1}\cap\overline{\mathrm{S1}}}(x)$

(*i*) $\mu_{\mathrm{S2}\cup\overline{\mathrm{S2}}}(x)$ (*j*) $\mu_{\mathrm{S2}\cup\overline{\mathrm{S2}}}(x)$

2.9. For flight simulator data the determination of certain changes in operating conditions of the aircraft is made on the basis of hard breakpoints in the mach region. Let us define a fuzzy set to represent the condition of "near" a mach number of 0.74. Further, define a second fuzzy set to represent the condition of "in the region of" a mach number of 0.74. In typical simulation data a mach number of 0.74 is a hard breakpoint.

$$\underset{\sim}{A} = \text{near mach } 0.74 = \left\{\frac{0}{0.730} + \frac{.8}{0.735} + \frac{1}{0.740} + \frac{.6}{0.745} + \frac{0}{0.750}\right\}$$

$$\underset{\sim}{B} = \text{in the region of mach } 0.74 = \left\{\frac{0}{0.730} + \frac{.4}{0.735} + \frac{0.8}{0.740} + \frac{1}{0.745} + \frac{0.6}{0.750}\right\}$$

For these two fuzzy sets find the following:

(*a*) $\underset{\sim}{A} \cup \underset{\sim}{B}$ (*b*) $\underset{\sim}{A} \cap \underset{\sim}{B}$

(*c*) $\overline{\underset{\sim}{A}}$ (*d*) $\underset{\sim}{A} \mid \underset{\sim}{B}$

(*e*) $\overline{\underset{\sim}{A} \cup \underset{\sim}{B}}$ (*f*) $\overline{\underset{\sim}{A} \cap \underset{\sim}{B}}$

2.10. A system component is tested on a drop table in the time domain, t, to shock loads of haversine pulses of various acceleration amplitudes, $\ddot{x}$, as shown in Fig. P2.10*a*. After the test the component is evaluated for damage. Define two fuzzy sets, "Passed" $= \underset{\sim}{P}$ and "Failed" $= \underset{\sim}{F}$. These sets are defined on the real line of $|\ddot{x}|$, which is the magnitude of the input pulse (see Fig. P2.10*b*). We define the following set operations:

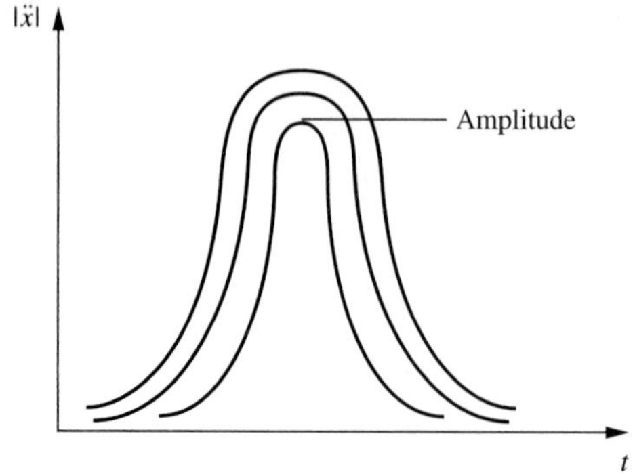

FIGURE P2.10*a*

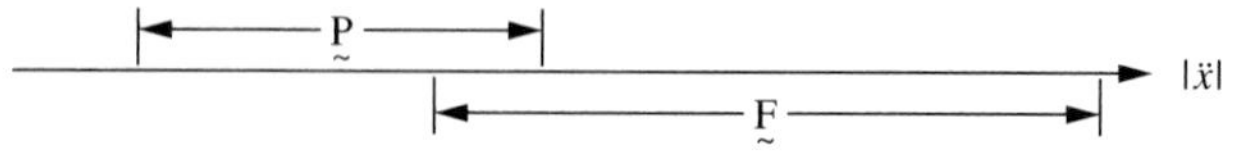

FIGURE P2.10*b*

(*a*) $\underset{\sim}{F} \cup \underset{\sim}{P} = \{|\ddot{x}|\}$: the universe of input shock level results.
(*b*) $\underset{\sim}{F} \cap \underset{\sim}{P}$: the portion of the universe where the component could either fail or pass.
(*c*) $\overline{\underset{\sim}{F}}$: portion of universe that definitely passed.
(*d*) $\overline{\underset{\sim}{P}}$: portion of universe that definitely failed.
(*e*) $\underset{\sim}{F} \mid \underset{\sim}{P}$: the portion of the failed set that definitely failed.

De Morgan's laws for

$$\left.\begin{aligned} \overline{\mathrm{F} \cap \mathrm{P}} &= \overline{\mathrm{F}} \cup \overline{\mathrm{P}} \\ \overline{\mathrm{F} \cup \mathrm{P}} &= \overline{\mathrm{F}} \cap \overline{\mathrm{P}} = \varnothing \end{aligned}\right\} \quad \text{(the portion of the results that either definitely passed or failed)}$$

$$\underset{\sim}{P} \cup \overline{\underset{\sim}{P}} \neq \{|\ddot{x}|\}$$

$$\underset{\sim}{P} \cap \overline{\underset{\sim}{P}} \neq \varnothing$$

Define suitable membership functions for the two fuzzy sets $\underset{\sim}{F}$ and $\underset{\sim}{P}$ and determine the operations just described.

2.11. You are assigned the task of identifying images in an overhead reconnaissance photograph. You design computer software to do image processing to locate objects within a scene. Define two fuzzy sets representing a car and a truck image:

$$\underset{\sim}{\mathrm{Car}} = \left\{ \frac{0.5}{\text{truck}} + \frac{0.4}{\text{motorcycle}} + \frac{0.3}{\text{boat}} + \frac{0.9}{\text{car}} + \frac{0.1}{\text{house}} \right\}$$

$$\underset{\sim}{\mathrm{Truck}} = \left\{ \frac{1}{\text{truck}} + \frac{0.1}{\text{motorcycle}} + \frac{0.4}{\text{boat}} + \frac{0.4}{\text{car}} + \frac{0.2}{\text{house}} \right\}$$

Find the following:

(*a*) $\underset{\sim}{\mathrm{Car}} \cup \underset{\sim}{\mathrm{Truck}}$ (*b*) $\underset{\sim}{\mathrm{Car}} \cap \underset{\sim}{\mathrm{Truck}}$
(*c*) $\overline{\underset{\sim}{\mathrm{Car}}}$ (*d*) $\overline{\underset{\sim}{\mathrm{Truck}}}$
(*e*) $\underset{\sim}{\mathrm{Car}} \mid \underset{\sim}{\mathrm{Truck}}$ (*f*) $\overline{\underset{\sim}{\mathrm{Car}} \cup \underset{\sim}{\mathrm{Truck}}}$
(*g*) $\overline{\underset{\sim}{\mathrm{Car}} \cap \underset{\sim}{\mathrm{Truck}}}$ (*h*) $\underset{\sim}{\mathrm{Car}} \cup \overline{\underset{\sim}{\mathrm{Car}}}$

(*i*) $\underset{\sim}{\text{Car}} \cap \overline{\underset{\sim}{\text{Car}}}$ (*j*) $\underset{\sim}{\text{Truck}} \cup \overline{\underset{\sim}{\text{Truck}}}$

(*k*) $\underset{\sim}{\text{Truck}} \cap \overline{\underset{\sim}{\text{Truck}}}$

2.12. Consider a local area network (LAN) of interconnected workstations that communicate using Ethernet protocols at a maximum rate of 10 Mbit/s. Traffic rates on the network can be expressed as the peak value of the total bandwidth (BW) used; and the two fuzzy variables, "Quiet" and "Congested," can be used to describe the perceived loading of the LAN. If the discrete universal set X = {0, 1, 2, 5, 7, 9, 10} represents bandwidth usage, then the membership grades of these elements in the fuzzy sets Quiet $\underset{\sim}{Q}$ and Congested $\underset{\sim}{C}$ are as shown in the table and in Fig. P2.12.

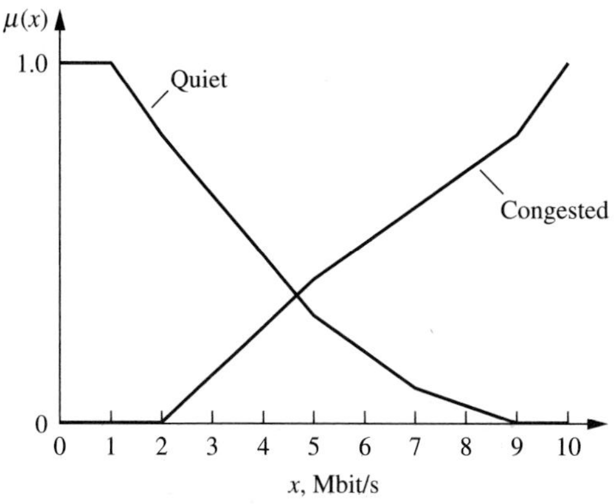

FIGURE P2.12

x(BW), Mbit/s	$\mu_{\underset{\sim}{Q}}(x)$	$\mu_{\underset{\sim}{C}}(x)$
0	1.0	0.0
1	1.0	0.0
2	0.8	0.0
5	0.3	0.4
7	0.1	0.6
9	0.0	0.8
10	0.0	1.0

For these two fuzzy sets, find the union, intersection, complement of each, difference $\underset{\sim}{Q} \mid \underset{\sim}{C}$, both De Morgan's laws, and both excluded middle laws, (*a*) graphically and (*b*) numerically.

2.13. You are asked to solve a problem in the area of security equipment, including surveillance cameras and monitors. You are to make a video camera that can run off either $120V_{ac}$/60Hz or $120V_{ac}$/50Hz. You will need a circuit that can detect the AC line frequency, to determine whether the camera should implement the European PAL video standard (for 50-Hz systems) or the U.S. NTSC video standard (for 60-Hz systems). You propose two fuzzy variables: Fuzzy set $\underset{\sim}{A}$ is "about 60 Hz" and fuzzy set $\underset{\sim}{B}$ is "about 50 Hz." The universe for both of these fuzzy variables is the crisp set of possible line frequencies, which theoretically includes DC (0 Hz) to infinite frequency. You choose a continuous, triangular membership function for both sets $\underset{\sim}{A}$ and $\underset{\sim}{B}$. Fuzzy variable $\underset{\sim}{A}$ has a prototypical value of 60 Hz, and fuzzy variable $\underset{\sim}{B}$ has a prototypical value of 50 Hz. You make the base widths for both $\underset{\sim}{A}$ and $\underset{\sim}{B}$ 16 Hz: this value makes the two membership functions overlap, since between 50 Hz and 60 Hz it is not crucial what the memberships are, and it makes the problem nontrivial. You also make the triangular membership functions symmetric about the prototypical value. Define the two membership functions as follows:

For fuzzy set $\underset{\sim}{A}$: $$\mu_{\underset{\sim}{A}}(x) = \frac{|60 - x|}{8} + 1$$

For fuzzy set $\underset{\sim}{B}$: $$\mu_{\underset{\sim}{B}}(x) = \frac{|50 - x|}{8} + 1$$

From these two sets, find the following:

(*a*) $\underset{\sim}{A} \cup \underset{\sim}{B}$ (*b*) $\underset{\sim}{A} \cap \underset{\sim}{B}$
(*c*) $\overline{\underset{\sim}{A}}$ (*d*) $\underset{\sim}{A} \mid \underset{\sim}{B}$
(*e*) $\overline{\underset{\sim}{A} \cup \underset{\sim}{B}}$ (*f*) $\overline{\underset{\sim}{A} \cap \underset{\sim}{B}}$

2.14. You have been given the task of developing a glass break detector/discriminator for use with residential alarm systems. The detector should be able to distinguish between the breaking of a pane of a glass (a window) and a drinking glass. From extensive study and analysis you have determined that the sound of a shattering window pane contains most of its energy at frequencies centered about 4 kHz whereas the sound of a shattering drinking glass contains most of its energy at frequencies centered about 8 kHz. The spectra of the two shattering sounds overlap. The membership functions for the window pane and the glass are given as $\mu_{\underset{\sim}{A}}(x)$ and $\mu_{\underset{\sim}{B}}(x)$, respectively. Illustrate the basic operations of union, intersection, complement, difference, De Morgan's laws, and the excluded middle laws for the following membership functions.

$$x = 0, 1, \ldots, 10 \qquad \sigma = 2 \qquad \mu_{\underset{\sim}{A}} = 4 \qquad \mu_{\underset{\sim}{B}} = 8$$

$$\mu_{\underset{\sim}{A}}(x) = \exp\left[\frac{-(x - \mu_{\underset{\sim}{A}})^2}{2\sigma^2}\right] \qquad \mu_{\underset{\sim}{B}}(x) = \exp\left[\frac{-(x - \mu_{\underset{\sim}{B}})^2}{2\sigma^2}\right]$$

2.15. Samples of a new microprocessor IC chip are to be sent to several customers for beta testing. The chips are sorted to meet certain maximum electrical characteristics, say frequency and temperature rating, so that the "best" chips are distributed to preferred customer 1. Suppose that each sample chip is screened and all chips are found to have a maximum operating frequency in the range 7–15 MHz at 20°C. Also, the maximum operating temperature range (20°C ± ΔT) at 8 MHz is determined. Suppose there are eight sample chips with the following electrical characteristics:

	Chip number							
	1	**2**	**3**	**4**	**5**	**6**	**7**	**8**
f_{max}, MHz	6	7	8	9	10	11	12	13
ΔT_{max}, °C	0	0	20	40	30	50	40	60

The following fuzzy sets are defined:

$$\underset{\sim}{A} = \text{set of "fast" chips} = \text{chips with } f_{max} \geq 12 \text{ MHz}$$
$$= \left\{\frac{0}{1} + \frac{0}{2} + \frac{0}{3} + \frac{0}{4} + \frac{0.2}{5} + \frac{0.6}{6} + \frac{1}{7} + \frac{1}{8}\right\}$$

$$\underset{\sim}{B} = \text{set of "slow" chips} = \text{chips with } f_{max} \geq 8 \text{ MHz}$$
$$= \left\{ \frac{0.1}{1} + \frac{0.5}{2} + \frac{1}{3} + \frac{1}{4} + \frac{1}{5} + \frac{1}{6} + \frac{1}{7} + \frac{1}{8} \right\}$$

$$\underset{\sim}{C} = \text{set of "cold" chips} = \text{chips with } \Delta T_{max} \geq 10°\text{C}$$
$$= \left\{ \frac{0}{1} + \frac{0}{2} + \frac{1}{3} + \frac{1}{4} + \frac{1}{5} + \frac{1}{6} + \frac{1}{7} + \frac{1}{8} \right\}$$

$$\underset{\sim}{D} = \text{set of "hot" chips} = \text{chips with } \Delta T_{max} \geq 50°\text{C}$$
$$= \left\{ \frac{0}{1} + \frac{0}{2} + \frac{0}{3} + \frac{0.5}{4} + \frac{0.1}{5} + \frac{1}{6} + \frac{0.5}{7} + \frac{1}{8} \right\}$$

Use these four fuzzy sets to illustrate various set operations. For example, the following operations relate the sets of "fast" and "hot" chips.

(*a*) $\underset{\sim}{A} \cup \underset{\sim}{D}$ (*b*) $\underset{\sim}{A} \cap \underset{\sim}{D}$
(*c*) $\overline{\underset{\sim}{A}}$ (*d*) $\underset{\sim}{A} \mid \underset{\sim}{D}$
(*e*) $\overline{\underset{\sim}{A} \cup \underset{\sim}{D}}$ (*f*) $\overline{\underset{\sim}{A} \cap \underset{\sim}{D}}$

2.16. We have a full test database containing articles from several popular news magazines like *Newsweek, U.S. News & World Report, Time,* etc. We want to use fuzzy sets to access linguistic information from our database. Let us say that the user enters four words, w_1, w_2, w_3, and w_4. Also, let us say we have partitioned our databases into two broad or general topics, $\underset{\sim}{T}_1$ and $\underset{\sim}{T}_2$. The following table illustrates the correlation between the words and the topics. Membership values close to 1 mean there is a strong correlation between a given word in a database and the associated topic; values close to 0 mean there is little correlation.

	$\underset{\sim}{T}_1$	$\underset{\sim}{T}_2$
w_1	0	0.9
w_2	0.3	0.7
w_3	0.5	0.1
w_4	0.8	0

For the two fuzzy sets ($\underset{\sim}{T}_1$ and $\underset{\sim}{T}_2$) given in the table determine the membership for various set operations.

2.17. Suppose you are to address a problem in the power control of a mobile cellular telephone transmitting to its base station. Let $\underset{\sim}{MP}$ be the medium-power fuzzy set, and $\underset{\sim}{HP}$ be the high-power set. Let the universe of discourse be comprised of discrete units of db · m, i.e., X = {0, 1, 2, ..., 10}. The membership functions for these two fuzzy sets are shown in Fig. P2.17. For these two fuzzy sets, demonstrate union, intersection, complement, difference, De Morgan's laws, and the excluded middle laws.

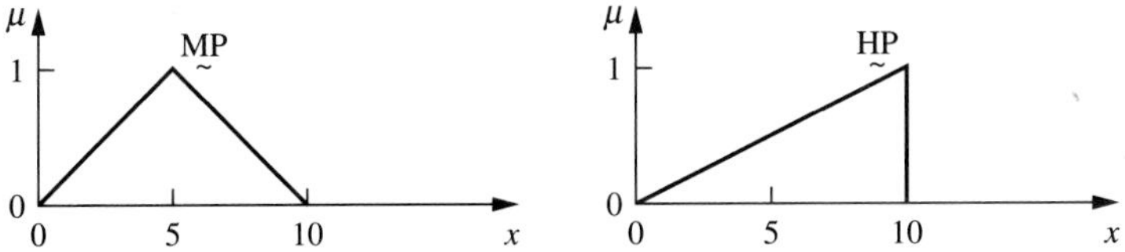

FIGURE P2.17

2.18. You are assigned a problem in communications protocols and control in embedded systems. In communications protocols two variables of interest are throughput (the number of bits of information transferred per second) and the number of nodes allowed on the communications bus. The following communications protocols are found in embedded systems and have the properties listed in the following table.

Protocol name	**Throughput, Kbit/s**	**Number of nodes on network**
CAN (controller automated network)	1,000	2,032
Standard modem	9.6	2
Echelon Lonworks	1,250	32,000
CEBus (Consumer Electronic bus)	10	32,000
SP50 (field bus)	500	31

Three fuzzy sets for throughput are defined as Fast, Medium, and Slow. Three fuzzy sets for the number of nodes on the network are defined as Few, Some, and Many. These fuzzy sets are shown graphically in Fig. P2.18. For these fuzzy sets determine various set operations graphically.

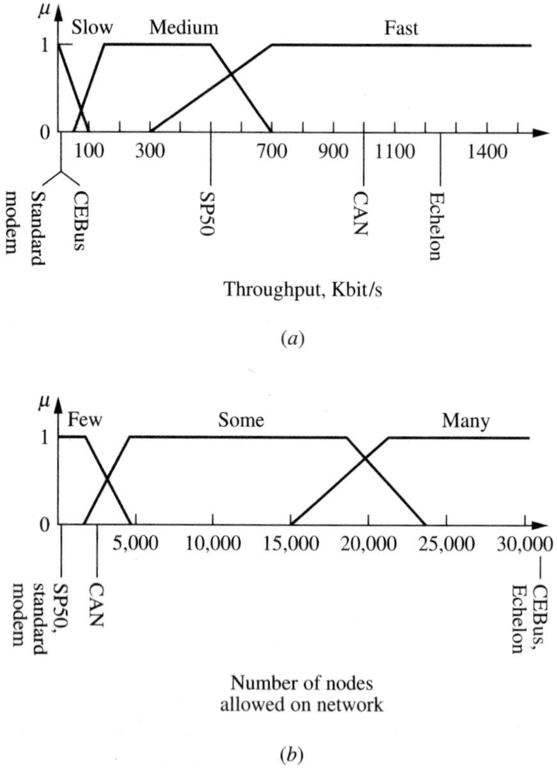

FIGURE P2.18

2.19. Let X be the universe of military aircraft of interest, as defined here:

$$\mathrm{X} = \{\mathrm{a10, b52, b117, c5, c130, f4, f14, f15, f16, f111, kc130}\}$$

Let $\underset{\sim}{\mathrm{A}}$ be the fuzzy set of bomber class aircraft:

$$\underset{\sim}{\mathrm{A}} = \left\{ \frac{0.2}{\mathrm{f16}} + \frac{0.4}{\mathrm{f4}} + \frac{0.5}{\mathrm{a10}} + \frac{0.5}{\mathrm{f14}} + \frac{0.6}{\mathrm{f15}} + \frac{0.8}{\mathrm{f111}} + \frac{1.0}{\mathrm{b117}} + \frac{1.0}{\mathrm{b52}} \right\}$$

Let $\underset{\sim}{\mathrm{B}}$ be the fuzzy set of fighter class aircraft:

$$\underset{\sim}{\mathrm{B}} = \left\{ \frac{0.1}{\mathrm{b117}} + \frac{0.3}{\mathrm{f111}} + \frac{0.5}{\mathrm{f4}} + \frac{0.8}{\mathrm{f15}} + \frac{0.9}{\mathrm{f14}} + \frac{1.0}{\mathrm{f16}} \right\}$$

Find the following various set combinations for these two sets:

(*a*) $\underset{\sim}{\mathrm{A}} \cup \underset{\sim}{\mathrm{B}}$ (*b*) $\underset{\sim}{\mathrm{A}} \cap \underset{\sim}{\mathrm{B}}$
(*c*) $\overline{\underset{\sim}{\mathrm{A}}}$ (*d*) $\overline{\underset{\sim}{\mathrm{B}}}$
(*e*) $\underset{\sim}{\mathrm{A}} \mid \underset{\sim}{\mathrm{B}}$ (*f*) $\underset{\sim}{\mathrm{B}} \mid \underset{\sim}{\mathrm{A}}$
(*g*) $\overline{\underset{\sim}{\mathrm{A}} \cup \underset{\sim}{\mathrm{B}}}$ (*h*) $\overline{\underset{\sim}{\mathrm{A}} \cap \underset{\sim}{\mathrm{B}}}$
(*i*) $\overline{\underset{\sim}{\mathrm{A}}} \cup \underset{\sim}{\mathrm{A}}$

2.20. Suppose the discretized membership functions (in nondimensional units) for a transistor and a resistor, shown in continuous form in Fig. P2.20, are as given in the following equations.

$$\mu_{\underset{\sim}{\mathrm{T}}} = \left\{ \frac{0}{0} + \frac{.3}{1} + \frac{.7}{2} + \frac{.8}{3} + \frac{.9}{4} + \frac{1}{5} \right\}$$

$$\mu_{\underset{\sim}{\mathrm{R}}} = \left\{ \frac{0}{0} + \frac{.1}{1} + \frac{.2}{2} + \frac{.3}{3} + \frac{.4}{4} + \frac{.5}{5} \right\}$$

For these two fuzzy sets perform the following calculations:

(*a*) $\mu_{\underset{\sim}{\mathrm{T}}} \vee \mu_{\underset{\sim}{\mathrm{R}}}$
(*b*) $\mu_{\underset{\sim}{\mathrm{T}}} \wedge \mu_{\underset{\sim}{\mathrm{R}}}$
(*c*) $\overline{\mu}_{\underset{\sim}{\mathrm{T}}} = 1 - \mu_{\underset{\sim}{\mathrm{T}}}$
(*d*) $\overline{\mu}_{\underset{\sim}{\mathrm{R}}} = 1 - \mu_{\underset{\sim}{\mathrm{R}}}$
(*e*) De Morgan's law: $\overline{\mu_{\underset{\sim}{\mathrm{T}}} \wedge \mu_{\underset{\sim}{\mathrm{R}}}} = \overline{\mu_{\underset{\sim}{\mathrm{T}}}} \vee \overline{\mu_{\underset{\sim}{\mathrm{R}}}}$

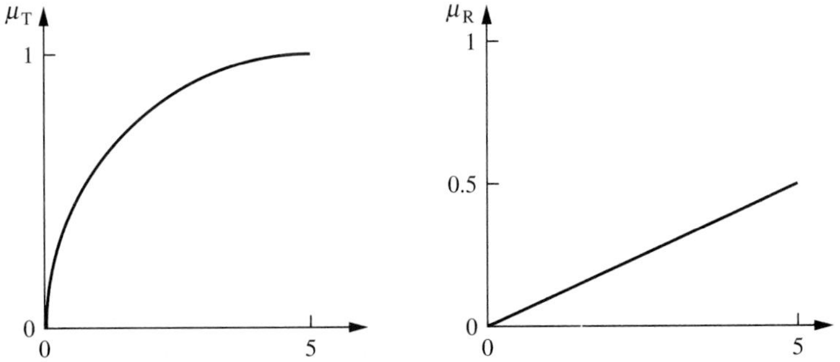

FIGURE P2.20

2.21. Enumerate the nonfuzzy subsets of the power set for a universe with $n = 4$ elements, i.e., $\mathrm{X} = \{x_1, x_2, x_3, x_4\}$, and indicate their coordinates as the vertices on a 4-cube.

CHAPTER 3

CLASSICAL RELATIONS AND FUZZY RELATIONS

It is the mark of an instructed mind to rest satisfied with that degree of precision which the nature of the subject admits, and not to seek exactness where only an approximation of the truth is possible.

Aristotle, 384–322 B.C.
Ancient Greek Philosopher

This chapter introduces the notion of a relation as the basic idea behind numerous operations on sets such as Cartesian products, composition of relations, and equivalence properties. Like a set, a relation is of fundamental importance in all engineering, science, and mathematically based fields. It is also associated with graph theory, a subject of wide impact in design and data manipulation.

Understanding relations is central to the understanding of a great many areas addressed in this textbook. Relations are intimately involved in logic, approximate reasoning, rule-based systems, nonlinear simulation, classification, pattern recognition, and control. Relations will be referred to repeatedly in this text in many different applications areas. Relations represent mappings for sets just as mathematical functions do; relations are also very useful in representing connectives in logic (see Chapter 7).

This chapter begins by describing Cartesian products as a means of producing ordered relationships among sets. Following this is an introduction to classical (crisp) relations—structures that represent the presence or absence of correlation, interaction, or propinquity between the elements of two or more crisp sets; in this case, a set could also be the universe. There are only two degrees of relationship between elements of the sets in a crisp relation: the relationships "completely related" and "not related," in a binary sense. Basic operations, properties, and the cardinality of relations are explained and illustrated. Two composition methods to relate elements of three or more universes are illustrated.

Fuzzy relations are then developed by allowing the relationship between elements of two or more sets to take on an infinite number of degrees of relationship between the extremes of "completely related" and "not related." In this sense, fuzzy relations are to crisp relations as fuzzy sets are to crisp sets; crisp sets and relations are constrained realizations of fuzzy sets and relations. Operations, properties, and cardinality of fuzzy relations are introduced and illustrated, as are Cartesian products and compositions of fuzzy relations. Some engineering examples are given to illustrate various issues associated with relations. The reader can consult the literature for more details on relations [for example, Gill, 1976; Dubois and Prade, 1980; Kandel, 1985; Klir and Folger, 1988; Zadeh, 1971].

This chapter contains a section on tolerance and equivalence relations—both classical and fuzzy—which is introduced for use in later chapters of the book. Both tolerance and equivalence relations are illustrated with some examples. Finally, the chapter concludes with a section on value assignments, which discusses various methods to develop the elements of relations. These various methods are discussed, and a few examples are given in the area of similarity methods.

CARTESIAN PRODUCT

An ordered sequence of r elements, written in the form $(a_1, a_2, a_3, \ldots, a_r)$, is called an ordered r-tuple; an unordered r-tuple is simply a collection of r elements without restrictions on order. In a ubiquitous special case where $r = 2$, the r-tuple is referred to as an ordered *pair.* For crisp sets $A_1, A_2, \ldots, A_r$, the set of all r-tuples $(a_1, a_2, a_3, \ldots, a_r)$, where $a_1 \in A_1$, $a_2 \in A_2$, and $a_r \in A_r$, is called the *Cartesian product* of $A_1, A_2, \ldots, A_r$, and is denoted by $A_1 \times A_2 \times \cdots \times A_r$. The Cartesian product of two or more sets is *not* the same thing as the arithmetic product of two or more sets. The latter will be dealt with in Chapter 6, when the extension principle is introduced.

When all the A_r are identical and equal to A, the Cartesian product $A_1 \times A_2 \times \cdots \times A_r$ can be denoted as A^r.

Example 3.1. The elements in two sets A and B are given as $A = \{0, 1\}$ and $B = \{a, b, c\}$. Various Cartesian products of these two sets can be written as shown:

$$A \times B = \{(0, a), (0, b), (0, c), (1, a), (1, b), (1, c)\}$$

$$B \times A = \{(a, 0), (a, 1), (b, 0), (b, 1), (c, 0), (c, 1)\}$$

$$A \times A = A^2 = \{(0, 0), (0, 1), (1, 0), (1, 1)\}$$

$$B \times B = B^2 = \{(a, a), (a, b), (a, c), (b, a), (b, b), (b, c), (c, a), (c, b), (c, c)\}$$

CRISP RELATIONS

A subset of the Cartesian product $A_1 \times A_2 \times \cdots \times A_r$ is called an *r-ary relation* over $A_1, A_2, \ldots, A_r$. Again, the most common case is for $r = 2$; in this situation the relation is a subset of the Cartesian product $A_1 \times A_2$ (i.e., a set of pairs, the first coordinate of which is from A_1 and the second from A_2). This subset of the full Cartesian product is called a *binary relation from* A_1 *into* A_2. If three, four, or five sets are involved in a subset of the full Cartesian product, the relations are called ternary, quaternary, and quinary, respectively. In this text, whenever the term *relation* is used without qualification, it is taken to mean a *binary relation.*

The Cartesian product of two universes X and Y is determined as

$$X \times Y = \{(x, y) \mid x \in X, y \in Y\} \tag{3.1}$$

which forms an ordered pair of every $x \in X$ with every $y \in Y$, forming *unconstrained* matches between X and Y. That is, every element in universe X is related completely to every element in universe Y. The *strength* of this relationship between ordered pairs of elements in each universe is measured by the characteristic function, denoted χ, where a value of unity is associated with *complete relationship* and a value of zero is associated with *no relationship;* i.e.,

$$\chi_{X\times Y}(x, y) = \begin{cases} 1, & (x, y) \in X \times Y \\ 0, & (x, y) \notin X \times Y \end{cases} \tag{3.2}$$

One can think of this strength of relation as a mapping from ordered pairs of the universe, or ordered pairs of sets defined on the universes, to the characteristic function. When the universes, or sets, are finite the relation can be conveniently represented by a matrix, called a *relation matrix.* An *r*-ary relation can be represented by an *r*-dimensional relation matrix. Hence, binary relations can be represented by two-dimensional matrices (used throughout this text).

An example of the strength of relation for the unconstrained case is given in the Sagittal diagram shown in Fig. 3.1 (a Sagittal diagram is simply a schematic depicting points as elements of universes, and lines as relationships between points). Lines in the Sagittal diagram and values of unity in the *relation matrix,*

$$\mathrm{R} = \begin{array}{c} \\ 1 \\ 2 \\ 3 \end{array} \begin{array}{c} \begin{matrix} a & b & c \end{matrix} \\ \begin{bmatrix} 1 & 1 & 1 \\ 1 & 1 & 1 \\ 1 & 1 & 1 \end{bmatrix} \end{array}$$

correspond to the ordered pairs of mappings in the relation. Here, the elements in the two universes are defined as $X = \{1, 2, 3\}$ and $Y = \{a, b, c\}$.

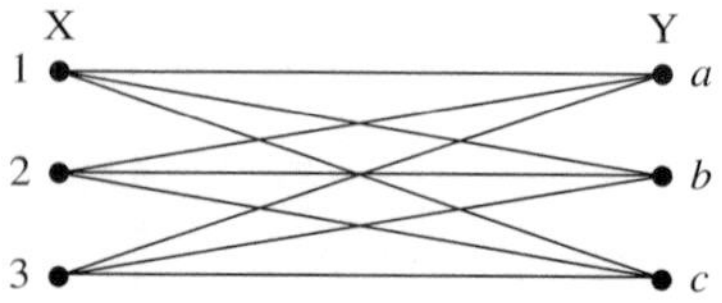

FIGURE 3.1
Sagittal diagram of an unconstrained relation.

A more general crisp relation, R, exists when matches between elements in two universes are *constrained.* Again, the characteristic function is used to assign values of relationship in the mapping of the Cartesian space X × Y to the binary values of (0, 1).

$$\chi_R(x, y) = \begin{cases} 1, & (x, y) \in R \\ 0, & (x, y) \notin R \end{cases} \tag{3.3}$$

Example 3.2. In many biological models, members of certain species can reproduce only with certain members of another species. Hence, only some elements in two or more universes have a relationship (nonzero) in the Cartesian product. An example is shown in Fig. 3.2 for two 2-member species, i.e., for $X = \{1, 2\}$ and for $Y = \{a, b\}$. In this case the locations of zeros in the relation matrix,

$$R = \{(1, a), (2, b)\} \qquad R \subset X \times Y$$

and the absence of lines in the Sagittal diagram correspond to pairs of elements between the two universes where there is "no relation"; i.e., the strength of the relationship is zero.

Special cases of the constrained and the unconstrained Cartesian product for sets where $r = 2$ (i.e., for A^2) are called the *identity relation* and the *universal relation,* respectively. For example, for $A = \{0, 1, 2\}$ the universal relation, denoted U_A, and the identity relation, denoted I_A, are found to be

$$U_A = \{(0, 0), (0, 1), (0, 2), (1, 0), (1, 1), (1, 2), (2, 0), (2, 1), (2, 2)\}$$
$$I_A = \{(0, 0), (1, 1), (2, 2)\}$$

Example 3.3. Relations can also be defined for continuous universes. Consider, for example, the continuous relation defined by the following expression:

$$R = \{(x, y) \mid y \geq 2x, x \in X, y \in Y\}$$

which is also given in function-theoretic form using the characteristic function as

$$\chi_R(x, y) = \begin{cases} 1, & y \geq 2x \\ 0, & y < 2x \end{cases}$$

Graphically, this relation is equivalent to the shaded region shown in Fig. 3.3.

Cardinality of Crisp Relations

Suppose n elements of the universe X are related (paired) to m elements of the universe Y. If the cardinality of X is n_X and the cardinality of Y is n_Y, then the cardinality of the relation, R, between these two universes is $n_{X\times Y} = n_X * n_Y$. The cardinality of the power set describing this relation, $P(X \times Y)$ is then $n_{P(X\times Y)} = 2^{(n_X n_Y)}$.

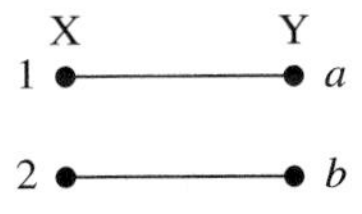

FIGURE 3.2
Relation matrix and Sagittal diagram for a constrained relation.

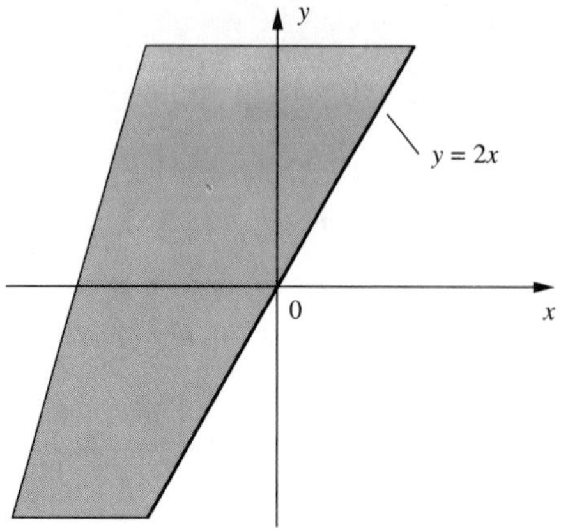

FIGURE 3.3
Relation corresponding to the expression $y \geq 2x$.

Operations on Crisp Relations

Define R and S as two separate relations on the Cartesian universe X × Y, and define the null relation and the complete relation as the relation matrices O and E, respectively. An example of a 4 × 4 form of the O and E matrices is given here:

$$\mathbf{O} = \begin{bmatrix} 0 & 0 & 0 & 0 \\ 0 & 0 & 0 & 0 \\ 0 & 0 & 0 & 0 \\ 0 & 0 & 0 & 0 \end{bmatrix} \qquad \mathbf{E} = \begin{bmatrix} 1 & 1 & 1 & 1 \\ 1 & 1 & 1 & 1 \\ 1 & 1 & 1 & 1 \\ 1 & 1 & 1 & 1 \end{bmatrix}$$

The following function-theoretic operations for the two crisp relations (R, S) can now be defined.

Union $\quad \mathrm{R} \cup \mathrm{S} \rightarrow \chi_{\mathrm{R}\cup\mathrm{S}}(x, y) : \chi_{\mathrm{R}\cup\mathrm{S}}(x, y) = \max[\chi_{\mathrm{R}}(x, y), \chi_{\mathrm{S}}(x, y)]$ (3.4)

Intersection $\quad \mathrm{R} \cap \mathrm{S} \rightarrow \chi_{\mathrm{R}\cap\mathrm{S}}(x, y) : \chi_{\mathrm{R}\cap\mathrm{S}}(x, y) = \min[\chi_{\mathrm{R}}(x, y), \chi_{\mathrm{S}}(x, y)]$ (3.5)

Complement $\quad \overline{\mathrm{R}} \rightarrow \chi_{\overline{\mathrm{R}}}(x, y) : \chi_{\overline{\mathrm{R}}}(x, y) = 1 - \chi_{\overline{\mathrm{R}}}(x, y)$ (3.6)

Containment $\quad \mathrm{R} \subset \mathrm{S} \rightarrow \chi_{\mathrm{R}}(x, y) : \chi_{\mathrm{R}}(x, y) \leq \chi_{\mathrm{S}}(x, y)$ (3.7)

Identity $\quad (\varnothing \rightarrow \mathbf{O} \text{ and } \mathrm{X} \rightarrow \mathbf{E})$ (3.8)

Properties of Crisp Relations

The properties of commutativity, associativity, distributivity, involution, and idempotency all hold for crisp relations just as they do for classical set operations. Moreover, *De Morgan's laws* and the *excluded middle laws* also hold for crisp (classical) relations just as they do for crisp (classical) sets. The null relation, **O**, and the complete relation, **E**, are analogous to the null set, ∅, and the whole set, X, in the set-theoretic case (see Chapter 2).

Composition

Let R be a relation that relates, or maps, elements from universe X to universe Y, and let S be a relation that relates, or maps, elements from universe Y to universe Z.

A useful question we seek to answer is whether we can find a relation, T, that relates the same elements in universe X that R contains to the same elements in universe Z that S contains. It turns out we can find such a relation using an operation known as *composition*. For the Sagittal diagram in Fig. 3.4, we see that the only "path" between relation R and relation S is the two routes that start at X_1 and end at Z_2 (i.e., X_1-Y_1-Z_2 and X_1-Y_3-Z_2). Hence, we wish to find a relation T that relates the ordered pair (X_1, Z_2); i.e., $(X_1, Z_2) \in T$. In this example,

$$R = \{(X_1, Y_1), (X_1, Y_3), (X_2, Y_4)\}$$
$$S = \{(Y_1, Z_2), (Y_3, Z_2)\}$$

There are two common forms of the composition operation; one is called the max-min composition and the other the max-product composition. (Several other forms of the composition operator are available for certain logic issues; these are described in Chapter 7.) The max-min composition is defined by the set-theoretic and membership function–theoretic expressions,

$$\begin{aligned} T &= R \circ S \\ \chi_T(x, z) &= \bigvee_{y \in Y} (\chi_R(x, y) \wedge \chi_S(y, z)) \end{aligned} \tag{3.9}$$

and the max-product (sometimes called max-dot) composition is defined by the set-theoretic and membership function–theoretic expressions,

$$\begin{aligned} T &= R \circ S \\ \chi_T(x, z) &= \bigvee_{y \in Y} (\chi_R(x, y) \bullet \chi_S(y, z)) \end{aligned} \tag{3.10}$$

There is a very interesting physical analogy for the max-min composition operator. Figure 3.5 illustrates a system comprising several chains placed together in a parallel fashion. In the system, each chain comprises a number of chain links. If we were to take one of the chains out of the system, place it in a tensile test machine, and exert a large tensile force on the chain, we would find that the chain would break at its weakest link. Hence, the strength of one chain is equal to the strength of its weakest link; in other words, the *minimum* ($\wedge$) strength of all the links in the chain governs the strength of the overall chain. Now, if we were to place the entire chain system in a tensile device and exert a tensile force on the chain system, we would find that the chain system would continue to carry increasing loads until the last chain in the system broke. That is, weaker chains would break with an increasing load until the strongest chain was left alone, and eventually it would break; in other

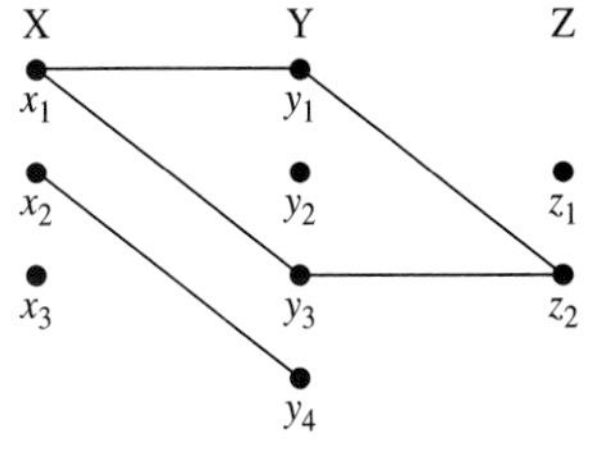

FIGURE 3.4
Sagittal diagram relating elements of three universes.

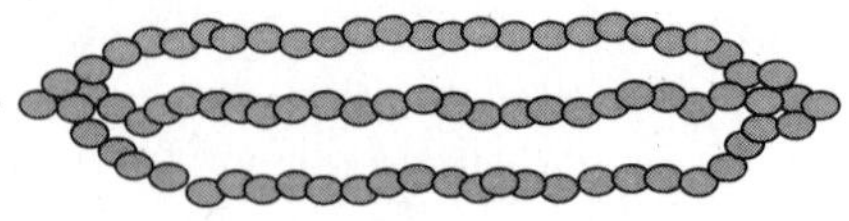

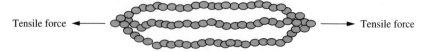

FIGURE 3.5
Chain strength analogy for max-min composition.

words, the *maximum* ($\vee$) strength of all the chains in the chain system would govern the overall strength of the chain system. Each chain in the system is analogous to the min operation in the max-min composition, and the overall chain system strength is analogous to the max operation in the max-min composition.

Example 3.4. The matrix expression for the crisp relations shown in Fig. 3.4 can be found using the max-min composition operation. Relation matrices for R and S would be expressed as

$$\mathrm{R} = \begin{array}{c} \\ x_1 \\ x_2 \\ x_3 \end{array} \begin{array}{c} \begin{array}{cccc} y_1 & y_2 & y_3 & y_4 \end{array} \\ \begin{bmatrix} 1 & 0 & 1 & 0 \\ 0 & 0 & 0 & 1 \\ 0 & 0 & 0 & 0 \end{bmatrix} \end{array} \quad \text{and} \quad \mathrm{S} = \begin{array}{c} \\ y_1 \\ y_2 \\ y_3 \\ y_4 \end{array} \begin{array}{c} \begin{array}{cc} z_1 & z_2 \end{array} \\ \begin{bmatrix} 0 & 1 \\ 0 & 0 \\ 0 & 1 \\ 0 & 0 \end{bmatrix} \end{array}$$

The resulting relation T would then be determined by max-min composition, Eq. (3.9), or max-product composition, Eq. (3.10). (In the crisp case these forms of the composi-ton operators produce identical results; other forms of this operator, such as those listed in Chapter 7, will not produce identical results.) For example,

$$\mu_{\mathrm{T}}(x_1, z_1) = \max[\min(1, 0), \min(0, 0), \min(1, 0), \min(0, 0)] = 0$$
$$\mu_{\mathrm{T}}(x_1, z_2) = \max[\min(1, 1), \min(0, 0), \min(1, 1), \min(0, 0)] = 1$$

and for the rest,

$$\mathrm{T} = \begin{array}{c} \\ x_1 \\ x_2 \\ x_3 \end{array} \begin{array}{c} \begin{array}{cc} z_1 & z_2 \end{array} \\ \begin{bmatrix} 0 & 1 \\ 0 & 0 \\ 0 & 0 \end{bmatrix} \end{array}$$

FUZZY RELATIONS

Fuzzy relations also map elements of one universe, say X, to those of another universe, say Y, through the Cartesian product of the two universes. However, the "strength" of the relation between ordered pairs of the two universes is not measured with the characteristic function, but rather with a membership function expressing various "degrees" of strength of the relation on the unit interval [0,1]. Hence, a fuzzy relation $\underset{\sim}{\mathrm{R}}$ is a mapping from the Cartesian space $\mathrm{X} \times \mathrm{Y}$ to the interval [0,1], where the strength of the mapping is expressed by the membership function of the relation for ordered pairs from the two universes, or $\mu_{\underset{\sim}{\mathrm{R}}}(x, y)$.

Cardinality of Fuzzy Relations

Since the cardinality of fuzzy sets on any universe is infinity, the cardinality of a fuzzy relation between two or more universes is also infinity.

Operations on Fuzzy Relations

Let $\underset{\sim}{R}$ and $\underset{\sim}{S}$ be fuzzy relations on the Cartesian space X × Y. Then the following operations apply for the membership values for various set operations:

Union $\mu_{\underset{\sim}{R}\cup\underset{\sim}{S}}(x, y) = \max(\mu_{\underset{\sim}{R}}(x, y), \mu_{\underset{\sim}{S}}(x, y))$ (3.11)

Intersection $\mu_{\underset{\sim}{R}\cap\underset{\sim}{S}}(x, y) = \min(\mu_{\underset{\sim}{R}}(x, y), \mu_{\underset{\sim}{S}}(x, y))$ (3.12)

Complement $\mu_{\overline{\underset{\sim}{R}}}(x, y) = 1 - \mu_{\underset{\sim}{R}}(x, y)$ (3.13)

Containment $\underset{\sim}{R} \subset \underset{\sim}{S} \Rightarrow \mu_{\underset{\sim}{R}}(x, y) \leq \mu_{\underset{\sim}{S}}(x, y)$ (3.14)

Properties of Fuzzy Relations

Just as for crisp relations, the properties of commutativity, associativity, distributivity, involution, and idempotency all hold for fuzzy relations. Moreover, *De Morgan's laws* hold for fuzzy relations just as they do for crisp (classical) relations, and the null relation, **O**, and the complete relation, **E**, are analogous to the null set and the whole set in set-theoretic form, respectively. The properties that do not hold for fuzzy relations, as is the case for fuzzy sets in general, are the *excluded middle laws*. Since a fuzzy relation $\underset{\sim}{R}$ is also a fuzzy set, there is overlap between a relation and its complement, hence,

$$\underset{\sim}{R} \cup \overline{\underset{\sim}{R}} \neq \mathbf{E}$$
$$\underset{\sim}{R} \cap \overline{\underset{\sim}{R}} \neq \mathbf{O}$$

As seen in the foregoing expressions, the *excluded middle laws* for relations do not result in the null relation, **O**, or the complete relation, **E**.

Fuzzy Cartesian Product and Composition

Because fuzzy relations in general are fuzzy sets, we can define the Cartesian product to be a relation between two or more fuzzy sets. Let $\underset{\sim}{A}$ be a fuzzy set on universe X and $\underset{\sim}{B}$ be a fuzzy set on universe Y; then the Cartesian product between fuzzy sets $\underset{\sim}{A}$ and $\underset{\sim}{B}$ will result in a fuzzy relation $\underset{\sim}{R}$, which is contained within the full Cartesian product space, or

$$\underset{\sim}{A} \times \underset{\sim}{B} = \underset{\sim}{R} \subset X \times Y \tag{3.15}$$

where the fuzzy relation $\underset{\sim}{R}$ has membership function

$$\mu_{\underset{\sim}{R}}(x, y) = \mu_{\underset{\sim}{A}\times\underset{\sim}{B}}(x, y) = \min(\mu_{\underset{\sim}{A}}(x), \mu_{\underset{\sim}{B}}(y)) \tag{3.16}$$

The Cartesian product defined by $\underset{\sim}{A} \times \underset{\sim}{B} = \underset{\sim}{R}$, Eq. (3.15), is implemented in the same fashion as is the cross product of two vectors. Again, the Cartesian product is *not* the same operation as the arithmetic product. In the case of two-dimensional relations ($r = 2$), the former employs the idea of pairing of elements among sets, whereas the latter uses actual arithmetic products between elements of sets. More will be given on arithmetic products in Chapter 6. Each of the fuzzy sets could be thought of as a vector of membership values; each value is associated with a particular element in each set. For example, for a fuzzy set (vector) $\underset{\sim}{A}$ that has four elements, hence column vector of size 4×1, and for a fuzzy set (vector) $\underset{\sim}{B}$ that has five elements, hence a row vector size of 1×5, the resulting fuzzy relation, $\underset{\sim}{R}$, will be represented by a matrix of size 4×5; i.e., $\underset{\sim}{R}$ will have four rows and five columns. This result is illustrated in the following example.

Example 3.5. Suppose we have two fuzzy sets, $\underset{\sim}{A}$ defined on a universe of three discrete temperatures, $X = \{x_1, x_2, x_3\}$, and $\underset{\sim}{B}$ defined on a universe of two discrete pressures, $Y = \{y_1, y_2\}$, and we want to find the fuzzy Cartesian product between them. Fuzzy set $\underset{\sim}{A}$ could represent the "ambient" temperature and fuzzy set $\underset{\sim}{B}$ the "near optimum" pressure for a certain heat exchanger, and the Cartesian product might represent the conditions (temperature-pressure pairs) of the exchanger that are associated with "efficient" operations. For example, let

$$\underset{\sim}{A} = \frac{0.2}{x_1} + \frac{0.5}{x_2} + \frac{1}{x_3} \quad \text{and} \quad \underset{\sim}{B} = \frac{0.3}{y_1} + \frac{0.9}{y_2}$$

Note that $\underset{\sim}{A}$ can be represented as a column vector of size 3×1 and $\underset{\sim}{B}$ can be represented by a row vector of 1×2. Then the fuzzy Cartesian product, using Eq. (3.16), results in a fuzzy relation $\underset{\sim}{R}$ (of size 3×2) representing "efficient"conditions, or

$$\underset{\sim}{A} \times \underset{\sim}{B} = \underset{\sim}{R} = \begin{matrix} x_1 \\ x_2 \\ x_3 \end{matrix} \overset{\begin{matrix} y_1 & \;\; y_2 \end{matrix}}{\begin{bmatrix} 0.2 & 0.2 \\ 0.3 & 0.5 \\ 0.3 & 0.9 \end{bmatrix}}$$

Fuzzy composition can be defined just as it is for crisp (binary) relations. Suppose $\underset{\sim}{R}$ is a fuzzy relation on the Cartesian space $X \times Y$, $\underset{\sim}{S}$ is a fuzzy relation on $Y \times Z$, and $\underset{\sim}{T}$ is a fuzzy relation on $X \times Z$; then fuzzy max-min composition is defined in terms of the set-theoretic notation and membership function–theoretic notation in the following manner:

$$\underset{\sim}{T} = \underset{\sim}{R} \circ \underset{\sim}{S}$$

$$\mu_{\underset{\sim}{T}}(x, z) = \bigvee_{y \in Y} (\mu_{\underset{\sim}{R}}(x, y) \wedge \mu_{\underset{\sim}{S}}(y, z)) \tag{3.17a}$$

and fuzzy max-product composition is defined in terms of the membership function–theoretic notation as

$$\mu_{\underset{\sim}{T}}(x, z) = \bigvee_{y \in Y} (\mu_{\underset{\sim}{R}}(x, y) \bullet \mu_{\underset{\sim}{S}}(y, z)) \tag{3.17b}$$

It should be pointed out that neither crisp nor fuzzy compositions have converses in general; that is,

$$\underset{\sim}{R} \circ \underset{\sim}{S} \neq \underset{\sim}{S} \circ \underset{\sim}{R} \tag{3.18}$$

Equation (3.18) is general for any matrix operation, fuzzy or otherwise, that must satisfy consistency between the cardinal counts of elements in respective universes. Even for the case of square matrices, the composition converse, represented by Eq. (3.18), is not guaranteed.

Example 3.6. Let us extend the information contained in the Sagittal diagram shown in Fig. 3.4 to include fuzzy relationships for X × Y (denoted by the fuzzy relation $\underset{\sim}{R}$) and Y × Z (denoted by the fuzzy relation $\underset{\sim}{S}$). In this case we have

$$X = \{x_1, x_2\}, \qquad Y = \{y_1, y_2\}, \qquad \text{and} \qquad Z = \{z_1, z_2, z_3\}$$

Consider the following fuzzy relations:

$$\underset{\sim}{R} = \begin{matrix} & y_1 & y_2 \\ x_1 & 0.7 & 0.5 \\ x_2 & 0.8 & 0.4 \end{matrix} \qquad \text{and} \qquad \underset{\sim}{S} = \begin{matrix} & z_1 & z_2 & z_3 \\ y_1 & 0.9 & 0.6 & 0.2 \\ y_2 & 0.1 & 0.7 & 0.5 \end{matrix}$$

Then the resulting relation, $\underset{\sim}{T}$, which relates elements of universe X to elements of universe Z, i.e., defined on Cartesian space X × Z, can be found by max-min composition, Eq. (3.17*a*), to be, for example,

$$\mu_{\underset{\sim}{T}}(x_1, z_1) = \max[\min(0.7, 0.9), \min(0.5, 0.1)] = 0.7$$

and the rest,

$$\underset{\sim}{T} = \begin{matrix} & z_1 & z_2 & z_3 \\ x_1 & 0.7 & 0.6 & 0.5 \\ x_2 & 0.8 & 0.6 & 0.4 \end{matrix}$$

and by max-product composition, Eq. (3.17*b*), to be, for example,

$$\mu_{\underset{\sim}{T}}(x_2, z_2) = \max[(0.8 \cdot 0.6), (0.4 \cdot 0.7)] = 0.48$$

and the rest,

$$\underset{\sim}{T} = \begin{matrix} & z_1 & z_2 & z_3 \\ x_1 & 0.63 & 0.42 & 0.25 \\ x_2 & 0.72 & 0.48 & 0.20 \end{matrix}$$

Noninteractive Fuzzy Sets

Later in the text, especially in Chapters 6 and 9, we will make reference to noninteractive fuzzy sets. Noninteractive sets in fuzzy set theory can be thought of as being analogous to independent events in probability theory. They always arise in the context of relations or in n-dimensional mappings [Zadeh, 1975; Bandemer and Näther, 1992]. A noninteractive fuzzy set can be defined as follows. Suppose we define a fuzzy set $\underset{\sim}{A}$ on the Cartesian space $X = X_1 \times X_2$. The set $\underset{\sim}{A}$ is separable into two *noninteractive* fuzzy sets, called its orthogonal projections, if and only if

$$\underset{\sim}{A} = \Pr_{X_1}(\underset{\sim}{A}) \times \Pr_{X_2}(\underset{\sim}{A}) \tag{3.19a}$$

where

$$\mu_{\Pr_{X_1}(\underset{\sim}{A})}(x_1) = \max_{x_2 \in X_2} \mu_{\underset{\sim}{A}}(x_1, x_2), \qquad \forall x_1 \in X_1 \tag{3.19b}$$

$$\mu_{\Pr_{X_2}(\underset{\sim}{A})}(x_2) = \max_{x_1 \in X_1} \mu_{\underset{\sim}{A}}(x_1, x_2), \qquad \forall x_2 \in X_2 \tag{3.19c}$$

are the membership functions for the projections of $\underset{\sim}{A}$ on universes X_1 and X_2, respectively. Hence, if Eq. (3.19*a*) holds for a fuzzy set, the membership functions

$\mu_{\text{Pr}_{X_1}(\underset{\sim}{A})}(x_1)$ and $\mu_{\text{Pr}_{X_2}(\underset{\sim}{A})}(x_2)$ describe noninteractive fuzzy sets, i.e., the projections are noninteractive fuzzy sets.

Separability or noninteractivity of fuzzy set $\underset{\sim}{A}$ describes a kind of independence of the components (x_1 and x_2): $\underset{\sim}{A}$ can be uniquely reconstructed by its projections; the components of the fuzzy set $\underset{\sim}{A}$ can vary without consideration of the other components. As an example, the two-dimensional planar fuzzy set shown in Fig. 2.24 in Chapter 2 comprises noninteractive fuzzy sets ($\underset{\sim}{F}$ and $\underset{\sim}{D}$), because it was constructed by the Cartesian product (intersection in this case) of the two fuzzy sets $\underset{\sim}{F}$ and $\underset{\sim}{D}$, whereas a two-dimensional fuzzy set comprising curved surfaces will be nonseparable; i.e., its components will be interactive. Interactive components are characterized by the fact that variation of one component depends on the values of the other components. See Fig. 3.6. We now illustrate the use of relations with fuzzy sets for three examples from the fields of electrical, computer, and industrial engineering.

Example 3.7. Suppose we are interested in understanding the speed control of the DC (direct current) shunt motor under no-load condition, as shown diagramatically in Fig. 3.7. Initially, the series resistance R_{se} in Fig. 3.7 should be kept in the cut-in position for the following reasons:

1. The back electromagnetic force, given by $E_b = kN\phi$, where k is a constant of proportionality, N is the motor speed, and ϕ is the flux (which is proportional to input voltage, V), is equal to zero because the motor speed is equal to zero initially.
2. We have $V = E_b + I_a(R_a + R_{se})$, therefore, $I_a = (V - E_b)/(R_a + R_{se})$, where I_a is the armature current and R_a is the armature resistance. Since E_b is equal to zero initially, the armature current will be $I_a = V/(R_a + R_{se})$, which is going to be quite large initially and may destroy the armature.

On the basis of both cases 1 and 2, keeping the series resistance R_{se} in the cut-in position will restrict the speed to a very low value. Hence, if the rated no-load speed of the motor is 1500 rpm, then the resistance in series with the armature, or the shunt resistance R_{sh}, has to be varied.

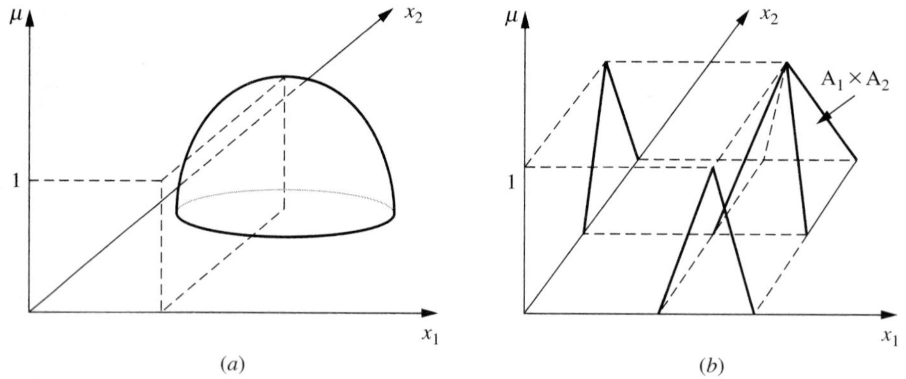

FIGURE 3.6
Fuzzy sets: (*a*) interactive and (*b*) noninteractive.

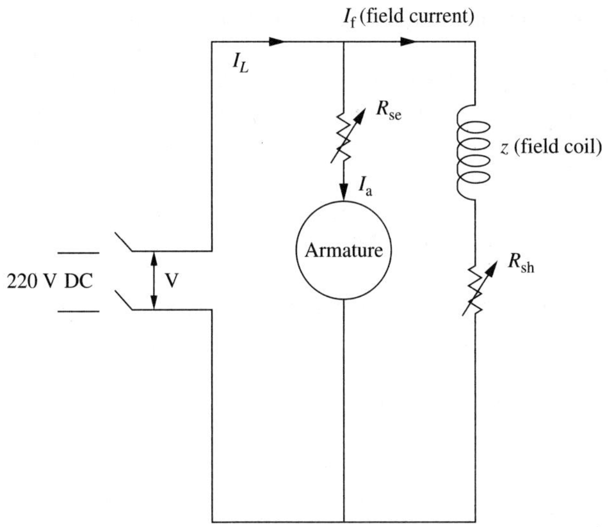

FIGURE 3.7
A DC shunt motor system.

Two methods provide this type of control: armature control and field control. For example, in armature control, suppose that ϕ (flux) is maintained at some constant value; then motor speed N is proportional to E_b.

If R_{se} is decreased step by step from its high value, I_a (armature current) increases. Hence, this method increases I_a. On the other hand, as I_a is increased the motor speed N increases. These two possible approaches to control could have been done manually or automatically. Either way, however, results in at least two problems, presuming we don't want to change the design of the armature:

- What should be the minimum and maximum level of R_{se}?
- What should be the minimum and maximum value of I_a?

Now let us suppose that load on the motor is taken into consideration. Then the problem of control becomes two-fold. First, owing to fluctuations in the load, the armature current may change, resulting in change in the motor speed. Second, as a result of changes in speed, the armature resistance control must be accomplished in order to maintain the motor's rated speed. Such control issues become very important in applications involving electric trains and a large number of consumer appliances making use of small batteries to run their motors.

We wish to use concepts of fuzzy sets to address this problem. Let $\underset{\sim}{R}_{se}$ be a fuzzy set representing a number of possible values for series resistance, say s_n values, given by

$$\underset{\sim}{R}_{s_1} = \{R_{s1}, R_{s2}, R_{s3}, \ldots, R_{s_n}\}$$

and let $\underset{\sim}{I}_a$ be a fuzzy set having a number of possible values of the armature current, say m values, given by

$$\underset{\sim}{I}_a = \{I_1, I_2, I_3, \ldots, I_m\}$$

The fuzzy sets $\underset{\sim}{\mathrm{R}}_{se}$ and $\underset{\sim}{\mathrm{I}}_a$ can be related through a fuzzy relation, say $\underset{\sim}{\mathrm{R}}$, which would allow for the establishment of various degrees of relationship between pairs of resistance and current. In this way, the resistance-current pairings could conform to the modeler's intuition about the trade-offs involved in control of the armature.

Let $\underset{\sim}{\mathrm{N}}$ be another fuzzy set having numerous values for the motor speed, say v values, given by

$$\underset{\sim}{\mathrm{N}} = \{N_1, N_2, N_3, \ldots, N_v\}$$

Now, we can determine another fuzzy relation, say $\underset{\sim}{\mathrm{S}}$, to relate current to motor speed, i.e., $\underset{\sim}{\mathrm{I}}_a$ to $\underset{\sim}{\mathrm{N}}$.

Using the operation of composition, we could then compute a relation, say $\underset{\sim}{\mathrm{T}}$, to be used to relate series resistance to motor speed, i.e., $\underset{\sim}{\mathrm{R}}_{se}$ to $\underset{\sim}{\mathrm{N}}$. The operations needed to develop these relations are as follows—two fuzzy cross products and one composition:

$$\underset{\sim}{\mathrm{R}} = \underset{\sim}{\mathrm{R}}_{se} \times \underset{\sim}{\mathrm{I}}_a$$
$$\underset{\sim}{\mathrm{S}} = \underset{\sim}{\mathrm{I}}_a \times \underset{\sim}{\mathrm{N}}$$
$$\underset{\sim}{\mathrm{T}} = \underset{\sim}{\mathrm{R}} \circ \underset{\sim}{\mathrm{S}}$$

Suppose the membership functions for both series resistance $\underset{\sim}{\mathrm{R}}_{se}$ and armature current $\underset{\sim}{\mathrm{I}}_a$ are given in terms of percentages *of their respective rated values,* i.e.,

$$\mu_{\mathrm{R_{se}}}(\%\mathrm{se}) = \frac{0.3}{30} + \frac{0.7}{60} + \frac{1.0}{100} + \frac{0.2}{120}$$

and

$$\mu_{\mathrm{I_a}}(\%\mathrm{a}) = \frac{0.2}{20} + \frac{0.4}{40} + \frac{0.6}{60} + \frac{0.8}{80} + \frac{1.0}{100} + \frac{0.1}{120}$$

and the membership value for $\underset{\sim}{\mathrm{N}}$ is given in units of motor speed in rpm,

$$\mu_{\mathrm{N}}(\mathrm{rpm}) = \frac{0.33}{500} + \frac{0.67}{1000} + \frac{1.0}{1500} + \frac{0.15}{1800}$$

The following relations then result from use of the Cartesian product to determine $\underset{\sim}{\mathrm{R}}$ and $\underset{\sim}{\mathrm{S}}$:

$$\underset{\sim}{\mathrm{R}} = \begin{array}{c} \\ 30 \\ 60 \\ 100 \\ 120 \end{array} \begin{array}{c} \begin{array}{cccccc} 20 & 40 & 60 & 80 & 100 & 120 \end{array} \\ \begin{bmatrix} 0.2 & 0.3 & 0.3 & 0.3 & 0.3 & 0.1 \\ 0.2 & 0.4 & 0.6 & 0.7 & 0.7 & 0.1 \\ 0.2 & 0.4 & 0.6 & 0.8 & 1 & 0.1 \\ 0.2 & 0.2 & 0.2 & 0.2 & 0.2 & 0.1 \end{bmatrix} \end{array}$$

and

$$\underset{\sim}{\mathrm{S}} = \begin{array}{c} \\ 20 \\ 40 \\ 60 \\ 80 \\ 100 \\ 120 \end{array} \begin{array}{c} \begin{array}{cccc} 500 & 1000 & 1500 & 1800 \end{array} \\ \begin{bmatrix} 0.2 & 0.2 & 0.2 & 0.15 \\ 0.33 & 0.4 & 0.4 & 0.15 \\ 0.33 & 0.6 & 0.6 & 0.15 \\ 0.33 & 0.67 & 0.8 & 0.15 \\ 0.33 & 0.67 & 1 & 0.15 \\ 0.1 & 0.1 & 0.1 & 0.1 \end{bmatrix} \end{array}$$

For example, $\mu_{\underset{\sim}{\mathrm{R}}}(60, 40) = \min(0.7, 0.4) = 0.4$, $\mu_{\underset{\sim}{\mathrm{R}}}(100, 80) = \min(1.0, 0.8) = 0.8$, and $\mu_{\underset{\sim}{\mathrm{S}}}(80, 1000) = \min(0.8, 0.67) = 0.67$.

The following relation results from a max-min composition for $\underset{\sim}{T}$:

$$\underset{\sim}{T} = \underset{\sim}{R} \circ \underset{\sim}{S} = \begin{array}{c} \\ 30 \\ 60 \\ 100 \\ 120 \end{array} \begin{array}{c} \begin{array}{cccc} 500 & 1000 & 1500 & 1800 \end{array} \\ \begin{bmatrix} 0.3 & 0.3 & 0.3 & 0.15 \\ 0.33 & 0.67 & 0.7 & 0.15 \\ 0.33 & 0.67 & 1 & 0.15 \\ 0.2 & 0.2 & 0.2 & 0.15 \end{bmatrix} \end{array}$$

For instance,

$$\begin{aligned} \mu_{\underset{\sim}{T}}(60, 1500) &= \max[\min(0.2, 0.2), \min(0.4, 0.4), \min(0.6, 0.6), \\ &\quad \min(0.7, 0.8), \min(0.7, 1.0), \min(0.1, 0.1)] \\ &= \max[0.2, 0.4, 0.6, 0.7, 0.7, 0.1] = 0.7 \end{aligned}$$

Example 3.8. In computer engineering, different logic families are often compared on the basis of their power-delay product. Consider the fuzzy set $\underset{\sim}{F}$ of logic families, the fuzzy set $\underset{\sim}{D}$ of delay times (in nanoseconds), and the fuzzy set $\underset{\sim}{P}$ of power dissipations (in milliwatts). We are interested in five possible logic families: $\underset{\sim}{F}$ = {NMOS, CMOS, TTL, ECL, JJ},

where NMOS = N-channel metal oxide semiconductors
CMOS = complementary metal oxide semiconductor
TTL = transistor-transistor logic
ECL = emitter-coupled logic
JJ = Josephson junction

The range of delay time is $\underset{\sim}{D} = \{0.1, 1, 10, 100\}$ (in ns) and of power dissipation is $\underset{\sim}{P} = \{0.01, 0.1, 1, 10, 100\}$ (in mW). Suppose various value assignments (see the last section in this chapter) from data allow us to develop the following relations between delay times and logic families ($\underset{\sim}{R}_1 = \underset{\sim}{D} \times \underset{\sim}{F}$), and between logic families and power dissipation ($\underset{\sim}{R}_2 = \underset{\sim}{F} \times \underset{\sim}{P}$):

$$\underset{\sim}{R}_1 = \begin{array}{c} \\ 0.1 \\ 1 \\ 10 \\ 100 \end{array} \begin{array}{c} \begin{array}{ccccc} N & C & T & E & J \end{array} \\ \begin{bmatrix} 0 & 0 & 0 & .6 & 1 \\ 0 & .1 & .5 & 1 & 0 \\ .4 & 1 & 1 & 0 & 0 \\ 1 & .2 & 0 & 0 & 0 \end{bmatrix} \end{array} \quad \text{and} \quad \underset{\sim}{R}_2 = \begin{array}{c} \\ N \\ C \\ T \\ E \\ J \end{array} \begin{array}{c} \begin{array}{ccccc} .01 & .1 & 1 & 10 & 100 \end{array} \\ \begin{bmatrix} 0 & .4 & 1 & .3 & 0 \\ .2 & 1 & 0 & 0 & 0 \\ 0 & 0 & .7 & 1 & 0 \\ 0 & 0 & 0 & 1 & .5 \\ 1 & .1 & 0 & 0 & 0 \end{bmatrix} \end{array}$$

We can use max-min composition to obtain a relation between delay times and power dissipation; i.e., we can compute $\underset{\sim}{R}_3 = \underset{\sim}{R}_1 \cup \underset{\sim}{R}_2$, in set notation, or in function-theoretic shorthand, $\mu_{\underset{\sim}{R}_3} = \bigvee (\mu_{\underset{\sim}{R}_1} \wedge \mu_{\underset{\sim}{R}_2})$, as

$$\underset{\sim}{R}_3 = \begin{array}{c} \\ 0.1 \\ 1 \\ 10 \\ 100 \end{array} \begin{array}{c} \begin{array}{ccccc} .01 & .1 & 1 & 10 & 100 \end{array} \\ \begin{bmatrix} 1 & .1 & 0 & .6 & .5 \\ .1 & .1 & .5 & 1 & .5 \\ .2 & 1 & .7 & 1 & 0 \\ .2 & .4 & 1 & .3 & 0 \end{bmatrix} \end{array}$$

Example 3.9. In the area of compensation for professional engineers many issues are important. Compensation for quality efforts in engineering does not have to be solely related to money. Public recognition through awards or notices can also be an effective reward because of the character of the professional practice. One important consideration in this endeavor is the correlation of performance, perceptions by management, and rewards. This example illustrates the utility of fuzzy relations in this area of engineering compensation.

The first problem to be addressed is the mapping from professional behavior to the perceptions by management of their professional staff. The second issue is the mapping from perception to monetary compensation. Suppose behavior is compressed into three attributes:

$$\underset{\sim}{B} = \{\text{hard work (HW), good manners (GM), high productivity (HP)}\}$$

Further, assume that management perceptions are reduced to two attributes:

$$\underset{\sim}{P} = \{\text{nice to have around (NHA), makes the boss look good (MLG)}\}$$

Moreover, the reward structure will be composed of two elements:

$$\underset{\sim}{R} = \{\text{public compliments (PC), good raise (GR)}\}$$

The membership functions for professional behavior and management perceptions can be established by engineering staff and management together through a combination of subjective judgment, consensus, and normalization of possible values on the interval [0,1]. The rewards are in the range of [0,1] where a membership value of 1.0 for a public compliment corresponds to the maximum number ever received by any nonmanagement employee, for example, 17 announcements during the year. A membership value of 1.0 for a good raise (GR) might correspond to 3.5 percent.

The following relation matrices are obtained through Cartesian products or from other value assignments as discussed later in this chapter. Suppose the relation matrix relating perception $\underset{\sim}{P}$ and behavior $\underset{\sim}{B}$ is postulated as

$$\underset{\sim}{R}_{P,B} = \begin{array}{c} \\ \text{NHA} \\ \text{MLG} \end{array} \begin{array}{c} \begin{array}{ccc} \text{NW} & \text{GM} & \text{HP} \end{array} \\ \begin{bmatrix} 0.2 & 0.8 & 0.7 \\ 0.4 & 0.5 & 0.8 \end{bmatrix} \end{array}$$

Further suppose the relation matrix relating reward $\underset{\sim}{R}$ and perception $\underset{\sim}{P}$ is postulated as

$$\underset{\sim}{R}_{R,P} = \begin{array}{c} \\ \text{PC} \\ \text{GR} \end{array} \begin{array}{c} \begin{array}{cc} \text{NHA} & \text{MLG} \end{array} \\ \begin{bmatrix} 0.5 & 0.9 \\ 0.7 & 0.8 \end{bmatrix} \end{array}$$

Then the relationship between reward $\underset{\sim}{R}$ and behavior $\underset{\sim}{B}$ can be computed through a composition. In this case, using max-min composition produces this relation,

$$\underset{\sim}{R}_{R,B} = \begin{array}{c} \\ \text{PC} \\ \text{GR} \end{array} \begin{array}{c} \begin{array}{ccc} \text{HW} & \text{GM} & \text{HP} \end{array} \\ \begin{bmatrix} 0.4 & 0.5 & 0.8 \\ 0.4 & 0.5 & 0.8 \end{bmatrix} \end{array}$$

This result shows that, for these data at any rate, it is equally possible to receive a reward through public compliments as through a good raise.

TOLERANCE AND EQUIVALENCE RELATIONS

Relations can exhibit various useful properties, a few of which will be discussed here. As mentioned in the introduction of this chapter, relations can be used in graph theory [Gill, 1976; Zadeh, 1971]. Consider the simple graphs in Fig. 3.8. This figure describes a universe of three elements, which are labeled as the vertices of this graph: 1, 2, and 3, or in set notation, X = {1, 2, 3}. The useful properties we wish to discuss are reflexivity, symmetry, and transitivity (there are other properties of relations that are the antonyms of these three, i.e., irreflexivity, asymmetry, and nontransitivity; these, and an additional property of antisymmetry, will not be discussed in this text). When a relation is reflexive every vertex in the graph originates a single loop, as shown in Fig. 3.8*a*. If a relation is symmetric, then in the graph for every edge pointing (the arrows on the edge lines in Fig. 3.8*b*) from vertex i to vertex j $(i, j = 1, 2, 3)$, there is an edge pointing in the opposite direction, i.e., from vertex j to vertex i. When a relation is transitive, then for every pair of edges in the graph, one pointing from vertex i to vertex j and the other from vertex j to vertex k $(i, j, k = 1, 2, 3)$, there is an edge pointing from vertex i directly to vertex k, as seen in Fig. 3.8*c* (e.g., an arrow from vertex 1 to vertex 2, an arrow from vertex 2 to vertex 3, and an arrow from vertex 1 to vertex 3).

Crisp Equivalence Relation

A relation R on a universe X can also be thought of as a relation from X to X. The relation R is an equivalence relation if it has the following three properties: (1) reflexivity, (2) symmetry, and (3) transitivity. For example, for a matrix relation the following properties will hold:

Reflexivity $\quad (x_i, x_i) \in \mathrm{R} \quad \text{or} \quad \chi_{\mathrm{R}}(x_i, x_i) = 1 \qquad (3.20a)$

Symmetry $\quad (x_i, x_j) \in \mathrm{R} \rightarrow (x_j, x_i) \in \mathrm{R}$
$\quad \text{or} \quad \chi_{\mathrm{R}}(x_i, x_j) = \chi_{\mathrm{R}}(x_j, x_i) \qquad (3.20b)$

Transitivity $\quad (x_i, x_j) \in \mathrm{R} \quad \text{and} \quad (x_j, x_k) \in \mathrm{R} \rightarrow (x_i, x_k) \in \mathrm{R}$
$\quad \text{or} \quad \chi_{\mathrm{R}}(x_i, x_j) \quad \text{and} \quad \chi_{\mathrm{R}}(x_j, x_k) = 1 \rightarrow \chi_{\mathrm{R}}(x_i, x_k) = 1 \qquad (3.20c)$

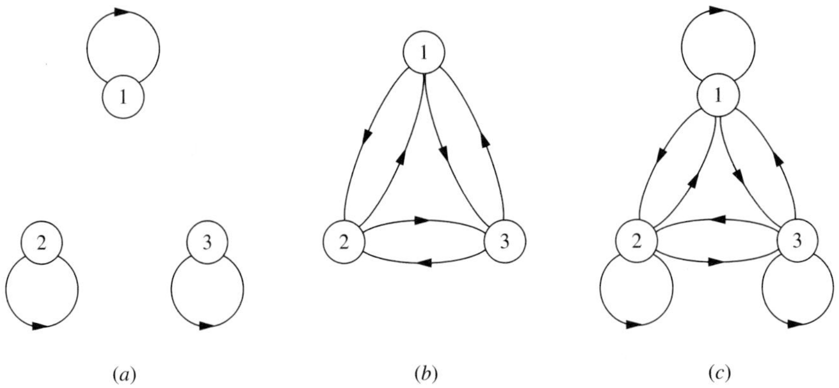

FIGURE 3.8
Three-vertex graphs for properties of (*a*) reflexivity, (*b*) symmetry, (*c*) transitivity [Gill, 1976].

The most familiar equivalence relation is that of equality among elements of a set. Other examples of equivalence relations include the relation of parallelism among lines in plane geometry, the relation of similarity among triangles, the relation "works in the same building as" among workers of a given city, and others.

Crisp Tolerance Relation

A tolerance relation R (also called a *proximity* relation) on a universe X is a relation that exhibits only the properties of reflexivity and symmetry. A tolerance relation, R, can be reformed into an equivalence relation by at most $(n-1)$ compositions with itself, where n is the cardinal number of the set defining R, in this case X; i.e.,

$$R_1^{n-1} = R_1 \circ R_1 \circ \cdots \circ R_1 = R \tag{3.21}$$

Example 3.10. Suppose in an airline transportation system we have a universe composed of five elements; the cities Omaha, Chicago, Rome, London, and Detroit. The airline is studying locations of potential hubs in various countries and must consider air mileage between cities and takeoff and landing policies in the various countries. These cities can be enumerated as the elements of a set, i.e.,

$$X = \{x_1, x_2, x_3, x_4, x_5\} = \{\text{Omaha, Chicago, Rome, London, Detroit}\}$$

Further, suppose we have a tolerance relation, R_1, that expresses relationships among these cities:

$$R_1 = \begin{bmatrix} 1 & 1 & 0 & 0 & 0 \\ 1 & 1 & 0 & 0 & 1 \\ 0 & 0 & 1 & 0 & 0 \\ 0 & 0 & 0 & 1 & 0 \\ 0 & 1 & 0 & 0 & 1 \end{bmatrix}$$

This relation is reflexive and symmetric. The graph for this tolerance relation would involve five vertices (five elements in the relation), as shown in Fig. 3.9. The property of reflexivity (diagonal elements equal unity) simply indicates that a city is totally related to itself. The property of symmetry might represent proximity; Omaha and Chicago (x_1 and x_2) are close (in a binary sense) geographically, and Chicago and Detroit (x_2 and

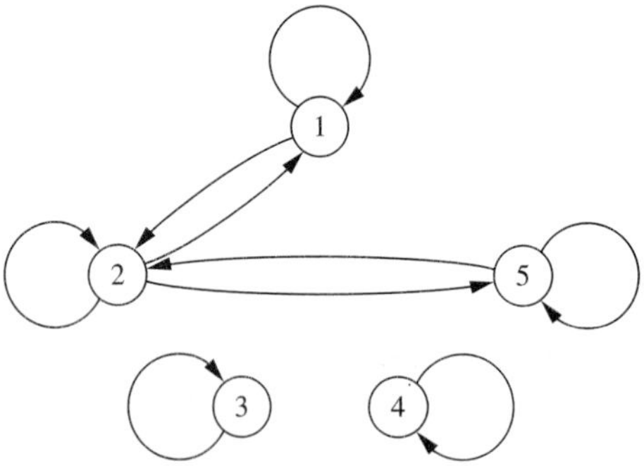

FIGURE 3.9
Five-vertex graph of tolerance relation (reflexive and symmetric) in Example 3.10.

x_5) are close geographically. This relation, R_1, does not have properties of transitivity; e.g.,

$$(x_1, x_2) \in R_1 \qquad (x_2, x_5) \in R_1 \qquad \text{but} \qquad (x_1, x_5) \notin R_1$$

R_1 can become an equivalence relation through two ($2 \leq n$, where $n = 5$) compositions. Using Eq. (3.21), we get

$$R_1 \circ R_1 = \begin{bmatrix} 1 & 1 & 0 & 0 & 1 \\ 1 & 1 & 0 & 0 & 1 \\ 0 & 0 & 1 & 0 & 0 \\ 0 & 0 & 0 & 1 & 0 \\ 1 & 1 & 0 & 0 & 1 \end{bmatrix} = R$$

Now, we see in this matrix that transitivity holds, i.e., $(x_1, x_5) \in R_1$, and R is an equivalence relation. Although the point is not important here, we will see in Chapter 11 that equivalence relations also have certain special properties useful in classification. For instance, in this example the equivalence relation expressed in the foregoing R matrix could represent cities in separate countries. Inspection of the matrix shows that columns 1, 2, and 5 are identical, that is, Omaha, Chicago, and Detroit are in the same class; and columns 3 and 4 are unique, indicating that Rome and London are cities each in their own class; these three different classes could represent distinct countries. The graph for this tolerance relation would involve five vertices (five elements in the relation), as shown in Fig. 3.10.

FUZZY TOLERANCE AND EQUIVALENCE RELATIONS

A fuzzy relation, $\underset{\sim}{R}$, on a single universe X is also a relation from X to X. It is a fuzzy equivalence relation if all three of the following properties for matrix relations define it; e.g.,

Reflexivity $\quad \mu_{\underset{\sim}{R}}(x_i, x_i) = 1 \qquad$ (3.22*a*)

Symmetry $\quad \mu_{\underset{\sim}{R}}(x_i, x_j) = \mu_{\underset{\sim}{R}}(x_j, x_i) \qquad$ (3.22*b*)

Transitivity $\quad \mu_{\underset{\sim}{R}}(x_i, x_j) = \lambda_1 \quad \text{and} \quad \mu_{\underset{\sim}{R}}(x_j, x_k) = \lambda_2 \rightarrow \mu_{\underset{\sim}{R}}(x_i, x_k) = \lambda \qquad$ (3.22*c*)

where $\quad \lambda \geq \min[\lambda_1, \lambda_2]$

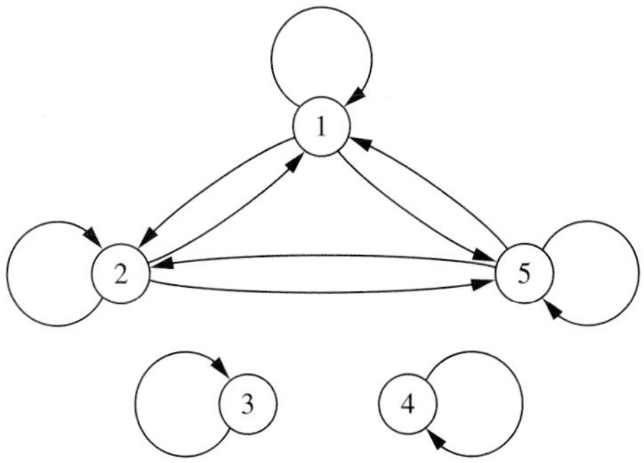

FIGURE 3.10
Five-vertex graph of equivalence relation (reflexive, symmetric, transitive) in Example 3.10.

Looking at the physical analog (see Fig. 3.5) of a composition operation, we see it comprises a parallel system of chains, where each chain represents a particular path through the chain system. The physical analogy behind transitivity is that the shorter the chain, the stronger the relation (the stronger is the chain system). In particular, the strength of the link between two elements must be greater than or equal to the strength of any indirect chain involving other elements, i.e., Eq. (3.22*c*) [Dubois and Prade, 1980].

It can be shown that any fuzzy tolerance relation, $\underset{\sim}{R}_1$, that has properties of relexivity and symmetry can be reformed into a fuzzy equivalence relation by at most $(n-1)$ compositions, just as a crisp tolerance relation can be reformed into a crisp equivalence relation. That is,

$$\underset{\sim}{R}_1^{n-1} = \underset{\sim}{R}_1 \circ \underset{\sim}{R}_1 \cdots \circ \underset{\sim}{R}_1 = \underset{\sim}{R} \tag{3.23}$$

Example 3.11. Suppose, in a biotechnology experiment, five potentially new strains of bacteria have been detected in the area around an anaerobic corrosion pit on a new aluminum-lithium alloy used in the fuel tanks of a new experimental aircraft. In order to propose methods to eliminate the biocorrosion caused by these bacteria, the five strains must first be categorized. One way to categorize them is to compare them to one another. In a pairwise comparison, the following "similarity" relation, $\underset{\sim}{R}_1$, is developed. For example, the first strain (column 1) has a strength of similarity to the second strain of 0.8, to the third strain a strength of 0 (i.e., no relation), to the fourth strain a strength of 0.1, and so on. Because the relation is for pairwise similarity it will be reflexive and symmetric. Hence,

$$\underset{\sim}{R}_1 = \begin{bmatrix} 1 & 0.8 & 0 & 0.1 & 0.2 \\ 0.8 & 1 & 0.4 & 0 & 0.9 \\ 0 & 0.4 & 1 & 0 & 0 \\ 0.1 & 0 & 0 & 1 & 0.5 \\ 0.2 & 0.9 & 0 & 0.5 & 1 \end{bmatrix}$$

is reflexive and symmetric. However, it is not transitive; e.g.,

$$\mu_{\underset{\sim}{R}}(x_1, x_2) = 0.8, \qquad \mu_{\underset{\sim}{R}}(x_2, x_5) = 0.9 \geq 0.8$$

but

$$\mu_{\underset{\sim}{R}}(x_1, x_5) = 0.2 \leq \min(.8, .9)$$

One composition results in the following relation:

$$\underset{\sim}{R}_1^2 = \underset{\sim}{R}_1 \circ \underset{\sim}{R}_1 = \begin{bmatrix} 1 & 0.8 & 0.4 & 0.2 & 0.8 \\ 0.8 & 1 & 0.4 & 0.5 & 0.9 \\ 0.4 & 0.4 & 1 & 0 & 0.4 \\ 0.2 & 0.5 & 0 & 1 & 0.5 \\ 0.8 & 0.9 & 0.4 & 0.5 & 1 \end{bmatrix}$$

where transitivity still does not result; for example,

$$\mu_{\underset{\sim}{R}^2}(x_1, x_2) = 0.8 \geq 0.5 \qquad \text{and} \qquad \mu_{\underset{\sim}{R}^2}(x_2, x_4) = 0.5$$

but

$$\mu_{\underset{\sim}{R}^2}(x_1, x_4) = 0.2 \leq \min(0.8, 0.5)$$

Finally, after one or two more compositions, transitivity results,

$$\underset{\sim}{R}_1^3 = \underset{\sim}{R}_1^4 = \underset{\sim}{R} = \begin{bmatrix} 1 & 0.8 & 0.4 & 0.5 & 0.8 \\ 0.8 & 1 & 0.4 & 0.5 & 0.9 \\ 0.4 & 0.4 & 1 & 0.4 & 0.4 \\ 0.5 & 0.5 & 0.4 & 1 & 0.5 \\ 0.8 & 0.9 & 0.4 & 0.5 & 1 \end{bmatrix}$$

$$\underset{\sim}{R}_1^3(x_1, x_2) = 0.8 \geq 0.5$$

$$\underset{\sim}{R}_1^3(x_2, x_4) = 0.5 \geq 0.5$$

$$\underset{\sim}{R}_1^3(x_1, x_4) = 0.5 \geq 0.5$$

Graphs can be drawn for fuzzy equivalence relations, but the arrows in the graphs between vertices will have various "strengths," i.e., values on the interval [0, 1]. Once the fuzzy relation $\underset{\sim}{R}$ in Example 3.11 is an equivalence relation, it can be used in categorizing the various bacteria according to preestablished levels of confidence. These levels of confidence will be illustrated with a method called "alpha cuts" in Chapter 5, and the categorization idea will be illustrated using classification in Chapter 11.

There is an interesting graphical analog for fuzzy equivalence relations. An inspection of a three-dimensional plot of the preceeding equivalence relation, $\underset{\sim}{R}_1^3$, is shown in Fig. 3.11. In this graph, which is a plot of the membership values of the equivalence relation, we can see that, if it were a watershed, there would be no location where water would *pool,* or be trapped. In fact, every equivalence relation will produce a surface on which water cannot be trapped; the converse is not true in general, however. That is, there can be relations that are not equivalence relations but

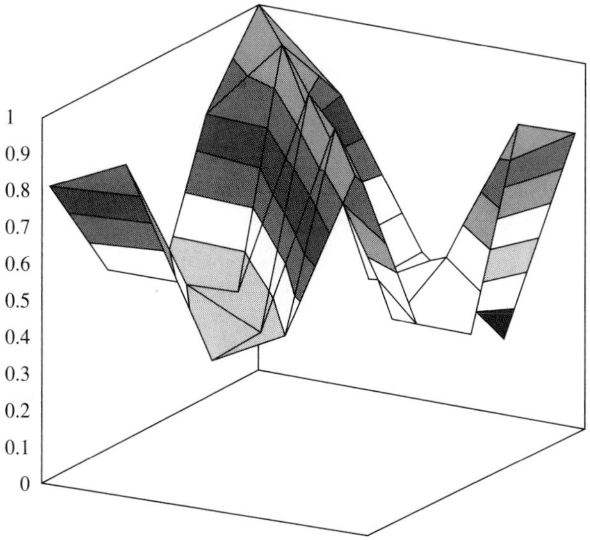

FIGURE 3.11
Three-dimensional representation of an equivalence relation.

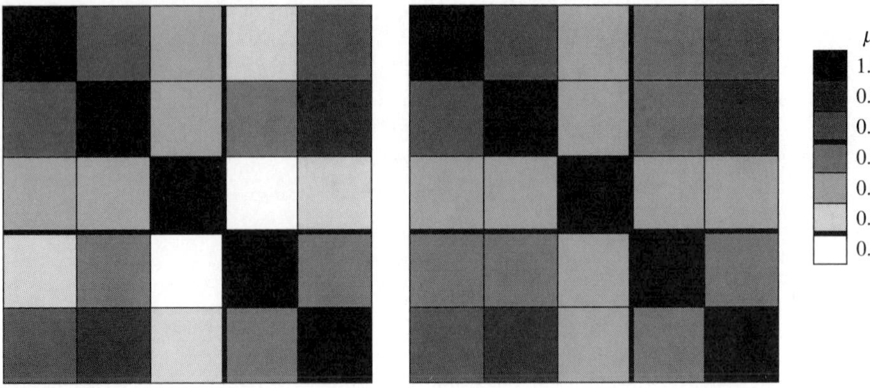

FIGURE 3.12
Two-dimensional contours of (*a*) tolerance relation $\underset{\sim}{R}_1^2$ and (*b*) equivalence relation $\underset{\sim}{R}_1^3$.

whose three-dimensional surface representations will not trap water. An example of the latter is given in the original tolerance relation, $\underset{\sim}{R}_1$, of Example 3.11.

Another way to show this same information is to construct a two-dimensional contour of the relation using various contour levels; these correspond to different degrees of membership. The graphic in Fig. 3.12 shows contour plots for the tolerance relation, $\underset{\sim}{R}_1^2$, and the equivalence level, $\underset{\sim}{R}_1^3$.

VALUE ASSIGNMENTS

An appropriate question regarding relations is, Where do the membership values that are contained in a relation come from? The answer to this question is that there are at least six different ways to develop the numerical values that characterize a relation:

1. Cartesian product
2. Closed-form expression
3. Lookup table
4. Linguistic rules of knowledge
5. Classification
6. Similarity methods in data manipulation

The first way is the one that has been illustrated so far in this chapter—to calculate relations from the Cartesian product of two or more fuzzy sets. A second way is through simple observation of a physical process. For a given set of inputs we observe a process yielding a set of outputs. If there is no variation between specific input-output pairs we may be led to model the process with a crisp relation. Moreover, if no variability exists, one might be able to express the relation as a closed-form algorithm of the form $\mathbf{Y} = f(\mathbf{X})$, where $\mathbf{X}$ is a vector of inputs and $\mathbf{Y}$ is a vector of outputs. If some variability exists, membership values on the interval [0, 1] may lead us to develop a fuzzy relation from a third approach—the use of a lookup table. Fuzzy relations can also be assembled from linguistic knowledge, expressed as if-then rules. Such knowledge may come from experts, from polls, or from consensus building. This fourth method is illustrated in more detail in Chapters 7, 8, and 9. Relations also arise from notions of classification where issues associated with similarity are central to determining relationships among patterns or clusters of data. The ability to develop relations in classification, the fifth method, is developed in more detail in Chapter 11.

One of the most prevalent forms of determining the values in relations is through manipulations of data, the sixth method mentioned. The more robust a data set, the more accurate the relational entities are in establishing relationships among elements of two or more data sets. This sixth way for determining value assignments for relations is actually a family of procedures termed similarity methods [see Zadeh, 1971; or Dubois and Prade, 1980]. All of these methods attempt to determine some sort of similar pattern or structure in data through various metrics. There are many of these methods available, but the most prevalent will be discussed here.

Cosine Amplitude

A useful method is the cosine amplitude method. As with all the following methods, this similarity metric makes use of a collection of data samples, n data samples in particular. If these data samples are collected they form a data array, X,

$$\mathbf{X} = \{\mathbf{x}_1, \mathbf{x}_2, \ldots, \mathbf{x}_n\}$$

Each of the elements, x_i, in the data array X is itself a vector of length m, that is,

$$\mathbf{x}_i = \{\mathbf{x}_{i_1}, \mathbf{x}_{i_2}, \ldots, \mathbf{x}_{i_m}\}$$

Hence, each of the data samples can be thought of as a point in m-dimensional space, where each point needs m coordinates for a complete description. Each element of a relation, r_{ij}, results from a pairwise comparison of two data samples, say $\mathbf{x}_i$ and $\mathbf{x}_j$, where the strength of the relationship between data sample $\mathbf{x}_i$ and data sample $\mathbf{x}_j$ is given by the membership value expressing that strength, that is, $r_{ij} = \mu_{\mathrm{R}}(x_i, y_j)$. The relation matrix will be of size $n \times n$ and, as will be the case for all similarity relations, the matrix will be reflexive and symmetric—hence a tolerance relation. The cosine amplitude method calculates r_{ij} in the following manner, and guarantees, as do all the similarity methods, that $0 \le r_{ij} \le 1$:

$$r_{ij} = \frac{\left|\sum_{k=1}^{m} x_{ik} x_{jk}\right|}{\sqrt{\left(\sum_{k=1}^{m} x_{ik}^2\right)\left(\sum_{k=1}^{m} x_{jk}^2\right)}}, \qquad \text{where } i, j = 1, 2, \ldots, n \qquad (3.24)$$

Close inspection of Eq. (3.24) reveals that this method is related to the dot product for the cosine function. When two vectors are colinear (most similar), their dot product is unity; when the two vectors are at right angles to one another (most dissimilar), their dot product is zero.

Example 3.12 [Dong, 1987]. Five separate regions along the San Andreas fault in California have suffered damage from a recent earthquake. For purposes of assessing payouts from insurance companies to building owners, the five regions must be classified as to their damage levels. Expression of the damage in terms of relations will prove helpful.

Surveys are conducted of the buildings in each region. All of the buildings in each region are described as being in one of three damage states: no damage, medium damage, and serious damage. Each region has each of these three damage states expressed as a percentage (ratio) of the total number of buildings. Hence, for this problem $n = 5$ and $m = 3$. The following table summarizes the findings of the survey team.

Regions	x_1	x_2	x_3	x_4	x_5
x_{i1}—Ratio with no damage	0.3	0.2	0.1	0.7	0.4
x_{i2}—Ratio with medium damage	0.6	0.4	0.6	0.2	0.6
x_{i3}—Ratio with serious damage	0.1	0.4	0.3	0.1	0

We wish to use the cosine amplitude method to express these data as a fuzzy relation. Equation (3.24) for an element in the fuzzy relation, r_{ij}, thus takes on the specific form

$$r_{ij} = \frac{\left|\sum_{k=1}^{3} x_{ik}x_{jk}\right|}{\sqrt{\left(\sum_{k=1}^{3} x_{ik}^2\right)\left(\sum_{k=1}^{3} x_{jk}^2\right)}}$$

For example, for $i = 1$ and $j = 2$ we get

$$r_{12} = \frac{0.3 \times 0.2 + 0.6 \times 0.4 + 0.1 \times 0.4}{\left[(0.3^2 + 0.6^2 + 0.1^2)(0.2^2 + 0.4^2 + 0.4^2)\right]^{1/2}} = \frac{0.34}{[0.46 \times 0.36]^{1/2}} = 0.836$$

Computing the other elements of the relation results in the following tolerance relation:

$$\underset{\sim}{\mathrm{R}}_1 = \begin{bmatrix} 1 & & & \text{sym} & \\ 0.836 & 1 & & & \\ 0.914 & 0.934 & 1 & & \\ 0.682 & 0.6 & 0.441 & 1 & \\ 0.982 & 0.74 & 0.818 & 0.774 & 1 \end{bmatrix}$$

and two compositions of $\underset{\sim}{\mathrm{R}}_1$ produce the equivalence relation, $\underset{\sim}{\mathrm{R}}$.

$$\underset{\sim}{\mathrm{R}} = \underset{\sim}{\mathrm{R}}_1^3 = \begin{bmatrix} 1 & & & \text{sym} & \\ 0.914 & 1 & & & \\ 0.914 & 0.934 & 1 & & \\ 0.774 & 0.774 & 0.774 & 1 & \\ 0.982 & 0.914 & 0.914 & 0.774 & 1 \end{bmatrix}$$

The tolerance relation, $\underset{\sim}{\mathrm{R}}_1$, expressed the pairwise similarity of damage for each of the regions; the equivalence relation, $\underset{\sim}{\mathrm{R}}$, also expresses this same information but additionally can be used to classify the regions into categories with *like properties* (see Chapter 11).

Max-min Method

Another popular method, which is computationally simpler than the cosine amplitude method, is known as the max-min method. Although the name sounds similar to the max-min composition method, this similarity method is different from composition. It is found through simple min and max operations on pairs of the data points, x_{ij}, and is given by

$$r_{ij} = \frac{\sum_{k=1}^{m} \min(x_{ik}, x_{jk})}{\sum_{k=1}^{m} \max(x_{ik}, x_{jk})}, \qquad \text{where } i, j = 1, 2, \ldots, n \tag{3.25}$$

Example 3.13. If we reconsider Example 3.12, the min-max method will produce the following result for $i = 1$, $j = 2$:

$$r_{12} = \frac{\sum_{k=1}^{3} (\min(0.3, 0.2), \min(0.6, 0.4), \min(0.1, 0.4))}{\sum_{k=1}^{3} (\max(0.3, 0.2), \max(0.6, 0.4), \max(0.1, 0.4))} = \frac{0.2 + 0.4 + 0.1}{0.3 + 0.6 + 0.4} = 0.538$$

Computing the other elements of the relation results in the following tolerance relation:

$$\underset{\sim}{\mathrm{R}}_1 = \begin{bmatrix} 1 & & & & \\ 0.538 & 1 & & \text{sym} & \\ 0.667 & 0.667 & 1 & & \\ 0.429 & 0.333 & 0.250 & 1 & \\ 0.818 & 0.429 & 0.538 & 0.429 & 1 \end{bmatrix}$$

Other Similarity Methods [Dong, 1987]

The list of other similarity methods is quite lengthy. For brevity, only a few of these are repeated here. For all of these, $i, j = 1, 2, \ldots, n$. Some of these methods estimate similarity through exponential functions:

Absolute exponential

$$r_{ij} = \exp\left(-\sum_{k=1}^{m} \left|x_{ik} - x_{jk}\right|\right) \tag{3.26}$$

Exponential similarity coefficient

$$r_{ij} = \frac{1}{m}\sum_{k=1}^{m} \exp\left(\frac{-3\left(x_{ik} - x_{jk}\right)^2}{4(\mathrm{S}_k)^2}\right) \tag{3.27}$$

where S_k = any general measure of variance for all the data, i.e., $(\mathrm{S}_k)^2 \geq 0$

Other methods produce scalar quantities which are similar to the cosine amplitude, such as the following:

Geometric average minimum

$$r_{ij} = \frac{\sum_{k=1}^{m} \min(x_{ik}, x_{jk})}{\sum_{k=1}^{m} [x_{ik} \cdot x_{jk}]^{1/2}} \tag{3.28}$$

Scalar product

$$r_{ij} = \begin{cases} 1 & i = j \\ \dfrac{1}{M}\displaystyle\sum_{k=1}^{m} x_{ik} \cdot x_{jk} & i \neq j \end{cases} \tag{3.29}$$

$$\text{where} \quad M \geq \max_{ij}\left(\sum_{k=1}^{m} x_{ik} \cdot x_{jk}\right)$$

Some methods are analogous to popular statistical quantities, such as

$$\textit{Correlation coefficient} \quad r_{ij} = \frac{\sum_{k=1}^{m} \left|x_{ik} - \overline{x}_i\right| \left|x_{jk} - \overline{x}_j\right|}{\left[\sum_{k=1}^{m} (x_{ik} - \overline{x}_i)^2\right]^{1/2} \cdot \left[\sum_{k=1}^{m} \left(x_{jk} - \overline{x}_j\right)^2\right]^{1/2}} \tag{3.30}$$

$$\text{where} \quad \overline{x}_i = \frac{1}{m}\sum_{k=1}^{m} x_{ik} \quad \text{and} \quad \overline{x}_j = \frac{1}{m}\sum_{k=1}^{m} x_{jk}$$

$$\textit{Arithmetic average-minimum} \quad r_{ij} = \frac{\sum_{k=1}^{m} \min(x_{ik}, x_{jk})}{\frac{1}{2}\sum_{k=1}^{m}(x_{ik} + x_{jk})} \tag{3.31}$$

Some methods are based on inverse relationships, for example,

$$\textit{Absolute reciprocal} \quad r_{ij} = \begin{cases} 1 & i = j \\ \dfrac{M}{\sum_{k=1}^{m} \left|x_{ik} - x_{jk}\right|} & i \neq j \end{cases} \tag{3.32}$$

where M is selected to make $0 \leq r_{ij} \leq 1$

$$\textit{Absolute subtrahend} \quad r_{ij} = \begin{cases} 1 & i = j \\ 1 - c\sum_{k=1}^{m} \left|x_{ik} - x_{jk}\right| & i \neq j \end{cases} \tag{3.33}$$

where c is selected to make $0 \leq r_{ij} \leq 1$

Other methods are nonparametric, such as

$$\textit{Nonparametric} \quad r_{ij} = \frac{\left|n^+ - n^-\right|}{n^+ + n^-} \tag{3.34}$$

where $x'_{ik} = x_{ik} - \overline{x}_i$ and $x'_{jk} = x_{jk} - \overline{x}_j$

n^+ = number of elements > 0 in $\{x'_{i1}x'_{j1}, x'_{i2}x'_{j2}, \ldots, x'_{im}x'_{jm}\}$

n^- = number of elements < 0 in $\{x'_{i1}x'_{j1}, x'_{i2}x'_{j2}, \ldots, x'_{im}x'_{jm}\}$

In Eq. 3.34, terms such as $x'_{i1}x'_{j1}$ are products of data elements.

SUMMARY

This chapter has shown some of the properties and operations of crisp and fuzzy relations. There are many more, but these will provide a sufficient foundation for the rest of the material in the text. The idea of a relation is most powerful; this modeling power will be shown in subsequent chapters dealing with such issues as logic, nonlinear simulation, classification, and control. The idea of composition was introduced, and it will be seen in Chapter 6 that the composition of a relation is similar to a method used to extend fuzziness into functions, called the *extension principle.* The principle of *non-interactivity* between sets was introduced as being analogous to the assumption of independence in probability modeling. Tolerant and equivalent relations hold some special properties, as will be illustrated in Chapter 11, when they are used in similarity applications and classification applications, respectively. There are some very interesting graphical analogies for relations as seen in some of the example problems (also see Exercise 3.9 in the problems at the end of the chapter). Finally, several similarity metrics were shown to be useful in developing the relational *strengths,* or distances, within fuzzy relations from data sets.

REFERENCES

Bandemer, H. and W. Näther. (1992). *Fuzzy data analysis,* Kluwer Academic, Dordrecht, Germany.

Dong, W. (1987). Personal notes.

Dubois, D., and H. Prade (1980). *Fuzzy sets and systems: Theory and applications,* Academic Press, New York.

Gill, A. (1976). *Applied algebra for the computer sciences,* Prentice Hall, Englewood Cliffs, NJ.

Kandel, A. (1985). *Fuzzy mathematical techniques with applications,* Addison-Wesley, Menlo Park, CA.

Klir, G., and T. Folger (1988). *Fuzzy sets, uncertainty, and information,* Prentice Hall, Englewood Cliffs, NJ.

Zadeh, L. (1971). "Similarity relations and fuzzy orderings," *Inf. Sci.,* vol. 3, pp. 177–200.

Zadeh, L. (1975). "The concept of a linguistic variable and its application to approximate reasoning, Parts 1, 2, and 3," *Inf. Sci.,* vol. 8, pp. 199–249, 301–357; vol. 9, pp. 43–80.

PROBLEMS

General Relations

3.1. Based on Example 3.9, calculate the relation between behavior and reward using max-product composition. Does the max product composition show that a raise is easier to get than a compliment?

3.2. For Example 3.8 on logic families show that

$$\underset{\sim}{R}_4 = \underset{\sim}{R}_1 \circ \underset{\sim}{R}_3$$

That is, using $\mu_{\underset{\sim}{R}_4} = \bigvee(\mu_{\underset{\sim}{R}_1} \circ \mu_{\underset{\sim}{R}_3})$ show that

$$\underset{\sim}{R}_4 = \begin{bmatrix} 1 & .1 & 0 & .6 & .3 \\ .02 & .1 & .35 & 1 & .5 \\ .2 & 1 & .7 & 1 & 0 \\ .04 & .4 & 1 & .3 & 0 \end{bmatrix}$$

3.3. A spray-on foam deposition process can be modeled by the following relationship:

$$T \propto (F)(P)$$

where T = thickness of deposited foam
F = Flow rate in mass per unit time
P = Period of time under spray gun

In general, each independent factor can have a tolerance (fuzziness) of ± 20 percent. Each factor's normalized fuzzy set is

$$\underset{\sim}{F} = \left\{ \frac{.9}{.8} + \frac{1}{1} + \frac{.9}{1.2} \right\}$$

$$\underset{\sim}{P} = \left\{ \frac{.9}{.8} + \frac{1}{1} + \frac{.9}{1.2} \right\}$$

(*a*) Find the fuzzy relation $\underset{\sim}{T} = \underset{\sim}{F} \times \underset{\sim}{P}$.

(*b*) Explain how this Cartesian product relation is interpreted differently from the simple arithmetic product $T = F \cdot P$.

3.4. A company sells a product called a video multiplexer, which multiplexes the video from 16 video cameras into a single video cassette recorder (VCR). The product has a motion detection feature that can increase the frequency with which a given camera's video is recorded to tape depending on the amount of motion that is present. It does this by recording more information from that camera at the expense of the amount of video that is recorded from the other 15 cameras. Define a universe X to be the speed of the objects that are present in the video of camera 1 (there are 16 cameras). For example, let X = {Low Speed, Medium Speed, High Speed} = {LS, MS, HS}. Now, define a universe Y to represent the frequency with which the video from camera 1 is recorded to a VCR tape, i.e., the record rate of camera 1. Suppose, Y = {Slow Record Rate, Medium Record Rate, Fast Record Rate} = {SRR, MRR,FRR}. Let us now define a fuzzy set $\underset{\sim}{A}$ on X and a fuzzy set $\underset{\sim}{B}$ on Y, where $\underset{\sim}{A}$ represents a fuzzy slow-moving object present in video camera 1, and $\underset{\sim}{B}$ represents a fuzzy slow record rate, biased to the slow side. For example,

$$\underset{\sim}{A} = \left\{ \frac{1}{\text{LS}} + \frac{.4}{\text{MS}} + \frac{.2}{\text{HS}} \right\}$$

$$\underset{\sim}{B} = \left\{ \frac{1}{\text{SRR}} + \frac{.5}{\text{MRR}} + \frac{.25}{\text{FRR}} \right\}$$

(*a*) Find the fuzzy relation for the Cartesian product of $\underset{\sim}{A}$ and $\underset{\sim}{B}$., i.e., find $\underset{\sim}{R} = \underset{\sim}{A} \times \underset{\sim}{B}$.

(*b*) Suppose we introduce another fuzzy set, $\underset{\sim}{C}$, which represents a fuzzy fast-moving object present in video camera 1, say for example, the following:

$$\underset{\sim}{C} = \left\{ \frac{.1}{\text{LS}} + \frac{.3}{\text{MS}} + \frac{1}{\text{HS}} \right\}$$

Find the relation between $\underset{\sim}{C}$ and $\underset{\sim}{B}$ using a Cartesian product, i.e., find $\underset{\sim}{S} = \underset{\sim}{C} \times \underset{\sim}{B}$.

(*c*) Find $\underset{\sim}{C} \circ \underset{\sim}{R}$ using max-min composition.

(*d*) Find $\underset{\sim}{C} \circ \underset{\sim}{S}$ using max-min composition.

(*e*) Comment on the meaning of parts (*c*) and (*d*) and on the differences between the results.

3.5. All new jet aircraft are subjected to intensive flight simulation studies before they are ever tested under actual flight conditions. In these studies an important relationship is that between the mach number (percent of the speed of sound) and the altitude of the aircraft. This relationship is important to the performance of the aircraft and has a definite impact in making flight plans over populated areas. If certain mach levels are reached, breaking the sound barrier (sonic booms) can result in human discomfort and light damage to glass enclosures on the earth's surface. Current rules of thumb establish crisp breakpoints for the conditions that cause performance changes (and sonic booms) in aircraft, but in reality these breakpoints are fuzzy, because other atmospheric conditions such as the humidity and temperature also affect breakpoints in performance. For this problem, suppose the flight test data can be characterized as "near" or "approximately" or "in the region of" the crisp database breakpoints.

Define a universe of aircraft speeds near the speed of sound as X = {0.730, 0.735, 0.740, 0.745, 0.750} and a fuzzy set on this universe for the speed "near mach 0.74" = $\underset{\sim}{M}$, where

$$\underset{\sim}{M} = \left\{ \frac{0}{.730} + \frac{.8}{.735} + \frac{1}{.740} + \frac{.8}{.745} + \frac{0}{.750} \right\}$$

and define a universe of altitudes as Y = {22.5, 23, 23.5, 24, 24.5, 25, 25.5} in k-feet, and a fuzzy set on this universe for the altitude fuzzy set "approximately 24,000 feet" = $\underset{\sim}{A}$, where

$$\underset{\sim}{A} = \left\{ \frac{0}{22.5\text{k}} + \frac{.2}{23\text{k}} + \frac{.7}{23.5\text{k}} + \frac{1}{24\text{k}} + \frac{.7}{24.5\text{k}} + \frac{.2}{25\text{k}} + \frac{0}{25.5\text{k}} \right\}$$

(*a*) Construct the relation $\underset{\sim}{R} = \underset{\sim}{M} \times \underset{\sim}{A}$

(*b*) For another aircraft speed, say $\underset{\sim}{M}_1$ = "in the region of mach 0.74," where

$$\underset{\sim}{M}_1 = \left\{ \frac{0}{.730} + \frac{.8}{.735} + \frac{1}{.740} + \frac{.6}{.745} + \frac{.2}{.750} \right\}$$

find the relation $\underset{\sim}{S} = \underset{\sim}{M}_1 \circ \underset{\sim}{R}$ using max-min composition.

3.6. Three variables of interest in power transistors are the amount of current that can be switched, the voltage that can be switched, and the cost. The following membership functions for power transistors were developed from a hypothetical components catalog:

$$\text{Average current (in amps)} = \underset{\sim}{I} = \left\{ \frac{.4}{.8} + \frac{.7}{.9} + \frac{1}{1} + \frac{.8}{1.1} + \frac{.6}{1.2} \right\}$$

$$\text{Average voltage (in volts)} = \underset{\sim}{V} = \left\{ \frac{.2}{30} + \frac{.8}{45} + \frac{1}{60} + \frac{.9}{75} + \frac{.7}{90} \right\}$$

Note how the membership values in each set taper off faster toward the lower voltage and currents. These two fuzzy sets are related to the "power" of the transistor. Power in electronics is defined by an algebraic operation, $P = VI$, but let us deal with a general Cartesian relationship between voltage and current, i.e., simply with $\underset{\sim}{P} = \underset{\sim}{V} \times \underset{\sim}{I}$. Keep in mind that the Cartesian product is different from the arithmetic product. The Cartesian product expresses the relationship between V_i and I_j, where V_i and I_j are individual elements in the fuzzy sets $\underset{\sim}{V}$ and $\underset{\sim}{I}$.

(*a*) Find the fuzzy Cartesian product $\underset{\sim}{P} = \underset{\sim}{V} \times \underset{\sim}{I}$.

Now let us define a fuzzy set for the cost $\underset{\sim}{C}$, in dollars, of a transistor, for example,

$$\underset{\sim}{C} = \left\{ \frac{.4}{.5} + \frac{1}{.6} + \frac{.5}{.7} \right\}$$

(*b*) Using a fuzzy Cartesian product, find $\underset{\sim}{T} = \underset{\sim}{I} \times \underset{\sim}{C}$. What would this relation, $\underset{\sim}{T}$, represent physically?

(*c*) Using max-min composition, find $\underset{\sim}{E} = \underset{\sim}{P} \circ \underset{\sim}{T}$. What would this relation, $\underset{\sim}{E}$, represent physically?

(*d*) Using max-product composition, find $\underset{\sim}{E} = \underset{\sim}{P} \circ \underset{\sim}{T}$ What would this relation, $\underset{\sim}{E}$, represent physically?

3.7. The relationship between temperature and maximum operating frequency R depends on various factors for a given electronic circuit. Let $\underset{\sim}{T}$ be a temperature fuzzy set (in degrees Fahrenheit) and $\underset{\sim}{F}$ represent a frequency fuzzy set (in MHz) on the following universes of discourse:

$$\underset{\sim}{T} = \{-100, -50, 0, 50, 100\} \quad \text{and} \quad \underset{\sim}{F} = \{8, 16, 25, 33\}$$

Suppose a Cartesian product between $\underset{\sim}{T}$ and $\underset{\sim}{F}$ is formed that results in the following relation:

$$\underset{\sim}{R} = \begin{array}{c} \\ 8 \\ 16 \\ 25 \\ 33 \end{array} \begin{array}{c} \begin{array}{ccccc} -100 & -50 & 0 & 50 & 100 \end{array} \\ \begin{bmatrix} .2 & .5 & .7 & 1 & .9 \\ .3 & .5 & .7 & 1 & .8 \\ .4 & .6 & .8 & .9 & .4 \\ .9 & 1 & .8 & .6 & .4 \end{bmatrix} \end{array}$$

The reliability of the electronic circuit is related to the maximum operating temperature. Such a relation $\underset{\sim}{S}$ can be expressed as a Cartesian product between the reliability index, $\underset{\sim}{M} = \{1, 2, 4, 8, 16\}$ (in dimensionless units), and the temperature, as in the following example:

$$\underset{\sim}{S} = \begin{array}{c} \\ -100 \\ -50 \\ 0 \\ 50 \\ 100 \end{array} \begin{array}{c} \begin{array}{ccccc} 1 & 2 & 4 & 8 & 16 \end{array} \\ \begin{bmatrix} 1 & .8 & .6 & .3 & .1 \\ .7 & 1 & .7 & .5 & .4 \\ .5 & .6 & 1 & .8 & .8 \\ .3 & .4 & .6 & 1 & .9 \\ .9 & .3 & .5 & .7 & 1 \end{bmatrix} \end{array}$$

Composition can be performed on any two or more relations with compatible row-column consistency. To find a relationship between frequency and the reliability index, use

(*a*) max-min composition

(*b*) max-product composition

3.8. Relating earthquake intensity to ground acceleration is an imprecise science. Suppose we have a universe of earthquake intensities (on the Mercalli scale), I = {5, 6, 7, 8, 9} and a universe of accelerations, A = {0.2, 0.4, 0.6, 0.8, 1.0, 1.2}, in g's. The following fuzzy relation, $\underset{\sim}{R}$, exists on the Cartesian space I × A.

$$\underset{\sim}{R} = \begin{array}{c} \\ 5 \\ 6 \\ 7 \\ 8 \\ 9 \end{array} \begin{array}{c} \begin{array}{cccccc} 0.2 & 0.4 & 0.6 & 0.8 & 1.0 & 1.2 \end{array} \\ \begin{bmatrix} 0.75 & 1 & 0.85 & 0.5 & 0.2 & 0 \\ 0.5 & 0.8 & 1 & 0.7 & 0.3 & 0 \\ 0.1 & 0.5 & 0.8 & 1 & 0.7 & 0.1 \\ 0 & 0.2 & 0.5 & 0.85 & 1 & 0.6 \\ 0 & 0 & 0.2 & 0.5 & 0.9 & 1 \end{bmatrix} \end{array}$$

If the fuzzy set "intensity about 7" is defined as

$$\underset{\sim}{I}_7 = \left\{ \frac{0.1}{5} + \frac{0.6}{6} + \frac{1}{7} + \frac{0.8}{8} + \frac{0.2}{9} \right\}$$

determine the fuzzy membership of $\underset{\sim}{I}_7$ on the universe of accelerations, A.

3.9. Given the continuous, noninteractive fuzzy sets $\underset{\sim}{A}$ and $\underset{\sim}{B}$ on universes X and Y, using Zadeh's notation for continuous fuzzy variables,

$$\underset{\sim}{A} = \left\{ \int \frac{1 - 0.1\,|x|}{x} \right\} \qquad \text{for } x \in [0, +10]$$

$$\underset{\sim}{B} = \left\{ \int \frac{0.2\,|y|}{y} \right\} \qquad \text{for } y \in [0, +5]$$

as seen in Figs. P3.9*a* and *b*,

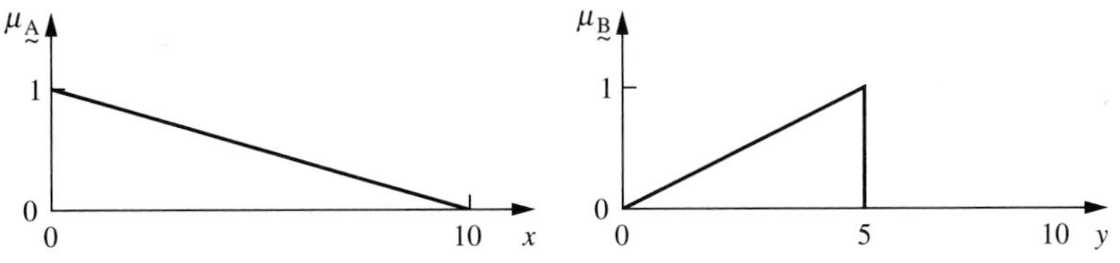

(*a*)

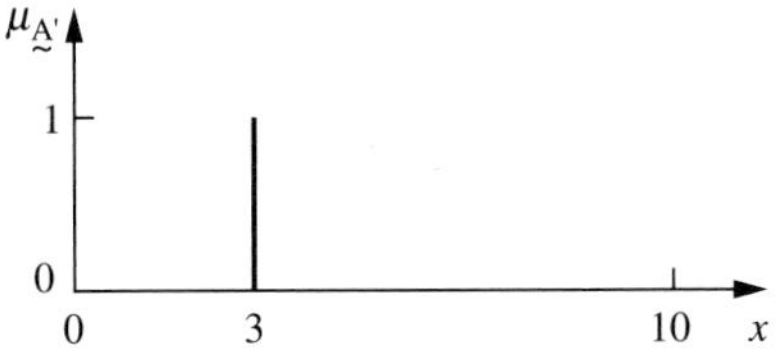

(*b*)

FIGURE P3.9
From [Dong, 1987].

(*a*) Construct a fuzzy relation R for the Cartesian product of $\underset{\sim}{A}$ and $\underset{\sim}{B}$.

(*b*) Use max-min composition to find $\underset{\sim}{B}'$, given the fuzzy singleton $\underset{\sim}{A}' = \frac{1}{3}$ (see Fig. P3.9*b*).

Hint: You can solve this problem graphically by segregating the Cartesian space into various regions according to the min and max operations, or you can approximate the continuous fuzzy variables as discrete variables and use matrix operations. In any case, sketch the solution.

3.10. The formation of algal and other biological colonies in surface waters is strongly dependent on such factors as the pH of the water, the temperature, and oxygen content. Relationships among these various factors enable environmental engineers to study issues involving bioremediation using the algae. Suppose we define a set T of water temperatures from a lake on the following discrete universe of temperatures in degrees Fahrenheit:

$$T = \{50, 55, 60\}$$

And suppose we define a universe O of oxygen content values in the water, as percent by volume:

$$O = \{1, 2, 6\}$$

Suppose a Cartesian product is performed between specific fuzzy sets $\underset{\sim}{T}$ and $\underset{\sim}{O}$ defined on T and Q to produce the following relation:

$$\underset{\sim}{R} = \underset{\sim}{T} \times \underset{\sim}{O} = \begin{matrix} \\ 50 \\ 55 \\ 60 \end{matrix} \begin{matrix} \begin{matrix} 1 & 2 & 6 \end{matrix} \\ \begin{bmatrix} 0.1 & 0.2 & 0.9 \\ 0.1 & 1 & 0.7 \\ 0.8 & 0.7 & 0.1 \end{bmatrix} \end{matrix}$$

Now suppose we define another fuzzy set of temperatures, "about 55°F," with the following membership values:

$$\underset{\sim}{I}_T = \left\{ \frac{0.5}{50} + \frac{1}{55} + \frac{0.7}{60} \right\}$$

(*a*) Using max-min composition, find $\underset{\sim}{S} = \underset{\sim}{I}_T \circ (\underset{\sim}{T} \times \underset{\sim}{O})$

(*b*) Using max-product composition, find $\underset{\sim}{S} = \underset{\sim}{I}_T \circ \underset{\sim}{R}$

3.11. In a typical database of documents, some documents are more useful than others to a given researcher using the database. If a document is too general for the researcher, she won't learn anything she doesn't already know, but if the document is too technical, it will be incomprehensible. So for a document to be useful, the document's *technicality* level must be well-matched to the researcher's *expertise* level. Define universe E of expertise levels,

$$E = \{\text{novice, student, master, expert}\}$$

and a universe T of technicality levels,

$$T = \{\text{general}(1), \text{medium}(2), \text{technical}(3)\}$$

The following relation, $\underset{\sim}{U}$, could represent the *usefulness* of a document, and it might be derived through a Cartesian product, defined on E × T,

$$\underset{\sim}{U} = \begin{bmatrix} \text{novice} \\ \text{student} \\ \text{master} \\ \text{expert} \end{bmatrix} \begin{matrix} [(1) & (2) & (3)] \\ \begin{bmatrix} 1 & .3 & 0 \\ .8 & .6 & .2 \\ .1 & .4 & .8 \\ 0 & .3 & .9 \end{bmatrix} \end{matrix}$$

Now suppose researcher R's expertise level is defined in terms of membership values as

$$\underset{\sim}{E}_R = \left\{ \frac{0}{\text{novice}} + \frac{0.7}{\text{student}} + \frac{0.4}{\text{master}} + \frac{0.3}{\text{expert}} \right\}$$

Determine the membership of the *usefulness* distribution of documents of the technicality levels in $\underset{\sim}{T}$ to researcher R; i.e., find $\underset{\sim}{S} = \underset{\sim}{E}_R \circ \underset{\sim}{U}$.

3.12. In photography it is important to relate reagent thickness to color balance on the film medium. Let Y be a universe of color balance, Y = [0, 1, 2, 3, 4], where 0 = yellow, 4 = blue, and 2 = neutral. Let X be a universe of the reagent thickness, X = [0, 1, 2, 3, 4], where 0 = thin, 4 = thick, and 2 = semi-thick. Now, suppose that a relation is obtained from a Cartesian product as follows:

$$\underset{\sim}{R} = \underset{\sim}{X} \times \underset{\sim}{Y} = \begin{array}{c} \\ 0 \\ 1 \\ 2 \\ 3 \\ 4 \end{array} \begin{array}{c} \begin{array}{ccccc} 0 & 1 & 2 & 3 & 4 \end{array} \\ \begin{bmatrix} 1 & .8 & .6 & .2 & 0 \\ .8 & 1 & .8 & .6 & 0 \\ .6 & .8 & 1 & .8 & .6 \\ .2 & .6 & .8 & 1 & .8 \\ 0 & .2 & .6 & .8 & 1 \end{bmatrix} \end{array}$$

Next, we wish to relate color balance on the film medium to the perceived quality of the picture. For this relation we need an additional universe of perceived picture quality, Z = [0, 1, 2, 3, 4], where 0 = bad; 4 = excellent; 2 = fair. Suppose a relation is obtained again from a Cartesian product,

$$\underset{\sim}{S} = \underset{\sim}{Y} \times \underset{\sim}{Z} = \begin{array}{c} \\ 0 \\ 1 \\ 2 \\ 3 \\ 4 \end{array} \begin{array}{c} \begin{array}{ccccc} 0 & 1 & 2 & 3 & 4 \end{array} \\ \begin{bmatrix} 1 & .6 & .4 & .2 & 0 \\ .6 & 1 & .6 & .4 & 0 \\ .4 & .6 & 1 & .6 & .4 \\ .2 & .4 & .6 & 1 & .6 \\ 0 & .2 & .4 & .6 & 1 \end{bmatrix} \end{array}$$

(*a*) Find $\underset{\sim}{T} = \underset{\sim}{R} \circ \underset{\sim}{S}$ using max-min composition. What would the relation $\underset{\sim}{T}$ represent?

(*b*) Find $\underset{\sim}{T} = \underset{\sim}{R} \circ \underset{\sim}{S}$ using max-product composition. What would the relation $\underset{\sim}{T}$ represent?

3.13. Risk assessment of hazardous waste situations requires the assimilation of a great deal of linguistic information. In this context, consider risk as being defined as "consequence of a hazard" multiplied by "possibility of the hazard" (instead of the conventional "probability of hazard"). Consequence is the result of an unintended action on humans, equipment or facilities, or the environment. Possibility is the estimate of the likelihood that the unintended action will occur. Consequence and possibility are dependent on several factors and therefore cannot be determined with precision. We will use composition, then, to define risk; hence risk = consequence ∘ possibility, or

$$\underset{\sim}{R} = \underset{\sim}{C} \circ \underset{\sim}{P}$$

We will consider that the consequence is the logical intersection of the hazard mitigation and the hazard source term (the source term defines the type of initiation, such as a smokestack emitting a toxic gas, or a truck spill of a toxic organic; hence we define consequence = mitigation ∩ source term, or

$$\underset{\sim}{C} = \underset{\sim}{M} \cap \underset{\sim}{ST}$$

Since humans and their systems are ultimately responsible for preventing or causing nonnatural hazards, we define the possibility of a hazard as the logical intersection of human errors and system vulnerabilities; hence possibility = human factors ∩ system reliabilities,

$$\underset{\sim}{P} = \underset{\sim}{H} \cap \underset{\sim}{S}$$

From these postulates, show that the membership form of the risk, $\underset{\sim}{R}$, is given by the expression

$$\mu_{\underset{\sim}{R}}(x, y) = \max\{\min(\min[\mu_{\underset{\sim}{M}}(x, y), \mu_{\underset{\sim}{ST}}(x, y)], \min[\mu_{\underset{\sim}{H}}(x, y), \mu_{\underset{\sim}{S}}(x, y)])\}$$

3.14. A new optical microscope camera uses a lookup table to relate voltage readings (which are related to illuminance) to exposure time. To aid in the creation of this lookup table, we need to determine how much time the camera should expose the pictures at a certain light level. Define a fuzzy set "around 3 volts" on a universe of voltage readings in volts

$$\underset{\sim}{V}_{1\times 5} = \left\{ \frac{.1}{2.98} + \frac{.3}{2.99} + \frac{.7}{3} + \frac{.4}{3.01} + \frac{.2}{3.02} \right\} \qquad \text{(volts)}$$

and a fuzzy set "around 1/10 second" on a universe of exposure time in seconds

$$\underset{\sim}{T}_{1\times 6} = \left\{ \frac{.1}{.05} + \frac{.3}{.06} + \frac{.3}{.07} + \frac{.4}{.08} + \frac{.5}{.09} + \frac{.2}{.1} \right\} \qquad \text{(seconds)}$$

(*a*) Find $\underset{\sim}{R} = \underset{\sim}{V} \times \underset{\sim}{T}$.

Now define a third universe of "stops." In photography, stops are related to making the picture some degree lighter or darker than the "average" exposed picture. Therefore, let Universe of Stops $= \{-2, -1.5, -1, 0, .5, 1, 1.5, 2\}$ (stops). We will define a fuzzy set on this universe as

$$\underset{\sim}{Z} = \text{a little bit lighter} = \left\{ \frac{.1}{0} + \frac{.7}{.5} + \frac{.3}{1} \right\}$$

(*b*) Find $\underset{\sim}{S} = \underset{\sim}{T} \times \underset{\sim}{Z}$.

(*c*) Find $\underset{\sim}{M} = \underset{\sim}{R} \circ \underset{\sim}{S}$ by max-min composition.

(*d*) Find $\underset{\sim}{M} = \underset{\sim}{R} \circ \underset{\sim}{S}$ by max-product composition.

3.15. In a digital code division multiple access (CDMA) cellular telephone system the carrier-to-interference (C/I) ratio of a given mobile unit is related to the number of mobile units in the sector. However, this ratio cannot be determined directly because it depends on other unknown factors such as voice quality, environment, distribution of mobile units within the sector, etc. So, we will define a universe of mobile units in a sector, N = {20, 25, 30, 35}, and a universe of C/I ratios in decibels, C = {4, 6, 8, 10, 12} (db). Suppose that a Cartesian product has been found such that

$$\underset{\sim}{R} = \begin{array}{c} \\ 20 \\ 25 \\ 30 \\ 35 \end{array} \begin{array}{c} \begin{array}{ccccc} 4 & 6 & 8 & 10 & 12 \end{array} \\ \begin{bmatrix} 0 & .3 & .7 & 1 & .8 \\ .1 & .5 & 1 & .9 & .7 \\ .6 & .1 & .8 & .6 & .1 \\ 1 & .8 & .6 & .2 & 0 \end{bmatrix} \end{array}$$

Define a linguistic fuzzy set on the universe of mobile units of "approximately $\underset{\sim}{30}$ mobile units" as

$$\underset{\sim}{N}_1 = \left\{ \frac{0}{20} + \frac{.7}{25} + \frac{1}{30} + \frac{.3}{35} \right\}$$

(*a*) Find $\underset{\sim}{C}_1 = \underset{\sim}{N}_1 \circ \underset{\sim}{R}$ using max-min composition.

(*b*) Find $\underset{\sim}{C}_1 = \underset{\sim}{N}_1 \circ \underset{\sim}{R}$ using max-product composition.

(*c*) Suppose now we wish to introduce a third variable into this problem, that of voice quality. We define a universe of voice qualities, say V = {good (G), medium (M), bad (B)}. Previous crisp approaches to this problem always segregated the three attributes of voice quality into three distinct C/I regions, i.e., V = G if $C > 8$, V = M if $6 < C \leq 8$, and V = B if $C \leq 6$.

How can we change this process to consider the voice quality as a fuzzy attribute instead of using C/I ratios?

3.16. A certain astronomy experiment uses a 200-inch telescope fitted with a photomultiplier and video camera to study faint objects in space. The telescope is capable of high magnification, but because of the "twinkle" effect of the earth's atmosphere, incoming photons do not focus into pinpoints, but instead tend to scatter around common centers, as shown in Fig. P3.16.

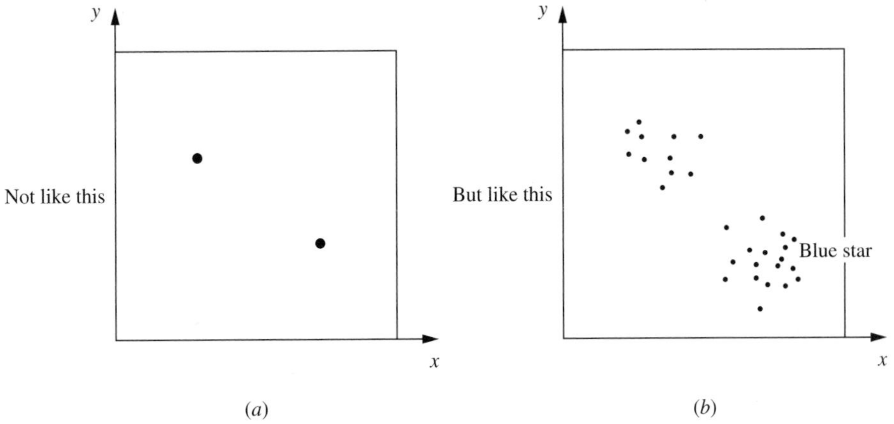

FIGURE P3.16

Astronomers are interested in tracking the motion of stars and need some way to determine velocity. The velocity of a point is easy to determine: Divide the distance traveled between telescope locations by the amount of time between these measurements. But how can astronomers measure the velocity of a star that appears, not as a point, but as a fuzzy mass of pinpoints as shown in the figure? Fuzzy relations might be useful. Consider for example the "blue star" cluster of pinpoints. Its location might be labeled as "about $x = 10$, and about $y = 6$." In fuzzy notation,

$$\text{“about } x = 10\text{”} = \underset{\sim}{X}_{10} = \left\{ \frac{0.1}{8} + \frac{0.5}{9} + \frac{1}{10} + \frac{0.3}{11} + \frac{0.1}{12} \right\}$$

$$\text{“about } y = 6\text{”} = \underset{\sim}{Y}_{6} = \left\{ \frac{0.3}{5} + \frac{1}{6} + \frac{0.9}{7} + \frac{0.2}{8} \right\}$$

(*a*) Find a fuzzy relation, $\underset{\sim}{R}$, relating to $\underset{\sim}{X}$ and $\underset{\sim}{Y}$.

Suppose the next day a new observation shows that the "blue star" has moved primarily in the x direction by an amount given by the fuzzy set

$$\text{“about } x = 9\text{”} = \underset{\sim}{X}_{9} = \left\{ \frac{0.5}{8} + \frac{1}{9} + \frac{0.9}{10} + \frac{0.4}{11} + \frac{0}{12} \right\}$$

The predicted $\underset{\sim}{Y}$ fuzzy number for this next day, which may or may not agree with the observed $\underset{\sim}{Y}$, could be determined using composition. Find the predicted $\underset{\sim}{Y}$

(*b*) Using max-min composition, $\underset{\sim}{X}_9 \circ \underset{\sim}{R}$.

(*c*) Using max-product composition.

3.17. Consider the problem of controlling the speed of a motor. Two variables related to this problem are speed (in rpm) and load (torque), resulting in the following two fuzzy membership functions:

$$\underset{\sim}{S} = \left\{ \frac{0.2}{x_1} + \frac{0.6}{x_2} + \frac{1.0}{x_3} + \frac{0.7}{x_4} + \frac{0.5}{x_5} \right\} \quad \text{where } \underset{\sim}{S} \text{ is on universe X}$$

$$\underset{\sim}{L} = \left\{ \frac{0.3}{y_1} + \frac{0.4}{y_2} + \frac{0.7}{y_3} + \frac{1.0}{y_4} + \frac{0.7}{y_5} + \frac{0.4}{y_6} + \frac{0.1}{y_7} \right\} \quad \text{where } \underset{\sim}{L} \text{ is on universe Y}$$

(*a*) Find a fuzzy relation that relates these two variables, $\underset{\sim}{R} = \underset{\sim}{S} \times \underset{\sim}{L}$.

Now define another relation, say the fuzzy current, $\underset{\sim}{I}$, that relates elements in the universe Y to elements in Z, as given here:

$$\underset{\sim}{I} = \begin{array}{c} \\ y_1 \\ y_2 \\ y_3 \\ y_4 \\ y_5 \\ y_6 \\ y_7 \end{array} \begin{array}{c} z_1 \\ \begin{bmatrix} 0.2 \\ 0.4 \\ 0.7 \\ 0.2 \\ 0.5 \\ 0.8 \\ 1.0 \end{bmatrix} \end{array}$$

(*b*) Find $\underset{\sim}{Q} = \underset{\sim}{I} \circ \underset{\sim}{R}$ using max-min composition to relate elements in X to elements in Z.

(*c*) Find $\underset{\sim}{Q} = \underset{\sim}{I} \circ \underset{\sim}{R}$ using max-product composition.

3.18. Music is not a precise science. Tactile movements by musicians on various instruments come from years of practice, and such movements are very subjective and imprecise. When a guitar player changes from an A chord to a C chord (major), his or her fingers have to move some distance, which can be measured in terms of frets (e.g., 1 fret = 0.1). This change in finger movement is described by the relation given in the following table. The table is for a six string guitar: x_i is the string number, for $i = 1, 2, \ldots, 6$. For example, −0.2 is 2 frets down and 0.3 is 3 frets up, where 0 is located at the top of the guitar fingerboard.

	C chord					
A chord	x_6	x_5	x_4	x_3	x_2	x_1

The finger positions on the guitar strings for the two chords can be given in terms of the following membership functions:

$$\underset{\sim}{C} \text{ chord} = \left\{ \frac{0}{x_6} + \frac{0.3}{x_5} + \frac{0.2}{x_4} + \frac{0}{x_3} + \frac{0.1}{x_2} + \frac{0}{x_1} \right\}$$

$$\underset{\sim}{A} \text{ chord} = \left\{ \frac{0}{x_6} + \frac{0}{x_5} + \frac{0.2}{x_4} + \frac{0.2}{x_3} + \frac{0.2}{x_2} + \frac{0}{x_1} \right\}$$

Suppose the placement of fingers on the six strings for a G chord is given as

$$\underset{\sim}{G} \text{ chord} = \left\{ \frac{0.3}{x_6} + \frac{0.2}{x_5} + \frac{0}{x_4} + \frac{0}{x_3} + \frac{0}{x_2} + \frac{0.3}{x_1} \right\}$$

(*a*) Find the relation that expresses moving from an A chord to a G chord; call this $\underset{\sim}{R}$.

(*b*) Use max-product composition to determine $\underset{\sim}{C} \circ \underset{\sim}{R}$.

3.19. In neuroscience research, it is often necessary to relate functional information to anatomical information. One source of anatomical information for a subject is a series of magnetic resonance imaging (MRI) pictures, composed of gray-level pixels, of the subject's head. For some applications, it is useful to segment (label) the brain into MRI slices (images along different virtual planes through the brain). This procedure can be difficult to do using gray-level values alone. A standard—or model—brain, combined with a distance criterion, can be used in conjunction with the gray-level information to improve the segmentation process. Define the following elements for the problem:

1. Normalized distance from the model ($\underset{\sim}{D}$)

$$\underset{\sim}{D} = \left\{ \frac{1}{0} + \frac{0.7}{1} + \frac{0.3}{2} \right\}$$

2. Intensity range for the cerebral cortex ($\underset{\sim}{I}_C$)

$$\underset{\sim}{I}_C = \left\{ \frac{0.5}{20} + \frac{1}{30} + \frac{0.6}{40} \right\}$$

3. Intensity range for the medulla ($\underset{\sim}{I}_M$)

$$\underset{\sim}{I}_M = \left\{ \frac{0.7}{20} + \frac{0.9}{30} + \frac{0.4}{40} \right\}$$

Based on these membership functions, find the following:

(*a*) $\underset{\sim}{R} = \underset{\sim}{D} \times \underset{\sim}{I}_C$

(*b*) Max-min composition of $\underset{\sim}{I}_M \circ \underset{\sim}{R}$

(*c*) Max-product composition of $\underset{\sim}{I}_M \circ \underset{\sim}{R}$

3.20. In the field of computer networking there is an imprecise relationship between the level of use of a network communication bandwidth and the latency experienced in peer-to-peer communications. Let $\underset{\sim}{X}$ be a fuzzy set of use levels (in terms of the percentage of full bandwidth used) and $\underset{\sim}{Y}$ be a fuzzy set of latencies (in milliseconds) with the following membership functions:

$$\underset{\sim}{X} = \left\{ \frac{0.2}{10} + \frac{0.5}{20} + \frac{0.8}{40} + \frac{1.0}{60} + \frac{0.6}{80} + \frac{0.1}{100} \right\}$$

$$\underset{\sim}{Y} = \left\{ \frac{0.3}{0.5} + \frac{0.6}{1} + \frac{0.9}{1.5} + \frac{1.0}{4} + \frac{0.6}{8} + \frac{0.3}{20} \right\}$$

(*a*) Find the Cartesian product represented by the relation $\underset{\sim}{R} = \underset{\sim}{X} \times \underset{\sim}{Y}$.

Now, suppose we have a second fuzzy set of bandwidth usage given by

$$\underset{\sim}{Z} = \left\{ \frac{0.3}{10} + \frac{0.6}{20} + \frac{0.7}{40} + \frac{0.9}{60} + \frac{1}{80} + \frac{0.5}{100} \right\}$$

Find $\underset{\sim}{S} = \underset{\sim}{Z}_{1\times 6} \circ \underset{\sim}{R}_{6\times 6}$

(*b*) Using max-min composition.

(*c*) Using max-product composition.

3.21. High-speed rail monitoring devices sometimes make use of sensitive sensors to measure the deflection of the earth when a rail car passes. These deflections are measured with respect to some distance from the rail car and, hence, are actually very small angles measured in microradians. Let a universe of deflections be A = {1, 2, 3, 4} where A is the angle in microradians, and let a universe of distances be D = {1, 2, 5, 7, } where D is distance in feet. Suppose a relation between these two parameters has been determined as follows:

$$\underset{\sim}{R} = \begin{array}{c} \\ A_1 \\ A_2 \\ A_3 \\ A_4 \end{array} \begin{array}{c} \begin{array}{cccc} D_1 & D_2 & D_3 & D_4 \end{array} \\ \begin{bmatrix} 1 & .3 & .1 & 0 \\ .2 & 1 & .3 & .1 \\ 0 & .7 & 1 & .2 \\ 0 & .1 & .4 & 1 \end{bmatrix} \end{array}$$

Now let a universe of rail car weights be W = {1, 2}, where W is the weight in units of 100,000 pounds. Suppose the fuzzy relation of W to A is given by

$$\underset{\sim}{S} = \begin{array}{c} \\ A_1 \\ A_2 \\ A_3 \\ A_4 \end{array} \begin{array}{c} \begin{array}{cc} W_1 & W_2 \end{array} \\ \begin{bmatrix} 1 & .4 \\ .5 & 1 \\ .3 & .1 \\ 0 & 0 \end{bmatrix} \end{array}$$

Using these two relations, find the relation, $\underset{\sim}{R}^T \circ \underset{\sim}{S} = \underset{\sim}{T}$ (note the matrix transposition here)

(*a*) Using max-min composition.

(*b*) Using max-product composition.

3.22. In engineering management it is often necessary to borrow money to finance projects of significant size. Engineering firms are therefore quite sensitive about their financing capacity, or their credit limits. Credit limits are often related to previous borrowing practices and to current credit account average balances. In turn, average balances are related to the profitability of the engineering venture. Suppose we have a fuzzy set $\underset{\sim}{C}$ for credit limits, a fuzzy set $\underset{\sim}{B}$ for average account balance, and a fuzzy set $\underset{\sim}{P}$ for profits, all in units of thousands of dollars; the universes for these fuzzy sets might be, for example,

$$\begin{aligned} \text{Credit limits} &= \{500, 1000, 1500, 2000\} \\ \text{Average account balance} &= \{20, 50, 100, 500, 1000, 1200\} \\ \text{Profits} &= \{-50, 0, 50, 100, 500\} \end{aligned}$$

Suppose that the following relation has been established between the credit limits of the firm and its average account balance,

$$\underset{\sim}{A} = \begin{bmatrix} .8 & 1 & .6 & .2 & 0 & 0 \\ .2 & .3 & .5 & .8 & .1 & 0 \\ .1 & .2 & .4 & .9 & .6 & .1 \\ .1 & .1 & .4 & .8 & .8 & .3 \end{bmatrix}$$

and that the following relation has been established between the average account balances of the firm and its profit margins:

$$\underset{\sim}{E} = \begin{bmatrix} .8 & .9 & .7 & .1 & 0 \\ .7 & 1 & .8 & 0 & 0 \\ .5 & .9 & .9 & .5 & .1 \\ .2 & .5 & .7 & .8 & .9 \\ .1 & .5 & .3 & .9 & .9 \\ 0 & .4 & .6 & .8 & .7 \end{bmatrix}$$

Find a relation between $\underset{\sim}{C}$ and $\underset{\sim}{P}$ by computing $\underset{\sim}{A} \circ \underset{\sim}{E}$,

(*a*) Using max-min composition.

(*b*) Using max-product composition.

3.23. In a ground transportation system, consider the relationship between number of people moved from point to point and the number of wheels used to accomplish this movement. For example, a motorcycle typically uses two wheels to move one person. Let X be a universe of speeds (miles per hour) at which these systems move people, i.e., X = {5, 30, 50, 100, 300+}. Let Y be the universe of the number of wheels that the system has, i.e., Y = {2, 4, 8, 10, 16}. We now define a "fast" system as a fuzzy set on X, or

$$\underset{\sim}{\text{Fast}} = \left\{ \frac{0}{5} + \frac{0.1}{30} + \frac{0.3}{50} + \frac{0.8}{100} + \frac{1.0}{300+} \right\}$$

and we define a "personally" owned system as a fuzzy set on Y, or

$$\underset{\sim}{\text{Personal}} = \left\{ \frac{0.7}{2} + \frac{1.0}{4} + \frac{0.1}{8} + \frac{0}{10} + \frac{0}{16} \right\}$$

(*a*) Find a fuzzy relation using the Cartesian product relating a "fast" system to a "personal" system.

Now, consider a "slow" system defined on X, given by the fuzzy set

$$\underset{\sim}{\text{Slow}} = \left\{ \frac{1.0}{5} + \frac{0.8}{30} + \frac{0}{50} + \frac{0}{100} + \frac{0}{300} \right\}$$

Find a relation between a "slow" system and the previously determined relation of part (*a*)

(*b*) Using max-min composition.

(*c*) Using max-product composition.

3.24. In a simple vision recognition system, we may want to locate specific objects within a scene containing many objects. Let X be a universe of general, well-known objects, such as

$$X = \{\text{car, boat, house, bike, tree, mountain}\}$$

and let Y be a universe of simple geometric shapes, such as

$$Y = \{\text{square, octagon, triangle, circle, ellipse}\}$$

We now define simple fuzzy sets of objects, such as those for "car" and "square:"

$$\underset{\sim}{A} = \text{car} = \left\{ \frac{1.0}{\text{car}} + \frac{0.4}{\text{boat}} + \frac{0.1}{\text{house}} + \frac{0.6}{\text{bike}} + \frac{0.1}{\text{tree}} + \frac{0}{\text{mountain}} \right\}$$

$$\underset{\sim}{B} = \text{square} = \left\{ \frac{1.0}{\text{square}} + \frac{0.5}{\text{octagon}} + \frac{0.4}{\text{triangle}} + \frac{0}{\text{circle}} + \frac{0.1}{\text{ellipse}} \right\}$$

(*a*) Find a relation between "car" and "square."

Now suppose we define another fuzzy set of objects, called "shapes with corners," as

$$\underset{\sim}{S} = \text{"shapes with corners"} = \left\{ \frac{0.6}{\text{square}} + \frac{0.9}{\text{octagon}} + \frac{0.4}{\text{triangle}} + \frac{0}{\text{circle}} + \frac{0.2}{\text{ellipse}} \right\}$$

Using an appropriate composition operation, find the fuzzy set $\underset{\sim}{T}$ = "objects with corners" using the following:

(*b*) Max-min composition.

(*c*) Max-product composition.

3.25. In computer engineering it is often desirable for management of memory to know the relationship between the number of "free" pages of memory on a machine and the number of pages on the modified page list of that machine. Another feature of the architecture that influences this memory management is the number of processors in the design. Suppose $\underset{\sim}{F}$ is a fuzzy set on the universe of possible free pages on a VAX, i.e.,

$$\underset{\sim}{F} = \{1500, 2000, 2500, 3000\}$$

and suppose $\underset{\sim}{M}$ is a fuzzy set on the universe of possible pages on the modified page list of that VAX, i.e.,

$$\underset{\sim}{M} = \{2000, 2500, 3000, 3500, 4000\}$$

Moreover, assume that a relation between $\underset{\sim}{M}$ and $\underset{\sim}{F}$ has been established as

$$\underset{\sim}{R} = \underset{\sim}{M} \times \underset{\sim}{F} = \begin{matrix} & 1500 & 2000 & 2500 & 3000 \\ 2000 & .4 & .4 & .4 & .3 \\ 2500 & .5 & .5 & .4 & .3 \\ 3000 & .6 & .5 & .4 & .3 \\ 3500 & .6 & .5 & .4 & 0 \\ 4000 & .1 & .1 & .1 & 0 \end{matrix}$$

Now suppose $\underset{\sim}{P}$ is a fuzzy set on the universe of the possible number of processors on the VAX,

$$\underset{\sim}{P} = \{40, 55, 70, 85\}$$

and a relation between $\underset{\sim}{F}$ and $\underset{\sim}{P}$ exists, such as

$$\underset{\sim}{S} = \underset{\sim}{F} \times \underset{\sim}{P} = \begin{matrix} & 40 & 55 & 70 & 85 \\ 1500 & .1 & .3 & .8 & .3 \\ 2000 & .3 & .5 & .6 & .7 \\ 2500 & .7 & .6 & .5 & .3 \\ 3000 & 1 & .8 & .3 & .1 \end{matrix}$$

Find the fuzzy relation $\underset{\sim}{T}$ that relates $\underset{\sim}{M}$ and $\underset{\sim}{P}$ using

(*a*) Max-min composition.

(*b*) Max-product composition.

3.26 In the field of soil mechanics new research methods involving vision recognition systems are being applied to soil masses to watch individual grains of soil as they translate and rotate under confining pressures. In tracking the motion of the soil particles some problems arise with the vision recognition software. One problem is called "occlusion," whereby a soil particle that is being tracked becomes partially or completely occluded from view by passing behind other soil particles. Occlusion can also occur when a tracked particle is behind a mark on the camera's lens, or the tracked particle is partly out of sight of the camera. In this problem we will consider only occlusions for the first two problems mentioned. Let us define a universe of parameters for particle occlusion, say

$$\mathrm{X} = \{x_1, x_2, x_3\}$$

and a universe of parameters for lens mark occlusion, say

$$\mathrm{Y} = \{y_1, y_2, y_3\}$$

Then define,

$$\underset{\sim}{A} = \left\{ \frac{0.1}{x_1} + \frac{0.9}{x_2} + \frac{0.0}{x_3} \right\}$$

as a specific fuzzy set for a tracked particle behind another particle, and let

$$\underset{\sim}{B} = \left\{ \frac{0}{y_1} + \frac{1}{y_2} + \frac{0}{y_3} \right\}$$

be a particular fuzzy set for an occlusion behind a lens mark.

(*a*) Find the relation, $\underset{\sim}{R} = \underset{\sim}{A} \times \underset{\sim}{B}$, using a Cartesian product.

Let $\underset{\sim}{C}$ be another fuzzy set in which a tracked particle is behind a particle, for example,

$$\underset{\sim}{C} = \left\{ \frac{0.3}{x_1} + \frac{1.0}{x_2} + \frac{0.0}{x_3} \right\}$$

(*b*) Using max-min composition, find $\underset{\sim}{S} = \underset{\sim}{C} \circ \underset{\sim}{R}$.

3.27. This problem deals with relating the speed of a motor to the voltage applied to it. The speed of the motor has a universe $\mathrm{S} = \{0, 1, 2, 3\}$ in degrees per second and the voltage has a universe $\mathrm{V} = \{0, 1, 2, 3, 4, 5, 6\}$ in volts.

Suppose we create a fuzzy set on each universe, e.g.,

$$\underset{\sim}{S}_2 = \left\{ \frac{1/3}{0} + \frac{2/3}{1} + \frac{1}{2} + \frac{2/3}{3} \right\} = \text{"speed about 2"}$$

$$\underset{\sim}{V}_0 = \left\{ \frac{1}{0} + \frac{3/4}{1} + \frac{1/2}{2} + \frac{1/4}{3} + \frac{0}{5} + \frac{0}{6} \right\} = \text{"voltage about 0"}$$

(*a*) Find the Cartesian product relation, $\underset{\sim}{R}$, between $\underset{\sim}{S}_2$ and $\underset{\sim}{V}_0$.

Creating another fuzzy set on universe $\underset{\sim}{V}$ for "voltage about 3" might give

$$\underset{\sim}{V}_3 = \left\{ \frac{0}{0} + \frac{1/4}{1} + \frac{1/2}{2} + \frac{1}{3} + \frac{1/2}{5} + \frac{0}{6} \right\}$$

(*b*) Use max-min composition to find $\underset{\sim}{V}_3 \circ \underset{\sim}{R}$.

(*c*) Use max-product composition to find $\underset{\sim}{V}_3 \circ \underset{\sim}{R}$.

Equivalence Relations

3.28. Which of the following are equivalence relations?

	Set	**Relation on the set**
(*a*)	People	Is the brother of
(*b*)	People	Has the same parents as
(*c*)	Points on a map	Is connected by a road to
(*d*)	Lines in plane geometry	Is perpendicular to
(*e*)	Positive integers	For some integer k, equals 10^k times

Draw graphs (similar to those in Figs. 3.8 and 3.9) of the equivalence relations with appropriate labels on the vertices.

3.29 The accompanying Sagittal diagrams (Fig. P3.29) show two relations on the universe, X = {1, 2, 3}. Are these relations equivalence relations [Gill, 1976]?

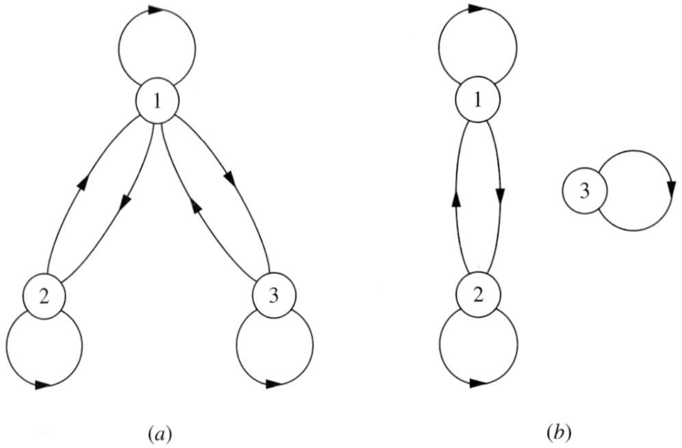

FIGURE P3.29
From [Gill, 1976].

CHAPTER 4

MEMBERSHIP FUNCTIONS

So far as the laws of mathematics refer to reality, they are not certain. And so far as they are certain, they do not refer to reality.

Albert Einstein,
Theoretical physicist and Nobel laureate
"Geometrie und Erfahrung," Lecture to Prussian Academy, 1921

The statement above, from Albert Einstein, attests to the fact that few things in real life are certain or can be conveniently reduced to the axioms of mathematical theories and models. A metaphorical expression that represents this idea is known as the "Law of Probable Dispersal," to wit, "Whatever it is that hits the fan will not be evenly distributed." As this enlightened law implies, most things in nature cannot be characterized with simple or convenient shapes or distributions. Membership functions characterize the fuzziness in a fuzzy set—whether the elements in the set are discrete or continuous—in a graphical form for eventual use in the mathematical formalisms of fuzzy set theory. But the shapes used to describe the fuzziness have very few restrictions indeed. It might be claimed that the rules used to describe fuzziness graphically are also fuzzy. Nevertheless, as with any formal mathematical structure, some standard terms related to the shape of membership functions have been developed over the years, and these terms are defined here. Just as there are an infinite number of ways to characterize fuzziness, there are an infinite number of ways to graphically depict the membership functions that describe fuzziness. This chapter describes only a very small number of possibilities for these membership functions.

Since the membership function essentially embodies all fuzziness for a particular fuzzy set, its description is the essence of a fuzzy property or operation. Because of the importance of the "shape" of the membership function, a great deal of attention has been focused on development of these functions. This chapter describes, then illustrates, seven procedures that have been used to build membership functions. There are many more. For example, fuzzy statistics has been used to develop membership functions; this method is not addressed in this text, because it requires a good deal of development and the limited space available would not provide an adequate introduction to this topic. References at the end of this chapter can be consulted on this topic.

FEATURES OF THE MEMBERSHIP FUNCTION

Since all information contained in a fuzzy set is described by its membership function, it is useful to develop a lexicon of terms to describe various special features of this function. For purposes of simplicity, the functions shown in the following figures will all be continuous, but the terms apply equally for both discrete and continuous fuzzy sets. Figure 4.1 assists in this description.

The *core* of a membership function for some fuzzy set $\underset{\sim}{A}$ is defined as that region of the universe that is characterized by complete and full membership in the set $\underset{\sim}{A}$. That is, the core comprises those elements x of the universe such that $\mu_{\underset{\sim}{A}}(x) = 1$.

The *support* of a membership function for some fuzzy set $\underset{\sim}{A}$ is defined as that region of the universe that is characterized by nonzero membership in the set $\underset{\sim}{A}$. That is, the support comprises those elements x of the universe such that $\mu_{\underset{\sim}{A}}(x) > 0$.

The *boundaries* of a membership function for some fuzzy set $\underset{\sim}{A}$ are defined as that region of the universe containing elements that have a nonzero membership but not complete membership. That is, the boundaries comprise those elements x of the universe such that $0 < \mu_{\underset{\sim}{A}}(x) < 1$. These elements of the universe are those with some *degree* of fuzziness, or only partial membership in the fuzzy set $\underset{\sim}{A}$. Figure 4.1 illustrates the regions in the universe comprising the core, support, and boundaries of a typical fuzzy set.

A *normal* fuzzy set is one whose membership function has at least one element x in the universe whose membership value is unity. For fuzzy sets where one and only one element has a membership equal to one, this element is typically referred to as the *prototype* of the set, or the prototypical element. Figure 4.2 illustrates typical normal and subnormal fuzzy sets.

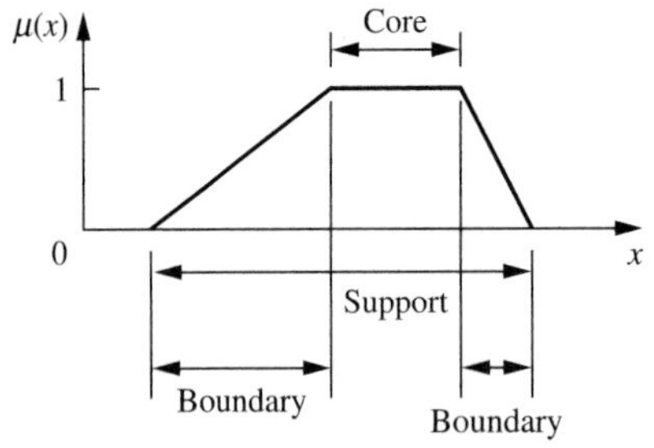

FIGURE 4.1
Core, support, and boundaries of a fuzzy set.

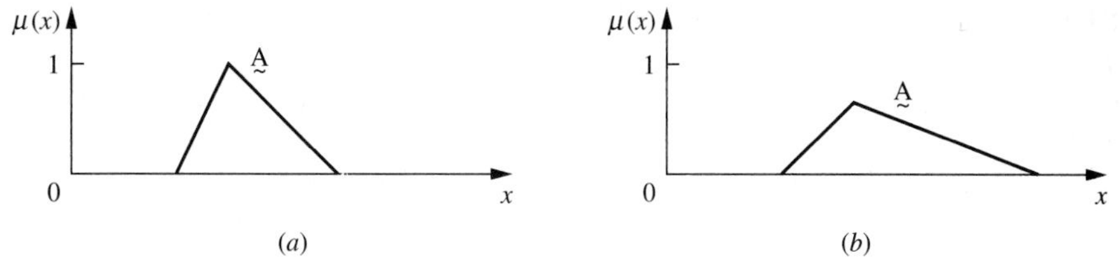

FIGURE 4.2
Fuzzy sets that are normal (*a*) and subnormal (*b*).

A *convex* fuzzy set is described by a membership function whose membership values are strictly monotonically increasing, or whose membership values are strictly monotonically decreasing, or whose membership values are strictly monotonically increasing then strictly monotonically decreasing with increasing values for elements in the universe. Said another way, if, for any elements x, y, and z in a fuzzy set $\underset{\sim}{A}$, the relation $x < y < z$ implies that

$$\mu_{\underset{\sim}{A}}(y) \geq \min[\mu_{\underset{\sim}{A}}(x), \mu_{\underset{\sim}{A}}(z)] \tag{4.1}$$

then $\underset{\sim}{A}$ is said to be a convex fuzzy set [Zadeh, 1965]. Figure 4.3 shows a typical convex fuzzy set and a typical nonconvex fuzzy set. It is important to remark here that this definition of convexity is *different* from some definitions of the same term in mathematics. In some areas of mathematics, convexity of shape has to do with whether a straight line through any part of the shape goes outside the boundaries of that shape. This definition of convexity is *not* used here; Eq. (4.1) succinctly summarizes our definition of convexity.

A special property of two convex fuzzy sets, say $\underset{\sim}{A}$ and $\underset{\sim}{B}$, is that the intersection of these two convex fuzzy sets is also a convex fuzzy set, as shown in Fig. 4.4. That is, for $\underset{\sim}{A}$ *and* $\underset{\sim}{B}$, which are both convex, $\underset{\sim}{A} \cap \underset{\sim}{B}$ is also convex.

The *crossover points* of a membership function are defined as the elements in the universe for which a particular fuzzy set $\underset{\sim}{A}$ has values equal to 0.5, i.e., for which $\mu_{\underset{\sim}{A}}(x) = 0.5$.

The *height* of a fuzzy set $\underset{\sim}{A}$ is the maximum value of the membership function, i.e., $\max\{\mu_{\underset{\sim}{A}}(x)\}$. If the height of a fuzzy set is less than unity, the fuzzy set is said to be subnormal.

If $\underset{\sim}{A}$ is a convex single-point normal fuzzy set defined on the real line, then $\underset{\sim}{A}$ is often termed a *fuzzy number.*

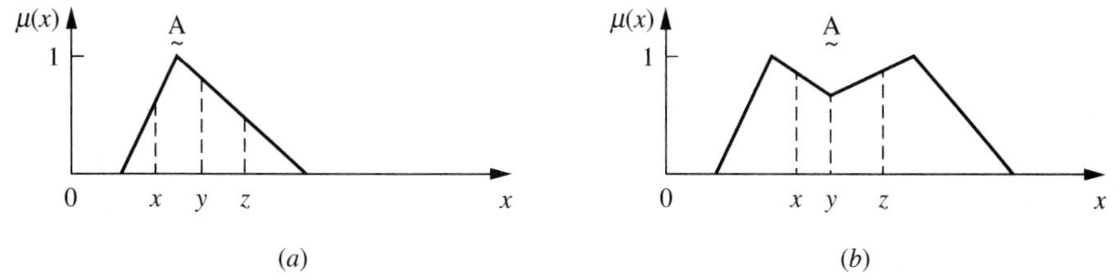

FIGURE 4.3
Convex, normal fuzzy set (*a*) and nonconvex, normal fuzzy set (*b*).

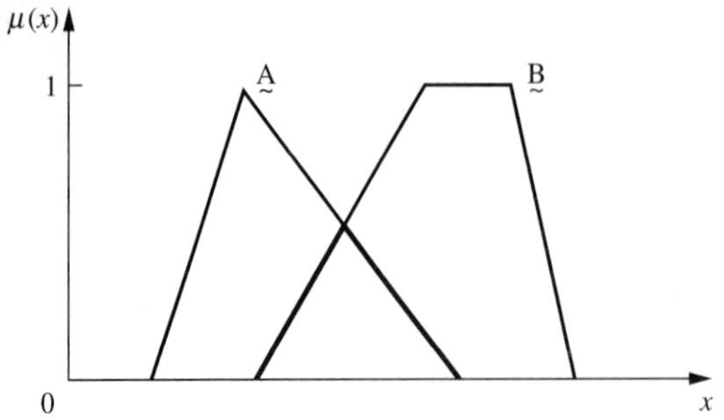

FIGURE 4.4
The intersection of two convex fuzzy sets produces a convex fuzzy set.

STANDARD FORMS AND BOUNDARIES

The most common forms of membership functions are those that are normal and convex. However, many operations on fuzzy sets, hence operations on membership functions, result in fuzzy sets that are subnormal and nonconvex. For example, the extension principle to be discussed in Chapter 6 and the union operator both can produce subnormal or nonconvex fuzzy sets.

Membership functions can be symmetrical or asymmetrical. They are typically defined on one-dimensional universes, but they certainly can be described on multidimensional (or n-dimensional) universes. For example, the membership functions shown in this chapter are one-dimensional curves. In two dimensions these curves become surfaces and for three or more dimensions these surfaces become hypersurfaces. These hypersurfaces, or curves, are simple mappings from combinations of the parameters in the n-dimensional space to a membership value on the interval [0, 1]. Again, this membership value expresses the degree of membership that the specific combination of parameters in the n-dimensional space has in a particular fuzzy set defined on the n-dimensional universe of discourse. The hypersurfaces for an n-dimensional universe are analogous to joint probability density functions; but, of course, the mapping for the membership function is to membership in a particular set and not to relative frequencies, as it is for probability density functions.

FUZZIFICATION

Fuzzification is the process of making a crisp quantity fuzzy. We do this by simply recognizing that many of the quantities that we consider to be crisp and deterministic are actually not deterministic at all: They carry considerable uncertainty. If the form of uncertainty happens to arise because of imprecision, ambiguity, or vagueness, then the variable is probably fuzzy and can be represented by a membership function.

In the real world, hardware such as a digital voltmeter generates crisp data, but these data are subject to experimental error. The information shown in Fig. 4.5 shows one possible range of errors for a typical voltage reading and the associated membership function that might represent such imprecision.

The representation of imprecise data as fuzzy sets is a useful but not mandatory step when those data are used in fuzzy systems. This idea is shown in Fig. 4.6, where we consider the data as a crisp reading, Fig. 4.6*a*, or as a fuzzy reading, as shown in

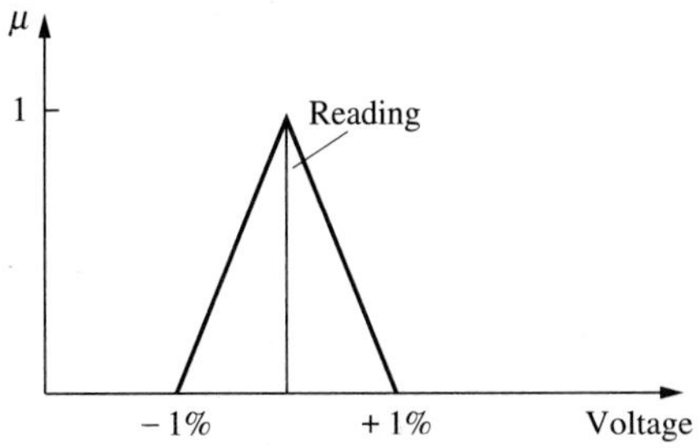

FIGURE 4.5
Membership function representing imprecision in "crisp voltage reading."

Fig. 4.6*b*. In Fig. 4.6*a* we might want to compare a crisp voltage reading to a fuzzy set, say "low voltage." In the figure we see that the crisp reading intersects the fuzzy set "low voltage" at a membership of 0.3; i.e., the fuzzy set and the reading can be said to agree at a membership value of 0.3. In Fig. 4.6*b* the intersection of the fuzzy set "medium voltage" and a fuzzified voltage reading occurs at a membership of 0.4. We can see in Fig. 4.6*b* that the set intersection of the two fuzzy sets is a small triangle, whose largest membership occurs at the membership value of 0.4. We will see more about the importance of fuzzification of crisp variables in Chapters 9 and 13 of this text.

MEMBERSHIP VALUE ASSIGNMENTS

There are possibly more ways to assign membership values or functions to fuzzy variables than there are to assign probability density functions to random variables

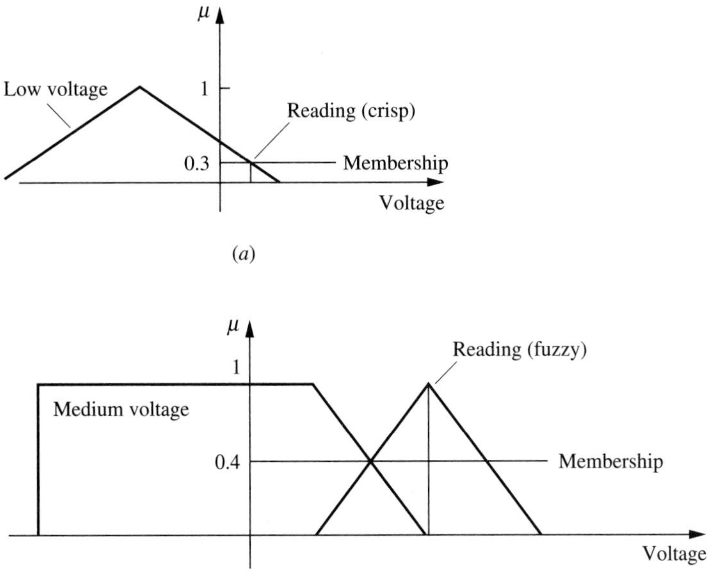

FIGURE 4.6
Comparisons of fuzzy sets and crisp or fuzzy readings: (*a*) fuzzy set and crisp reading; (*b*) fuzzy set and fuzzy reading.

[see Dubois and Prade, 1980]. This assignment process can be intuitive or it can be based on some algorithmic or logical operations. The following list provides some of the methods described in the literature to assign membership values or functions to fuzzy variables. Only the first seven of these methods will be illustrated in simple examples in this chapter. The eighth method (soft partitioning) is considered separately in Chapter 11, "Classification," because of its importance to other disciplines such as decision making and pattern recognition. The literature on this topic is rich with references, and a short list of those consulted is provided in the summary of this chapter.

- Intuition
- Inference
- Rank ordering
- Angular fuzzy sets
- Neural networks
- Genetic algorithms
- Inductive reasoning
- Soft partitioning
- Meta rules
- Fuzzy statistics

The last two methods listed are not addressed in this book because of the wealth of information available in other texts and because of the relative complexity and depth required to address them adequately.

Intuition

This method needs little or no introduction. It is simply derived from the capacity of humans to develop membership functions through their own innate intelligence and understanding. Intuition involves contextual and semantic knowledge about an issue; it can also involve linguistic truth values about this knowledge [see Zadeh, 1972]. As an example, consider the membership functions for the fuzzy variable temperature. Figure 4.7 shows various shapes on the universe of temperature as measured in units of degrees Celsius. Each curve is a membership function corresponding to various fuzzy variables, such as very cold, cold, normal, hot, and very hot. Of course, these curves are a function of context and the analyst developing them. For example, if the temperatures are referred to the range of human comfort we get one set of curves, and

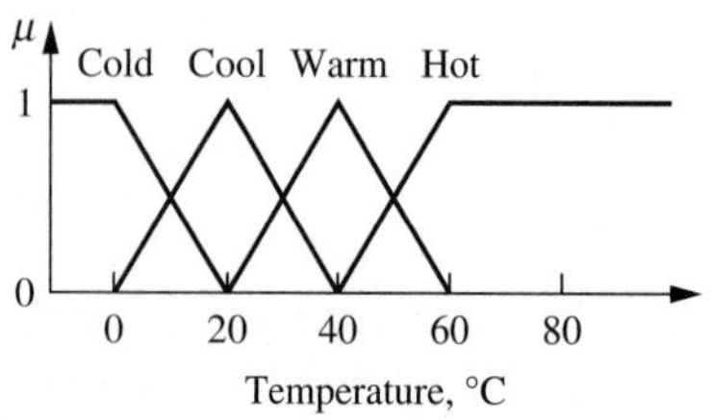

FIGURE 4.7
Membership functions for the fuzzy variable "temperature."

if they are referred to the range of safe operating temperatures for a steam turbine we get another set. However, the important character of these curves for purposes of use in fuzzy operations is the fact that they overlap. In numerous examples throughout the rest of this text we shall see that the precise shapes of these curves are not so important in their utility. Rather, it is the approximate placement of the curves on the universe of discourse, the number of curves (partitions) used, and the overlapping character that are the most important ideas.

Inference

In the inference method we use knowledge to perform deductive reasoning. That is, we wish to deduce or infer a conclusion, given a body of facts and knowledge. There are many forms of this method documented in the literature, but the one we will illustrate here relates to our formal knowledge of geometry and geometric shapes, similar to ideas posed in Chapter 1.

In the identification of a triangle, let A, B, and C be the inner angles of a triangle, in the order $A \geq B \geq C \geq 0$, and let U be the universe of triangles; i.e.,

$$\mathrm{U} = \{(A, B, C) \mid A \geq B \geq C \geq 0;\ A + B + C = 180^\circ\} \tag{4.2}$$

We define a number of geometric shapes that we wish to be able to identify for any collection of angles fulfilling the constraints given in Eq. (4.2). For this purpose we will define the following five types of triangles:

$\underset{\sim}{\mathrm{I}}$	Approximate isosceles triangle
$\underset{\sim}{\mathrm{R}}$	Approximate right triangle
$\underset{\sim}{\mathrm{IR}}$	Approximate isosceles *and* right triangle
$\underset{\sim}{\mathrm{E}}$	Approximate equilateral triangle
$\underset{\sim}{\mathrm{T}}$	Other triangles

We can infer membership values for all of these triangle types through the method of inference, because we possess knowledge about geometry that helps us to make the membership assignments. So we shall list this knowledge here to develop an algorithm to assist us in making these membership assignments for any collection of angles meeting the constraints of Eq. (4.2).

For the approximate isosceles triangle we have the following algorithm for the membership, again for the situation of $A \geq B \geq C \geq 0$ and $A + B + C = 180^\circ$:

$$\mu_{\underset{\sim}{\mathrm{I}}}(A, B, C) = 1 - \frac{1}{60^\circ} \min(A - B, B - C) \tag{4.3}$$

So, for example, if $A = B$ or $B = C$, the membership value in the approximate isosceles triangle is $\mu_{\underset{\sim}{\mathrm{I}}} = 1$; if $A = 120^\circ$, $B = 60^\circ$, and $C = 0$, then $\mu_{\underset{\sim}{\mathrm{I}}} = 0$. For a fuzzy right triangle, we have

$$\mu_{\underset{\sim}{\mathrm{R}}}(A, B, C) = 1 - \frac{1}{90^\circ} \left|A - 90^\circ\right| \tag{4.4}$$

For instance, when $A = 90^\circ$, the membership value in the fuzzy right triangle, $\mu_{\underset{\sim}{\mathrm{R}}} = 1$, or when $A = 180^\circ$, this membership vanishes, i.e., $\mu_{\underset{\sim}{\mathrm{R}}} = 0$. For the case of an approximate isosceles *and* right triangle (there is only one of these in the crisp domain), we can find this membership function by taking the logical intersection (*and* operator) of the isosceles and right triangle membership functions, or

$$\underset{\sim}{\mathrm{IR}} = \underset{\sim}{\mathrm{I}} \cap \underset{\sim}{\mathrm{R}}$$

which results in

$$\begin{aligned}\mu_{\underset{\sim}{\mathrm{IR}}}(A, B, C) &= \min[\mu_{\underset{\sim}{\mathrm{I}}}(A, B, C), \mu_{\underset{\sim}{\mathrm{R}}}(A, B, C)] \\ &= 1 - \max\left[\tfrac{1}{60^\circ}\min(A - B, B - C), \tfrac{1}{90^\circ}\left|A - 90^\circ\right|\right]\end{aligned} \tag{4.5}$$

For the case of a fuzzy equilateral triangle, the membership function is given by

$$\mu_{\underset{\sim}{\mathrm{E}}}(A, B, C) = 1 - \tfrac{1}{180^\circ}(A - C) \tag{4.6}$$

For example, when $A = B = C$, the membership value is $\mu_{\underset{\sim}{\mathrm{E}}}(A, B, C) = 1$; when $A = 180^\circ$, the membership value vanishes, or $\mu_{\underset{\sim}{\mathrm{E}}} = 0$. Finally, for the set of "all other triangles" (all triangular shapes other than I, R, and E) we simply invoke the complement of the logical union of the three previous cases [or, from De Morgan's laws (Eq. (2.13)), the intersection of the complements of the triangular shapes],

$$\underset{\sim}{\mathrm{T}} = = \overline{(\underset{\sim}{\mathrm{I}} \cup \underset{\sim}{\mathrm{R}} \cup \underset{\sim}{\mathrm{E}})} = \overline{\underset{\sim}{\mathrm{I}}} \cap \overline{\underset{\sim}{\mathrm{R}}} \cap \overline{\underset{\sim}{\mathrm{E}}}$$

which results in

$$\begin{aligned}\mu_{\underset{\sim}{\mathrm{T}}}(A, B, C) &= \min\{1 - \mu_{\underset{\sim}{\mathrm{I}}}(A, B, C), 1 - \mu_{\underset{\sim}{\mathrm{E}}}(A, B, C), 1 - \mu_{\underset{\sim}{\mathrm{R}}}(A, B, C)\} \\ &= \tfrac{1}{180^\circ}\min\{3(A - B), 3(B - C), 2\left|A - 90^\circ\right|, A - C\}\end{aligned} \tag{4.7}$$

Example 4.1. [Dong, 1987]. Define a specific triangle, as shown in Fig. 4.8, with these three ordered angles:

$$\{\mathrm{X} : A = 85^\circ \geq B = 50^\circ \geq C = 45^\circ, \text{ where } A + B + C = 180^\circ\}$$

The membership values for the fuzzy triangle shown in Fig. 4.8 for each of the fuzzy triangles types are determined from Eqs. (4.3)–(4.7), as listed here:

$$\begin{aligned}\mu_{\underset{\sim}{\mathrm{R}}}(x) &= 0.94 \\ \mu_{\underset{\sim}{\mathrm{I}}}(x) &= 0.916 \\ \mu_{\underset{\sim}{\mathrm{IR}}}(x) &= 0.916 \\ \mu_{\underset{\sim}{\mathrm{E}}}(x) &= 0.7 \\ \mu_{\underset{\sim}{\mathrm{T}}}(x) &= 0.05\end{aligned}$$

Hence, it appears that the triangle given in Fig. 4.8 has the highest membership in the set of fuzzy right triangles, i.e., in $\underset{\sim}{\mathrm{R}}$. Notice, however, that the triangle in Fig. 4.8 also has high membership in the isosceles triangle fuzzy set, and reasonably high membership in the equilateral fuzzy triangle set.

Rank Ordering

Assessing preferences by a single individual, a committee, a poll, and other opinion methods can be used to assign membership values to a fuzzy variable. Prefer-

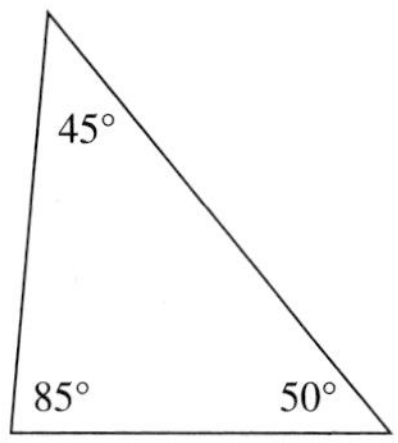

FIGURE 4.8
A specific triangle.

ence is determined by pairwise comparisons, and these determine the ordering of the membership. More is provided on the area of rank ordering in Chapter 10, "Fuzzy Decision Making." This method is very similar to a relative preferences method developed by Saaty [1974].

Example 4.2. [Dong, 1987]. Suppose 1000 people respond to a questionnaire about their pairwise preferences among five colors, X = {red, orange, yellow, green, blue}. Define a fuzzy set as $\underset{\sim}{A}$ on the universe of colors "best color." Table 4.1 is a summary of the opinion survey. In this table, for example, out of 1000 people 517 preferred the color red to the color orange, 841 preferred the color orange to the color yellow, etc. Note that the color columns in the table represent an "antisymmetric" matrix. Such a matrix will be seen to relate to a *reciprocal relation,* which is introduced in Chapter 10. The total number of responses is 10,000 (10 comparisons). If the sum of the preferences of each color (row sum) is normalized to the total number of responses, a rank ordering can be determined as shown in the last two columns of the table.

If the percentage preference (the percentage column of Table 4.1) of these colors is plotted to a normalized scale on the universe of colors in an ascending order on the color universe, the membership function for "best color" shown in Fig. 4.9 would result. Alternatively, the membership function could be formed based on the rank order developed (last column of Table 4.1).

Angular Fuzzy Sets

Angular fuzzy sets were first introduced by Hadipriono and Sun [1990]. Angular fuzzy sets differ from standard fuzzy sets only in their coordinate description. Angular fuzzy sets are defined on a universe of angles, hence are repeating shapes

TABLE 4.1
Example in rank ordering

	Number who preferred							
	Red	**Orange**	**Yellow**	**Green**	**Blue**	**Total**	**Percentage**	**Rank order**
Red	—	517	525	545	661	2,248	22.5	2
Orange	483	—	841	477	576	2,377	23.8	1
Yellow	475	159	—	534	614	1,782	17.8	4
Green	455	523	466	—	643	2,087	20.9	3
Blue	339	524	386	357	—	1,506	15	5
Total						10,000		

every 2π cycles. In most applications angular fuzzy sets are used in the quantitative description of linguistic variables known as truth values. These will be discussed further in Chapter 8, but for now we can simply suggest that the variable "truth" is no different from any other linguistic variable in that it can be described by a fuzzy set. We will see in Chapter 7 that logical propositions are associated with a degree of truth, which will be equated to a membership value in the fuzzy set "truth." When a certain proposition has a membership value of 1 it is said to be true, and when the proposition has a membership value of 0 it is said to be false; values in between 0 and 1 correspond to a proposition being partially true (or partially false).

We will explain angular fuzzy sets with an example in environmental engineering. Suppose we measure the pH value of some water samples collected from a polluted pond and give these pH readings linguistic labels, such as very basic, fairly acidic, etc., to communicate qualitative information about the pollution. It is well known that a neutral solution has a pH of 7. The linguistic terms can be built in such a way that a "neutral" (N) solution corresponds to $\theta = 0$ rad, "absolutely basic" (AB) corresponds to $\theta = \pi/2$ rad, and "absolutely acidic" (AA) corresponds to $\theta = -\pi/2$. Levels of pH between 14 and 7 can be labeled as "very basic" (VB), "basic" (B), "fairly basic" (FB), and so on, and are represented graphically between $\theta = 0$ and $\theta = \pi/2$. Levels of pH between 7 and 0 are labeled as "very acidic" (VA), "acidic" (A), "fairly acidic" (FA), and so on and are represented between $\theta = 0$ and $\theta = -\pi/2$. Using these linguistic labels for pH, the model of the angular fuzzy set is shown in Fig. 4.10.

The linguistic values vary with θ, and their membership values are on the $\mu(\theta)$ axis. The membership value of the linguistic term can be obtained from the equation

$$\mu_t(\theta) = t \cdot \tan(\theta) \tag{4.8}$$

where t is the horizontal projection of a radial vector (see Fig. 4.10). Angular fuzzy sets are useful in situations that have a natural basis in polar coordinates, or in situations where the value of a variable is cyclic. Truth qualification (Chapter 8) examples can be described in an angular sense; a proposition can move from true through ambivalence (undecided) to false, back through ambivalence and return to true, just as one could travel around a circle.

Example 4.3. AC motors have a special advantage over DC motors in the rotation of armatures. Where DC motors rotate unidirectionally, AC motors can rotate in both directions. Among AC motors, let us take the case of stepper motors. There are two types of stepper motors: (*i*) a variable reluctance (resistance to magnetic flux) motor

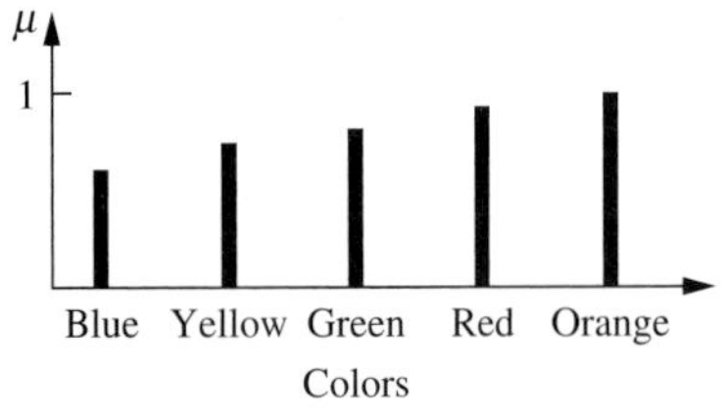

FIGURE 4.9
Membership function for best color.

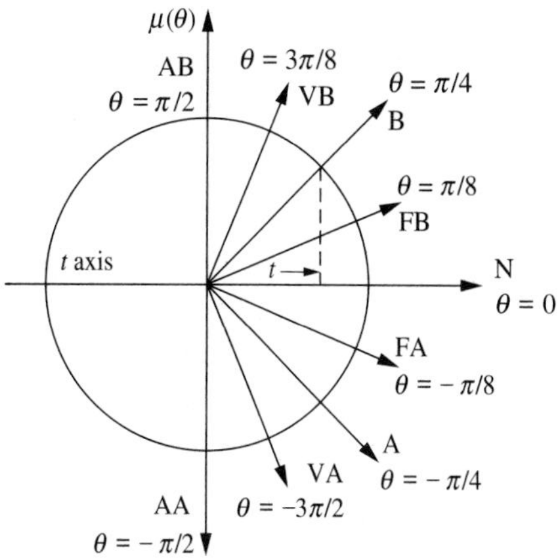

FIGURE 4.10
Angular fuzzy sets defined on the unit circle.

and (*ii*) a permanent magnet motor. In the case of variable reluctance motors, stators (housing for a rotor on a shaft) have a common frame and rotors have a common shaft. Stators are pulse excited and rotors are unexcited. Variable reluctance motors are used in incremental motion control systems, such as in computer peripherals like printers, tape drivers, capstan drives, etc.

The direction of rotation of the variable reluctance motor can be regulated by the kind of pulse applied. By defining the following linguistic terms related to the motion of the motor, we can form an angular fuzzy set:

Completely anticlockwise (CA)	$\theta = \pi/2$
Partially anticlockwise (PA)	$\theta = \pi/4$
No motion in the motor (NM)	$\theta = 0$
Partially clockwise (PC)	$\theta = -\pi/4$
Completely clockwise (CC)	$\theta = -\pi/2$

The angular fuzzy set in Fig. 4.11 can be interpreted alternatively as follows. The circle shown in Fig. 4.11 is of unit radius, and the membership value of the directional motions is defined by $\mu_t(z) = z(\tan\theta)$ where $z = |\cos\theta|$ here. Using this equation, we can calculate the membership values for the motions as given in Table 4.2.

Using the values of θ and $z(\tan\theta)$, one can obtain the membership function shown in Fig. 4.12. Note the nonconvex form of this shape. If convexity is desired, the θ axis values can be shifted by $\pi/2$ or, alternatively, the data can be transformed by the expression $1 - |\sin\theta|$.

Example 4.4. In many optical experiments, we need to differentiate colors from one another. For the sake of this problem, let us say we are trying to differentiate between two colored regions, namely violet and blue. To show these differences, we use an angular fuzzy set model. To form the set, first we have to define the following linguistic variables:

Absolutely violet (AV)	$\alpha = \pi/2$
Very violet (VV)	$\alpha = 3\pi/8$
Violet (V)	$\alpha = \pi/4$

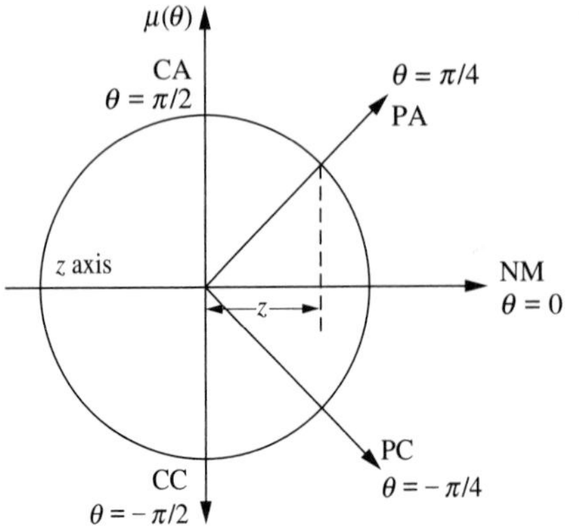

FIGURE 4.11
Angular fuzzy set for the motion of a motor.

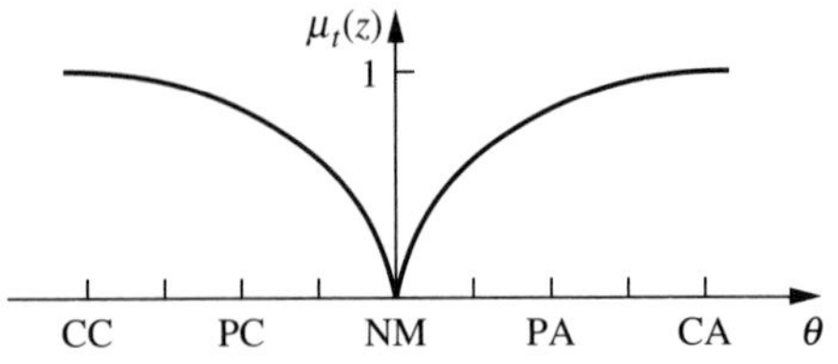

FIGURE 4.12
Angular fuzzy set membership function obtained from Eq. (4.8).

Fairly violet (FV)	$\alpha = \pi/8$
Undecided (UN)	$\alpha = 0$
Fairly blue (FB)	$\alpha = -\pi/8$
Blue (B)	$\alpha = -\pi/4$
Very blue (VB)	$\alpha = -3\pi/8$
Absolutely blue (AB)	$\alpha = -\pi/2$

These angular variables can be plotted in polar coordinate form as shown in Fig. 4.13, where violet is in the positive angular region ($+\alpha$) and blue is in the negative ($-\alpha$) angular region.

The above set can also be interpreted as follows. The circle shown in Fig. 4.13 is of unit radius, and the membership values of the colors are defined by Eq. (4.8) and are shown in Table 4.3. By using the values of α and $z(\tan\alpha)$ in Table 4.3, the membership function shown in Fig. 4.14 is obtained. Again, convexity for this shape can be obtained with a suitable shift in the axis or other transformation.

Neural Networks

In this section we explain how a neural network can be used to determine membership functions. We first present a brief introduction to neural networks and then show how they can be used to determine membership functions.

TABLE 4.2
Angular fuzzy membership values

θ	$\tan\theta$	z	$\mu_t(z) = \|(z)\tan\theta\|$
0°	0	1	0
45°	1	0.707	0.707
90°	∞	0	1

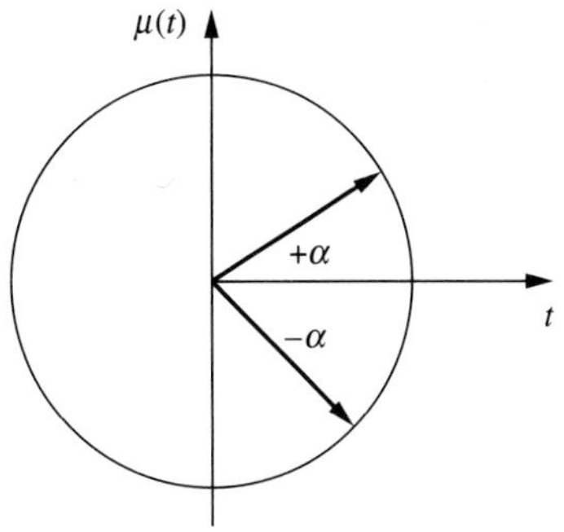

FIGURE 4.13
Polar coordinate form for angular fuzzy sets.

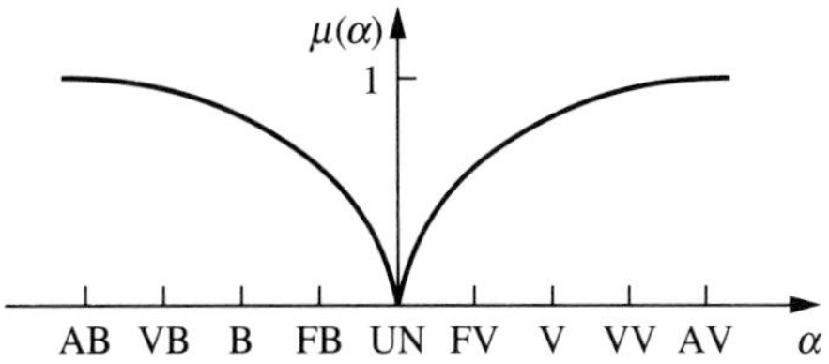

FIGURE 4.14
Membership values for colors between violet and blue.

A neural network is a technique that seeks to build an intelligent program (to implement intelligence) using models that simulate the working network of the neurons in the human brain [Yamakawa, 1992; Hopfield, 1982, 1986]. A neuron, Fig. 4.15, is made up of several protrusions called dendrites and a long branch called the axon. A neuron is joined to other neurons through the dendrites. The dendrites of different neurons meet to form synapses, the areas where messages pass. The neurons receive the impulses via the synapses. If the total of the impulses received exceeds a certain threshold value, then the neuron sends an impulse down the axon where the axon is connected to other neurons through more synapses. The synapses may be excitatory or inhibitory in nature. An excitatory synapse adds to the total of the impulses reaching the neuron, whereas an inhibitory neuron reduces the total of the impulses reaching the neuron. In a global sense, a neuron receives a set of input pulses and sends out another pulse that is a function of the input pulses.

This concept of how neurons work in the human brain is utilized in performing computations on computers. Researchers have long felt that the neurons are responsible for the human capacity to learn, and it is in this sense that the physical structure is being emulated by a neural network to accomplish machine learning. Each computational unit computes some function of its inputs and passes the result to connected units in the network. The knowledge of the system comes out of the entire network of the neurons.

TABLE 4.3
Angular membership values for colors

α, degrees	$\tan\alpha$	z	$\mu(\alpha) = z\tan\alpha$
0	0	1	0
22.5	0.414	0.923	0.382
45	1	0.707	0.707
67.5	2.41	0.382	0.923
90	∞	0	1

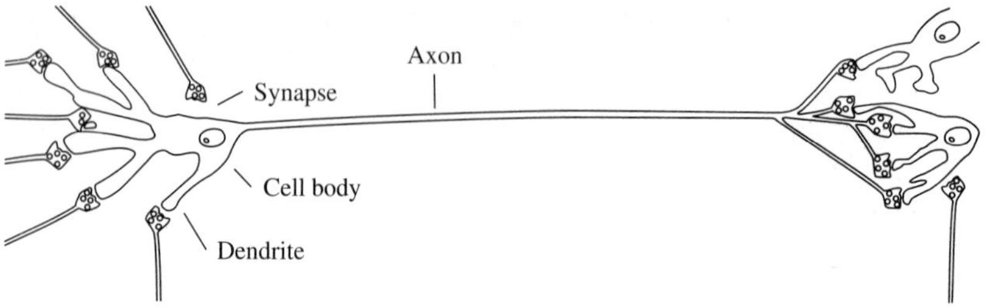

FIGURE 4.15
A simple schematic of a human neuron.

Figure 4.16 shows the analog of a neuron as a threshold element. The variables $x_1, x_2, \ldots, x_i, \ldots, x_n$ are the n inputs to the threshold element. These are analogous to impulses arriving from several different neurons to one neuron. The variables $w_1, w_2, \ldots, w_i, \ldots, w_n$ are the weights associated with the impulses/inputs, signifying the relative importance that is associated with the path from which the input is coming. When w_i is positive, input x_i acts as an excitatory signal for the element. When w_i is negative, input x_i acts as an inhibitory signal for the element. The threshold element sums the product of these inputs and their associated weights ($\sum w_i x_i$), compares it to a prescribed threshold value and, if the summation is greater than the threshold value, computes an output using a nonlinear function (F). The signal output y (Fig. 4.16) is a nonlinear function (F) of the difference between the preceding computed summation and the threshold value and is expressed as

$$y = F\left(\sum w_i x_i - t\right) \tag{4.9}$$

where
x_i = signal input ($i = 1, 2, \ldots, n$)
w_i = weight associated with the signal input x_i
t = threshold level prescribed by user
$F(s)$ = a nonlinear function; e.g., a sigmoid function $F(s) = 1 + \dfrac{1}{e^{-s}}$

The nonlinear function, F, is a modeling choice and is a function of the type of output signal desired in the neural network model. Popular choices for this function are a sigmoid function, a step function, and a ramp function.

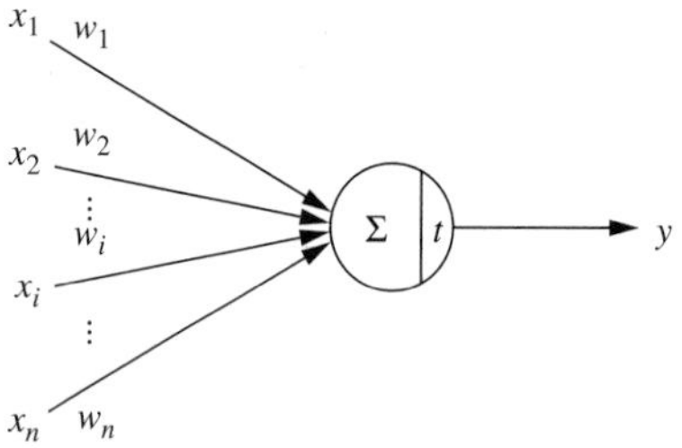

FIGURE 4.16
A threshold element as an analog to a neuron.

Figure 4.17 shows a simple neural network for a system with single-input signal x and a corresponding single-output signal $f(x)$. Layer number one has only one element that has a single input, but the element sends its output to four other elements in the second layer. Elements shown in the second layer are all single-input single-output elements. The third layer has only one element that has four inputs, and it computes the output for the system. This neural network is termed a $(1 \times 4 \times 1)$ neural network. The numbers represent the number of elements in each layer of the network. The layers other than the first (input layer) and the last (output layer) layers constitute the set of hidden layers. (Systems can have more than three layers, in which case we would have more than one hidden layer.)

Neural systems solve problems by adapting to the nature of the data (signals) they receive. One of the ways to accomplish this is to use a training-data set and a checking-data set of input and output data/signals (x, y) (for a multiple-input, multiple-output system using a neural network, we may use input-output sets comprised of vectors $(\mathbf{x_1}, \mathbf{x_2}, \ldots, \mathbf{x_n}, \mathbf{y_1}, \mathbf{y_2}, \ldots, \mathbf{y_n})$). We start with a random assignment of weights w^i_{jk} to the paths joining the elements in the different layers (Fig. 4.17). Then an input x from the training-data set is passed through the neural network. The neural network computes a value $(f(x)_{\text{output}})$, which is compared with the actual value $(f(x)_{\text{actual}} = y)$. The error measure δ is computed from these two output values as

$$\delta = f(x)_{\text{actual}} - f(x)_{\text{output}} \tag{4.10}$$

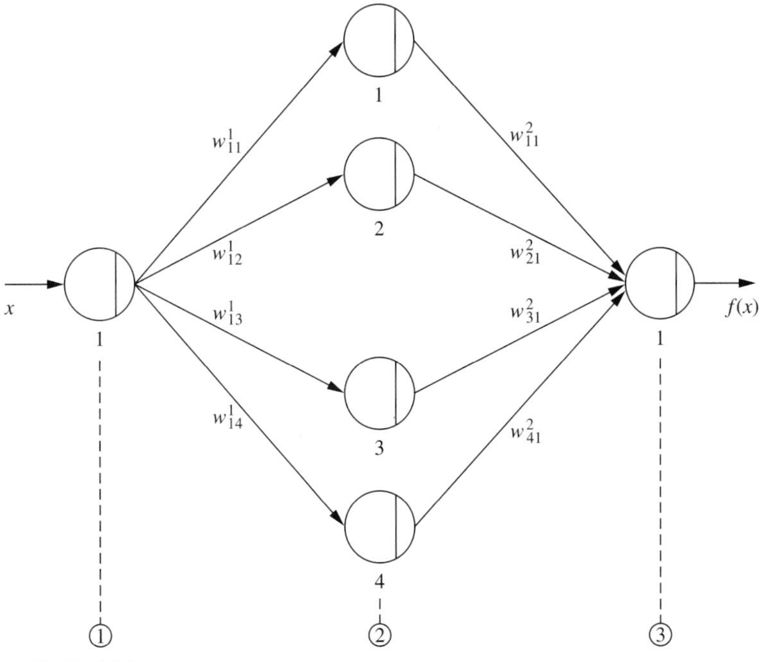

FIGURE 4.17
A simple $1 \times 4 \times 1$ neural network, where w^i_{jk} represents the weight associated with the path connecting the jth element of the ith layer to the kth element of the $(i + 1)$th layer.

This is the error measure associated with the last layer of the neural network (for Fig. 4.17); in this case the error measure δ would be associated with the third layer in the neural network. Next we try to distribute this error to the elements in the hidden layers using a technique called back-propagation.

The error measure associated with the different elements in the hidden layers is computed as follows. Let δ_j be the error associated with the jth element (Fig. 4.18). Let w_{nj} be the weight associated with the line from element n to element j and let I be the input to unit n. The error for element n is computed as

$$\delta_n = F'(I)w_{nj}\delta_j \tag{4.11}$$

where, for $F(I) = 1/(1 + e^{-I})$, the sigmoid function, we have

$$F'(I) = F(I)(1 - F(I)) \tag{4.12}$$

Next the different weights w^i_{jk} connecting different elements in the network are corrected so that they can approximate the final output more closely. For updating the weights, the error measure on the elements is used to update the weights on the lines joining the elements.

For an element with an error δ associated with it, as shown in Fig. 4.19, the associated weights may be updated as

$$w_i\text{ (new)} = w_i\text{ (old)} + \alpha\delta x_i \tag{4.13}$$

where α = learning constant
δ = associated error measure
x_i = input to the element

The input value x_i is passed through the neural network (now having the updated weights) again, and the errors, if any, are computed again. This technique is iterated until the error value of the final output is within some user-prescribed limits.

The neural network then uses the next set of input-output data. This method is continued for all data in the training-data set. This technique makes the neural network simulate the nonlinear relation between the input-output data sets. Finally a checking-data set is used to verify how well the neural network can simulate the nonlinear relationship.

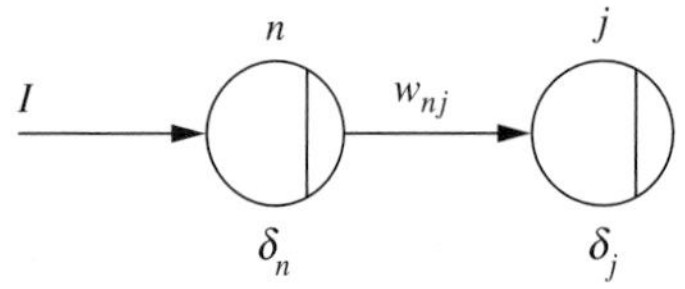

FIGURE 4.18
Distribution of error to different elements.

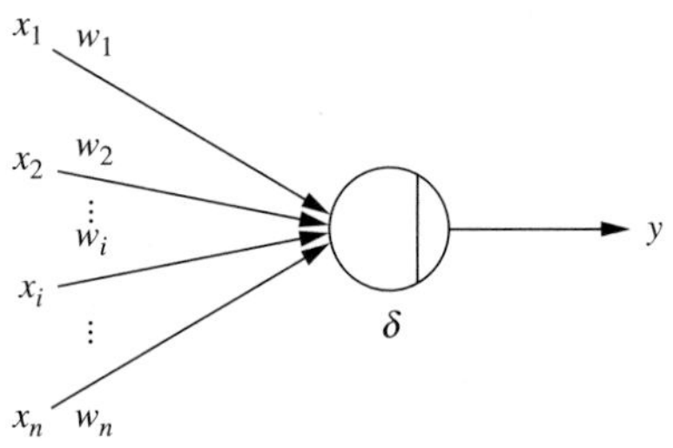

FIGURE 4.19
A threshold element with an error δ associated with it.

For systems where we may have data sets of inputs and corresponding outputs, and where the relationship between the input and output may be highly nonlinear or not known at all, we may want to use fuzzy logic to classify the input and the output data sets broadly into different fuzzy classes. Furthermore, for systems that are dynamic in nature (the system parameters may change in a nondeterministic fashion) the fuzzy membership functions would have to be repeatedly updated. For these types of systems it is advantageous to use a neural network since the network can modify itself (by changing the weight assignments in the neural network) to accommodate the changes. Unlike symbolic learning algorithms, e.g., expert systems written in PROLOG and LISP [Luger and Stubblefield, 1989], neural networks do not learn by adding new rules to their knowledge base; they learn by modifying their overall structure.

GENERATION OF MEMBERSHIP FUNCTIONS USING A NEURAL NETWORK. We consider here a method by which fuzzy membership functions may be created for fuzzy classes of an input data set [Takagi and Hayashi, 1991]. We select a number of input data values and divide them into a training-data set and a checking-data set. The training-data set is used to train the neural network. Let us consider an input training-data set as shown in Fig. 4.20*a*. Table 4.4 shows the coordinate values of the different data points considered (e.g., crosses in Fig. 4.20*a*). The data points are expressed with two coordinates each, since the data shown in Fig. 4.20*a* represent a two-dimensional problem. The data points are first divided into different classes (Fig. 4.20*a*) by conventional clustering techniques (these are explained in Chapter 11).

As shown in Fig. 4.20*a* the data points have been divided into three regions, or classes, R^1, R^2, and R^3. Let us consider data point 1, which has input coordinate values of $x_1 = 0.7$ and $x_2 = 0.8$ (Fig. 4.20*d*). As this is in region R_2, we assign to it a complete membership of one in class R_2 and zero membership in classes R_1 and R_3 (Fig. 4.20*f*). Similarly, the other data points are assigned membership values of unity for the classes they belong to initially. A neural network is created (Fig. 4.20*b, e, h*) that uses the data point marked 1 and the corresponding membership values in different classes for training itself to simulate the relationship between coordinate locations and the membership values. Figure 4.20*c* represents the output of the neural network, which classifies data points into one of the three regions. The neural network then uses the next set of data values (e.g., point 2) and membership values to train itself further as seen in Fig. 4.20*d*. This repetitive process is continued until the neural network can simulate the entire set of input-output (coordinate location–membership value) values. The performance of the neural network is then checked using the checking-data set. Once the neural network is ready, its final version (Fig. 4.20*h*) can be used to determine the membership values (function) of any input data (Fig. 4.20*g*) in the different regions (Fig. 4.20*i*).

Notice that the points shown in the table in Fig. 4.20*i* are actually the membership values in each region for the data point shown in Fig. 4.20*g*. These could be plotted as a membership function, as shown in Fig. 4.21. A complete mapping of the membership of different data points in the different fuzzy classes can be derived to

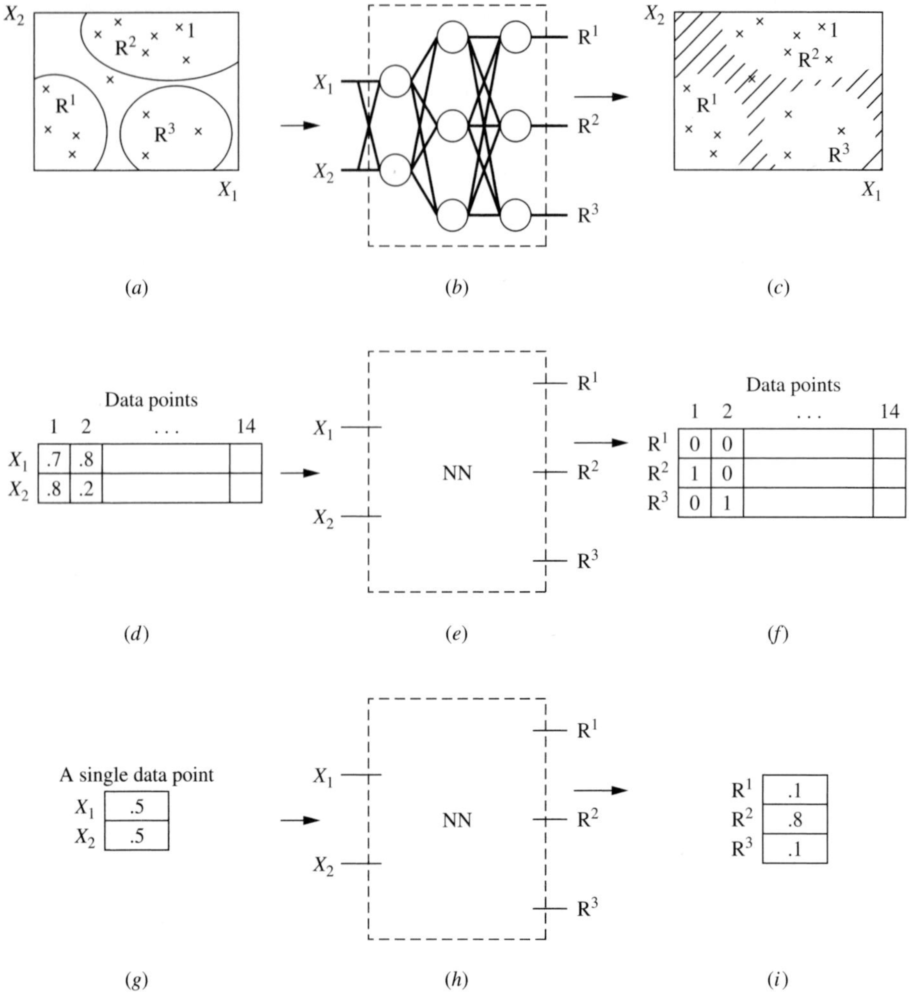

FIGURE 4.20
Using a neural network to determine membership functions [Takagi and Hayashi, 1991].

determine the overlap of the different classes (the crosshatched portion in Fig. 4.20*c* shows the overlap of the three fuzzy classes). These steps will become clearer as we go through the computations in the following example.

Example 4.5. Let us consider a system that has 20 data points described in two-dimensional format (two variables) as shown in Tables 4.4 and 4.5. We have placed these data points in two fuzzy classes, R_1 and R_2, using a clustering technique (see Chapter 11). We would like to form a neural network that can determine the membership values of any data point in the two classes. We would use the data points in Table 4.4 to train the neural network and the data points in Table 4.5 to check its performance. The membership values in Table 4.6 are to be used to train and check the performance of the neural network. The data points that are to be used for training and checking the

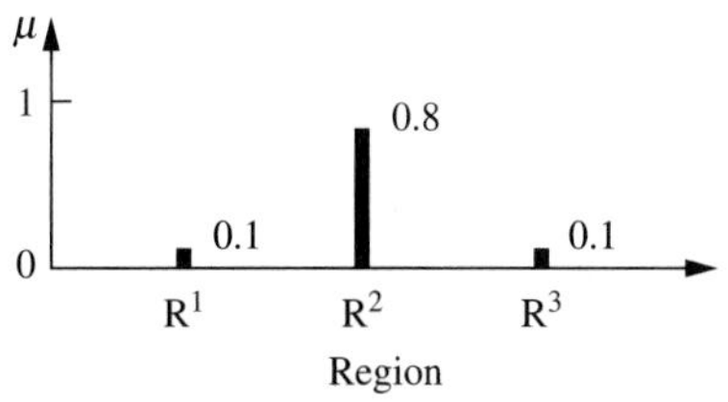

FIGURE 4.21
Membership function for data point ($X_1, X_2 = (0.5, 0.5)$).

performance of the neural network have been assigned membership values of unity for the classes into which they have been originally assigned, as seen in Table 4.6.

We select a 2 × 3 × 3 × 2 neural network to simulate the relationship between the data points and their membership in the two fuzzy sets, R_1 and R_2 (Fig. 4.22). The coordinates x_1 and x_2 for each data point are used as the input values, and the corresponding membership values in the two fuzzy classes for each data point are the output values for the neural network.

Table 4.7 shows the initial quasi-random values that have been assigned to the different weights connecting the paths between the elements in the layers in the net-

TABLE 4.4
Variables describing the data points to be used as a training data set

Data point	1	2	3	4	5	6	7	8	9	10
x_1	.05	.09	.12	.15	.20	.75	.80	.82	.90	.95
x_2	.02	.11	.20	.22	.25	.75	.83	.80	.89	.89

TABLE 4.5
Variables describing the data points to be used as a checking data set

Data point	11	12	13	14	15	16	17	18	19	20
x_1	.09	.10	.14	.18	.22	.77	.79	.84	.94	.98
x_2	.04	.10	.21	.24	.28	.78	.81	.82	.93	.99

TABLE 4.6
Membership values of the data points in the training and checking data sets to be used for training and checking the performance of the neural network

Data points	1 & 11	2 & 12	3 & 13	4 & 14	5 & 15	6 & 16	7 & 17	8 & 18	9 & 19	10 & 20
R_1	1.0	1.0	1.0	1.0	1.0	0.0	0.0	0.0	0.0	0.0
R_2	0.0	0.0	0.0	0.0	0.0	1.0	1.0	1.0	1.0	1.0

TABLE 4.7
The initial quasi-random values that have been assigned to the different weights connecting the paths between the elements in the layers in the network of Fig. 4.22.

$w_{11}^1 = 0.5$	$w_{11}^2 = 0.10$	$w_{11}^3 = 0.30$
$w_{12}^1 = 0.4$	$w_{12}^2 = 0.55$	$w_{12}^3 = 0.35$
$w_{13}^1 = 0.1$	$w_{13}^2 = 0.35$	$w_{21}^3 = 0.35$
$w_{21}^1 = 0.2$	$w_{21}^2 = 0.20$	$w_{22}^3 = 0.25$
$w_{22}^1 = 0.6$	$w_{22}^2 = 0.45$	$w_{31}^3 = 0.45$
$w_{23}^1 = 0.2$	$w_{23}^2 = 0.35$	$w_{32}^3 = 0.30$
	$w_{31}^2 = 0.25$	
	$w_{32}^2 = 0.15$	
	$w_{33}^2 = 0.60$	

work shown in Fig. 4.22. We take the first data point ($x_1 = 0.05, x_2 = 0.02$) as the input to the neural network. We will use Eq. (4.9) in the form

$$O = \frac{1}{1 + \exp - (\sum x_i w_i - t)} \tag{4.14}$$

where O = output of the threshold element computed using the sigmoidal function
x_i = inputs to the threshold element ($i = 1, 2, \ldots, n$)
w_i = weights attached to the inputs
t = threshold for the element

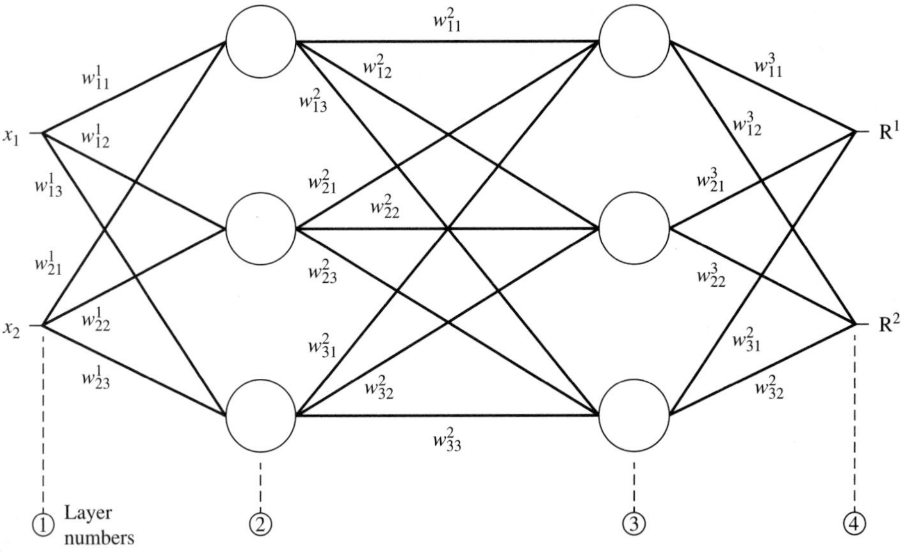

FIGURE 4.22
The [2 × 3 × 3 × 2] neural network to be trained for the data set of Example 4.5.

First iteration. We start off with the first iteration in training the neural network using Eq. (4.14) to determine the outputs of the different elements by calculating the outputs for each of the neural network layers. We select a threshold value of $t = 0$.

Outputs for the second layer.

$$O_1^2 = \frac{1}{1 + \exp -[(0.05 \times 0.50) + (0.02 \times 0.20) - 0.0]} = 0.507249$$

$$O_2^2 = \frac{1}{1 + \exp -[(0.05 \times 0.40) + (0.02 \times 0.60) - 0.0]} = 0.507999$$

$$O_3^2 = \frac{1}{1 + \exp -[(0.05 \times 0.10) + (0.02 \times 0.20) - 0.0]} = 0.502250$$

Outputs for the third layer.

$$O_1^3 = \frac{1}{1 + \exp -[(0.507249 \times 0.10) + (0.507999 \times 0.20) + (0.502250 \times 0.25) - 0.0]}$$
$$= 0.569028$$

$$O_2^3 = \frac{1}{1 + \exp -[(0.507249 \times 0.55) + (0.507999 \times 0.45) + (0.502250 \times 0.15) - 0.0]}$$
$$= 0.641740$$

$$O_3^3 = \frac{1}{1 + \exp -[(0.507249 \times 0.35) + (0.507999 \times 0.35) + (0.502250 \times 0.60) - 0.0]}$$
$$= 0.658516$$

Outputs for the fourth layer.

$$O_1^4 = \frac{1}{1 + \exp -[(0.569028 \times 0.30) + (0.641740 \times 0.35) + (0.658516 \times 0.45) - 0.0]}$$
$$= 0.666334$$

$$O_2^4 = \frac{1}{1 + \exp -[(0.569028 \times 0.35) + (0.641740 \times 0.25) + (0.658516 \times 0.30) - 0.0]}$$
$$= 0.635793$$

Determining errors. Next we compare the outputs of the fourth layer (which is the output layer) to the correct outputs (previously known membership values listed in Table 4.6) to determine the final error of the neural network, Eq. (4.10).

$$\mathrm{R}_1 : E_1^4 = O_1^4 - O_{1,\ \mathrm{actual}}^4 = 0.666334 - 1.0 = -0.333666$$

$$\mathrm{R}_2 : E_2^4 = O_2^4 - O_{2,\ \mathrm{actual}}^4 = 0.635793 - 0.0 = 0.635793$$

Now that we know the final errors for the neural network for the first iteration, we distribute this error to the other nodes (elements) in the network using Eqs. (4.11)–(4.12) in the form

$$E_n = O_n(1 - O_n)\sum_j w_{nj}E_j \tag{4.15}$$

Assigning errors. First, we assign errors to the elements in the third layer,

$$E_1^3 = 0.569028(1.0 - 0.569028)$$
$$\times [(0.30 \times (-0.333666)) + (0.35 \times 0.635793)] = 0.030024$$
$$E_2^3 = 0.641740(1.0 - 0.641740)$$
$$\times [(0.35 \times (-0.333666)) + (0.25 \times 0.635793)] = 0.009694$$
$$E_3^3 = 0.658516(1.0 - 0.658516)$$
$$\times [(0.45 \times (-0.333666)) + (0.30 \times 0.635793)] = 0.009127$$

and then assign errors to the elements in the second layer,

$$E_1^2 = 0.507249(1.0 - 0.507249)$$
$$\times [(0.10 \times 0.030024) + (0.55 \times 0.009694) + (0.35 \times 0.009127)] = 0.002882$$
$$E_2^2 = 0.507999(1.0 - 0.507999)$$
$$\times [(0.20 \times 0.030024) + (0.45 \times 0.009694) + (0.35 \times 0.009127)] = 0.003390$$
$$E_3^2 = 0.502250(1.0 - 0.502250)$$
$$\times [(0.25 \times 0.030024) + (0.15 \times 0.009694) + (0.60 \times 0.009127)] = 0.003609$$

Now that we know the errors associated with each element in the network we can update the weights associated with these elements so that the network approximates the output more closely. To update the weights we use Eq. (4.13) in the form

$$w_{jk}^i(\text{new}) = w_{jk}^i(\text{old}) + \alpha E_k^{(i+1)} x_{jk} \tag{4.16}$$

where w_{jk}^i = weight associated with the path connecting the jth element of the ith layer to the kth element of the $(i+1)$th layer

α = learning constant, taken as 0.3 for this example

E_k^{i+1} = error associated with the kth element of the $(i+1)$th layer

x_{jk} = input from the jth element in the ith layer to the kth element in the $(i+1)$th layer (O_j^i)

Updating weights. We will update the weights connecting elements in the third and the fourth layers,

$$w_{11}^3 = 0.30 + 0.3 \times (-0.333666) \times 0.569028 = 0.243040$$
$$w_{21}^3 = 0.35 + 0.3 \times (-0.333666) \times 0.641740 = 0.285762$$
$$w_{31}^3 = 0.45 + 0.3 \times (-0.333666) \times 0.658516 = 0.384083$$

$$w_{12}^3 = 0.35 + 0.3 \times 0.635793 \times 0.569028 = 0.458535$$
$$w_{22}^3 = 0.25 + 0.3 \times 0.635793 \times 0.641740 = 0.372404$$
$$w_{32}^3 = 0.30 + 0.3 \times 0.635793 \times 0.658516 = 0.425604$$

then update weights connecting elements in the second and the third layers,

$$w_{11}^2 = 0.10 + 0.3 \times 0.030024 \times 0.507249 = 0.104968$$
$$w_{21}^2 = 0.20 + 0.3 \times 0.030024 \times 0.507999 = 0.204576$$
$$w_{31}^2 = 0.25 + 0.3 \times 0.030024 \times 0.502250 = 0.254524$$

$$w_{12}^2 = 0.55 + 0.3 \times 0.009694 \times 0.507249 = 0.551475$$
$$w_{22}^2 = 0.45 + 0.3 \times 0.009694 \times 0.507999 = 0.451477$$
$$w_{32}^2 = 0.15 + 0.3 \times 0.009694 \times 0.502250 = 0.151461$$

$$w_{13}^2 = 0.35 + 0.3 \times 0.009127 \times 0.507249 = 0.351389$$
$$w_{23}^2 = 0.35 + 0.3 \times 0.009127 \times 0.507999 = 0.351391$$
$$w_{33}^2 = 0.60 + 0.3 \times 0.009127 \times 0.502250 = 0.601375$$

and then, finally, update weights connecting elements in the first and the second layers,

$$w_{11}^1 = 0.50 + 0.3 \times 0.002882 \times 0.05 = 0.500043$$
$$w_{12}^1 = 0.40 + 0.3 \times 0.003390 \times 0.05 = 0.400051$$
$$w_{13}^1 = 0.10 + 0.3 \times 0.003609 \times 0.05 = 0.100054$$

$$w_{21}^1 = 0.20 + 0.3 \times 0.002882 \times 0.02 = 0.200017$$
$$w_{22}^1 = 0.60 + 0.3 \times 0.003390 \times 0.02 = 0.600020$$
$$w_{23}^1 = 0.20 + 0.3 \times 0.003609 \times 0.02 = 0.200022$$

Now that all the weights in the neural network have been updated, the input data point ($x_1 = 0.05, x_2 = 0.02$) is again passed through the neural network. The errors in approximating the output are computed again and redistributed as before. This process is continued until the errors are within acceptable limits. Next, the second data point ($x_1 = 0.09, x_2 = 0.11$, Table 4.4) and the corresponding membership values ($R^1 = 1, R^2 = 0$, Table 4.6) are used to train the network. This process is continued until all the data points in the *training*-data set (Table 4.4) are used. The performance of the neural network (how closely it can predict the value of the membership of the data point) is then checked using the data points in the *checking*-data set (Table 4.5).

Once the neural network is trained and verified to be performing satisfactorily, it can be used to find the membership of any other data points in the two fuzzy classes. A complete mapping of the membership of different data points in the different fuzzy classes can be derived to determine the overlap of the different classes (R_1 and R_2).

Genetic Algorithms

As in the previous section we will first provide a brief introduction to genetic algorithms and then show how these can be used to determine membership functions. In the previous section we introduced the concept of a neural network. In implementing a neural network algorithm, we try to recreate the working of neurons in the human brain. In this section we introduce another class of algorithms, which use the concept of Darwin's theory of evolution. Darwin's theory basically stressed the fact that the existence of all living things is based on the rule of "survival of the fittest." Darwin also postulated that new breeds or classes of living things come into existence through the processes of reproduction, crossover, and mutation among existing organisms [Forrest, 1993].

These concepts in the theory of evolution have been translated into algorithms to search for solutions to problems in a more "natural" way. First, different possible solutions to a problem are created. These solutions are then tested for their performance (i.e., how good a solution they provide). Among all possible solutions, a fraction of the good solutions is selected, and the others are eliminated (survival of the fittest). The selected solutions undergo the processes of reproduction, crossover, and mutation to create a new generation of possible solutions (which are expected to perform better than the previous generation). This process of production of a new generation and its evaluation is repeated until there is convergence within a generation. The benefit of this technique is that it searches for a solution from a broad spectrum of possible solutions, rather than restrict the search to a narrow domain where the results would be normally expected. Genetic algorithms try to perform an intelligent search for a solution from a nearly infinite number of possible solutions.

In the following material we show how the concepts of genetics are translated into a search algorithm [Goldberg, 1989]. In a genetic algorithm, the parameter set of the problem is coded as a finite string of bits. For example, given a set of two-dimensional data ((x, y) data points), we want to fit a linear curve (straight line) through the data. To get a linear fit, we encode the parameter set for a line ($y = C_1x + C_2$) by creating independent bit strings for the two unknown constants C_1 and C_2 (parameter set describing the line) and then join them (concatenate the strings). The bit strings are combinations of 0s and 1s, which represent the value of a number in binary form. An n-bit string can accommodate all integers up to the value $2^n - 1$. For example, the number 7 requires a three-bit string, i.e., $2^3 - 1 = 7$, and the bit string would look like "111," where the first unit-digit is in the 2^2 place ($= 4$), the second unit-digit is in the 2^1 place ($= 2$), and the last unit-digit is in the 2^0 place ($= 1$); hence, $4 + 2 + 1 = 7$. The number 10 would look like "1010," i.e., $2^3 + 2^1 = 10$, from a 4-bit string. This bit string may be mapped to the value of a parameter, say $C_i, i = 1, 2$, by the mapping

$$C_i = C_{\min} + \frac{b}{2^L - 1}\left(C_{\max_i} - C_{\min_i}\right) \tag{4.17}$$

where "b" is the number in decimal form that is being represented in binary form (e.g., 152 may be represented in binary form as 10011000), L is the length of the bit string (i.e., the number of bits in each string), and $C_{\max}$ and $C_{\min}$ are user-defined constants between which C_1 and C_2 vary linearly. The parameters C_1 and C_2 depend on the problem. The length of the bit strings is based on the handling capacity of the computer being used, i.e., on how long a string (strings of each parameter are concatenated to make one long string representing the whole parameter set) the computer can manipulate at an optimum speed.

All genetic algorithms contain three basic operators: reproduction, crossover, and mutation, where all three are analogous to their namesakes in genetics. Let us consider the overall process of a genetic algorithm before trying to understand the basic processes.

First, an initial population of n strings (for n parameters) of length L is created. The strings are created in a random fashion, i.e., the values of the parameters that are coded in the strings are random values (created by randomly placing the 0s and

1s in the strings). Each of the strings is decoded into a set of parameters that it represents. This set of parameters is passed through a numerical model of the problem space. The numerical model gives out a solution based on the input set of parameters. On the basis of the quality of this solution, the string is assigned a fitness value. The fitness values are determined for each string in the entire population of strings. With these fitness values, the three genetic operators are used to create a new generation of strings, which is expected to perform better than the previous generations (better fitness values). The new set of strings is again decoded and evaluated, and a new generation is created using the three basic genetic operators. This process is continued until convergence is achieved within a population.

Among the three genetic operators, reproduction is the process by which strings with better fitness values receive correspondingly *better copies* in the new generation, i.e., we try to ensure that better solutions persist and contribute to better offsprings (new strings) during successive generations. This is a way of ensuring the "survival of the fittest" strings. Because the total number of strings in each generation is kept a constant (for computational economy and efficiency), strings with lower fitness values are eliminated.

The second operator, crossover, is the process in which the strings are able to mix and match their desirable qualities in a random fashion. After reproduction, crossover proceeds in three simple steps. First, two new strings are selected at random (Fig. 4.23*a*). Second, a random location in both strings is selected (Fig. 4.23*b*). Third, the portions of the strings to the right of the randomly selected location in the two strings are exchanged (Fig. 4.23*c*). In this way information is exchanged between strings, and portions of high-quality solutions are exchanged and combined.

Reproduction and crossover together give genetic algorithms most of their searching power. The third genetic operator, mutation, helps to increase the search-

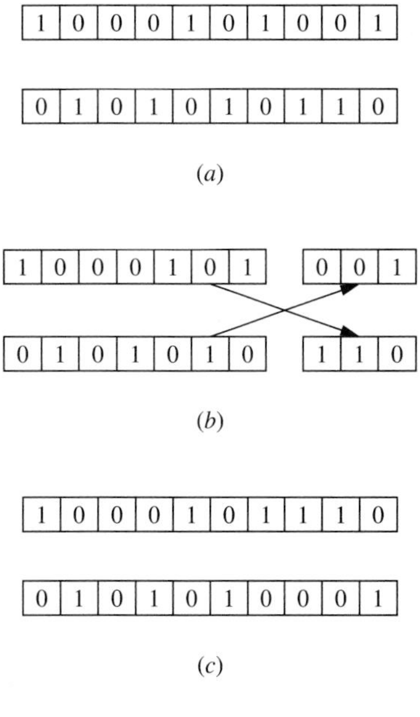

FIGURE 4.23
Crossover in strings. (*a*) Two strings are selected at random to be mated; (*b*) a random location in the strings is located (here the location is before the last three bit locations); and (*c*) the string portions following the selected location are exchanged.

ing power. In order to understand the need for mutation, let us consider the case where reproduction or crossover may not be able to find an optimum solution to a problem. During the creation of a generation it is possible that the entire population of strings is missing a vital bit of information (e.g., none of the strings has a 1 at the fourth location) that is important for determining the correct or the most nearly optimum solution. Future generations that would be created using reproduction and crossover would not be able to alleviate this problem. Here mutation becomes important. Occasionally, the value at a certain string location is changed, i.e., if there is a 1 originally at a location in the bit string, it is changed to a 0, or vice versa. Mutation thus ensures that the vital bit of information is introduced into the generation. Mutation, as it does in nature, takes place very rarely, on the order of once in a thousand bit string locations (a suggested mutation rate is 0.005/bit/generation [Forrest, 1993]).

Let us now consider an example that shows how a line may be fit through a given data set using a genetic algorithm.

Example 4.6. Let us consider the data set in Table 4.8. For performing a line ($y = C_1x + C_2$) fit, as mentioned earlier, we first encode the parameter set (C_1, C_2) in the form of bit strings. Bit strings are created with random assignment of 1s and 0s at different bit locations. We start with an initial population of four strings (Table 4.9*a*, column 2). The strings are 12 bits in length. The first six bits encode the parameter C_1, and the next six bits encode the parameter C_2. Table 4.9*a*, columns 3 and 5, shows the decimal equivalent of their binary coding. These binary values for C_1 and C_2 are then mapped into values relevant to the problem using Eq. (4.17). We assume that the minimum value to which we would expect C_1 or C_2 to go would be -2 and the maximum would be 5 (these are arbitrary values—any other values could just as easily have been chosen). Therefore, for Eq. (4.17), $C_{\min\, i} = -2$ and $C_{\max\, i} = 5$. Using these values, we compute C_1 and C_2 (Table 4.9*a*, columns 4 and 6). The values shown in Table 4.9*a*, columns 7, 8, 9, and 10, are the values computed using the equation $y = C_1x + C_2$, using the values of C_1 and C_2 from columns 4 and 6, respectively, for different values of x as given in Table 4.8. These computed values for y's are compared with the correct values (Table 4.8), and the square of the errors in estimating the y's is estimated for each string. This summation is subtracted from a large number (400 in this problem) (Table 4.9*a*, column 11) to convert the problem into a maximization problem. The values in Table 4.9*a*, column 11 are the fitness values for the four strings. These fitness values are added. Their average is also computed. The fitness value of each string is divided by the average fitness value of the whole population of strings to give an estimate of the relative fitness of each string (Table 4.9*a*, column 12). This measure also acts as a

TABLE 4.8
Data set through which a line fit is required

Data number	x	y'
1	1.0	1.0
2	2.0	2.0
3	3.0	3.0
4	6.0	6.0

TABLE 4.9*a*
First iteration using a genetic algorithm, Example 4.6

(1) String number	(2) String	(3) C_1 (bin.)	(4) C_1	(5) C_2 (bin.)	(6) C_2	(7) y_1	(8) y_2	(9) y_3	(10) y_4	(11) $f(x) = 400 - \sum(y_1 - y_{i'})^2$	(12) Expected count = f/average	(13) Actual count
1	000111 010100	7	−1.22	20	0.22	−1.00	−2.22	−4.67	−7.11	147.49	0.48	0
2	010010 001100	18	0.00	12	−0.67	−0.67	−0.67	−0.67	−0.67	332.22	1.08	1
3	010101 101010	21	0.33	42	2.67	3.00	3.33	4.00	4.67	391.44	1.27	2
4	100100 001001	36	2.00	9	−1.00	1.00	3.00	7.00	11.00	358.00	1.17	1
									Sum	1229.15		
									Average	307.29		
									Maximum	391.44		

TABLE 4.9*b*
Second iteration using a genetic algorithm, Example 4.6

Selected strings	New strings	C_1 (bin.)	C_1	C_2 (bin.)	C_2	y_1	y_2	y_3	y_4	$f(x) = 400 - \sum(y_1 - y_{i'})^2$	Expected count = f/average	Actual count
0101|01 101010	010110 001100	22	0.44	12	−0.67	−0.22	0.22	1.11	2.00	375.78	1.15	1
0100|10 001100	010001 101010	17	−0.11	42	2.67	2.56	2.44	2.22	2.00	380.78	1.17	2
010101 101|010	010101 101001	21	0.33	41	2.56	2.89	3.22	3.89	4.56	292.06	0.90	1
100100 001|001	100100 001010	36	2.0	10	−0.89	1.11	3.11	7.11	11.11	255.73	0.78	0
									Sum	1304.35		
									Average	326.09		
									Maximum	380.78		

guide as to which strings are eliminated from consideration for the next generation and which string "gets reproduced" in the next generation. In this problem a cutoff value of 0.80 (relative fitness) has been used for the acceptability of a string succeeding into the next generation. Table 4.9*a*, column 13 shows the number of copies of each of the four strings that would be used to create the next generation of strings.

Table 4.9*b* is a continuation of Table 4.9*a*. The first column in Table 4.9*b* shows the four strings selected from the previous generation aligned for crossover at the locations shown in the strings in the column. After crossover, the new strings generated are shown in Table 4.9*b*, column 2. These strings undergo the same process of decoding and evaluation as the previous generation. This process is shown in Table 4.9*b*, columns 3 through 13. We notice that the average fitness of the second generation is greater than that of the first generation of strings.

The process of generation of strings and their evaluation is continued until we get a convergence to the solution within a generation.

COMPUTING MEMBERSHIP FUNCTIONS USING GENETIC ALGORITHMS. Genetic algorithms as just described can be used to compute membership functions [Karr and Gentry, 1993]. Given some functional mapping for a system, some membership functions and their shapes are assumed for the various fuzzy variables defined for a problem. These membership functions are then coded as bit strings that are then concatenated. An evaluation (fitness) function is used to evaluate the fitness of each set of membership functions (parameters that define the functional mapping). This procedure is illustrated for a simple problem in the next example.

Example 4.7. Let us consider that we have a single input (x)–single output (y) system with input-output values as shown in Table 4.10. Table 4.11 shows a functional mapping for this system between the input (x) and the output (y).

In Table 4.11 we see that each of the variables x and y makes use of two fuzzy classes [x uses S (small) and L (large); y uses L (large) and VL (very large)]. The functional mapping tells us that a *small* x maps to a *small* y, and a *large* x maps to a *very large* y. We assume that the range of the variable x is [0, 5] and that that of y is [0, 25]. We assume that each membership function has the shape of a right triangle, as shown in Fig. 4.24.

The membership function on the right side of Fig. 4.24 is constrained to have the right-angle wedge at the upper limit of the range of the fuzzy variable. The membership function on the left side is constrained to have the right-angle wedge on the lower limit of the range of the fuzzy variable. It is intuitively obvious that under the foregoing constraints the only thing needed to describe the shape and position of the membership function fully is the length of the base of the right-triangle membership functions. We use this fact in encoding the membership functions as bit strings.

The unknown variables in this problem are the lengths of the bases of the four membership functions (x(S, L) and y(S, VL)). We use six-bit binary strings to define

TABLE 4.10
Data for a single input–single output system

x	1	2	3	4	5
y	1	4	9	16	25

TABLE 4.11
Functional mapping for the system

x	S	L
y	S	VL

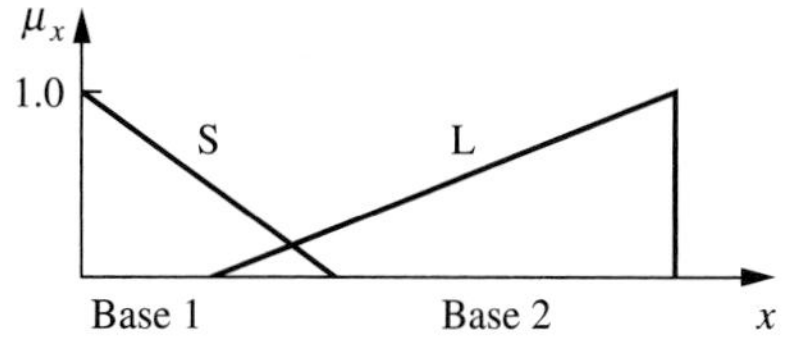

FIGURE 4.24
Membership functions for the input variables are assumed to be right triangles.

the base of each of the membership functions. (The binary values are later mapped to decimal values using Eq. (4.17).) These strings are then concatenated to give us a 24-bit (6 × 4) string. As shown in Table 4.12*a*, column 1, we start with an initial population of four strings. These are decoded to the binary values of the variables as shown in Table 4.12*a*, columns 2, 3, 4, and 5. The binary values are mapped to decimal values for the fuzzy variables using Eq. (4.17) (Table 4.12*a*, columns 6, 7, 8, and 9). For the fuzzy variable x (range $x = 0, 5$) we use $C_{min} = 0$ and $C_{max} = 5$ for both the membership functions S (Small) and L (Large). For the fuzzy variable y (range $y = 0, 25$) we use $C_{min} = 0$ and $C_{max} = 25$.

The physical representation of the first string is shown in Fig. 4.25. In this figure the base values are obtained from Table 4.12*a*, columns 6,7,8, and 9. So, for example, the base values for the x variable for string number 1 are 0.56 and $5 - 1.59 = 3.41$, and the base values for the y variable are 8.73 and $25 - 20.24 = 4.76$. To determine the fitness of the combination of membership functions in each of the strings, we want a measure of the square of the errors that are produced in estimating the value of the outputs y, given the inputs x from Table 4.10. Figure 4.25 shows how the value of the output y can be computed graphically from the membership functions for string number 1 in Table 4.12*a*. For example, for $x = 4$ we see that the membership of x in the fuzzy class Large is 0.37. Referring to the rules in Table 4.11, we see that if x is Large then y is Very Large. Therefore, we look for the value in the fuzzy class Very Large (VL) of fuzzy variable y that has a membership of 0.37. We determine this to be equal to 12.25. The corresponding actual value for y is 16 (Table 4.10). Therefore, the squared error is $(16 - 12.25)^2 = 14.06$. Columns 10, 11, 12, 13, and 14 of Table 4.12*a* show the values computed for y using the respective membership functions. Table 4.12*a*, column 15 shows the sum of the squared errors subtracted from 1000 (this is done to convert the fitness function from a minimization problem to a maximization problem). Table 4.12*a*, column 15 thus shows the fitness values for the four strings. We find the sum of all the fitness values in the generation

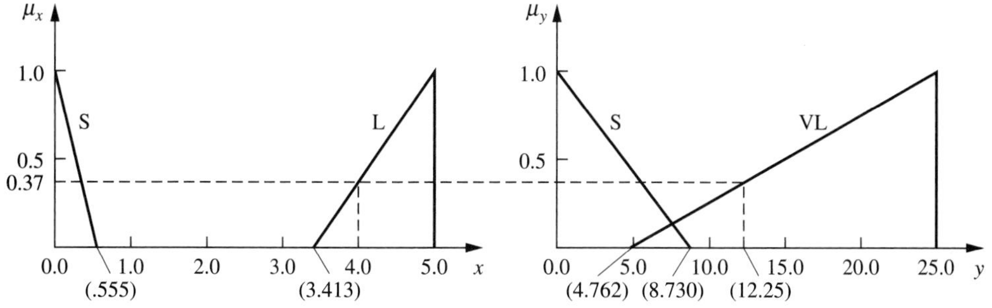

FIGURE 4.25
Physical representation of the first string in Table 4.12*a* and the graphical determination of y for a given x.

TABLE 4.12*a*
First iteration using a genetic algorithm for determining optimal membership functions

String number	(1) String	(2) base 1 (bin.)	(3) base 2 (bin.)	(4) base 3 (bin.)	(5) base 4 (bin.)	(6) base 1	(7) base 2	(8) base 3
1	000111 010100 010110 110011	7	20	22	51	0.56	1.59	8.73
2	010010 001100 101100 100110	18	12	44	38	1.43	0.95	17.46
3	010101 101010 001101 101000	21	42	13	40	1.67	3.33	5.16
4	100100 001001 101100 100011	36	9	44	35	2.86	0.71	17.46

and the average fitness of the generation. The average fitness of the generation is used to determine the relative fitness of the strings in the generation, as seen in Table 4.12*a*, column 16. These relative fitness values are used to determine which strings are to be eliminated and which string gets how many copies to make the next generation of strings. In this problem a cutoff value of 0.75 (relative fitness) has been used for the acceptability of a string propagating into the next generation. Table 4.12*a*, column 17 shows the number of copies of each of the four strings that would be used to create the next generation of strings.

Table 4.12*b* is a continuation of Table 4.12*a*. The first column in Table 4.12*b* shows the four strings selected from the previous generation aligned for crossover at the locations shown in the strings in the column. After crossover, the new strings generated are shown in Table 4.12*b*, column 2. These strings undergo the same process of decoding and evaluation as the previous generation. This process is shown in Table 4.12*b*, columns 3 through 18. We notice that the average fitness of the second generation is greater than that of the first generation of strings. Also, the fitness of the best string in the second generation is greater than the fitness of the best string in the first generation. Figure 4.26 shows the physical mapping of the best string in the first generation. Figure 4.27 shows the physical mapping of the best string in the second generation;

TABLE 4.12*b*
Second iteration using a genetic algorithm for determining optimal membership functions

(1) Selected strings	(2) New strings	(3) base 1 (bin.)	(4) base 2 (bin.)	(5) base 3 (bin.)	(6) base 4 (bin.)	(7) base 1
000111 0101|00 010110 110011	000111 010110 001101 101000	7	22	13	40	0.56
010101 1010|10 001101 101000	010101 101000 010110 110011	21	40	22	51	1.67
010101 101010 001101 1|01000	010101 101010 001101 100011	21	42	13	35	1.67
100100 001001 101100 1|00011	100100 001001 101100 101000	36	9	44	40	2.86

(9) base 4	(10) $y'(x = 1)$	(11) $y'(x = 2)$	(12) $y'(x = 3)$	(13) $y'(x = 4)$	(14) $y'(x = 5)$	(15) $\sum$error	(16) Expected count $= f/f_{av.}$	(17) Actual count
20.24	0	0	0	7.49	25	829.58	1.18	1
15.08	12.22	0	0	0	25	521.11	0.74	0
15.87	3.1	10.72	15.48	20.24	25	890.46	1.27	2
13.89	6.98	12.22	0	0	25	559.67	0.80	1
					Sum	2800.82		
					Average	700.20		
					Maximum	890.46		

notice that the membership values for the y variable in Fig. 4.27 show overlap, which is a very desirable property of membership functions.

The process of generating and evaluating strings is continued until we get a convergence to the solution within a generation, i.e., we get the membership functions with the best fitness value.

Inductive Reasoning

An automatic generation of membership functions can also be accommodated by using the essential characteristic of *inductive reasoning,* which derives a general consensus from the particular (derives the generic from the specific). The induction is performed by the entropy minimization principle, which clusters most optimally the parameters corresponding to the output classes [De Luca and Termini, 1972].

This method is based on an ideal scheme that describes the input and output relationships for a well-established database, i.e., the method generates membership functions based solely on the data provided. The method can be quite useful for complex systems where the data are abundant and static. In situations where the data are dynamic, the method may not be useful, since the membership functions will continually change with time (see the chapter summary for a discussion on the merits of this method).

(8) base 2	(9) base 3	(10) base 4	(11) y' $(x = 1)$	(12) y' $(x = 2)$	(13) y' $(x = 3)$	(14) y' $(x = 4)$	(15) y' $(x = 5)$	(16) $\sum$error	(17) Expected count $= f/f_{av}$.	(18) Actual count
1.75	5.16	15.87	0	0	0	14.99	25	900.98	1.10	1
3.17	8.73	20.24	5.24	5.85	12.23	18.62	25	961.30	1.18	2
3.33	5.16	13.89	3.1	12.51	16.68	20.84	25	840.78	1.03	1
0.71	17.46	15.87	6.11	12.22	0	0	25	569.32	0.70	0
							Sum	3272.36		
							Average	818.09		
							Maximum	961.30		

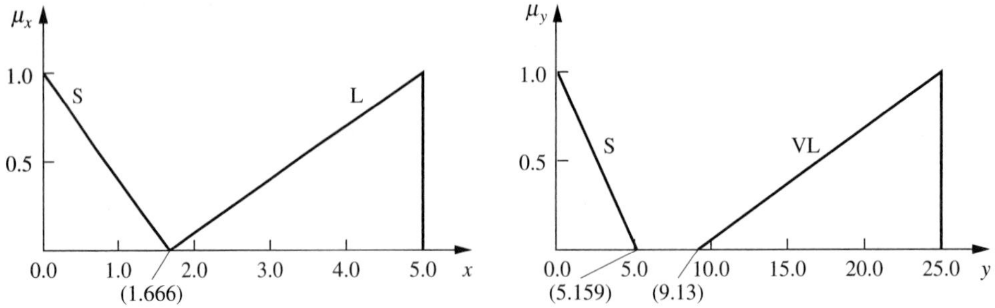

FIGURE 4.26
Physical mapping of the best string in the first generation of strings in the GA.

The intent of induction is to discover a law having objective validity and universal application. Beginning with the particular, induction concludes with the general. The essential principles of induction have been known for centuries. Three laws of induction are summarized here [Christensen, 1980]:

1. Given a set of irreducible outcomes of an experiment, the induced probabilities are those probabilities consistent with all available information that maximize the entropy of the set.
2. The induced probability of a set of independent observations is proportional to the probability density of the induced probability of a single observation.
3. The induced rule is that rule consistent with all available information of which the entropy is minimum.

Among the three laws above, the third one is appropriate for classification (or, for our purposes, membership function development) and the second one for calculating the mean probability of each step of separation (or partitioning). In classification, the probability aspects of the problem are completely disregarded since the issue is simply a binary one; a data point is either in a class or not.

A key goal of entropy minimization analysis is to determine the quantity of information in a given data set. The entropy of a probability distribution is a measure of the uncertainty of the distribution [Yager and Filev, 1994]. This information

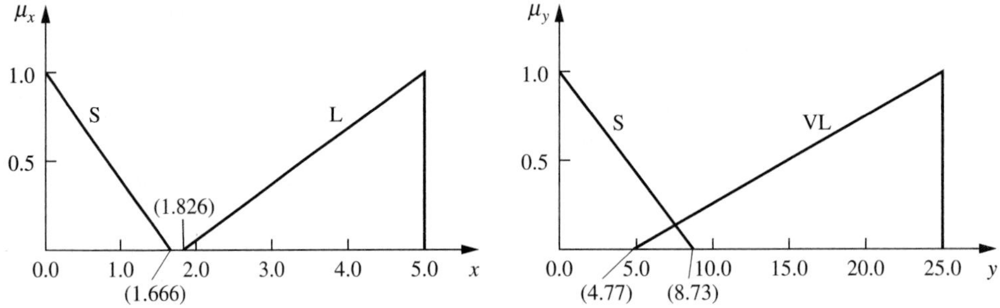

FIGURE 4.27
Physical mapping of the best string in the second generation of strings in the GA.

measure compares the contents of data to a prior probability for the same data. The higher the prior estimate of the probability for an outcome to occur, the lower will be the information gained by observing it to occur. The entropy on a set of possible outcomes of a trial where one and only one outcome is true is defined by the summation of probability and the logarithm of the probability for all outcomes. In other words, the entropy is the expected value of information.

For a simple one-dimensional (one uncertain variable) case, let us assume that the probability of the ith sample w_i to be true is $\{p(w_i)\}$. If we actually observe the sample w_i in the future and discover that it is true, then we gain the following information, $I(w_i)$:

$$I(w_i) = -k \ln p(w_i) \tag{4.18}$$

where k is a normalizing parameter. If we discover that it is false, we still gain this information:

$$I(\overline{w}_i) = -k \ln[1 - p(w_i)] \tag{4.19}$$

Then the entropy of the inner product of all the samples (N) is

$$S = -k \sum_{i-1}^{N} [p_i \ln p_i + (1 - p_i) \ln(1 - p_i)] \tag{4.20}$$

where $p_i = p(w_i)$. The minus sign before parameter k in Eq. (4.20) ensures that $S \geq 0$, because $\ln x \leq 0$ for $0 \leq x \leq 1$.

The third law of induction, which is typical in pattern classification, says that the entropy of a rule should be minimized. Minimum entropy (S) is associated with all p_i's being as close to 1s or 0s as possible, which in turn implies that they have a very high probability of either happening or not happening, respectively. Note in Eq. (4.20) that if $p_i = 1$ then $S = 0$. This result makes sense since p_i is the probability measure of whether a value belongs to a partition or not.

MEMBERSHIP FUNCTION GENERATION. To subdivide our data set into membership functions we need some procedure to establish fuzzy thresholds between classes of data. We can determine a threshold line with an entropy minimization screening method, then start the segmentation process, first into two classes. By partitioning the first two classes one more time, we can have three different classes. Therefore, a repeated partitioning with threshold value calculations will allow us to partition the data set into a number of classes, or fuzzy sets, depending on the shape used to describe membership in each set.

Membership function generation is based on a partitioning or analog screening concept, which draws a threshold line between two classes of sample data. The main idea behind drawing the threshold line is to classify the samples while minimizing the entropy for an optimum partitioning. The following is a brief review of the threshold value calculation using the induction principle for a two-class problem. First, we assume that we are seeking a threshold value for a sample in the range between x_1 and x_2. Considering this sample alone, we write an entropy equation for the regions $[x_1, x]$ and $[x, x_2]$. We denote the first region p and the second region q,

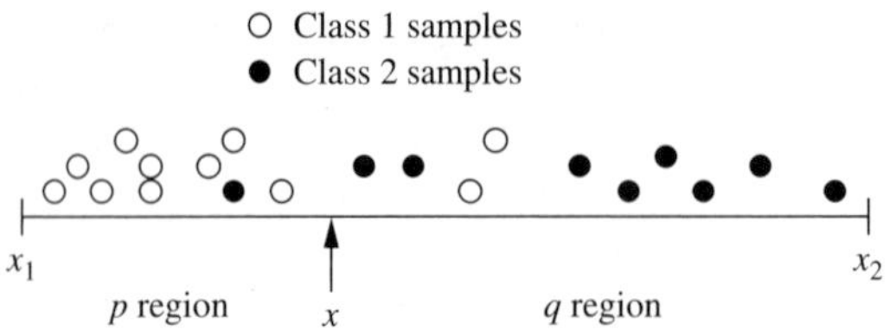

FIGURE 4.28
Illustration of threshold value idea.

as is shown in Fig. 4.28. By moving an imaginary threshold value x between x_1 and x_2, we calculate entropy for each value of x.

An entropy with each value of x in the region x_1 and x_2 is expressed by Christensen [1980] as

$$S(x) = p(x)S_p(x) + q(x)S_q(x) \tag{4.21}$$

where

$$S_p(x) = -[p_1(x)\ln p_1(x) + p_2(x)\ln p_2(x)] \tag{4.22}$$

$$S_q(x) = -[q_1(x)\ln q_1(x) + q_2(x)\ln q_2(x)] \tag{4.23}$$

and where $p_k(x)$ and $q_k(x)$ = conditional probabilities that the class k sample is in the region $[x_1, x_l + x]$ and $[x_l + x, x_2]$, respectively

$p(x)$ and $q(x)$ = probabilities that all samples are in the region $[x_1, x_l + x]$ and $[x_l + x, x_2]$, respectively

$p(x) + q(x) = 1$

A value of x that gives the minimum entropy is the optimum threshold value. We calculate entropy estimates of $p_k(x)$, $q_k(x)$, $p(x)$, and $q(x)$, as follows [Christensen, 1980]:

$$p_k(x) = \frac{n_k(x) + 1}{n(x) + 1} \tag{4.24}$$

$$q_k(x) = \frac{N_k(x) + 1}{N(x) + 1} \tag{4.25}$$

$$p(x) = \frac{n(x)}{n} \tag{4.26}$$

$$q(x) = 1 - p(x) \tag{4.27}$$

where $n_k(x)$ = number of class k samples located in $[x_1, x_l + x]$

$n(x)$ = the total number of samples located in $[x_1, x_l + x]$

$N_k(x)$ = number of class k samples located in $[x_l + x, x_2]$

$N(x)$ = the total number of samples located in $[x_l + x, x_2]$

n = total number of samples in $[x_1, x_2]$

While moving x in the region $[x_1, x_2]$ we calculate the values of entropy for each position of x. The value of x that holds the minimum entropy we will call the primary threshold (PRI) value. With this PRI value, we divide the region $[x_1, x_2]$ in two. We may say that the left side of the primary threshold is the *negative* side and the right, *positive* side; these labels are purely arbitrary but should hold some contextual meaning for the particular problem. With this first PRI value we can choose a shape for the two membership functions; one such shape uses two trapezoids, as seen in Fig. 4.29*a*. But the particular choice of shape is arbitrary; we could just as well have chosen to make the threshhold crisp and use two rectangles as membership functions. However, we do want to employ some amount of overlap since this develops the power of a membership function. As we get more and more subdivisions of the region $[x_1, x_2]$, the choice of shape for the membership function becomes less and less important as long as there is overlap between sets. Therefore, selection of simple shapes like triangles and trapezoids, which exhibit some degree of overlap, is judicious. In the next sequence we conduct the segmentation again, on each of the regions shown in Fig. 4.29*a;* this process will determine *secondary* threshold values. The same procedure is applied to calculate these secondary threshold values. If we denote a secondary threshold in the negative area as SEC1 and the other secondary threshold in the positive area SEC2, we now have three threshold lines in the sample space. The thresholds SEC1 and SEC2 are the minimum entropy points that divide the respective areas into two classes. Then we can use three labels of PO (positive), ZE (zero), and NG (negative) for each of the classes, and the three threshold values (PRI, SEC1, SEC2) are used as the *toes* of the three separate membership shapes shown in Fig. 4.29*b*. In fuzzy logic applications we often use an odd number of membership functions to partition a region, say five labels or seven. To develop seven partitions we would need *tertiary* threshold values in each of the three classes of Fig. 4.29*b*. Each threshold level, in turn, gradually separates the region into more and more classes. We have four tertiary threshold values: TER1, TER2, TER3, and TER4. Two of the tertiary thresholds lie between primary and secondary thresholds, and the other two lie between secondary thresholds and the ends of the sample space;

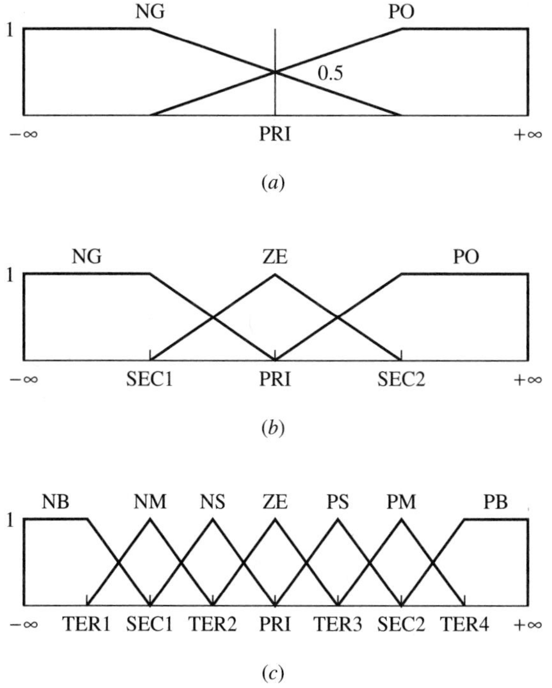

FIGURE 4.29
Repeated partitions and corresponding fuzzy set labels; (*a*) the first partition, (*b*) the second partition, and (*c*) the third partition.

this arrangement is shown in Fig. 4.29*c*. In this figure we use labels such as NB, NM, NS, ZE, PS, PM, and PB.

Example 4.8. The shape of an ellipse may be characterized by the ratio of the length of two chords a and b, as shown in Fig. 4.30 (a similar problem was originally posed in Chapter 1; see Fig. 1.5).

Let $x = a/b$; then as the ratio $a/b \to \infty$, the shape of the ellipse tends to a horizontal line, whereas as $a/b \to 0$, the shape tends to a vertical line. For $a/b = 1$ the shape is a circle. Given a set of a/b values that have been classified into two classes (class division is not necessarily based on the value of x alone; other properties like line thickness, shading of the ellipse, etc. may also be criteria), divide the variable $x = a/b$ into fuzzy partitions, as illustrated in Table 4.13.

First we determine the entropy for different values of x. The value of x is selected as approximately the midvalue between any two adjacent values. Equations (4.21)–(4.27) are then used to compute p_1, p_2, q_1, q_2, $p(x)$, $q(x)$, $S_p(x)$, $S_q(x)$, and S; and the results are displayed in Table 4.14. The value of x that gives the minimum value of the entropy (S) is selected as the first threshold partition point, PRI. From Table 4.14 (see checkmark at $S = 0.4$) we see that the first partition point is selected at $x = 1.5$, and its location for determining membership function selection is shown in Fig. 4.31.

The same process as displayed in Table 4.14 is repeated for the negative and positive partitions for different values of x. For example, in determining the threshold value to partition the negative (NE) side of Fig. 4.31, Table 4.15 displays the appropriate calculations.

Table 4.16 illustrates the calculations to determine the threshold value to partition the positive side of Fig. 4.29.

The partitions are selected based on the minimum entropy principle; the S values with a checkmark in Tables 4.15 and 4.16 are those selected. The resulting fuzzy partitions are as shown in Fig. 4.32. If required, these partitions can be further subdivided into more fuzzy subpartitions of the variable x.

SUMMARY

This chapter attempts to summarize several methods—classical and modern—that have been and are being used to develop membership functions. This field is rapidly developing, and this chapter is simply an introduction. Many methods for developing membership functions have not been discussed in this chapter. Ideas like deformable prototypes [Bremermann, 1976], implicit analytical definition [Kochen and Badre,

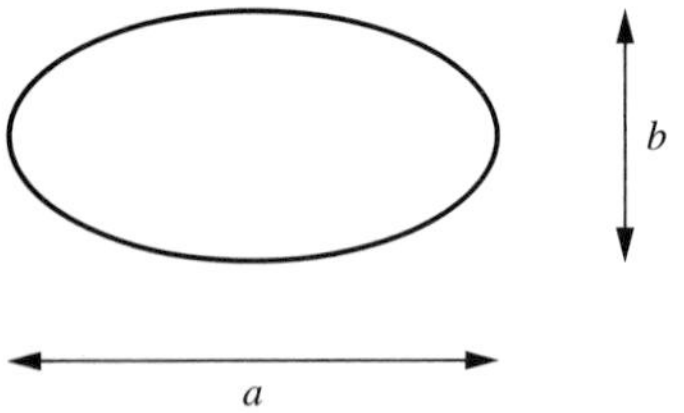

FIGURE 4.30
Geometry of an ellipse.

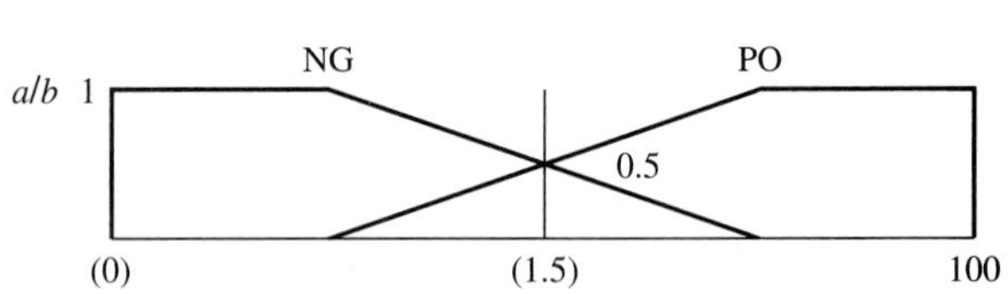

FIGURE 4.31
Partitioning of the variable $x = a/b$ into positive (PO) and negative (NE) partitions.

TABLE 4.13
Segmentation of x into two arbitrary classes (from raw data)

$x = a/b$	0	0.1	0.15	0.2	0.2	0.5	0.9	1.1	1.9	5	50	100
Class	1	1	1	1	1	2	1	1	2	2	2	2

TABLE 4.14
Calculations for selection of partition point PRI

x	**0.7**	**1.0**	**1.5**	**3.45**
p_1	$\frac{5+1}{6+1}=\frac{6}{7}$	$\frac{6+1}{7+1}=\frac{7}{8}$	$\frac{7+1}{8+1}=\frac{8}{9}$	$\frac{7+1}{9+1}=\frac{8}{10}$
p_2	$\frac{1+1}{6+1}=\frac{2}{7}$	$\frac{1+1}{7+1}=\frac{2}{8}$	$\frac{1+1}{8+1}=\frac{2}{9}$	$\frac{2+1}{9+1}=\frac{3}{10}$
q_1	$\frac{2+1}{6+1}=\frac{3}{7}$	$\frac{1+1}{5+1}=\frac{2}{6}$	$\frac{0+1}{4+1}=\frac{1}{5}$	$\frac{0+1}{3+1}=\frac{1}{4}$
q_2	$\frac{4+1}{6+1}=\frac{5}{7}$	$\frac{4+1}{5+1}=\frac{5}{6}$	$\frac{4+1}{4+1}=1.0$	$\frac{3+1}{3+1}=1.0$
$p(x)$	$\frac{6}{12}$	$\frac{7}{12}$	$\frac{8}{12}$	$\frac{9}{12}$
$q(x)$	$\frac{6}{12}$	$\frac{5}{12}$	$\frac{4}{12}$	$\frac{3}{12}$
$Sp(x)$	0.49	0.463	0.439	0.54
$Sq(x)$	0.603	0.518	0.32	0.347
S	0.547	0.486	0.4 √	0.49

TABLE 4.15
Calculations to determine secondary threshold value: NE side

x	**0.175**	**0.35**	**0.7**
p_1	$\frac{3+1}{3+1}=1.0$	$\frac{5+1}{5+1}=1.0$	$\frac{5+1}{6+1}=\frac{6}{7}$
p_2	$\frac{0+1}{3+1}=\frac{1}{4}$	$\frac{0+1}{5+1}=\frac{1}{6}$	$\frac{1+1}{6+1}=\frac{2}{7}$
q_1	$\frac{4+1}{5+1}=\frac{5}{6}$	$\frac{2+1}{3+1}=\frac{3}{4}$	$\frac{2+1}{2+1}=1.0$
q_2	$\frac{1+1}{5+1}=\frac{2}{6}$	$\frac{1+1}{3+1}=\frac{2}{4}$	$\frac{0+1}{2+1}=\frac{1}{3}$
$p(x)$	$\frac{3}{8}$	$\frac{5}{8}$	$\frac{6}{8}$
$q(x)$	$\frac{5}{8}$	$\frac{3}{8}$	$\frac{2}{8}$
$Sp(x)$	0.347	0.299	0.49
$Sq(x)$	0.518	0.562	0.366
S	0.454	0.398 √	0.459

TABLE 4.16
Calculations to determine secondary threshold value: PO side

x	27.5
p_1	$\frac{0+1}{2+1} = \frac{1}{3}$
p_2	$\frac{2+1}{2+1} = 1.0$
q_1	$\frac{0+1}{2+1} = \frac{1}{3}$
q_2	$\frac{2+1}{2+1} = 1.0$
$p(x)$	$\frac{2}{4}$
$q(x)$	$\frac{2}{4}$
$Sp(x)$	0.366
$Sq(x)$	0.366
S	0.366 √

1976], relative preferences [Saaty, 1974], and various uses of statistics [Dubois and Prade, 1980] are just a few of the many omitted here for brevity.

This chapter has dealt at length with only seven of the methods currently used in developing membership functions. There is a growing number of papers in the area of cognitive systems, where learning methods like neural networks and reasoning systems like fuzzy systems are being combined to form powerful problem solvers. In these cases, the membership functions are generally tuned in a cyclic fashion and are inextricably tied to their associated rule structure [for example, see Hayashi, et al., 1992].

In the case of genetic algorithms a number of works have appeared recently [see Karr and Gentry, 1993; Lee and Takagi, 1993]. Vast improvements have been made in the ability of genetic algorithms to find optimum solutions, for example, the *best* shape for a membership function. One of these improvements makes use of gray codes in solving a traditional binary coding problem, where sometimes all the bits used to map a decimal number had to be changed to increase that number by 1 [Forrest, 1993]. This problem had made it difficult for some algorithms to find an optimum solution from a point in the solution space that was already close to the optimum. Both neural network and genetic algorithm approaches to determining membership functions generally make use of associated rules in the knowledge base.

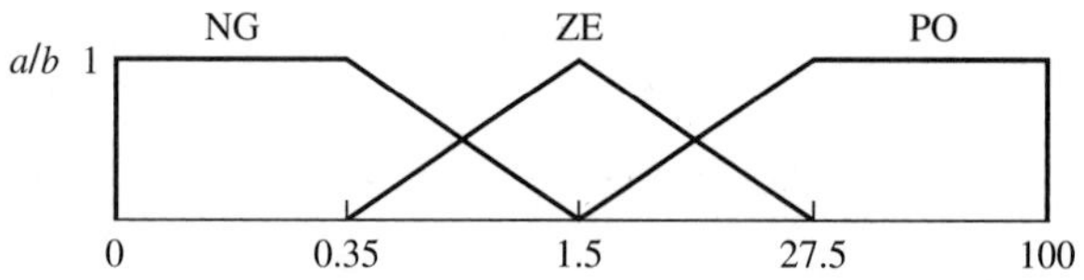

FIGURE 4.32
Secondary partitioning for Example 4.8.

In inductive reasoning, as long as the database is not dynamic the method will produce good results; when the database changes, the partitioning must be reaccomplished. Compared to neural networks and genetic algorithms, inductive reasoning has an advantage in the fact that the method may not require a convergence analysis, which in the case of genetic algorithms and neural networks is computationally very expensive. On the other hand, the inductive reasoning method uses the entire database to formulate rules and membership functions and, if the database is large, this method can also be computationally expensive. The choice of which of the three methods to use depends entirely on the problem size and problem type.

REFERENCES

Bremermann, H. (1976). "Pattern recognition," in H. Bossel, S. Klaszko, and N. Müller (eds.), *Systems theory in the social sciences,* Birkhaeuser, Basel, pp. 116–159.

Christensen, R. (1980). *Entropy minimax sourcebook,* vols. 1–4, and *Fundamentals of inductive reasoning,* Entropy Ltd., Lincoln, MA.

De Luca, A., and S. Termini (1972). "A definition of a non-probabilistic entropy in the setting of fuzzy sets theory," *Inf. Control,* vol. 20, pp. 301–312.

Dong, W. (1987). Personal notes.

Dubois, D., and H. Prade (1980). *Fuzzy sets and systems: Theory and applications,* Academic, New York.

Einstein, A. (1922). "Geometry and experience," in *Sidelights of relativity,* Methuem, London (English translation of 1921 speech to Prussian Academy).

Forrest, S. (1993). "Genetic algorithms: principles of natural selection applied to computation," *Science,* vol. 261, pp. 872–878.

Goldberg, D. (1989). *Genetic algorithms,* Addison-Wesley, New York.

Hadipriono, F., and K. Sun (1990). "Angular fuzzy set models for linguistic values," *Civ. Eng. Syst.,* vol. 7, no. 3, pp. 148–156.

Hayashi, I., H. Nomura, H. Yamasaki, and N. Wakami (1992). "Construction of fuzzy inference rules by NDF and NDFL," *Int. J. Approximate Reasoning,* vol. 6, pp. 241–266.

Hopfield, J. (1982). "Neural networks and physical systems with emergent collective computational abilities," *Proc. National Academy of Sciences USA,* vol. 79, pp. 2554–2558.

Hopfield, J., and D. Tank (1986). "Computing with neural circuits: a model," *Science,* vol. 233, pp. 625–633.

Karr, C. L., and E. J. Gentry (1993). "Fuzzy control of pH using genetic algorithms," *IEEE Trans. Fuzzy Syst.,* vol. 1, no. 1, pp. 46–53.

Kim, C. J., and B. D. Russel (1993). "Automatic generation of membership function and fuzzy rule using inductive reasoning," IEEE Trans. Paper 0-7803-1485-9/93.

Kochen, M., and A. Badre (1976). "On the precision of adjectives which denote fuzzy sets," *J. Cybern.,* vol. 4, no. 1, pp. 49–59.

Lee, M., and H. Takagi (1993). "Integrating design stages of fuzzy systems using genetic algorithms," IEEE Trans. Paper 0-7803-0614-7/93.

Luger, G., and W. Stubblefield (1989). *Artificial intelligence and the design of expert systems,* Benjamin-Cummings, Redwood City, CA.

Saaty, T. (1974). "Measuring the fuzziness of sets," *J. Cybern.,* vol. 4, no. 4, pp. 53–61.

Takagi, H., and I. Hayashi (1991). "NN-driven fuzzy reasoning," *Int. J. Approximate Reasoning,* vol. 5, pp. 191–212.

Yager, R., and D. Filev (1994). "Template-based fuzzy systems modeling," *Intelligent and Fuzzy Sys.,* vol. 2, no. 1, pp. 39–54.

Yamakawa, T. (1992). "A fuzzy logic controller," *J. Biotechnol.,* vol. 24, pp. 1–32.

Zadeh, L. (1972). "A rationale for fuzzy control," *J. Dyn. Syst. Meas. Control Trans. ASME,* vol. 94, pp. 3–4.

Zadeh, L. (1965). "Fuzzy sets," *Inf. Control,* vol. 8, pp. 338–353.

PROBLEMS

4.1. Using your own intuition, develop fuzzy membership functions on the real line for the fuzzy number 3, using the following function shapes:

(*a*) Symmetric triangle

(*b*) Trapezoid

(*c*) Gaussian function

4.2. Using your own intuition, develop fuzzy membership functions on the real line for the fuzzy number "approximately 2 *or* approximately 8" using the following function shapes:

(*a*) Symmetric triangles

(*b*) Trapezoids

(*c*) Gaussian functions

4.3. Using your own intuition, develop fuzzy membership functions on the real line for the fuzzy number "approximately 2 *and* approximately 8" using the following function shapes:

(*a*) Symmetric triangles

(*b*) Trapezoids

(*c*) Gaussian functions

4.4. Using your own intuition, develop fuzzy membership functions on the real line for the fuzzy number "approximately 6 *to* approximately 8" using the following function shapes:

(*a*) Symmetric triangles

(*b*) Trapezoids

(*c*) Gaussian functions

4.5. Using your own intuition and your own definitions of the universe of discourse, plot fuzzy membership functions for the following variables:

(*a*) Weight of people
- (*i*) Very light
- (*ii*) Light
- (*iii*) Average
- (*iv*) Heavy
- (*v*) Very heavy

(*b*) Age of people
- (*i*) Very young
- (*ii*) Young
- (*iii*) Middle-aged
- (*iv*) Old
- (*v*) Very old

(*c*) Education of people
- (*i*) Fairly educated
- (*ii*) Educated
- (*iii*) Highly educated
- (*iv*) Not highly educated
- (*v*) More or less educated

4.6. Using the inference approach outlined in this chapter, find the membership values for each of the triangular shapes ($\underset{\sim}{I}$, $\underset{\sim}{R}$, $\underset{\sim}{IR}$, $\underset{\sim}{E}$, $\underset{\sim}{T}$) for each of the following triangles:

(*a*) 80, 75, 25 degrees
(*b*) 55, 65, 60 degrees
(*c*) 50, 50, 80 degrees
(*d*) 45, 45, 90 degrees
(*e*) 45, 75, 60 degrees

4.7. Using the inference approach outlined in the chapter, plot the membership function for the following triangles (see appropriate expressions in the text) on a three-dimensional plot. Since only two of the angles in a triangle are independent (the third is equal to 180 minus the sum of the other two) the plots will have two independent axes (x, y) and one dependent axis (z), which is the membership value, μ. In plotting, the two largest angles can be taken as the independent values. For each of the following triangles plot just enough points to recognize the general shape of each membership function (a surface in this case). The plots can be three-dimensional, or they can be two-dimensional images in the x-z and y-z planes.

(*a*) Isosceles ($\underset{\sim}{\mathrm{I}}$)
(*b*) Right ($\underset{\sim}{\mathrm{R}}$)
(*c*) Isosceles and right ($\underset{\sim}{\mathrm{IR}}$)
(*d*) Equilateral ($\underset{\sim}{\mathrm{E}}$)
(*e*) Other ($\underset{\sim}{\mathrm{T}}$)

4.8. Develop a membership function for rectangles that is similar to the algorithm on triangles in this chapter. This function should have two independent variables; hence, it can be plotted.

4.9. The following raw data were determined in a pairwise comparison of new premium car preferences in a poll of 100 people. When it was compared with a Porsche (P), 79 of those polled preferred a BMW (B), 85 preferred a Mercedes (M), 59 preferred a Lexus (L), and 67 preferred an Infinity (I). When a BMW was compared, the preferences were 21-P, 23-M, 37-L, and 45-I. When a Mercedes was compared, the preferences were 15-P, 77-B, 35-L, and 48-I. When a Lexus was compared, the preferences were 41-P, 63-B, 65-M, and 51-I. Finally, when an Infinity was compared, the preferences were 33-P, 55-B, 52-M, and 49-L. Using rank ordering, plot the membership function for "most preferred car."

4.10. The energy E of a particle spinning in a magnetic field B is given by the equation

$$E = \mu B \sin\theta$$

where μ = magnetic moment of the spinning particle
θ = complement angle of magnetic moment with respect to the direction of the magnetic field

Assuming the magnetic field B and magnetic moment μ to be constants, we propose linguistic terms for the complement angle of magnetic moment as follows:

High moment (H)	$\theta = \pi/2$
Slightly high moment (SH)	$\theta = \pi/4$
No moment (Z)	$\theta = 0$
Slightly low moment (SL)	$\theta = -\pi/4$
Low moment (L)	$\theta = -\pi/2$

Find the membership values using the angular fuzzy set approach for these linguistic labels for the complement angle, and plot these values versus θ.

4.11. A small mining company signs a lease for a mineral quarry. Management must solve some problems before they can proceed with the work properly. For example, whenever water springs up from natural springs or rain, work will be stopped until the water is removed. Another problem that may be encountered is loose soil on the walls of the quarry. After the completion of some initial work, miners can be faced with soil falling freely from the walls and must take time to remove this soil. The lost time associated with these problems is a negative process in terms of maximizing profits. Only when mining work goes smoothly and the ore is being extracted is progress toward profits being made. Phrase some linguistic terms for these activities and build an angular fuzzy set model.

4.12. For the intersection of Central and Stanford in Albuquerque (Fig. P4.12), the traffic data shown in the following table are determined through traffic counters. Sample volumes of the traffic are taken at three intervals, each lasting one hour. The data consist of vehicles turning left, turning right, and going straight through the intersection. Making your own set of assumptions, how could you use these traffic data to develop crude membership functions using the angular fuzzy set method?

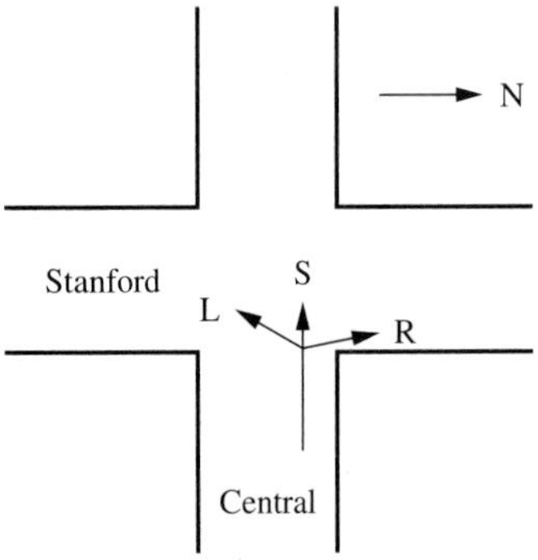

FIGURE P4.12

Sample number	**Vehicles turning left**	**Vehicles turning right**	**Vehicles moving straight**
1	48	50	201
2	15	38	168
3	42	60	220

4.13. For the data shown in the accomanying table, show the first iteration in trying to compute the membership values for the input variables x_1, x_2, and x_3 in the output regions R^1 and R^2. Assume a random set of weights for your neural network.

x_1	x_2	x_3	$\mathbf{R}^1$	$\mathbf{R}^2$
1.0	0.5	2.3	1.0	0.0

(*a*) Use a $3 \times 3 \times 1$ neural network.
(*b*) Use a $3 \times 3 \times 2$ neural network.
(*c*) Explain the difference in results when using (*a*) and (*b*).

4.14. For the data shown in the following table, show the first iteration in trying to compute the membership values for the input variables x_1, x_2, x_3, and x_4 in the regions R^1, R^2, and R^3.

x_1	x_2	x_3	x_4	R^1	R^2	R^3
10	0	−4	2	0	1	0

Use a 4 × 3 × 3 neural network with a random set of weights.

4.15. For the data shown in the accompanying Table A, show the first two iterations using a genetic algorithm in trying to find the optimum membership functions (use right-triangle functions) for the input variable x and output variable y in the rule table, Table B.

TABLE A—DATA

x	0	45	90
y	0	0.71	1

TABLE B—RULES

x	SM	MD
y	SM	LG

For the rule table, the symbols SM, MD, and LG mean small, medium, and large, respectively.

4.16. For the data shown in the following table (Table A), show the first two iterations using a genetic algorithm in trying to find the optimum membership functions (use right-triangle functions) for the input variable x and output variable y in the rule table, Table B. For the rule table, Table B, the symbols ZE, S, and LG mean zero, small, and large, respectively.

TABLE A—DATA

x	0	0.3	0.6	1.0	100
y	1	0.74	0.55	0.37	0

TABLE B—RULES

x	LG	S
y	ZE	S

4.17. The results of a price survey for 30 automobiles is presented here.

Class	Automobile prices (in units of $1000)
Economy	5.5, 5.8, 7.5, 7.9, 8.2, 8.5, 9.2, 10.4, 11.2, 13.5
Midsize	11.9, 12.5, 13.2, 14.9, 15.6, 17.8, 18.2, 19.5, 20.5, 24.0
Luxury	22.0, 23.5, 25.0, 26.0, 27.5, 29.0, 32.0, 37.0, 43.0, 47.5

Consider the car prices as a variable and the classes as economy, midsize, and luxury cars. Develop three membership function envelopes for car prices using the method of inductive reasoning as shown in Example 4.8 in the text.

CHAPTER 5

FUZZY-TO-CRISP CONVERSIONS

"Let's consider your age, to begin with—how old are you?"
"I'm seven and a half, exactly."
"You needn't say 'exactually,'" the Queen remarked;
"I can believe it without that. Now I'll give you *something to believe. I'm just one hundred and one, five months, and a day."*
"I can't believe that*!" said Alice.*
"Can't you?" the Queen said in a pitying tone. "Try again; draw a long breath, and shut your eyes."
Alice laughed. "There's no use trying," she said; "one can't *believe impossible things."*

Lewis Carroll
Through the Looking Glass, *1871*

It is one thing to compute, to reason, and to model with fuzzy information; it is another to apply the fuzzy results to the world around us. Despite the fact that the bulk of the information we assimilate every day is fuzzy, like the age of people in the Lewis Carroll example above, most of the actions or decisions implemented by humans or machines are crisp or binary. The decisions we make are binary, the hardware we use is binary, and certainly the computers we use are based on binary digital instructions. For example, in making a decision about developing a new engineering product the eventual decision is to go forward with development or not; the fuzzy choice to "maybe go forward" might be acceptable in planning stages, but eventually funds are released for development or they are not. In giving instructions to an electric motor, it is not possible to increase the voltage "slightly"; a machine does not understand the natural language of a human. We have to increase the voltage by 3.4 volts, for example, a crisp number. An electrical circuit typically is either on or off, not partially on.

The bulk of this textbook illustrates procedures to "fuzzify" the mathematical and engineering principles we have so long considered to be deterministic. But in various applications and engineering scenarios there will be a need to "defuzzify" the fuzzy results we generate through a fuzzy set analysis. In other words, we may eventually find a need to convert the fuzzy results to crisp results. For example, in classification (see Chapter 11) we may want to transform a fuzzy partition matrix into a crisp partition; in pattern recognition (see Chapter 12) we may want to compare a fuzzy pattern to a crisp pattern; in control (see Chapter 13) we may want to give a single-valued input to a semiconductor device instead of a fuzzy input command. This "defuzzification" has the result of reducing a fuzzy set to a crisp single-valued quantity, or to a crisp set; of converting a fuzzy matrix to a crisp matrix; or of making a fuzzy number crisp.

Mathematically, the defuzzification of a fuzzy set is the process of "rounding it off" from its location in the unit hypercube to the nearest (in a geometric sense) vertex (see Chapter 1). If one thinks of a fuzzy set as a collection of membership values, or a vector of values on the unit interval, defuzzification reduces this vector to a single scalar quantity— presumably to the most typical (prototype) or representative value. Various popular forms of converting fuzzy sets to crisp sets or to single scalar values are introduced in the following sections.

LAMBDA-CUTS FOR FUZZY SETS

We begin by considering a fuzzy set $\underset{\sim}{A}$; then, define a lambda-cut set, A_λ, where $0 \leq \lambda \leq 1$. The set A_λ is a crisp set called the lambda (λ)-cut (or alpha-cut) set of the fuzzy set $\underset{\sim}{A}$, where $A_\lambda = \{x|\mu_{\underset{\sim}{A}}(x) \geq \lambda\}$. Note that the λ-cut set A_λ does not have a tilde underscore; it is a crisp set derived from its parent fuzzy set, $\underset{\sim}{A}$. Any particular fuzzy set $\underset{\sim}{A}$ can be transformed into an infinite number of λ-cut sets, because there are an infinite number of values λ on the interval [0, 1].

Any element $x \in A_\lambda$ belongs to $\underset{\sim}{A}$ with a grade of membership that is greater than or equal to the value λ. The following example illustrates this idea.

Example 5.1. Let us consider the discrete fuzzy set, using Zadeh's notation, defined on universe $X = \{a, b, c, d, e, f\}$,

$$\underset{\sim}{A} = \left\{ \frac{1}{a} + \frac{.9}{b} + \frac{.6}{c} + \frac{.3}{d} + \frac{.01}{e} + \frac{0}{f} \right\}$$

This fuzzy set is shown schematically in Fig. 5.1. We can reduce this fuzzy set into several λ-cut sets, all of which are crisp. For example, we can define λ-cut sets for the values of $\lambda = 1, 0.9, 0.6, 0.3, 0^+$, and 0.

$$A_1 = \{a\} \qquad A_{0.9} = \{a, b\}$$

$$A_{0.6} = \{a, b, c\} \qquad A_{0.3} = \{a, b, c, d\}$$

$$A_{0^+} = \{a, b, c, d, e\} \qquad A_0 = X$$

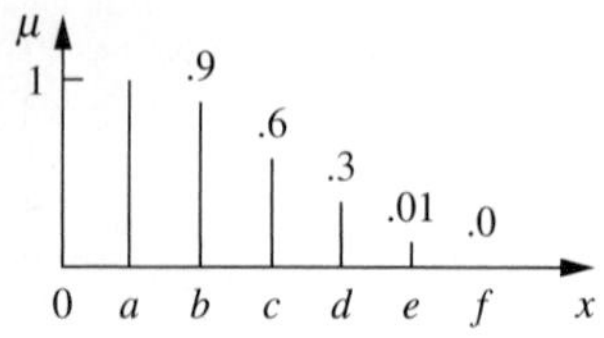

FIGURE 5.1
A discrete fuzzy set $\underset{\sim}{A}$.

The quantity $\lambda = 0^+$ is defined as a small "δ" value > 0, i.e., a value just greater than zero. By definition, $\lambda = 0$ produces the universe X, since all elements in the universe have at least a 0 membership value in any set on the universe. Since all A_λ are crisp sets, all the elements just shown in the example λ-cut sets have unit membership in the particular λ-cut set. For example, for $\lambda = 0.3$, the elements a, b, c, and d of the universe have membership $= 1$ in the λ-cut set, $A_{0.3}$, and the elements e and f of the universe have membership $= 0$ in the λ-cut set, $A_{0.3}$. Figure 5.2 shows schematically the crisp λ-cut sets for the values $\lambda = 1, 0.9, 0.6, 0.3, 0^+$, and 0. Notice in these plots of membership value versus the universe X that the effect of a λ-cut is to rescale the membership values: to one for all elements of the fuzzy set $\underset{\sim}{A}$ having membership values greater than or equal to λ, and to zero for all elements of the fuzzy set $\underset{\sim}{A}$ having membership values less than λ.

We can express λ-cut sets using Zadeh's notation. For the example in Fig. 5.1, λ-cut sets for the values $\lambda = 0.9$ and 0.25 are given here:

$$A_{0.9} = \left\{\frac{1}{a} + \frac{1}{b} + \frac{0}{c} + \frac{0}{d} + \frac{0}{e} + \frac{0}{f}\right\} \qquad A_{0.25} = \left\{\frac{1}{a} + \frac{1}{b} + \frac{1}{c} + \frac{1}{d} + \frac{0}{e} + \frac{0}{f}\right\}$$

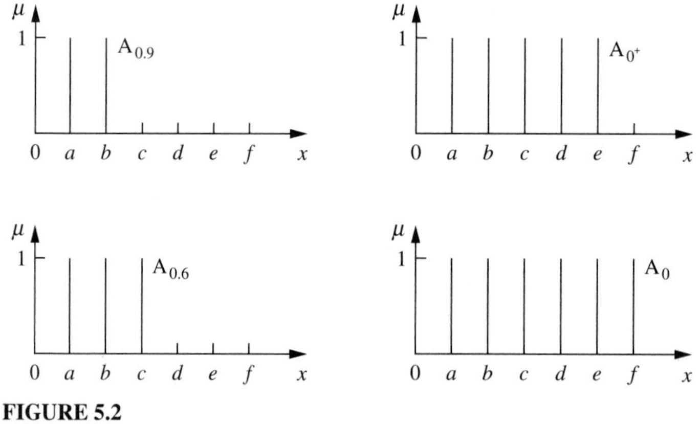

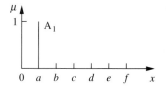

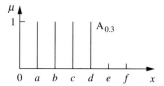

FIGURE 5.2
Lambda-cut sets for $\lambda = 1, 0.9, 0.6, 0.3, 0^+, 0$.

Lambda-cut sets obey the following four very special properties:

1. $(\underset{\sim}{A} \cup \underset{\sim}{B})_\lambda = A_\lambda \cup B_\lambda$ (5.1*a*)
2. $(\underset{\sim}{A} \cap \underset{\sim}{B})_\lambda = A_\lambda \cap B_\lambda$ (5.1*b*)
3. $(\overline{\underset{\sim}{A}})_\lambda \neq (\overline{A_\lambda})$ except for a value of $\lambda = 0.5$ (5.1*c*)
4. For any $\lambda \leq \alpha$, where $0 \leq \alpha \leq 1$, it is true that $A_\alpha \subseteq A_\lambda$, where $A_0 = X$ (5.1*d*)

These properties show that λ-cuts on standard operations on fuzzy sets are equivalent with standard set operations on λ-cut sets. The last operation, Eq. (5.1*d*), can be shown more conveniently using graphics. Figure 5.3 shows a continuous-valued fuzzy set with two λ-cut values. Notice in the graphic that for $\lambda = 0.3$ and $\alpha = 0.6$ $A_{0.3}$ has a greater domain than $A_{0.6}$, i.e., for $\lambda \leq \alpha (0.3 \leq 0.6)$, $A_{0.6} \subseteq A_{0.3}$.

In Chapter 4, various definitions of a membership function were discussed and illustrated. Many of these same definitions arise through the use of λ-cut sets. As seen in Fig. 4.1, we can provide the following definitions for a convex fuzzy set $\underset{\sim}{A}$. The core of $\underset{\sim}{A}$ is the $\lambda = 1$ cut set, A_1. The support of $\underset{\sim}{A}$ is the λ-cut set A_{0^+}, where $\lambda = 0^+$, or symbolically, $A_{0^+} = \{x \mid \mu_{\underset{\sim}{A}(x)} > 0\}$. The intervals $[A_{0^+}, A_1]$ form the boundaries of the fuzzy set $\underset{\sim}{A}$, i.e., those regions that have membership values between 0 and 1 (exclusive of 0 and 1), that is, for $0 < \lambda < 1$.

LAMBDA-CUTS FOR FUZZY RELATIONS

In Chapter 3, a biotechnology example, Example 3.11, was developed using a fuzzy relation that was reflexive and symmetric. Recall this matrix,

$$\underset{\sim}{R} = \begin{bmatrix} 1 & 0.8 & 0 & 0.1 & 0.2 \\ 0.8 & 1 & 0.4 & 0 & 0.9 \\ 0 & 0.4 & 1 & 0 & 0 \\ 0.1 & 0 & 0 & 1 & 0.5 \\ 0.2 & 0.9 & 0 & 0.5 & 1 \end{bmatrix}$$

We can define a λ-cut procedure for relations similar to the one developed for sets. Consider a fuzzy relation $\underset{\sim}{R}$, where each row of the relational matrix is consid-

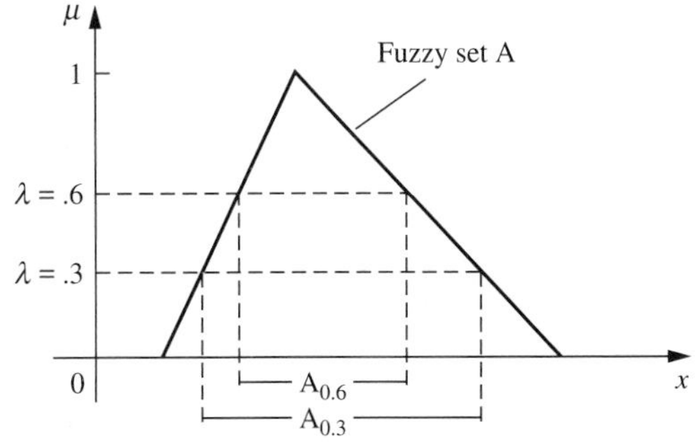

FIGURE 5.3
Two different λ-cut sets for a continuous-valued fuzzy set.

ered a fuzzy set; i.e., the jth row in $\underset{\sim}{R}$ represents a discrete membership function for a fuzzy set, $\underset{\sim}{R}_j$. Hence, a fuzzy relation can be converted to a crisp relation in the following manner. Let us define $R_\lambda = \{x, y) \mid \mu_{\underset{\sim}{R}}(x, y) \geq \lambda\}$ as a λ-cut relation of the fuzzy relation, $\underset{\sim}{R}$. Since in this case $\underset{\sim}{R}$ is a two-dimensional array defined on the universes X and Y, then any pair $(x, y) \in R_\lambda$ belongs to $\underset{\sim}{R}$ with a "strength" of relation greater than or equal to λ. These ideas for relations can be illustrated with an example.

Example 5.2. Suppose we take the fuzzy relation from the biotechnology example in Chapter 3 (Example 3.11), and perform λ-cut operations for the values of $\lambda = 1, .9, 0$. These crisp relations are given below:

$$\lambda = 1 \qquad R_1 = \begin{bmatrix} 1 & 0 & 0 & 0 & 0 \\ 0 & 1 & 0 & 0 & 0 \\ 0 & 0 & 1 & 0 & 0 \\ 0 & 0 & 0 & 1 & 0 \\ 0 & 0 & 0 & 0 & 1 \end{bmatrix}$$

$$\lambda = .9 \qquad R_{.9} = \begin{bmatrix} 1 & 0 & 0 & 0 & 0 \\ 0 & 1 & 0 & 0 & 1 \\ 0 & 0 & 1 & 0 & 0 \\ 0 & 0 & 0 & 1 & 0 \\ 0 & 1 & 0 & 0 & 1 \end{bmatrix}$$

$$\lambda = 0 \qquad R_0 = \text{E (whole relation; see Chapter 3)}$$

Lambda-cuts on fuzzy relations obey certain properties, just as lambda-cuts on fuzzy sets do [see Eqs. (5.1)], as given in Eqs. (5.2):

1. $(\underset{\sim}{R} \cup \underset{\sim}{S})_\lambda = R_\lambda \cup S_\lambda$ (5.2*a*)
2. $(\underset{\sim}{R} \cap \underset{\sim}{S})_\lambda = R_\lambda \cap S_\lambda$ (5.2*b*)
3. $(\overline{\underset{\sim}{R}})_\lambda \neq \overline{R}_\lambda$ (5.2*c*)
4. For any $\lambda \leq \alpha, 0 \leq \alpha \leq 1$, then $R_\alpha \subseteq R_\lambda$ (5.2*d*)

DEFUZZIFICATION METHODS

As mentioned in the introduction, there may be situations where the output of a fuzzy process needs to be a single scalar quantity as opposed to a fuzzy set. Defuzzification is the conversion of a fuzzy quantity to a precise quantity, just as fuzzification is the conversion of a precise quantity to a fuzzy quantity. The output of a fuzzy process can be the logical union of two or more fuzzy membership functions defined on the universe of discourse of the output variable. For example, suppose a fuzzy output is comprised of two parts: the first part, $\underset{\sim}{C}_1$, a trapezoidal shape, shown in Fig. 5.4*a*, and the second part, $\underset{\sim}{C}_2$, a triangular membership shape, shown in Fig. 5.4*b*. The union of these two membership functions, i.e., $\underset{\sim}{C} = \underset{\sim}{C}_1 \cup \underset{\sim}{C}_2$, involves the max-operator, which graphically is the outer envelope of the two shapes shown in Figs. 5.4*a* and *b;* the resulting shape is shown in Fig. 5.4*c*. Of course, a general fuzzy output process can involve many output parts (more than two), and the membership

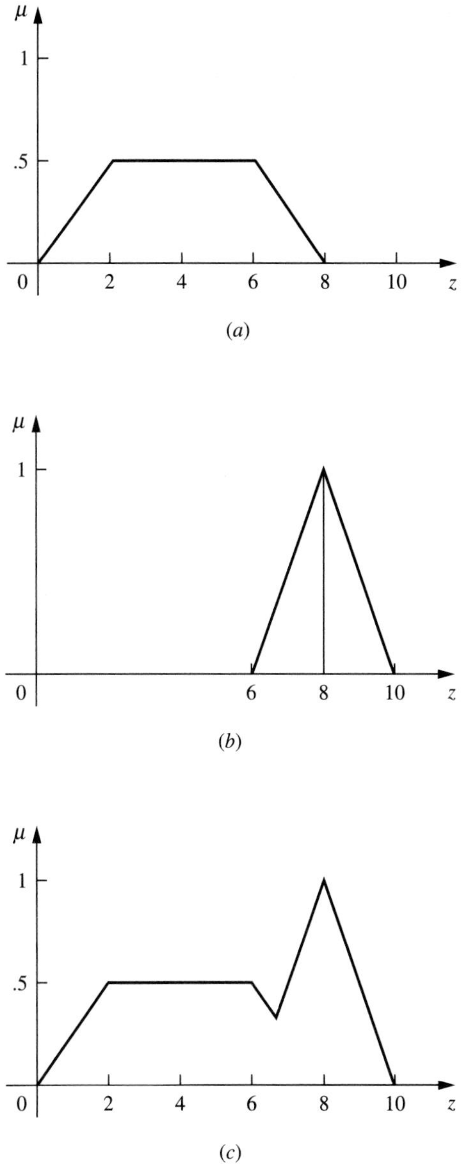

FIGURE 5.4
Typical fuzzy process output: (*a*) first part of fuzzy output; (*b*) second part of fuzzy output; (*c*) union of both parts.

function representing each part of the output can have shapes other than triangles and trapezoids. Further, as Fig. 5.4*a* shows, the membership functions may not always be normal. In general, we can have

$$\underset{\sim}{C}_k = \bigcup_{i=1}^{k} \underset{\sim}{C}_i = \underset{\sim}{C} \tag{5.3}$$

At least seven methods in the literature, among the many that have been proposed by investigators in recent years, are popular for defuzzifying fuzzy output functions (membership functions) [Hellendoorn and Thomas, 1993]. Four of these methods are summarized here and then are illustrated in two examples.

1. *Max-membership principle:* Also known as the *height* method, this scheme is limited to peaked output functions. This method is given by the algebraic expression

$$\mu_{\underset{\sim}{C}}(z^*) \geq \mu_{\underset{\sim}{C}}(z) \qquad \text{for all } z \in Z \tag{5.4}$$

and is shown graphically in Fig. 5.5.

2. *Centroid method:* This procedure (also called center of area, center of gravity) is the most prevalent and physically appealing of all the defuzzification methods [Sugeno, 1985; Lee, 1990]; it is given by the algebraic expression

$$z^* = \frac{\int \mu_{\underset{\sim}{C}}(z) \cdot z\, dz}{\int \mu_{\underset{\sim}{C}}(z)\, dz} \tag{5.5}$$

where $\int$ denotes an algebraic integration. This method is shown in Fig. 5.6.

3. *Weighted average method:* This method is only valid for symmetrical output membership functions. It is given by the algebraic expression

$$z^* = \frac{\sum \mu_{\underset{\sim}{C}}(\overline{z}) \cdot \overline{z}}{\sum \mu_{\underset{\sim}{C}}(\overline{z})} \tag{5.6}$$

where $\sum$ denotes an algebraic sum. This method is shown in Fig. 5.7. The weighted average method is formed by weighting each membership function in the output by its respective maximum membership value. As an example, the two functions shown in Fig. 5.7 would result in the following general form for the defuzzified value:

$$z^* = \frac{a(.5) + b(.9)}{.5 + .9}$$

Since the method is limited to symmetrical membership functions, the values a and b are the means of their respective shapes.

4. *Mean-max membership:* This method (also called middle-of-maxima) is closely related to the first method, except that the locations of the maximum membership can be non-unique (i.e., the maximum membership can be a plateau rather than a

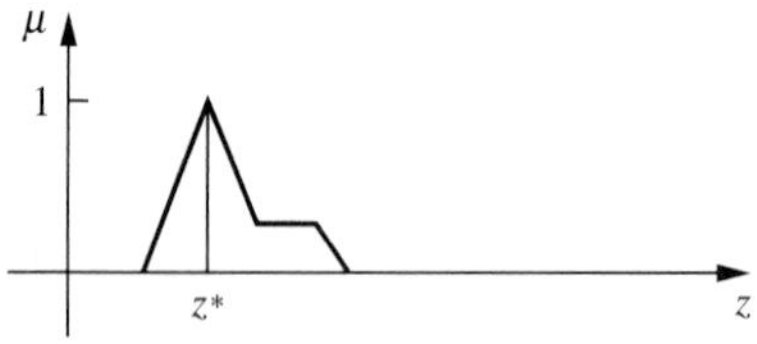

FIGURE 5.5
Max-membership defuzzification method.

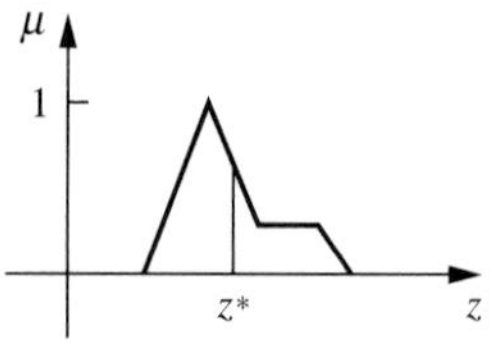

FIGURE 5.6
Centroid defuzzification method.

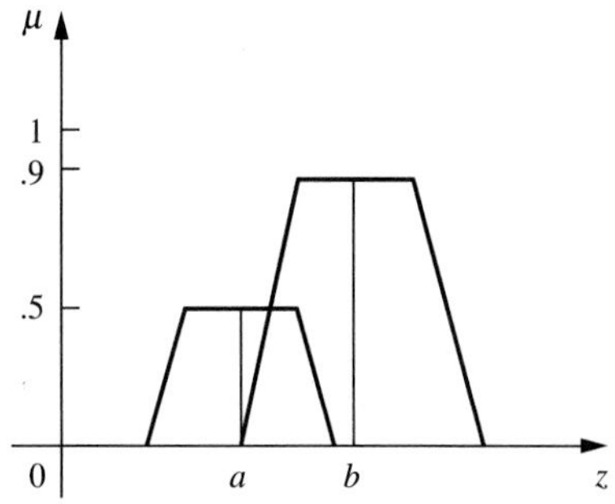

FIGURE 5.7
Weighted average method of defuzzification.

single point). This method is given by the expression [Sugeno, 1985; Lee, 1990]

$$z^* = \frac{a + b}{2} \tag{5.7}$$

where a and b are as defined in Fig. 5.8.

Example 5.3. A railroad company intends to lay a new rail line in a particular part of a county. The whole area through which the new line is passing must be purchased for right-of-way considerations. It is surveyed in three stretches, and the data are collected for analysis. The surveyed data for the road are given by the sets, $\underset{\sim}{B}_1$, $\underset{\sim}{B}_2$, and $\underset{\sim}{B}_3$, where the sets are defined on the universe of right-of-way widths, in meters. For the railroad to purchase the land, it must have an assessment of the amount of land to be bought. The three surveys on right-of-way width are ambiguous, however, because some of the land along the proposed railway route is already public domain and will not need to be purchased. Additionally, the original surveys are so old (circa 1860) that some ambiguity exists on boundaries and public right-of-way for old utility lines and old roads. The three fuzzy sets, $\underset{\sim}{B}_1$, $\underset{\sim}{B}_2$, and $\underset{\sim}{B}_3$, shown in Figs. 5.9, 5.10, and 5.11, respectively, represent the uncertainty in each survey as to the membership of right-of-way width, in meters, in privately owned land.

We now want to aggregate these three survey results to find the single most nearly representative right-of-way width (z) to allow the railroad to make its initial estimate

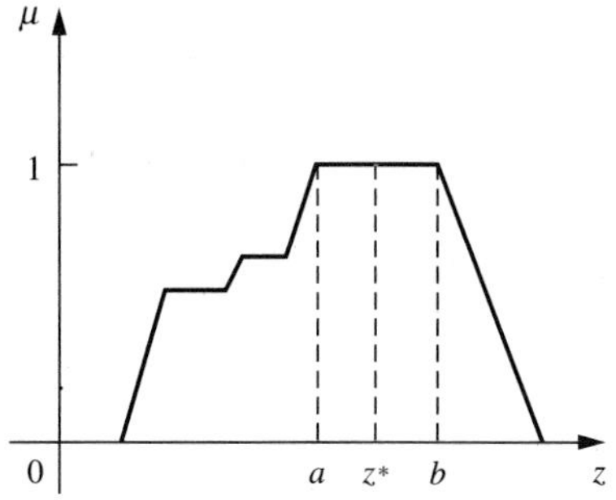

FIGURE 5.8
Mean-max membership defuzzification method.

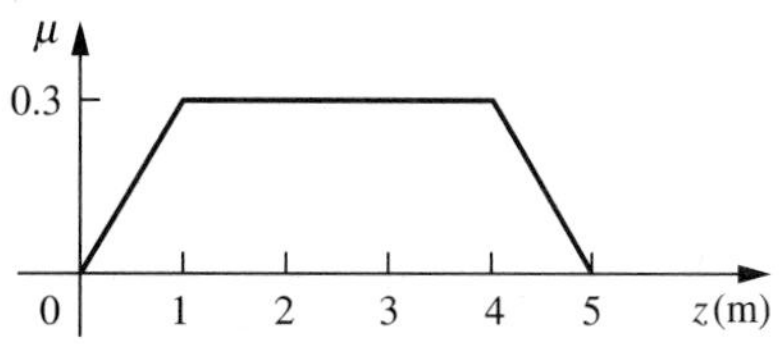

FIGURE 5.9
Fuzzy set $\underset{\sim}{B}_1$: Public right-of-way width (z) for survey 1.

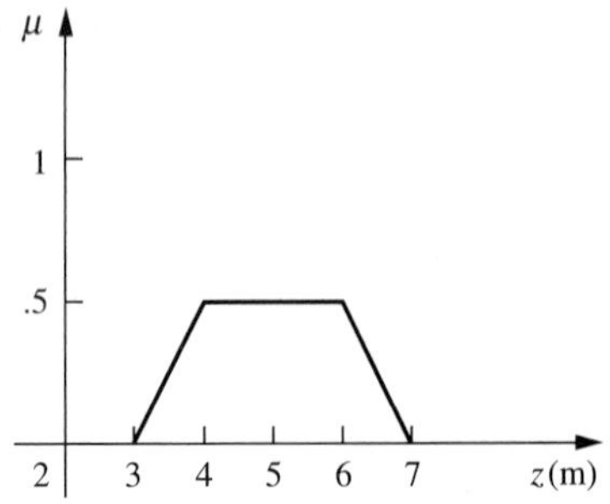

FIGURE 5.10
Fuzzy set $\underset{\sim}{B}_2$: Public right-of-way width (z) for survey 2.

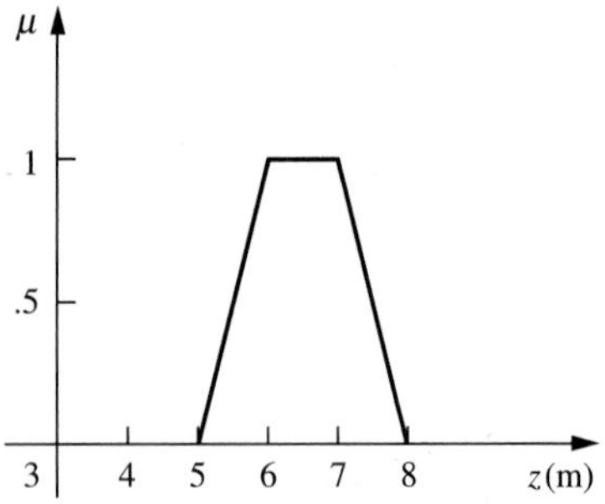

FIGURE 5.11
Fuzzy set $\underset{\sim}{B}_3$: Public right-of-way width (z) for survey 3.

of the right-of-way purchasing cost. Using Eqs. (5.5) through (5.7) and the preceding three fuzzy sets, we want to find z^*.

According to the centroid method, Eq. (5.5), z^* can be found using

$$
\begin{aligned}
z^* &= \frac{\int \mu_{\underset{\sim}{B}}(z)\cdot z\,dz}{\int \mu_{\underset{\sim}{B}}(z)\,dz} = \\
&\left[\int_0^1 (.3z)z\,dz + \int_1^{3.6} (.3z)\,dz + \int_{3.6}^{4}\left(\frac{z-3}{2}\right)z\,dz + \int_4^{5.5}(.5)z\,dz\right. \\
&\left. + \int_{5.5}^{6}(z-5)z\,dz + \int_6^7 z\,dz + \int_7^8 (8-z)z\,dz\right] \\
&\div\left[\int_0^1 (.3z)\,dz + \int_1^{3.6}(.3)\,dz + \int_{3.6}^{4}\left(\frac{z-3}{2}\right)dz + \int_4^{5.5}(.5)\,dz\right. \\
&\left. + \int_{5.5}^{6}(z-5)\,dz + \int_6^7 dz + \int_7^8 (8-z)\,dz\right] \\
&= 4.9 \text{ meters}
\end{aligned}
$$

where z^* is shown in Fig. 5.12. According to the weighted average method, Eq. (5.6),

$$
z^* = \frac{(.3\times 2.5) + (.5\times 5) + (1\times 6.5)}{.3 + .5 + 1} = 5.41 \text{ meters}
$$

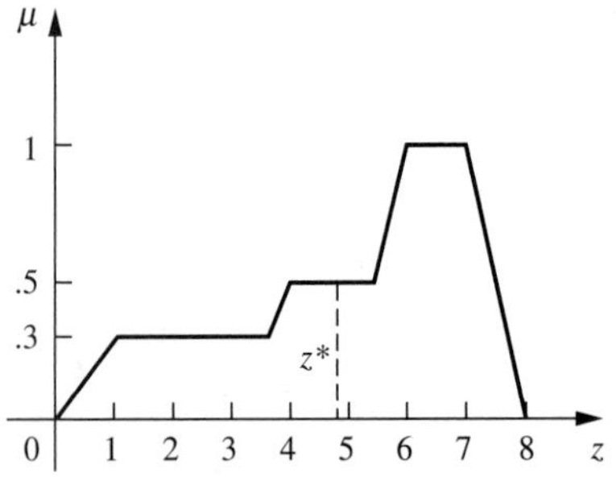

FIGURE 5.12
The centroid method for finding z^*.

and is shown in Fig. 5.13. According to the mean-max membership method, Eq. (5.7), z^* is given by $(6 + 7)/2 = 6.5$ meters, and is shown in Fig. 5.14.

Example 5.4. Many products, such as tar, petroleum jelly, and petroleum, are extracted from crude oil. In a newly drilled oil well, three sets of oil samples are taken and tested for their viscosity. The results are given in the form of the three fuzzy sets $\underset{\sim}{B}_1$, $\underset{\sim}{B}_2$, and $\underset{\sim}{B}_3$, all defined on a universe of normalized viscosity, as shown in Figs. 5.15–5.17. Using Eqs. (5.4) through (5.6), we want to find the most nearly representative viscosity value for all three oil samples, and hence find z^* for the three fuzzy viscosity sets.

To find z^* using the centroid method, we first need to find the logical union of the three fuzzy sets. This is shown in Fig. 5.18. Also shown in Fig. 5.18 is the result of the max-membership method, Eq. (5.4). For this method, we see that $\mu_{\underset{\sim}{B}}(z^*)$ has three locations where the membership equals unity. This result is ambiguous and, in this case, the selection of the intermediate point is arbitrary, but it is closer to the centroid of the area shown in Fig. 5.18. There could be other compelling reasons to select another value in this case; perhaps max-membership is not a good metric for this problem.

According to the centroid method, Eq. (5.5),

$$z^* = \frac{\int \mu_{\underset{\sim}{B}}(z)z\,dz}{\int \mu_{\underset{\sim}{B}}(z)\,dz} =$$

$$\left[\int_0^{1.5} (.67z)z\,dz + \int_{1.5}^{1.8} (2 - .67z)z\,dz + \int_{1.8}^{2} (z - 1)z\,dz + \int_{2}^{2.33} (3 - z)z\,dz \right.$$
$$\left. + \int_{2.33}^{3} (.5z - .5)z\,dz + \int_{3}^{5} (2.5 - .5z)z\,dz\right]$$
$$\div \left[\int_0^{1.5} (.67z)\,dz + \int_{1.5}^{1.8} (2 - .67z)\,dz + \int_{1.8}^{2} (z - 1)\,dz + \int_{2}^{2.33} (3 - z)\,dz \right.$$
$$\left. + \int_{2.33}^{3} (.5z - .5)\,dz + \int_{3}^{5} (2.5 - .5z)\,dz\right]$$
$$= 2.5$$

The centroid value obtained, z^*, is shown in Fig. 5.19.

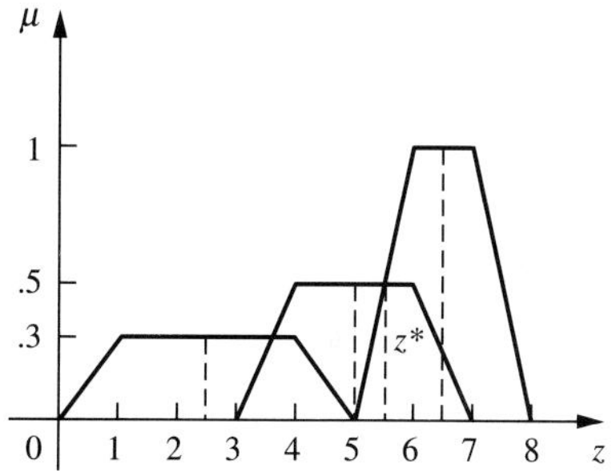

FIGURE 5.13
The weighted average method for finding z^*.

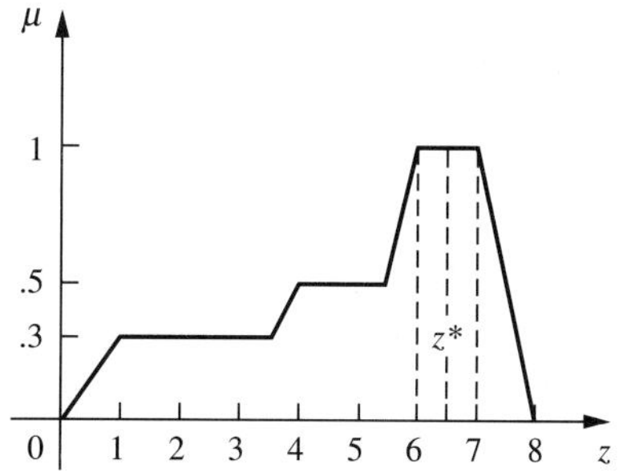

FIGURE 5.14
The mean-max membership method for finding z^*.

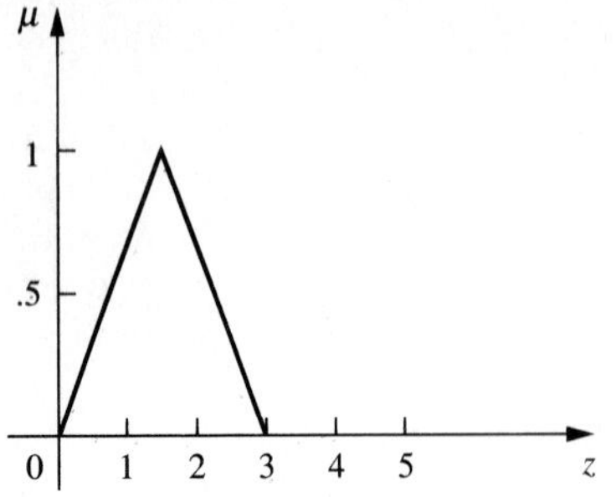

FIGURE 5.15
Membership in viscosity of oil sample 1, $\underset{\sim}{B}_1$.

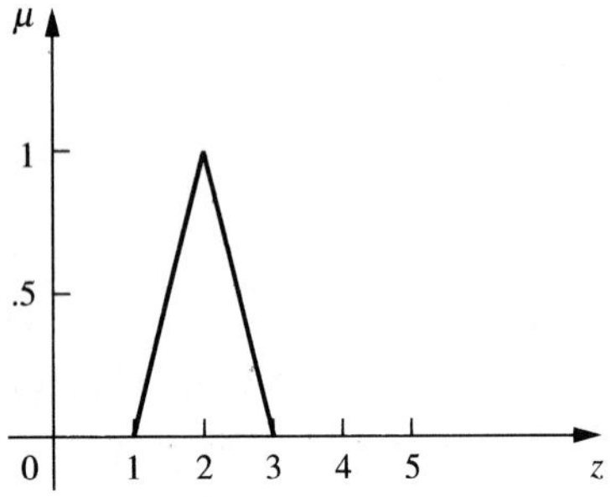

FIGURE 5.16
Membership in viscosity of oil sample 2, $\underset{\sim}{B}_2$.

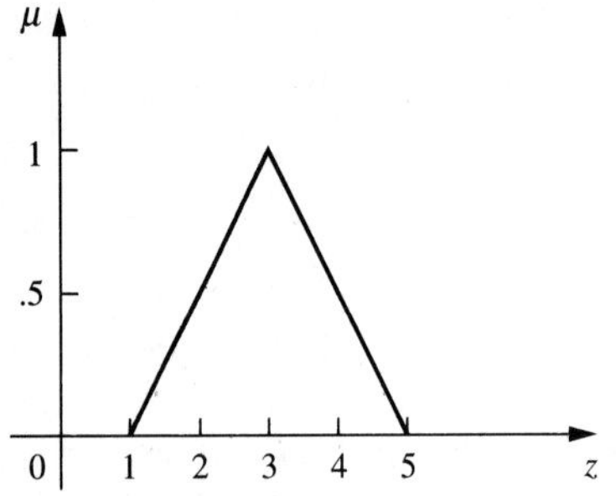

FIGURE 5.17
Membership in viscosity of oil sample 3, $\underset{\sim}{B}_3$.

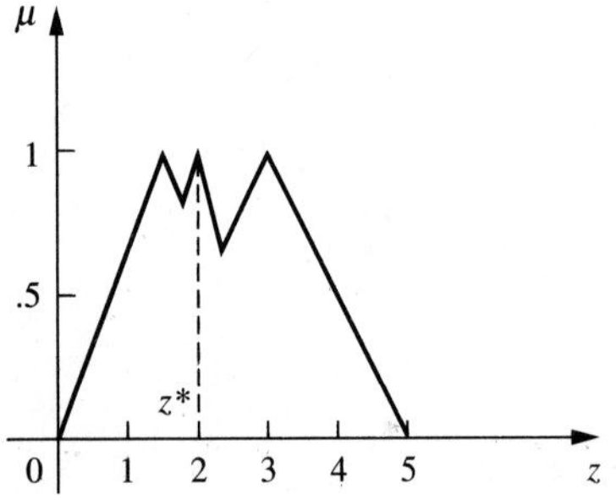

FIGURE 5.18
Logical union of three fuzzy sets $\underset{\sim}{B}_1$, $\underset{\sim}{B}_2$, and $\underset{\sim}{B}_3$.

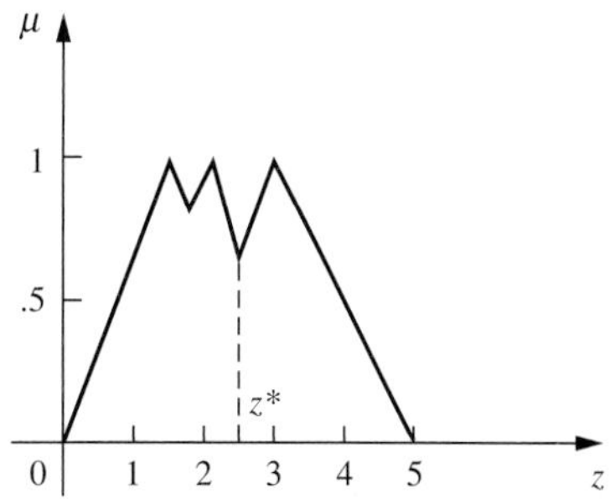

FIGURE 5.19
Centroid value z^* for three fuzzy oil samples.

According to the weighted average method, Eq. (5.6),

$$z^* = \frac{(1 \times 1.5) + (1 \times 2) + (1 \times 3)}{1 + 1 + 1} = 2.25$$

and is shown in Fig. 5.20.

Three other popular methods, which are worthy of discussion because of their appearance in some applications, are the center of sums, center of largest area, and first of maxima methods [Hellendoorn and Thomas, 1993]. These methods are now developed.

5. *Center of sums:* This is faster than many defuzzification methods that are presently in use. This process involves the algebraic sum of individual output fuzzy sets, say $\underset{\sim}{C}_1$ and $\underset{\sim}{C}_2$, instead of their union. One drawback to this method is that the intersecting areas are added twice. The defuzzified value z^* is given by the following equation:

$$z^* = \frac{\int_Z z \sum_{k=1}^{n} \mu_{\underset{\sim}{C}_k}(z)\, dz}{\int_z \sum_{k=1}^{n} \mu_{\underset{\sim}{C}_k}(z)\, dz} \tag{5.8}$$

This method is similar to the weighted average method, Eq. (5.6), except in the center of sums method the weights are the areas of the respective membership functions whereas in the weighted average method the weights are individual membership values. Figure 5.21 is an illustration of the center of sums method.

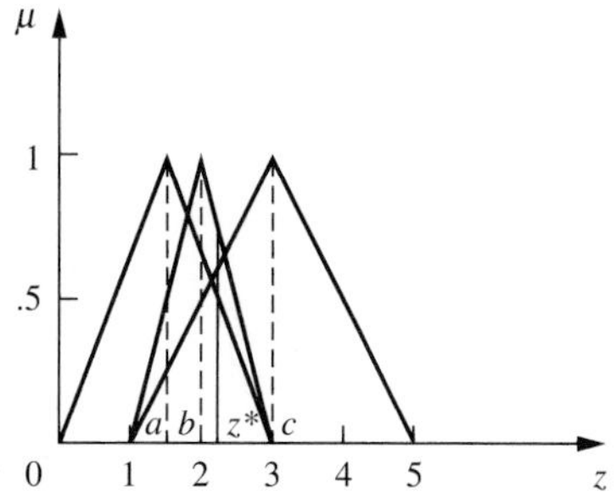

FIGURE 5.20
Weighted average method for z^*.

6. *Center of largest area:* If the output fuzzy set has at least two convex subregions, then the center of gravity [i.e., z^* is calculated using the centroid method, Eq. (5.5)] of the convex fuzzy subregion with the largest area is used to obtain the defuzzified value z^* of the output. This is shown graphically in Fig. 5.22, and given algebraically here:

$$z^* = \frac{\int \mu_{\underset{\sim}{C}_m}(z) z \, dz}{\int \mu_{\underset{\sim}{C}_m}(z) \, dz} \tag{5.9}$$

where $\underset{\sim}{C}_m$ is the convex subregion that has the largest area making up $\underset{\sim}{C}_k$. This condition applies in the case when the overall output $\underset{\sim}{C}_k$ is nonconvex; and in the case when $\underset{\sim}{C}_k$ is convex, z^* is the same quantity as determined by the centroid method or the center of largest area method (because then there is only one convex region).

7. *First (or last) of maxima:* This method uses the overall output or union of all individual output fuzzy sets $\underset{\sim}{C}_k$ to determine the smallest value of the domain with maximized membership degree in $\underset{\sim}{C}_k$. The equations for z^* are as follows.

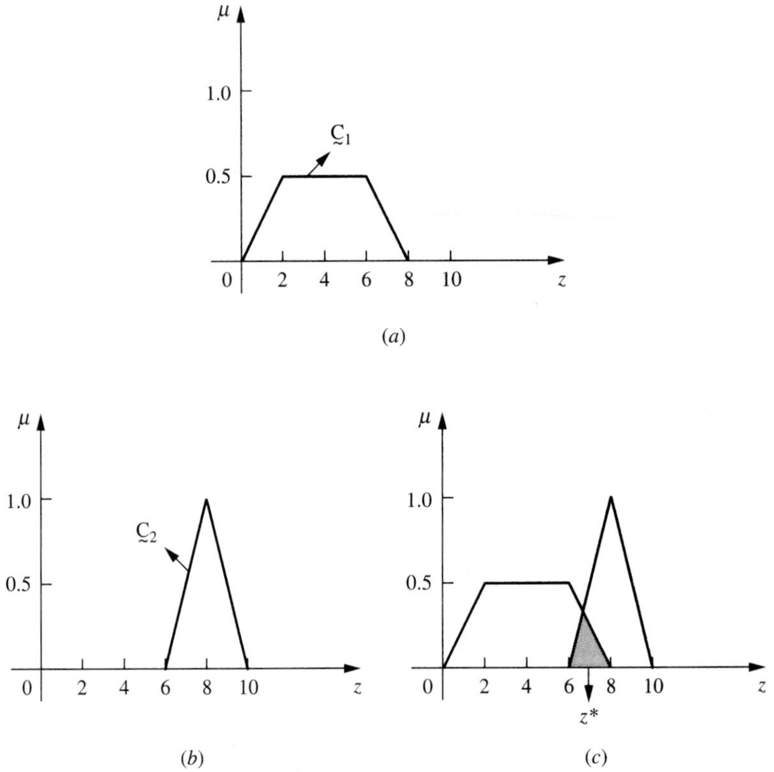

FIGURE 5.21
Center of sums method: (*a*) first membership function; (*b*) second membership function; and (*c*) defuzzification step.

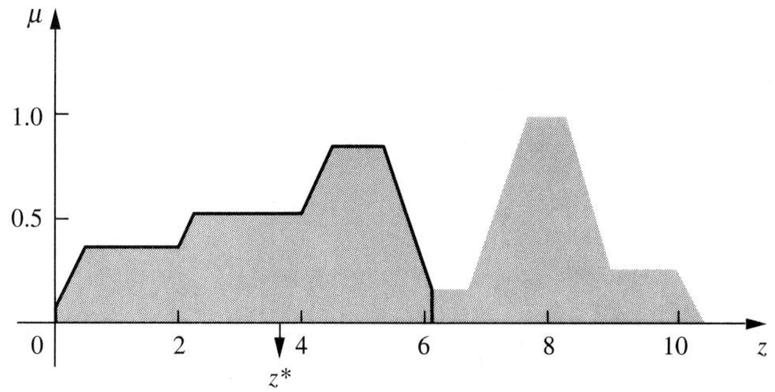

FIGURE 5.22
Center of largest area method (outlined with bold lines), shown for a nonconvex $\underset{\sim}{C}_k$.

First, the largest height in the union [denoted hgt($\underset{\sim}{C}_k$)] is determined,

$$\text{hgt}(\underset{\sim}{C}_k) = \sup_{z \in Z} \mu_{\underset{\sim}{C}_k}(z) \tag{5.10}$$

Then the first of the maxima is found,

$$z* = \inf_{z \in Z} \{ z \in Z \mid \mu_{\underset{\sim}{C}_k}(z) = \text{hgt}(\underset{\sim}{C}_k) \} \tag{5.11}$$

An alternative to this method is called the last of maxima, and it is given by

$$z^* = \sup_{z \in Z} \{ z \in Z \mid \mu_{\underset{\sim}{C}_k}(z) = \text{hgt}(\underset{\sim}{C}_k) \} \tag{5.12}$$

In Eqs. (5.10)–(5.12) the supremum (sup) is the least upper bound and the infimum (inf) is the greatest lower bound. Graphically, this method is shown in Fig. 5.23, where, in the case illustrated in the figure, the first max is also the last max and, because it is a distinct max, is also the mean-max. Hence, the methods presented in Eqs. (5.4) (max or height), (5.7) (mean-max), (5.11) (first-max), and (5.12) (last-max) all provide the same defuzzified value, z^*, for the particular situation illustrated in Fig. 5.23.

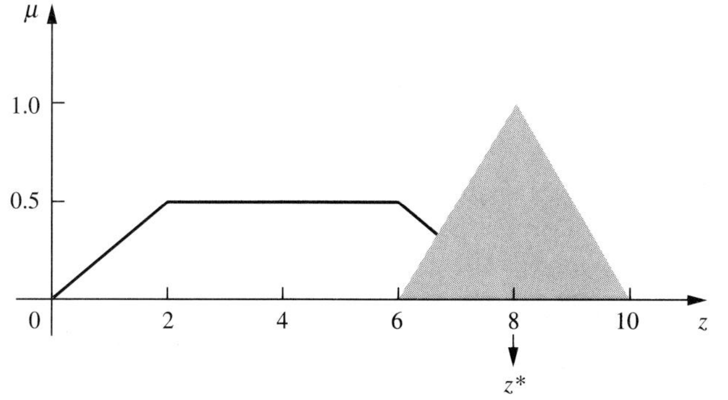

FIGURE 5.23
First of max (and last of max) method.

The problems illustrated in Examples 5.3 and 5.4 are now continued, to illustrate the last three methods presented.

Example 5.5. Continuing with Example 5.3 on the railroad company planning to lay a new rail line, we will calculate the defuzzified values using the (*i*) center of sums method, (*ii*) center of largest area, and (*iii*) first maxima and last maxima.

According to the center of sums method, Eq. (5.8), z^* will be as follows:

$$z^* = \frac{\int_0^8 [2.5 \times 0.5 \times 0.3(3+5) + 5 \times 0.5 \times 0.5(2+4) + 6.5 \times 0.5 \times 1(3+1)]\,dz}{\int_0^8 [0.5 \times 0.3(3+5) + 0.5 \times 0.5(2+4) + 0.5 \times 1(3+1)]\,dz}$$

$$= 5.042 \text{ m}$$

with the result shown in Fig. 5.24. The center of largest area method, Eq. (5.9), provides the same result (i.e., $z^* = 4.9$) as the centroid method, Eq. (5.5), because the complete output fuzzy set is convex, as seen in Fig. 5.25. According to the first of maxima and last of maxima methods, Eqs. (5.11) and (5.12), z^* is shown as z_1^* and z_2^*, respectively, in Fig. 5.26.

Example 5.6. Continuing with Example 5.4 on the crude oil problem, the center of sums method, Eq. (5.8), produces a defuzzified value for z^* of

$$z^* = \frac{\int_0^5 (0.5 \times 3 \times 1 \times 1.5 + 0.5 \times 2 \times 1 \times 2 + 0.5 \times 4 \times 1 \times 3)\,dz}{\int_0^5 (0.5 \times 3 \times 1 + 0.5 \times 2 \times 1 + 0.5 \times 4 \times 1)\,dz} = 2.3 \text{ m}$$

which is shown in Fig. 5.27. In the center of largest area method we first determine the areas of the three individual convex fuzzy output sets, as seen in Fig. 5.28. These areas are 1.02, 0.46, and 1.56 square units, respectively. Among them, the third area is largest, so the centroid of that area will be the center of the largest area. The defuzzified value is calculated to be $z^* = 3.5$.

$$z^* = \frac{\int_{2.33}^{3} \left[\left(\frac{0.67}{2} + 2.33\right)(0.5 \times 0.67(1+0.67))\right] dz + \int_3^5 3.66(0.5 \times 2 \times 1)\,dz}{\int_{2.33}^{3} [0.5 \times 0.67(1+0.67)]\,dz + \int_3^5 (0.5 \times 2 \times 1)\,dz} = 3.5 \text{ m}$$

Finally, one can see graphically in Fig. 5.29 that the first of maxima and last of maxima, Eqs. (5.11)–(5.12), give different values for z^*, namely, $z^* = 1.5$ and 3.0, respectively.

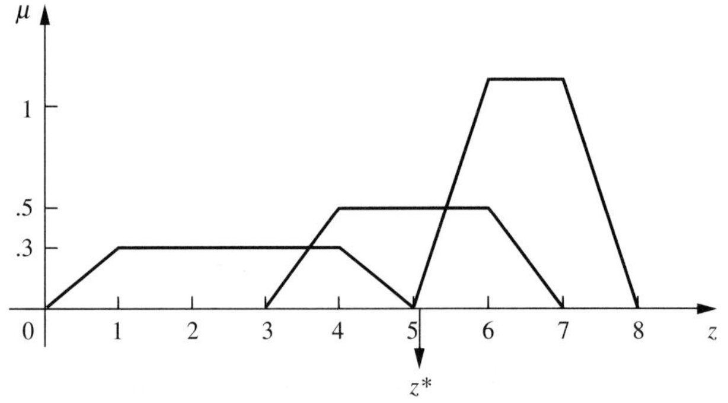

FIGURE 5.24
Center of sums result for Example 5.5.

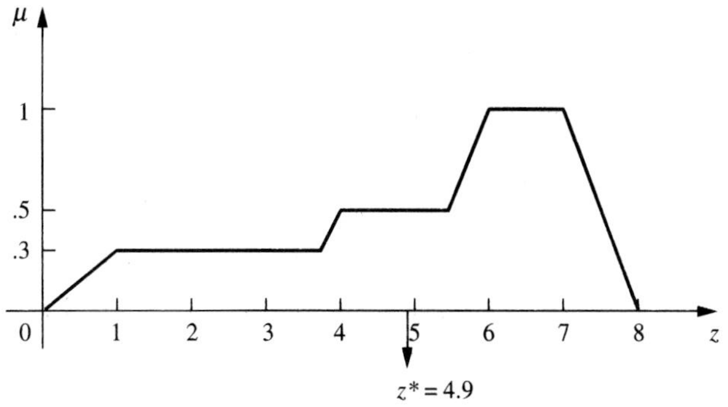

FIGURE 5.25
Output fuzzy set for Example 5.5 is convex.

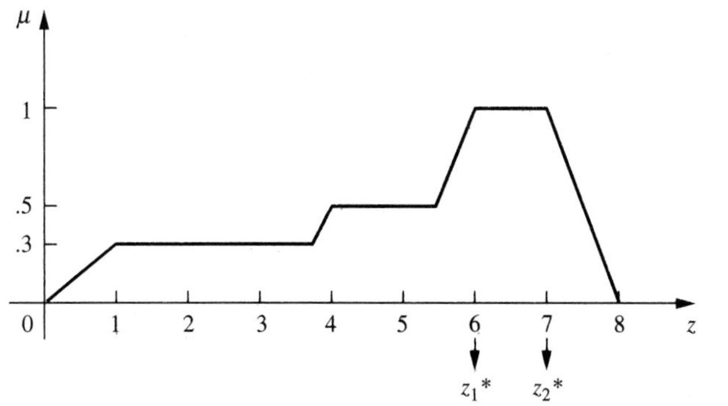

FIGURE 5.26
First of maxima solution ($z_1^* = 6$) and last of maxima solution ($z_2^* = 7$).

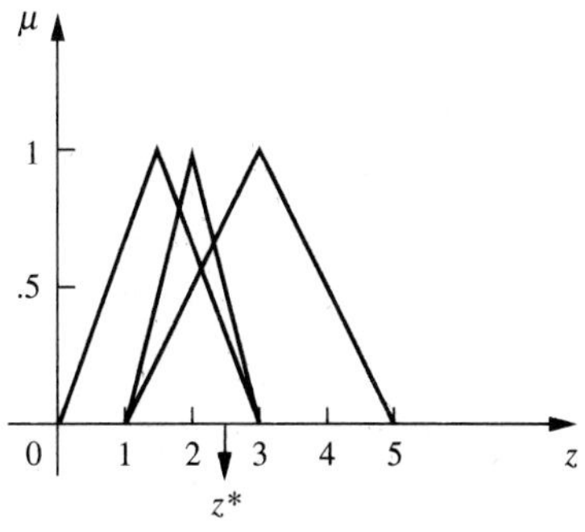

FIGURE 5.27
Center of sums solution for Example 5.6.

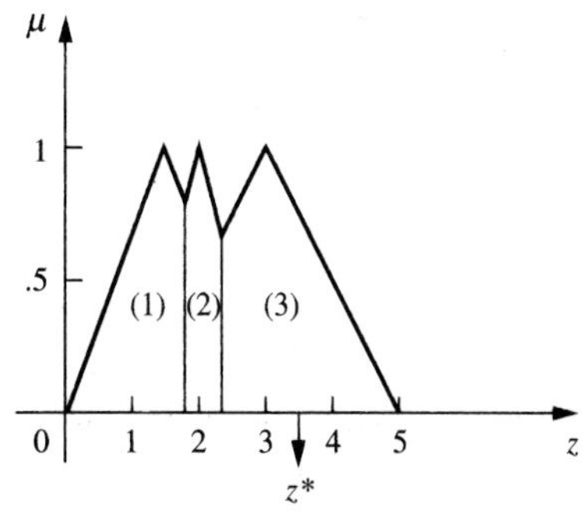

FIGURE 5.28
Center of largest area method for Example 5.6.

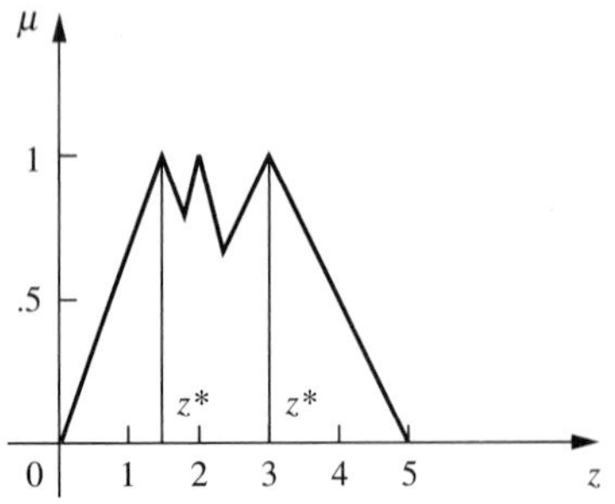

FIGURE 5.29
First of maxima gives $z^* = 1.5$ and last of maxima gives $z^* = 3$.

SUMMARY

This chapter has introduced the notion of converting from fuzzy forms to crisp forms—a process called defuzzification. Defuzzification is necessary because, for example, we cannot instruct the voltage going into a machine to increase "slightly," even if this instruction comes from a fuzzy controller—we must alter its voltage by a specific amount. Defuzzification is a natural and necessary process. In fact, there is an analogous form of defuzzification in mathematics where we solve a complicated problem in the complex plane, find the real and imaginary parts of the solution, then *decomplexify* the imaginary solution back to the real numbers space [Bezdek, 1993]. There are numerous other methods for defuzzification that have not been presented here. A review of the literature will provide the details on some of these [see for example, Filev and Yager, 1991; and Yager and Filev, 1993].

A natural question to ask is, "Of the seven defuzzification methods presented, which is the best?" One obvious answer to the question is that it is context- or problem-dependent. To answer this question in more depth, Hellendoorn and Thomas [1993] have specified five criteria against which to measure the methods. These criteria will be repeated here for the benefit of the reader who also ponders the question just given in terms of the advantages and disadvantages of the various methods. The first criterion is *continuity.* A small change in the input of a fuzzy process should not produce a large change in the output. Second, a criterion known as *disambiguity* simply points out that a defuzzification method should always result in a unique value for z^*, i.e., no ambiguity in the defuzzified value. This criterion is not satisfied by the center of largest area method, Eq. (5.9), because, as seen in Fig. 5.24, when the largest membership functions have equal area, there is ambiguity in selecting a z^*. The third criterion is called *plausibility.* To be plausible, z^* should lie approximately in the middle of the support region of $\underset{\sim}{C}_k$ and have a high degree of membership in $\underset{\sim}{C}_k$. The centroid method, Eq. (5.5), does not exhibit plausibility in the situation illustrated in Fig. 5.24 because, although z^* lies in the middle of the support of $\underset{\sim}{C}_k$, it does not have a high degree of membership (also seen in the darkened area of Fig. 5.21*c*). The fourth criterion is that of *computational simplicity,* which suggests that the more time-consuming a method is, the less value it should have in a computation system. The height method, Eq. (5.4), the mean-max method, Eq. (5.7), and the first of maxima method are faster than the centroid, Eq. (5.5), or center of sum, Eq. (5.8), methods, for example. The fifth criterion is called the *weighting method,* which weights the output fuzzy sets. This criterion constitutes the difference between the centroid method, Eq. (5.5), the weighted average method, Eq. (5.6), and center of sum methods, Eq. (5.8). The problem with the fifth criterion is that it is problem-dependent, as there is little by which to judge the best weighting method; the weighted average method involves less computation than the center of sums, but that attribute falls under the fourth criterion, computational simplicity.

As with many issues in fuzzy logic, the method of defuzzification should be assessed in terms of the goodness of the answer in the context of the data available. Other methods are available that purport to be superior to the simple methods presented here [Hellendoorn and Thomas, 1993].

REFERENCES

Bezdek, J. (1993). "Editorial: Fuzzy models—what are they, and why?" *IEEE Trans. Fuzzy Syst.,* vol. 1, pp. 1–5.

Filev, D., and R. Yager (1991). "A generalized defuzzification method under BAD distributions," *Int. J. Intelligent Syst.,* vol. 6, pp. 689–697.

Hellendoorn, H., and C. Thomas (1993). "Defuzzification in fuzzy controllers," *Intelligent and Fuzzy Systems,* vol. 1, pp. 109–123.

Klir, G., and T. Folger (1988). *Fuzzy sets, uncertainty, and information,* Prentice Hall, Englewood Cliffs, NJ.

Lee, C. (1990). "Fuzzy logic in control systems: fuzzy logic controller, Parts I and II," *IEEE Trans. Syst., Man & Cybern.,* vol. 20, pp. 404–435.

Sugeno, M. (1985). "An introductory survey of fuzzy control," *Inf. Sci.,* vol. 36, pp. 59–83.

Yager, R., and D. Filev (1993). "SLIDE: A simple adaptive defuzzification method," *IEEE Trans. Fuzzy Syst.,* vol. 1, pp. 69–78.

PROBLEMS

5.1. Two fuzzy sets $\underset{\sim}{A}$ and $\underset{\sim}{B}$, both defined on X, are as follows:

$\mu(x_i)$	x_1	x_2	x_3	x_4	x_5	x_6
$\underset{\sim}{A}$	0.1	0.6	0.8	0.9	0.7	0.1
$\underset{\sim}{B}$	0.9	0.7	0.5	0.2	0.1	0

Express the following λ-cut sets using Zadeh's notation:

(*a*) $(\overline{\underset{\sim}{A}})_{0.7}$ (*e*) $(\underset{\sim}{A} \cup \overline{\underset{\sim}{A}})_{0.7}$

(*b*) $(\underset{\sim}{B})_{0.4}$ (*f*) $(\underset{\sim}{B} \cap \overline{\underset{\sim}{B}})_{0.5}$

(*c*) $(\underset{\sim}{A} \cup \underset{\sim}{B})_{0.7}$ (*g*) $(\overline{\underset{\sim}{A} \cap \underset{\sim}{B}})_{0.7}$

(*d*) $(\underset{\sim}{A} \cap \underset{\sim}{B})_{0.6}$ (*h*) $(\overline{\underset{\sim}{A}} \cup \overline{\underset{\sim}{B}})_{0.7}$

5.2. [Klir and Folger, 1988] Show that all λ-cuts of any fuzzy set $\underset{\sim}{A}$ defined in R^n space $(n \geq 1)$ are convex if and only if

$$\mu_{\underset{\sim}{A}}[\lambda r + (1-\lambda)s] \geq \min[\mu_{\underset{\sim}{A}}(r), \mu_{\underset{\sim}{A}}(s)]$$

for all $r, s \in R^n$, and all $\lambda \in [0, 1]$.

5.3. The fuzzy sets $\underset{\sim}{A}$, $\underset{\sim}{B}$, and $\underset{\sim}{C}$ are all defined on the universe X = [0, 5] with the following membership functions:

$$\mu_{\underset{\sim}{A}}(x) = \frac{1}{1 + 5(x - 5)^2} \qquad \mu_{\underset{\sim}{B}}(x) = 2^{-x} \qquad \mu_{\underset{\sim}{C}}(x) = \frac{2x}{x + 5}$$

(*a*) Sketch the membership functions.

(*b*) Define the intervals along the x axis corresponding to the λ-cut sets for each of the fuzzy sets $\underset{\sim}{A}$, $\underset{\sim}{B}$, and $\underset{\sim}{C}$ for the following values of λ:

(*i*) $\lambda = 0.2$
(*ii*) $\lambda = 0.4$
(*iii*) $\lambda = 0.7$
(*iv*) $\lambda = 0.9$
(*v*) $\lambda = 1.0$

5.4. Sketch λ-cut sets for the four set operations (*i–iv*) in Problem 2.1 using $\lambda = 0.2$ and 0.8.

5.5. Sketch λ-cut sets for the five set operations in Problem 2.5 using $\lambda = 0.3$ and 0.9.

5.6. Determine the intervals on the real line for the λ-cut sets for the six set operations (*a–f*) in Problem 2.14 using $\lambda = 0.5$ and 0.7.

5.7. Using Zadeh's notation, determine the λ-cut sets for the first six set operations (*a–f*) in Problem 2.8 using $\lambda = 0.5$.

5.8. Using Zadeh's notation, determine the λ-cut sets for the six set operations (*a–f*) in Problem 2.9 using $\lambda = 0.2$ and 0.8.

5.9. Determine the crisp λ-cut relations for $\lambda = 0.1j$, for $j = 0, 1, \ldots, 10$ for the following fuzzy relation matrix $\underset{\sim}{R}$:

$$\underset{\sim}{R} = \begin{bmatrix} 0.2 & 0.7 & 0.8 & 1 \\ 1 & 0.9 & 0.5 & 0.1 \\ 0 & 0.8 & 1 & 0.6 \\ 0.2 & 0.4 & 1 & 0.3 \end{bmatrix}$$

5.10. For the fuzzy relation $\underset{\sim}{R}_4$ in Exercise 3.2 find the λ-cut relations for the following values of λ:

(*a*) $\lambda = 0^+$
(*b*) $\lambda = 0.1$
(*c*) $\lambda = 0.4$
(*d*) $\lambda = 0.7$

5.11. For the fuzzy relation $\underset{\sim}{R}$ in Problem 3.7 find the λ-cut relations for the following values of λ:

(*a*) $\lambda = 0.2$
(*b*) $\lambda = 0.4$
(*c*) $\lambda = 0.7$
(*d*) $\lambda = 0.9$

5.12. For the fuzzy relation $\underset{\sim}{R}$ in Problem 3.9*i* sketch (in 3D) the λ-cut relations for the following values of λ:

(*a*) $\lambda = 0^+$
(*b*) $\lambda = 0.3$
(*c*) $\lambda = 0.5$

(*d*) $\lambda = 0.9$

(*e*) $\lambda = 1$

5.13. For the individual fuzzy relations $\underset{\sim}{A}$ and $\underset{\sim}{E}$ in Problem 3.22 determine the λ-cut relations for the following values of λ:

(*a*) $\lambda = 0^+$

(*b*) $\lambda = 0.5$

(*c*) $\lambda = 0.9$

(*d*) $\lambda = 1$

5.14. Show that any λ-cut relation of a fuzzy tolerance relation results in a crisp tolerance relation.

5.15. Show that any λ-cut relation of a fuzzy equivalence relation results in a crisp equivalence relation.

5.16. In metallurgy materials are made with mixtures of various metals and other elements to achieve certain desirable properties. In a particular preparation of steel, three elements, namely iron, manganese, and carbon, are mixed in two different proportions. The samples obtained from these two different proportions are placed on a normalized scale, as shown in Fig. P5.16 and are represented as fuzzy sets $\underset{\sim}{A}_1$ and $\underset{\sim}{A}_2$. You are interested in finding some sort of "average" steel proportion. For the logical union of the membership functions shown, find the defuzzified value, z^*, using each of the seven methods discussed in the text. Comment on the differences in results.

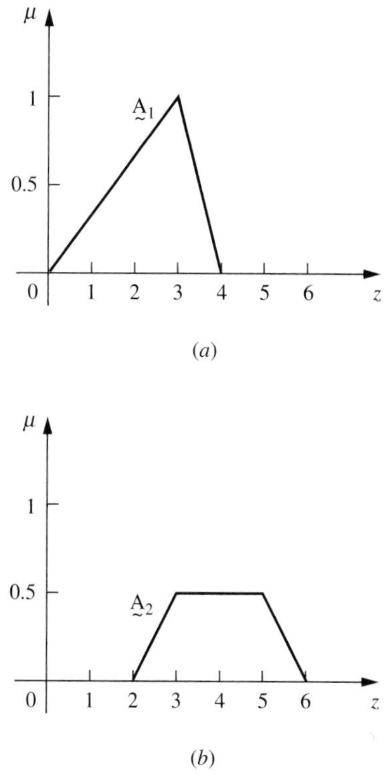

FIGURE P5.16

5.17. Two companies bid for a contract. A committee has to review the estimates of those companies and give reports to its chairperson. The reviewed reports are evaluated on a nondimensional scale and assigned a weighted score that is represented by a fuzzy membership function, as illustrated by the two fuzzy sets, $\underset{\sim}{B}_1$ and $\underset{\sim}{B}_2$, in Fig. P 5.17. The chairperson is interested in the lowest bid, as well as a metric to measure the combined "best" score. For the logical union of the membership functions shown here, find the defuzzified value, z^*, using each of the seven methods discussed in the text. Comment on the differences of the methods in terms of the criteria provided in the summary section of the chapter.

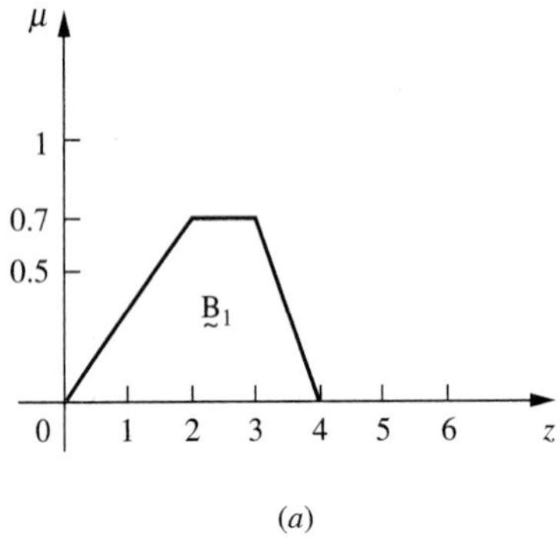

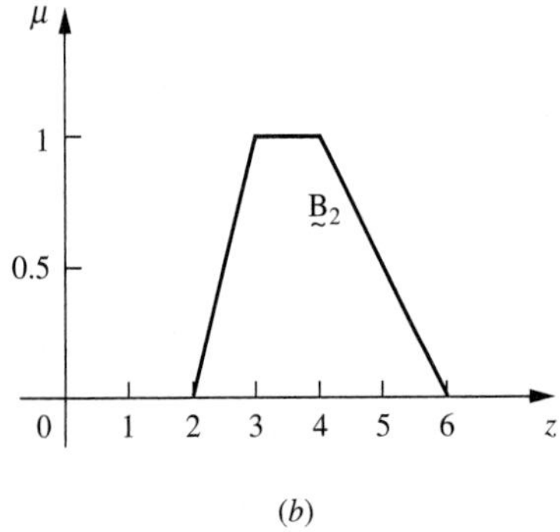

FIGURE P5.17

CHAPTER 6

FUZZY ARITHMETIC, NUMBERS, VECTORS, AND THE EXTENSION PRINCIPLE

Said the Mock Turtle with a sigh, "I only took the regular course." "What was that?" inquired Alice. "Reeling and Writhing, of course, to begin with," the Mock Turtle replied; "and the different branches of Arithmetic—Ambition, Distraction, Uglification, and Derision."

Lewis Carroll
Alice in Wonderland, *1865*

As Lewis Carroll so cleverly implied as early as 1865 (he was, by the way, a brilliant mathematician), there possibly could be other elements of arithmetic—consider those of ambition, distraction, uglification, and derision. Certainly fuzzy logic has been described in worse terms by many people over the last three decades! Perhaps Mr. Carroll had a presage of fuzzy set theory exactly one hundred years before Dr. Zadeh; perhaps, possibly.

In this chapter we see that standard arithmetic and algebraic operations, which are based after all on the foundations of classical set theory, can be extended to fuzzy arithmetic and fuzzy algebraic operations. This extension is accomplished with Zadeh's extension principle [Zadeh, 1975]. Fuzzy numbers, briefly described in Chapter 4, are used here because such numbers are the basis for fuzzy arithmetic. In this context the arithmetic operations are not fuzzy; the numbers on which the operations are performed are fuzzy and, hence, so too are the results of these operations. Conventional interval analysis is reviewed as a prelude to some improvements and approximations to the extension principle, most notably the fuzzy vertex method and its alternative forms.

The chapter concludes with a description of fuzzy vectors. Since vectors are simple one-dimensional arrays of numbers, properties developed in the extension principle apply directly to vectors of fuzzy numbers. This attribute for vectors is not exploited in this chapter, however. What is described are operations on fuzzy vectors that contain simple scalar quantities as their elements (instead of fuzzy numbers as elements of the vector). Despite the omission in this chapter of fuzzy vector operations on vectors of fuzzy numbers, the operations should be transparent for manipulations on fuzzy numbers as well. The material on fuzzy vectors will become quite useful in the study of fuzzy pattern recognition (see Chapter 12).

EXTENSION PRINCIPLE

In engineering, mathematics, and the sciences, functions are ubiquitous elements in modeling. Consider a simple relationship between one independent variable and one dependent variable as shown in Fig. 6.1. This relationship is a single-input, single-output process where the transfer function (the box in Fig. 6.1) represents the mapping provided by the general function f. In the typical case, f is of analytic form, for example $y = f(x)$, the input, x, is deterministic, and the resulting output, y, is also deterministic.

How can we extend this mapping to the case where the input, x, is a fuzzy variable or a fuzzy set, and where the function itself could be fuzzy? That is, how can we determine the fuzziness in the output, y, based on either a fuzzy input or a fuzzy function or both (mapping)? An extension principle developed by Zadeh [1975] and later elaborated by Yager [1986] enables us to extend the domain of a function on fuzzy sets.

The material of the next several sections introduces the extension principle by first reviewing theoretical issues of classical (crisp) transforms, mappings, and relations. The theoretical material then moves to the case where the input is fuzzy but the function itself is crisp, then to the case where the input and the function both are fuzzy. Simple examples to illustrate the ideas are provided. The next section serves as a more practical guide to the implementation of the extension principle, with several numerical examples. The extension principle is a very powerful idea that, in many situations, provides the capabilities of a "fuzzy calculator."

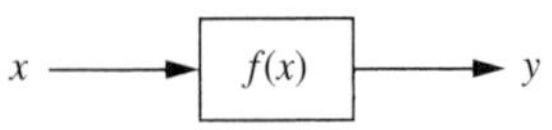

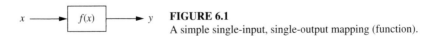

FIGURE 6.1
A simple single-input, single-output mapping (function).

Crisp Functions, Mapping, and Relations

Functions (also called transforms), such as the logarithmic function, $y = \log(x)$, or the linear function $y = ax + b$, are mappings from one universe, X, to another universe, Y. Symbolically, this mapping (function, f) is sometimes denoted, $f : \mathrm{X} \rightarrow \mathrm{Y}$. Other terminology calls the mapping $y = f(x)$ the *image of x under f*, and the inverse mapping, $x = f^{-1}(y)$, is termed the *original image of y*. A mapping can also be expressed by a relation R (as described in Chapter 3), on the Cartesian space $\mathrm{X} \times \mathrm{Y}$. Such a relation (crisp) can be described symbolically as $\mathrm{R} = \{(x, y) \mid y = f(x)\}$, with the characteristic function describing membership of specific x, y pairs to the relation R as

$$\chi_{\mathrm{R}}(x, y) = \begin{cases} 1, & y = f(x) \\ 0, & y \neq f(x) \end{cases} \tag{6.1}$$

Now, since we can define transform functions, or mappings, for specific elements of one universe (x) to specific elements of another universe (y), we can also do the same thing for collections of elements in X mapped to collections of elements in Y. Such collections have been referred to in this text as sets. Presumably, then, all possible sets in the power set of X can be mapped in some fashion (there may be null mapping for many of the combinations) to the sets in the power set of Y; i.e., $f : \mathrm{P(X)} \rightarrow \mathrm{P(Y)}$. For a set A defined on universe X, its image, set B on the universe Y, is found from the mapping, $\mathrm{B} = f(\mathrm{A}) = \{y \mid \text{ for all } x \in \mathrm{A}, y = f(x)\}$, where B will be defined by its characteristic value

$$\chi_{\mathrm{B}}(y) = \chi_{f(\mathrm{A})}(y) = \bigvee_{y=f(x)} \chi_{\mathrm{A}}(x) \tag{6.2}$$

Example 6.1. Suppose we have a crisp set $\mathrm{A} = \{0, 1\}$, or, using Zadeh's notation,

$$\mathrm{A} = \left\{ \frac{0}{-2} + \frac{0}{-1} + \frac{1}{0} + \frac{1}{1} + \frac{0}{2} \right\}$$

defined on the universe $\mathrm{X} = \{-2, -1, 0, 1, 2\}$ and a simple mapping $y = |4x| + 2$. We wish to find the resulting crisp set B on an output universe Y using the extension principle. From the mapping we can see that the universe Y will be $\mathrm{Y} = \{2, 6, 10\}$. The mapping described by Eq. (6.2) will yield the following calculations for the membership values of each of the elements in universe Y:

$$\begin{aligned} \chi_{\mathrm{B}}(2) &= \vee\{\chi_{\mathrm{A}}(0)\} = 1 \\ \chi_{\mathrm{B}}(6) &= \vee\{\chi_{\mathrm{A}}(-1), \chi_{\mathrm{A}}(1)\} = \vee\{0, 1\} = 1 \\ \chi_{\mathrm{B}}(10) &= \vee\{\chi_{\mathrm{A}}(-2), \chi_{\mathrm{A}}(2)\} = \vee\{0, 0\} = 0 \end{aligned}$$

Notice there is only one way to get the element 2 in the universe Y, but there are two ways to get the elements 6 and 10 in Y. Written in Zadeh's notation this mapping results in the output

$$\mathrm{B} = \left\{ \frac{1}{2} + \frac{1}{6} + \frac{0}{10} \right\}$$

or, alternatively, $\mathrm{B} = \{2, 6\}$.

Suppose we want to find the image B on universe Y using a relation that expresses the mapping. This transform can be accomplished by using the composition operation described in Chapter 3 for finite universe relations, where the mapping $y = f(x)$ is a general relation. Again, for X = {−2, −1, 0, 1, 2} and a generalized universe Y = {0, 1, ..., 9, 10}, the crisp relation describing this mapping ($y = |4x| + 2$) is

$$R = \begin{array}{c} \\ -2 \\ -1 \\ 0 \\ 1 \\ 2 \end{array} \begin{array}{c} \begin{array}{ccccccccccc} 0 & 1 & 2 & 3 & 4 & 5 & 6 & 7 & 8 & 9 & 10 \end{array} \\ \begin{bmatrix} 0 & 0 & 0 & 0 & 0 & 0 & 0 & 0 & 0 & 0 & 1 \\ 0 & 0 & 0 & 0 & 0 & 0 & 1 & 0 & 0 & 0 & 0 \\ 0 & 0 & 1 & 0 & 0 & 0 & 0 & 0 & 0 & 0 & 0 \\ 0 & 0 & 0 & 0 & 0 & 0 & 1 & 0 & 0 & 0 & 0 \\ 0 & 0 & 0 & 0 & 0 & 0 & 0 & 0 & 0 & 0 & 1 \end{bmatrix} \end{array}$$

The image B can be found through composition (since X and Y are finite), that is, B = A ∘ R (we note here that any set, say A, can be regarded as a one-dimensional relation), where, again using Zadeh's notation,

$$A = \left\{ \frac{0}{-2} + \frac{0}{-1} + \frac{1}{0} + \frac{1}{1} + \frac{0}{2} \right\}$$

and B is found by means of Eq. (3.9) to be

$$\chi_B(y) = \bigvee_{x \in X} (\chi_A(x) \wedge \chi_R(x, y)) = \begin{cases} 1 & \text{for } y = 2, 6 \\ 0 & \text{otherwise} \end{cases}$$

or in Zadeh's notation on Y,

$$B = \left\{ \frac{0}{0} + \frac{0}{1} + \frac{1}{2} + \frac{0}{3} + \frac{0}{4} + \frac{0}{5} + \frac{1}{6} + \frac{0}{7} + \frac{0}{8} + \frac{0}{9} + \frac{0}{10} \right\}$$

Functions of Fuzzy Sets—Extension Principle

Again we start with two universes of discourse, X and Y, and a functional transform (mapping) of the form $y = f(x)$. Now suppose that we have a collection of elements in universe x that form a fuzzy set $\underset{\sim}{A}$. What is the image of fuzzy set $\underset{\sim}{A}$ on X under the mapping f? This image will also be fuzzy, say we denote it fuzzy set $\underset{\sim}{B}$; and it will be found through the same mapping, i.e., $\underset{\sim}{B} = f(\underset{\sim}{A})$.

The membership functions describing $\underset{\sim}{A}$ and $\underset{\sim}{B}$ will now be defined on the universe of a unit interval [0, 1], and for the fuzzy case Eq. (6.2) becomes

$$\mu_{\underset{\sim}{B}}(y) = \bigvee_{f(x)=y} \mu_{\underset{\sim}{A}}(x) \tag{6.3}$$

A convenient shorthand for many fuzzy calculations that utilize matrix relations involves the *fuzzy vector*. Basically, a fuzzy vector is a vector containing fuzzy membership values. Suppose the fuzzy set $\underset{\sim}{A}$ is defined on n elements in X, for instance on $x_1, x_2, \ldots, x_n$, and fuzzy set $\underset{\sim}{B}$ is defined on m elements in Y, say on $y_1, y_2, \ldots, y_m$. The array of membership functions for each of the fuzzy sets $\underset{\sim}{A}$ and $\underset{\sim}{B}$ can then be reduced to fuzzy vectors by the following substitutions:

$$\underset{\sim}{a} = \{a_1, \ldots, a_n\} = \{\mu_{\underset{\sim}{A}}(x_1), \ldots, \mu_{\underset{\sim}{A}}(x_n)\} = \{\mu_{\underset{\sim}{A}}(x_i)\}, \qquad \text{for } i = 1, 2, \ldots, n \tag{6.4}$$

$$\underset{\sim}{b} = \{b_1, \ldots, b_m\} = \{\mu_{\underset{\sim}{B}}(y_1), \ldots, \mu_{\underset{\sim}{B}}(y_m)\} = \{\mu_{\underset{\sim}{B}}(y_j)\}, \qquad \text{for } j = 1, 2, \ldots, m \tag{6.5}$$

Now, the image of fuzzy set $\underset{\sim}{A}$ can be determined through the use of the composition operation, or $\underset{\sim}{B} = \underset{\sim}{A} \circ \underset{\sim}{R}$, or when using the fuzzy vector form, $\underset{\sim}{b} = \underset{\sim}{a} \circ \underset{\sim}{R}$ where $\underset{\sim}{R}$ is an $n \times m$ fuzzy relation matrix.

More generally, suppose our input universe comprises the Cartesian product of many universes. Then the mapping f is defined on the power sets of this Cartesian input space and the output space, or,

$$f : \text{P}(\text{X}_1 \times \text{X}_2 \times \cdots \times \text{X}_\text{n}) \rightarrow \text{P}(\text{Y}) \tag{6.6}$$

Let fuzzy sets $\underset{\sim}{A}_1, \underset{\sim}{A}_2, \ldots, \underset{\sim}{A}_n$ be defined on the universes $\text{X}_1, \text{X}_2, \ldots, \text{X}_n$. The mapping for these particular input sets can now be defined as $\underset{\sim}{B} = f(\underset{\sim}{A}_1, \underset{\sim}{A}_2, \ldots, \underset{\sim}{A}_n)$, where the membership function of the image $\underset{\sim}{B}$ is given by

$$\mu_{\underset{\sim}{B}}(y) = \max_{y = f(x_1, x_2, \ldots, x_n)} \left\{\min[\mu_{\underset{\sim}{A}_1}(x_1), \mu_{\underset{\sim}{A}_2}(x_2), \ldots, \mu_{\underset{\sim}{A}_n}(x_n)]\right\} \tag{6.7}$$

In the literature Eq. (6.7) is generally called Zadeh's *extension principle.* Equation (6.7) is expressed for a discrete-valued function, f. If the function, f, is a continuous-valued expression, the max operator is replaced by the sup (supremum) operator (the supremum is the least upper bound).

Fuzzy Transform (Mapping)

The material presented in the preceding two sections is associated with the issue of "extending" fuzziness in an input set to an output set. In this case, the input is fuzzy, the output is fuzzy, but the transform (mapping) is crisp, or $f : \underset{\sim}{A} \rightarrow \underset{\sim}{B}$. What happens in a more restricted case where the input is a single element (a nonfuzzy singleton) and this single element maps to a fuzzy set in the output universe? In this case the transform, or mapping, is termed a fuzzy transform.

Formally, let a mapping exist from an element x in universe X ($x \in \text{X}$) to a fuzzy set $\underset{\sim}{B}$ in the power set of universe Y, P(Y). Such a mapping is called a fuzzy mapping, $\underset{\sim}{f}$, where the output is no longer a single element, y, but a fuzzy set $\underset{\sim}{B}$; i.e.,

$$\underset{\sim}{B} = \underset{\sim}{f}(x) \tag{6.8}$$

If X and Y are finite universes, the fuzzy mapping expressed in Eq. (6.8) can be described as a fuzzy relation, $\underset{\sim}{R}$, or, in matrix form,

$$\underset{\sim}{R} = \begin{array}{c} \\ x_1 \\ x_2 \\ x_i \\ x_n \end{array} \begin{array}{c} \begin{array}{ccccccc} y_1 & y_2 & \cdots & y_j & \cdots & y_m \end{array} \\ \begin{bmatrix} r_{11} & r_{12} & \cdots & r_{1j} & \cdots & r_{1m} \\ r_{21} & r_{22} & \cdots & r_{2j} & \cdots & r_{2m} \\ r_{i1} & r_{i2} & \cdots & r_{ij} & \cdots & r_{im} \\ r_{n1} & r_{n2} & \cdots & r_{nj} & \cdots & r_{nm} \end{bmatrix} \end{array} \tag{6.9}$$

For a particular single element of the input universe, say x_i, its fuzzy image, $\underset{\sim}{B}_i = \underset{\sim}{f}(x_i)$, is given in a general symbolic form as

$$\mu_{\underset{\sim}{B}_i}(y_j) = r_{ij} \tag{6.10}$$

or, in fuzzy vector notation,

$$\underset{\sim}{\mathbf{b}}_i = \{r_{i1}, r_{i2}, \ldots, r_{im}\} \tag{6.11}$$

Hence, the fuzzy image of the element x_i is given by the elements in the ith row of the fuzzy relation, $\underset{\sim}{R}$, defining the fuzzy mapping, Eq. (6.9).

Suppose we now further generalize the situation where a fuzzy input set, say $\underset{\sim}{A}$, maps to a fuzzy output through a fuzzy mapping, or

$$\underset{\sim}{B} = \underset{\sim}{f}(\underset{\sim}{A}) \tag{6.12}$$

The extension principle again can be used to find this fuzzy image, $\underset{\sim}{B}$, by the following expression:

$$\mu_{\underset{\sim}{B}}(y) = \bigvee_{x \in X} (\mu_{\underset{\sim}{A}}(x) \wedge \mu_{\underset{\sim}{R}}(x, y)) \tag{6.13}$$

The preceding expression is analogous to a fuzzy composition performed on fuzzy vectors, or $\underset{\sim}{\mathbf{b}} = \underset{\sim}{\mathbf{a}} \circ \underset{\sim}{R}$, or in vector form,

$$\underset{\sim}{\mathbf{b}}_j = \max_i(\min(a_i, r_{ij})) \tag{6.14}$$

where $\underset{\sim}{\mathbf{b}}_j$ is the jth element of the fuzzy image $\underset{\sim}{B}$.

Example 6.2. Suppose we have a fuzzy mapping, $\underset{\sim}{f}$, given by the following fuzzy relation, $\underset{\sim}{R}$:

$$\begin{array}{c} \\ \underset{\sim}{R} = \end{array} \begin{array}{c} \begin{matrix} 1.4 & 1.5 & 1.6 & 1.7 & 1.8 & (\text{m}) \end{matrix} \\ \begin{bmatrix} 1 & 0.8 & 0.2 & 0.1 & 0 \\ 0.8 & 1 & 0.8 & 0.2 & 0.1 \\ 0.2 & 0.8 & 1 & 0.8 & 0.2 \\ 0.1 & 0.2 & 0.8 & 1 & 0.8 \\ 0 & 0.1 & 0.2 & 0.8 & 1 \end{bmatrix} \begin{matrix} 40 \ (\text{kg}) \\ 50 \\ 60 \\ 70 \\ 80 \end{matrix} \end{array}$$

which represents a fuzzy mapping between the length and mass of test articles scheduled for flight in a space experiment. The mapping is fuzzy because of the complicated relationship between mass and the cost to send the mass into space, the constraints on length of the test articles fitted into the cargo section of the spacecraft, and the scientific value of the experiment. Suppose a particular experiment is being planned for flight, but specific mass requirements have not been determined. For planning purposes the mass (in kilograms) is presumed to be a fuzzy quantity described by the following membership function:

$$\underset{\sim}{A} = \left\{ \frac{0.8}{40} + \frac{1}{50} + \frac{0.6}{60} + \frac{0.2}{70} + \frac{0}{80} \right\} \text{kg}$$

or as a fuzzy vector $\underset{\sim}{\mathbf{a}} = \{0.8, 1, 0.6, 0.2, 0\}$ kg.

The fuzzy image $\underset{\sim}{B}$ can be found using the extension principle (or, equivalently, composition for this fuzzy mapping), $\underset{\sim}{\mathbf{b}} = \underset{\sim}{\mathbf{a}} \circ \underset{\sim}{R}$ (recall that a set is also a one-dimensional relation). This composition results in a fuzzy output vector describing the fuzziness in the length of the experimental object (in meters), to be used for planning purposes, or $\underset{\sim}{\mathbf{b}} = \{0.8, 1, 0.8, 0.6, 0.2\}$ m.

Practical Considerations

Heretofore we have discussed features of fuzzy sets on certain universes of discourse. Suppose there is a mapping between elements, u, of one universe, U, onto elements, v, of another universe, V, through a function f. Let this mapping be described by $f : u \to v$. Define $\underset{\sim}{A}$ to be a fuzzy set on universe U; that is, $\underset{\sim}{A} \subset$ U. This relation is described by the membership function

$$\underset{\sim}{A} = \left\{ \frac{\mu_1}{u_1} + \frac{\mu_2}{u_2} + \cdots + \frac{\mu_n}{u_n} \right\} \tag{6.15}$$

Then the extension principle, as manifested in Eq. (6.3), asserts that, for a function f that performs a one-to-one mapping (i.e., maps one element in universe U to one element in universe V), an obvious consequence of Eq. (6.3) is

$$\begin{aligned} f(\underset{\sim}{A}) &= f\left(\frac{\mu_1}{u_1} + \frac{\mu_2}{u_2} + \cdots + \frac{\mu_n}{u_n} \right) \\ &= \left\{ \frac{\mu_1}{f(u_1)} + \frac{\mu_2}{f(u_2)} + \cdots + \frac{\mu_n}{f(u_n)} \right\} \end{aligned} \tag{6.16}$$

The mapping in Eq. (6.16) is said to be *one-to-one.*

Example 6.3. Let a fuzzy set $\underset{\sim}{A}$ be defined on the universe U $= \{1, 2, 3\}$. We wish to map elements of this fuzzy set to another universe, V, under the function

$$v = f(u) = 2u - 1$$

We see that the elements of V are V $= \{1, 3, 5\}$. Suppose the fuzzy set $\underset{\sim}{A}$ is given by

$$\underset{\sim}{A} = \left\{ \frac{0.6}{1} + \frac{1}{2} + \frac{0.8}{3} \right\}$$

Then the fuzzy membership function for $v = f(u) = 2u - 1$ would be

$$f(\underset{\sim}{A}) = \left\{ \frac{0.6}{1} + \frac{1}{3} + \frac{0.8}{5} \right\}$$

For cases where this functional mapping f maps products of elements from two universes, say U_1 and U_2, to another universe V, and we define $\underset{\sim}{A}$ as a fuzzy set on the Cartesian space $U_1 \times U_2$, then

$$f(\underset{\sim}{A}) = \left\{ \sum \frac{\min[\mu_1(i), \mu_2(j)]}{f(i, j)} \mid i \in U_1, j \in U_2 \right\} \tag{6.17}$$

where $\mu_1(i)$ and $\mu_2(j)$ are the separable membership projections of $\mu(i, j)$ from the Cartesian space $U_1 \times U_2$ when $\mu(i, j)$ cannot be determined. This projection involves the invocation of a condition known as *noninteraction* (see Chapter 3) between the separate universes. It is analogous to the assumption of independence employed in probability theory, which reduces a joint probability density function to the product of its separate marginal density functions. In the fuzzy noninteraction case we are doing a kind of intersection; hence, we use the minimum operator (some logics use operators other than the minimum operator) as opposed to the product operator used in probability theory.

Example 6.4. Suppose we have the integers 1 to 10 as the elements of two identical but different universes; let

$$U_1 = U_2 = \{1, 2, 3, \ldots, 10\}$$

Then define two fuzzy numbers $\underset{\sim}{A}$ and $\underset{\sim}{B}$ on universe U_1 and U_2, respectively:

$$\text{Define } \underset{\sim}{A} = \underset{\sim}{2} = \text{“approximately 2”} = \left\{ \frac{0.6}{1} + \frac{1}{2} + \frac{0.8}{3} \right\}$$

$$\text{Define } \underset{\sim}{B} = \underset{\sim}{6} = \text{“approximately 6”} = \left\{ \frac{0.8}{5} + \frac{1}{6} + \frac{0.7}{7} \right\}$$

The product of (“approximately 2”) × (“approximately 6”) should map to a fuzzy number “approximately 12,” which is a fuzzy set defined on a universe, say V, of integers, $V = \{5, 6, \ldots, 18, 21\}$, as determined by the extension principle, Eq. (6.7), or

$$\underset{\sim}{2} \times \underset{\sim}{6} = \left(\frac{0.6}{1} + \frac{1}{2} + \frac{0.8}{3} \right) \times \left(\frac{0.8}{5} + \frac{1}{6} + \frac{0.7}{7} \right)$$

$$= \left\{ \frac{\min(0.6, 0.8)}{5} + \frac{\min(0.6, 1)}{6} + \cdots + \frac{\min(0.8, 1)}{18} + \frac{\min(0.8, 0.7)}{21} \right\}$$

$$= \left\{ \frac{0.6}{5} + \frac{0.6}{6} + \frac{0.6}{7} + \frac{0.8}{10} + \frac{1}{12} + \frac{0.7}{14} + \frac{0.8}{15} + \frac{0.8}{18} + \frac{0.7}{21} \right\}$$

In this example each of the elements in the universe, V, is determined by a unique mapping of the input variables. For example, $1 \times 5 = 5$, $2 \times 6 = 12$, etc. Hence, the maximum operation expressed in Eq. (6.7) is not necessary. It should also be noted that the result of this arithmetic product is not convex, hence does not appear to be a fuzzy number (i.e., normal and convex). However, the nonconvexity arises from numerical aberrations from the discretization of the two fuzzy numbers, $\underset{\sim}{2}$ and $\underset{\sim}{6}$, and not from any inherent problems in the extension principle. This issue is discussed at length later in this chapter in Example 6.14.

The complexity of the extension principle increases when we consider if more than one of the combinations of the input variables, U_1 and U_2, are mapped to the same variable in the output space, V; i.e., if the mapping is not one-to-one. In this case we take the maximum membership grades of the combinations mapping to the same output variable, or, for the following mapping, we get

$$\mu_{\underset{\sim}{A}}(u_1, u_2) = \max_{v=f(u_1,u_2)} [\min\{\mu_1(u_1), \mu_2(u_2)\}] \tag{6.18}$$

Example 6.5. We have two fuzzy sets $\underset{\sim}{A}$ and $\underset{\sim}{B}$, each defined on its own universe as follows:

$$\underset{\sim}{A} = \left\{\frac{0.2}{1} + \frac{1}{2} + \frac{0.7}{4}\right\} \quad \text{and} \quad \underset{\sim}{B} = \left\{\frac{0.5}{1} + \frac{1}{2}\right\}$$

We wish to determine the membership values for the algebraic product mapping

$$\begin{aligned} f(\underset{\sim}{A}, \underset{\sim}{B}) &= \underset{\sim}{A} \times \underset{\sim}{B} \text{ (arithmetic product)} \\ &= \left\{\frac{\min(0.2, 0.5)}{1} + \frac{\max[\min(0.2, 1), \min(0.5, 1)]}{2}\right. \\ &\quad \left. + \frac{\max[\min(0.7, 0.5), \min(1, 1)]}{4} + \frac{\min(0.7, 1)}{8}\right\} \\ &= \left\{\frac{0.2}{1} + \frac{0.5}{2} + \frac{1}{4} + \frac{0.7}{8}\right\} \end{aligned}$$

In this case, the mapping involves two ways to produce a 2 (1×2 and 2×1) and two ways to produce a 4 (4×1 and 2×2), hence the maximum operation expressed in Eq. (6.7) is necessary.

The extension principle can also be useful in propagating fuzziness through generalized relations that are discrete mappings of ordered pairs of elements from input universes to ordered pairs of elements in an output universe.

Example 6.6. We want to map ordered pairs from input universes $X_1 = \{a, b\}$ and $X_2 = \{1, 2, 3\}$ to an output universe, $Y = \{x, y, z\}$. The mapping is given by the crisp relation, R,

$$\mathrm{R} = \begin{array}{c} \\ a \\ b \end{array} \begin{array}{c} \begin{array}{ccc} 1 & 2 & 3 \end{array} \\ \begin{bmatrix} x & z & x \\ x & y & z \end{bmatrix} \end{array}$$

We note that this relation represents a mapping, and it does not contain membership values. We define a fuzzy set $\underset{\sim}{A}$ on universe X_1 and a fuzzy set $\underset{\sim}{B}$ on universe X_2 as

$$\underset{\sim}{A} = \left\{\frac{0.6}{a} + \frac{1}{b}\right\} \quad \text{and} \quad \underset{\sim}{B} = \left\{\frac{0.2}{1} + \frac{0.8}{2} + \frac{0.4}{3}\right\}$$

We wish to determine the membership function of the output, $\underset{\sim}{C} = f(\underset{\sim}{A}, \underset{\sim}{B})$, whose relational mapping, f, is described by R. This is accomplished with the extension principle, Eq. (6.7), as follows:

$$\begin{aligned} \mu_{\underset{\sim}{C}}(x) &= \max[\min(0.2, 0.6), \min(0.2, 1), \min(0.4, 0.6)] = 0.4 \\ \mu_{\underset{\sim}{C}}(y) &= \max[\min(0.8, 1)] = 0.8 \\ \mu_{\underset{\sim}{C}}(z) &= \max[\min(0.8, 0.6), \min(0.4, 1)] = 0.6 \end{aligned}$$

Hence,

$$\underset{\sim}{C} = \left\{\frac{0.4}{x} + \frac{0.8}{y} + \frac{0.6}{z}\right\}$$

The extension principle is also useful in mapping fuzzy inputs through continuous-valued functions. The process is the same as for a discrete-valued function, but the effort involved in the computations is more rigorous.

Example 6.7 [Wong and Ross, 1985]. Suppose we have a nonlinear system given by the harmonic function, $\underset{\sim}{x} = \cos(\underset{\sim}{\omega} t)$, where the frequency of excitation, $\underset{\sim}{\omega}$, is a fuzzy variable described by the membership function shown in Fig. 6.2*a*. The output variable, $\underset{\sim}{x}$, will be fuzzy because of the fuzziness provided in the mapping from the input variable, $\underset{\sim}{\omega}$. This function represents a one-to-one mapping in two stages, $\underset{\sim}{\omega} \to \underset{\sim}{\omega} t \to x$. The membership function of $\underset{\sim}{x}$ will be determined through the use of the extension principle, which for this example will take on the following form:

$$\mu_{\underset{\sim}{x}}(x) = \bigvee_{x=\cos(\omega t)} [\mu_{\underset{\sim}{\omega}}(\omega)]$$

To show the development of this expression, we will take several time points, such as $t = 0, 1, \ldots$. For $t = 0$, all values of $\underset{\sim}{\omega}$ map into a single point in the $\underset{\sim}{\omega} t$ domain, viz., $\underset{\sim}{\omega} t = 0$, and into a single point in the x universe, viz., $x = 1$. Hence, the membership

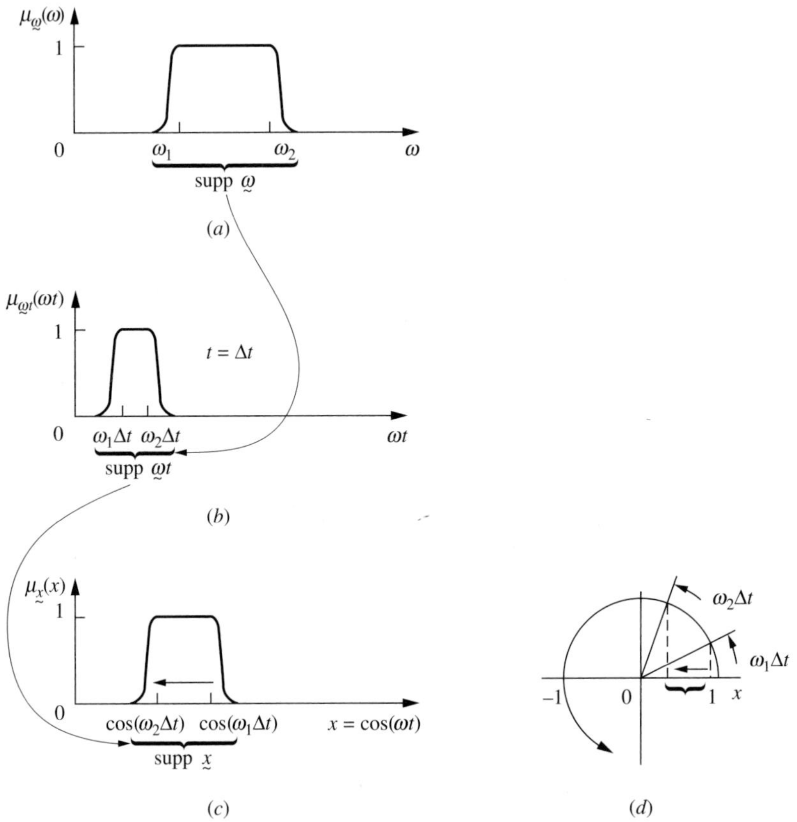

FIGURE 6.2
Extension principle applied to $\underset{\sim}{x} = \cos(\underset{\sim}{\omega} t)$, at $t = \Delta t$.

of $\underset{\sim}{x}$ is simply a singleton at $x = 1$; i.e.,

$$\mu_{\underset{\sim}{x}}(x) = \begin{cases} 1 & \text{if } x = 1 \\ 0 & \text{otherwise} \end{cases}$$

For a nonzero but small t, say $t = \Delta t$, the support of $\underset{\sim}{\omega}$, denoted in Fig. 6.2*a* as supp $\underset{\sim}{\omega}$, is mapped into a small but finite support of $\underset{\sim}{x}$, denoted in Fig. 6.2*c* as supp $\underset{\sim}{x}$ (the support of a fuzzy set was defined in Chapter 4 as the interval corresponding to a λ-cut of $\lambda = 0^+$). The membership value for each $\underset{\sim}{x}$ in this interval is determined directly from the membership of ω in a one-to-one mapping. As can be seen in Fig. 6.2, as t increases, the support of $\underset{\sim}{x}$ increases, and the fuzziness in the response spreads with time. Eventually, there will be a value of t when the support of $\underset{\sim}{x}$ folds partly onto itself, i.e., we have multi-ω-to-single-x mapping. In this event, the maximum of all candidate membership values of $\underset{\sim}{\omega}$ is used as the membership value of x according to the extension principle, Eq. (6.3), as shown in Fig. 6.3.

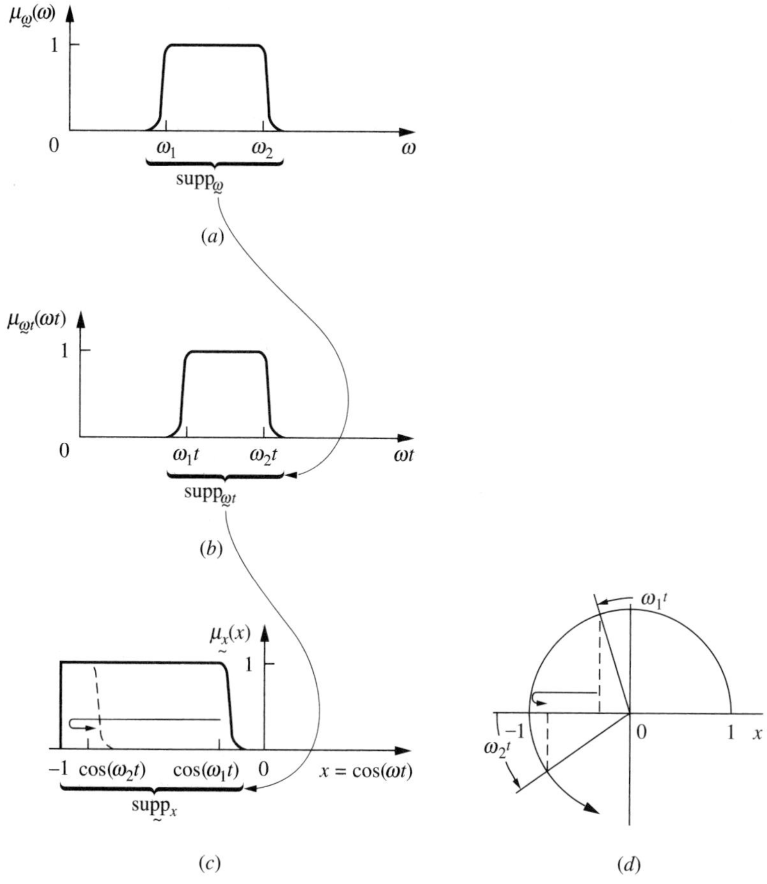

FIGURE 6.3
Extension principle applied to $\underset{\sim}{x} = \cos(\underset{\sim}{\omega}t)$ when t causes overlap in support of $\underset{\sim}{x}$.

When t is of such magnitude that the support of x occupies the interval $[-1, 1]$ completely, the largest support possible, the membership $\mu_{\underset{\sim}{x}}(x)$ will be unity for all x within this interval. This is the state of complete fuzziness, as illustrated in Fig. 6.4. In the equation $\underset{\sim}{x} = \cos(\underset{\sim}{\omega}t)$, the output can have any value in the interval $[-1, 1]$ with equal and complete membership. Once this state is reached the system remains there for all future time.

FUZZY NUMBERS

Chapter 4 defines a fuzzy number as being described by a normal, convex membership function on the real line. In this chapter we wish to use the extension principle to perform algebraic operations on fuzzy numbers (as illustrated in previous examples in this chapter). We define a normal, convex fuzzy set on the real line to be a fuzzy number, and denote it $\underset{\sim}{I}$.

Let $\underset{\sim}{I}$ and $\underset{\sim}{J}$ be two fuzzy numbers, with $\underset{\sim}{I}$ defined on the real line in universe X and $\underset{\sim}{J}$ defined on the real line in universe Y, and let the symbol $*$ denote a general

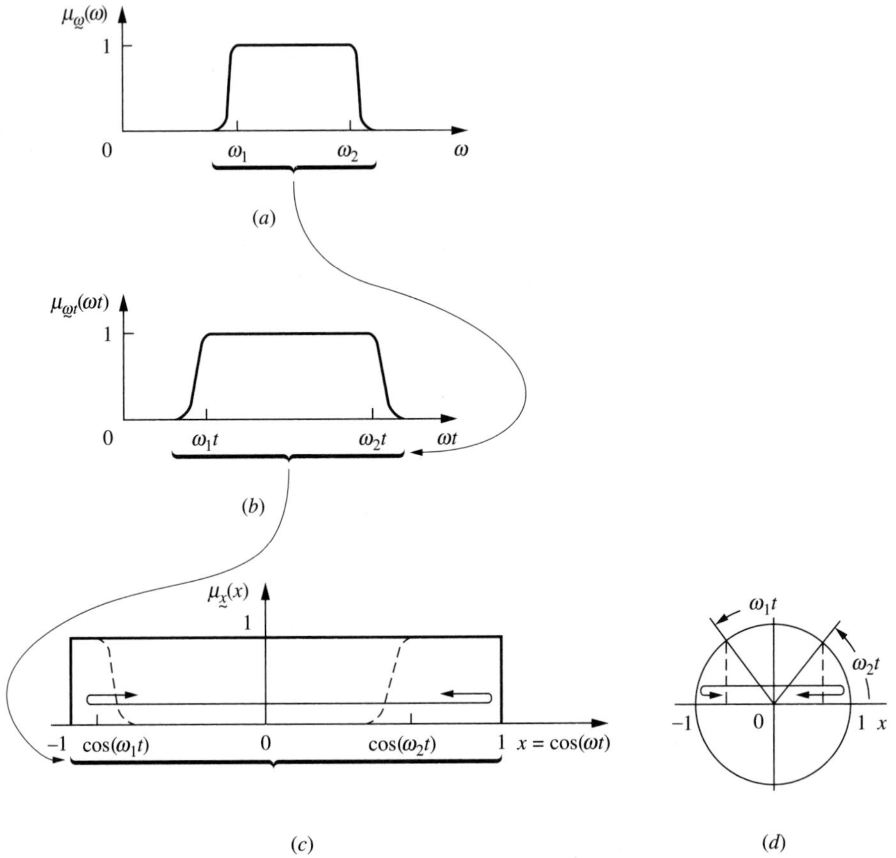

FIGURE 6.4
Extension principle applied to $\underset{\sim}{x} = \cos(\underset{\sim}{\omega}t)$ when t causes complete fuzziness.

arithmetic operation, i.e., $* \equiv \{+, -, \times, \div\}$. An arithmetic operation (mapping) between these two fuzzy numbers, denoted $\underset{\sim}{I} * \underset{\sim}{J}$, will be defined on universe Z, and can be accomplished using the extension principle, by

$$\mu_{\underset{\sim}{I}*\underset{\sim}{J}}(z) = \bigvee_{x*y=z} (\mu_{\underset{\sim}{I}}(x) \wedge \mu_{\underset{\sim}{J}}(y)) \tag{6.19}$$

Equation (6.19) results in another fuzzy set, the fuzzy number resulting from the arithmetic operation on fuzzy numbers $\underset{\sim}{I}$ and $\underset{\sim}{J}$.

Example 6.8. We want to perform a simple addition ($* \equiv +$) of two fuzzy numbers. Define a fuzzy *one* by the normal, convex membership function defined on the integers,

$$\underset{\sim}{1} = \left\{\frac{.2}{0} + \frac{1}{1} + \frac{.2}{2}\right\}$$

Now, we want to add "fuzzy one" plus "fuzzy one," using the extension principle, Eq. (6.19), to get

$$\begin{aligned}
\underset{\sim}{1} + \underset{\sim}{1} = \underset{\sim}{2} &= \left(\frac{.2}{0} + \frac{1}{1} + \frac{.2}{2}\right) + \left(\frac{.2}{0} + \frac{1}{1} + \frac{.2}{2}\right) \\
&= \left\{\frac{\min(0.2, 0.2)}{0} + \frac{\max[\min(0.2, 1), \min(1, 0.2)]}{1}\right. \\
&\quad + \frac{\max[\min(0.2, 0.2), \min(1, 1), \min(0.2, 0.2)]}{2} \\
&\quad \left. + \frac{\max[\min(1, 0.2), \min(0.2, 1)]}{3} + \frac{\min(0.2, 0.2)}{4}\right\} \\
&= \left\{\frac{.2}{0} + \frac{.2}{1} + \frac{1}{2} + \frac{.2}{3} + \frac{.2}{4}\right\}
\end{aligned}$$

Note that there are two ways to get the resulting membership value for a 1 (0 + 1 and 1 + 0), three ways to get a 2 (0 + 2, 1 + 1, 2 + 0), and two ways to get a 3 (1 + 2 and 2 + 1). These are accounted for in the implementation of the extension principle.

The support for a fuzzy number, say $\underset{\sim}{I}$ (see Chapter 4), is given by

$$\operatorname{supp} \underset{\sim}{I} = \{x \mid \mu_{\underset{\sim}{I}}(x) > 0\} \tag{6.20}$$

which is an interval on the real line, denoted symbolically as I. Since applying Eq. (6.19) to arithmetic operations on fuzzy numbers results in a quantity that is also a fuzzy number, we can find the support of the fuzzy number resulting from the arithmetic operation, $\underset{\sim}{I} * \underset{\sim}{J}$, i.e.,

$$\operatorname*{supp}_{\underset{\sim}{I}*\underset{\sim}{J}}(z) = \mathrm{I} * \mathrm{J} \tag{6.21}$$

which is seen to be the arithmetic operation on the two individual supports (crisp intervals), I and J, for fuzzy numbers $\underset{\sim}{I}$ and $\underset{\sim}{J}$, respectively.

Chapter 5 revealed that the support of a fuzzy set is equal to its lambda-cut value at $\lambda = 0^+$. In general, we can perform λ-cut operations on fuzzy numbers for any value of λ. A result we saw in Chapter 5 for set operations [Eq. (5.1)] is also valid for general arithmetic operations:

$$(\underset{\sim}{I} * \underset{\sim}{J})_\lambda = I_\lambda * J_\lambda \tag{6.22}$$

Equation 6.22 shows that the λ-cut on a general arithmetic operation ($* \equiv \{+, -, \times, \div\}$) on two fuzzy numbers is equivalent to the arithmetic operation on the respective λ-cuts of the two fuzzy numbers. Both $(\underset{\sim}{I} * \underset{\sim}{J})_\lambda$ and $I_\lambda * J_\lambda$ are interval quantities; and manipulations of these quantities can make use of classical interval analysis, the subject of the next section.

INTERVAL ANALYSIS IN ARITHMETIC

As alluded to in Chapter 2, a fuzzy set can be thought of as a crisp set with moving boundaries. In this sense, as Chapter 5 illustrated, a convex membership function defining a fuzzy set can be described by the intervals associated with different levels of lambda-cuts. Let I_1 and I_2 be two interval numbers defined by ordered pairs of real numbers with lower and upper bounds:

$$I_1 = [a, b], \qquad \text{where } a \leq b$$
$$I_2 = [c, d], \qquad \text{where } c \leq d$$

When $a = b$ and $c = d$, these interval numbers degenerate to a scalar real number. We again define a general arithmetic property with the symbol $*$, where $* \equiv \{+, -, \times, \div\}$. Symbolically, the operation

$$I_1 * I_2 = [a, b] * [c, d] \tag{6.23}$$

represents another interval. This interval calculation depends on the magnitudes and signs of the elements *a, b, c,* and *d.* Table 6.1 shows the various combinations of set-theoretic intersection ($\cap$) and set-theoretic union ($\cup$) for the six possible combinations of these elements ($a < b$ and $c < d$ still hold). Based on the information in Table 6.1, the four arithmetic interval operations associated with Eq. (6.23) are given as follows:

$$[a, b] + [c, d] = [a + c, b + d] \tag{6.24}$$
$$[a, b] - [c, d] = [a - d, b - c] \tag{6.25}$$

TABLE 6.1
Set operations on intervals

Cases	Intersection—$\cap$	Union—$\cup$
$a > d$	$\varnothing$	$[c, d] \cup [a, b]$
$c > b$	$\varnothing$	$[a, b] \cup [c, d]$
$a > c, b < d$	$[a, b]$	$[c, d]$
$c > a, d < b$	$[c, d]$	$[a, b]$
$a < c < b < d$	$[c, b]$	$[a, d]$
$c < a < d < b$	$[a, d]$	$[c, b]$

$$[a, b] \cdot [c, d] = [\min(ac, ad, bc, bd), \max(ac, ad, bc, bd)] \tag{6.26}$$

$$[a, b] \div [c, d] = [a, b] \cdot \left[\frac{1}{d}, \frac{1}{c}\right] \quad \text{provided that } 0 \notin [c, d] \tag{6.27}$$

$$\alpha[a, b] = \begin{cases} [\alpha a, \alpha b] & \text{for } \alpha > 0 \\ [\alpha b, \alpha a] & \text{for } \alpha < 0 \end{cases} \tag{6.28}$$

where *ac, ad, bc,* and *bd* are arithmetic products and $1/d$ and $1/c$ are quotients.

The caveat applied to Eq. (6.27) is that the equivalence stated is not valid for the case when $c \le 0$ and $d \ge 0$ (obviously the constraint $c < d$ still holds); i.e., zero cannot be contained within the interval $[c, d]$. Interval arithmetic follows properties of associativity and commutativity for both summations and products, but it does not follow the property of distributivity. Rather, intervals do follow a special subclass of distributivity known as *subdistributivity,* i.e., for three intervals, I, J, and K,

$$\mathrm{I} \cdot (\mathrm{J} + \mathrm{K}) \subset \mathrm{I} \cdot \mathrm{J} + \mathrm{I} \cdot \mathrm{K} \tag{6.29}$$

The failure of distributivity to hold for intervals is due to the treatment of two occurrences of identical interval numbers (i.e., I) as two independent interval numbers [Dong and Shah, 1987].

Example 6.9.

$$\begin{aligned} -3 \cdot [1, 2] &= [-6, -3] \\ [0, 1] - [0, 1] &= [-1, 1] \\ [1, 3] \cdot [2, 4] &= [\min(2, 4, 6, 12), \max(2, 4, 6, 12)] = [2, 12] \\ [1, 2] \div [1, 2] &= [1, 2] \cdot [\tfrac{1}{2}, 1] = [\tfrac{1}{2}, 2] \end{aligned}$$

Consider the following example of subdistributivity. For $\mathrm{I} = [1, 2], \mathrm{J} = [2, 3], \mathrm{K} = [1, 4]$, then

$$\begin{aligned} \mathrm{I} \cdot (\mathrm{J} - \mathrm{K}) &= [1, 2] \cdot ([2, 3] - [1, 4]) = [1, 2] \cdot [-2, 2] = [-4, 4] \\ \mathrm{I} \cdot \mathrm{J} - \mathrm{I} \cdot \mathrm{K} &= [1, 2] \cdot [2, 3] - [1, 2] \cdot [1, 4] = [2, 6] - [1, 8] = [-6, 5] \end{aligned}$$

Now, $[-4, 4] \neq [-6, 5]$, but $[-4, 4] \subset [-6, 5]$.

Interval arithmetic can be thought of in the following way. When we add or multiply two crisp numbers, the result is a crisp singleton. When we add or multiply two intervals we are essentially performing these operations on the infinite number of combinations of pairs of crisp singletons from each of the two intervals; hence, in this sense, an interval is expected as the result. In the simplest case, when we multiply two intervals containing only positive real numbers, it is easy conceptually to see that the interval comprising the solution is found by taking the product of the two lowest values from each of the intervals to form the solution's lower bound, and by taking the product of the two highest values from each of the intervals to form the solution's upper bound. Even though we can see conceptually that an infinite number of combinations of products between these two intervals exist, we need only the endpoints of the intervals to find the endpoints of the solution.

APPROXIMATE METHODS OF EXTENSION

A serious disadvantage of the discretized form of the extension principle in propagating fuzziness for continuous-valued mappings is the irregular and erroneous membership functions determined for the output variable if the membership functions of the input variables are discretized for numerical convenience (this problem is demonstrated in Example 6.14). The reason for this anomaly is that the solution to the extension principle, as expressed in Eq. (6.7), is really a nonlinear programming problem for continuous-valued functions. It is well-known that, in any optimization process, discretization of any variables can lead to an erroneous optimum solution because portions of the solution space are omitted in the calculations. For example, try to plot a tenth-order curve with a series of equally spaced points; some local minimum and maximum points on the curve are going to be missed if the discretization is not small enough. Again, these problems do not arise because of any inherent problems in the extension principle itself; they arise when continuous-valued functions are discretized, then allowed to propagate from the input domain to the output domain using the extension principle.

Other methods have been proposed to ease the computational burden in implementing the extension principle for continuous-valued functions and mappings. Among the alternative methods proposed in the literature to avoid this disadvantage for continuous fuzzy variables are three approaches that are summarized here along with illustrative numerical examples. All of these approximate methods make use of intervals, at various λ-cut levels, in defining membership functions.

Vertex Method

A procedure known as the vertex method [Dong and Shah, 1987] greatly simplifies manipulations of the extension principle for continuous-valued fuzzy variables, such as fuzzy numbers defined on the real line. The method is based on a combination of the λ-cut concept and standard interval analysis. The vertex method can prevent abnormality in the output membership function due to application of the discretization technique on the fuzzy variables' domain, and it can prevent the widening of the resulting function value set due to multiple occurrences of variables in the functional expression by conventional interval analysis methods. The algorithm is very easy to implement and can be computationally efficient.

The algorithm works as follows. Any continuous membership function can be represented by a continuous sweep of λ-cut intervals from $\lambda = 0^+$ to $\lambda = 1$. Figure 6.5 shows a typical membership function with an interval associated with a specific value of λ. Suppose we have a single-input mapping given by $y = f(x)$ that is to be extended for fuzzy sets, or $\underset{\sim}{\mathrm{B}} = f(\underset{\sim}{\mathrm{A}})$, and we want to decompose $\underset{\sim}{\mathrm{A}}$ into a series of λ-cut intervals, say I_λ.

When the function $f(x)$ is continuous and monotonic on $\mathrm{I}_\lambda = [a, b]$, the interval representing $\underset{\sim}{\mathrm{B}}$ at a particular value of λ, say B_λ, can be obtained by

$$\mathrm{B}_\lambda = f(\mathrm{I}_\lambda) = [\min(f(a), f(b)), \max(f(a), f(b))] \tag{6.30}$$

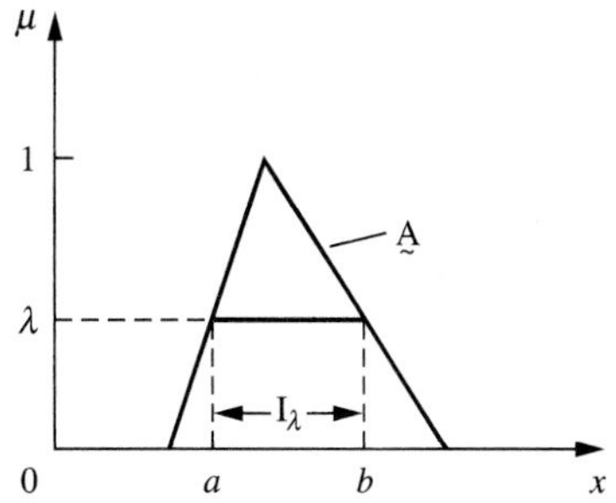

FIGURE 6.5
Interval corresponding to a λ-cut level on fuzzy set $\underset{\sim}{A}$.

Equation (6.30) has reduced the interval analysis problem for a functional mapping to a simple procedure dealing only with the endpoints of the interval. When the mapping is given by n inputs, i.e., $y = f(x_1, x_2, x_3, \ldots, x_n)$, then the input space can be represented by an n-dimensional Cartesian region; a 3D Cartesian region is shown in Fig. 6.6. Each of the input variables can be described by an interval, say $\mathrm{I}_{i\lambda}$, at a specific λ-cut, where

$$\mathrm{I}_{i\lambda} = [a_i, b_i] \qquad i = 1, 2, \ldots, n \tag{6.31}$$

As seen in Fig. 6.6, the endpoint pairs of each interval given in Eq. (6.31) intersect in the 3D space and form the vertices (corners) of the Cartesian space. The coordinates of these vertices are the values used in the vertex method when determining the output interval for each λ-cut. The number of vertices, N, is a quantity equal to $N = 2^n$, where n is the number of fuzzy input variables. When the mapping $y = f(x_1, x_2, x_3, \ldots, x_n)$ is continuous in the n-dimensional Cartesian region

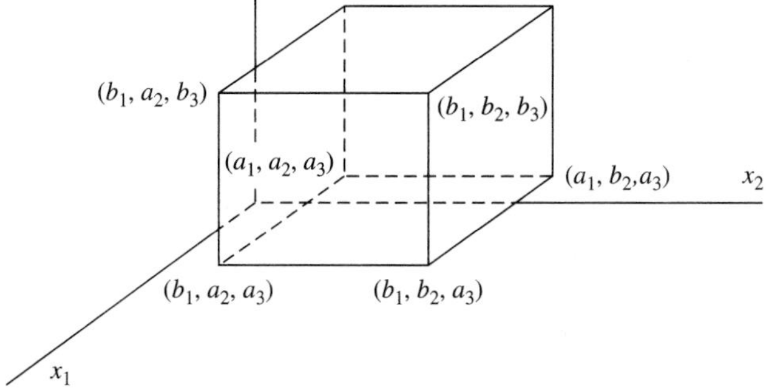

FIGURE 6.6
Three-dimensional Cartesian region involving intervals for three input variables, x_1, x_2, and x_3.

and when also there is no extreme point in this region (or along the boundaries), the value of the interval function for a particular λ-cut can be obtained by

$$\mathrm{B}_\lambda = f(\mathrm{I}_{1\lambda}, \mathrm{I}_{2\lambda}, \mathrm{I}_{3\lambda}, \ldots, \mathrm{I}_{n\lambda}) = \left[\min_j(f(c_j)), \max_j(f(c_j))\right] \quad j = 1, 2, \ldots, N \tag{6.32}$$

where c_j is the coordinate of the jth vertex representing the n-dimensional Cartesian region.

The vertex method is accurate only when the conditions of continuity and no extreme point are satisfied. When extreme points of the function $y = f(x_1, x_2, x_3, \ldots, x_n)$ exist in the n-dimensional Cartesian region of the input parameters, the vertex method will miss certain parts of the interval that should be included in the output interval value, B_λ. Extreme points can be missed, for example, in certain mappings that are not one-to-one. If the extreme points can be identified, they are simply treated as additional vertices, E_k, in the Cartesian space and Eq. (6.32) becomes, because the continuity property still holds,

$$\mathrm{B}_\lambda = \left[\min_{j,k}(f(c_j), f(E_k)), \max_{j,k}(f(c_j), f(E_k))\right] \tag{6.33}$$

where $j = 1, 2, \ldots, N$ and $k = 1, 2, \ldots, m$ for m extreme points in the region.

Example 6.10. We wish to determine the fuzziness in the output of a simple nonlinear mapping given by the expression, $y = f(x) = x(2 - x)$, seen in Fig. 6.7*a*, where the fuzzy input variable, x, has the membership function shown in Fig. 6.7*b*.

We shall solve this problem using the fuzzy vertex method at three λ-cut levels, for $\lambda = 0^+, 0.5, 1$. As seen in Fig. 6.7*b*, the intervals corresponding to these λ-cuts are $\mathrm{I}_{0^+} = [0.5, 2]$, $\mathrm{I}_{.5} = [0.75, 1.5]$, $\mathrm{I}_1 = [1, 1]$ (a single point). Since the problem is one-dimensional, the vertices, c_j, are described by a single coordinate; there are $N = 2^1 = 2$ vertices ($j = 1, 2$). In addition, an extreme point does exist within the region of the

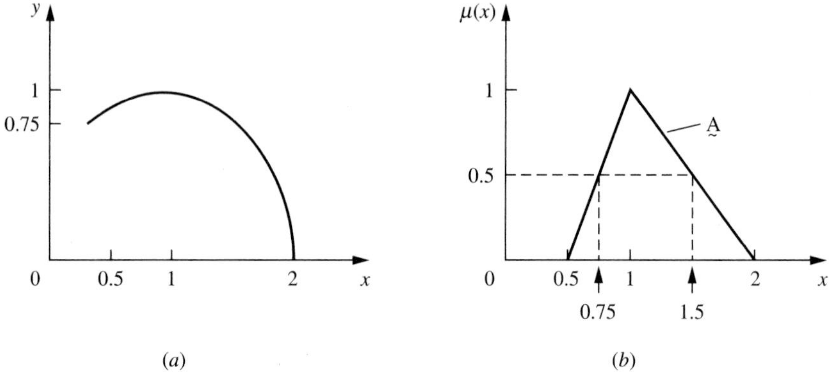

FIGURE 6.7
Nonlinear function and fuzzy input membership.

membership function and is determined using a derivative of the function, $df(x)/dx = 2 - 2x = 0$, $x_0 = E_1 = 1$ (E_k, where $k = 1$). This extreme point is within each of the three λ-cut intervals, so will be involved in all the following calculations for B_λ:

$$I_{0+} = [0.5, 2]$$
$$c_1 = 0.5, \quad c_2 = 2, \quad E_1 = 1$$
$$f(c_1) = 0.5(2 - 0.5) = 0.75, \quad f(c_2) = 2(2 - 2) = 0,$$
$$f(E_1) = 1(2 - 1) = 1$$
$$B_{0+} = [\min(0.75, 0, 1), \max(0.75, 0, 1)] = [0, 1]$$

$$I_{0.5} = [0.75, 1.5]$$
$$c_1 = 0.75, \quad c_2 = 1.5, \quad E_1 = 1$$
$$f(c_1) = 0.75(2 - 0.75) = 0.9375, \quad f(c_2) = 1.5(2 - 1.5) = 0.75,$$
$$f(E_1) = 1(2 - 1) = 1$$
$$B_{0.5} = [\min(0.9375, 0.75, 1), \max(0.9375, 0.75, 1)] = [0.75, 1]$$

$$I_1 = [1, 1]$$
$$c_1 = 1, \quad c_2 = 1, \quad E_1 = 1$$
$$f(c_1) = f(c_2) = f(E_1) = 1(2 - 1) = 1$$
$$B_1 = [\min(1, 1, 1), \max(1, 1, 1)] = [1, 1] = 1$$

Figure 6.8 provides a plot of the intervals B_{0+}, $B_{0.5}$, and B_1 to form the fuzzy output, y.

DSW Algorithm

The DSW algorithm [Dong, Shah, and Wong, 1985] also makes use of the λ-cut representation of fuzzy sets, but, unlike the vertex method, it uses the full λ-cut intervals in a standard interval analysis. The DSW algorithm consists of the following steps:

1. Select a λ value where $0 \le \lambda \le 1$.
2. Find the interval(s) in the input membership function(s) that correspond to this λ.
3. Using standard binary interval operations, compute the interval for the output membership function for the selected λ-cut level.

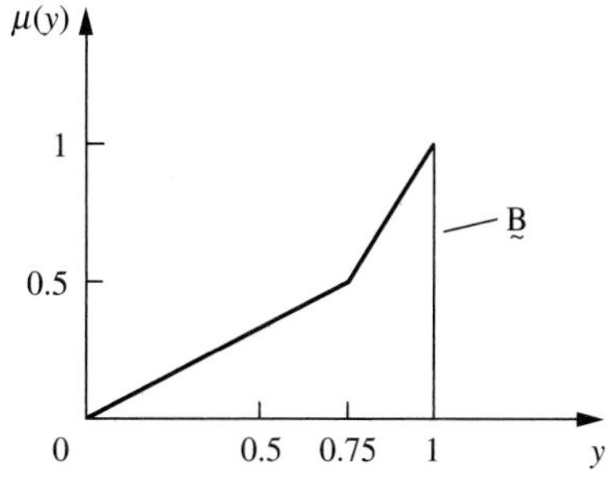

FIGURE 6.8
Fuzzy membership function for the output to $y = x(2 - x)$.

4. Repeat steps 1–3 for different values of λ to complete a λ-cut representation of the solution.

Example 6.11. Let us consider a nonlinear, one-dimensional expression similar to the previous example, or $y = x(2 + x) = 2x + x^2$, where we again use the fuzzy input variable shown in Fig. 6.7. The new function is shown in Fig. 6.9*a*, along with the fuzzy input in Fig. 6.9*b*. Again, if we decompose the membership function for the input into three λ-cut intervals, for $\lambda = 0^+, 0.5$, and 1 we get the intervals $\mathrm{I}_{0^+} = [0.5, 2]$, $I_{0.5} = [0.75, 1.5]$, and $\mathrm{I}_1 = [1, 1]$ (a single point). In terms of binary interval operations, the functional mapping on the intervals would take place as follows for each λ-cut level:

$$\mathrm{I}_{0^+} = [0.5, 2]$$
$$\mathrm{B}_{0^+} = 2[0.5, 2] + [0.5^2, 2^2] = [1, 4] + [0.25, 4] = [1.25, 8]$$

$$\mathrm{I}_{0.5} = [0.75, 1.5]$$
$$\mathrm{B}_{0.5} = 2[0.75, 1.5] + [0.75^2, 1.5^2] = [1.5, 3] + [0.5625, 2.25] = [2.0625, 5.25]$$

$$\mathrm{I}_1 = [1, 1]$$
$$\mathrm{B}_1 = 2[1, 1] + [1^2, 1^2] = [2, 2] + [1, 1] = [3, 3] = 3$$

Figure 6.10 provides a plot of the intervals B_{0^+}, $\mathrm{B}_{0.5}$, and B_1 to form the fuzzy output, y.

The previous example worked with a fuzzy input that was defined on the positive side of the real line, hence DSW operations were conducted on positive quantities. Suppose we want to conduct the same DSW operations, but on a fuzzy input that is defined on both the positive and negative side of the real line. The user of the DSW algorithm must be careful in this case. If the lower bound of an interval is negative and the upper bound is positive (i.e., if the interval contains zero) and if the function involves a square or an even-power operation, then the lower bound of the result should be zero. This feature of an interval analysis, like the DSW method, is demonstrated in the following example.

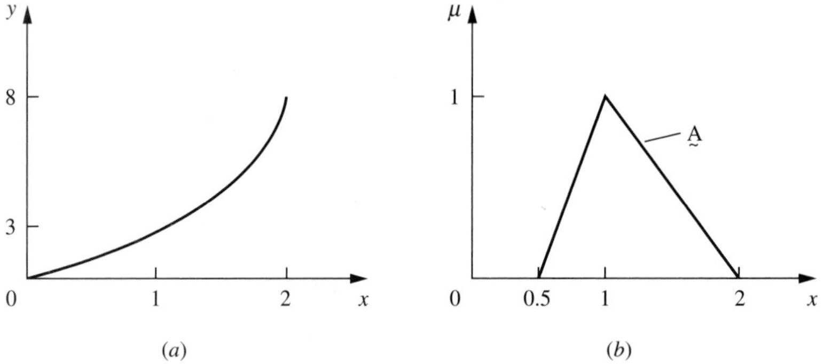

FIGURE 6.9
Nonlinear function and fuzzy input membership.

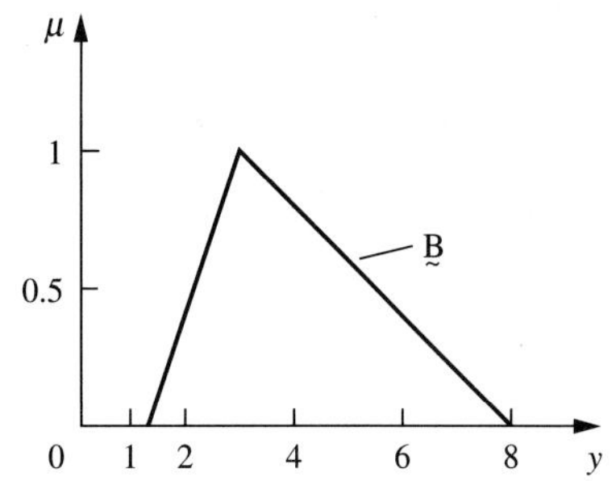

FIGURE 6.10
Fuzzy membership function for the output to $y = x(2+x)$.

Example 6.12. Let us consider the nonlinear, one-dimensional expression from Example 6.11, i.e., $y = x(2 + x) = 2x + x^2$, which is shown in Fig. 6.9*a* and is repeated in Fig. 6.11*a*. Suppose we change the domain of the input variable, *x*, to include negative numbers, as shown in Fig. 6.11*b*. Again, if we decompose the membership function for the input into three λ-cut intervals, for $\lambda = 0^+$, 0.5, and 1, we get the intervals, $I_{0+} = [-0.5, 1]$, $I_{0.5} = [-0.25, 0.5]$, and $I_1 = [0, 0]$ (a single point). In terms of binary interval operations, the functional mapping on the intervals would take place as follows for each λ-cut level:

$$I_{0+} = [-0.5, 1]$$
$$B_{0+} = 2[-0.5, 1] + [\mathbf{0}, 1^2] = [-1, 2] + [0, 1] = [-1, 3]$$

(*Note:* The boldface zero is taken as the minimum, since $(-0.5)^2 > 0$; because zero is contained in the interval $[-0.5, 1]$ the minimum of squares of any number in the interval will be zero.)

$$I_{0.5} = [-0.25, 0.5]$$
$$B_{0.5} = 2[-0.25, 0.5] + [\mathbf{0}, 0.5^2] = [-0.5, 1] + [0, 0.25] = [-0.5, 1.25]$$

$$I_1 = [0, 0]$$
$$B_1 = 2[0, 0] + [0, 0] = [0, 0]$$

Figure 6.12 is a plot of the intervals B_{0+}, $B_{0.5}$, and B_1 that form the fuzzy output, *y*.

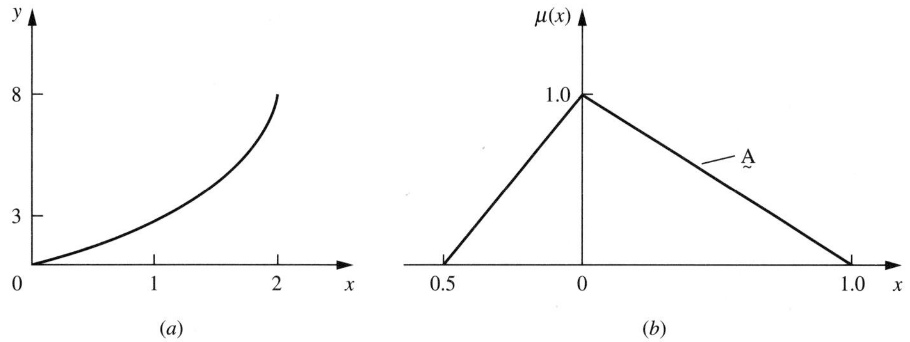

FIGURE 6.11
Nonlinear function and fuzzy input membership.

Restricted DSW Algorithm

This method, proposed by Givens and Tahani [1987], is a slight restriction of the original DSW algorithm. Suppose we have two interval numbers, I = [a, b] and J = [c, d]. For the special case where neither of these intervals contains negative numbers, i.e., $a, b, c, d \geq 0$, and none of the calculations using these intervals involves subtraction, the definitions of interval multiplication, Eq. (6.26), and interval division, Eq. (6.27), can be simplified as follows:

$$\mathrm{I} \cdot \mathrm{J} = [a, b] \cdot [c, d] = [ac, bd] \tag{6.34}$$

$$\mathrm{I}/\mathrm{J} = [a, b] \div [c, d] = \left[\frac{a}{d}, \frac{b}{c}\right] \tag{6.35}$$

These definitions of interval multiplication and division require only one-fourth the number of multiplications (or divisions), and there is no need for the min or max operations, unlike the previous definitions, Eqs. (6.26, 6.27).

Example 6.13. Let us consider the function in Example 6.12, $y = x(2 + x)$, and another nonlinear, one-dimensional expression of the form $y = x/(2 + x)$, where we again use the fuzzy input variable shown in Fig. 6.7 in both functions. In interval calculations we can represent the scalar value 2 by the interval [2, 2]. The λ-cut interval calculations using the restricted DSW calculations are now as follows:

$\boldsymbol{y = x(2 + x)}$

$$\mathrm{I}_{0+} = [0.5, 2]$$
$$\mathrm{B}_{0+} = [0.5, 2] \cdot \{[2, 2] + [0.5, 2]\} = [0.5, 2] \cdot [2.5, 4] = [1.25, 8]$$

$$\mathrm{I}_{0.5} = [0.75, 1.5]$$
$$\mathrm{B}_{0.5} = [0.75, 1.5] \cdot \{[2, 2] + [0.75, 1.5]\} = [0.75, 1.5] \cdot [2.75, 3.5] = [2.0625, 5.25]$$

$$\mathrm{I}_1 = [1, 1]$$
$$\mathrm{B}_1 = [1, 1] \cdot \{[2, 2] + [1, 1]\} = [1, 1] \cdot [3, 3] = 3$$

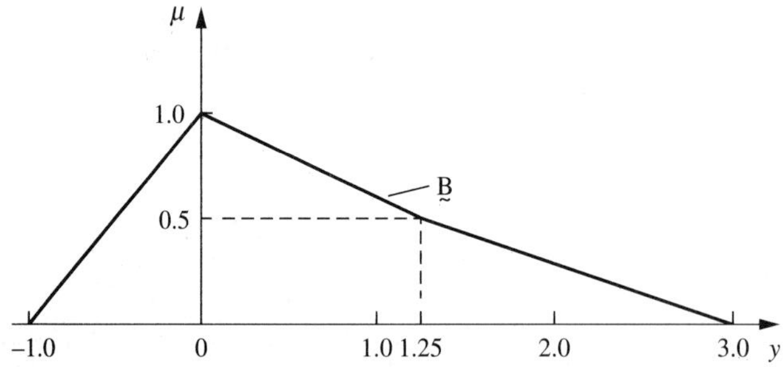

FIGURE 6.12
Fuzzy membership function for the output to $y = x(2 + x)$.

Note that these three intervals for the output $\underset{\sim}{B}$ are identical to those in the previous example.

$y = x/(2 + x)$

$$I_{0+} = [0.5, 2]$$
$$B_{0+} = [0.5, 2] \div \{[2, 2] + [0.5, 2]\} = [0.5, 2] \div [2.5, 4] = [0.125, 0.8]$$

$$I_{0.5} = [0.75, 1.5]$$
$$B_{0.5} = [0.75, 1.5] \div \{[2, 2] + [0.75, 1.5]\} = [0.75, 1.5] \div [2.75, 3.5] = [0.2143, 0.5455]$$

$$I_1 = [1, 1]$$
$$B_1 = [1, 1] \div \{[2, 2] + [1, 1]\} = [1, 1] \div [3, 3] = 0.3333$$

Comparisons

It will be useful at this point to compare the three methods discussed so far—the extension principle, the vertex method, and the DSW algorithm—by applying them to the same problem. This comparison will illustrate the problems faced with using the extension principle on discretized membership functions, as compared to the other two methods.

Example 6.14. We define fuzzy sets $\underset{\sim}{X}$ and $\underset{\sim}{Y}$ with the membership functions as shown in Fig. 6.13. We will use the following methods to compute $\underset{\sim}{X} * \underset{\sim}{Y}$ and to demonstrate the similarity of results:

- The extension principle
- The vertex method
- The DSW algorithm

Extension principle using discretized fuzzy sets. The fuzzy variables may be discretized at seven points as

$$\underset{\sim}{X} = \left\{\frac{0}{1} + \frac{0.33}{2} + \frac{0.66}{3} + \frac{1.0}{4} + \frac{0.66}{5} + \frac{0.33}{5} + \frac{0}{7}\right\}$$

and

$$\underset{\sim}{Y} = \left\{\frac{0}{2} + \frac{0.33}{3} + \frac{0.66}{4} + \frac{1.0}{5} + \frac{0.66}{6} + \frac{0.33}{7} + \frac{0}{8}\right\}$$

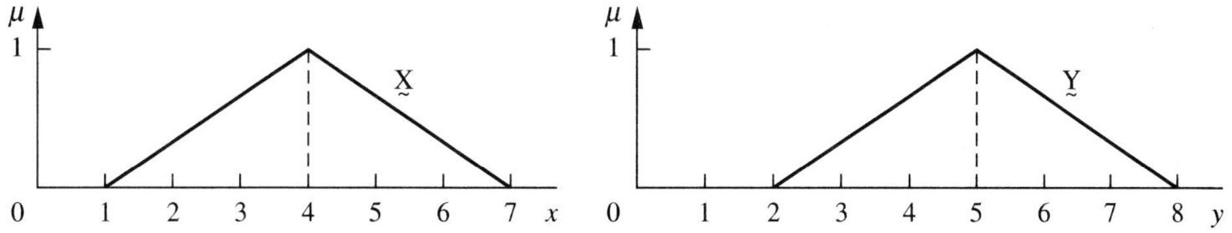

FIGURE 6.13
Fuzzy sets $\underset{\sim}{X}$ and $\underset{\sim}{Y}$.

Their product would then give us

$$\underset{\sim}{X} \times \underset{\sim}{Y} = \left\{ \frac{0}{2} + \frac{0}{3} + \frac{0}{4} + \frac{0}{5} + \frac{0.33}{6} + \frac{0}{7} + \frac{0.33}{8} + \frac{0.33}{9} + \frac{0.33}{10} + \frac{0.66}{12} \right.$$
$$+ \frac{0.33}{14} + \frac{0.66}{15} + \frac{0.33}{16} + \frac{0.66}{18} + \frac{1.0}{20} + \frac{0.33}{21} + \frac{0.66}{24} + \frac{0.66}{25} + \frac{0.33}{28}$$
$$\left. + \frac{0.66}{30} + \frac{0}{32} + \frac{0.33}{35} + \frac{0.33}{36} + \frac{0.0}{40} + \frac{0.33}{42} + \frac{0.0}{48} + \frac{0.0}{49} + \frac{0.0}{56} \right\}$$

The result of the operation $\underset{\sim}{X} \times \underset{\sim}{Y}$ for a discretization level of seven points is plotted in Fig. 6.14*a*. Figures 6.14*b*, *c*, and *d* show the product function $\underset{\sim}{X} \times \underset{\sim}{Y}$ for greater discretization levels of the fuzzy variables $\underset{\sim}{X}$ and $\underset{\sim}{Y}$.

Vertex method.

I_{0^+}: Support for X is the interval [1, 7] and support for Y is the interval [2, 8].

$(a)\ x = 1, \quad y = 2, \quad f(a) = 2$
$(b)\ x = 1, \quad y = 8, \quad f(b) = 8$
$(c)\ x = 7, \quad y = 2, \quad f(c) = 14$
$(d)\ x = 7, \quad y = 8, \quad f(d) = 56$

Therefore, min = 2, max = 56, and $B_{0^+} = [2, 56]$.

$I_{0.33}$: X[2, 6], Y[3, 7].

$(a)\ x = 2, \quad y = 3, \quad f(a) = 6$
$(b)\ x = 2, \quad y = 7, \quad f(b) = 14$
$(c)\ x = 6, \quad y = 3, \quad f(c) = 18$
$(d)\ x = 6, \quad y = 7, \quad f(d) = 42$

Therefore, min = 6, max = 42, and $B_{0.33} = [6, 42]$.

$I_{0.66}$: X[3, 5], Y[4, 6].

$(a)\ x = 3, \quad y = 4, \quad f(a) = 12$
$(b)\ x = 3, \quad y = 6, \quad f(b) = 18$
$(c)\ x = 5, \quad y = 4, \quad f(c) = 20$
$(d)\ x = 5, \quad y = 6, \quad f(d) = 30$

Therefore, min = 12, max = 30, and $B_{0.33} = [12, 30]$.

$I_{1.0}$: X[4, 4], Y[5, 5].

(a) x = 4, y = 5, f(a) = 20

Therefore, min = 20, max = 20, and $B_{1.0} = [20, 20]$.

The results of plotting the four λ-cut levels is shown in Fig. 6.15.

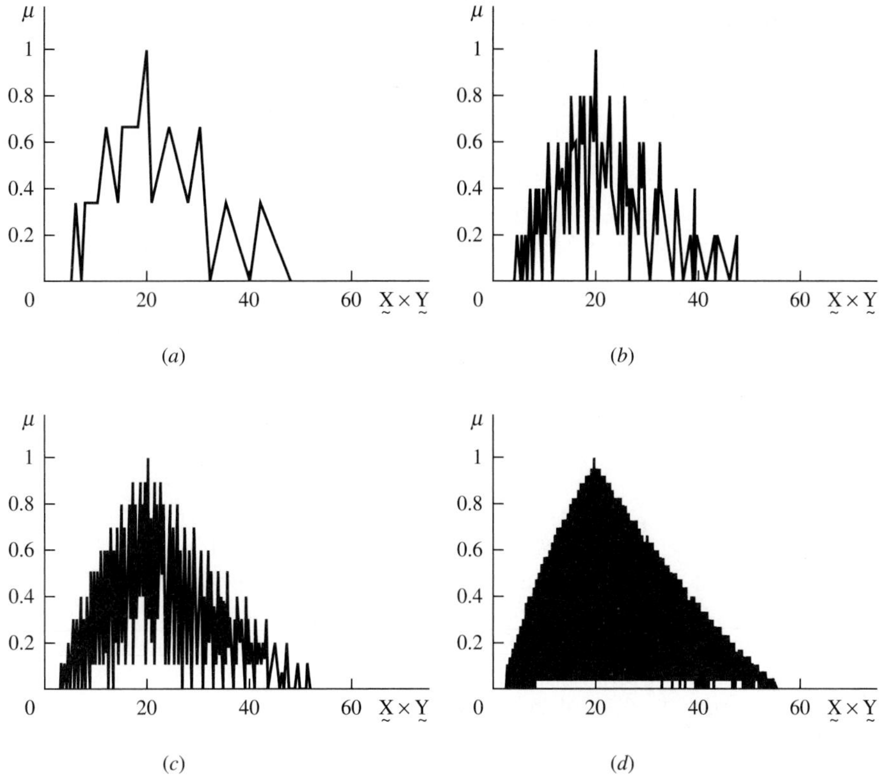

FIGURE 6.14
$\underset{\sim}{X} \times \underset{\sim}{Y}$ for increasing discretization of both X and Y (both variables are discretized for the same number of points): (*a*) 7 points; (*b*) 13 points; (*c*) 23 points; (*d*) 63 points.

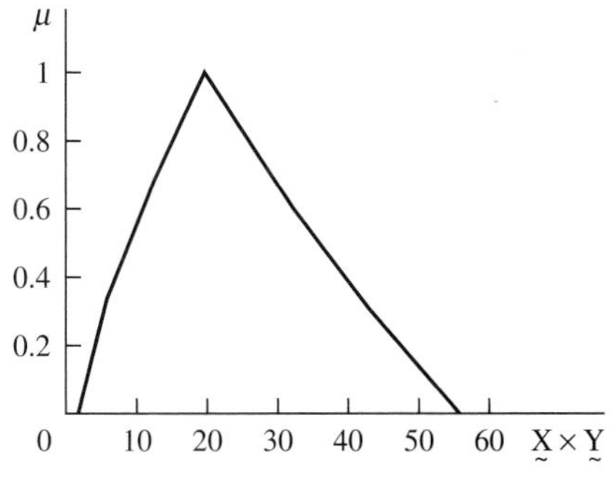

FIGURE 6.15
Output profile of $\underset{\sim}{X} \times \underset{\sim}{Y}$ determined using the vertex method.

DSW method.

$$\begin{aligned} I_{0^+} &: [1,7] \bullet [2,8] = [\min(2,14,8,56), \max(2,14,8,56)] = [2,56] \\ I_{.33} &: [2,6] \bullet [3,7] = [\min(6,18,14,42), \max(6,18,14,42)] = [6,42] \\ I_{.66} &: [3,5] \bullet [4,6] = [\min(12,20,28,30), \max(12,20,28,30)] = [12,30] \\ I_1 &: [4,4] \bullet [5,5] = [\min(20,20,20,20), \max(20,20,20,20)] = [20,20] \end{aligned}$$

The results of plotting the four λ-cut levels are shown in Fig. 6.16.

Comparing the results from the foregoing three methods, we see that the results are the same. The discretization method was performed for increasing levels of discretization. In each case the outer envelope of the curves due to the discretized method gave the correct results. It should be intuitively obvious that as the discretization is increased (the equation is exactly simulated) the resulting curve approaches the true values of membership functions. Also note that the discretization technique is computationally expensive for complex problems.

FUZZY VECTORS

Fuzzy vectors were introduced earlier in this chapter as a one-dimensional array of membership values. This section introduces some interesting features and operations on fuzzy vectors that will become quite useful in Chapter 12 in the discipline of fuzzy pattern recognition [Dong, 1986]. Formally, a vector, $\underset{\sim}{\mathbf{a}} = (a_1, a_2, \ldots, a_n)$, is called a fuzzy vector if for any element we have $0 \le a_i \le 1$ for $i = 1, 2, \ldots, n$. Similarly, the transpose of the fuzzy vector $\underset{\sim}{\mathbf{a}}$, denoted $\underset{\sim}{\mathbf{a}}^{\mathrm{T}}$, is a column vector if $\underset{\sim}{\mathbf{a}}$ is a row vector, i.e.,

$$\underset{\sim}{\mathbf{a}}^{\mathrm{T}} = \begin{bmatrix} a_1 \\ a_2 \\ \vdots \\ a_n \end{bmatrix}$$

Let us define $\underset{\sim}{\mathbf{a}}$ and $\underset{\sim}{\mathbf{b}}$ as fuzzy vectors of length n, and define

$$\underset{\sim}{\mathbf{a}} \bullet \underset{\sim}{\mathbf{b}}^{\mathrm{T}} = \bigvee_{i=1}^{n} (a_i \wedge b_i) \tag{6.36}$$

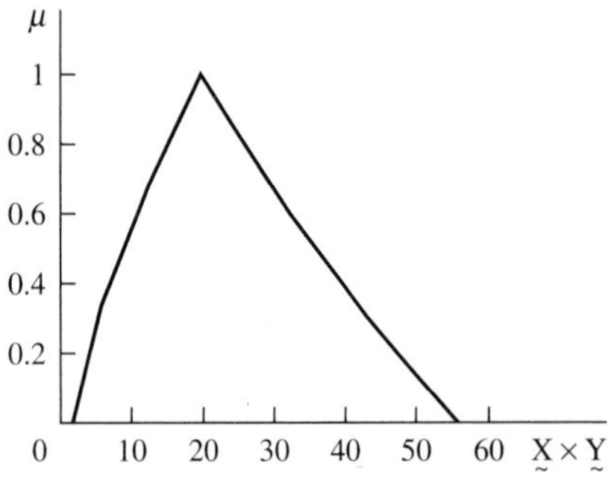

FIGURE 6.16
Output profile of $\underset{\sim}{X} \times \underset{\sim}{Y}$ determined using the DSW algorithm.

as the fuzzy *inner product* of $\underset{\sim}{\mathbf{a}}$ and $\underset{\sim}{\mathbf{b}}$, and

$$\underset{\sim}{\mathbf{a}} \oplus \underset{\sim}{\mathbf{b}}^{\mathrm{T}} = \bigwedge_{i=1}^{n} (a_i \vee b_i) \tag{6.37}$$

as the fuzzy *outer product* of $\underset{\sim}{\mathbf{a}}$ and $\underset{\sim}{\mathbf{b}}$.

Example 6.15. We have two fuzzy vectors of length 4 as defined here, and want to find the inner product and the outer product for these two fuzzy vectors:

$$\underset{\sim}{\mathbf{a}} = (0.3, 0.7, 1, 0.4)$$
$$\underset{\sim}{\mathbf{b}} = (0.5, 0.9, 0.3, 0.1)$$

$$\underset{\sim}{\mathbf{a}} \bullet \underset{\sim}{\mathbf{b}}^{\mathrm{T}} = (3, .7, .1, .4)\begin{pmatrix} .5 \\ .9 \\ .3 \\ .1 \end{pmatrix}$$

$$= (.3 \wedge .5) \vee (.7 \wedge .9) \vee (1 \wedge .3) \vee (.4 \wedge .1) = .3 \vee .7 \vee .3 \vee .1 = 0.7$$

$$\underset{\sim}{\mathbf{a}} \oplus \underset{\sim}{\mathbf{b}}^{\mathrm{T}} = (.3 \vee .5) \wedge (.7 \vee .9) \wedge (1 \vee .3) \wedge (.4 \vee .1) = .5 \wedge .9 \wedge 1 \wedge .4 = 0.4$$

The symbol $\oplus$ has also been used in the literature to describe the Boolean outer product. In this context we will use this symbol to refer to the outer product of two fuzzy vectors. An interesting feature of these products is found in comparing them to standard algebraic operations on vectors in physics. Whereas the inner and outer products on fuzzy vectors result in scalar quantities, only the algebraic inner product on vectors in physics produces a scalar; the outer product on two vectors in physics produces another vector, whose direction is orthogonal to the plane containing the original two vectors.

We now define the complement of the fuzzy vector, or fuzzy complement vector, as

$$\overline{\underset{\sim}{\mathbf{a}}} = (1 - a_1, 1 - a_2, \ldots, 1 - a_n) = (\overline{a}_1, \overline{a}_2, \ldots, \overline{a}_n) \tag{6.38}$$

It should be obvious that since $\overline{\underset{\sim}{\mathbf{a}}}$ is subject to the constraint, $0 \le \overline{a}_i \le 1$ for $i = 1, 2, \ldots, n$, the fuzzy complement vector is also another fuzzy vector. Moreover, we define the largest element $\hat{a}$ in the fuzzy vector $\underset{\sim}{\mathbf{a}}$ as its *upper bound,* i.e.,

$$\hat{a} = \max_i(a_i) \tag{6.39}$$

and the smallest element $\underset{\wedge}{a}$ in the fuzzy vector $\underset{\sim}{\mathbf{a}}$ as its *lower bound,* i.e.,

$$\underset{\wedge}{a} = \min_i(a_i) \tag{6.40}$$

Some properties of fuzzy vectors that will become quite useful in the area of pattern recognition will be summarized here. For two fuzzy vectors, $\underset{\sim}{\mathbf{a}}$ and $\underset{\sim}{\mathbf{b}}$, both of length n, the following properties hold:

$$\overline{\underset{\sim}{\mathbf{a}} \bullet \underset{\sim}{\mathbf{b}}^{\mathrm{T}}} = \overline{\underset{\sim}{\mathbf{a}}} \oplus \overline{\underset{\sim}{\mathbf{b}}}^{\mathrm{T}} \quad \text{and alternatively} \quad \overline{\underset{\sim}{\mathbf{a}} \oplus \underset{\sim}{\mathbf{b}}^{\mathrm{T}}} = \overline{\underset{\sim}{\mathbf{a}}} \bullet \overline{\underset{\sim}{\mathbf{b}}}^{\mathrm{T}} \tag{6.41}$$

$$\underset{\sim}{\mathbf{a}} \bullet \underset{\sim}{\mathbf{b}}^{\mathrm{T}} \le (\hat{a} \wedge \hat{b}) \quad \text{and alternatively} \quad \underset{\sim}{\mathbf{a}} \oplus \underset{\sim}{\mathbf{b}}^{\mathrm{T}} \ge (\underset{\wedge}{a} \vee \underset{\wedge}{b}) \tag{6.42}$$

$$\underset{\sim}{\mathbf{a}} \bullet \underset{\sim}{\mathbf{a}}^{\mathrm{T}} = \hat{a} \quad \text{and} \quad \underset{\sim}{\mathbf{a}} \oplus \underset{\sim}{\mathbf{a}}^{\mathrm{T}} \geq \underset{\wedge}{a} \tag{6.43}$$

$$\text{For } \underset{\sim}{\mathbf{a}} \subseteq \underset{\sim}{\mathbf{b}} \text{ then } \underset{\sim}{\mathbf{a}} \bullet \underset{\sim}{\mathbf{b}}^{\mathrm{T}} = \hat{a}; \text{ and for } \underset{\sim}{\mathbf{b}} \subseteq \underset{\sim}{\mathbf{a}} \text{ then } \underset{\sim}{\mathbf{a}} \oplus \underset{\sim}{\mathbf{b}}^{\mathrm{T}} = \underset{\wedge}{a} \tag{6.44}$$

$$\underset{\sim}{\mathbf{a}} \bullet \underset{\sim}{\overline{\mathbf{a}}} \leq \tfrac{1}{2} \quad \text{and} \quad \underset{\sim}{\mathbf{a}} \oplus \underset{\sim}{\overline{\mathbf{a}}} \geq \tfrac{1}{2} \tag{6.45}$$

From the fuzzy vector properties given in Eqs. (6.42)–(6.45), one can show (see Problem 6.13) that when two separate fuzzy vectors are identical, i.e., $\underset{\sim}{\mathbf{a}} = \underset{\sim}{\mathbf{b}}$, the inner product $\underset{\sim}{\mathbf{a}} \bullet \underset{\sim}{\mathbf{b}}^{\mathrm{T}}$ reaches a maximum while the outer product $\underset{\sim}{\mathbf{a}} \oplus \underset{\sim}{\mathbf{b}}^{\mathrm{T}}$ reaches a minimum. This result is extremely powerful when used in any problem requiring a metric of similarity between two vectors (see Chapter 12). If two vectors are identical, the inner product metric will yield a maximum value, and if the two vectors are completely dissimilar the inner product will yield a minimum value. Chapter 12 makes use of the inverse duality between the inner product and the outer product for fuzzy vectors and fuzzy sets in developing an algorithm for pattern recognition.

SUMMARY

The extension principle is one of the most basic ideas in fuzzy set theory. It provides a general method for extending crisp mathematical concepts to address fuzzy quantities, such as real algebra operations on fuzzy numbers. These operations are computationally effective generalizations of interval analysis. Several methods to convert extended fuzzy operations into efficient computational algorithms have been presented in this chapter. All of these approximations make use of the decomposition of a membership function into a series of λ-cut intervals. The employment of the extension principle on discretized fuzzy numbers can lead to counterintuitive results, unless sufficient resolution in the discretization is maintained. This statement is simply a caution to potential users of some of the simpler ideas in the extension principle. Although the set of real fuzzy numbers equipped with an extended addition or multiplication is no longer a group, many structural properties of the resulting fuzzy numbers are preserved in the process [Dubois and Prade, 1980]. Finally, operations on, and properties of, fuzzy vectors were developed in this chapter for eventual use with some similarity metrics in Chapter 12.

REFERENCES

Dong, W. (1986). "Applications of fuzzy sets theory in structural and earthquake engineering," Ph.D. dissertation, Department of Civil Engineering, Stanford University, Stanford, CA.

Dong, W., and H. Shah (1987). "Vertex method for computing functions of fuzzy variables," *Fuzzy Sets Syst.,* vol. 24, pp. 65–78.

Dong, W., H. Shah, and F. Wong (1985). "Fuzzy computations in risk and decision analysis," *Civ. Eng. Syst.,* vol. 2, pp. 201–208.

Dubois, D., and H. Prade (1980). *Fuzzy sets and systems: Theory and applications,* Academic, New York.

Givens, J., and H. Tahani (1987). "An improved method of performing fuzzy arithmetic for computer vision," Proceedings of North American Information Processing Society (NAFIPS), Purdue University, West Lafayette, IN, pp. 275–280.

Wong, F., and T. Ross (1985). "Treatment of uncertainties in structural dynamics models," in F. Deyi and L. Xihui (eds.), *Proc. Int. Symp. on Fuzzy Math. in Earthquake Res.*, Seismological Press, Beijing, China.

Yager, R. R. (1986). "A characterization of the extension principle," *Fuzzy Sets Syst.*, vol. 18, pp. 205–217.

Zadeh, L. (1975). "The concept of a linguistic variable and its application to approximate reasoning, Part I," *Inf. Sci.*, vol. 8, pp. 199–249.

PROBLEMS

6.1. Perform the following operations on intervals:

(*a*) $[2, 3] + [3, 4]$

(*b*) $[1, 2] \times [1, 3]$

(*c*) $[4, 6] \div [1, 2]$

(*d*) $[3, 5] - [4, 5]$

6.2. Given the following fuzzy numbers and using Zadeh's extension principle, calculate $\underset{\sim}{K} = \underset{\sim}{I} \cdot \underset{\sim}{J}$ and explain (or show) why $\underset{\sim}{6}$ is nonconvex.

$$\underset{\sim}{I} = \underset{\sim}{3} = \frac{0.2}{2} + \frac{1}{3} + \frac{0.1}{4}$$

$$\underset{\sim}{J} = \underset{\sim}{2} = \frac{0.1}{1} + \frac{1}{2} + \frac{0.3}{3}$$

6.3. These exercises use Zadeh's extension principle. You are given the fuzzy sets $\underset{\sim}{A}$ and $\underset{\sim}{B}$ on the real line as follows:

$\mu(x_i)$	**0**	**1**	**2**	**3**	**4**	**5**	**6**	**7**
$\underset{\sim}{A}$	0.0	0.1	0.6	0.8	0.9	0.7	0.1	0.0
$\underset{\sim}{B}$	0.0	1.0	0.7	0.5	0.2	0.1	0.0	0.0

If x and y are real numbers defined by sets $\underset{\sim}{A}$ and $\underset{\sim}{B}$, respectively, calculate the fuzzy set $\underset{\sim}{C}$ representing the real numbers z given by

(*a*) $z = 3x - 2$

(*b*) $z = 4x^2 + 3$

(*c*) $z = x^2 + y^2$

(*d*) $z = x - x$

(*e*) $z = \min(x, y)$

6.4. For the function $y = x_1^2 + x_2^2 - 4x_1 + 4$ and the membership functions for fuzzy variables x_1 and x_2 shown in Fig. P6.4, find and plot the membership function for the fuzzy output variable, y, using

(*a*) A discretized form of the extension principle

(*b*) The vertex method

(*c*) The DSW algorithm

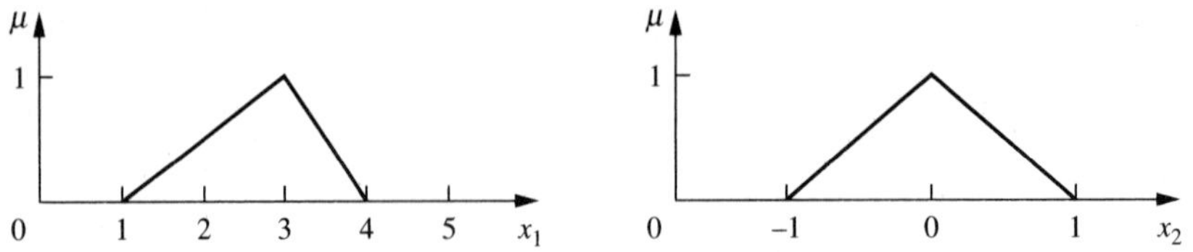

FIGURE P6.4

6.5. The voltage drop across an element in a series circuit is equal to the series current multiplied by the element's impedance. The current, $\underset{\sim}{I}$, impedance, $\underset{\sim}{R}$, and voltage, $\underset{\sim}{V}$, are presumed to be fuzzy variables. Membership functions for the current and impedance are as follows:

$$\underset{\sim}{I} = \left\{ \frac{0}{0} + \frac{.8}{.5} + \frac{1}{1} + \frac{.8}{1.5} + \frac{0}{2} \right\}$$

$$\underset{\sim}{R} = \left\{ \frac{.5}{500} + \frac{.9}{750} + \frac{1}{1000} + \frac{.9}{1250} + \frac{.5}{1500} \right\}$$

Find the arithmetic product for $\underset{\sim}{V} = \underset{\sim}{I} \cdot \underset{\sim}{R}$ using the extension principle.

6.6. Determine equivalent resistance of the circuity shown in Fig. P6.6, where $\underset{\sim}{R}_1$ and $\underset{\sim}{R}_2$ are fuzzy sets describing the resistance of resistors R_1 and R_2, repectively, expressed in ohms. Since the resistors are in series they can be added arithmetically. Using the extension principle, find the equivalent resistance,

$$\underset{\sim}{R}_{eq} = \underset{\sim}{R}_1 + \underset{\sim}{R}_2$$

The membership functions for the two resistors are

$$\underset{\sim}{R}_1 = \left\{ \frac{0.5}{3} + \frac{0.8}{4} + \frac{0.6}{5} \right\} \quad \text{and} \quad \underset{\sim}{R}_2 = \left\{ \frac{0.3}{8} + \frac{1.0}{9} + \frac{0.4}{10} \right\}$$

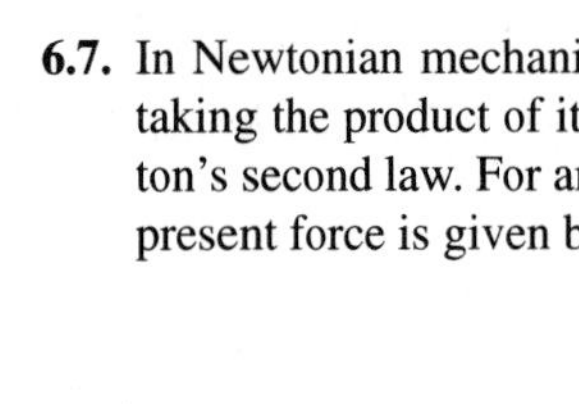

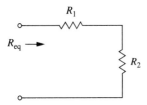

FIGURE P6.6

6.7. In Newtonian mechanics the equivalent force on a body in motion can be found by taking the product of its mass and acceleration; this is commonly referred to as Newton's second law. For an object in a particular state, suppose the acceleration under the present force is given by the fuzzy set

$$\underset{\sim}{A} = \left\{ \frac{0}{0} + \frac{.2}{1} + \frac{.7}{2} + \frac{1}{3} + \frac{0}{4} \right\}$$

and the mass is given by the fuzzy set

$$\underset{\sim}{M} = \left\{ \frac{0}{1} + \frac{.5}{2} + \frac{1}{3} + \frac{.5}{4} + \frac{0}{5} \right\}$$

Assume both sets are in nondimensionalized units.

(*a*) Find the fuzzy set representing the force on the object using the extension principle.

(*b*) Develop analogous continuous membership functions, and plot them, for the fuzzy acceleration and mass and solve for the fuzzy force using (*i*) the vertex method and (*ii*) the restricted DSW algorithm.

6.8. For fluids, the product of the pressure (P) and the volume (V) of the fluid is a constant for a given temperature, i.e.,

$$PV = \text{constant}$$

Assume that at a given temperature a fluid of fuzzy volume

$$\underset{\sim}{V}_1 = \left\{ \frac{0.0}{0.5} + \frac{0.5}{0.75} + \frac{1.0}{1.0} + \frac{0.5}{1.25} + \frac{0.0}{1.5} \right\}$$

is under a fuzzy pressure

$$\underset{\sim}{P}_1 = \left\{ \frac{0.0}{0.5} + \frac{0.5}{1.75} + \frac{1.0}{2.0} + \frac{0.5}{2.25} + \frac{0.0}{2.5} \right\}$$

(*a*) Using the extension principle, determine the pressure $\underset{\sim}{P}_2$ if the volume is reduced to

$$\underset{\sim}{V}_2 = \left\{ \frac{0.0}{0.4} + \frac{0.5}{0.45} + \frac{1.0}{0.5} + \frac{0.5}{0.55} + \frac{0.0}{0.6} \right\}$$

(*b*) Develop analogous continuous membership functions for the fuzzy pressure $\underset{\sim}{P}_1$ and volume $\underset{\sim}{V}_1$ and solve for the pressure $\underset{\sim}{P}_2$ using (*i*) the vertex method and (*ii*) the DSW algorithm. Plot the resulting membership function.

(*c*) Explain why $\underset{\sim}{P}_2 \cdot \underset{\sim}{V}_2$ would not be the same as $\underset{\sim}{P}_1 \cdot \underset{\sim}{V}_1$.

6.9. A circle is governed by the equation $x^2 + y^2 = 8$. Its fuzzy x coordinate is defined by the fuzzy set

$$\underset{\sim}{x} = \left\{ \frac{0}{0} + \frac{.6}{2} + \frac{.65}{3} + \frac{.7}{4} + \frac{.75}{5} + \frac{.8}{6} \right\}$$

Find the fuzzy y coordinate, and plot its membership function for the equation of a circle.

(*a*) Use the DSW algorithm.

(*b*) Perform the same calculation using the restricted DSW algorithm.

(*c*) Comment on the nature of the results using a fuzzy x that is nonnormal.

6.10. For the function $y = x_1^2 \cdot x_2 - 3x_2$, where the membership functions of x_1 and x_2 are given in Fig. P6.10, find and plot the fuzzy membership function for y using

(*a*) The vertex method

(*i*) Ignoring any extreme points

(*ii*) Including any extreme points

(*b*) The restricted DSW algorithm

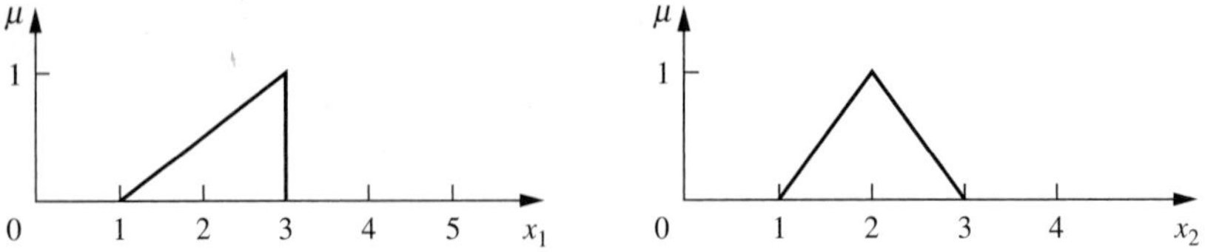

FIGURE P6.10

6.11. Define a fuzzy set $\underset{\sim}{X}$ with the membership function

$$\underset{\sim}{x} = \left\{ \frac{0.1}{1} + \frac{1}{2} + \frac{0.4}{3} \right\}$$

Using the extension principle, determine the membership function for $\underset{\sim}{z}$, written in two different forms, i.e., for

(*a*) $\underset{\sim}{z} = \underset{\sim}{x}^2$

(*b*) $\underset{\sim}{z} = \underset{\sim}{x} \cdot \underset{\sim}{x}$

For parts (*a*) and (*b*) produce analogous continuous membership functions for $\underset{\sim}{x}$ (i.e., a function that passes through the three points given).

(*c*) $\underset{\sim}{z} = \underset{\sim}{x}^2$ and $\underset{\sim}{z} = \underset{\sim}{x} \cdot \underset{\sim}{x}$ using the vertex method

(*d*) $\underset{\sim}{z} = \underset{\sim}{x}^2$ and $\underset{\sim}{z} = \underset{\sim}{x} \cdot \underset{\sim}{x}$ using the DSW algorithm

(*e*) Discuss your answers from the different forms and methods.

6.12. Now suppose $\underset{\sim}{x}$ has membership function

$$\underset{\sim}{x} = \left\{ \frac{0.1}{-3} + \frac{0.3}{-2} + \frac{0.7}{-1} + \frac{1}{0} + \frac{0.7}{1} + \frac{0.3}{2} + \frac{0.1}{3} \right\}$$

Repeat steps (*a*), (*b*), (*c*), and (*d*) of Problem 6.11 and (*e*) comment on any differences or similarities.

6.13. Show that when two separate fuzzy vectors are identical, i.e., $\underset{\sim}{\mathbf{a}} = \underset{\sim}{\mathbf{b}}$, the inner product $\underset{\sim}{\mathbf{a}} \bullet \underset{\sim}{\mathbf{b}}^{\mathrm{T}}$ reaches a maximum as the outer product $\underset{\sim}{\mathbf{a}} \oplus \underset{\sim}{\mathbf{b}}^{\mathrm{T}}$ reaches a minimum.

6.14. For two fuzzy vectors $\underset{\sim}{\mathbf{a}}$ and $\underset{\sim}{\mathbf{b}}$ and the particular case where $\hat{a} = \hat{b} = 1$ and $\underset{\wedge}{a} = \underset{\wedge}{b} = 0$, show that when $\underset{\sim}{\mathbf{a}} = \underset{\sim}{\mathbf{b}}$, then the inner product $\underset{\sim}{\mathbf{a}} \bullet \underset{\sim}{\mathbf{b}}^{\mathrm{T}} = 1$ and the outer product $\underset{\sim}{\mathbf{a}} \oplus \underset{\sim}{\mathbf{b}}^{\mathrm{T}} = 0$.

CHAPTER 7

CLASSICAL LOGIC AND FUZZY LOGIC

"I know what you're thinking about," said Tweedledum; "but it isn't so, nohow." "Contrariwise," continued Tweedledee, "if it was so, it might be; and if it were so, it would be; but as it isn't, it ain't. That's logic."

Lewis Carroll
Through the Looking Glass, *1871*

Logic is but a small part of the human capacity to reason. Logic can be a means to compel us to infer correct answers, but it cannot by itself be responsible for our creativity or for our ability to remember. In other words, logic can assist us in organizing words to make clear sentences, but it cannot help us determine what sentences to use in various contexts. Consider the passage above from the nineteenth-century mathematician Lewis Carroll in his classic *Through the Looking Glass.* How many of us can see the logical context in the discourse of these fictional characters? Logic for humans is a way quantitatively to develop a reasoning process that can be replicated and manipulated with mathematical precepts. The interest in logic is the study of truth in logical propositions; in classical predicate logic this truth is binary—a proposition is either true or false.

From this perspective, fuzzy logic is a method to formalize the human capacity of imprecise reasoning, or—later in this chapter—approximate reasoning. Such reasoning represents the human ability to reason approximately and judge under uncertainty. In fuzzy logic all truths are partial or approximate. In this sense this reasoning has also been termed interpolative reasoning, where the process of interpolating between the binary extremes of true and false is represented by the ability of fuzzy logic to encapsulate partial truths.

This chapter introduces the reader to fuzzy logic with a review of classical predicate logic and its operations, logical implications, and certain classical inference mechanisms such as tautologies. The concept of a proposition is introduced as are associated concepts of truth sets, tautologies, and contradictions. The operations of disjunction, conjunction, and negation are introduced as well as classical implication and equivalence; all of these are useful tools to construct compound propositions from single propositions. Operations on propositions are shown to be isomorphic with operations on sets, hence an algebra of propositions is developed by using the algebra of sets discussed in Chapter 2. Fuzzy logic is then shown to be an extension of classical logic when partial truths are included to extend bi-valued logic (true or false) to a multivalued logic (degrees of truth between true and not-true).

CLASSICAL PREDICATE LOGIC

In classical predicate logic, a simple proposition P is a linguistic, or declarative, statement contained within a universe of elements, say X, that can be identified as being a collection of elements in X that are strictly true or strictly false. Hence, a proposition P is a collection of elements, that is, a set, where the truth values for all elements in the set are either all true or all false. The veracity (truth) of an element in the proposition P can be assigned a binary truth value, called $T(\mathrm{P})$, just as an element in a universe is assigned a binary quantity to measure its membership in a particular set. For binary (Boolean) predicate logic, $T(\mathrm{P})$ is assigned a value of 1 (truth) or 0 (false). If U is the universe of all propositions, then T is a mapping of the elements, u, in these propositions (sets) to the binary quantities (0, 1), or

$$T : u \in \mathrm{U} \rightarrow [0, 1]$$

All elements u in the universe U that are true for proposition P are called the truth set of P, denoted T(P). Those elements u in the universe U that are false for proposition P are called the falsity set of P.

In logic we need to postulate the boundary conditions of truth values just as we do for sets; that is, in function-theoretic terms we need to define the truth value of a universe of discourse. For a universe Y and the null set $\varnothing$, we define the following truth values:

$$T(\mathrm{Y}) = 1 \qquad \text{and} \qquad T(\varnothing) = 0$$

Now let P and Q be two simple propositions on the same universe of discourse that can be combined using the following five logical connectives,

(*a*) Disjunction ($\vee$)
(*b*) Conjunction ($\wedge$)
(*c*) Negation ($-$)
(*d*) Implication ($\rightarrow$)
(*e*) Equivalence ($\leftrightarrow$)

to form logical expressions involving the two simple propositions. These connectives can be used to form new propositions from simple propositions.

The disjunction connective, the logical *or,* is the term used to represent what is commonly referred to as the *inclusive or.* The natural language term *or* and the logical *or* differ in that the former implies exclusion (denoted in the literature as the *exclusive or;* further details are given in this chapter). For example, "soup or salad" on a restaurant menu implies the choice of one or the other option, but not both. The *inclusive or* is the one most often employed in logic; the inclusive or (*logical or* as used here) implies that a compound proposition is true if either of the simple propositions is true or both are true.

The equivalence connective arises from dual implication, that is, for some propositions P and Q, if $P \rightarrow Q$ and $Q \rightarrow P$, then $P \leftrightarrow Q$.

Now define sets A and B from universe X (universe X is isomorphic with universe U), where these sets might represent linguistic ideas or thoughts. A *propositional calculus* (sometimes called the *algebra of propositions*) will exist for the case where proposition P measures the truth of the statement that an element, x, from the universe X is contained in set A and the truth of the statement Q that this element, x, is contained in set B, or more conventionally,

$$\begin{aligned} &\mathrm{P} : \text{truth that } x \in \mathrm{A} \\ &\mathrm{Q} : \text{truth that } x \in \mathrm{B} \end{aligned}$$

where truth is measured in terms of the truth value, i.e.,

$$\begin{aligned} &\text{If } x \in \mathrm{A}, T(\mathrm{P}) = 1; \text{ otherwise, } T(\mathrm{P}) = 0 \\ &\text{If } x \in \mathrm{B}, T(\mathrm{Q}) = 1; \text{ otherwise, } T(\mathrm{Q}) = 0 \end{aligned}$$

or, using the characteristic function to represent truth (1) and falsity (0), the following notation results:

$$\chi_{\mathrm{A}}(x) = \begin{cases} 1, & x \in \mathrm{A} \\ 0, & x \notin \mathrm{A} \end{cases}$$

A notion of *mutual exclusivity* arises in this calculus. For the situation involving two propositions P and Q, where $T(\mathrm{P}) \cap T(\mathrm{Q}) = \varnothing$, we have that the truth of P always implies the falsity of Q and vice versa; hence, P and Q are mutually exclusive propositions.

Example 7.1. Let P be the proposition "The structural beam is an 18WF45" and let Q be the proposition "The structural beam is made of steel." Let X be the universe of structural members comprised of girders, beams, and columns; x is an element (beam), A is the set of all wide-flange (WF) beams, and B is the set of all steel beams. Hence,

$$\begin{aligned} &\mathrm{P} : x \text{ is in A} \\ &\mathrm{Q} : x \text{ is in B} \end{aligned}$$

The five logical connectives already defined can be used to create compound propositions, where a compound proposition is defined as a logical proposition formed by logically connecting two or more simple propositions. Just as we are interested in the truth of a simple proposition, predicate logic also involves the assessment of the truth of compound propositions. For the case of two simple propositions, the resulting compound propositions are defined next in terms of their binary truth values,

Given a proposition $\mathrm{P} : x \in \mathrm{A}, \overline{\mathrm{P}} : x \notin \mathrm{A}$, we have the following for the logical connectives:

Disjunction

$$\begin{aligned} \mathrm{P} \vee \mathrm{Q} &: x \in \mathrm{A} \text{ or } x \in \mathrm{B} \\ \text{Hence, } T(\mathrm{P} \vee \mathrm{Q}) &= \max(T(\mathrm{P}), T(\mathrm{Q})) \end{aligned} \tag{7.1a}$$

Conjunction

$$\begin{aligned} \mathrm{P} \wedge \mathrm{Q} &: x \in \mathrm{A} \text{ and } x \in \mathrm{B} \\ \text{Hence, } T(\mathrm{P} \wedge \mathrm{Q}) &= \min(T(\mathrm{P}), T(\mathrm{Q})) \end{aligned} \tag{7.1b}$$

Negation

$$\text{If } T(\mathrm{P}) = 1, \text{ then } T(\overline{\mathrm{P}}) = 0; \text{ if } T(\mathrm{P}) = 0, \text{ then } T(\overline{\mathrm{P}}) = 1. \tag{7.1c}$$

Implication

$$\begin{aligned} (\mathrm{P} \rightarrow \mathrm{Q}) &: x \notin \mathrm{A} \text{ or } x \in \mathrm{B} \\ \text{Hence, } T(\mathrm{P} \rightarrow \mathrm{Q}) &= T(\overline{\mathrm{P}} \cup \mathrm{Q}) \end{aligned} \tag{7.1d}$$

Equivalence

$$(\mathrm{P} \leftrightarrow \mathrm{Q}) : T(\mathrm{P} \leftrightarrow \mathrm{Q}) = \begin{cases} 1, & \text{for } T(\mathrm{P}) = T(\mathrm{Q}) \\ 0, & \text{for } T(\mathrm{P}) \neq T(\mathrm{Q}) \end{cases} \tag{7.1e}$$

The logical connective *implication,* i.e., $\mathrm{P} \rightarrow \mathrm{Q}$ (P implies Q) presented here is also known as the classical implication, to distinguish it from an alternative form devised in the 1930s by Lukasiewicz, a Polish mathematician, who was first credited with exploring logics other than Aristotelian (classical or binary logic) [Rescher, 1969], and from several other forms (see end of this chapter). In this implication the proposition P is also referred to as the *hypothesis* or the *antecedent,* and the proposition Q is also referred to as the *conclusion* or the *consequent.* The compound proposition $\mathrm{P} \rightarrow \mathrm{Q}$ is true in all cases except where a true antecedent P appears with a false consequent, Q; i.e., a true hypothesis cannot imply a false conclusion.

Example 7.2 [Similar to Gill, 1976]. Consider the following four propositions:

1. If $1 + 1 = 2$, then $4 > 0$.
2. If $1 + 1 = 3$, then $4 > 0$.
3. If $1 + 1 = 3$, then $4 < 0$.
4. If $1 + 1 = 2$, then $4 < 0$.

The first three propositions are all true; the fourth is false. In the first two, the conclusion $4 > 0$ is true regardless of the truth of the hypothesis; in the third case both propositions are false, but this does not disprove the implication; finally, in the fourth case, a true hypothesis cannot produce a false conclusion.

Hence, the classical form of the implication is true for all propositions of P and Q except for those propositions that are in both the truth set of P and the false set of Q, i.e.,

$$\mathrm{T(P \rightarrow Q)} = \overline{\mathrm{T(P)} \cap \mathrm{T(\overline{Q})}} \tag{7.2}$$

This classical form of the implication operation requires some explanation. For a proposition P defined on set A and a proposition Q defined on set B, the implication "P implies Q" is equivalent to taking the union of elements in the complement of set A with the elements in the set B [this result can also be derived by using De Morgan's laws on Eq. (7.2)]. That is, the logical implication is analogous to the set-theoretic form,

$$(\mathrm{P} \rightarrow \mathrm{Q}) \equiv (\overline{A} \cup \mathrm{B} \text{ is true}) \equiv (\text{either "not in A" or "in B"})$$

so that

$$T(\mathrm{P} \rightarrow \mathrm{Q}) = T(\overline{\mathrm{P}} \vee \mathrm{Q}) = \max(T(\overline{\mathrm{P}}), T(\mathrm{Q})) \tag{7.3}$$

This expression is linguistically equivalent to the statement, "P → Q is true" when either "not A" or "B" is true (logical or). Graphically, this implication and the analogous set operation are represented by the Venn diagram in Fig. 7.1. As noted in the diagram, the region represented by the difference A | B is the set region where the implication P → Q is false (the implication "fails"). The shaded region in Fig. 7.1 represents the collection of elements in the universe where the implication is true, that is, the set

$$\overline{A \mid B} = \overline{A} \cup \overline{\overline{B}} = \overline{A \cap \overline{B}}$$

If x is in A *and* x is not in B, then

$$\mathrm{A} \rightarrow \mathrm{B} \text{ fails} \equiv \mathrm{A} \mid \mathrm{B} \text{ (difference)}$$

Now, with two propositions (P and Q) each being able to take on one of two truth values (true or false, 1 or 0), there will be a total of $2^2 = 4$ propositional situations. These situations are illustrated, along with the appropriate truth values, for the propositions P and Q and the various logical connectives between them in Table 7.1. The values in the last five columns of the table are calculated using

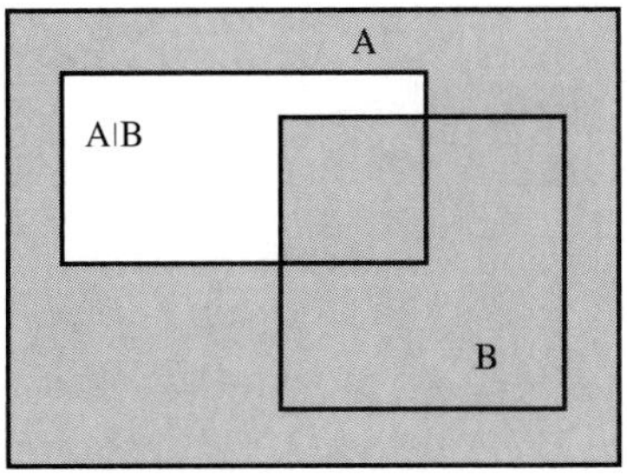

FIGURE 7.1
Graphical analog of the classical implication operation; gray area is where implication holds.

the expressions in Eqs. (7.1) and (7.3). In Table 7.1 T (or 1) denotes true and F (or 0) denotes false.

Suppose the implication operation involves two different universes of discourse; P is a proposition described by set A, which is defined on universe X, and Q is a proposition described by set B, which is defined on universe Y. Then the implication $P \rightarrow Q$ can be represented in set-theoretic terms by the relation R, where R is defined by

$$\begin{array}{ll} R = (A \times B) \cup (\overline{A} \times Y) \equiv \text{IF A, THEN B} & \\ \text{IF } x \in A & \text{where } x \in X \text{ and } A \subset X \\ \text{THEN } y \in B & \text{where } y \in Y \text{ and } B \subset Y \end{array} \tag{7.4}$$

This implication, Eq. (7.4), is also equivalent to the linguistic rule form, IF A, THEN B. The graphic shown in Fig. 7.2 represents the space of the Cartesian product $X \times Y$, showing typical sets A and B; and superposed on this space is the set-theoretic equivalent of the implication. That is,

$$P \rightarrow Q: \text{ If } x \in A, \text{ Then } y \in B, \qquad \text{or} \qquad P \rightarrow Q \equiv \overline{A} \cup B$$

The shaded regions of the compound Venn diagram in Fig. 7.2 represent the truth domain of the implication, IF A, THEN B ($P \rightarrow Q$).

Another compound proposition in linguistic rule form is the expression

IF A, THEN B, ELSE C.

Linguistically, this compound proposition could be expressed as

IF A, THEN B, or IF $\overline{\text{A}}$, THEN C.

In predicate logic this rule has the form

$$(P \rightarrow Q) \vee (\overline{P} \rightarrow S) \tag{7.5}$$

where $P: x \in A, A \subset X$
$Q: y \in B, B \subset Y$
$S: y \in C, C \subset Y$

The set-theoretic equivalent of this compound proposition is given by

$$\text{IF A, THEN B, ELSE C} \equiv (A \times B) \cup (\overline{A} \times C) = R = \text{relation on } X \times Y \tag{7.6}$$

TABLE 7.1
Truth table for various compound propositions

P	Q	$\overline{P}$	$P \vee Q$	$P \wedge Q$	$P \rightarrow Q$	$P \leftrightarrow Q$
T (1)	T (1)	F (0)	T (1)	T (1)	T (1)	T (1)
T (1)	F (0)	F (0)	T (1)	F (0)	F (0)	F (0)
F (0)	T (1)	T (1)	T (1)	F (0)	T (1)	F (0)
F (0)	F (0)	T (1)	F (0)	F (0)	T (1)	T (1)

The graphic in Fig. 7.3 illustrates the shaded region representing the truth domain for this compound proposition for the particular case where $B \cap C = \varnothing$.

Tautologies

In predicate logic it is useful to consider compound propositions that are always true, irrespective of the truth values of the individual simple propositions. Classical logical compound propositions with this property are called *tautologies*. Tautologies are useful for deductive reasoning, for proving theorems, and for making deductive inferences. So, if a compound proposition can be expressed in the form of a tautology, the truth value of that compound proposition is known to be true. Inference schemes in expert systems often employ tautologies because tautologies are formulas that are true on logical grounds alone. For example, if A is the set of all prime numbers ($A_1 = 1, A_2 = 2, A_3 = 3, A_4 = 5, \ldots$) on the real line universe, X, then the proposition "A_i is not divisible by 6" is a tautology.

One tautology, known as *modus ponens* deduction, is a very common inference scheme used in forward-chaining rule-based expert systems. It is an operation whose task is to find the truth value of a consequent in a production rule, given the truth value of the antecedent in the rule. *Modus ponens* deduction concludes that, given two propositions, P and $P \rightarrow Q$, both of which are true, then the truth of the simple proposition Q is automatically inferred. Another useful tautology is the *modus tollens* inference, which is used in backward-chaining expert systems. In *modus tollens* an implication between two propositions is combined with a second proposition and both are used to imply a third proposition. Some common tautologies follow.

$$\overline{B} \cup B \leftrightarrow X$$

$$A \cup X; \qquad \overline{A} \cup X \leftrightarrow X$$

$$A \leftrightarrow B$$

$$(A \wedge (A \rightarrow B)) \rightarrow B \qquad (\textit{modus ponens}) \tag{7.7}$$

$$(\overline{B} \wedge (A \rightarrow B)) \rightarrow \overline{A} \qquad (\textit{modus tollens}) \tag{7.8}$$

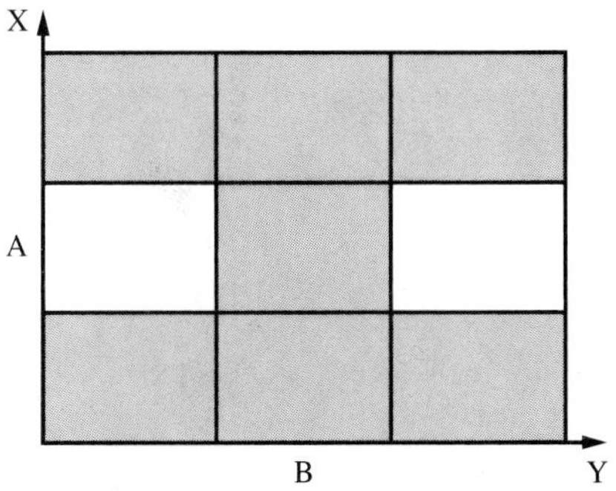

FIGURE 7.2
The Cartesian space showing the implication IF A, THEN B.

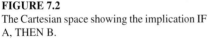

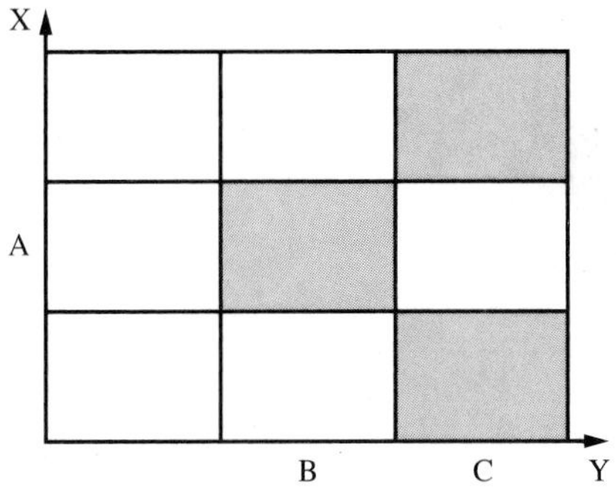

FIGURE 7.3
Truth domain for IF A, THEN B, ELSE C.

A simple proof of the truth value of the *modus ponens* deduction is provided here, along with the various properties for each step of the proof, for purposes of illustrating the utility of a tautology in classical reasoning.

Proof

$$
\begin{aligned}
(A \wedge (A \rightarrow B)) &\rightarrow B && \\
(A \wedge (\overline{A} \cup B)) &\rightarrow B && \text{Implication} \\
((A \wedge \overline{A}) \cup (A \wedge B)) &\rightarrow B && \text{Distributivity} \\
(\varnothing \cup (A \wedge B)) &\rightarrow B && \text{Excluded middle laws} \\
(A \wedge B) &\rightarrow B && \text{Identity} \\
\overline{(A \wedge B)} &\cup B && \text{Implication} \\
(\overline{A} \vee \overline{B}) &\cup B && \text{De Morgan's laws} \\
\overline{A} \vee (\overline{B} &\cup B) && \text{Associativity} \\
\overline{A} &\cup X && \text{Excluded middle laws} \\
X \Rightarrow T(X) &= 1 && \text{Identity; } QED
\end{aligned}
$$

A simpler manifestation of the truth value of this tautology is shown in Table 7.2 in truth table form, where a column of all ones for the result shows a tautology.

Similarly, a simple proof of the truth value of the *modus tollens* inference is listed here.

Proof

$$
\begin{aligned}
(\overline{B} \wedge (A \rightarrow B)) &\rightarrow \overline{A} \\
(\overline{B} \wedge (\overline{A} \cup B)) &\rightarrow \overline{A} \\
((\overline{B} \wedge \overline{A}) \cup (\overline{B} \wedge B)) &\rightarrow \overline{A} \\
((\overline{B} \wedge \overline{A}) \cup \varnothing) &\rightarrow \overline{A} \\
(\overline{B} \wedge \overline{A}) &\rightarrow \overline{A} \\
\overline{(\overline{B} \wedge \overline{A})} &\cup \overline{A} \\
(\overline{\overline{B}} \vee \overline{\overline{A}}) &\cup \overline{A} \\
B \cup (A &\cup \overline{A}) \\
B \cup X = X \Rightarrow T(X) &= 1 \qquad QED
\end{aligned}
$$

The truth table form of this result is shown in Table 7.3.

TABLE 7.2
Truth table (*modus ponens*)

A	B	A → B	(A ∧ (A → B))	(A ∧ (A → B)) → B	
0	0	1	0	1	
0	1	1	0	1	Tautology
1	0	0	0	1	
1	1	1	1	1	

TABLE 7.3
Truth table (*modus tollens*)

A	B	$\overline{A}$	$\overline{B}$	$A \rightarrow B$	$(\overline{B} \wedge (A \rightarrow B))$	$(\overline{B} \wedge (A \rightarrow B)) \rightarrow \overline{A}$	
0	0	1	1	1	1	1	
0	1	1	0	1	0	1	Tautology
1	0	0	1	0	0	1	
1	1	0	0	1	0	1	

Contradictions

Compound propositions that are always false, regardless of the truth value of the individual simple propositions constituting the compound proposition, are called contradictions. For example, if A is the set of all prime numbers ($A_1 = 1, A_2 = 2, A_3 = 3, A_4 = 5, \ldots$) on the real line universe, X, then the proposition "A_i is a multiple of 2" is a contradiction. Some simple contradictions are listed here:

$$\overline{B} \cap B$$
$$A \cap \varnothing; \qquad \overline{A} \cap \varnothing$$

Equivalence

As mentioned, propositions P and Q are equivalent, i.e., $P \leftrightarrow Q$, is true only when both P and Q are true or when both P and Q are false. For example, the propositions P: "triangle is equilateral" and Q: "triangle is equiangular" are equivalent because they are either both true or both false for some triangle. This condition of equivalence is shown in Fig. 7.4, where the shaded region is the region of equivalence.

It can be easily proved that the statement $P \leftrightarrow Q$ is a tautology if P is identical to Q, i.e., if and only if $T(P) = T(Q)$.

Example 7.3. Suppose we consider the universe of positive integers, $X = \{1 \leq n \leq 8\}$. Let P = "n is an even number" and let Q = "$(3 \leq n \leq 7) \wedge (n \neq 6)$." Then $T(P) = \{2, 4, 6, 8\}$ and $T(Q) = \{3, 4, 5, 7\}$. The equivalence $P \leftrightarrow Q$ has the truth set

$$T(P \leftrightarrow Q) = (T(P) \cap T(Q)) \cup (\overline{T(P)} \cap \overline{T(Q)}) = \{4\} \cup \{1\} = \{1, 4\}$$

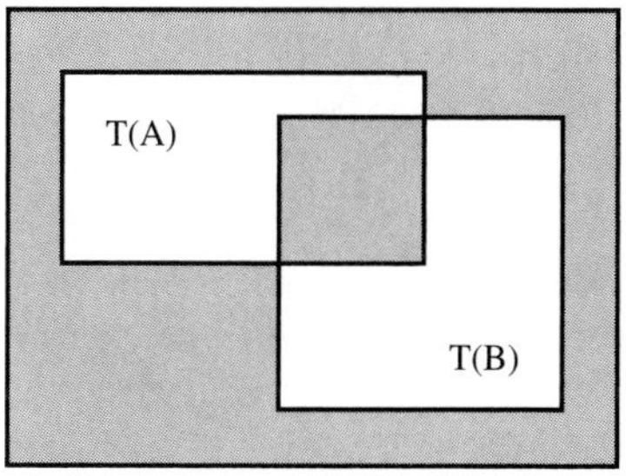

FIGURE 7.4
Venn diagram for equivalence (darkened areas), i.e., for $T(A \leftrightarrow B)$.

One can see that "1 is an even number" and "$(3 \le 1 < 7) \wedge (1 \ne 6)$" are both false, and "4 is an even number" and "$(3 \le 4 < 7) \wedge (4 \ne 6)$" are both true.

Example 7.4. Prove that $P \leftrightarrow Q$ if P = "n is an integer power of 2 less than 7 and greater than zero" and Q = "$n^2 - 6n + 8 = 0$." Since T(P) = {2, 4} and T(Q) = {2, 4}, it follows that $P \leftrightarrow Q$ is an equivalence.

Suppose a proposition R has the form $P \rightarrow Q$. Then the proposition $\overline{Q} \rightarrow \overline{P}$ is called the *contrapositive* of R; the proposition $Q \rightarrow P$ is called the *converse* of R; and the proposition $\overline{P} \rightarrow \overline{Q}$ is called the *inverse* of R. Interesting properties of these propositions can be shown (see Problem 7.3 at the end of this chapter).

The *dual* of a compound proposition that does not involve implication is the same proposition with false (0) replacing true (1), true replacing false, conjunction ($\wedge$) replacing disjunction ($\vee$), and disjunction replacing conjunction. If a proposition is true, then its *dual* is also true (see Problems 7.4 and 7.5).

Exclusive Or and Exclusive Nor

Two more interesting compound propositions are worthy of discussion. These are the *exclusive or* and the *exclusive nor.* The exclusive or is of interest because it arises in many situations involving natural language and human reasoning. For example, when you are going to travel by plane or boat to some destination, the implication is that you can travel by air or sea, but not both; i.e., one or the other. This situation involves the exclusive or; it does not involve the intersection, as does the logical or [union in Eq. (2.1) and Fig. 2.2 and disjunction in Eq. (7.1*a*)]. For two propositions, P and Q, the exclusive or, denoted here as "$\vee$", is given in Table 7.4 and Fig. 7.5.

The *exclusive nor* is the complement of the *exclusive or* [Mano, 1988]. A look at its truth table, Table 7.5, shows that it is an equivalence operation; i.e.,

$$\overline{P \text{``}\vee\text{''} Q} \leftrightarrow (P \leftrightarrow Q)$$

and, hence, it is graphically equivalent to the Venn diagram in Fig. 7.4.

Logical Proofs

Logic involves the use of inference in everyday life, as well as in mathematics. In the latter, we often want to prove theorems to form foundations for solution procedures. In natural language, if we are given some hypotheses it is often useful to make certain conclusions from them—the so-called process of inference (inferring new facts from established facts). In the terminology we have been using, we want to know if the proposition, $(P_1 \wedge P_2 \wedge \cdots \wedge P_n) \rightarrow Q$, is true. That is, is the statement a tautology?

The process works as follows. First, the linguistic statement (compound proposition) is made. Second, the statement is decomposed into its respective single propositions. Third, the statement is expressed algebraically with all pertinent logical connectives in place. Fourth, a truth table is used to establish the veracity of the statement.

TABLE 7.4
Truth table for exclusive or, "$\bigvee$"

P	Q	P "$\bigvee$" Q
1	1	0
1	0	1
0	1	1
0	0	0

TABLE 7.5
Truth table for exclusive nor

P	Q	$\overline{\text{P "}\bigvee\text{" Q}}$
1	1	1
1	0	0
0	1	0
0	0	1

Example 7.5.

Hypotheses: Engineers are mathematicians. Logical thinkers do not believe in magic. Mathematicians are logical thinkers.

Conclusion: Engineers do not believe in magic.

Let us decompose this information into individual propositions.

$$\begin{aligned} &\text{P : a person is an engineer}\\ &\text{Q : a person is a mathematician}\\ &\text{R : a person is a logical thinker}\\ &\text{S : a person believes in magic}\end{aligned}$$

The statements can now be expressed as algebraic propositions as

$$((P \rightarrow Q) \wedge (R \rightarrow \overline{S}) \wedge (Q \rightarrow R)) \rightarrow (P \rightarrow \overline{S})$$

It can be shown that this compound proposition is a tautology (see Problem 7.6).

Sometimes it might be difficult to prove a proposition by a direct proof (i.e., verify that it is true), so an alternative is to use an indirect proof. For example, the popular *proof by contradiction* (reductio ad absurdum) exploits the fact that $P \rightarrow Q$ is true if and only if $P \wedge \overline{Q}$ is false. Hence, if we want to prove that the compound statement, $(P_1 \wedge P_2 \wedge \cdots \wedge P_n) \rightarrow Q$ is a tautology, we can alternatively show that the alternative statement, $P_1 \wedge P_2 \wedge \cdots \wedge P_n \wedge \overline{Q}$ is a contradiction.

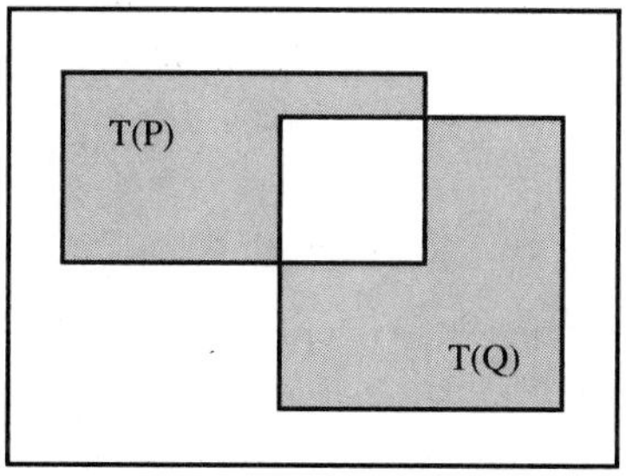

FIGURE 7.5
Exclusive or shown in gray areas.

Example 7.6.

Hypotheses: If an arch-dam fails, the failure is due to a poor subgrade. An arch-dam fails.

Conclusion: The arch-dam failed because of a poor subgrade.

This statement can be shown to be algebraically equivalent to the expression,

$$((P \rightarrow Q) \wedge P) \rightarrow Q$$

To prove this by contradiction, we need to show that the algebraic expression,

$$((P \rightarrow Q) \wedge P \wedge \overline{Q})$$

is a contradiction. We can do this by constructing the truth table in Table 7.6. Recall that a contradiction is indicated when the last column of a truth table is filled with zeros.

Deductive Inferences

The *modus ponens* deduction is used as a tool for making inferences in rule-based systems. A typical if-then rule is used to determine whether an antecedent (cause or action) infers a consequent (effect or reaction). Suppose we have a rule of the form IF A, THEN B, where A is a set defined on universe X and B is a set defined on universe Y. As discussed before, this rule can be translated into a relation between sets A and B, that is, recalling Eq. (7.4), $R = (A \times B) \cup (\overline{A} \times Y)$. Now suppose a new antecedent, say A′, is known. Can we use *modus ponens* deduction, Eq. (7.7), to infer a new consequent, say B′, resulting from the new antecedent? That is, can we deduce, in rule form IF A′, THEN B′ ? The answer, of course, is yes, through the use of the composition operation (defined initially in Chapter 3). Since "A implies B" is defined on the Cartesian space $X \times Y$, B′ can be found through the following set-theoretic formulation, again from Eq. (7.4):

$$B' = A' \circ R = A' \circ ((A \times B) \cup (\overline{A} \times Y))$$

where the symbol ∘ denotes the composition operation. *Modus ponens* deduction can also be used for the compound rule, IF A, THEN B, ELSE C, where this compound rule is equivalent to the relation defined in Eq. (7.8) as, $R = (A \times B) \cup (\overline{A} \times C)$. For this compound rule, if we define another antecedent A′, the following possibilities exist, depending on (*i*) whether A′ is fully contained in the original antecedent A,

TABLE 7.6
Truth table for dam failure problem

P	Q	$\overline{P}$	$\overline{Q}$	$\overline{P} \vee Q$	$(\overline{P} \vee Q) \wedge P \wedge \overline{Q}$
0	0	1	1	1	0
0	1	1	0	1	0
1	0	0	1	0	0
1	1	0	0	1	0

(*ii*) whether A′ is contained only in the complement of A, or (*iii*) whether A′ and A overlap to some extent as described next:

$$\text{IF A}' \subset \text{A, THEN } y = \text{B}$$
$$\text{IF A}' \subset \overline{\text{A}}\text{, THEN } y = \text{C}$$
$$\text{IF A}' \cap \text{A} \neq \varnothing,\ \text{A}' \cap \overline{\text{A}} \neq \varnothing,\ \text{THEN } y = \text{B} \cup \text{C}$$

The rule IF A, THEN B (proposition P is defined on set A in universe X, and proposition Q is defined on set B in universe Y), i.e., $(\text{P} \to \text{Q}) = \text{R} = (\text{A} \times \text{B} \cup (\overline{\text{A}} \times \text{Y}))$, is then defined in function-theoretic terms as

$$\chi_{\text{R}}(x, y) = \max\left[(\chi_{\text{A}}(x) \wedge \chi_{\text{B}}(y)), ((1 - \chi_{\text{A}}(x)) \wedge 1)\right] \tag{7.9}$$

where $\chi(\,)$ is the characteristic function as defined before.

Example 7.7. Suppose we have two universes of discourse for a heat exchanger problem described by the following collection of elements, $\text{X} = \{1, 2, 3, 4\}$ and $\text{Y} = \{1, 2, 3, 4, 5, 6\}$. Suppose X is a universe of normalized temperatures and Y is a universe of normalized pressures. Define crisp set A on universe X and crisp set B on universe Y as follows: $\text{A} = \{2, 3\}$ and $\text{B} = \{3, 4\}$. The deductive inference IF A, THEN B (i.e., IF temperature is A, THEN pressure is B) will yield a matrix describing the membership values of the relation R, i.e., $\chi_{\text{R}}(x, y)$ through the use of Eq. (7.9). That is, the matrix R represents the rule IF A, THEN B as a matrix of characteristic (crisp membership) values.

Crisp sets A and B can be written using Zadeh's notation,

$$\text{A} = \left\{\frac{0}{1} + \frac{1}{2} + \frac{1}{3} + \frac{0}{4}\right\}$$

$$\text{B} = \left\{\frac{0}{1} + \frac{0}{2} + \frac{1}{3} + \frac{1}{4} + \frac{0}{5} + \frac{0}{6}\right\}$$

If we treat set A as a column vector and set B as a row vector, the following matrix results from the Cartesian product of A × B, using Eq. (3.16):

$$\text{A} \times \text{B} = \begin{bmatrix} 0 & 0 & 0 & 0 & 0 & 0 \\ 0 & 0 & 1 & 1 & 0 & 0 \\ 0 & 0 & 1 & 1 & 0 & 0 \\ 0 & 0 & 0 & 0 & 0 & 0 \end{bmatrix}$$

The Cartesian product $\overline{\text{A}} \times \text{Y}$ can be determined using Eq. (3.16) by arranging $\overline{\text{A}}$ as a column vector and the universe Y as a row vector (sets $\overline{\text{A}}$ and Y can be written using Zadeh's notation),

$$\overline{\text{A}} = \left\{\frac{1}{1} + \frac{0}{2} + \frac{0}{3} + \frac{1}{4}\right\}$$

$$\text{Y} = \left\{\frac{1}{1} + \frac{1}{2} + \frac{1}{3} + \frac{1}{4} + \frac{1}{5} + \frac{1}{6}\right\}$$

$$\overline{\text{A}} \times \text{Y} = \begin{bmatrix} 1 & 1 & 1 & 1 & 1 & 1 \\ 0 & 0 & 0 & 0 & 0 & 0 \\ 0 & 0 & 0 & 0 & 0 & 0 \\ 1 & 1 & 1 & 1 & 1 & 1 \end{bmatrix}$$

Then the full relation R describing the implication IF A, THEN B is the maximum of the two matrices A × B and $\overline{A}$ × Y, or, using Eq. (7.9),

$$R = \begin{array}{c} \\ 1 \\ 2 \\ 3 \\ 4 \end{array} \begin{array}{c} \begin{array}{cccccc} 1 & 2 & 3 & 4 & 5 & 6 \end{array} \\ \begin{bmatrix} 1 & 1 & 1 & 1 & 1 & 1 \\ 0 & 0 & 1 & 1 & 0 & 0 \\ 0 & 0 & 1 & 1 & 0 & 0 \\ 1 & 1 & 1 & 1 & 1 & 1 \end{bmatrix} \end{array}$$

The compound rule IF A, THEN B, ELSE C can also be defined in terms of a matrix relation as, $R = (A \times B) \cup (\overline{A} \times C) \Rightarrow (P \to Q) \vee (\overline{P} \to S)$, as given by Eqs. (7.5) and (7.6), where the membership function is determined as

$$\chi_R(x, y) = \max [(\chi_A(x) \wedge \chi_B(y)), ((1 - \chi_A(x)) \wedge \chi_C(y))] \qquad (7.10)$$

Example 7.8. Continuing with the previous heat exchanger example, suppose we define a crisp set C on the universe of normalized temperatures Y as C = {5, 6}, or using Zadeh's notation,

$$C = \left\{ \frac{0}{1} + \frac{0}{2} + \frac{0}{3} + \frac{0}{4} + \frac{1}{5} + \frac{1}{6} \right\}$$

The deductive inference IF A, THEN B, ELSE C (i.e., IF pressure is A, THEN temperature is B, ELSE temperature is C) will yield a relational matrix R, with characteristic values $\chi_R(x, y)$ obtained using Eq. (7.10). The first half of the expression in Eq. (7.10) (i.e., A × B) has already been determined in the previous example. The Cartesian product $\overline{A}$ × C can be determined using Eq. (3.16) by arranging the set $\overline{A}$ as a column vector and the set C as a row vector (see set $\overline{A}$ in Example 7.7), or

$$\overline{A} \times C = \begin{bmatrix} 0 & 0 & 0 & 0 & 1 & 1 \\ 0 & 0 & 0 & 0 & 0 & 0 \\ 0 & 0 & 0 & 0 & 0 & 0 \\ 0 & 0 & 0 & 0 & 1 & 1 \end{bmatrix}$$

Then the full relation R describing the implication IF A, THEN B, ELSE C is the maximum of the two matrices A × B and $\overline{A}$ × C [see Eq. (7.10)],

$$R = \begin{array}{c} \\ 1 \\ 2 \\ 3 \\ 4 \end{array} \begin{array}{c} \begin{array}{cccccc} 1 & 2 & 3 & 4 & 5 & 6 \end{array} \\ \begin{bmatrix} 0 & 0 & 0 & 0 & 1 & 1 \\ 0 & 0 & 1 & 1 & 0 & 0 \\ 0 & 0 & 1 & 1 & 0 & 0 \\ 0 & 0 & 0 & 0 & 1 & 1 \end{bmatrix} \end{array}$$

FUZZY LOGIC

The restriction of classical propositional calculus to a two-valued logic has created many interesting paradoxes over the ages. For example, the Barber of Seville is a classic paradox (also termed Russell's barber). In the small Spanish town of Seville, there is a rule that all and only those men who do not shave themselves are shaved by the barber. Who shaves the barber? Another example comes from ancient Greece. Does the liar from Crete lie when he claims, "All Cretians are liars"? If he is telling the truth, his statement is false. But if his statement is false, he is not telling the truth. A simpler form of this paradox is the two-word proposition, "I lie." The statement can't be both true and false.

Returning to the Barber of Seville, we conclude that the only way for this paradox (or any classic paradox for that matter) to work is if the statement is both true and false simultaneously. This can be shown, using set notation [Kosko, 1992]. Let S be the proposition that the barber shaves himself and $\overline{\mathrm{S}}$ (not-S) that he does not. Then since $\mathrm{S} \to \overline{\mathrm{S}}$ (S implies not-S), and $\overline{\mathrm{S}} \to \mathrm{S}$, the two propositions are logically equivalent: $\mathrm{S} \leftrightarrow \overline{\mathrm{S}}$. Equivalent propositions have the same truth value, hence,

$$T(\mathrm{S}) = T(\overline{\mathrm{S}}) = 1 - T(\mathrm{S})$$

which yields the expression,

$$T(\mathrm{S}) = \tfrac{1}{2}$$

As seen, paradoxes reduce to half-truths (or half-falsities) mathematically. In classical binary (bivalued) logic, however, such conditions are not allowed; i.e., only $T(\mathrm{S}) = 1$ or 0 is valid.

A more subtle form of paradox can also be addressed by a multivalued logic. Consider the paradoxes represented by the classical *sorites* (literally, a heap of syllogisms) (see also Problem 7.24), for example, the case of a liter-full glass of water. Often this example is called the optimist's conclusion (is the glass half-full or half-empty when the volume is at 500 milliliters?). Is the liter-full glass still full if we remove one milliliter of water? Is the glass still full if we remove two milliliters of water, three, four, or 100 milliliters? If we continue to answer yes, then eventually we will have removed all the water, and an empty glass will still be characterized as full! At what point did the liter-full glass of water become empty? Perhaps at 500 milliliters full? Unfortunately no single milliliter of liquid provides for a transition between full and empty. This transition is gradual, so that as each milliliter of water is removed, the truth value of the glass being full gradually diminishes from a value of 1 at 1000 milliliters to 0 at 0 milliliters. Hence, for many problems we have need for a multivalued logic other than the classic binary logic that is so prevalent today.

A fuzzy logic proposition, $\underset{\sim}{\mathrm{P}}$, is a statement involving some concept without clearly defined boundaries. Linguistic statements that tend to express subjective ideas and that can be interpreted slightly differently by various individuals typically involve fuzzy propositions. Most natural language is fuzzy, in that it involves vague and imprecise terms. Statements describing a person's height or weight or assessments of people's preferences about colors or menus can be used as examples of fuzzy propositions. The truth value assigned to $\underset{\sim}{\mathrm{P}}$ can be any value on the interval [0, 1]. The assignment of the truth value to a proposition is actually a mapping from the interval [0, 1] to the universe U of truth values, T, as indicated in Eq. (7.11),

$$T : u \in \mathrm{U} \to \{0, 1\} \tag{7.11}$$

As in classical binary logic, we assign a logical proposition to a set in the universe of discourse. Fuzzy propositions are assigned to fuzzy sets. Suppose proposition $\underset{\sim}{\mathrm{P}}$ is assigned to fuzzy set $\underset{\sim}{\mathrm{A}}$; then the truth value of a proposition, denoted $T(\underset{\sim}{\mathrm{P}})$, is given by

$$T(\underset{\sim}{P}) = \mu_{\underset{\sim}{A}}(x) \qquad \text{where } 0 \leq \mu_{\underset{\sim}{A}} \leq 1 \tag{7.12}$$

Equation (7.12) indicates that the degree of truth for the proposition $\underset{\sim}{P} : x \in \underset{\sim}{A}$ is equal to the membership grade of x in the fuzzy set $\underset{\sim}{A}$.

The logical connectives of negation, disjunction, conjunction, and implication are also defined for a fuzzy logic. These connectives are given in Eqs. (7.13)–(7.16) for two simple propositions: proposition $\underset{\sim}{P}$ defined on fuzzy set $\underset{\sim}{A}$ and proposition $\underset{\sim}{Q}$ defined on fuzzy set $\underset{\sim}{B}$.

Negation

$$T(\overline{\underset{\sim}{P}}) = 1 - T(\underset{\sim}{P}) \tag{7.13}$$

Disjunction

$$\underset{\sim}{P} \vee \underset{\sim}{Q} : x \text{ is } \underset{\sim}{A} \text{ or } \underset{\sim}{B} \qquad T(\underset{\sim}{P} \vee \underset{\sim}{Q}) = \max(T(\underset{\sim}{P}), T(\underset{\sim}{Q})) \tag{7.14}$$

Conjunction

$$\underset{\sim}{P} \wedge \underset{\sim}{Q} : x \text{ is } \underset{\sim}{A} \text{ and } \underset{\sim}{B} \qquad T(\underset{\sim}{P} \wedge \underset{\sim}{Q}) = \min(T(\underset{\sim}{P}), T(\underset{\sim}{Q})) \tag{7.15}$$

Implication [Zadeh, 1973]

$$\begin{aligned} &\underset{\sim}{P} \rightarrow \underset{\sim}{Q} : x \text{ is } \underset{\sim}{A}, \text{ then } x \text{ is } \underset{\sim}{B} \\ &T(\underset{\sim}{P} \rightarrow \underset{\sim}{Q}) = T(\overline{\underset{\sim}{P}} \vee \underset{\sim}{Q}) = \max(T(\overline{\underset{\sim}{P}}), T(\underset{\sim}{Q})) \end{aligned} \tag{7.16}$$

As before in binary logic, the implication connective can be modeled in rule-based form; $\underset{\sim}{P} \rightarrow \underset{\sim}{Q}$ is, IF x is $\underset{\sim}{A}$, THEN y is $\underset{\sim}{B}$ and it is equivalent to the following fuzzy relation, $\underset{\sim}{R} = (\underset{\sim}{A} \times \underset{\sim}{B}) \cup (\overline{\underset{\sim}{A}} \times Y)$ [recall Eq. (7.4)], just as it is in classical logic. The membership function of $\underset{\sim}{R}$ is expressed by the following formula:

$$\mu_{\underset{\sim}{R}}(x, y) = \max\left[(\mu_{\underset{\sim}{A}}(x) \wedge \mu_{\underset{\sim}{B}}(y)), (1 - \mu_{\underset{\sim}{A}}(x))\right] \tag{7.17}$$

Example 7.9. Suppose we are evaluating a new invention to determine its commercial potential. We will use two metrics to make our decisions regarding the innovation of the idea. Our metrics are the "uniqueness" of the invention, denoted by a universe of novelty scales, $X = \{1, 2, 3, 4\}$, and the "market size" of the invention's commercial market, denoted on a universe of scaled market sizes, $Y = \{1, 2, 3, 4, 5, 6\}$. In both universes the lowest numbers are the "highest uniqueness" and the "largest market," respectively. A new invention in your group, say a compressible liquid of very useful temperature and viscosity conditions, has just received scores of "medium uniqueness," denoted by fuzzy set $\underset{\sim}{A}$, and "medium market size," denoted fuzzy set $\underset{\sim}{B}$. We wish to determine the implication of such a result, i.e., IF $\underset{\sim}{A}$, THEN $\underset{\sim}{B}$. We assign the invention the following fuzzy sets to represent its ratings:

$$\underset{\sim}{A} = \text{medium uniqueness} = \left\{\frac{0.6}{2} + \frac{1}{3} + \frac{0.2}{4}\right\}$$

$$\underset{\sim}{B} = \text{medium market size} = \left\{\frac{0.4}{2} + \frac{1}{3} + \frac{0.8}{4} + \frac{0.3}{5}\right\}$$

The following matrices are then determined in developing the membership function of the implication, $\mu_{\underset{\sim}{R}}(x, y)$, illustrated in Eq. (7.17),

$$\underset{\sim}{A} \times \underset{\sim}{B} = \begin{array}{c} \\ 1 \\ 2 \\ 3 \\ 4 \end{array} \begin{array}{c} \begin{array}{cccccc} 1 & 2 & 3 & 4 & 5 & 6 \end{array} \\ \left[\begin{array}{cccccc} 0 & 0 & 0 & 0 & 0 & 0 \\ 0 & 0.4 & 0.6 & 0.6 & 0.3 & 0 \\ 0 & 0.4 & 1 & 0.8 & 0.3 & 0 \\ 0 & 0.2 & 0.2 & 0.2 & 0.2 & 0 \end{array}\right] \end{array}$$

$$\overline{\underset{\sim}{A}} \times Y = \begin{array}{c} \\ 1 \\ 2 \\ 3 \\ 4 \end{array} \begin{array}{c} \begin{array}{cccccc} 1 & 2 & 3 & 4 & 5 & 6 \end{array} \\ \left[\begin{array}{cccccc} 1 & 1 & 1 & 1 & 1 & 1 \\ 0.4 & 0.4 & 0.4 & 0.4 & 0.4 & 0.4 \\ 0 & 0 & 0 & 0 & 0 & 0 \\ 0.8 & 0.8 & 0.8 & 0.8 & 0.8 & 0.8 \end{array}\right] \end{array}$$

and finally, $\underset{\sim}{R} = \max(\underset{\sim}{A} \times \underset{\sim}{B}, \overline{\underset{\sim}{A}} \times Y)$

$$\underset{\sim}{R} = \begin{array}{c} \\ 1 \\ 2 \\ 3 \\ 4 \end{array} \begin{array}{c} \begin{array}{cccccc} 1 & 2 & 3 & 4 & 5 & 6 \end{array} \\ \left[\begin{array}{cccccc} 1 & 1 & 1 & 1 & 1 & 1 \\ 0.4 & 0.4 & 0.6 & 0.6 & 0.4 & 0.4 \\ 0 & 0.4 & 1 & 0.8 & 0.3 & 0 \\ 0.8 & 0.8 & 0.8 & 0.8 & 0.8 & 0.8 \end{array}\right] \end{array}$$

When the logical conditional implication is of the compound form,

$$\text{IF } x \text{ is } \underset{\sim}{A}, \text{ THEN } y \text{ is } \underset{\sim}{B}, \text{ ELSE } y \text{ is } \underset{\sim}{C}$$

then the equivalent fuzzy relation, $\underset{\sim}{R}$, is expressed as, $\underset{\sim}{R} = (\underset{\sim}{A} \times \underset{\sim}{B}) \cup (\overline{\underset{\sim}{A}} \times \underset{\sim}{C})$, in a form just as Eq. (7.5), whose membership function is expressed by the following formula:

$$\mu_{\underset{\sim}{R}}(x, y) = \max\left[(\mu_{\underset{\sim}{A}}(x) \wedge \mu_{\underset{\sim}{B}}(y)), ((1 - \mu_{\underset{\sim}{A}}(x)) \wedge \mu_{\underset{\sim}{C}}(y))\right] \qquad (7.18)$$

APPROXIMATE REASONING

The ultimate goal of fuzzy logic is to form the theoretical foundation for reasoning about imprecise propositions; such reasoning has been referred to as approximate reasoning [Zadeh, 1976, 1979]. Approximate reasoning is analogous to predicate logic for reasoning with precise propositions, and hence is an extension of classical propositional calculus that deals with partial truths.

Suppose we have a rule-based format to represent fuzzy information. These rules are expressed in conventional antecedent-consequent form, such as

Rule 1: IF x is $\underset{\sim}{A}$, THEN y is $\underset{\sim}{B}$, where $\underset{\sim}{A}$ and $\underset{\sim}{B}$ represent fuzzy propositions (sets).

Now suppose we introduce a new antecedent, say $\underset{\sim}{A}'$, and we consider the following rule:

Rule 2: IF x is $\underset{\sim}{A}'$, THEN y is $\underset{\sim}{B}'$

From information derived from Rule 1, is it possible to derive the consequent in Rule 2, $\underset{\sim}{B}'$? The answer is yes, and the procedure is fuzzy composition. The consequent $\underset{\sim}{B}'$ can be found from the composition operation, $\underset{\sim}{B}' = \underset{\sim}{A}' \circ \underset{\sim}{R}$.

The two most common forms of the composition operator are the max-min and the max-product compositions, as initially defined in Chapter 3. There are other forms of the composition operation, however, and these are summarized at the end of this chapter.

Example 7.10. Continuing with the invention example, Example 7.9, suppose that the fuzzy relation just developed, i.e., $\underset{\sim}{R}$, describes the invention's commercial potential. We wish to know what market size would be associated with a uniqueness score of "almost high uniqueness." That is, with a new antecedent, $\underset{\sim}{A}'$, the following consequent, $\underset{\sim}{B}'$, can be determined using composition. Let

$$\underset{\sim}{A}' = \text{almost high uniqueness} = \left\{ \frac{0.5}{1} + \frac{1}{2} + \frac{0.3}{3} + \frac{0}{4} \right\}$$

Then, using the following max-min composition,

$$\underset{\sim}{B}' = \underset{\sim}{A}' \circ \underset{\sim}{R} = \left\{ \frac{0.5}{1} + \frac{0.5}{2} + \frac{0.6}{3} + \frac{0.6}{4} + \frac{0.5}{5} + \frac{0.5}{6} \right\}$$

we get the fuzzy set describing the associated market size. In other words, the consequent is fairly diffuse, where there is no strong (or weak) membership value for any of the market size scores (i.e., no membership values near 0 or 1).

This power of fuzzy logic and approximate reasoning to assess qualitative knowledge can be illustrated in more familiar terms to engineers in the context of the following two examples in the fields of electrical engineering and biophysics.

Example 7.11. In the design of a fuzzy logic-based train, a controller is introduced whose inputs and outputs are shown in Fig. 7.6. The input operation is called the *desired operation* (described shortly). The output variable is either a (1) clockwise torque or a (2) counterclockwise torque.

In Fig. 7.7 the letters A–H are different stations separated by different distances. Directions of the train can be from A to H or H to A; for the train to run from station A to station H we will let the train motor run clockwise, and for the train to run from station H to station A we let the motor run counterclockwise. Motor torques can vary depending on the number of people boarding the train at different stops.

The control system for this train will be a computer-based system having different station distances. We can sense the arrival of a train with a weight sensor. The controller can easily calculate the speed of the train by dividing the distance between stations by the elapsed time between stations. The speed and the direction of the arriving train will be used as the inputs to the motor controller.

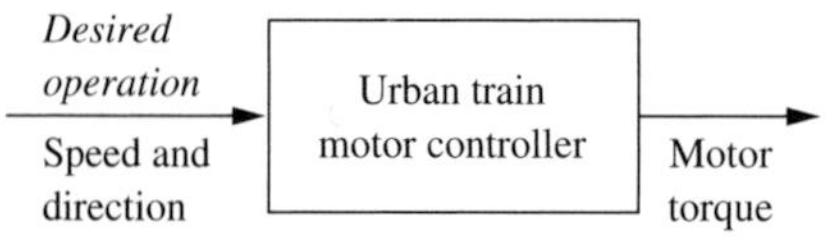

FIGURE 7.6
Schematic of urban train controller.

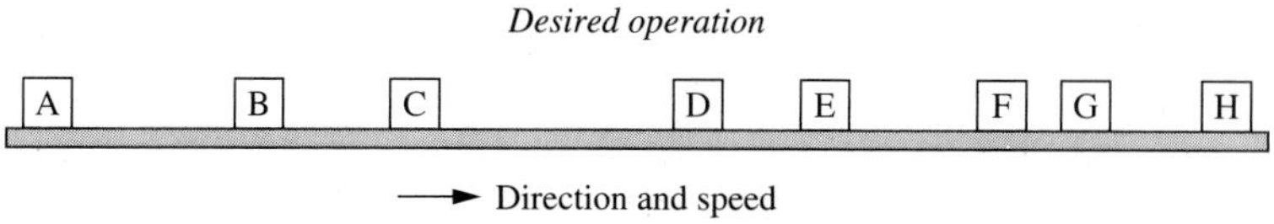

FIGURE 7.7
Typical train station separation distances.

An urban transit system will have a number of stops at close distances. To save energy and to maintain efficiency, we consider the graph shown in Fig. 7.8. The speed of the train motor N depends not only on the distance to the next station but also on the torque put on the train. In Fig. 7.6 we have called the input to the controller the *desired operation.* In this input we are interested in making adjustments (called the *difference*) to the operation of the train by making changes in the direction and speed of the motor controlling the train. Let *positive* indicate clockwise direction; *negative* indicates counterclockwise direction. For example, if we need to get to the next station, and it is a long distance away but in the same direction that the train is presently going, then the adjustment to the motor torque might be a "clockwise increase in motor speed (torque)."

Fuzzy inference rules may be developed as follows:

IF desired operation is positive high difference, THEN clockwise motor speed is high;

IF desired operation is positive medium difference, THEN clockwise motor speed is medium;

IF desired operation is positive small difference, THEN clockwise motor speed is low;

IF desired operation is positive zero difference, THEN clockwise motor speed is zero;

IF desired operation is negative zero difference, THEN counterclockwise motor speed is zero;

IF desired operation is negative small difference, THEN counterclockwise motor speed is low;

IF desired operation is negative medium difference, THEN counterclockwise motor speed is medium;

IF desired operation is negative high difference, THEN counterclockwise motor speed is high.

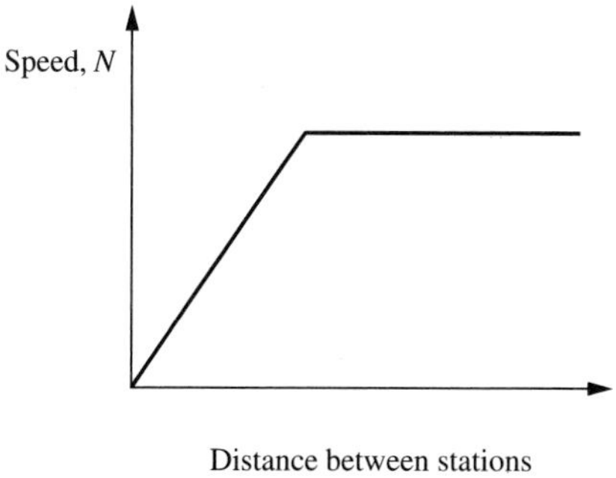

FIGURE 7.8
Graph of individual speed vs. distance between stations to get best results.

Positive high ↓ Negative high	} Difference;	Clockwise / Counterclockwise	} Direction; Speed: medium, high, . . .	} Torque rate

FIGURE 7.9
Symbolic form of the implication for the train controller.

Now, suppose we define a family of propositions $\underset{\sim}{P}$ of fuzzy sets on the desired operation differences, e.g., positive high difference ($\underset{\sim}{A}$), and so on. Let a family of propositions of fuzzy sets $\underset{\sim}{Q}$ be defined for the motor speeds and directions (torques), e.g., high, medium, low, and so on ($\underset{\sim}{B}$). The implication $\underset{\sim}{P} \rightarrow \underset{\sim}{Q}$ is illustrated in symbolic form in Fig. 7.9.

To set up our problem in the form of relations, we will say that the *desired operation* (input, x) is the speed difference ($\underset{\sim}{A}$) (see Fig. 7.9) and that the combination of motor direction and motor speed (output, y) is the torque rate ($\underset{\sim}{B}$). This is equivalent to the fuzzy relation

$$\underset{\sim}{R} = (\underset{\sim}{A} \times \underset{\sim}{B}) \cup (\overline{\underset{\sim}{A}} \times Y)$$

whose membership function is described by the expression

$$\mu_{\underset{\sim}{R}}(x, y) = \max\{[\mu_{\underset{\sim}{A}}(x) \wedge \mu_{\underset{\sim}{B}}(y)], (1 - \mu_{\underset{\sim}{A}}(x))\}$$

(Y is the output universe—motor speed and direction—on which $\underset{\sim}{B}$ is defined.) If we now introduce passenger *load* as an additional variable, then we can see that even if distance is low, whatever the motor direction is, speed will be dependent on passenger *load*. So, depending on *load*, we can form new rules as follows:

IF desired operation is positive high difference, THEN clockwise motor speed is high; ELSE clockwise motor speed is less [high load].
IF desired operation is positive medium difference, THEN clockwise motor speed is high; ELSE clockwise motor speed is high [low load],
and so on.

Denoting the kth load effect as Rate_k, we can have, for example, IF desired operation (x) is speed difference ($\underset{\sim}{A}$), THEN clockwise motor speed (y) is rate ($\underset{\sim}{R}$), ELSE clockwise motor speed (y) is Rate_k ($\underset{\sim}{C}$). The equivalent fuzzy relation is, $\underset{\sim}{R} = (\underset{\sim}{A} \times \underset{\sim}{B}) \cup (\overline{\underset{\sim}{A}} \times \underset{\sim}{C})$ whose membership function is expressed by the following formula:

$$\mu_{\underset{\sim}{R}}(x, y) = \max\{[\mu_{\underset{\sim}{A}}(x) \wedge \mu_{\underset{\sim}{B}}(y)], (1 - \mu_{\underset{\sim}{A}}(x) \wedge \mu_{\underset{\sim}{C}}(y))\}$$

Now, define fuzzy sets on speed of the motor, $\underset{\sim}{A}$, in dimensionless units on a range of $[1, 5]$; on direction of the motor motion, $\underset{\sim}{B}$, from 0° to 180° (clockwise to counterclockwise); and on another direction of motor motion, say $\underset{\sim}{C}$. Specific examples of these are as follows:

$$\underset{\sim}{A} = \frac{1}{1} + \frac{0.5}{2} + \frac{0}{3} + \frac{0}{4} + \frac{0}{5} = \text{“slow speed”}$$

$$\underset{\sim}{B} = \frac{0}{0^\circ} + \frac{0}{45^\circ} + \frac{0}{90^\circ} + \frac{0.5}{135^\circ} + \frac{1}{180^\circ} = \text{“mostly counterclockwise”}$$

$$\underset{\sim}{C} = \frac{0}{0^\circ} + \frac{0.5}{45^\circ} + \frac{1}{90^\circ} + \frac{0.5}{135^\circ} + \frac{0}{180^\circ} = \text{“around neutral direction”}$$

With these examples, suppose we have a rule: IF "slow speed," THEN "mostly counterclockwise." The Cartesian product of $\underset{\sim}{A}$ and $\underset{\sim}{B}$ is given by

$$\underset{\sim}{A} \times \underset{\sim}{B} = \begin{bmatrix} 1 \\ .5 \\ 0 \\ 0 \\ 0 \end{bmatrix} \times [0\,0\,0\,.5\,1] = \begin{bmatrix} 0 & 0 & 0 & .5 & 1 \\ 0 & 0 & 0 & .5 & .5 \\ 0 & 0 & 0 & 0 & 0 \\ 0 & 0 & 0 & 0 & 0 \\ 0 & 0 & 0 & 0 & 0 \end{bmatrix}$$

and the Cartesian product of $\overline{\underset{\sim}{A}}$ and Y is given by

$$\overline{\underset{\sim}{A}} \times \underset{\sim}{B} = \begin{bmatrix} 0 \\ .5 \\ 1 \\ 1 \\ 1 \end{bmatrix} \times [1\,1\,1\,1\,1] = \begin{bmatrix} 0 & 0 & 0 & 0 & 0 \\ .5 & .5 & .5 & .5 & .5 \\ 1 & 1 & 1 & 1 & 1 \\ 1 & 1 & 1 & 1 & 1 \\ 1 & 1 & 1 & 1 & 1 \end{bmatrix}$$

Finally, the rule is represented by the relation

$$\underset{\sim}{R} = \underset{\sim}{A} \rightarrow \underset{\sim}{B} = [\underset{\sim}{A} \times \underset{\sim}{B}] \cup [\overline{\underset{\sim}{A}} \times Y] = \begin{bmatrix} 0 & 0 & 0 & .5 & 1 \\ .5 & .5 & .5 & .5 & .5 \\ 1 & 1 & 1 & 1 & 1 \\ 1 & 1 & 1 & 1 & 1 \\ 1 & 1 & 1 & 1 & 1 \end{bmatrix}$$

This relation, $\underset{\sim}{R}$, expresses the relationship between a "slow speed" and a "mostly counterclockwise direction."

With the introduction of a new motor speed, say fuzzy set $\underset{\sim}{A}' = [1, .4, .2, 0, 0] =$ "a little slow" on the universe of dimensionless speeds [1, 5], with approximate reasoning we can find the direction, $\underset{\sim}{B}'$ associated with this new motor speed, using a max-min composition,

$$\underset{\sim}{B}' = \underset{\sim}{A}' \circ \underset{\sim}{R} = [1\,.4\,.2\,0\,0] \circ \begin{bmatrix} 0 & 0 & 0 & .5 & 1 \\ .5 & .5 & .5 & .5 & .5 \\ 1 & 1 & 1 & 1 & 1 \\ 1 & 1 & 1 & 1 & 1 \\ 1 & 1 & 1 & 1 & 1 \end{bmatrix} = [.4\,.4\,.4\,.5\,1]$$

The result, $\underset{\sim}{B}'$, might be given a linguistic label of "not so counterclockwise."

Now, suppose we have another rule of knowledge of the form $\underset{\sim}{R}_1$ = IF $\underset{\sim}{A}$ THEN $\underset{\sim}{B}$ ELSE $\underset{\sim}{C}$; i.e., IF "slow speed," THEN "mostly counterclockwise" direction, ELSE "around neutral" direction. First we calculate $\overline{\underset{\sim}{A}} \times \underset{\sim}{C}$,

$$\overline{\underset{\sim}{A}} \times \underset{\sim}{C} = \begin{bmatrix} 0 \\ .5 \\ 1 \\ 1 \\ 1 \end{bmatrix} [0 \quad .5 \quad 1 \quad .5 \quad 0] = \begin{bmatrix} 0 & 0 & 0 & 0 & 0 \\ 0 & .5 & .5 & .5 & 0 \\ 0 & .5 & 1 & .5 & 0 \\ 0 & .5 & 1 & .5 & 0 \\ 0 & .5 & 1 & .5 & 0 \end{bmatrix}$$

Then, using the preceding results, the new relation is

$$\underset{\sim}{R}_1 = (\underset{\sim}{A} \times \underset{\sim}{B}) \cup (\overline{\underset{\sim}{A}} \times \underset{\sim}{C}) = \begin{bmatrix} 0 & 0 & 0 & .5 & 1 \\ 0 & 0 & 0 & .5 & .5 \\ 0 & 0 & 0 & 0 & 0 \\ 0 & 0 & 0 & 0 & 0 \\ 0 & 0 & 0 & 0 & 0 \end{bmatrix} \cup \begin{bmatrix} 0 & 0 & 0 & 0 & 0 \\ 0 & .5 & .5 & .5 & 0 \\ 0 & .5 & 1 & .5 & 0 \\ 0 & .5 & 1 & .5 & 0 \\ 0 & .5 & 1 & .5 & 0 \end{bmatrix}$$

$$= \begin{bmatrix} 0 & 0 & 0 & .5 & 1 \\ 0 & .5 & .5 & .5 & .5 \\ 0 & .5 & 1 & .5 & 0 \\ 0 & .5 & 1 & .5 & 0 \\ 0 & .5 & 1 & .5 & 0 \end{bmatrix}$$

This relation, $\underset{\sim}{R}_1$, expresses the knowledge in the IF, THEN, ELSE clause just given.

Example 7.12. For research on the human visual system, it is sometimes necessary to characterize the strength of response to a visual stimulus based on a magnetic field measurement or on an electrical potential measurement. When using magnetic field measurements, a typical experiment will require nearly 100 off/on presentations of the stimulus at one location to obtain useful data. If the researcher is attempting to map the visual cortex of the brain, several stimulus locations must be used in the experiments. When working with a new subject, a researcher will make preliminary measurements to determine if the type of stimulus being used evokes a good response in the subject. The magnetic measurements are in units of femtotesla (10^{-15} tesla). Therefore, the inputs and outputs are both measured in terms of magnetic units.

We will define inputs on the universe X = [0, 50, 100, 150, 200] femtotesla, and outputs on the universe Y = [0, 50, 100, 150, 200] femtotesla. We will define two fuzzy sets, two different stimuli, on universe X:

$$\underset{\sim}{W} = \text{"weak stimulus"} = \left\{ \frac{1}{0} + \frac{.9}{50} + \frac{.3}{100} + \frac{0}{150} + \frac{0}{200} \right\} \subset X$$

$$\underset{\sim}{M} = \text{"medium stimulus"} = \left\{ \frac{0}{0} + \frac{.4}{50} + \frac{1}{100} + \frac{.4}{150} + \frac{0}{200} \right\} \subset X$$

and one fuzzy set on the output universe Y,

$$\underset{\sim}{S} = \text{"severe response"} = \left\{ \frac{0}{0} + \frac{0}{50} + \frac{0.5}{100} + \frac{0.9}{150} + \frac{1}{200} \right\} \subset \text{Y}$$

The complement of $\underset{\sim}{S}$ will then be

$$\overline{\underset{\sim}{S}} = \left\{ \frac{1}{0} + \frac{1}{50} + \frac{.5}{100} + \frac{.1}{150} + \frac{0}{200} \right\}$$

We will construct the proposition : IF "weak stimulus" THEN not "severe response," using classical implication.

$$\text{IF } \underset{\sim}{W} \text{ THEN } \overline{\underset{\sim}{S}} = \underset{\sim}{W} \rightarrow \overline{\underset{\sim}{S}} = (\underset{\sim}{W} \times \overline{\underset{\sim}{S}}) \cup (\overline{\underset{\sim}{W}} \times Y)$$

$$\underset{\sim}{W} \times \overline{\underset{\sim}{S}} = \begin{bmatrix} 1 \\ .9 \\ .3 \\ 0 \\ 0 \end{bmatrix} [1\ 1\ .5\ .1\ 0] = \begin{array}{c} \\ 0 \\ 50 \\ 100 \\ 150 \\ 200 \end{array} \begin{array}{c} \begin{array}{ccccc} 0 & 50 & 100 & 150 & 200 \end{array} \\ \begin{bmatrix} 1 & 1 & .5 & .1 & 0 \\ .9 & .9 & .5 & .1 & 0 \\ .3 & .3 & .3 & .1 & 0 \\ 0 & 0 & 0 & 0 & 0 \\ 0 & 0 & 0 & 0 & 0 \end{bmatrix} \end{array}$$

$$\overline{\underset{\sim}{W}} \times Y = \begin{bmatrix} 0 \\ .1 \\ .7 \\ 1 \\ 1 \end{bmatrix} [1\ 1\ 1\ 1\ 1] = \begin{array}{c} \\ 0 \\ 50 \\ 100 \\ 150 \\ 200 \end{array} \begin{array}{c} \begin{array}{ccccc} 0 & 50 & 100 & 150 & 200 \end{array} \\ \begin{bmatrix} 0 & 0 & 0 & 0 & 0 \\ .1 & .1 & .1 & .1 & .1 \\ .7 & .7 & .7 & .7 & .7 \\ 1 & 1 & 1 & 1 & 1 \\ 1 & 1 & 1 & 1 & 1 \end{bmatrix} \end{array}$$

$$\underset{\sim}{R} = (\underset{\sim}{W} \times \overline{\underset{\sim}{S}}) \cup (\overline{\underset{\sim}{W}} \times Y) = \begin{array}{c} \\ 0 \\ 50 \\ 100 \\ 150 \\ 200 \end{array} \begin{array}{c} \begin{array}{ccccc} 0 & 50 & 100 & 150 & 200 \end{array} \\ \begin{bmatrix} 1 & 1 & .5 & .1 & 0 \\ .9 & .9 & .5 & .1 & .1 \\ .7 & .7 & .7 & .7 & .7 \\ 1 & 1 & 1 & 1 & 1 \\ 1 & 1 & 1 & 1 & 1 \end{bmatrix} \end{array}$$

This relation $\underset{\sim}{R}$, then, expresses the knowledge embedded in the rule: IF "weak stimuli" THEN not "severe response." Now, using a new antecedent (IF part) for the input, $\underset{\sim}{M}$ = "medium stimuli," and a max-min composition we can find another response on the Y universe to relate approximately to the new stimulus $\underset{\sim}{M}$, i.e., to find $\underset{\sim}{M} \circ \underset{\sim}{R}$;

$$\underset{\sim}{M} \circ \underset{\sim}{R} = [0\quad .4\quad 1\quad .4\quad 0] \begin{array}{c} \begin{array}{ccccc} 0 & 50 & 100 & 150 & 200 \end{array} \\ \begin{bmatrix} 1 & 1 & .5 & .1 & 0 \\ .9 & .9 & .5 & .1 & .1 \\ .7 & .7 & .7 & .7 & .7 \\ 1 & 1 & 1 & 1 & 1 \\ 1 & 1 & 1 & 1 & 1 \end{bmatrix} \end{array} = [.7\quad .7\quad .7\quad .7\quad .7]$$

This result might be labeled linguistically as "no measurable response."

An interesting issue in approximate reasoning is the idea of an inverse relationship between fuzzy antecedents and fuzzy consequences arising from the composition operation. Consider the following problem. Suppose we use the original antecedent, $\underset{\sim}{A}$, in the fuzzy composition. Do we get the original fuzzy consequent, $\underset{\sim}{B}$, as a result of the operation? That is, does the composition operation have a unique inverse, i.e., $\underset{\sim}{B} = \underset{\sim}{A} \circ \underset{\sim}{R}$? The answer is an unqualified no, and one should not expect an inverse to exist for fuzzy composition.

Example 7.13. Again, continuing with the invention example, Examples 7.9 and 7.10, suppose that $\underset{\sim}{A}' = \underset{\sim}{A}$ = "medium uniqueness," then

$$B' = A' \circ R = A \circ R = \left\{ \frac{0.4}{1} + \frac{0.4}{2} + \frac{1}{3} + \frac{0.8}{4} + \frac{0.4}{5} + \frac{0.4}{6} \right\} \neq \underset{\sim}{B}$$

That is, the new consequent does not yield the original consequent ($\underset{\sim}{B}$ = medium market size) because the inverse is not guaranteed with fuzzy composition.

In classical binary logic this inverse does exist, that is, crisp *modus ponens* would give

$$B' = A' \circ R = A \circ R = B$$

where the sets A and B are crisp, and the relation R is also crisp. In the case of approximate reasoning, the fuzzy inference is not precise but rather is approximate. However, the inference does represent an approximate linguistic characteristic of the relation between two universes of discourse, X and Y.

Example 7.14. Suppose you are a soils engineer and you wish to track the movement of soil particles under applied loading in an experimental apparatus that allows viewing of the soil motion. You are building pattern recognition software to enable a computer to monitor and detect the motions. However, there are some difficulties in "teaching" your software to view the motion. The tracked particle can be occluded by another particle. The occlusion can occur when a tracked particle is behind another particle, behind a mark on the camera's lens, or partially out of sight of the camera. We want to establish a relationship between particle occlusion, which is a poorly known phenomenon, and lens occlusion, which is quite well-known in photography. Let these membership functions,

$$\underset{\sim}{A} = \left\{ \frac{0.1}{x_1} + \frac{0.9}{x_2} + \frac{0.0}{x_3} \right\} \quad \text{and} \quad \underset{\sim}{B} = \left\{ \frac{0}{y_1} + \frac{1}{y_2} + \frac{0}{y_3} \right\}$$

describe fuzzy sets for a *tracked particle moderately occluded* behind another particle and a *lens mark associated with moderate image quality,* respectively. Fuzzy set $\underset{\sim}{A}$ is defined on a universe $X = \{x_1, x_2, x_3\}$ of tracked particle indicators, and fuzzy set $\underset{\sim}{B}$ (note in this case that $\underset{\sim}{B}$ is a crisp singleton) is defined on a universe $Y = \{y_1, y_2, y_3\}$ of lens obstruction indices. A typical rule might be

IF occlusion due to particle occlusion is moderate,
THEN image quality will be similar to a moderate lens obstruction,

or symbolically,

$$\text{If } x \text{ is } \underset{\sim}{A}, \text{ then } y \text{ is } \underset{\sim}{B} \text{ or } (\underset{\sim}{A} \times \underset{\sim}{B}) \cup (\overline{\underset{\sim}{A}} \times Y) = \underset{\sim}{R}$$

We can find the relation, $\underset{\sim}{R}$, as follows:

$$\underset{\sim}{A} \times \underset{\sim}{B} = \begin{matrix} & y_1 & y_2 & y_3 \\ x_1 & 0 & .1 & 0 \\ x_2 & 0 & .9 & 0 \\ x_3 & 0 & 0 & 0 \end{matrix}$$

$$\overline{\underset{\sim}{A}} \times \underset{\sim}{Y} = \begin{array}{c} \\ x_1 \\ x_2 \\ x_3 \end{array} \begin{array}{c} \begin{array}{ccc} y_1 & y_2 & y_3 \end{array} \\ \begin{bmatrix} .9 & .9 & .9 \\ .1 & .1 & .1 \\ 1 & 1 & 1 \end{bmatrix} \end{array}$$

$$\underset{\sim}{R} = (\underset{\sim}{A} \times \underset{\sim}{B}) \cup (\overline{\underset{\sim}{A}} \times \underset{\sim}{Y}) = \begin{bmatrix} .9 & .9 & .9 \\ .1 & .9 & .1 \\ 1 & 1 & 1 \end{bmatrix}$$

This relation expresses in matrix form all the knowledge embedded in the implication. Let $\underset{\sim}{A}'$ be a fuzzy set, in which a tracked particle is behind a particle with *a slight bit more occlusion* than the particle expressed in the original antecedent $\underset{\sim}{A}$, given by

$$\underset{\sim}{A}' = \left\{ \frac{0.3}{x_1} + \frac{1.0}{x_2} + \frac{0.0}{x_3} \right\}$$

We can find the associated membership of the image quality using max-min composition. For example, approximate reasoning will provide

$$\text{IF } x \text{ is } \underset{\sim}{A}', \text{ THEN } \underset{\sim}{B}' = \underset{\sim}{A}' \circ \underset{\sim}{R}$$

and we get

$$\underset{\sim}{B}' = [.3\ 1\ 0] \circ \begin{bmatrix} .9 & .9 & .9 \\ .1 & .9 & .1 \\ 1 & 1 & 1 \end{bmatrix} = \left\{ \frac{.3}{y_1} + \frac{.9}{y_2} + \frac{.3}{y_3} \right\}$$

This image quality, $\underset{\sim}{B}'$, is more fuzzy than $\underset{\sim}{B}$, as indicated by the former's membership function.

FUZZY TAUTOLOGIES, CONTRADICTIONS, EQUIVALENCE, AND LOGICAL PROOFS

The extension of truth operations for tautologies, contradictions, equivalence, and logical proofs is no different for fuzzy sets; the results, however, can differ considerably from those in classical logic. If the truth values for the simple propositions of a fuzzy logic compound proposition are strictly true (1) or false (0), the results follow identically those in classical logic. However, the use of partially true (or partially false) simple propositions in compound propositional statements results in new ideas termed quasi tautologies, quasi contradictions, and quasi equivalence. Moreover, the idea of a logical proof is altered because now a proof can be shown only to a "matter of degree." Some examples of these will be useful.

Suppose we want to verify "approximate *modus ponens,*" where the truth value of the propositions is neither strictly true nor strictly false. Notice in the truth table, Table 7.7*a,* that the last column contains values other than unity—this represents a quasi tautology. How can these numbers be interpreted? There are a number of ways to look at this. One would be to take the lower and upper bounds of these numbers to represent the "range" of truth of the quasi tautology for the propositions considered. Of course, as the truth values of the simple propositions change, so too do the truth

TABLE 7.7*a*
Truth table (approximate *modus ponens*)

A	B	A → B	(A ∧ (A → B))	(A ∧ (A → B)) → B	
.3	.2	.7	.3	.7	
.3	.8	.8	.3	.8	Quasi tautology
.7	.2	.3	.3	.7	
.7	.8	.8	.7	.8	

values in the quasi tautology. If we use the same truth table and change the truth values, we see a different result (Table 7.7*b*). Note also that if all the propositions are half-truths (truth values all equal 0.5) the resulting truth value of the tautology is also a half-truth; i.e., the last column in the truth table is comprised of values equal to 0.5.

Consider the classical contradiction, $\overline{B} \cap B$. Suppose the truth value, T(B) is equal to 0.6, then the contradiction results in a value of min(0.4, 0.6) = 0.4, which we call a quasi contradiction (it is not zero as the classical contradiction would be).

Quasi equivalence follows from the results of applying the implication connective to propositions that are not strictly true or false. Here again, equivalence can only be thought of in an approximate sense. For example, the statements "the angle between two orthogonal lines in the same plane" and "right angle" are only approximately equivalent if the terms "orthogonal" and "right angle" are interpreted approximately, i.e., as fuzzy quantities or fuzzy numbers. Equivalence in the fuzzy realm is a function of the degree of membership. Equivalence between two propositions is manifested "if the truth values of the two propositions are identical." So, if the truth values of two propositions are equal to 0.3, then they are equivalent at that degree of membership.

Theorem proving takes on a whole new context in the fuzzy domain. What does it mean to prove a theorem approximately? Not many mathematicians would base their theories on approximate theorems because theorems, by definition, have to be tautologies. Hence, the proof that a theorem is a quasi tautology would not produce a result that is true under any circumstance. It seems that fuzzy theorem proving might not be useful in developing irrefutable corollaries, lemmas, and so forth; but it certainly would be useful in showing that any theorem based on vague natural language constructs could be challenged in a classical sense.

TABLE 7.7*b*
Truth table (approximate *modus ponens*)

A	B	A → B	(A ∧ (A → B))	(A ∧ (A → B)) → B	
.4	.1	.6	.4	.6	
.4	.9	.9	.4	.9	Quasi tautology
.6	.1	.4	.4	.6	
.6	.9	.9	.6	.9	

OTHER FORMS OF THE IMPLICATION OPERATION

There are other techniques for obtaining the fuzzy relation $\underset{\sim}{R}$ based on the IF $\underset{\sim}{A}$, THEN $\underset{\sim}{B}$, or $\underset{\sim}{R} = \underset{\sim}{A} \to \underset{\sim}{B}$. These are known as fuzzy implication operations, and they are valid for all values of $x \in X$ and $y \in Y$. The following forms of the implication operator show different techniques for obtaining the membership function values of fuzzy relation $\underset{\sim}{R}$ defined on the Cartesian product space $X \times Y$:

$$\mu_{\underset{\sim}{R}}(x, y) = \max\{\min[\mu_{\underset{\sim}{A}}(x), \mu_{\underset{\sim}{B}}(y)], 1 - \mu_{\underset{\sim}{A}}(x)\} \tag{7.19}$$

$$\mu_{\underset{\sim}{R}}(x, y) = \max\{\mu_{\underset{\sim}{B}}(y), 1 - \mu_{\underset{\sim}{A}}(x)\} \tag{7.20}$$

$$\mu_{\underset{\sim}{R}}(x, y) = \min[\mu_{\underset{\sim}{A}}(x), \mu_{\underset{\sim}{B}}(y)] \tag{7.21}$$

$$\mu_{\underset{\sim}{R}}(x, y) = \min\{1, [1 - \mu_{\underset{\sim}{A}}(x) + \mu_{\underset{\sim}{B}}(y)]\} \tag{7.22}$$

$$\mu_{\underset{\sim}{R}}(x, y) = \min\{1, [\mu_{\underset{\sim}{A}}(x) + \mu_{\underset{\sim}{B}}(y)]\} \tag{7.23}$$

$$\mu_{\underset{\sim}{R}}(x, y) = \min\left\{1, \left[\frac{\mu_{\underset{\sim}{B}}(y)}{\mu_{\underset{\sim}{A}}(x)}\right]\right\}, \qquad \mu_{\underset{\sim}{A}}(x) > 0 \tag{7.24}$$

$$\mu_{\underset{\sim}{R}}(x, y) = \max\{\mu_{\underset{\sim}{A}}(x) \cdot \mu_{\underset{\sim}{B}}(y), [1 - \mu_{\underset{\sim}{A}}(x)]\} \tag{7.25}$$

$$\mu_{\underset{\sim}{R}}(x, y) = \mu_{\underset{\sim}{A}}(x) \cdot \mu_{\underset{\sim}{B}}(y) \tag{7.26}$$

$$\mu_{\underset{\sim}{R}}(x, y) = \begin{cases} 1, & \text{for } \mu_{\underset{\sim}{A}}(x) \leq \mu_{\underset{\sim}{B}}(y) \\ \mu_{\underset{\sim}{B}}(y), & \text{otherwise} \end{cases} \tag{7.27}$$

$$\mu_{\underset{\sim}{R}}(x, y) = \begin{cases} 1, & \text{for } \mu_{\underset{\sim}{A}}(x) \leq \mu_{\underset{\sim}{B}}(y) \\ 0, & \text{otherwise} \end{cases} \tag{7.28}$$

In situations where the universes are represented by discrete elements the fuzzy relation $\underset{\sim}{R}$ is a matrix.

Equation (7.19) was developed earlier in this chapter and is known as classical implication; this form was considered by Zadeh [1973]. Equation (7.20) is equivalent to Equation (7.19) for $\mu_{\underset{\sim}{B}}(y) \leq \mu_{\underset{\sim}{A}}(x)$. Equation (7.21) has been given various terms in the literature; it has been referred to as *correlation-minimum* and as *Mamdani's implication,* after British Prof. Mamdani's work in the area of system control [Mamdani, 1976]. This formulation for the implication is also equivalent to the fuzzy cross product of fuzzy sets $\underset{\sim}{A}$ and $\underset{\sim}{B}$, i.e., $\underset{\sim}{R} = \underset{\sim}{A} \times \underset{\sim}{B}$. For $\mu_{\underset{\sim}{A}}(x) \geq 0.5$ and $\mu_{\underset{\sim}{B}}(y) \geq 0.5$ classical implication reduces to Mamdani's implication. The implication defined by Eq. (7.22) is known as *Lukasiewicz' implication,* after the Polish logician Jan Lukasiewicz [Rescher, 1969]. The fuzzy implication relation defined by Eq. (7.23) is commonly referred to as the *bounded sum implication.* Equation (7.24) is due to Goguen [1969]. Equations (7.25) and (7.26) describe two forms of *correlation-product implication* and are based on the notions of conditioning and reinforcement. Both of these product forms tend to dilute the influence of joint membership values that are small and, as such, are related to Hebbian-type learning algorithms in neuropsychology when used in artificial neural network computations. Equation (7.25) is a recently suggested form by Vadiee [1993] and is equally valid for crisp and fuzzy cases. Equation (7.27) is sometimes called *Brouwerian* implication and is discussed in Sanchez [1976]. Equation (7.28) gives a very simple form of Eq. (7.27) that has been termed in the literature as *R-SEQ (standard sequence logic) implication* [Maydole, 1975]. Although the classical implication continues to be the most popular and is valid for fuzzy and crisp applications, these other methods have been introduced as computationally effective under certain conditions of the membership values, $\mu_{\underset{\sim}{A}}(x)$ and $\mu_{\underset{\sim}{B}}(y)$. The appropriate choice of an implication operator is a matter left to the analyst, since it is typically context-dependent (see Problems 7.45 and 7.46 for comparisons).

OTHER FORMS OF THE COMPOSITION OPERATION

Max-min and max-product (also referred to as max-dot) methods of composition of fuzzy relations are the two most commonly used techniques. Many other techniques are mentioned in the literature. Each method of composition of fuzzy relations reflects a special inference machine and has its own significance and applications. The max-min method is the one used by Zadeh in his original paper on approximate reasoning using natural language IF-THEN rules. Many have claimed, since Zadeh's introduction, that this method of composition effectively expresses the approximate and interpolative reasoning used by humans when they employ linguistic propositions for deductive reasoning [Vadiee, 1993].

The following common methods are among those proposed in the literature for the composition operation $\underset{\sim}{B} = \underset{\sim}{A} \circ \underset{\sim}{R}$, where $\underset{\sim}{A}$ is the input, or antecedent defined on the universe X, $\underset{\sim}{B}$ is the output, or consequent defined on universe Y, and $\underset{\sim}{R}$ is a fuzzy relation characterizing the relationship between specific inputs (x) and specific outputs (y):

$$\text{max-min} \qquad \mu_{\underset{\sim}{B}}(y) = \max_{x \in X}\{\min[\mu_{\underset{\sim}{A}}(x), \mu_{\underset{\sim}{R}}(x, y)]\} \tag{7.29}$$

$$\text{max-product} \qquad \mu_{\underset{\sim}{B}}(y) = \max_{x \in X}[\mu_{\underset{\sim}{A}}(x) \cdot \mu_{\underset{\sim}{R}}(x, y)] \tag{7.30}$$

$$\text{min-max} \qquad \mu_{\underset{\sim}{B}}(y) = \min_{x \in X}\{\max[\mu_{\underset{\sim}{A}}(x), \mu_{\underset{\sim}{R}}(x, y)]\} \tag{7.31}$$

$$\text{max-max} \qquad \mu_{\underset{\sim}{B}}(y) = \max_{x \in X}\{\max[\mu_{\underset{\sim}{A}}(x), \mu_{\underset{\sim}{R}}(x, y)]\} \tag{7.32}$$

$$\text{min-min} \qquad \mu_{\underset{\sim}{B}}(y) = \min_{x \in X}\{\min[\mu_{\underset{\sim}{A}}(x), \mu_{\underset{\sim}{R}}(x, y)]\} \tag{7.33}$$

$$\text{max-average} \qquad \mu_{\underset{\sim}{B}}(y) = \tfrac{1}{2}\max_{x \in X}[\mu_{\underset{\sim}{A}}(x) + \mu_{\underset{\sim}{R}}(x, y)] \tag{7.34}$$

$$\text{sum-product} \qquad \mu_{\underset{\sim}{B}}(y) = f\left\{\sum_{x \in X}[\mu_{\underset{\sim}{A}}(x) \cdot \mu_{\underset{\sim}{R}}(x, y)]\right\} \tag{7.35}$$

where $f(\cdot)$ is a logistic function (like a sigmoid or a step function) that limits the value of the function within the interval [0, 1]. This composition method is commonly used in applications of artificial neural networks for mapping between parallel layers in a multi-layer network.

It is left as an exercise for the reader (see Problems 7.45 and 7.47) to determine the relationship among these alternative forms of the composition operator for various combinations of the membership values for $\mu_{\underset{\sim}{A}}(x)$ and $\mu_{\underset{\sim}{R}}(x, y)$.

SUMMARY

This chapter has presented the basic axioms, operations, and properties of binary logic and fuzzy logic. Just as in Chapter 2, we find that the only significant difference between a binary logic and a fuzzy logic stems from the logical equivalent of the excluded middle laws. Examples are provided that illustrate the various operations of a fuzzy logic. The notion that a truth table in fuzzy logic has an infinite number of possibilities is introduced. An approximate reasoning, proposed by Zadeh [1976, 1979], is presented to illustrate the power of using fuzzy sets in the reasoning process. Other works in the area of fuzzy reasoning and approximate reasoning have been helpful in explaining the theory; for example, a useful comparison study [Mizumoto and Zimmerman, 1982] and a work defining the mathematical foundations [Yager, 1985]. From a general point of view, other multivalued logics have been developed [DuBois and Prade, 1980; Klir and Folger, 1988], and these other logics may be viewed as *fuzzy logics* in the sense that they represent more than just the crisp truth values of 0 and 1. In fact, Gaines [1976] has shown that some forms of multivalued logics result from fuzzifying, in the sense of the extension principle, the standard propositional calculus. The illustration of approximate reasoning given here is conducted using fuzzy relations to represent the rules of inference. The chapter concludes by pointing out the rich variety in reasoning possible with fuzzy logic when one considers the vast array of implication and composition operations; an example of this can be found in Yager [1983]. The implications can be interpreted as specific chains of reasoning. Giles [1976] gives a very nice interpretation of these chains of reasoning in terms of risk: every chain of reasoning is analogous to a dialogue between speakers whose assertions entail a commitment about their truth.

REFERENCES

Dubois, D., and H. Prade (1980). *Fuzzy sets and systems: Theory and applications,* Academic, New York.

Gaines, B. (1976). "Foundations of fuzzy reasoning," *Int. J. Man Mach. Stud.,* vol. 8, pp. 623–688.

Giles, R. (1976). "Lukasiewicz logic and fuzzy theory," *Int. J. Man Mach. Stud.,* vol. 8, pp. 313–327.

Gill, A. (1976). *Applied algebra for the computer sciences,* Prentice Hall, Englewood Cliffs, N.J.

Goguen, J. (1969). "The logic of inexact concepts," *Synthese,* vol. 19, pp. 325–373.

Klir, G., and T. Folger (1988). *Fuzzy sets, uncertainty, and information,* Prentice Hall, Englewood Cliffs, N.J.

Kosko, B. (1992). *Neural networks and fuzzy systems,* Prentice Hall, Englewood Cliffs, N.J.

Mamdani, E. H. (1976),. "Advances in linguistic synthesis of fuzzy controllers," *Int. J. Man Mach. Stud.,* vol. 8, pp. 669–678.

Mano, M. (1988). *Computer engineering: Hardware design,* Prentice Hall, Englewood Cliffs, N.J., p. 59.

Maydole, R. (1975). "Many-valued logic as a basis for set theory," *J. Philos. Logic,* vol. 4, pp. 269–291.

Mizumoto, M., and H.-J. Zimmerman (1982). "Comparison of fuzzy reasoning methods," *Fuzzy Sets Syst.,* vol. 8, pp. 253–283.

Rescher, N. (1969). *Many-valued logic,* McGraw-Hill, New York.

Sanchez, E. (1976). "Resolution of composite fuzzy relation equations," *Inf. Control,* vol. 30, pp. 38–48.

Vadiee, N. (1993). "Fuzzy rule based expert systems–I," chap. 4 in M. Jamshidi, N. Vadiee, and T. Ross (eds.), *Fuzzy logic and control: Software and hardware applications,* Prentice Hall, Englewood Cliffs, N.J.

Yager, R. R. (1983). "On the implication operator in fuzzy logic," *Inf. Sci.,* vol. 31, pp. 141–164.

Yager, R. R. (1985). "Strong truth and rules of inference in fuzzy logic and approximate reasoning," *Cybern. Syst.*, vol. 16, pp. 23–63.

Zadeh, L. (1973). "Outline of a new approach to the analysis of complex systems and decision processes," *IEEE Trans. Syst. Man Cybern.*, vol. 3, pp. 28-44.

Zadeh, L. (1976). "The concept of a linguistic variable and its application to approximate reasoning—Part 3," *Inf. Sci.*, vol. 9, pp. 43–80.

Zadeh, L. (1979). "A theory of approximate reasoning," in J. Hayes, D. Michie, and L. Mikulich (eds.), *Machine Intelligence*, Halstead Press, New York, pp. 149–194.

PROBLEMS

7.1. Under what conditions of P and Q is the implication $\mathrm{P} \to \mathrm{Q}$ a tautology?

7.2. The exclusive-or is given by the expression, P"$\vee$"Q $= (\overline{\mathrm{P}} \wedge \mathrm{Q}) \vee (\mathrm{P} \wedge \overline{\mathrm{Q}})$. Show that the logical-or, given by $\mathrm{P} \vee \mathrm{Q}$, gives a different result from the exclusive-or and comment on this difference using an example in your own field.

7.3. For a proposition R of the form $\mathrm{P} \to \mathrm{Q}$, show the following:

(*a*) R and its contrapositive are equivalent, i.e., prove that $(\mathrm{P} \to \mathrm{Q}) \leftrightarrow (\overline{\mathrm{Q}} \to \overline{\mathrm{P}})$.

(*b*) The converse of R and the inverse of R are equivalent, i.e., prove that $(\mathrm{Q} \to \mathrm{P}) \leftrightarrow (\overline{\mathrm{P}} \to \overline{\mathrm{Q}})$.

7.4. Show that the dual of the equivalence $((\mathrm{P} \vee \mathrm{Q}) \vee ((\overline{\mathrm{P}}) \wedge (\overline{\mathrm{Q}}))) \leftrightarrow \mathrm{X}$ is also true.

7.5. Show that De Morgan's laws are duals.

7.6. Show that the compound proposition $((\mathrm{P} \to \mathrm{Q}) \wedge (\mathrm{R} \to \overline{\mathrm{S}}) \wedge (\mathrm{Q} \to \mathrm{R})) \to (\mathrm{P} \to \overline{\mathrm{S}})$ is a tautology.

7.7. Show that the following propositions from Lewis Carroll are tautologies [Gill, 1976]:

(*a*) No ducks waltz; no officers ever decline to waltz; all my poultry are ducks. Therefore, none of my poultry are officers.

(*b*) Babies are illogical; despised persons cannot manage crocodiles; illogical persons are despised; therefore, babies cannot manage crocodiles.

(*c*) Promise-breakers are untrustworthy; wine-drinkers are very communicative; a man who keeps his promise is honest; all pawnbrokers are wine-drinkers; we can always trust a very communicative person; therefore, all pawnbrokers are honest. (This problem requires $2^6 = 64$ lines of a truth table; perhaps it should be tackled with a computer).

7.8. Prove the following statements by contradiction.

(*a*) $((\mathrm{P} \to \mathrm{Q}) \wedge \mathrm{P}) \to \mathrm{Q}$

(*b*) $((\mathrm{P} \to \overline{\mathrm{Q}}) \wedge (\mathrm{Q} \vee \overline{\mathrm{R}}) \wedge (\mathrm{R} \wedge \overline{\mathrm{S}})) \to \overline{\mathrm{P}}$

(*c*) $\sqrt{3}$ is not a rational number; i.e., show that it cannot be the ratio of two even integers.

7.9. Prove that $((\mathrm{P} \to \overline{\mathrm{Q}}) \wedge (\mathrm{R} \to \overline{\mathrm{Q}}) \wedge (\mathrm{P} \vee \mathrm{R})) \to \mathrm{R}$ is not a tautology (i.e., a fallacy) by developing a counterexample.

7.10. Prove that the following statements are tautologies.

(*a*) $((\mathrm{P} \to \mathrm{Q}) \wedge \mathrm{P}) \to \mathrm{Q}$

(*b*) $\mathrm{P} \to (\mathrm{P} \vee \mathrm{Q})$

(*c*) $(\mathrm{P} \wedge \mathrm{Q}) \to \mathrm{P}$

(*d*) $((\mathrm{P} \to \mathrm{Q}) \wedge (\mathrm{Q} \to \mathrm{R})) \to (\mathrm{P} \to \mathrm{R})$

(*e*) $((\mathrm{P} \vee \mathrm{Q}) \wedge \overline{\mathrm{P}}) \to \mathrm{Q}$

7.11. For this inference rule,

$$[(A \to B) \wedge (B \to C)] \to (A \to C)$$

(*a*) Prove that the rule is a tautology for crisp sets.

(*b*) Show that the rule is a quasi tautology for fuzzy sets.

7.12. Consider the following two discrete fuzzy sets, which are defined on universe X = $\{-5, 5\}$:

$$\underset{\sim}{A} = \text{"zero"} = \left\{ \frac{0}{-2} + \frac{0.5}{-1} + \frac{1.0}{0} + \frac{0.5}{1} + \frac{0}{2} \right\}$$

$$\underset{\sim}{B} = \text{"positive medium"} = \left\{ \frac{0}{0} + \frac{0.5}{1} + \frac{1.0}{2} + \frac{0.5}{3} + \frac{0}{4} \right\}$$

(*a*) Construct the relation for the rule IF $\underset{\sim}{A}$, THEN $\underset{\sim}{B}$ (i.e., IF x is "zero" THEN y is "positive medium") using the Mamdani implication, Eq. (7.21), and the product implication, Eq. (7.26), or

$$\mu_R(x, y) = \min[\mu_A(x), \mu_B(y)]$$

and

$$\mu_R(x, y) = \mu_A(x) \cdot \mu_B(y)$$

(*b*) If we introduce a new antecedent,

$$\underset{\sim}{A}' = \text{"positive small"} = \left\{ \frac{0}{-1} + \frac{0.5}{0} + \frac{1.0}{1} + \frac{0.5}{2} + \frac{0}{3} \right\}$$

find the new consequent $\underset{\sim}{B}'$, using max-min composition, i.e., $\underset{\sim}{B}' = \underset{\sim}{A}' \circ \underset{\sim}{R}$, for both relations from part (*a*).

7.13. Given the fuzzy sets $\underset{\sim}{A}$ and $\underset{\sim}{B}$ on X and Y, respectively,

$$\underset{\sim}{A} = \int \left\{ \frac{1 - 0.1x}{x} \right\}, \qquad \text{for } x \in [0, +10]$$

$$\underset{\sim}{B} = \int \left\{ \frac{0.2y}{y} \right\}, \qquad \text{for } y \in [0, +5]$$

$$\mu_A(x) = 0 \qquad \text{outside the } [0, 10] \text{ interval}$$

$$\mu_B(y) = 0 \qquad \text{outside the } [0, 5] \text{ interval}$$

(*a*) Construct a fuzzy relation $\underset{\sim}{R}$ for the implication $\underset{\sim}{A} \to \underset{\sim}{B}$ using the classical implication operation, i.e., construct $\underset{\sim}{R} = (\underset{\sim}{A} \times \underset{\sim}{B}) \cup (\overline{\underset{\sim}{A}} \times Y)$

(*b*) Use max-min composition to find $\underset{\sim}{B}'$, given

$$\underset{\sim}{A}' = \left\{ \frac{1}{3} \right\}$$

Note: $\underset{\sim}{A}'$ is a crisp singleton, i.e., the number 3 has a membership of 1, and all other numbers in the universe X have a membership of 0.

Hint: You can solve this problem graphically by segregating the Cartesian space into various regions according to the min and max operations, or you can approximate the continuous fuzzy variables as discrete variables and use matrix operations. In either case, "sketch" the solutions for part (*a*) in 3D space (x, y, μ) and (*b*) in 2D space (y, μ).

7.14. Suppose we have a distillation process where the objective is to separate components of a mixture in the input stream. The process is pictured in Fig. P7.14. The relationship between the input variable, temperature, and the output variable, distillate fractions, is not precise but the human operator of this process has developed an intuitive understanding of this relationship. The universe for each of these variables is

$$\mathrm{X} = \text{universe of temperatures } (^{\circ}\mathrm{F}) = \{160, 165, 170, 175, 180, 185, 190, 195\}$$

$$\mathrm{Y} = \text{universe of distillate fractions (percentages)} = \{77, 80, 83, 86, 89, 92, 95, 98\}$$

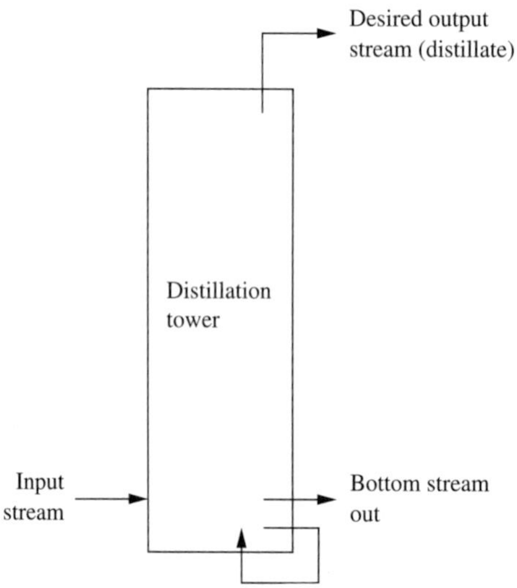

FIGURE P7.14

Now we define fuzzy sets $\underset{\sim}{A}$ and $\underset{\sim}{B}$ on X and Y, respectively:

$$\underset{\sim}{A} = \text{temperature of input steam is hot} = \left\{\frac{0}{175} + \frac{.7}{180} + \frac{1}{185} + \frac{.4}{190}\right\}$$

$$\underset{\sim}{B} = \text{separation of mixture is good} = \left\{\frac{0}{89} + \frac{.5}{92} + \frac{.8}{95} + \frac{1}{98}\right\}$$

We wish to determine the proposition, IF "temperature is hot" THEN "separation of mixture is good," or symbolically, $\underset{\sim}{A} \rightarrow \underset{\sim}{B}$. From this,

(*a*) Find $\underset{\sim}{R} = (\underset{\sim}{A} \times \underset{\sim}{B}) \cup (\overline{\underset{\sim}{A}} \times \mathrm{Y})$

(*b*) Now define another fuzzy linguistic variable as

$$\underset{\sim}{A}' = \left\{\frac{1}{170} + \frac{.8}{175} + \frac{.5}{180} + \frac{.2}{185}\right\}$$

and for the "new" rule IF $\underset{\sim}{A}'$ THEN $\underset{\sim}{B}'$, find $\underset{\sim}{B}'$ using max-min composition, i.e., find $\underset{\sim}{B}' = \underset{\sim}{A}' \circ \underset{\sim}{R}$.

7.15. The calculation of the vibration of an elastic structure depends on knowing the material properties of the structure as well as its support conditions. Suppose we have an elastic structure, such as a bar of known material, with properties like wave speed (C), modulus of elasticity (E), and cross-sectional area (A). However, the support stiffness is not well-known, hence the fundamental natural frequency of the system is not precise either. A relationship does exist between them, though, as illustrated in Fig. P7.15.

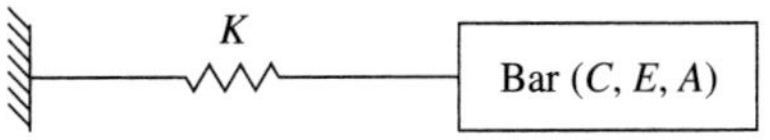

FIGURE P7.15

Define two fuzzy sets,

$$\underset{\sim}{\mathrm{K}} = \text{"support stiffness," in pounds per square inch}$$

$$\underset{\sim}{\mathrm{f}}_1 = \text{"first natural frequency of the system," in hertz}$$

with membership functions

$$\underset{\sim}{\mathrm{K}} = \left\{ \frac{0}{1e+3} + \frac{.2}{1e+4} + \frac{.5}{1e+5} + \frac{.8}{5e+5} + \frac{1}{1e+6} + \frac{.8}{5e+6} + \frac{.2}{1e+7} \right\}$$

$$\underset{\sim}{\mathrm{f}}_1 = \left\{ \frac{0}{100} + \frac{0}{200} + \frac{.2}{500} + \frac{.5}{800} + \frac{1}{1000} + \frac{.8}{2000} + \frac{.2}{5000} \right\}$$

(*a*) Using the proposition, IF x is $\underset{\sim}{\mathrm{K}}$, THEN y is $\underset{\sim}{\mathrm{f}}_1$, find this relation using the following forms of the implication $\underset{\sim}{\mathrm{K}} \rightarrow \underset{\sim}{\mathrm{f}}_1$:
  (*i*) Classical $\mu_\mathrm{R} = \max[\min(\mu_\mathrm{K}, \mu_{\mathrm{f}_1}), (1 - \mu_\mathrm{K})]$
  (*ii*) Mamdani $\mu_\mathrm{R} = \min(\mu_\mathrm{K}, \mu_{\mathrm{f}_1})$
  (*iii*) Product $\mu_\mathrm{R} = \mu_\mathrm{K} \cdot \mu_{\mathrm{f}_1}$

(*b*) Now define another antecedent, say $\underset{\sim}{\mathrm{K}}' =$ "damaged support,"

$$\underset{\sim}{\mathrm{K}}' = \left\{ \frac{0}{1e+3} + \frac{.8}{1e+4} + \frac{.1}{1e+5} \right\}$$

Find the system's fundamental (first) natural frequency due to the change in the support conditions; i.e., find $\underset{\sim}{\mathrm{f}}_1 =$ "first natural frequency due to damaged support" using classical implication from part (*a*), section (*i*) preceding, and
  (*i*) max-min composition
  (*ii*) max-product composition

7.16. We want to set up a database of documents that cover a certain broad scientific subject to various degrees of depth. We are interested in the relationship, for each *individual* document, between a researcher's amount of expertise in this subject and the relevance of that document to the researcher's scientific endeavors. A "relevant" document is defined as

$$\text{"Relevant"} = \underset{\sim}{\mathrm{V}} = \left\{ \frac{0.0}{\text{irrelevant}} + \frac{0.2}{\text{tangential}} + \frac{1.0}{\text{relevant}} + \frac{1.0}{\text{crucial}} \right\}$$

A "knowledgeable" researcher is defined as

$$\text{"Knowledgeable"} = \underset{\sim}{\mathrm{K}} = \left\{ \frac{0.0}{\text{novice}} + \frac{0.3}{\text{student}} + \frac{0.8}{\text{graduate}} + \frac{1.0}{\text{expert}} \right\}$$

Consider a certain very in-depth document. A knowledgeable researcher will find this document relevant to the research at hand (whereas it might be too complex for a less knowledgeable researcher). There is a relationship ($\underset{\sim}{\mathrm{R}}$) between $\underset{\sim}{\mathrm{K}}$ and $\underset{\sim}{\mathrm{V}}$.

(*a*) Find the relation IF $\underset{\sim}{\mathrm{K}}$, THEN $\underset{\sim}{\mathrm{V}}$ for this in-depth document, using classical implication.

(*b*) A "freshman" researcher would have a different view of the same document. A freshman could be described by another fuzzy set on the universe of researchers, say

$$\underset{\sim}{F} = \text{"freshman"} = \left\{ \frac{0.5}{\text{novice}} + \frac{0.7}{\text{student}} + \frac{0.0}{\text{graduate}} + \frac{0.0}{\text{expert}} \right\}$$

Find the relevancy of the in-depth document to this freshman using max-min composition on the relation found in part (*a*).

7.17. A company sells a product called a video multiplexer that multiplexes the video signal from 16 video cameras into a single video cassette recorder (VCR). The product has a unit motion detection feature that can increase the frequency with which a given camera's video is recorded to tape depending on the amount of motion that is present. It does this by recording more information from that camera at the expense of the amount of the video that is recorded from the other 15 cameras. Define the universe of X to be the discrete speed of the objects that are present in the video of camera 1 (there are 16 cameras); i.e., X = {low speed, medium speed, high speed} = {LS, MS, HS}.

Define the universe of Y to represent the discrete frequency with which the video from camera 1 is recorded to VCR tape, i.e., the recording rate of camera 1: or Y = {slow recording rate, medium recording rate, fast recording rate} = {SRR, MRR, FRR}. Two linguistic variables are defined to develop a relationship between these two universes. Define fuzzy variable $\underset{\sim}{A}$ to be "slow-moving object" present in camera 1, and fuzzy variable $\underset{\sim}{B}$ to be "slow recording rate," which is biased to the slow side in the new multiplexer product. The two membership functions are as follows:

$$\underset{\sim}{A} = \left\{ \frac{1}{\text{LS}} + \frac{.4}{\text{MS}} + \frac{.2}{\text{HS}} \right\}$$

$$\underset{\sim}{B} = \left\{ \frac{1}{\text{SRR}} + \frac{.5}{\text{MRR}} + \frac{.25}{\text{FRR}} \right\}$$

(*a*) Calculate the associated fuzzy logic relation for the compound proposition, "If *x* is $\underset{\sim}{A}$, then *y* is $\underset{\sim}{B}$" using classical implication.

(*b*) We pose another fuzzy linguistic variable $\underset{\sim}{A}'$ = "fast-moving object" present in video camera 1, which is also defined on universe X, with membership function,

$$\underset{\sim}{A}' = \left\{ \frac{.1}{\text{LS}} + \frac{.3}{\text{MS}} + \frac{1}{\text{HS}} \right\}$$

Using max-min composition, find the multiplexer's recording rate associated with the relation developed in part (*a*).

7.18. In a problem concerning character recognition, we would like to establish an approximate relationship between occurrences of special characters in the Czech language and readability of a word by nonnative Czech speakers. Let's take three special symbols creating the universe of special characters, namely M = { ˇ , ´ , ° }. Readability, X, will be defined over the universe of the following measures:

e—eye movement

s—silent self-pronunciation

d—distinction (apparent distinction of characters allowing guessing at the first glance rather than detailed decoding)

Let us consider the Czech word *žížaliččca,* which maps to our symbol universe M as

$$\underset{\sim}{A} = \left\{ \frac{.9}{\text{ˇ}} + \frac{.3}{\text{´}} + \frac{0}{\text{˚}} \right\}$$

and readability is

$$\underset{\sim}{B} = \left\{ \frac{.3}{e} + \frac{.9}{s} + \frac{.7}{d} \right\}$$

The readability $\underset{\sim}{B}$ is linguistically equivalent to "pretty darn difficult," and $\underset{\sim}{A}$ is "common" use of the special characters for the Czech word *žížaliččca.*

(*a*) Establish a relation, $\underset{\sim}{R}$, for the classical implication operation IF $\underset{\sim}{A}$, THEN $\underset{\sim}{B}$.

(*b*) Take a less common Czech word, say *úl*, with membership function

$$\underset{\sim}{A}' = \left\{ \frac{0}{\text{ˇ}} + \frac{.3}{\text{´}} + \frac{0}{\text{˚}} \right\}$$

and obtain the associated readability using max-min composition. What might the linguistic meaning of this readability be?

(*c*) On the basis of this approximate reasoning, would you say that it's "more wrong" to have "rather difficult readability" words, rather than "easily readable" words in the Czech language for native speakers and native readers?

7.19. In computer systems there is a relationship between CPU board temperatures and power supply voltage. Let us consider the following relation: IF the temperature (in degrees Fahrenheit) is high, THEN the power supply voltage (in volts) will drop or become low. Let

$$\underset{\sim}{A} = \text{temperature is high.}$$
$$\underset{\sim}{B} = \text{voltage is low.}$$
$$\underset{\sim}{A} \rightarrow \underset{\sim}{B} = \text{IF the temperature is high, THEN voltage will be low.}$$

The following membership functions might be appropriate for these two variables:

$$\underset{\sim}{A} = \left\{ \frac{.1}{50} + \frac{.5}{75} + \frac{.7}{100} + \frac{.9}{125} + \frac{1}{150} \right\}$$

$$\underset{\sim}{B} = \left\{ \frac{1}{4.0} + \frac{.8}{4.25} + \frac{.5}{4.5} + \frac{.2}{4.75} + \frac{0}{5.0} \right\}$$

(*a*) Using classical implication, find the relation $\underset{\sim}{A} \rightarrow \underset{\sim}{B}$.

(*b*) Suppose we consider another temperature; say, let $\underset{\sim}{A}$ = temperature is very high.

$$\underset{\sim}{A} = \left\{ \frac{0}{50} + \frac{.2}{75} + \frac{.4}{100} + \frac{.6}{125} + \frac{1}{150} \right\}$$

Find the appropriate voltage for this temperature using max-product composition with the relation found in part (*a*).

7.20. Application of a spray-on foam insulation to the large external tank for the space shuttle is implemented in a special manufacturing spray cell. The tank is rotated on a turntable as foam is sprayed on by a spray gun moving vertically up the cell wall on an elevator. Erratic motion of the turntable servodrive due to worn or unlubricated bearings will cause an expensive spray abort. We can employ two process variables to characterize the health of the bearing system: motor armature drive current and the variability of the current. We define these variables as fuzzy sets on the universe of currents [1, 5] amps, and on the universe of variability in current [1, 5] in amps per second, as follows:

$$\underset{\sim}{A} = \text{drive current} = \left\{ \frac{.3}{3} + \frac{1}{4} + \frac{.2}{5} \right\}$$

$$\underset{\sim}{B} = \text{variability} = \left\{ \frac{.5}{2} + \frac{1}{3} + \frac{.6}{4} \right\}$$

Using Mamdani implication, find a fuzzy relation, $\underset{\sim}{R}$, for these variables.

7.21. In distribution circuits, fault indicators (FI) are used to indicate problems in the circuit. Consider a very sophisticated FI that indicated the degree of fault according to a color band. If the fault indicator shows the color bright red, this would be an indication of a fault, which means some fault current passed through the circuit. If the FI shows the color green, then a fault does not occur within the indicator device. But what if the color band indicates orange, or reddish-orange? How would an intelligent system assess this situation? We can express this as, IF FI is "sort-of red," THEN fault is medium, or $\underset{\sim}{A} \rightarrow \underset{\sim}{B}$.

Let the fuzzy set $\underset{\sim}{A} \subset X$, where X is described in terms of discrete color frequency units on the nondimensional interval [0, 4]. Let the fuzzy set $\underset{\sim}{B} \subset Y$, where Y is also a discrete universe of degrees of fault on the nondimensional interval [0, 4]. Define the following membership functions for the variables "sort of red" and "medium fault:"

$$\underset{\sim}{A} = \left\{ \frac{.5}{1} + \frac{.6}{2} + \frac{.1}{3} + \frac{.8}{4} \right\}$$

$$\underset{\sim}{B} = \left\{ \frac{0}{2} + \frac{1}{3} + \frac{.5}{4} \right\}$$

Notice in this case that fuzzy set $\underset{\sim}{A}$ is nonnormal and nonconvex, which indicates that the color "sort of red" does not have unit membership in any of the discrete frequencies, and it results from a "mixture" of standard colors. This result is entirely possible in this situation where the color spectrum has been discretized and we want to accommodate nonstandard colors. In addition, we have the capability of fuzzifying a typical binary output: fault or no-fault. In this problem we can consider degrees of fault between no-fault [0] and fault [4].

(*a*) Using Mamdani implication, find a fuzzy relation, $\underset{\sim}{R}$, for $\underset{\sim}{A} \rightarrow \underset{\sim}{B}$.

(*b*) Suppose we want to consider another FI color, say

$$\underset{\sim}{A}' = \left\{ \frac{.4}{1} + \frac{.5}{2} + \frac{0}{3} + \frac{1}{4} \right\}$$

Find the degree of fault associated with this color using max-product composition on the relation, $\underset{\sim}{R}$, found in part (*a*).

7.22. In astrophysics it is well known that most asteroids are confined to the asteroid belt, a region extending roughly from Mars to Jupiter, well out of the way of Earth's orbit. But every few years, someone discovers an asteroid that has an orbit passing dangerously close to Earth. These asteroids are sometimes not even discovered until they are within striking distance. This makes it important to determine quickly whether a newly found heavenly object is a threat. Suppose an astronomer devises a way of identifying dangerous objects, as expressed in the following implication:

IF the object is moving quickly when it is discovered, THEN it is a potentially dangerous asteroid.

Based on past occurrences, the astrophysicist then assigns speeds to the objects, where "speed" is on the universe X = [0, 5] of 10^6 mph, with

$$\underset{\sim}{S} = \text{high speed} = \left\{ \frac{.6}{4} + \frac{1}{5} \right\}$$

and an encounter distance,

$$\underset{\sim}{T} = \text{too close} = \left\{ \frac{.3}{1} + \frac{1}{2} + \frac{.5}{3} \right\}$$

on the universe Y of log(distance in 10^8 miles), where Y = [1, 6].

(*a*) Express the foregoing implication by the relation $\underset{\sim}{R}$, using correlation-product implication, (Eq. 7.25) i.e.,

$$\mu_{\underset{\sim}{R}}(x, y) = \max[\mu_{\underset{\sim}{S}}(x)\mu_{\underset{\sim}{T}}(y), 1 - \mu_{\underset{\sim}{S}}(x)]$$

(*b*) One night a new moving object is discovered, and it is found to have velocity

$$\underset{\sim}{S}' = \text{"might be fast"} = \left\{ \frac{.5}{3} + \frac{1}{4} + \frac{.2}{5} \right\}$$

Using max-product composition, find the fuzzy distance associated with this velocity.

7.23. In a CDMA cellular telephone system, average voice quality of a mobile unit is related to how many active mobiles are in the same sector and on the same frequency. To model the system using fuzzy logic, one could relate the number of mobile units to voice quality. Suppose the following holds: IF "Number of active cells is about 30," THEN "bad voice quality." Let us define fuzzy sets for "about 30" on a universe of integers and for "bad voice quality:"

$$\underset{\sim}{A} = \text{"}\sim 30\text{"} = \left\{ \frac{0.0}{10} + \frac{0.0}{15} + \frac{0.1}{20} + \frac{0.6}{25} + \frac{1.0}{30} + \frac{0.3}{35} \right\}$$

$$\underset{\sim}{B} = \text{"bad voice quality"} = \left\{ \frac{1.0}{0} + \frac{0.4}{1} + \frac{0.1}{2} + \frac{0}{3} \right\}$$

(0 being bad and 3 being good).

(*a*) Find the relation for $\underset{\sim}{A} \rightarrow \underset{\sim}{B}$ using Mamdani implication.

(*b*) For another antecedent, say "about 20 active cells" find the associated voice quality using classical max-min composition.

$$\underset{\sim}{A}' = \text{"about 20 active cells"} = \left\{ \frac{0.1}{10} + \frac{0.4}{15} + \frac{1.0}{20} + \frac{0.7}{25} + \frac{0.1}{30} + \frac{0.0}{35} \right\}$$

7.24. The question as to when a man is bald as determined by the number of hairs on his head relates to the classic sorites. To address this question, let us define two linguistic variables: one is "old" and the another is "bald," $\underset{\sim}{O}$ and $\underset{\sim}{B}$, respectively.

$$\underset{\sim}{B} = \left\{ \frac{1}{< 500} + \frac{1}{< 1000} + \frac{0.8}{< 2000} + \frac{0.6}{< 3000} + \frac{0.2}{< 4000} + \frac{0}{> 4000} \right\}$$

where the denominators are numbers of strands of head-hair, and

$$\underset{\sim}{\mathrm{O}} = \left\{ \frac{0}{<30} + \frac{0.2}{<40} + \frac{0.4}{<50} + \frac{0.6}{<60} + \frac{0.8}{<70} + \frac{1}{>70} \right\}$$

where the denominators are years.

(*a*) We want to determine if the variables are related; do we think $\underset{\sim}{\mathrm{O}} \rightarrow \underset{\sim}{\mathrm{B}}$? Construct the relationship between $\underset{\sim}{\mathrm{O}}$ and $\underset{\sim}{\mathrm{B}}$ for $\underset{\sim}{\mathrm{O}} \rightarrow \underset{\sim}{\mathrm{B}}$ using

$$\underset{\sim}{\mathrm{R}} = (\underset{\sim}{\mathrm{O}} \times \underset{\sim}{\mathrm{B}}) \cup (\overline{\underset{\sim}{\mathrm{O}}} \times \mathrm{Y})$$

(*b*) Let us use *modus ponens* to find a new fuzzy logic relation for a "very old" man. Suppose $\underset{\sim}{\mathrm{O}}'$ now represents very old,

$$\underset{\sim}{\mathrm{O}}' = \left\{ \frac{0}{<30} + \frac{0.04}{<40} + \frac{0.16}{<50} + \frac{0.36}{<60} + \frac{0.64}{<70} + \frac{1}{>70} \right\}$$

How "bald" is this man?

(*c*) Notice that the relation $\underset{\sim}{\mathrm{R}}$ in part (*a*) is constructed from the domain of "old" and "bald." What would result if we use this relation in an approximate reasoning sense for the antecedent "young" (not old)? Show the result using *modus ponens,* then discuss the results for the following membership function for "young."

$$\underset{\sim}{\mathrm{Y}} = \left\{ \frac{1}{<30} + \frac{0.8}{<40} + \frac{0.6}{<50} + \frac{0.3}{<60} + \frac{0.1}{<70} + \frac{0}{>70} \right\}$$

7.25. Simulator data for flight testing of high-performance jets usually are in tables composed of hard breakpoints. By using fuzzy logic, the flight test speed can be characterized approximately, such as "near" or "in the region of" the database breakpoints, and their acceptability as being useful to the simulation can also be characterized in a non-crisp manner. Consider a relation that states, "IF the data are near mach 0.74, THEN the data quality is acceptable." Define the two fuzzy sets "near mach 0.74" and "acceptable" as

$$\underset{\sim}{\mathrm{M}} = \text{“near mach 0.74”} = \left\{ \frac{0}{.730} + \frac{0.8}{.735} + \frac{1}{.74} + \frac{0.8}{.745} + \frac{0}{.75} \right\}$$

$$\underset{\sim}{\mathrm{Q}} = \text{“acceptable”} = \left\{ \frac{0}{0} + \frac{0}{.25} + \frac{.25}{.5} + \frac{.8}{.75} + \frac{1}{1} \right\}$$

(*a*) Using Mamdani implication, find this relation.

(*b*) Using max-min composition, find the acceptability of another flight test speed, "in the region of mach 0.74," where

$$\underset{\sim}{\mathrm{M}} = \text{“in the region of mach 0.74”} = \left\{ \frac{0}{.730} + \frac{0.4}{.735} + \frac{0.8}{.74} + \frac{1}{.745} + \frac{0.6}{.75} \right\}$$

7.26. When gyros are calibrated for axis bias, they are matched with a temperature. Thus, we can have a relation of gyro bias (GB) vs. temperature ($\underset{\sim}{\mathrm{T}}$). Suppose we have fuzzy sets for a given gyro bias and a given Fahrenheit temperature, as follows:

$$\mu_{\underset{\sim}{\mathrm{GB}}}(x) = \left\{ \frac{0.2}{3} + \frac{0.4}{4} + \frac{1}{5} + \frac{0.4}{6} + \frac{0.2}{7} \right\} \quad \text{bias in degrees Fahrenheit per hour}$$

$$\mu_{\underset{\sim}{\mathrm{T}}}(y) = \left\{ \frac{0.4}{66} + \frac{0.6}{68} + \frac{1}{70} + \frac{0.6}{72} + \frac{0.4}{74} \right\} \quad \text{temperature in degrees Fahrenheit}$$

(*a*) Use a Mamdani implication to find the relation IF gyro bias, THEN temperature.

(*b*) Suppose we are given a new gyro bias ($\underset{\sim}{\text{GB}}'$) as follows:

$$\mu_{\underset{\sim}{\text{GB}}'}(x) = \left\{ \frac{.6}{3} + \frac{1}{4} + \frac{.6}{5} \right\}$$

Using max-min composition, find the temperature associated with this new bias.

7.27. A laser imaging printer used for nuclear medicine prints an 8-bit gray-scale image on special carbon-based media. After printing the image on the media, lamination is required to ensure that image integrity is maintained. This process requires applying a protective adhesive at high pressure, at high temperature, and at a speed that is not excessive in relation to print cycle-time. Lamination quality is evaluated on a metric called "haze." Define the two universes of discourse for the variables image haze and lamination speed as

Universe Y = image haze {1, 2, 3, 4, 5}; The lowest member is a haze-free image

Universe X = lamination speed {1, 2, 3, 4, 5}; Highest number is the fastest speed

(*a*) If we define two particular fuzzy sets on these universes as

$$\underset{\sim}{A} = \text{“fast lamination speed”} = \left\{ \frac{.6}{2} + \frac{1}{3} + \frac{.7}{4} \right\}$$

$$\underset{\sim}{B} = \text{“very desirable haze”} = \left\{ \frac{.2}{3} + \frac{1}{4} + \frac{.2}{5} \right\}$$

then find the relation, IF "fast lamination speed" is required, THEN "very desirable haze" is observed on the printed image, using classical implication.

(*b*) If lamination is done at "low speed,"

$$A' = \left\{ \frac{.5}{1} + \frac{1}{2} + \frac{.3}{3} \right\}$$

then find the fuzzy consequent "acceptable haze" artifacts are observed, using max-min composition and the relation found in part (*a*).

7.28. Suppose we are trying to maintain a consistent average outgoing quality level in a manufacturing process. Attribute imperfections are assigned a severity of S = {0, 2, 3, 4, 5} and the universe of sampling rate (on a frequency basis) is F = {200, 250, 300, 350, 400}. A severity of zero means "no imperfection." A severity of 5 means "severe imperfection." A sample is evaluated and the result is denoted by two fuzzy sets $\underset{\sim}{A}$ for "medium severity" and $\underset{\sim}{B}$ for "moderate sampling rate." We wish to construct a compound proposition, IF $\underset{\sim}{A}$ THEN $\underset{\sim}{B}$. The evaluator has assigned the following fuzzy sets to represent the linguistic assignments:

$$\underset{\sim}{A} = \text{“medium severity”} = \left\{ \frac{.4}{2} + \frac{1}{3} + \frac{.5}{4} \right\}$$

$$\underset{\sim}{B} = \text{“moderate sampling rate”} = \left\{ \frac{0.6}{200} + \frac{0.9}{250} + \frac{0.5}{300} + \frac{0.2}{350} \right\}$$

(*a*) Find the relation expressing this compound proposition using classical implication.

(*b*) Suppose the imperfection severity for a new batch of samples changes to "severe imperfection," i.e., the membership function for the new antecedent is

$$\underset{\sim}{A}' = \left\{ \frac{0.6}{3} + \frac{1}{4} + \frac{0.6}{5} \right\}$$

Find the sampling rate associated with this imperfection.

7.29. Consider the possible types of electronic designs: the universe X is defined to be {RF analog, LF analog, digital, mixed}, where

RF = radio/high frequency analog design
LF = low frequency analog design
digital = digital logic design
mixed = a combination of two or more of the above

The universe Y is the complexity of the design and is defined to be Y = {simple, moderate, complicated}. We define two fuzzy sets on these universes:

$$\underset{\sim}{T} = \text{type of my work} = \left\{ \frac{0}{\text{RF}} + \frac{.05}{\text{LF}} + \frac{1.0}{\text{digital}} + + \frac{.1}{\text{mixed}} \right\}$$

$$\underset{\sim}{D} = \text{depth of my work} = \left\{ \frac{.1}{\text{simple}} + + \frac{1.0}{\text{moderate}} + + \frac{.5}{\text{complicated}} \right\}$$

(*a*) Consider the implication that if we know the type of the next design job we can calculate the depth, that is, $\underset{\sim}{T} \rightarrow \underset{\sim}{D}$. Find this relation using classical implication.

(*b*) Consider your next job to be the following:

$$\underset{\sim}{J} = \text{job type} = \left\{ \frac{0}{\text{RF}} + \frac{0}{\text{LF}} + \frac{1}{\text{digital}} + \frac{0}{\text{mixed}} \right\}$$

What is the value of job complexity associated with this next job using max-min composition?

7.30. Suppose we are doing the preliminary design of a new video system. In the design we have a video camera mounted on a positioning device known as a pan and tilt motor. We want to relate the speed of the pan and tilt motor to the distance of this motor from its desired position. The speed universe is X = {0, 1, 2, 3, 4, 5} in degrees per second and the distance universe is Y = {0, 45, 90, 135, 180} in degrees. We define two fuzzy sets on these universes, namely,

$$\underset{\sim}{S} = \text{"slow"} = \frac{1/4}{0} + \frac{1}{1} + \frac{1/4}{2}$$

$$\underset{\sim}{D} = \text{"not near and not far"} = \frac{1/4}{90} + \frac{1/2}{135} + \frac{1/4}{180}$$

Find a fuzzy relation expressing the implication IF $\underset{\sim}{S}$, THEN $\underset{\sim}{D}$.

7.31. In a radar system, the pulse repetition rate (frequency-PRF) is chosen on some function range. Let ranges be defined on {10, 20, 40, 80, 160} km, and let PRFs be defined on {1000, 2000, 4000, 8000} Hz. Suppose a medium range is

$$\underset{\sim}{A} = \left\{ \frac{0}{10} + \frac{.5}{20} + \frac{1}{40} + \frac{.2}{80} + \frac{0}{160} \right\}$$

and a medium PRF is

$$\underset{\sim}{B} = \left\{ \frac{0}{1000} + \frac{.3}{2000} + \frac{1}{4000} + \frac{.6}{8000} \right\}$$

Using classical implication find a fuzzy relation expressing $\underset{\sim}{A} \rightarrow \underset{\sim}{B}$.

7.32. In the production of a photographic plate there are two key variables: exposure time and development time. Suppose we represent each of these variables as specific fuzzy sets,

$$\underset{\sim}{A} = \text{“exposure”} = \left\{ \frac{0}{0} + \frac{1}{1} + \frac{.7}{2} + \frac{.1}{3} \right\}$$

which is defined on the universe X = [0, 3] seconds, and

$$\underset{\sim}{B} = \text{“ development time”} = \left\{ \frac{0}{0} + \frac{.2}{1} + \frac{.3}{2} + \frac{.5}{3} + \frac{.7}{4} + \frac{1}{5} + \frac{.9}{6} + \frac{.6}{7} \right\}$$

which is defined on the universe Y = [0, 7] minutes.

(*a*) Construct a relation for the compound proposition IF $\underset{\sim}{A}$, THEN $\underset{\sim}{B}$ based on Mamdani implication.

(*b*) Let a new exposure time be represented by

$$\underset{\sim}{A}' = \text{“about 1”} = \left\{ \frac{0}{0} + \frac{1}{1} + \frac{.4}{2} + \frac{0}{3} \right\}$$

Using max-min composition, find the fuzzy development time associated with the new exposure time.

7.33. You are asked to develop a controller to regulate the temperature of a room. Knowledge of the system allows you to construct a simple rule of thumb: When the temperature is HOT then cool room down by turning the fan at the fast speed, or, expressed in rule form, IF temperature is HOT, THEN fan should turn FAST. Fuzzy sets for hot temperature and fast fan speed can be developed, for example,

$$\underset{\sim}{H} = \text{“hot”} = \left\{ \frac{0}{60} + \frac{.1}{70} + \frac{.7}{80} + \frac{.9}{90} + \frac{1}{100} \right\}$$

represents universe X in °F.

$$\underset{\sim}{F} = \text{“fast”} = \left\{ \frac{0}{0} + \frac{.2}{1} + \frac{.5}{2} + \frac{.9}{3} + \frac{1}{4} \right\}$$

represents universe Y in 1000 rpm.

(*a*) From these two fuzzy sets construct a relation for the rule using classical implication.

(*b*) Suppose a new rule uses a slightly different temperature, say “moderately hot,” and is expressed by the fuzzy membership function for “moderately hot,” or

$$\underset{\sim}{H}' = \left\{ \frac{0}{60} + \frac{.2}{70} + \frac{1}{80} + \frac{1}{90} + \frac{1}{100} \right\}$$

Using max-product composition, find the resulting fuzzy fan speed.

7.34. Let $\underset{\sim}{A}$ be a fuzzy set of power supply voltage for given current source and let $\underset{\sim}{B}$ be a fuzzy set of reference currents being output from the circuit. For this problem the fuzzy sets can be fuzzy numbers, or they can be fuzzy linguistic expressions. Suppose we consider fuzzy numbers for the voltage and current,

$$\underset{\sim}{\mathrm{A}} = \text{“about 14 volts”} = \left\{ \frac{.5}{12} + \frac{.8}{13} + \frac{1}{14} + \frac{.8}{15} \right\}$$

$$\underset{\sim}{\mathrm{B}} = \text{“very nearly 5 amperes”} = \left\{ \frac{.7}{4.9} + \frac{1}{5.0} + \frac{.8}{5.1} \right\}$$

(*a*) Construct a relation using classical implication for the rule, IF the voltage being supplied is "about 14 volts," THEN the current output of the motor is "very nearly 5 amperes."

(*b*) Now, suppose we want to represent the voltage by a fuzzy expression, such as

$$\underset{\sim}{\mathrm{A}}' = \text{“around 14 volts”} = \left\{ \frac{.2}{13} + \frac{.8}{14} + \frac{1}{15} + \frac{.8}{16} \right\}$$

Find the fuzzy current associated with this new voltage. What would you call this current linguistically?

7.35. You are faced with the problem of controlling a motor that is subjected to a variable load. The motor must maintain a constant speed, regardless of the load placed on it; therefore, the voltage applied to the motor must change to compensate for changes in load. Define fuzzy sets for motor speed (rpm) and motor voltage (volts) as follows:

$$\underset{\sim}{\mathrm{A}} = \text{“motor speed OK”} = \left\{ \frac{.3}{20} + \frac{.6}{30} + \frac{.8}{40} + \frac{1}{50} + \frac{.7}{60} + \frac{.4}{70} \right\}$$

$$\underset{\sim}{\mathrm{B}} = \text{“motor voltage nominal”} = \left\{ \frac{.1}{1} + \frac{.3}{2} + \frac{.8}{3} + \frac{1}{4} + \frac{.7}{5} + \frac{.4}{6} + \frac{.2}{7} \right\}$$

(*a*) Using classical implication, find a relation for the following compound proposition: IF motor speed is OK, THEN motor voltage is nominal.

(*b*) Now specify a new antecedent $\underset{\sim}{\mathrm{A}}'$ = "motor speed a little slow," where

$$\underset{\sim}{\mathrm{A}}' = \left\{ \frac{.4}{20} + \frac{.7}{30} + \frac{1}{40} + \frac{.6}{50} + \frac{.3}{60} + \frac{.1}{70} \right\}$$

Using max-min composition (i.e., $\mathrm{B}' = \mathrm{A}' \circ \mathrm{R}$), find the new consequent.

7.36. In the art of navigation, gyros are often used along with accelerometers to navigate. Biases are calculated on gyros. But as the cost of the gyro goes down, the turn-on-to-turn-off reliability of the bias goes down; i.e., decreased gyro cost implies decreased reliability. Let us pose a bias, called *x*-gyro bias, as denoted by the symbol x_{gb}. This problem is actually continuous, but we will discretize it. We have uncertainty about the nominal value of the bias, and we construct a fuzzy set about the nominal x_{gb}, as follows:

$$\underset{\sim}{\mathrm{A}} = \left\{ \frac{.2}{x_{\text{gb}} - 3\delta_x} + \frac{.4}{x_{\text{gb}} - 2\delta_x} + \frac{.6}{x_{\text{gb}} - \delta_x} + \frac{.8}{x_{\text{gb}}} + \frac{.6}{x_{\text{gb}} + \delta_x} + \frac{.4}{x_{\text{gb}} + 2\delta_x} + \frac{.2}{x_{\text{gb}} + 3\delta_x} \right\}$$

For this problem, let $x_{\text{gb}} = 2°/\text{hour}$ and $\delta_x = 0.1$. We get

$$x\text{-gyro bias} = \underset{\sim}{\mathrm{A}} = \left\{ \frac{.2}{1.7} + \frac{.4}{1.8} + \frac{.6}{1.9} + \frac{.8}{2.0} + \frac{.6}{2.1} + \frac{.4}{2.2} + \frac{.2}{2.3} \right\}$$

Let $\underset{\sim}{B}$ be a fuzzy set describing an accelerator bias in the x direction, where the normal bias $= 0.3g$, where g is the acceleration due to gravity. The membership function might look like

$$\underset{\sim}{B} = \left\{ \frac{.1}{.25} + \frac{.4}{.27} + \frac{.9}{.3} + \frac{.4}{.33} + \frac{.1}{.35} \right\}$$

(*a*) Using Mamdani implication, find a relation for IF $\underset{\sim}{A}$, THEN $\underset{\sim}{B}$.

(*b*) Say we have to change x-gyros and the new gyro has the following fuzzy bias:

$$\underset{\sim}{A}' = \left\{ \frac{0}{1.7} + \frac{.5}{1.8} + \frac{.7}{1.9} + \frac{.95}{2.0} + \frac{.7}{2.1} + \frac{.5}{2.2} + \frac{0}{2.3} \right\}$$

Calculate the associated accelerometer bias using

(*i*) max-min composition, $\underset{\sim}{T} = \underset{\sim}{A}' \circ \underset{\sim}{R}$

(*b*) max-product composition, $\underset{\sim}{T} = \underset{\sim}{A}' \circ \underset{\sim}{R}$

(*c*) In a similar fashion we define a fuzzy set for a gyro bias in the y direction of the gyro, denoted y_{gb}, and for $y_{gb} = 3.5°/\text{hour}$, $\delta_y = 0.15$, we get

$$y\text{-gyro bias} = \underset{\sim}{C} = \left\{ \frac{.3}{3.05} + \frac{.5}{3.2} + \frac{.6}{3.35} + \frac{.75}{3.5} + \frac{.6}{3.65} + \frac{.5}{3.8} + \frac{.3}{3.95} \right\}$$

Repeat steps (*a*) and (*b*) using the y-gyro bias in place of an x-gyro bias and using the accelerometer bias in the x-direction as given.

7.37. In network computing many applications involve communication between two separate systems interconnected via a network. Two metrics of interest during the interaction of these client-and-server systems are the "response" on the system where the user resides (client) and the "load" on the remote system (server). Let X represent the universe of response, X = {1, 2, 3, 4, 5}, and Y, the universe of load, Y = {1, 2, 3, 4}, where lower numbers correspond to "faster response" and "lighter load," respectively. Now let us define two fuzzy variables $\underset{\sim}{A}$ and $\underset{\sim}{B}$ representing "average response" and "medium load" where

$$\underset{\sim}{A} = \left\{ \frac{.2}{1} + \frac{.5}{2} + \frac{1}{3} + \frac{.3}{4} + \frac{0}{5} \right\} \quad \text{and} \quad \underset{\sim}{B} = \left\{ \frac{0}{1} + \frac{.7}{2} + \frac{.4}{3} + \frac{.1}{4} \right\}$$

(*a*) Find the implication $\underset{\sim}{A} \rightarrow \underset{\sim}{B}$, using the classical approach, i.e.,

$$\underset{\sim}{R} = (\underset{\sim}{A} \times \underset{\sim}{B}) \cup (\overline{\underset{\sim}{A}} \times \text{Y})$$

or

$$\mu_{\underset{\sim}{R}}(x, y) = \max\{[\mu_{\underset{\sim}{A}}(x) \wedge \mu_{\underset{\sim}{B}}(y)], (1 - \mu_{\underset{\sim}{A}}(x))\}$$

(*b*) Now, what "degree of load" would be associated with a new fuzzy set $\underset{\sim}{A}'$ denoting "quick response?" Let

$$\underset{\sim}{A}' = \left\{ \frac{.9}{1} + \frac{.7}{2} + \frac{.3}{3} + \frac{.1}{4} + \frac{0}{5} \right\}$$

Using max-product composition, find $\underset{\sim}{B}' = \underset{\sim}{A}' \circ \underset{\sim}{R}$. What might this "degree of load" be called?

7.38. A camera system is judged on the basis of picture quality. Determining whether a picture is good is very subjective. The output can be affected by the quality of the camera as well as the quality of film. A possible universe of camera ratings is X = {1, 2, 3, 4, 5}, where 1 represents the highest camera rating. A possible universe of picture ratings is Y = {1, 2, 3, 4, 5}, where once again, 1 is the highest rating for pictures. We now define two fuzzy sets,

$$\underset{\sim}{A} = \text{“above average camera”} = \left\{ \frac{.7}{1} + \frac{.9}{2} + \frac{.2}{3} + \frac{0}{4} + \frac{0}{5} \right\}$$

$$\underset{\sim}{B} = \text{“above average picture quality”} = \left\{ \frac{.6}{1} + \frac{.8}{2} + \frac{.5}{3} + \frac{.1}{4} + \frac{0}{5} \right\}$$

(*a*) From the proposition, IF $\underset{\sim}{A}$, THEN $\underset{\sim}{B}$, find the relation using Mamdani implication.

(*b*) Suppose the camera manufacturer wants to improve camera and film sales by improving the quality of the camera. A new camera is rated as follows:

$$\underset{\sim}{A}' = \text{“new and improved camera”} = \left\{ \frac{.8}{1} + \frac{.8}{2} + \frac{.1}{3} + \frac{0}{4} + \frac{0}{5} \right\}$$

What might be the resulting picture rating from this new camera?

7.39. In public transportation systems there often is a significant need for speed control. For subway systems, for example, the train speed cannot go too far beyond a certain target speed or the trains will have trouble stopping at a desired location in the station. Set up a fuzzy set

$$\underset{\sim}{A} = \text{“speed way over target”} = \left\{ \frac{0}{T_0} + \frac{.8}{T_0 + 5} + \frac{1}{T_0 + 10} + \frac{.8}{T_0 + 15} \right\}$$

on a universe of target speeds, say T = $[T_0, T_0 + 15]$, where T_0 is a lower bound on speed. Define another fuzzy set,

$$\underset{\sim}{B} = \text{“apply brakes with high strength”} = \left\{ \frac{.3}{10} + \frac{.8}{20} + \frac{.9}{30} + \frac{1}{40} \right\}$$

on a universe of braking pressures, say S = [10, 40].

(*a*) For the compound proposition, IF speed is “way over target,” THEN “apply brakes with high strength,” find a fuzzy relation using classical implication.

(*b*) For a new antecedent,

$$\underset{\sim}{A}' = \text{“speed moderately over target”} = \left\{ \frac{.2}{T_0} + \frac{.6}{T_0 + 5} + \frac{.8}{T_0 + 10} + \frac{.3}{T_0 + 15} \right\}$$

find the fuzzy brake pressure using max-min composition.

7.40. We want to consider the engineering of amplifiers. Here, the amplifier is a simple voltage-measuring input and current output, as shown in Fig. P7.40. We define two fuzzy linguistic variables for a fuzzy relation: $\underset{\sim}{V}_{\text{in}}$, the input voltage and $\underset{\sim}{I}_{\text{out}}$, the output current:

$$\underset{\sim}{V}_{\text{in}} = \text{“small”} = \left\{ \frac{.5}{.10} + \frac{1}{.20} + \frac{.8}{.30} + \frac{.2}{.40} \right\} \quad \text{volts}$$

$$\underset{\sim}{I}_{\text{out}} = \text{“big”} = \left\{ \frac{.3}{.6} + \frac{1}{1} + \frac{.5}{1.4} \right\} \quad \text{amperes}$$

where $\underset{\sim}{V}_{\text{in}}$ is defined on a universe of voltages, and $\underset{\sim}{I}_{\text{out}}$ is defined on a universe of currents.

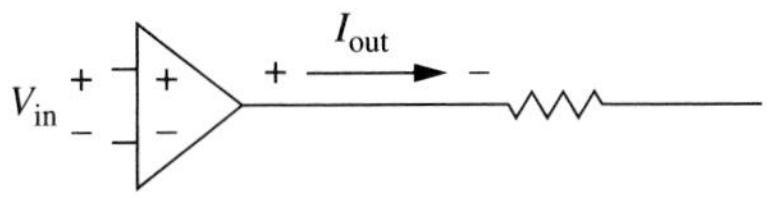

FIGURE P7.40

(*a*) Find the relation, IF $\underset{\sim}{V}_{in}$, THEN $\underset{\sim}{I}_{out}$ using classical implication.

(*b*) Another fuzzy linguistic variable in this problem is input impedance, $\underset{\sim}{Z}$. The higher the impedance, generally the better the amplifier. For the following impedance defined on a universe of resistances,

$$\underset{\sim}{Z} = \text{“high impedance”} = \left\{ \frac{0}{10^4} + \frac{.3}{10^5} + \frac{1}{10^6} + \frac{.6}{10^7} \right\} \quad \text{ohms}$$

find the relation, IF $\underset{\sim}{Z}$, THEN $\underset{\sim}{I}_{out}$, using Mamdani implication.

7.41. People use a subjective relationship between picture quality and the deviation from correct focus when setting their own focus levels as they take pictures with their camera. This process can be quantified using fuzzy logic. Let us define a fuzzy set $\underset{\sim}{A}$ = set of people who set “nearly correct” focus, as measured on the universe X = measure of deviation from correct focus, X = {−10, −5, 0, 5, 10} meters in focal length and a fuzzy set $\underset{\sim}{B}$ = set of “acceptable” picture quality, as measured on the universe Y = measure of picture quality {1 to 5} or {very bad to excellent}. Let

$$\underset{\sim}{A} = \left\{ \frac{.1}{-10} + \frac{.3}{-5} + \frac{1}{0} + \frac{.3}{5} + \frac{.1}{10} \right\}$$

$$\underset{\sim}{B} = \left\{ \frac{.1}{1} + \frac{.2}{2} + \frac{1}{3} + \frac{.2}{4} + \frac{.1}{5} \right\}$$

(*a*) For the compound proposition, IF the camera has a “nearly correct” focus, THEN the pictures will be “acceptable,” find a fuzzy relation using Mamdani implication.

(*b*) Suppose another camera operator uses an “incorrect” focus, say

$$\underset{\sim}{A}' = \left\{ \frac{0.9}{-10} + \frac{0.7}{-5} + \frac{0}{0} + \frac{0.7}{5} + \frac{0.9}{10} \right\}$$

Find the associated picture quality using max-min composition.

7.42. A drive system that rotates a large turntable consists of a motor, drive unit, gears, and bearings. The motor draws a nominal amount of current with respect to the turntable speed. However, if motor current increases while the revolutions per minute (RPM) are constant, then it is reasonable to assume that a binding problem exists somewhere in the gears and bearings. Let us define a fuzzy set $\underset{\sim}{I}$ on a universe of currents in amps, and a fuzzy set $\underset{\sim}{F}$ on a universe of bearing frictions (tangent of the friction angle) as

$$\underset{\sim}{I} = \text{“large motor current at 3 RPM”}$$
$$= \left\{ \frac{.3}{10} + \frac{.4}{11} + \frac{.5}{12} + \frac{.6}{13} + \frac{.7}{14} + \frac{.8}{15} + \frac{.9}{16} + \frac{1}{17} \right\}$$

$$\underset{\sim}{F} = \text{“high bearing friction”} = \left\{ \frac{0}{.1} + \frac{.2}{.2} + \frac{.4}{.3} + \frac{.6}{.4} + \frac{.8}{.5} + \frac{1}{.6} \right\}$$

(*a*) For the compound proposition: IF current is large, THEN bearing friction is high, find a fuzzy relation using Mamdani implication.

(*b*) Suppose another fuzzy set for currents is available, say

$$\underset{\sim}{I}' = \left\{ \frac{.7}{10} + \frac{.8}{11} + \frac{.9}{12} + \frac{1}{13} + \frac{.9}{14} + \frac{.8}{15} + \frac{.7}{16} + \frac{.6}{17} \right\}$$

Using fuzzy *modus ponens,* find the associated bearing friction using max-min composition.

7.43. Approximate reasoning can be very useful in a digital circuit design problem of sizing a CMOS (complementary metal oxide semiconductor) inverter to drive an output load capacitance. As the load capacitance increases, the W/L (width/length) ratio of both the N (*n* channel) MOS and P (*p* channel) MOS devices at the CMOS driver must increase proportionally in order to maintain the same delay time. However, other considerations such as area or power consumption must also be included in the design. A general design rule could be developed as

IF the output capacitance is large (in picofarads), THEN the W/L ratio should be high.

Let $\underset{\sim}{A}$ be a fuzzy set for "output capacitance is large": using Zadeh's notation,

$$\underset{\sim}{A} = \left\{ \frac{0}{.25} + \frac{.2}{.5} + \frac{.4}{1} + \frac{.6}{2} + \frac{.8}{4} + \frac{1}{8} \right\}$$

defined on universe X in picofarads, and

$$\underset{\sim}{B} = \text{"W/L ratio is high"} = \left\{ \frac{0}{2} + \frac{.1}{4} + \frac{.5}{8} + \frac{.8}{16} + \frac{.9}{32} + \frac{1}{64} \right\}$$

on universe Y.

(*a*) Find a fuzzy relation for the design rule using classical implication.

(*b*) Now suppose that we have a new output capacitance,

$$\underset{\sim}{A}' = \text{"output capacitance is very large"}$$

$$= \left\{ \sum \frac{(\mu_{\underset{\sim}{A}}(x))^2}{\underset{\sim}{A}} \right\} \approx \left\{ \frac{0}{.25} + \frac{0}{.5} + \frac{.1}{1} + \frac{.3}{2} + \frac{.6}{4} + \frac{1}{8} \right\}$$

Find the fuzzy W/L ratio associated with the new output capacitance.

7.44. Fuzzy logic is useful in the process of ranking hazardous waste consequences. We will define two fuzzy linguistic variables: gas concentration and high hazard. Now, consider the following compound propositions:

1. If gas concentration is above the lower flammability limit and gas concentration is equal to or below the upper flammability limit, then a high hazard exists.
2. If gas concentration is below the lower flammability limit, then a high hazard does not exist.
3. If gas concentration is above the upper flammability limit, then a high hazard does not exist.

These propositions are represented in the fuzzy membership function shown in Fig. P7.44*a*.

Now consider a relationship between gas concentration and the threshold limit concentration through the following propositions:

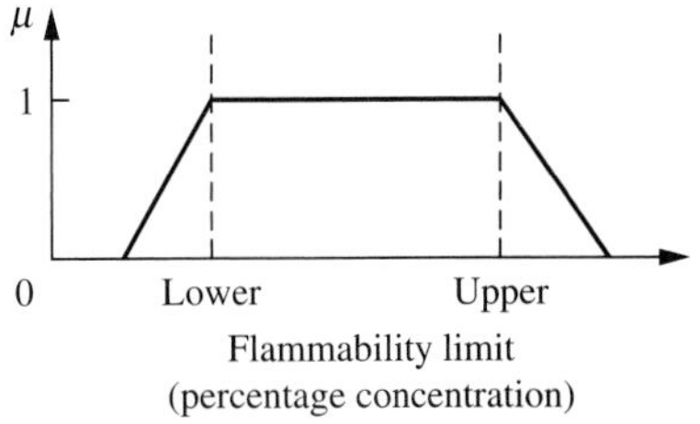

FIGURE P7.44*a*

1. If gas concentration is above the threshold limit value (TLV), then a hazard exists.
2. If gas concentration is equal to or greater than the immediately-dangerous-to-life-and-health value (IDLH), then a high hazard exists.

These propositions are represented in the fuzzy membership function shown in Fig. P7.44*b*.

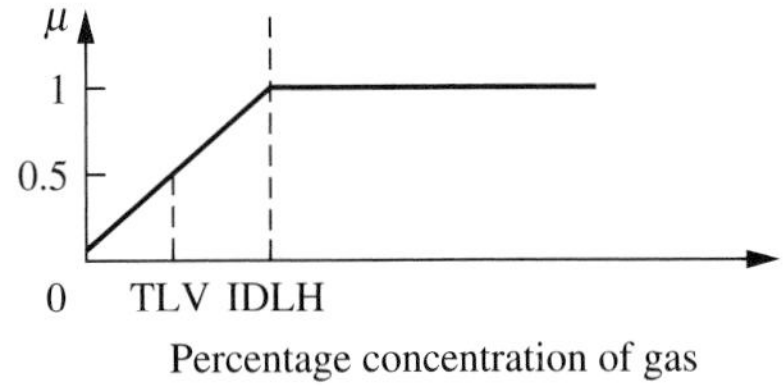

FIGURE P7.44*b*

Suppose we now want to combine the IDLH and the flammability limits in some way to relate to the fuzzy term "high hazard." Develop a procedure, using those we have described in this chapter, to relate the two preceding membership functions to a single membership function describing "high hazard." For purposes of this problem define the hazard to be the universe of numbers on the interval [0, 40], where 0 is no hazard and 40 is the highest hazard. Figure P7.44*c* provides a hint in getting started with this problem.

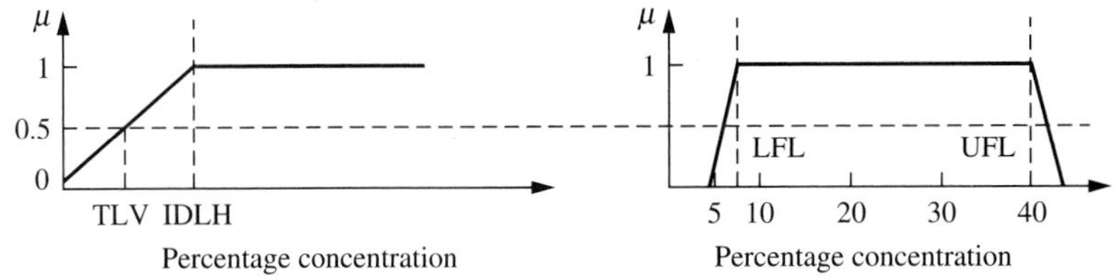

FIGURE P7.44*c*

7.45 For Example 7.9 in this chapter,

(*a*) Recalculate the fuzzy relation $\underset{\sim}{R}$ using

(*i*) Equation 7.20
(*ii*) Equation 7.22
(*iii*) Equation 7.23
(*iv*) Equation 7.24
(*v*) Equation 7.25

(*vi*) Equation 7.26
(*vii*) Equation 7.27
(*viii*) Equation 7.28

(*b*) Recalculate the fuzzy consequent $\underset{\sim}{B}'$ using
(*i*) Equation 7.31
(*ii*) Equation 7.32
(*iii*) Equation 7.33
(*iv*) Equation 7.34
(*v*) Equation 7.35, where $f(\bullet) = \begin{cases} 1 - e^{-x}, & \text{for } x \geq 0 \\ e^{x} - 1, & \text{for } x \leq 0 \end{cases}$

7.46. Fill in the following table using Eqs. (7.19) through (7.28) to determine the values of the implication $\underset{\sim}{A} \rightarrow \underset{\sim}{B}$. Comment on the similarities and dissimilarities of the various implication methods with respect to the various values for $\underset{\sim}{A}$ and $\underset{\sim}{B}$.

$\underset{\sim}{A}$	$\underset{\sim}{B}$	$\underset{\sim}{A} \rightarrow \underset{\sim}{B}$
0	0	
0	1	
1	0	
1	1	
0.2	0.3	
0.2	0.7	
0.8	0.3	
0.8	0.7	

7.47 Fill in the following table using Eqs. (7.29) through (7.35) to determine values of the composition $\underset{\sim}{B} = \underset{\sim}{A} \circ \underset{\sim}{R}$ for the fuzzy relation,

$$\underset{\sim}{R} = \begin{matrix} & y_1 & y_2 & y_3 \\ x_1 & 0.1 & 0.2 & 0.3 \\ x_2 & 0.4 & 0.5 & 0.6 \\ x_3 & 0.7 & 0.8 & 0.9 \end{matrix}$$

Comment on the similarities and dissimilarities of the various composition methods with respect to the various antecedents, $\underset{\sim}{A}$.

$\underset{\sim}{A}$	$\underset{\sim}{B}$
[0.1 0.5 1.0]	
[1.0 0.6 0.1]	
[0.2 0.6 0.4]	
[0.7 0.9 0.8]	

7.48 A sensor system is capable of detecting the infrared heat generated by an object. The system that processes the data is also capable of tracking the object. Define a fuzzy set called detectability,

$$\underset{\sim}{D} = \text{“detectability”} = \left\{ \frac{0}{98} + \frac{.1}{99} + \frac{.3}{100} + \frac{.5}{101} + \frac{1}{102} \right\}$$

on the universe X = [98, 102], which is the range of temperatures (degrees Fahrenheit) that can be detected. Assume another fuzzy set for target velocity,

$$\underset{\sim}{V} = \text{“typical target velocity”} = \left\{ \frac{.8}{5} + \frac{.6}{10} + \frac{.4}{15} + \frac{.9}{20} + \frac{.1}{25} \right\}$$

defined on the universe Y = [5, 25] in units of miles per hour (mph) of typical targets.

(*a*) Using classical implication, determine, IF a target moves, THEN is it detectable; i.e., $\underset{\sim}{V} \rightarrow \underset{\sim}{D}$.

(*b*) We propose a third fuzzy set called “target size.” Assume that there is a relation between “target size” and detectability. Using Mamdani implication, find this relation for the fuzzy set

$$\underset{\sim}{T} = \text{“target size”} = \left\{ \frac{.7}{1} + \frac{.9}{2} + \frac{.3}{3} + \frac{.1}{4} + \frac{0}{5} \right\}$$

defined on the universe C = [1, 5] in square meters (m^2).

7.49. Illustrate four separate situations in your own field where the exclusive or proposition is appropriate.

7.50. Illustrate four separate situations in your own field where the exclusive nor proposition is appropriate, and discuss why the issue of *equivalence* could address the same situation.

CHAPTER 8

FUZZY RULE-BASED SYSTEMS

It was the best of times, it was the worst of times, it was the age of wisdom, it was the age of foolishness, it was the epoch of belief, it was the epoch of incredulity, it was the season of Light, it was the season of Darkness, it was the spring of hope, it was the winter of despair, we had everything before us, we had nothing before us, we were all going direct to Heaven, we were all going direct the other way—in short, the period was so far like the present period, that some of its noisiest authorities insisted on its being received, for good or for evil, in the superlative degree of comparison only.

Charles Dickens
A Tale of Two Cities, *Chapter 1, 1859*

Natural language is perhaps the most powerful form of conveying information that humans possess for any given problem or situation that requires solving or reasoning. This power has largely remained untapped in today's mathematical paradigms; not so anymore with the utility of fuzzy logic. Consider the information contained in the passage above from Charles Dickens' *A Tale of Two Cities.* Imagine reducing this passage to more precise form such that it could be assimilated by a binary computer. First, we will have to remove the fuzziness inherent in the passage, limiting the statements to precise, either-or, Aristotelian logic. Consider the following crisp version of the first few words of the Dickens passage:

> The time interval x was the period exhibiting a 100 percent maximum of possible values as measured along some arbitrary social scale, [and] the interval x was also the period

> of time exhibiting a 100 percent minimum of these values as measured along the same scale. [Clark, 1992]

The crisp version of this passage has established an untenable paradox, identical to that posed by the excluded middle laws in probability theory. Another example is available from the same classic, the last sentence in Dickens' *A Tale of Two Cities:* "It is a far, far better thing that I do, than I have ever done; it is a far, far better rest that I go to, than I have ever known." Suppose this original "fuzzy" phrase is replaced by a defuzzified version more understandable to an "intelligent machine" using crisp logic, to wit:

> *t* is a thing which exceeds by 98 to 99.5 percent the value of each member of the set of all other things I have done prior to this precise moment in time; it is a rest to which I go that exceeds by the same range of values any other I have known prior to the same temporal reference point. [Clark, 1992]

Both of these examples demonstrate the power of communication inherent in natural language, and they demonstrate how far we are from enabling intelligent machines to reason the way humans do—a long way!

In this chapter we introduce the use of fuzzy sets as a calculus for the interpretation of natural language. Natural language, despite its vagueness and ambiguity, is the vehicle for human communication, and it seems appropriate that a mathematical theory that deals with fuzziness and ambiguity is also the same tool used to express and interpret the linguistic character of our language. The chapter continues with the use of natural language in the expression of a knowledge form known as rule-based systems. The decomposition of compound rules into canonical forms and the treatment of canonical rule forms as logical propositions is addressed. The characterization of the confidence in a particular rule is addressed using truth qualifications. The expression of rules as a collection of logical implications being manipulated by some inferencing scheme is introduced and illustrated with examples. The chapter concludes with a simple graphical interpretation of inference, which is illustrated with some examples.

NATURAL LANGUAGE

Cognitive scientists tell us that humans base their thinking primarily on conceptual patterns and mental images rather than on any numerical quantities. In fact the expert system paradigm known as "frames" is based on the notion of a cognitive picture in one's mind. Furthermore, humans communicate with their own natural language by referring to previous mental images with rather vague but simple terms. Despite the vagueness and ambiguity in natural language, humans communicating in a common language have very little trouble in basic understanding. Our language has been termed the *shell of our thoughts* [Zadeh, 1975*a*]. Hence, any attempts to model the human thought process as expressed in our communications with one another must be preceded by models that attempt to emulate our natural language.

Since a vast amount of the information involved in human communication involves natural language terms that, by their very nature, are often vague, imprecise, ambiguous, and fuzzy, we will propose the use of fuzzy sets as the mathematical foundation of our natural language. This language can be broken down into fundamental terms and certain linguistic connectors of those terms. Fuzzy sets will be employed in the numerical description of these terms and in the prescription of the connection of terms into strings of intelligible expressions. This task will be accomplished in what is termed a set description of our natural language.

Our natural language consists of fundamental terms characterized as atoms in the literature. A collection of these atoms will form the molecules, or phrases, of our natural language. The fundamental terms can be called *atomic* terms. Examples of some atomic terms are *slow, medium, young, beautiful,* etc. A collection of atomic terms is called a composite, or simply a set of terms. Examples of composite terms are *very slow horse, medium-weight female, young tree, fairly beautiful painting,* etc. Suppose we define the atomic terms and sets of atomic terms to exist as elements and sets on a universe of natural language terms, say universe X. Furthermore, let us define another universe, called Y, as a universe of cognitive interpretations, or meanings. Although it may seem straightforward to envision a universe of terms, it may be difficult to ponder a universe of *interpretations.* Consider this universe, however, to be a collection of individual elements and sets that represent the cognitive patterns and mental images referred to earlier in this chapter. Clearly, then, these interpretations would be rather vague, and they might best be represented as fuzzy sets. Hence, an atomic term, or as Zadeh [1975*a*] defines it, a linguistic variable, can be interpreted using fuzzy sets.

The need for expressing linguistic variables using the precepts of mathematics is quite well established. Leibnitz, who was an early developer of calculus, once claimed, "If we could find characters or signs appropriate for expressing all our thoughts as definitely and as exactly as arithmetic expresses numbers or geometric analysis expresses lines, we could in all subjects, in so far as they are amenable to reasoning, accomplish what is done in arithmetic and geometry." Fuzzy sets are a relatively new quantitative method to accomplish just what Leibnitz had suggested.

With these definitions and foundations, we are now in a position to establish a formal model of linguistics using fuzzy sets. Suppose we define a specific atomic term in the universe of natural language, X, as element α, and we define a fuzzy set $\underset{\sim}{A}$ in the universe of interpretations, or meanings, Y, as a specific meaning for the term α. Then natural language can be expressed as a mapping, $\underset{\sim}{M}$ from a set of atomic terms in X to a corresponding set of interpretations defined on universe Y. Each atomic term α in X corresponds to a fuzzy set $\underset{\sim}{A}$ in Y, which is the "interpretation" of α. This mapping, which can be denoted $\underset{\sim}{M}(\alpha, \underset{\sim}{A})$, is shown schematically in Fig. 8.1.

The fuzzy set $\underset{\sim}{A}$ represents the fuzziness in the mapping between an atomic term and its interpretation, and can be denoted by the membership function $\mu_{\underset{\sim}{M}}(\alpha, y)$, or more simply by

$$\mu_{\underset{\sim}{M}}(\alpha, y) = \mu_{\underset{\sim}{A}}(y) \tag{8.1}$$

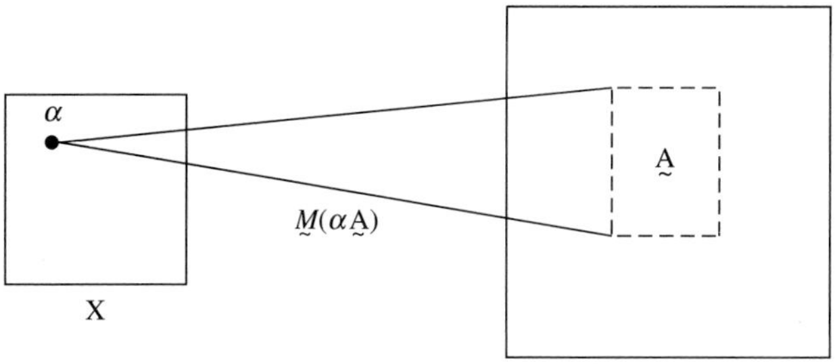

FIGURE 8.1
Mapping of a linguistic atom, α, to a cognitive interpretation, $\underset{\sim}{A}$.

As an example, suppose we have the atomic term "young" (α) and we want to interpret this linguistic atom in terms of age, y, by a membership function that expresses the term "young." The membership function given here in the notation of Zadeh [1975b], and labeled $\underset{\sim}{A}$, might be one interpretation of the term young expressed as a function of age,

$$\underset{\sim}{A} = \text{"young"} = \int_0^{25} \frac{1}{y} + \int_{25}^{100} \frac{1}{y}\left(1 + \left(\frac{y-25}{5}\right)^2\right)^{-1}$$

or alternatively,

$$\mu_{\underset{\sim}{M}}(\text{young}, y) = \begin{cases} \left(1 + \left(\frac{y-25}{5}\right)^2\right)^{-1} & y > 25 \text{ years} \\ 1 & y \leq 25 \text{ years} \end{cases}$$

Similarly, the atomic term "old" might be expressed as another fuzzy set, $\underset{\sim}{O}$, on the universe of interpretation, Y, as

$$\mu_{\underset{\sim}{M}}(\text{old}, y) = 1 - \left(1 + \left(\frac{y-50}{5}\right)^2\right)^{-1} \qquad \text{for } 50 \leq y \leq 100$$

On the basis of the foregoing, we can call α, a natural language variable whose "value" is defined by the fuzzy set $\mu_\alpha(y)$. Hereinafter, the "value" of a linguistic variable will be synonymous with its *interpretation.*

As suggested before, a composite is a collection, or set, of atomic terms combined by various linguistic connectives such as *and, or,* and *not.* Define two atomic terms, α and β, on the universe X. The *interpretation* of the composite, defined on universe Y, can be defined by the following set-theoretic operations [Zadeh, 1975*b*],

$$\begin{aligned} \alpha \text{ or } \beta : \mu_{\alpha \text{ or } \beta}(y) &= \max(\mu_\alpha(y), \mu_\beta(y)) \\ \alpha \text{ and } \beta : \mu_{\alpha \text{ and } \beta}(y) &= \min(\mu_\alpha(y), \mu_\beta(y)) \\ \text{Not } \alpha = \overline{\alpha} : \mu_{\overline{\alpha}}(y) &= 1 - \mu_\alpha(y) \end{aligned} \tag{8.2}$$

These operations are analogous to those proposed in Chapter 7, where the natural language connectives *and, or,* and *not* were logical connectives.

LINGUISTIC HEDGES

In linguistics, fundamental atomic terms are often modified with adjectives (nouns) or adverbs (verbs) like *very, low, slight, more-or-less, fairly, slightly, almost, barely, mostly, roughly, approximately,* and so many more that it would be difficult to list them all. We will call these modifiers "linguistic hedges," that is, the singular meaning of an atomic term is modified, or hedged, from its original interpretation. Using fuzzy sets as the calculus of interpretation, these linguistic hedges have the effect of modifying the membership function for a basic atomic term [Zadeh, 1972]. As an example, let's look at the basic linguistic atom, α, and subject it to some hedges. Define $\alpha = \int_Y \mu_\alpha(y)/y$, then

$$\text{“Very” } \alpha = \alpha^2 = \int_Y \frac{[\mu_\alpha(y)]^2}{y} \tag{8.3}$$

$$\text{“Very, very” } \alpha = \alpha^4 \tag{8.4}$$

$$\text{“Plus” } \alpha = \alpha^{1.25} \tag{8.5}$$

$$\text{“Slightly” } \alpha = \sqrt{\alpha} = \int_Y \frac{[\mu_\alpha(y)]^{0.5}}{y} \tag{8.6}$$

$$\text{“Minus” } \alpha = \alpha^{0.75} \tag{8.7}$$

The expressions shown in Eqs.(8.3)–(8.5) are linguistic hedges known as *concentrations* [Zadeh, 1972]. Concentrations tend to concentrate the elements of a fuzzy set by reducing the degree of membership of all elements that are only "partly" in the set. The less an element is in a set (i.e., the lower its original membership value), the more it is reduced in membership through concentration. For example, by using Eq.(8.3) for the hedge *very,* a membership value of 0.9 is reduced by 10 percent to a value of 0.81, but a membership value of 0.1 is reduced by an order-of-magnitude to 0.01. This decrease is simply a manifestation of the properties of the membership value itself; for $0 \leq \mu \leq 1$, then $\mu \geq \mu^2$. Alternatively, the expressions given in Eqs.(8.6) and (8.7) are linguistic hedges known as *dilations* (or dilutions in some publications). Dilations stretch or dilate a fuzzy set by increasing the membership of elements that are "partly" in the set [Zadeh, 1972]. For example, using Eq.(8.6) for the hedge *slightly,* a membership value of 0.81 is increased by 11 percent to a value of 0.9, whereas a membership value of 0.01 is increased by an order of magnitude to 0.1.

Another operation on linguistic fuzzy sets is known as *intensification.* This operation acts in a combination of concentration and dilation. It increases the degree of membership of those elements in the set with original membership values greater than 0.5, and it decreases the degree of membership of those elements in the set with original membership values less than 0.5. This also has the effect of making the boundaries of the membership function (see Chapter 4) steeper. *Intensification* can be expressed by numerous algorithms, one of which, proposed by Zadeh [1972], is

$$\text{“intensify” } \alpha = = \begin{cases} 2\mu_\alpha{}^2(y) & \text{for } 0 \le \mu_\alpha(y) \le 0.5 \\ 1 - 2\,[1 - \mu_\alpha(y)]^2 & \text{for } 0.5 \le \mu_\alpha(y) \le 1 \end{cases} \tag{8.8}$$

Intensification increases the contrast between the elements of the set that have more than half-membership and those that have less than half-membership. Figures 8.2, 8.3, and 8.4 illustrate the operations of concentration, dilation, and intensification, respectively, for fuzzy linguistic hedges on a typical fuzzy set $\underset{\sim}{A}$.

Composite terms can be formed from one or more combinations of atomic terms, logical connectives, and linguistic hedges. Since an atomic term is essentially a fuzzy mapping from the universe of terms to a universe of fuzzy sets represented by membership functions, the implementation of linguistic hedges and logical connectives is manifested as function-theoretic operations on the values of the membership functions. In order to conduct the function-theoretic operations, a precedence order must be established. For example, suppose we have two atomic terms "small" and "red," and their associated membership functions, and we pose the following linguistic expression: a "not-small" *and* "very red" fruit. Which of the operations—i.e., not, and, very—would we perform first, which would we perform second, and so on? In the literature, the following preference table (Table 8.1) has been suggested for standard Boolean operations.

Parentheses may be used to change the precedence order and ambiguities may be resolved by the use of association-to-the-right. For example, "plus very minus very small" should be interpreted as

plus (very (minus (very (small))))

Every atomic term and every composite term has a syntax represented by its linguistic label and a semantics, or meaning (interpretation), which is given by a membership function. The use of a membership function gives the flexibility of an elastic meaning to a linguistic term. On the basis of this elasticity and flexibility, it is possible to incorporate subjectivity and bias into the meaning of a linguistic term. These are some of the most important benefits of using fuzzy mathematics in the modeling of linguistic variables. This capability allows us to encode and automate human knowledge, which is often expressed in natural language propositions.

In our example, a "not small" *and* "very red" fruit, we would perform the hedges "not small" and "very red" first, then we would perform the logical operation *and* on the two phrases as suggested in Table 8.1. To further illustrate Table 8.1 consider the following numerical example.

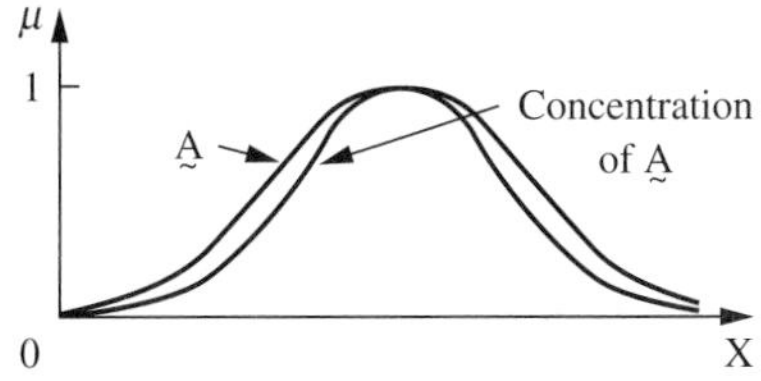

FIGURE 8.2
Fuzzy concentration.

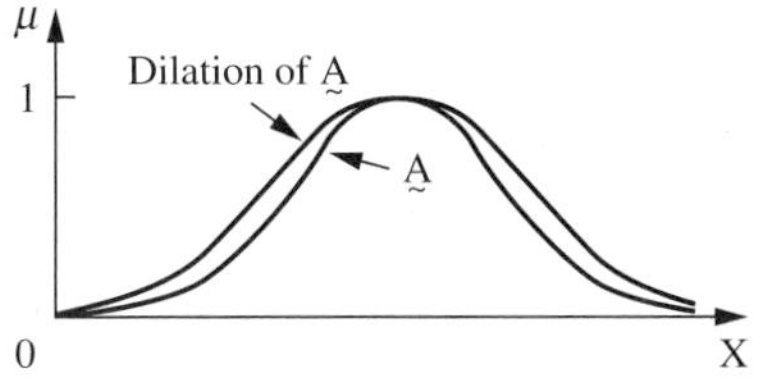

FIGURE 8.3
Fuzzy dilation.

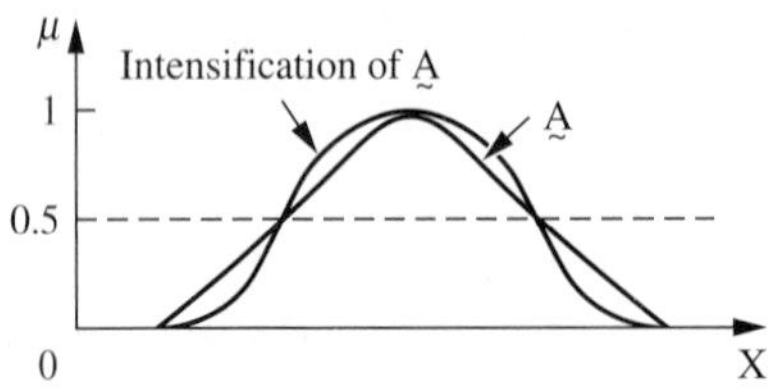

FIGURE 8.4
Fuzzy intensification.

Example 8.1. Suppose we have a universe of integers, Y = {1, 2, 3, 4, 5}. We define the following linguistic terms as a mapping onto Y:

$$\text{“Small”} = \left\{\frac{1}{1} + \frac{.8}{2} + \frac{.6}{3} + \frac{.4}{4} + \frac{.2}{5}\right\}$$

$$\text{“Large”} = \left\{\frac{.2}{1} + \frac{.4}{2} + \frac{.6}{3} + \frac{.8}{4} + \frac{1}{5}\right\}$$

Now we modify these two linguistic terms with hedges,

$$\text{“Very small”} = \text{“small”}^2 \text{ [Eq. (8.3)]} = \left\{\frac{1}{1} + \frac{.64}{2} + \frac{.36}{3} + \frac{.16}{4} + \frac{.04}{5}\right\}$$

$$\text{“Not very small”} = 1 - \text{“very small”} = \left\{\frac{0}{1} + \frac{0.36}{2} + \frac{.64}{3} + \frac{.84}{4} + \frac{.96}{5}\right\}$$

Then we construct a phrase, or a composite term:

$$\alpha = \text{“not very small and not very, very large”}$$

which involves the following set-theoretic operations:

$$\alpha = \left(\frac{.36}{2} + \frac{.64}{3} + \frac{.84}{4} + \frac{.96}{5}\right) \cap \left(\frac{1}{1} + \frac{1}{2} + \frac{.9}{3} + \frac{.6}{4}\right) = \left(\frac{.36}{2} + \frac{.64}{3} + \frac{.6}{4}\right)$$

Suppose we want to construct a linguistic variable “intensely small” (extremely small); we will make use of Eq.(8.8) to modify “small” as follows:

$$\text{“Intensely small”} = \left\{\frac{1-2[1-1]^2}{1} + \frac{1-2[1-0.8]^2}{2} + \frac{1-2[1-0.6]^2}{3} + \frac{2[0.4]^2}{4} + \frac{2[0.2]^2}{5}\right\}$$

$$= \left\{\frac{1}{1} + \frac{0.92}{2} + \frac{0.68}{3} + \frac{0.32}{4} + \frac{0.08}{5}\right\}$$

TABLE 8.1
Precedence for linguistic hedges and logical operations

Precedence	Operation
First	Hedge, not
Second	And
Third	Or

Source: Zadeh, 1973

In summary, the foregoing material introduces the idea of a *linguistic variable* (atomic term), which is a variable whose values (interpretation) are natural language expressions referring to the contextual semantics of the variable. Lotfi Zadeh [1975*b*] described this notion quite well:

> A linguistic variable differs from a numerical variable in that its values are not numbers but words or sentences in a natural or artificial language. Since words, in general, are less precise than numbers, the concept of a linguistic variable serves the purpose of providing a means of approximate characterization of phenomena which are too complex or too ill-defined to be amenable to description in conventional quantitative terms. More specifically, the fuzzy sets which represent the restrictions associated with the values of a linguistic variable may be viewed as summaries of various subclasses of elements in a universe of discourse. This, of course, is analogous to the role played by words and sentences in a natural language. For example, the adjective handsome is a summary of a complex of characteristics of the appearance of an individual. It may also be viewed as a label for a fuzzy set which represents a restriction imposed by a fuzzy variable named handsome. From this point of view, then, the terms *very handsome, not handsome, extremely handsome, quite handsome,* etc., are names of fuzzy sets which result from operating on the fuzzy set handsome with the modifiers named *very, not, extremely, quite,* etc. In effect, these fuzzy sets, together with the fuzzy set labeled *handsome,* play the role of values of the linguistic variable *Appearance.*

RULE-BASED SYSTEMS

In the field of artificial intelligence (machine intelligence) there are various ways to represent knowledge. Perhaps the most common way to represent human knowledge is to form it into natural language expressions of the type,

IF premise (antecedent), THEN conclusion (consequent) (8.9)

The form in Expression (8.9) is commonly referred to as the IF-THEN *rule-based* form. It typically expresses an inference such that if we know a fact (premise, hypothesis, antecedent), then we can infer, or derive, another fact called a conclusion (consequent). This form of knowledge representation, characterized as *shallow knowledge,* is quite appropriate in the context of linguistics because it expresses human empirical and heuristic knowledge in our own language of communication. It does not, however, capture the *deeper* forms of knowledge usually associated with intuition, structure, function, and behavior of the objects around us simply because these latter forms of knowledge are not readily reduced to linguistic phrases or representations. The rule-based system is distinguished from classical expert systems in the sense that the rules comprising a rule-based system might derive from sources other that human experts and, in this context, are distinguished from expert systems. This chapter confines itself to the broader category of fuzzy rule-based systems (of which expert systems could be seen as a subset) because of their prevalence and popularity in the literature, because of their preponderant use in engineering practice, and because the rule-based form makes use of linguistic variables as its antecedents and consequents. As illustrated earlier in this chapter, these linguistic variables can be naturally represented by fuzzy sets and logical connectives of these sets.

Canonical Rule Forms

In general, three general forms exist for any linguistic variable [Vadiee, 1993]. These forms are the (*i*) assignment statements, (*ii*) conditional statements, and (*iii*) unconditional statements. Examples of each of these are given below.

Assignment statements

$x =$ large
Banana's color = yellow
$x \approx s$
x is not large and not very small
Season = winter
Outside temperature = hot

Conditional statements

IF the tomato is red THEN the tomato is ripe
IF x is very hot THEN stop
IF x is large THEN y is small ELSE y is not small

Unconditional statements

Go to 9
Stop
Divide by x
Turn the pressure higher

The assignment statements restrict the value of a variable to a specific quantity. The unconditional statements may be thought of as conditional restrictions with their IF clause conditions being the universe of discourse of the input conditions, which is always true (see Chapter 7 on tautologies). An unconditional restriction such as "output is low" could equivalently be written as

IF any conditions THEN output is low, or
IF anything THEN low.

Hence, the rule-base under consideration could be described using a collection of conditional restrictive statements. These statements may also be modeled as fuzzy conditional statements, such as

IF condition C^1 THEN restriction R^1.

The unconditional restrictions might be in the form

R^1: The output is B^1
AND
R^2: The output is B^2
AND
etc.

where $B^1, B^2, \ldots$ are fuzzy consequents.

Table 8.2 is the rule-based system composed of a set of conditional rules. Hence, the canonical rule set may be put in the form shown in Table 8.2.

In general, unconditional as well as conditional statements place some restrictions on the consequent of the rule-based process because of certain immediate or past conditions. These restrictions are usually manifested in terms of vague natural language words that can be modeled using fuzzy mathematics.

Consider the problem of the control of an industrial furnace with some restrictive statements:

If the temperature is hot, then the pressure is rather high.
If the temperature is cold, then the pressure is very low.
If the temperature is warm, then the pressure is medium and not high.
Etc.

The vague term "rather high" in the first statement places a fuzzy restriction on the pressure, based on a fuzzy "hot" temperature condition in the antecedent.

In summary, the fuzzy level of understanding and describing a complex system is expressed in the form of a set of restrictions on the output based on certain conditions of the input. Restrictions are generally modeled by fuzzy sets and relations. These restriction statements, unconditional as well as conditional, are usually connected by linguistic connectives such as "and," "or," or "else." The restrictions $R^1, R^2, \ldots, R^r$ apply to the output actions, or consequents of the rules.

Decomposition of Compound Rules

A linguistic statement expressed by a human might involve compound rule structures. As an example, consider a rule-base for a simple home temperature control problem, which might contain the following rules [Vadiee, 1993]:

```
IF it is raining hard
    THEN close the window.
IF the room temperature is very hot,
    THEN
        IF the heat is on
            THEN turn the heat lower
        ELSE
            IF (the window is closed) AND (the air conditioner is off)
                THEN (turn on the air conditioner)
    AND (it is not raining hard)
            THEN open the window
        ELSE
            IF (the window is closed) AND (the air conditioner is on)
                THEN open the window; etc.
```

By using the basic properties and operations defined for fuzzy sets (see Chapter 2), any compound rule structure may be decomposed and reduced to a number of simple canonical rules as given in Table 8.2. These rules are based on natural language representations and models, which are themselves based on fuzzy sets and

TABLE 8.2
The canonical form for a fuzzy rule-based system

Rule 1:	IF condition C^1, THEN restriction R^1
Rule 2:	IF condition C^2, THEN restriction R^2
⋮	⋮
Rule *r*:	IF condition C^r, THEN restriction R^r

fuzzy logic. The following illustrates a number of the most common techniques [Vadiee, 1993] for decomposition of compound linguistic rules into simple canonical forms:

Multiple conjunctive antecedents

$$\text{IF } x \text{ is } \underset{\sim}{A}^1 \text{ and } \underset{\sim}{A}^2 \ldots \text{ and } \underset{\sim}{A}^L \text{ THEN } y \text{ is } \underset{\sim}{B}^s \tag{8.10}$$

Assuming a new fuzzy subset A^s as

$$\underset{\sim}{A}^s = \underset{\sim}{A}^1 \cap \underset{\sim}{A}^2 \cap \cdots \cap \underset{\sim}{A}^L$$

expressed by means of membership function

$$\mu_{\underset{\sim}{A}^s}(x) = \min[\mu_{\underset{\sim}{A}^1}(x), \mu_{\underset{\sim}{A}^2}(x), \ldots, \mu_{\underset{\sim}{A}^L}(x)]$$

based on the definition of the fuzzy intersection operation, the compound rule may be rewritten as

$$\text{IF } \underset{\sim}{A}^s \text{ THEN } \underset{\sim}{B}^s$$

Multiple disjunctive antecedents

$$\text{IF } x \text{ is } \underset{\sim}{A}^1 \text{ OR } x \text{ is } \underset{\sim}{A}^2 \ldots \text{OR } x \text{ is } \underset{\sim}{A}^L \text{ THEN } y \text{ is } \underset{\sim}{B}^s \tag{8.11}$$

could be rewritten as

$$\text{IF } x \text{ is } \underset{\sim}{A}^s \text{ THEN } y \text{ is } \underset{\sim}{B}^s$$

where the fuzzy set $\underset{\sim}{A}^s$ is defined as

$$\underset{\sim}{A}^s = \underset{\sim}{A}^1 \cup \underset{\sim}{A}^2 \cup \cdots \cup \underset{\sim}{A}^L$$

$$\mu_{\underset{\sim}{A}^s}(x) = \max[\mu_{\underset{\sim}{A}^1}(x), \mu_{\underset{\sim}{A}^2}(x), \ldots, \mu_{\underset{\sim}{A}^L}(x)]$$

which is based on the definition of the fuzzy union operation.

Conditional statements with ELSE and UNLESS

(*a*)

$$\text{IF } \underset{\sim}{A}^1 \text{ THEN } (\underset{\sim}{B}^1 \text{ ELSE } \underset{\sim}{B}^2) \tag{8.12}$$

may be decomposed into two simple canonical form rules connected by "OR:"

$$\text{IF } \underset{\sim}{A}^1 \text{ THEN } \underset{\sim}{B}^1$$

$$\text{OR}$$

$$\text{IF NOT } \underset{\sim}{A}^1 \text{ THEN } \underset{\sim}{B}^2$$

(*b*) $$\text{IF } \underset{\sim}{A}^1 \text{ (THEN } \underset{\sim}{B}^1\text{) UNLESS } \underset{\sim}{A}^2 \tag{8.13}$$

could be decomposed as

$$\text{IF } \underset{\sim}{A}^1 \text{ THEN } \underset{\sim}{B}^1$$
$$\text{OR}$$
$$\text{IF } \underset{\sim}{A}^2 \text{ THEN NOT } \underset{\sim}{B}^1$$

(*c*) $$\text{IF } \underset{\sim}{A}^1 \text{ THEN (} \underset{\sim}{B}^1 \text{ ELSE IF } \underset{\sim}{A}^2 \text{ THEN (} \underset{\sim}{B}^2\text{))} \tag{8.14}$$

may be put into the following form:

$$\text{IF } \underset{\sim}{A}^1 \text{ THEN } \underset{\sim}{B}^1$$
$$\text{OR}$$
$$\text{IF NOT } \underset{\sim}{A}^1 \text{ AND } \underset{\sim}{A}^2 \text{ THEN } \underset{\sim}{B}^2$$

Nested IF-THEN rules

$$\text{IF } \underset{\sim}{A}^1 \text{ THEN (IF } \underset{\sim}{A}^2 \text{ THEN (} \underset{\sim}{B}^1\text{))} \tag{8.15}$$

may be put into the form

$$\text{IF } \underset{\sim}{A}^1 \text{ AND } \underset{\sim}{A}^2 \text{ THEN } \underset{\sim}{B}^1$$

When rules are decomposed into a series of canonical forms, each of these forms is an implication, and we can then reduce the rules to a series of relations.

Likelihood and Truth Qualification

Primary (atomic) and composite terms using linguistic hedges may also be followed by linguistic variables connoting likelihood, such as "likely," "very likely," "highly likely," "unlikely," or they might be modified semantically by truth qualification statements such as "true," "fairly true," "very true," "false," "fairly false," and "very false." These likelihood labels are based on notions of probability. The primary terms, as well as the rules, may also be restricted by linguistic variables associated with certainty, such as "indefinite," "unknown," and "definite." An example of a linguistic variable connoting likelihood is illustrated next.

Example 8.2 [Zadeh, 1973]. We assume that the universe of discourse is given by a normalized scale,

$$\text{U} = \{0, 0.1, 0.2, 0.3, 0.4, 0.5, 0.6, 0.7, 0.8, 0.9, 1.0\}$$

in which the elements of U represent probabilities. Suppose we wish to compute the meaning (value) of the linguistic variable x,

$$x = \text{"highly unlikely"}$$

in which "highly" and "unlikely" are defined as

$$\text{"highly"} = \text{"minus very very"} = (\text{very very})^{0.75}$$

and

$$\text{"unlikely"} = \text{"not likely"}$$

With the meaning of the primary term "likely" defined on U and given by

$$\text{"likely"} = \left\{\frac{1}{1} + \frac{1}{0.9} + \frac{1}{0.8} + \frac{0.8}{0.7} + \frac{0.6}{0.6} + \frac{0.5}{0.5} + \frac{0.3}{0.4} + \frac{0.2}{0.3}\right\}$$

we obtain

$$\text{"unlikely"} = \text{"1}-\text{likely"} = \left\{\frac{1}{0} + \frac{1}{0.1} + \frac{1}{0.2} + \frac{0.8}{0.3} + \frac{0.7}{0.4} + \frac{0.5}{0.5} + \frac{0.4}{0.6} + \frac{0.2}{0.7}\right\}$$

and hence

$$\text{"very very unlikely"} = \text{"(unlikely)}^4\text{"} = \left\{\frac{1}{0} + \frac{1}{0.1} + \frac{1}{0.2} + \frac{0.4}{0.3} + \frac{0.2}{0.4}\right\}$$

where terms with membership values less than 0.1 have been deleted. Finally,

$$\text{"highly unlikely"} = \text{"minus very very unlikely"} = \left(\frac{1}{0} + \frac{1}{0.1} + \frac{1}{0.2} + \frac{0.4}{0.3} + \frac{0.2}{0.4}\right)^{0.75}$$

$$= \left\{\frac{1}{0} + \frac{1}{0.1} + \frac{1}{0.2} + \frac{0.5}{0.3} + \frac{0.3}{0.4}\right\}$$

The primary terms "yes," "maybe," and "no" may also be assigned meanings based on membership functions given by "very very likely," "likely," and "very very unlikely." Note also that the atomic term "anything" is equivalent to the universe of discourse and given by

$$\mu_{\text{anything}}(x) = 1 \qquad \text{for all } x \in X \tag{8.16}$$

Suppose we are interested in the quantification of the truth value of an antecedent or consequent in a rule [Hadipriono and Sun, 1990]. Let τ be a fuzzy truth value, for example, "very true," "true," "fairly true," "fairly false," "false," etc. Such a truth value may be regarded as a fuzzy element on the unit interval that is characterized by its own membership function. A truth qualification proposition can be expressed as "x is $\underset{\sim}{A}$ is τ." The transformation for such propositions can be given by

$$x \text{ is } \underset{\sim}{A} \text{ is } \tau = \mu_{\underset{\sim}{A}}(x_\tau) \tag{8.17}$$

Equation (8.17) has the effect of reducing the membership values of the antecedent as qualified by the truth value, τ. Figure 8.5 illustrates the meaning of true (T), fairly true (FT), very true (VT), false (F), fairly false (FF), and very false (VF). The fuzzy assignment statements such as

$$x \text{ is } \underset{\sim}{A} \text{ is very true}$$

or

$$y \text{ is } \underset{\sim}{B} \text{ is very false}$$

and fuzzy conditional statements such as

$$\text{It is very true that IF } \underset{\sim}{A} \text{ THEN } \underset{\sim}{B}$$

are transformed to a new meaning using Eq. (8.17). In the case of conditional statements, the fuzzy set describing the fuzzy relation $\underset{\sim}{A} \rightarrow \underset{\sim}{B}$ will be transformed to a new fuzzy relation, i.e., $\underset{\sim}{R} = \overline{\underset{\sim}{A}} \cup \underset{\sim}{B}$.

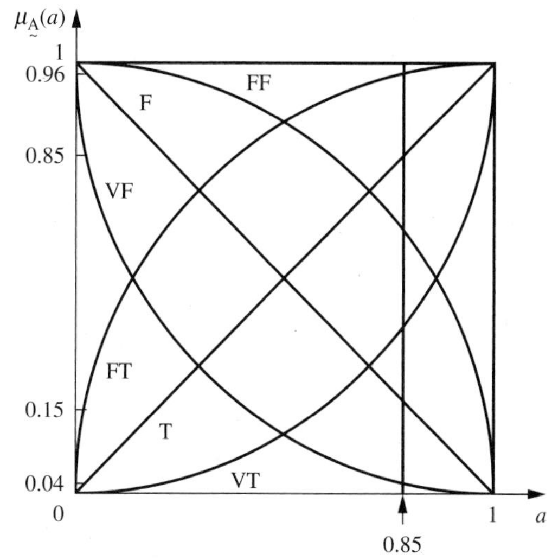

FIGURE 8.5
Truth value qualification graphs.

Example 8.3. If a fuzzy variable x has a membership value equal to 0.85 in the fuzzy set $\underset{\sim}{A}$, i.e., $\mu_{\underset{\sim}{A}}(x) = 0.85$, as shown in Fig. 8.6, then its membership values for the following truth qualification statements are determined from Fig. 8.5:

τ: x is $\underset{\sim}{A}$ is true	$\mu_{\underset{\sim}{A}}(x_\tau) = 0.85$
τ: x is $\underset{\sim}{A}$ is false	$\mu_{\underset{\sim}{A}}(x_\tau) = 0.15$
τ: x is $\underset{\sim}{A}$ is fairly true	$\mu_{\underset{\sim}{A}}(x_\tau) = 0.96$
τ: x is $\underset{\sim}{A}$ is very false	$\mu_{\underset{\sim}{A}}(x_\tau) = 0.04$

Aggregation of Fuzzy Rules

Most rule-based systems involve more than one rule. The process of obtaining the overall consequent (conclusion) from the individual consequents contributed by each rule in the rule-base is known as aggregation of rules. In determining an aggregation strategy, two simple extreme cases exist [Vadiee, 1993]:

(a) *Conjunctive system of rules.* In the case of a system of rules that must be jointly satisfied, the rules are connected by "and" connectives. In this case the aggregated output (consequent), y, is found by the fuzzy intersection of all individual rule consequents, y^i, where $i = 1, 2, \ldots r$, as

$$y = y^1 \text{ and } y^2 \text{ and } \ldots \text{ and } y^r$$

or (8.18)

$$y = y^1 \cap y^2 \cap \cdots \cap y^r$$

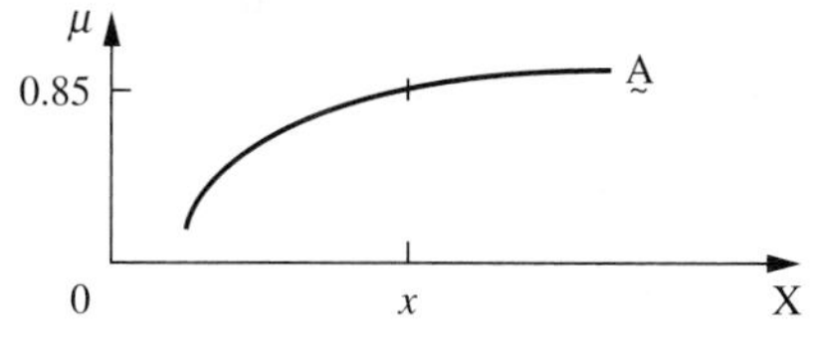

FIGURE 8.6
Point x has membership 0.85 in fuzzy set $\underset{\sim}{A}$ when statement is "true."

which is defined by the membership function

$$\mu_y(y) = \min(\mu_{y^1}(y),\ \mu_{y^2}(y),\ \ldots,\ \mu_{y^r}(y)) \text{ for } y \in Y \tag{8.19}$$

(*b*) *Disjunctive system of rules.* For the case of a disjunctive system of rules where the satisfaction of at least one rule is required, the rules are connected by the "or" connectives. In this case the aggregated output is found by the fuzzy union of all individual rule contributions, as

$$y = y^1 \text{ or } y^2 \text{ or } \ldots \text{ or } y^r$$

or (8.20)

$$y = y^1 \cup y^2 \cup \cdots \cup y^r$$

which is defined by the membership function

$$\mu_y(y) = \max(\mu_{y^1}(y), \mu_{y^2}(y), \ldots, \mu_{y^r}(y)) \text{ for } y \in \mathrm{Y} \tag{8.21}$$

GRAPHICAL TECHNIQUES OF INFERENCE

Chapter 7 illustrated mathematical procedures to conduct inferencing of IF-THEN rules. These procedures can be implemented on a computer for processing speed. Sometimes, however, it is useful to be able to conduct the inference computation manually with a few rules to check computer programs or to verify the inference operations. Conducting the matrix operations illustrated in Chapter 7 for a few rule sets can quickly become quite onerous. Graphical methods that emulate the inference process and that make manual computations involving a few simple rules straightforward have been proposed [see Jamshidi et al., 1993]. To illustrate this idea, we consider a simple two-rule system where each rule comprises two antecedents and one consequent. This is analogous to a dual-input and single-output fuzzy system. The graphical procedures illustrated here can be easily extended and will hold for fuzzy rule-bases (or fuzzy systems) with any number of antecedents (inputs) and consequents (outputs). A fuzzy system with two noninteractive inputs x_1 and x_2 (antecedents) and a single output y (consequent) is described by a collection of r linguistic IF-THEN propositions:

$$\text{IF } x_1 \text{ is } \underset{\sim}{\mathrm{A}}_1{}^k \text{ and } x_2 \text{ is } \underset{\sim}{\mathrm{A}}_2{}^k \text{ THEN } y^k \text{ is } \underset{\sim}{\mathrm{B}}^k \qquad \text{for } k = 1, 2, \ldots, r \tag{8.22}$$

where $\underset{\sim}{\mathrm{A}}_1{}^k$ and $\underset{\sim}{\mathrm{A}}_2{}^k$ are the fuzzy sets representing the kth-antecedent pairs, and $\underset{\sim}{\mathrm{B}}^k$ are the fuzzy sets representing the kth-consequent.

In the following presentation, we consider four different cases of two-input systems: (1) the inputs to the system are crisp values, and we use a max-min inference method; (2) the inputs to the system are crisp values, and we use a max-product inference method; (3) the inputs to the system are represented by fuzzy sets, and we use a max-min inference method; and (4) the inputs to the system are represented by fuzzy sets and we use a max-product inference method.

Case 1. Inputs x_1 and x_2 are crisp values, i.e., delta functions. The rule-based system is described by Eq. (8.22), so membership for the inputs x_1 and x_2 will be described by

$$\mu(x_1) = \delta(x_1 - \text{input}\ (i)) = \begin{cases} 1, & x_1 = \text{input}(i) \\ 0, & \text{otherwise} \end{cases} \quad (8.23)$$

$$\mu(x_2) = \delta(x_2 - \text{input}\ (j)) = \begin{cases} 1, & x_2 = \text{input}(j) \\ 0, & \text{otherwise} \end{cases} \quad (8.24)$$

Based on the Mamdani implication method of inference given in Chapter 7, Eq. (7.21), and for a set of disjunctive rules, the aggregated output for the r rules will be given by

$$\mu_{\underset{\sim}{B}^k}(y) = \max_k \left[\min \left[\mu_{\underset{\sim}{A}_1^k}(\text{input}(i)), \mu_{\underset{\sim}{A}_2^k}(\text{input}(j)) \right] \right] \qquad k = 1, 2, \ldots, r \quad (8.25)$$

Equation (8.25) has a very simple graphical interpretation, as seen in Fig. 8.7. Figure 8.7 illustrates the graphical analysis of two rules, where the symbols A_{11} and A_{12} refer to the first and second fuzzy antecedents of the first rule, respectively, and the symbol B_1 refers to the fuzzy consequent of the first rule; the symbols A_{21} and A_{22} refer to the first and second fuzzy antecedents, respectively, of the second rule, and the symbol B_2 refers to the fuzzy consequent of the second rule. The minimum function in Eq. (8.25) is illustrated in Fig. 8.7 and arises because the antecedent pairs given in the general rule structure for this system are connected by a logical "and" connective, as seen in Eq. (8.22). The minimum membership value for the antecedents propagates through to the consequent and truncates the membership

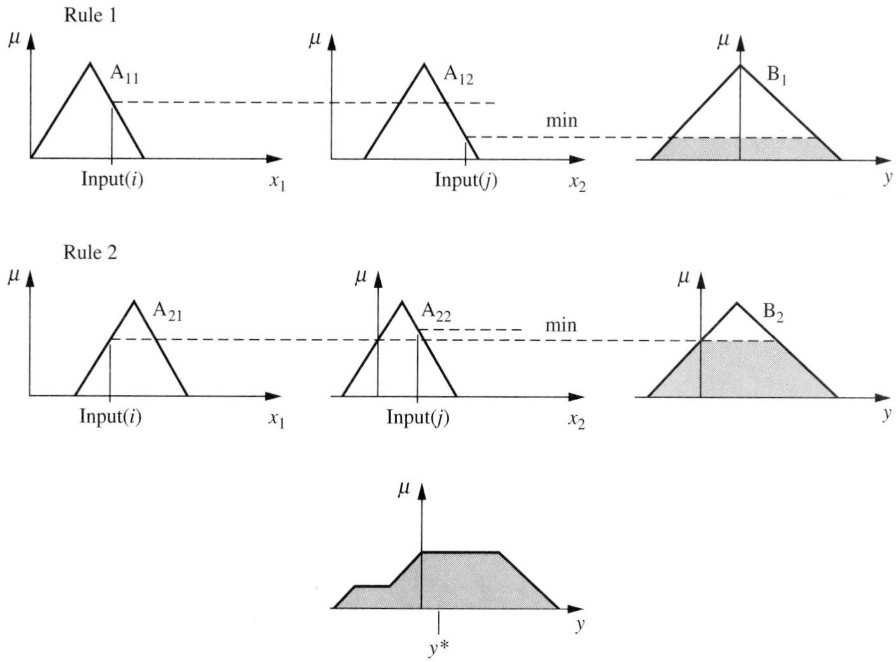

FIGURE 8.7
Graphical Mamdani (max-min) inference method with crisp inputs.

function for the consequent of each rule. This graphical inference is done for each rule. Then the truncated membership functions for each rule are aggregated, using the graphical equivalent of either Eq. (8.19), for conjunction rules, or Eq. (8.21), for disjunctive rules; in Fig. 8.7 the rules are disjunctive, so the aggregation operation *max* results in an aggregated membership function comprised of the outer envelope of the individual truncated membership forms from each rule. If one wishes to find a crisp value for the aggregated output, some appropriate defuzzification technique (see Chapter 5) could be employed to the aggregated membership function, and a value such as y^* shown in Fig. 8.7 would result.

Case 2. In the preceding example, if we were to use a max-product (or correlation-product) implication technique [see Eq. (7.26)] for a set of disjunctive rules, the aggregated output for the r rules would be given by

$$\mu_{\underset{\sim}{B}^k}(y) = \max_k \left[\mu_{\underset{\sim}{A}_1^k}(\text{input}(i)) \cdot \mu_{\underset{\sim}{A}_2^k}(\text{input}(j)) \right] \qquad k = 1, 2, \ldots, r \qquad (8.26)$$

and the resulting graphical equivalent of Eq. (8.26) would be as shown in Fig. 8.8. In Fig. 8.8 the effect of the max-product implication is shown by the consequent membership functions remaining as scaled triangles (instead of truncated triangles as in case 1). Again, Fig. 8.8 shows the aggregated consequent resulting from a disjunctive set of rules (the outer envelope of the individual scaled-consequents) and a defuzzified value, y^*, resulting from some defuzzification method (see Chapter 5).

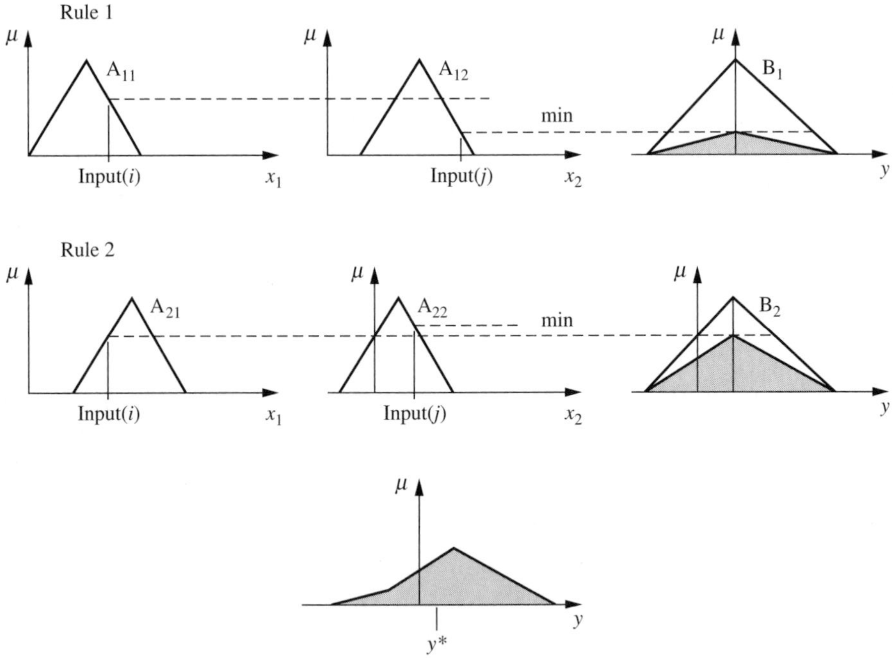

FIGURE 8.8
Graphical max-product implication method with crisp inputs.

Case 3. Inputs input(i) and input(j) are fuzzy variables described by fuzzy membership functions. The rule-based system is described by Eq. (8.22), so for a set of disjunctive rules, where $k = 1, 2, \ldots, r$, the aggregated output using a Mamdani implication, Eq. (7.21) will be given by

$$\mu_{\underset{\sim}{B}^k}(y) = \max_k \left[\min\{\max\left[\mu_{\underset{\sim}{A}_1^k}(x) \wedge \mu(x_1)\right], \max\left[\mu_{\underset{\sim}{A}_2^k}(x) \wedge \mu(x_2)\right]\}\right] \qquad (8.27)$$

Equation (8.27) has a very simple graphical interpretation, which is illustrated in Fig. 8.9. In this figure the fuzzy inputs are represented by triangular membership functions (input(i) and input(j) in the figure). The intersection of these inputs and the stored membership functions for the antecedents (A_{11}, A_{12} for the first rule, and A_{21}, A_{22} for the second rule) results in triangles (see Fig. 2.16 for example). The maximum value of each of these intersection triangles results in a membership value, the minimum of which is propagated for each rule [because of the "and" connective between the antecedents of each rule, Eq. (8.22)]. Figure 8.9 shows the aggregated consequent resulting from a disjunctive set of rules (the outer envelope of the individual truncated-consequents) and a defuzzified value, y^*, resulting from some defuzzification method (see Chapter 5).

Case 4. Inputs input(i) and input(j) are fuzzy variables described by fuzzy membership functions, and the inference method is a correlation-product method, Eq. (7.26). The resulting expression for this inference for the r disjunctive rules would be

$$\mu_{\underset{\sim}{B}^k}(y) = \max_k \left[\max\left[\mu_{\underset{\sim}{A}_1^k}(x) \wedge \mu(x_1)\right] \cdot \max\left[\mu_{\underset{\sim}{A}_2^k}(x) \wedge \mu(x_2)\right]\right] \qquad (8.28)$$

Equation (8.28) also has a simple graphical interpretation, which is illustrated in Fig. 8.10. In this figure the fuzzy inputs are represented by triangular membership functions [input(i) and input(j) in the figure]. As before, the intersection of these inputs and the stored membership functions for the antecedents (A_{11}, A_{12} for the first rule, and A_{21}, A_{22} for the second rule) results in other triangles (see Fig. 2.16, for example). The maximum value of each of these intersection triangles results in a membership value, the minimum of which is propagated for each rule (because of the "and" connective between the antecedents of each rule [Equation (8.22)]. Figure 8.10 shows the aggregated consequent resulting from a disjunctive set of rules (the outer envelope of the individual scaled-consequents) and a defuzzified value, y^*, resulting from some defuzzification method.

Example 8.4. In mechanics, the energy of a moving body is called kinetic energy. If an object of mass m (kilograms) is moving with a velocity v (meters per second), then the kinetic energy k (in joules) is given by the equation $k = \frac{1}{2}mv^2$. Suppose we model the mass and velocity as inputs to a system (moving body) and the energy as output, then observe the system for a while and deduce the following two disjunctive rules of inference based on our observations:

Rule 1: IF x_1 is $\underset{\sim}{A}_1^1$(small mass) *and* x_2 is $\underset{\sim}{A}_2^1$ (high velocity),
THEN y is $\underset{\sim}{B}^1$ (medium energy).

Rule 2: IF x_1 is $\underset{\sim}{A}_1^2$ (large mass) *or* x_2 is $\underset{\sim}{A}_2^2$ (medium velocity),
THEN y is $\underset{\sim}{B}^2$ (high energy).

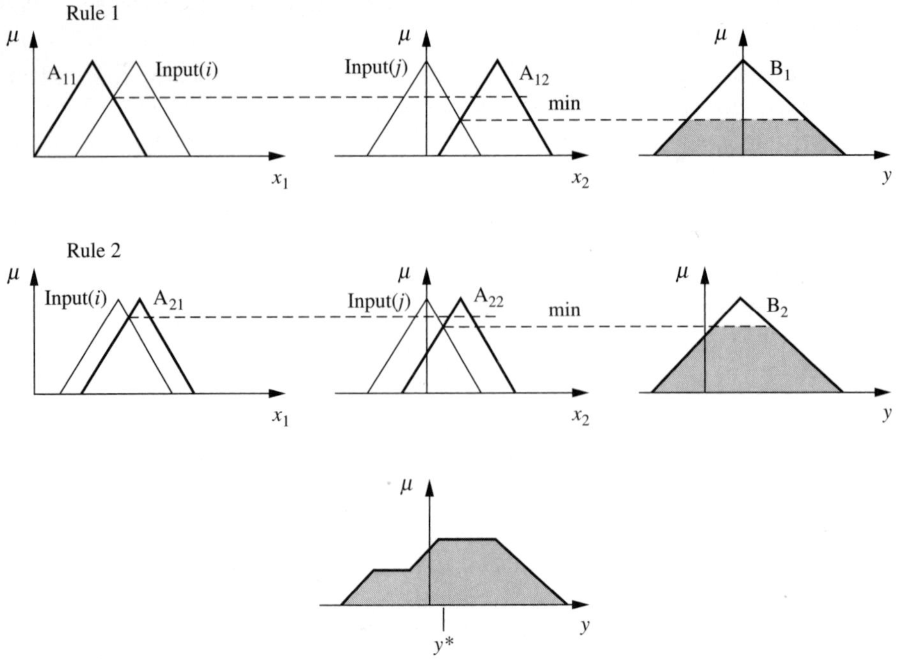

FIGURE 8.9
Graphical Mamdani (max-min) implication method with fuzzy inputs.

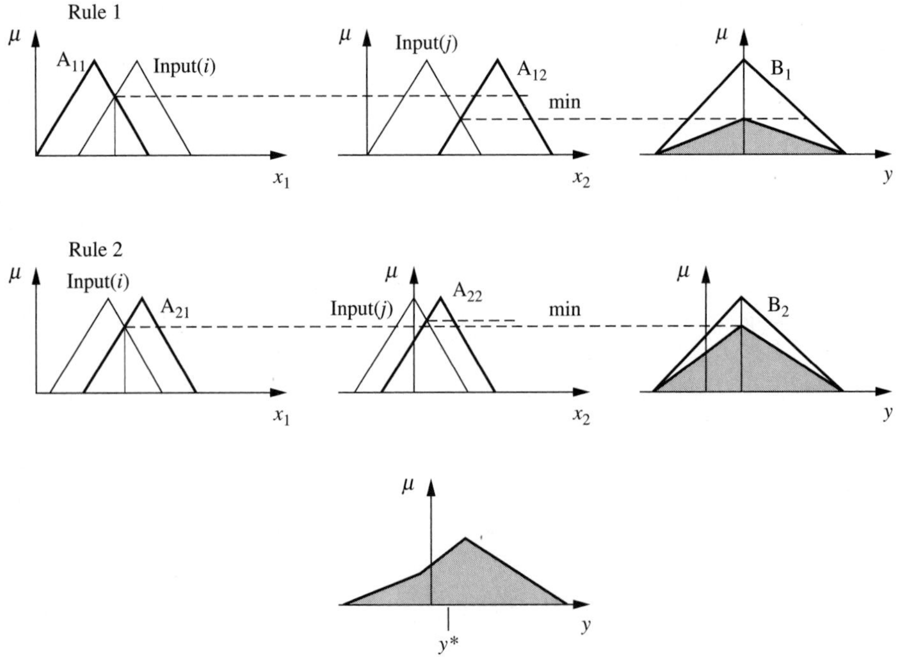

FIGURE 8.10
Graphical correlation-product inference using fuzzy inputs.

We now proceed to describe these two rules in a graphical form and illustrate the four cases of graphical inference presented earlier in this section.

Suppose we have made some observations of the system (moving body) and we estimate the values of the two inputs, mass and velocity, as crisp values. For example, let input(i) = 0.35 kg (mass) and input(j) = 55 m/s (velocity). Case 1 models the inputs as delta functions, Eqs. (8.23)–(8.24) and uses a Mamdani implication, Eq. (8.25). Graphically, this is illustrated in Fig. 8.11, where the output fuzzy membership function is defuzzified using a centroid method.

In Figs. 8.11–8.14, the two rules governing the behavior of the moving body system are illustrated graphically. The antecedents, mass (kg) and velocity (m/s), for each rule are shown as fuzzy membership functions corresponding to the linguistic values for each antecedent. Moreover, the consequent, energy (joules), for each rule is also shown as a fuzzy membership function corresponding to the linguistic label for that consequent. The inputs for mass and velocity intersect the antecedent membership functions at some membership level. The minimum or maximum of the two membership values is propagated to the consequent depending on whether the "and" or "or" connective, respectively, is used between the two antecendents in the rule. The propagated membership value from operations on the antecedents then truncates (for Mamdani implication) or scales (for max-product implication) the membership function for the consequent for that rule. This truncation or scaling is conducted for each rule, and then the truncated or scaled membership functions from each rule are aggregated according to Eq. (8.19) (conjunctive) or (8.21) (disjunctive). In this example we are using two disjunctive rules.

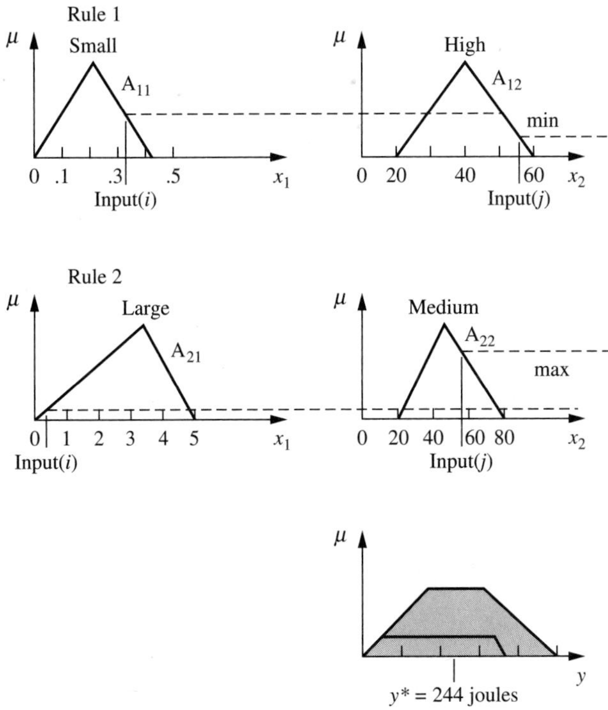

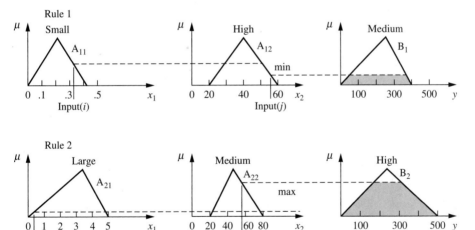

FIGURE 8.11
Fuzzy inference method using the case 1 graphical approach.

In case 2 we only change the method of implication from the first case. Now using a max-product implication method, Eq. (8.26), and a centroidal defuzzification method, the graphical result is shown in Figure 8.12.

Suppose that when we made the observations of the system (moving body) we estimated the values of the two inputs, mass and velocity, to be fuzzy values. For example, we observe that input(i) = "mass is approximately equal to 0.35 kg" and that input(j) = "velocity is approximately equal to 55 m/s." Case 3 models the inputs as fuzzy sets and uses a Mamdani implication, Eq. (8.27). Graphically, this is illustrated in Fig. 8.13, where the output fuzzy membership function is defuzzified using a centroid method.

In case 4 we only change the method of implication from case 3. Now using a max-product implication method, Eq. (8.28), and a centroidal defuzzification method, the graphical result is shown in Fig. 8.14. In this figure the second consequent membership function completely envelops the membership function for the consequent for the first rule, so the membership shape for the first rule is not seen in the final membership function of the aggregated consequent for the two rules.

As the foregoing example illustrates, the four methods yield different shapes for the aggregated fuzzy consequents for the two rules used. However, the defuzzified values for the output energy are all fairly consistent; they vary from 240 joules to 260 joules. The power of fuzzy rule-based systems is their ability to yield "good" results with reasonably simple mathematical operations.

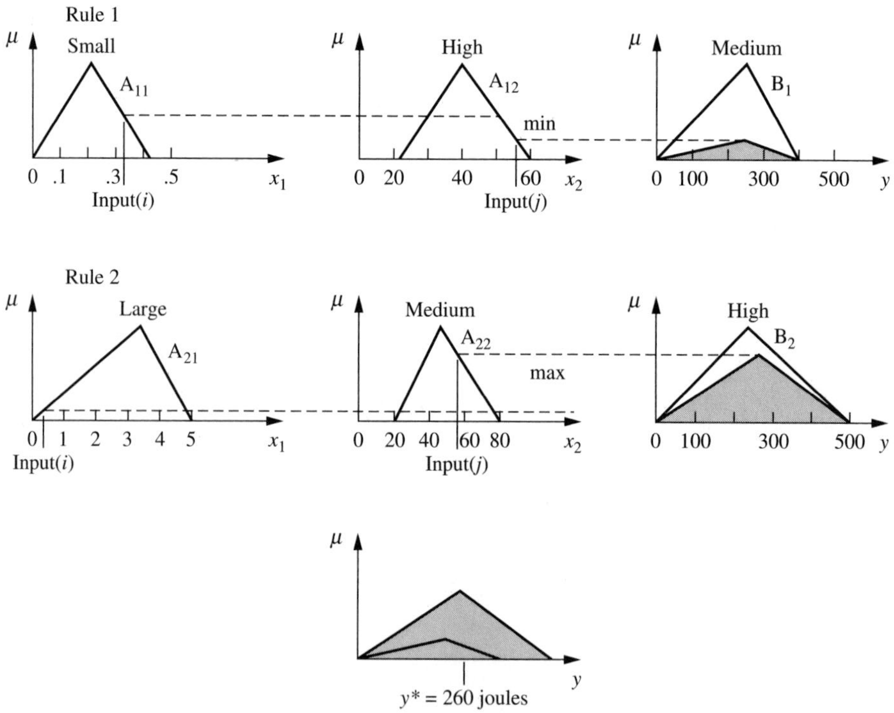

FIGURE 8.12
Fuzzy inference method using the case 2 graphical approach.

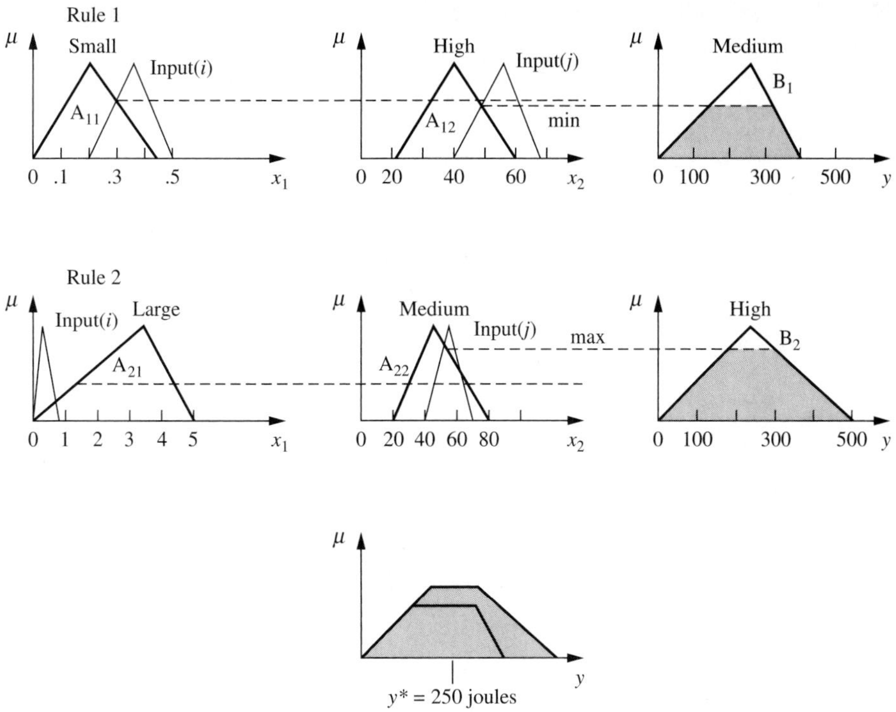

FIGURE 8.13
Fuzzy inference method using Case (*c*) graphical approach.

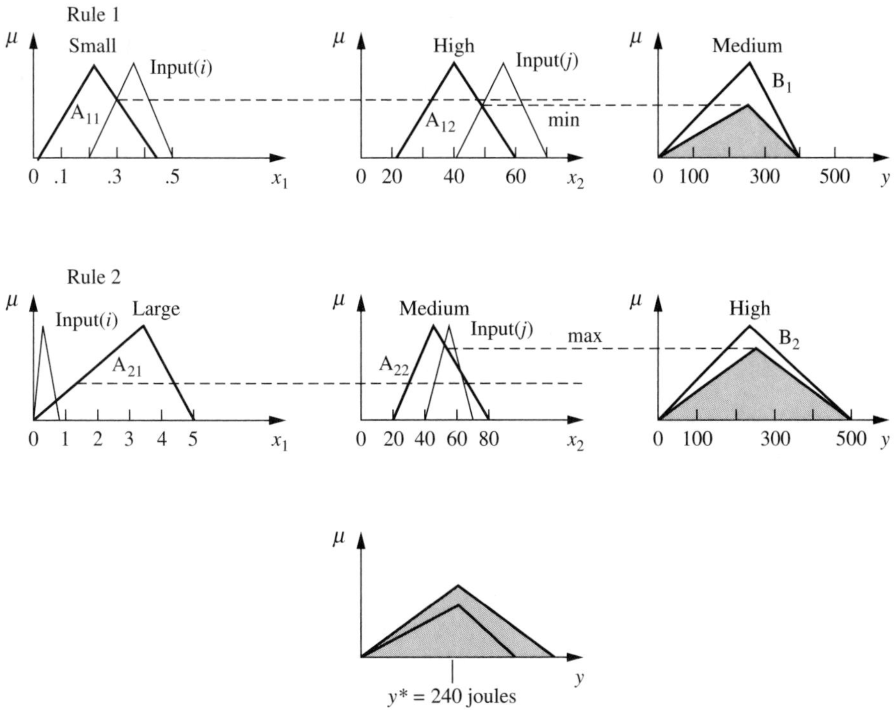

FIGURE 8.14
Fuzzy inference method using Case (*d*) graphical approach.

SUMMARY

The subjectivity that exists in fuzzy modeling is a blessing rather than a curse. The vagueness present in the definition of terms is consistent with the information contained in the conditional rules developed by the engineer when observing some complex process. Even though the set of linguistic variables and their meanings is compatible and consistent with the set of conditional rules used, the overall outcome of the qualitative process is translated into objective and quantifiable results. Fuzzy mathematical tools and the calculus of fuzzy IF-THEN rules provide a most useful paradigm for the automation and implementation of an extensive body of human knowledge heretofore not embodied in the quantitative modeling process. These mathematical tools provide a means of sharing, communicating, and transferring this human subjective knowledge of systems and processes.

This chapter has summarized the seminal works of Zadeh [1972, 1973, 1975*a,b*] in the area of linguistic modeling and truth valuation. Many investigators later extended his methods in truth qualification. For example, see the works in the civil engineering field by Baldwin [1978], Blockley [1980], and Hadipriono and Ross [1991], where methods such as *truth functional modification* and *inverse truth functional modification* were developed to model complex problems using *modus ponens* and *modus tollens* deductive techniques. Modeling in the area of linguistics has reached far beyond the boundaries of engineering. For example, Kickert [1979] used fuzzy linguistic modeling to adapt a factual prototype of Mulder's power theory to a numerical simulation. This is a marvelous illustration of the power of fuzzy sets in a situation where ideals of the *rational man* run contrary to the satisfaction gained simply through the exercise of power. The material developed in this chapter provides a good foundation for discussions in Chapter 9 on nonlinear simulation and in Chapter 13 on fuzzy control.

REFERENCES

Baldwin, J. (1978). "A new approach to approximate reasoning using a fuzzy logic," University of Bristol, Eng. Math. Dept., Research Report EM/FS3, February.

Blockley, D. (1980). *The nature of structural design and safety,* John Wiley & Sons, New York.

Clark, D. W. (1992). "Computer illogic...," *Mirage Magazine,* University of New Mexico Alumni Association, Fall, pp. 12–13.

Hadipriono, F., and K. Sun (1990). "Angular fuzzy set models for linguistic variables," *Civ. Eng. Sys.,* vol. 2, pp. 148–156.

Hadipriono, F., and T. Ross (1991). "A rule based fuzzy logic deduction technique for damage assessment of protective structures," Fuzzy Sets and Systems, vol. 44, no. 3, pp. 459–468.

Jamshidi, M., N. Vadiee, and T. Ross (eds.) (1993). *Fuzzy logic and control: Software and hardware applications,* Prentice Hall, Englewood Cliffs, N.J.

Kickert, W. (1979). "An example of linguistic modelling: The case of Mulder's theory of power," in M. Gupta, R. Ragade, and R. Yager (eds.), *Advances in fuzzy set theory and applications,* pp. 519–540. Elsevier Publishers, Amsterdam, the Netherlands.

Vadiee, N. (1993). "Fuzzy rule-based expert systems—I," in M. Jamshidi, N. Vadiee, and T. Ross (eds.), *Fuzzy logic and control: software and hardware applications,* Prentice Hall, Englewood Cliffs, N.J., pp. 51–85.

Zadeh, L. (1972) "A fuzzy-set-theoretic interpretation of linguistic hedges," *J. Cybern.,* vol. 2, no. 2, pp. 4–34.

Zadeh, L. (1973). "Outline of a new approach to the analysis of complex systems and decision processes," *IEEE Trans. Sys., Man Cybern.*, vol. SMC-3, pp. 28–44.

Zadeh, L. (1975*a*). "The concept of a linguistic variable and its application to approximate reasoning—I," *Inf. Sci.*, vol. 8, pp. 199–249.

Zadeh, L. (1975*b*). "The concept of a linguistic variable and its application to approximate reasoning—II," *Inf. Sci.*, vol. 8, pp. 301–357.

PROBLEMS

8.1. A factory process control operation involves two linguistic (atomic) parameters consisting of pressure and temperature in a fluid delivery system. Nominal pressure limits range from 400 psi minimum to 1000 psi maximum. Nominal temperature limits are 130°F to 140°F. We characterize each parameter in fuzzy linguistic terms as follows:

$$\text{“Low” temperature} = \left\{\frac{1}{131} + \frac{.8}{132} + \frac{.6}{133} + \frac{.4}{134} + \frac{.2}{135} + \frac{0}{136}\right\}$$

$$\text{“High” temperature} = \left\{\frac{0}{134} + \frac{.2}{135} + \frac{.4}{136} + \frac{.6}{137} + \frac{.8}{138} + \frac{1}{139}\right\}$$

$$\text{“High” pressure} = \left\{\frac{0}{400} + \frac{.2}{600} + \frac{.4}{700} + \frac{.6}{800} + \frac{.8}{900} + \frac{1}{1000}\right\}$$

$$\text{“Low” pressure} = \left\{\frac{1}{400} + \frac{.8}{600} + \frac{.6}{700} + \frac{.4}{800} + \frac{.2}{900} + \frac{0}{1000}\right\}$$

(*a*) Find the following membership functions:
  (*i*) Temperature not very low
  (*ii*) Temperature not very high
  (*iii*) Temperature not very low and not very high

(*b*) Find the following membership functions:
  (*i*) Pressure slightly high
  (*ii*) Pressure fairly high ([high]$^{2/3}$)
  (*iii*) Pressure not very low or fairly low

8.2. In information retrieval, having fuzzy information about the size of a document helps when trying to scale the word frequencies of the document (i.e., how often a word occurs in a document is important when determining relevance). So on the universe of document sizes, we define two fuzzy sets:

$$\text{“Small” document} = \begin{cases} 1 - e^{-k(a-x)} & \text{for } x \le a \\ 0 & \text{for } x > a \end{cases}$$

$$\text{“Large” document} = \begin{cases} 1 - e^{-k(x-b)} & \text{for } x \ge b \\ 0 & \text{for } x < b \end{cases}$$

where the parameters k, a, and b change from database to database. Graphically the parameters a and b look as shown in Fig. P8.2.

Develop a graphical solution to the following linguistic phrases, for the specific values of $a = 2$, $b = 4$, and $k = 0.5$:

(*a*) "Not very large" document

(*b*) "Large and small" documents

(*c*) "Not very large or small" documents

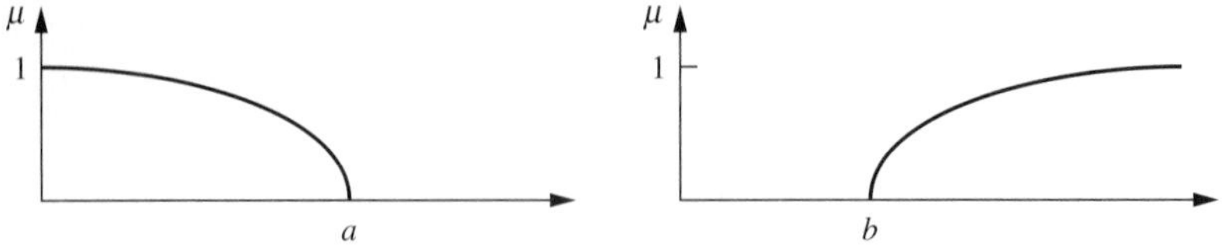

FIGURE P8.2

8.3. One way to implement a document retrieval system is to ask the user for a list of terms likely to appear in the sort of document he or she wants to see. The higher the number of such terms appearing in a particular document, the more *relevant* that document is to the user's topic of research. If a document contains only a few of the listed terms, it is only *tangentially* related to the topic. Define "relevant" and "tangential" as follows, where the denominator is the ratio of search terms found in a particular document to the total number of search terms provided by the user, rounded to the nearest fifth.

$$\text{“Relevant”} = \left\{\frac{0}{0/5} + \frac{.1}{1/5} + \frac{.3}{2/5} + \frac{.8}{3/5} + \frac{.9}{4/5} + \frac{1}{5/5}\right\}$$

$$\text{“Tangential”} = \left\{\frac{0}{0/5} + \frac{1}{1/5} + \frac{.6}{2/5} + \frac{.1}{3/5} + \frac{0}{4/5} + \frac{0}{5/5}\right\}$$

For these atomic terms, find membership functions for the following phrases:

(*a*) Very relevant and not tangential

(*b*) Tangential or slightly relevant

8.4. In the field of hydrology the study of rainfall patterns is most important. The rate of rainfall, in units of mm/h, falling in a particular geographic region could be described linguistically. Suppose we define membership functions for the linguistic variables "heavy" and "light" as follows:

$$\text{“Heavy”} = \left\{\frac{.2}{5} + \frac{.4}{8} + \frac{.6}{12} + \frac{.8}{20} + \frac{1}{30}\right\}$$

$$\text{“Light”} = \left\{\frac{0}{30} + \frac{.1}{20} + \frac{.5}{12} + \frac{.8}{8} + \frac{1}{5}\right\}$$

Develop membership functions for the following linguistic phrases:

(*a*) Very heavy

(*b*) Fairly heavy ($=$ [heavy]$^{2/3}$)

(*c*) Not very light

8.5. In the design of robust control systems, uncertainty in the system models must be estimated and used in the specifications of frequency-dependent weights to be used in the control design. Uncertainty at a particular frequency can be described by a number from 0 to 1 (e.g., 0 = completely certain and 1 = completely uncertain). For these membership functions,

$$\text{“Uncertain”} = \left\{\frac{0}{0} + \frac{0}{.1} + \frac{.2}{.2} + \frac{.6}{.3} + \frac{1}{.4} + \frac{1}{.5} + \frac{1}{.6} + \frac{1}{.7} + \frac{1}{.8} + \frac{1}{.9} + \frac{1}{1}\right\}$$

$$\text{“Certain”} = \left\{\frac{1}{0} + \frac{.8}{.1} + \frac{.5}{.2} + \frac{.2}{.3} + \frac{.1}{.4} + \frac{0}{.5} + \frac{0}{.6} + \frac{0}{.7} + \frac{0}{.8} + \frac{0}{.9} + \frac{0}{1}\right\}$$

find the following linguistic phrases:

(*a*) Not very certain

(*b*) Not very certain and uncertain

(*c*) Not very certain and not very uncertain

8.6. In a problem related to the computer tracking of soil particles as they move under stress, the program displays desired particles on the screen. Particles can be small and large. Because of segmentation problems in computer imaging, the particles can become too large and obscure particles of interest or become too small and be obscured. To solve this problem linguistically, suppose we define the following atomic terms on a scale of sizes [0, 50] in units of mm^2.

$$\text{“Large”} = \left\{\frac{0}{0} + \frac{.1}{10} + \frac{.3}{20} + \frac{.5}{30} + \frac{.6}{40} + \frac{.7}{50}\right\}$$

$$\text{“Small”} = \left\{\frac{1}{0} + \frac{.8}{10} + \frac{.5}{20} + \frac{.3}{30} + \frac{.1}{40} + \frac{0}{50}\right\}$$

For these atomic terms find membership functions for the following phrases:

(*a*) Very small or very large

(*b*) Not small and not large

(*c*) Large or not small

8.7. In a computer system, performance depends to a large extent on the relative speed of the components making up the system. The "speeds" of the CPU and memory, for example, are important factors in defining the limits of operating speed in terms of instructions executed per unit time. Let's say we have system "speeds" referenced by the following linguistic variables:

$$\text{“Fast”} = \left\{\frac{0}{0} + \frac{0}{1} + \frac{.1}{4} + \frac{.3}{8} + \frac{.5}{20} + \frac{.7}{45} + \frac{1}{100}\right\}$$

$$\text{“Slow”} = \left\{\frac{1}{0} + \frac{.9}{1} + \frac{.8}{4} + \frac{.5}{8} + \frac{.2}{20} + \frac{.1}{45} + \frac{0}{100}\right\}$$

where the "speeds" are in units of mips (million instructions per second). Now, let's say we are evaluating the performance of two computer systems for some set of applications. The evaluation of the performance of these systems may be expressed in terms of the following linguistic phrases. Calculate the membership functions for these phrases.

(*a*) Not very fast and slightly slow

(*b*) Very, very fast and not slow

(*c*) Very slow or not fast

8.8. Suppose we had an image processing application whereby we were interested in describing the relative "brightness" or "darkness" of an image. In talking about the brightness or darkness of an image we refer to its gray-level. In this example we will assume that we have a universe of gray-levels, G = {0, 50, 100, 150, 200, 250}, where a gray-level of 0 signifies maximum "darkness," or black, and a gray-level of 250 corresponds to a maximum "brightness," or white. Define the atomic terms "bright" and "dark" as

$$\text{“Bright”} = \left\{\frac{0}{0} + \frac{.2}{50} + \frac{.3}{100} + \frac{.4}{150} + \frac{.8}{200} + \frac{1}{250}\right\}$$

$$\text{“Dark”} = \left\{\frac{1}{0} + \frac{.9}{50} + \frac{.7}{100} + \frac{.4}{150} + \frac{.2}{200} + \frac{0}{250}\right\}$$

From these atomic terms determine the membership functions for the following linguistic phrases:

(*a*) Not bright

(*b*) Dark or not bright

(*c*) Not dark and bright

8.9. A certain software product developer tests their products for user-friendliness by trying their programs out on new, potential users. The test participants are each given a user's manual and a task to perform with the software. Meanwhile, the developer watches the users' reactions while they are using the software. The developer keeps a record of each user's performance, counting the number of times each user looks at the manual, and recording each user's opinion of how user-friendly the software was. The developer then creates two fuzzy variables, "user-friendly" and "easy to learn," relating the users' reactions to how many times they had to look at the manuals; i.e., [0, 5] times.

$$\text{"User-friendly"} = \left\{ \frac{1}{0} + \frac{.8}{1} + \frac{.65}{2} + \frac{.4}{3} + \frac{.15}{4} + \frac{.05}{5} \right\}$$

$$\text{"Easy to learn"} = \left\{ \frac{1}{0} + \frac{.9}{1} + \frac{.75}{2} + \frac{.6}{3} + \frac{.3}{4} + \frac{.1}{5} \right\}$$

Suppose that when the users express their feelings about the new software they often use the term "but" really to mean "and." On the basis of this observation, find membership functions for the following expressions:

(*a*) User-friendly but not very user-friendly

(*b*) User-friendly or easy to learn

8.10. We have a situation involving a fluid pump where the flow through the pump must be controlled in a fairly accurate manner. We can assess the flow with two atomic terms, "high" and "low" flow, which are given as follows:

$$\text{"High"} = \left\{ \frac{0}{0} + \frac{0}{50} + \frac{.3}{100} + \frac{.6}{150} + \frac{.7}{200} + \frac{.8}{250} + \frac{.9}{300} + \frac{.95}{350} + \frac{1}{400} \right\}$$

$$\text{"Low"} = \left\{ \frac{1}{0} + \frac{.9}{50} + \frac{.8}{100} + \frac{.5}{150} + \frac{.2}{200} + \frac{.1}{250} + \frac{0}{300} + \frac{0}{350} + \frac{0}{400} \right\}$$

Find membership functions for other kinds of fluid flow for the following phrases:

(*a*) Slightly low and not very low

(*b*) Slightly low and not very high

(*c*) Not very high or not very low

8.11. With respect to imaging radars when we speak of "resolution," we are referring to the "fineness" of our ability to distinguish closely spaced targets. We may have "high" resolution radars or "low" resolution radars, or some other fuzzy description of their resolution ability. Define the following fuzzy sets on the universe of radar image resolution as measured in meters:

$$\text{"High resolution"} = \left\{ \frac{1}{0.1} + \frac{.9}{0.3} + \frac{.5}{1} + \frac{.2}{3} + \frac{.1}{10} + \frac{0}{30} \right\}$$

$$\text{"Low resolution"} = \left\{ \frac{0}{0.1} + \frac{.1}{0.3} + \frac{.3}{1} + \frac{.7}{3} + \frac{.9}{10} + \frac{1}{30} \right\}$$

Find membership functions for the following linguistic phrases:

(*a*) Not high resolution and not low resolution

(*b*) Low resolution or not very high resolution

(*c*) High resolution and not very, very high resolution

8.12. In engineering it is sometimes easier to use a linguistic term than a difficult algorithm. But many numerical models cannot handle linguistic information. For example, in the ship maintenance industry, when you want to study the influence of "ship age" on maritime operational economy, you can define two atomic terms like "old" and "young."

$$\text{"Old"} = \left\{ \frac{0}{0} + \frac{.1}{5} + \frac{.3}{10} + \frac{.5}{15} + \frac{.7}{20} + \frac{.9}{25} + \frac{1}{30} \right\}$$

$$\text{"Young"} = \left\{ \frac{1}{0} + \frac{.9}{5} + \frac{.7}{10} + \frac{.5}{15} + \frac{.3}{20} + \frac{.1}{25} + \frac{0}{30} \right\}$$

For these ship ages, in years, find the membership functions for the following expressions:

(*a*) Very old

(*b*) Very old or very young

(*c*) Not very old and fairly young ($= [\text{young}]^{2/3}$)

(*d*) Young or slightly old

8.13. Let's say we build a temperature controller, but instead of building a thermostatic system that controls temperature to a desired specific value of, say, 72°F or 85°F, we would like to indicate linguistic comfort ranges such as "slightly cold" and "not too hot." Using these membership functions for the atomic terms defined on a universe of temperatures in °F,

$$\text{"Hot"} = \left\{ \frac{0}{50} + \frac{0}{60} + \frac{.1}{70} + \frac{.5}{80} + \frac{.9}{90} + \frac{1}{100} \right\}$$

$$\text{"Cold"} = \left\{ \frac{1}{50} + \frac{.9}{60} + \frac{.3}{70} + \frac{0}{80} + \frac{0}{90} + \frac{0}{100} \right\}$$

find the membership functions for

(*a*) Not very hot

(*b*) Slightly cold or slightly hot

(*c*) Not very cold and not very hot

8.14. Amplifier capacity on a normalized universe, say [0, 100], can be described linguistically by fuzzy variables like these:

$$\text{"Powerful"} = \left\{ \frac{0}{1} + \frac{.4}{10} + \frac{.8}{50} + \frac{1}{100} \right\}$$

$$\text{"Weak"} = \left\{ \frac{1}{1} + \frac{.9}{10} + \frac{.3}{50} + \frac{0}{100} \right\}$$

Find the membership functions for the following linguistic phrases used to describe the capacity of various amplifiers:

(*a*) Powerful and not weak

(*b*) Very powerful or very weak

(*c*) Very, very powerful and not weak

8.15. Consider the domain of attitude control of a spin-stabilized space vehicle. In order to change the attitude of the vehicle, the roll orientation of the vehicle, say Φ, has to be in a specific position, and the roll rate has to be within a certain bound, say a slow rate and a fast rate. Let these two rates be defined as linguistic variables on a universe of degrees per second:

$$\text{“Fast”} = \left\{\frac{0}{1} + \frac{.2}{100} + \frac{.4}{200} + \frac{.6}{300} + \frac{.8}{400} + \frac{1}{500}\right\}$$

$$\text{“Slow”} = \left\{\frac{1}{1} + \frac{.8}{100} + \frac{.6}{200} + \frac{.4}{300} + \frac{.2}{400} + \frac{0}{500}\right\}$$

Find membership functions for the following natural language phrases:

(*a*) Slightly fast or very fast
(*b*) Fairly slow ($=$ [slow]$^{2/3}$)
(*c*) Slightly fast and fairly slow
(*d*) Not slightly fast or slow
(*e*) Not very fast or fairly slow

8.16. Suppose we have an image-processing problem in associating a group of pixels with a single object. If we have a universe of pixel clusters, assume that a large object is defined by

$$\text{“Large object”} = \left\{\frac{.3}{4} + \frac{.5}{6} + \frac{.7}{8} + \frac{.9}{9}\right\}$$

where the number of pixels is 4, 6, 8, and 9, and that a small object is defined as

$$\text{“Small object”} = \left\{\frac{1}{4} + \frac{.7}{6} + \frac{.2}{8} + \frac{.1}{9}\right\}$$

Find the membership functions for the following definitions of objects:

(*a*) Very small object and not large object
(*b*) Very, very small object or very, very large object
(*c*) Slightly small object and very large object

8.17. In magnetoencephalography (MEG), human skull thickness plays a role in determining how well magnetic fields emanating from the brain can be measured. Since the skull acts as a conductivity barrier, a thicker skull can make it more difficult to analyze magnetic measurements of brain activity. One criterion for choosing subjects for MEG experiments could be the thickness or thinness of their skulls. For the following linguistic atomic terms, as defined on a universe of thicknesses in millimeters,

$$\text{“Thick”} = \left\{\frac{0}{1} + \frac{.2}{2} + \frac{.5}{3} + \frac{.8}{4} + \frac{1}{5}\right\}$$

$$\text{“Thin”} = \left\{\frac{1}{1} + \frac{.9}{2} + \frac{.5}{3} + \frac{.3}{4} + \frac{0}{5}\right\}$$

Find the membership functions for the following phrases:

(*a*) Thick or very thick
(*b*) Not very thick and not very thin
(*c*) Thick and thin

8.18. Suppose you are a system manager for a large VAX cluster. You may ask your users, “How is the response time today?” As a response you might get some variation on "good" or "bad." You define membership functions for "good" and "bad" on a universe of response times in seconds as follows:

$$\text{"Good"} = \left\{ \frac{.9}{.25} + \frac{.7}{.75} + \frac{.5}{1.25} + \frac{.3}{2} \right\}$$

$$\text{"Bad"} = \left\{ \frac{.1}{.25} + \frac{.3}{.75} + \frac{.7}{1.25} + \frac{.9}{2} \right\}$$

Find the membership functions for these phrases:

(*a*) Pretty $[(\alpha)^{0.9}]$ good

(*b*) Not good and very bad

(*c*) Very good and not bad

8.19. Assume that you are in an engineering discipline that deals with scene illumination. The video cameras you design are specified in terms of their response to scene illumination. Scene illumination is measured in units of Lux and typically ranges from 0 to 10,000 Lux. Two atomic terms that can be used to describe scene illumination are bright and dark, i.e.,

$$\text{"Bright"} = \left\{ \frac{0}{0} + \frac{0}{1} + \frac{.1}{10} + \frac{.3}{100} + \frac{.8}{1000} + \frac{1}{10{,}000} \right\}$$

$$\text{"Dark"} = \left\{ \frac{1}{0} + \frac{.8}{1} + \frac{.3}{10} + \frac{.1}{100} + \frac{0}{1000} + \frac{0}{10{,}000} \right\}$$

For these atomic terms calculate membership functions for the following:

(*a*) Not very bright

(*b*) Slightly dark

(*c*) Slightly dark or dark

8.20. The sensitivity of typical CCD (charge-coupled device) cameras is defined in terms of the minimum usable face-plate illumination as measured in footcandles (fc). This is the illumination that produces a specific peak-to-peak video level on an oscilloscope. The universe of discourse is defined as fc = face-plate illumination in footcandles $= \{0, .01, .02, .03, .04, .05, .06, .07, .08, .09, .1\}$. Two atomic variables are defined on the universe fc: (*i*) "Sensitive" represents a camera that has high sensitivity, hence a small minimum usable face-plate illumination, so it has high membership values at low illuminations and low membership values at high illuminations; (*ii*) "Insensitive" represents a camera that has low sensitivity, hence a high minimum usable face-plate illumination, so it has low membership values at low illuminations and high membership values at high illuminations. The membership function for these two variables is given here, where the summation sign denotes set union:

$$\text{"Sensitive"} = \left\{ \sum_i \frac{1 - 0.1 \cdot i}{0.01 \cdot i} \right\} \qquad \text{for } i = 1, 2, \ldots, 10$$

$$\text{"Insensitive"} = \left\{ \sum_i \frac{0.1 \cdot i}{0.01 \cdot i} \right\} \qquad \text{for } i = 1, 2, \ldots, 10$$

Define the membership functions for the following three linguistic expressions:

(*a*) Not very sensitive

(*b*) Very, very sensitive or very insensitive

(*c*) Not highly insensitive and not very sensitive

8.21. In color perception, blue and yellow are complements of one another. The membership functions for these two colors are given here on a normalized universe of discourse, [0, 100], with 0 indicating absolute yellow (complete absence of blue) and 100 indicating absolute (i.e., completely saturated) blue:

$$\text{“Blue”} = \left\{ \frac{0}{0} + \frac{.1}{10} + \frac{.2}{20} + \frac{.3}{30} + \frac{.4}{40} + \frac{.5}{50} + \frac{.6}{60} + \frac{.7}{70} + \frac{.8}{80} + \frac{.9}{90} + \frac{1}{100} \right\}$$

$$\text{“Yellow”} = \left\{ \frac{1}{0} + \frac{.9}{10} + \frac{.8}{20} + \frac{.7}{30} + \frac{.6}{40} + \frac{.5}{50} + \frac{.4}{60} + \frac{.3}{70} + \frac{.2}{80} + \frac{.1}{90} + \frac{0}{100} \right\}$$

Calculate the membership functions for the following mix of colors:

(*a*) Not very blue

(*b*) Blue or fairly yellow (= $[\text{yellow}]^{2/3}$)

(*c*) Very blue and not very yellow

8.22. Using image processing techniques, suppose you are trying to locate objects or shapes within an image field. An object is big or small according to whether it has a value above or below a predefined threshold based on the number of consecutive pixel fields in a row-column image matrix. Define a universe of discourse of the number of adjacent pixels above a certain threshold on the interval [50, 300], then, for the membership functions for “big” and “small,”

$$\text{“Big”} = \left\{ \frac{0}{50} + \frac{.2}{100} + \frac{.3}{150} + \frac{.8}{200} + \frac{.9}{250} + \frac{1}{300} \right\}$$

$$\text{“Small”} = \left\{ \frac{1}{50} + \frac{.5}{100} + \frac{.1}{150} + \frac{0}{200} + \frac{0}{250} + \frac{0}{300} \right\}$$

define the membership functions for the following three linguistic expressions:

(*a*) Not big and very small

(*b*) Very, very big or not small

(*c*) Not very, very big

8.23. In vehicle navigation the mapping source of information uses *shape points* to define the curvature of a turning maneuver. A segment is a length of road between two points. If the segment is linear, it has no or very few shape points. If the road is winding or circular, the segment can have many shape points. Figure P8.23 shows the relationship of curvature and shape points. Assume that up to nine shape points can define any curvature in a typical road segment. The universe of discourse of shape points then varies from 0 (linear road) to 9 (extremely curved). Define the following membership functions:

$$\text{“Somewhat straight”} = \left\{ \frac{1}{0} + \frac{.9}{1} + \frac{.8}{2} + \frac{.7}{3} + \frac{.6}{4} + \frac{.5}{5} + \frac{.4}{6} + \frac{.3}{7} + \frac{.2}{8} + \frac{.1}{9} \right\}$$

$$\text{“Curved”} = \left\{ \frac{0}{0} + \frac{.1}{1} + \frac{.2}{2} + \frac{.3}{3} + \frac{.4}{4} + \frac{.5}{5} + \frac{.6}{6} + \frac{.7}{7} + \frac{.8}{8} + \frac{.9}{9} \right\}$$

Calculate the membership functions for the following phrases:

(*a*) Very curved

(*b*) Fairly curved (= $[\text{curved}]^{2/3}$)

(*c*) Very, very somewhat straight

(*d*) Not fairly curved and very, very somewhat straight

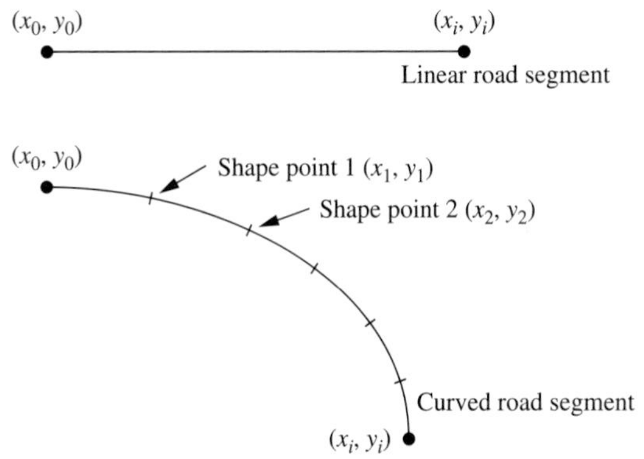

FIGURE P8.23

8.24. This problems deals with the voltages generated internally in switching power supplies. Embedded systems are often supplied 120 V AC for power. A "power supply" is required to convert this to a useful voltage (quite often +5 V DC). Some power supply designs employ a technique called "switching." This technique generates the appropriate voltages by storing and releasing the energy between inductors and capacitors. This problem characterizes two linguistic variables, high and low voltage, on the voltage range of 0 to 200 V AC.

$$\text{“High”} = \left\{ \frac{0}{0} + \frac{0}{25} + \frac{0}{50} + \frac{.1}{75} + \frac{.2}{100} + \frac{.4}{125} + \frac{.6}{150} + \frac{.8}{175} + \frac{1}{200} \right\}$$

$$\text{“Medium”} = \left\{ \frac{.2}{0} + \frac{.4}{25} + \frac{.6}{50} + \frac{.8}{75} + \frac{1}{100} + \frac{.8}{125} + \frac{.6}{150} + \frac{.4}{175} + \frac{.2}{200} \right\}$$

Find the membership functions for the following phrases:

(*a*) Not very high

(*b*) Slightly medium and very high

(*c*) Very, very high or very, very medium

8.25. In risk assessment we deal with characterizing uncertainty in assessing the hazard to human health posed by various toxic chemicals. Because the pharmacokinetics of the human body are very difficult to explain for long-term chemical hazards, such as chronic exposure to lead or to cigarette smoke, hazards can sometimes be uncertain because of scarce data or uncertainty in the exposure patterns. Let us characterize hazard linguistically with two terms: "low" hazard and "high" hazard:

$$\text{“Low” hazard} = \left\{ \frac{0}{1} + \frac{.3}{2} + \frac{.8}{3} + \frac{.1}{4} + \frac{0}{5} \right\}$$

$$\text{“High” hazard} = \left\{ \frac{0}{1} + \frac{.1}{2} + \frac{.2}{3} + \frac{.8}{4} + \frac{0}{5} \right\}$$

Find the membership functions for the following linguistic expressions:

(*a*) Low hazard and not high hazard

(*b*) Very high hazard and not low hazard

(*c*) Low hazard or high hazard

8.26. In reference to car speeds we have the linguistic variables "fast" and "slow" for speed:

$$\text{"Fast"} = \left\{ \frac{0}{0} + \frac{0.1}{10} + \frac{0.2}{20} + \frac{0.3}{30} + \frac{0.4}{40} + \frac{0.5}{50} + \frac{0.6}{60} + \frac{0.7}{70} + \frac{0.8}{80} + \frac{0.9}{90} + \frac{1}{100} \right\}$$

$$\text{"Slow"} = \left\{ \frac{1}{0} + \frac{0.9}{10} + \frac{0.8}{20} + \frac{0.7}{30} + \frac{0.6}{40} + \frac{0.5}{50} + \frac{0.4}{60} + \frac{0.3}{70} + \frac{0.2}{80} + \frac{0.1}{90} + \frac{0}{100} \right\}$$

Using these variables, compute the membership function for the following linguistic terms:

(*a*) Very fast
(*b*) Very, very fast
(*c*) Highly fast (= minus very, very fast)
(*d*) Plus very fast
(*e*) Fairly fast (= $[\text{fast}]^{2/3}$)
(*f*) Not very slow and not very fast
(*g*) Slow or not very slow

8.27. For finding the volume of a cylinder, we need two parameters, namely, radius and height of the cylinder. When the radius is 7 centimeters and height is 12 centimeters, then the volume equals 1847.26 cubic centimeters (using volume $= \pi r^2 h$). Reduce the following rule to canonical form: IF x_1 is radius AND x_2 is height, THEN y is volume.

8.28. According to Boyle's law, for an ideal gas at constant temperature t, pressure is inversely proportional to volume, or volume is inversely proportional to pressure. When we consider different sets of pressures and volumes under the same temperature, we can apply the following rule: IF x_1 is $p_1 v_1$ AND x_2 is $p_2 v_2$, THEN t is a constant. Here p is pressure and v is volume of the gas considered. Reduce this rule to canonical form.

8.29. In mechanics of materials, the moments are calculated by multiplying forces with the distances. Distances are the spaces between the assumed datum and the points at which the forces are acting. Decompose the following rule to canonical form: IF A^1 is large moment THEN (B^1 is large force ELSE B^2 is small force).

8.30. The normal stress for any material is defined as the force per unit area. Decompose the following rule to a canonical form: IF A^1 is low stress (THEN B^1 is high area) UNLESS A^2 is high stress.

8.31. The normal strain for any material is defined as the ratio of change in length to the original length under the influence of some force. Decompose the following rule to canonical form: IF A^1 is a small change in length (THEN B^1 is low strain ELSE IF A^2 is a high change in length (B^2 is a high change in length)).

8.32. Charles' law says that at constant pressure, the temperature of an ideal gas is proportional to the volume, or the volume is proportional to the temperature. For a set of volumes (v) and temperatures (t), decompose the following rule into canonical form: IF A^1 is $v_1 \propto t_1$ THEN (IF A^2 is $v_2 \propto t_2$ THEN B^1 is $v_1 t_2 = v_2 t_1$).

8.33. The formula

$$\frac{1}{u} + \frac{1}{v} = \frac{1}{f}$$

is used extensively in optics. The variables u, v, and f are the distance from the center of the lens to the center of the object, the distance from center of lens to the center of the image, and the focal length, respectively. Change the form of the following rule to canonical form: IF A^1 is u THEN (IF A^2 is v THEN B^1 is f).

8.34. The summation of pressure head, velocity head, and gravitational head is total energy in a fluid flow. Change the form of the following rule to canonical form: IF A^1 is pressure head (THEN (IF A^2 is gravitational head (THEN IF A^3 is velocity head THEN

(B^1 is total energy in the fluid flow ELSE B^2 is hydrostatic head))) ELSE B^3 is sum of gravitational and velocity heads).

8.35. Change the form of the following symbolic rule to canonical form: IF A^1 is R_1 (THEN B^1 AND B^2 (IF A^2 is R_2 (THEN B^3 (IF A^3 is R_3 THEN B^4)))).

8.36. Given the discretized form of the fuzzy variables $\underset{\sim}{X}$, $\underset{\sim}{Y}$, $\underset{\sim}{Z}_1$, $\underset{\sim}{Z}_2$, and $\underset{\sim}{Z}_3$,

$$\underset{\sim}{X} = \left\{\frac{0.0}{0.0} + \frac{0.5}{1.0} + \frac{1.0}{2.0} + \frac{0.5}{3.0} + \frac{0.0}{4.0}\right\}$$

$$\underset{\sim}{Y} = \left\{\frac{0.0}{2.0} + \frac{0.5}{3.0} + \frac{1.0}{4.0} + \frac{0.5}{5.0} + \frac{0.0}{6.0}\right\}$$

$$\underset{\sim}{Z}_1 = \left\{\frac{0.0}{5.0} + \frac{0.5}{6.0} + \frac{1.0}{7.0} + \frac{0.5}{8.0} + \frac{0.0}{9.0}\right\}$$

$$\underset{\sim}{Z}_2 = \left\{\frac{0.0}{10.0} + \frac{0.5}{11.0} + \frac{1.0}{12.0} + \frac{0.5}{13.0} + \frac{0.0}{14.0}\right\}$$

$$\underset{\sim}{Z}_3 = \left\{\frac{0.0}{20.0} + \frac{0.5}{21.0} + \frac{1.0}{22.0} + \frac{0.5}{23.0} + \frac{0.0}{24.0}\right\}$$

(*a*) Form analogous continuous membership functions for $\underset{\sim}{X}$, $\underset{\sim}{Y}$, $\underset{\sim}{Z}_1$, $\underset{\sim}{Z}_2$, and $\underset{\sim}{Z}_3$.

(*b*) A system is described by a set of three rules, using the foregoing fuzzy variables. All the rules have to be satisfied simultaneously for the system to work. The rules are these:

1. If $\underset{\sim}{X}$ and $\underset{\sim}{Y}$ then $\underset{\sim}{Z}_1$.
2. If $\underset{\sim}{X}$ or $\underset{\sim}{Y}$ then $\underset{\sim}{Z}_2$.
3. If $\underset{\sim}{X}^2$ or $\underset{\sim}{Y}^2$ then $\underset{\sim}{Z}_3$.

Determine the output of the system by graphical inference, using the max-min technique, if $x = 3$ and $y = 4$. Use the centroid method for defuzzification.

(*c*) What would be the output of the system if, for the system to work, either of the rules just described may be satisfied.

8.37. Repeat Problem 8.36 using the max-product technique instead of the max-min technique.

8.38. Repeat Problem 8.36 if the inputs (antecedents) to the system are as follows:

$$\underset{\sim}{X} = \left\{\frac{0.0}{2.6} + \frac{0.5}{2.8} + \frac{1.0}{3.0} + \frac{0.5}{3.2} + \frac{0.0}{3.4}\right\}$$

$$\underset{\sim}{Y} = \left\{\frac{0.0}{3.6} + \frac{0.5}{3.8} + \frac{1.0}{4.0} + \frac{0.5}{4.2} + \frac{0.0}{4.4}\right\}$$

8.39. Repeat Problem 8.38 using the max-product technique.

8.40. Compare and comment on the differences and similarities in the outputs obtained in Problems 8.36, 8.37, 8.38, and 8.39.

CHAPTER 9

FUZZY NONLINEAR SIMULATION

There is no idea or proposition in the field, which cannot be put into mathematical language, although the utility of doing so can very well be doubted.

H. W. Brand
Mathematician, 1961

Virtually all physical processes in the real world are nonlinear. It is our abstraction of the real world that leads us to the use of linear systems in modeling these processes. The linear systems are simple and understandable, and, in many situations, they provide acceptable simulations of the actual processes that we observe. Unfortunately, only the simplest of systems can be modeled with linear system theory and only a very small fraction of the nonlinear systems have verifiable solutions. The bulk of the physical processes that we must address are too complex to be reduced to algorithmic form—linear or nonlinear. Most observable processes have only a small amount of information available with which to develop an algorithmic understanding. The vast majority of information we have on most processes tends to be nonnumeric and nonalgorithmic. Most of the information is fuzzy and linguistic in form.

The quotation above from H. W. Brand is an appropriate introduction to this chapter. We can always reduce a complicated process to simple mathematical form. And, for a while at least, we may feel comfortable that our model is a useful replicate of the process we seek to understand. However, reliance on simple linear, or nonlinear, models of the algorithmic form can lead to quite disastrous results, as many engineers have found in documented failures of the past.

A classic example in mechanics serves to illustrate the problems encountered in overlooking the simplest of assumptions. In most beginning textbooks in mechanics, Newton's second law is described by the following equation:

$$\sum F = m \cdot a \tag{9.1}$$

which states that the motion (acceleration) of a body under an imbalance of external forces acting on the body is equal to the sum of the forces ($\sum F$) divided by the body's mass (m). Specifically, the forces and acceleration of the body are vectors containing magnitude and direction. Unfortunately, Eq. (9.1) is not specifically Newton's second law. Newton hypothesized that the imbalance of forces was equivalent to the rate of change in the momentum ($m \cdot v$) of the body; i.e.,

$$\sum F = \frac{d(m \cdot v)}{dt} = m \cdot \frac{dv}{dt} + v \cdot \frac{dm}{dt} \tag{9.2}$$

where v is the velocity of the body and t is time. As one can see, Eqs. (9.1) and (9.2) are not equivalent unless the body's mass does not change with time. In many mechanics applications the mass does not change with time, but in other applications, such as in the flight of spacecraft or aircraft, where fuel consumption reduces total system mass, mass most certainly changes over time. It may be asserted that such an oversight has nothing to do with the fact that Newton's second law is not a valid algorithmic model, but rather it is a model that must be applied against an appropriate physical phenomenon. The point is this: Algorithmic models are useful only when we understand and can observe all the underlying physics of the process. In the airplane example, fuel consumption may not have been an observable phenomenon, and Eq. (9.1) could have been applied to the model. Most complex problems have only a few observables, and an understanding of all the pertinent physics is usually not available. As another example, Newton's first and second laws are not very useful in quantum mechanics applications.

If a process can be described algorithmically, we can describe the solution set for a given input set. If the process is not reducible to algorithmic form, perhaps the input-output features of the system are at least observable or measurable. This chapter deals with systems that cannot be simulated with conventional crisp or algorithmic approaches but that can be simulated because of the presence of other information—observed or linguistic—using fuzzy nonlinear simulation methods.

This chapter proposes to use fuzzy rule-based systems as suitable representations of simple and complex physical systems. For this purpose, a fuzzy rule-based system consists of (*i*) a set of rules that represent the engineer's understanding of the behavior of the system, (*ii*) a set of input data observed going into the system, and (*iii*) a set of output data coming from the system. The input and output data can be numerical, or they can be nonnumeric observations. Figure 9.1 shows a general static physical system, which could be a simple mapping from the input space to the

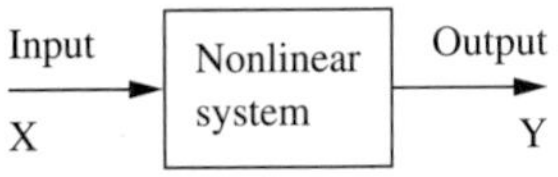

FIGURE 9.1
A general static physical system with observed inputs and outputs.

output space, an industrial control system, a system identification problem, a pattern recognition process, or a decision-making process.

The system inputs and outputs can be vector quantities. Consider an n-input and m-output system. Let X be a Cartesian product of n universes X_i, for $i = 1, 2, \ldots, n$, i.e., $X = X_1 \times X_2 \times \cdots \times X_n$ and let Y be a Cartesian product of m universes Y_j for $j = 1, 2, \ldots, m$; i.e., $Y = Y_1 \times Y_2 \times \cdots \times Y_m$. The vector $\mathbf{X} = (X_1, X_2, \ldots, X_n)$ is the input vector to the system defined on real space R^n, and $\mathbf{Y} = (Y_1, Y_2, \ldots, Y_m)$ is the output vector of the system defined on real space R^m. The input data, rules, and output actions or consequences are generally fuzzy sets expressed by means of appropriate membership functions defined on an appropriate universe of discourse. The method of evaluation of rules is known as *approximate reasoning* or *interpolative reasoning* and is commonly represented by the composition of the fuzzy relations that are formed by the IF-THEN rules (see Chapter 7).

Three spaces are present in the general system posed in Fig. 9.1 [Vadiee, 1993]:

1. The space of possible conditions of the inputs to the system, which, in general, can be represented by a collection of fuzzy subsets $\underset{\sim}{A}^k$, for $k = 1, 2, \ldots$, which are fuzzy partitions of space X, expressed by means of membership functions,

$$\mu_{\underset{\sim}{A}^k}(x) \qquad \text{where } k = 1, 2, \ldots \tag{9.3}$$

2. The space of possible output consequences, based on some specific conditions of the inputs, which can be represented by a collection of fuzzy subsets $\underset{\sim}{B}^p$, for $p = 1, 2, \ldots$, which are fuzzy partitions of space Y, expressed by means of membership functions,

$$\mu_{\underset{\sim}{B}^p}(y) \qquad \text{where } p = 1, 2, \ldots \tag{9.4}$$

3. The space of possible mapping relations from the input space X onto the output space Y. The mapping relations are, in general, represented by fuzzy relations $\underset{\sim}{R}^q$, for $q = 1, 2, \ldots$, and expressed by means of membership functions

$$\mu_{\underset{\sim}{R}^q}(x, y) \qquad \text{where } q = 1, 2, \ldots \tag{9.5}$$

A human perception of the system shown in Fig. 9.1 is based on experience and expertise, empirical observation, intuition, a knowledge of the physics of the system, or a set of subjective preferences and goals. The human observer usually puts this type of knowledge in the form of a set of unconditional as well as conditional propositions in natural language. Our understanding of complex systems is at a qualitative and declarative level, based on vague linguistic terms; this is our so-called *fuzzy* level of understanding of the physical system.

FUZZY RELATIONAL EQUATIONS

Consider a typical crisp nonlinear function relating elements of a single input variable, say x, to the elements of a single output variable, say y, as shown in Fig. 9.2. Notice in Fig. 9.2 that every x in the domain of the independent variable (each x')

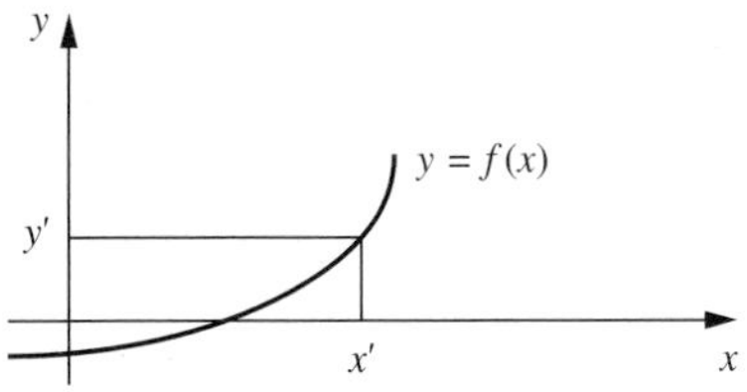

FIGURE 9.2
A crisp relation represented as a nonlinear function.

is "related" to a y (y') in the dependent variable (we called this relation a mapping in Chapter 6). The curve in Fig. 9.2 represents a transfer function, which, in generic terms, is a relation. In fact, any continuous-valued function, such as the curve in Fig. 9.2, can be discretized and reformulated as a matrix relation [see Eqs. (6.1) and (6.2)].

Example 9.1. For the nonlinear function, $y = x^2$, we can formulate a matrix relation to model the mapping imposed by the function. Discretize the independent variable x (the input variable) on the domain, $x = -2, -1, 0, 1, 2$. We find that the mapping provides for the dependent variable y (the output variable) to take on the values $y = 0, 1, 4$. This mapping can be represented by a matrix relation, R, or

$$\mathrm{R} = \begin{array}{c} \\ -2 \\ -1 \\ 0 \\ 1 \\ 2 \end{array} \begin{array}{c} \begin{array}{ccc} 0 & 1 & 4 \end{array} \\ \begin{bmatrix} 0 & 0 & 1 \\ 0 & 1 & 0 \\ 1 & 0 & 0 \\ 0 & 1 & 0 \\ 0 & 0 & 1 \end{bmatrix} \end{array}$$

The elements in the crisp relation, R, are the indicator values as given by Eq. (6.2).

We saw in Chapter 7 that a fuzzy relation can also represent a logical inference. The fuzzy implication, IF $\underset{\sim}{A}$ THEN $\underset{\sim}{B}$, is known as the *generalized modus ponens* form of inference. There are numerous techniques for obtaining a fuzzy relation $\underset{\sim}{R}$ that will represent this inference in the form of a fuzzy relational equation given by

$$\underset{\sim}{B} = \underset{\sim}{A} \circ \underset{\sim}{R} \qquad (9.6)$$

where $\circ$ represents a general method for composition of fuzzy relations (see Chapter 7 for various forms of composition). Equation (9.6) appeared previously in Chapter 7 as the generalized form of approximate reasoning, where Eqs. (7.4) and (7.5) provided two of the most common forms of determining the fuzzy relation $\underset{\sim}{R}$ from a single rule of the form IF $\underset{\sim}{A}$ THEN $\underset{\sim}{B}$.

Suppose our knowledge concerning a certain nonlinear process is not algorithmic, like the algorithm $y = x^2$ in Example 9.1, but rather is in some other more complex form. This more complex form could be data observations of measured inputs and measured outputs. Relations can be developed from these data that are analogous to a look up table, and methods for this step have been given in Chapter 3. Alternatively, the complex form of the knowledge of a nonlinear process could be described with some linguistic rules of the form IF $\underset{\sim}{A}$ THEN $\underset{\sim}{B}$. For example, suppose we are monitoring a thermodynamic process involving an input heat, measured by temperature, and an output variable, pressure. We observe that when we use a "low" temperature, we get out of the process a "low" pressure; when we input a "moderate" temperature, we see a "high" pressure in the system; when we input "high" temperature into the thermodynamics of the system, the output pressure reaches an "extremely high" value; and so on. This process is shown in Fig. 9.3, where the inputs are now *not* points in the input universe (heat) and the output universe (pressure), but *patches* of the variables in each universe. These patches represent the fuzziness in describing the variables linguistically. Obviously, the mapping describing this relationship between heat and pressure is fuzzy. That is, patches from the input space map, or relate, to patches in the output space; and the relations R^1, R^2, and R^3 in Fig. 9.3 represent the fuzziness in this mapping. In general, all the patches, including those representing the relations, overlap because of the ambiguity in their definitions.

Each of the patches in the input space shown in Fig. 9.3 could represent a fuzzy set, say $\underset{\sim}{A}$, defined on the input variable, say x; each of the patches in the output space could be represented by a fuzzy set, say $\underset{\sim}{B}$, defined on the output variable, say y; and each of the patches lying on the general nonlinear function path could be represented by a fuzzy relation, say $\underset{\sim}{R}^k$, where $k = 1, 2, \ldots, r$ represents r possible linguistic relationships between input and output. Suppose we have a situation where a fuzzy input, say x, results in a series of fuzzy outputs, say y^k, depending on which fuzzy relation, $\underset{\sim}{R}^k$, is used to determine the mapping. Each of these relationships, as listed in Table 9.1, could be described by what is called a *fuzzy relational equation,* where y^k is the output of the system contributed by the kth rule, and whose membership function is given by $\mu_{y^k}(y)$. Both x and y^k ($k = 1, 2, \ldots, r$) can be written as single-variable fuzzy relations of dimensions $1 \times n$ and $1 \times m$, respectively. The unary relations, in this case, are actually similarity relations between the elements of the fuzzy set and a most typical or prototype element, usually with membership value equal to unity.

The system of fuzzy relational equations given in Table 9.1 describes a general fuzzy nonlinear system. If the fuzzy system is described by a system of conjunctive rules, we could decompose the rules into a single aggregated fuzzy relational equation by making use of Eqs. (8.18)–(8.19) for each input, x, as follows:

$$y = (x \circ \underset{\sim}{R}^1) \text{ AND } (x \circ \underset{\sim}{R}^2) \text{ AND } \ldots \text{ AND } (x \circ \underset{\sim}{R}^r)$$

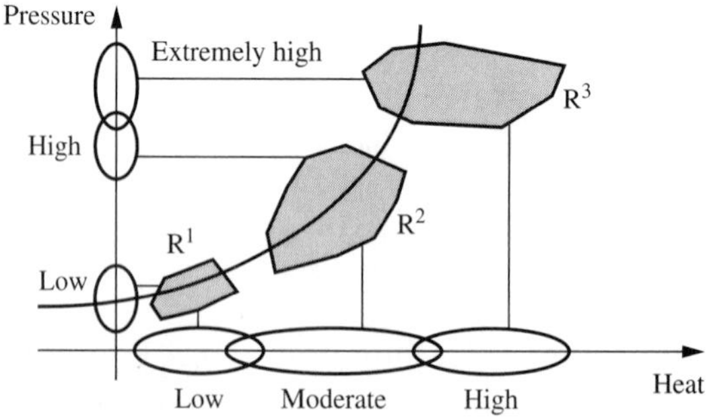

FIGURE 9.3
A fuzzy nonlinear relation matching patches in the input space to patches in the output space.

TABLE 9.1
System of fuzzy relational equations

$\underset{\sim}{R}^1$:	$y^1 = x \circ \underset{\sim}{R}^1$
$\underset{\sim}{R}^2$:	$y^2 = x \circ \underset{\sim}{R}^2$
⋮	⋮
$\underset{\sim}{R}^r$:	$y^r = x \circ \underset{\sim}{R}^r$

and equivalently,

$$y = x \circ (\underset{\sim}{R}^1 \text{ AND } \underset{\sim}{R}^2 \text{ AND } \ldots \text{ AND } \underset{\sim}{R}^r)$$

and finally

$$y = x \circ \underset{\sim}{R} \tag{9.7}$$

where $\underset{\sim}{R}$ is defined as

$$\underset{\sim}{R} = \underset{\sim}{R}^1 \cap \underset{\sim}{R}^2 \cap \cdots \cap \underset{\sim}{R}^r \tag{9.8}$$

The aggregated fuzzy relation $\underset{\sim}{R}$ in Eq. (9.8) is called the *fuzzy system transfer relation* for a single input, x. For the case of a system with n noninteractive fuzzy inputs (see Chapter 3), x_i, and a single output, y, described in Eq. (8.10), the fuzzy relational Eq. (9.7) can be written in the form

$$y = x_1 \circ x_2 \circ \cdots \circ x_n \circ \underset{\sim}{R} \tag{9.9}$$

If the fuzzy system is described by a system of disjunctive rules, we could decompose the rules into a single aggregated fuzzy relational equation by making use of Eqs. (8.20)–(8.21) as follows:

$$y = (x \circ \underset{\sim}{R}^1) \text{ OR } (x \circ \underset{\sim}{R}^2) \text{ OR } \ldots \text{ OR } (x \circ \underset{\sim}{R}^r)$$

and equivalently,

$$y = x \circ (\underset{\sim}{R}^1 \text{ OR } \underset{\sim}{R}^2 \text{ OR} \ldots \text{ OR } \underset{\sim}{R}^r)$$

and finally

$$y = x \circ \underset{\sim}{R} \tag{9.10}$$

where $\underset{\sim}{R}$ is defined as

$$\underset{\sim}{R} = \underset{\sim}{R}^1 \cup \underset{\sim}{R}^2 \cup \cdots \cup \underset{\sim}{R}^r \tag{9.11}$$

The aggregated fuzzy relation, i.e., $\underset{\sim}{R}$, again is called the *fuzzy system transfer relation.*

For the case of a system with n noninteractive (see Chapter 3) fuzzy inputs, x_i, and single output, y, described as in Eq. (8.10), the fuzzy relational Eq. (9.10) can be written in the same form as Eq. (9.9).

There is an interesting interpretation for the way that the fuzzy system transfer relation $\underset{\sim}{R}$ is described. The fuzzy system transfer relation $\underset{\sim}{R}$ is given by a collection of fuzzy relations representing the rules,

$$\underset{\sim}{R} = \{\underset{\sim}{R}^1, \underset{\sim}{R}^2, \ldots, \underset{\sim}{R}^r\}$$

and in its aggregated form $\underset{\sim}{R}$ is given by Eqs. (9.8) and (9.11). Each individual relation $\underset{\sim}{R}^k$, for $k = 1, 2, \ldots, r$, represents a fuzzy data point (a *patch*) in the Cartesian product space X × Y. The fuzzy system transfer relation, i.e., $\underset{\sim}{R}$, is being approximated by r fuzzy input-output data points $\underset{\sim}{R}^k$, for $k = 1, 2, \ldots, r$. It is analogous to the case where a crisp function $y = f(x)$ is approximately described by r numerical input-output values. Each of the k rules is an implication, $\underset{\sim}{A}^k \rightarrow \underset{\sim}{B}^k$, which gives a fuzzy data point (or single *patch*) for approximating the overall fuzzy system transfer relation $\underset{\sim}{R}$. The larger the number of these fuzzy data points with overlapping supports, the better is the approximation of the system input-output mapping.

The fuzzy relations making up the complete set $\underset{\sim}{R}^k$ for $k = 1, 2, \ldots, r$, are fuzzy restrictions, describing a real-world dynamic system or process; they represent the perceptual or cognitive understanding of the system. Chapter 3 summarizes some of the methods used in developing fuzzy relations. In the context of this chapter for nonlinear simulation, these fuzzy restrictions, $\underset{\sim}{R}^k$, can be found through the use of one of the following methods [Vadiee, 1993]:

1. Linguistic knowledge expressed in the form of IF-THEN rules. These rules can be extracted from interviews conducted or protocols obtained when gathering the human knowledge concerning identification or control of a physical process or system.
2. Common sense and intuitive knowledge of the design engineer about the physical process or system under investigation.
3. Use of the general physical principles and laws governing the dynamics of the process or the system under study.
4. Use of pattern classification, clustering, and statistical analysis of some available set of input-output numerical results obtained from measurements of a physical system or process (see Chapters 11 and 12).
5. Use of available closed-form analytical equations that describe the process or the system with Zadeh's extension principle [Zadeh, 1975] to derive a set of fuzzy restrictions on the input-output mapping of the process or the system.

The two most common techniques for obtaining the fuzzy relation $\underset{\sim}{R}$ are the implication of a fuzzy antecedent and a fuzzy consequent (see Chapter 7) and Zadeh's extension principle (see Chapter 6) to propagate a fuzzy input to a fuzzy output through a crisp mapping. The general solution of a system of fuzzy relational equations [Eqs. (9.7) and (9.10)] makes use of the composition operation as discussed in Chapter 7.

PARTITIONING

The fuzzy relation $\underset{\sim}{R}$ is very context-dependent and therefore has local properties with respect to the Cartesian space of the input and output universes. This fuzzy relation results from the Cartesian product of the fuzzy sets representing the inputs and outputs of the fuzzy nonlinear system. However, before the relation $\underset{\sim}{R}$ can be determined, one must consider the more fundamental question of how to *partition* the input and output spaces (universes of discourse) into meaningful fuzzy sets. There are three frequently used methods for the definition of the membership functions (*partitioning*) of the input fuzzy sets (antecedents) as well as the output fuzzy sets (consequents): (1) prototype categorization, (2) degree of similarity, and (3) "similarity as distance."

All three of these methods employ the assumption that the fuzzy system under consideration has n noninteractive inputs (see Chapter 3), where knowledge regarding the functional relation of the system is determined through observation or is extracted from interviews with a human. The knowledge can be put in the form of linguistic IF-THEN rules such as, "If the input is medium-negative then the output is positive-low," or "if the input is high the output is medium," and so on. In this way, the human observer is simply giving the functional relation between pairs of prototype points "medium-negative" and "high" in the input space and the prototype points "positive-low" and "medium" in the output space. In this instance, a crisp binary relation between prototype input points and prototype output points is all we are able to extract from the observer. The exact numerical values of these prototype variables and the corresponding membership functions that these prototype variables represent are unclear to us. These values are highly subjective and intuitive for a human, even though he or she may not know the exact values. Some intuitive methods for assigning membership functions to linguistic values, used by humans in their propositions about a system, will be described by means of two different situations [Vadiee, 1993].

1. There are situations where the ranges of the input and output variables are known in some way. In this case the interval from the lower limit to the higher limit is divided into n equal partitions and the midpoint of each partition is taken to represent the prototype point for that partition, i.e., a membership equal to unity. For the two extreme partitions the minimum and maximum endpoints are assumed to be the corresponding prototype points. It is further assumed that the value of membership function corresponding to a typical prototype point is equal to zero at all other prototype points, and the value of any point in a particular fuzzy set is proportional to the distance of that point from the prototype point in that set. Based on this assumption, n triangular-shaped membership functions will be defined on the interval for each of the input and output variables. Figure 9.4 is an illustration of this procedure for some input variable range defined on the interval $[x_{\min}, x_{\max}]$. Some intuitive linguistic values, such as negative-big or positive-medium, might be given to the fuzzy sets $\underset{\sim}{P}^1$(partition^1), $\underset{\sim}{P}^2$(partition^2), and so on for n partitions. Once the membership functions are determined for the input and output variables, the fuzzy system relational matrices, i.e., $\underset{\sim}{R}$'s, are found based on methods discussed in Chapter 8.
2. It is also possible to ask the observer of the nonlinear system to give the value(s) of the most typical point, points, or intervals (prototypes) that correctly represent a linguistic category used in their knowledge of the system. A similar procedure to

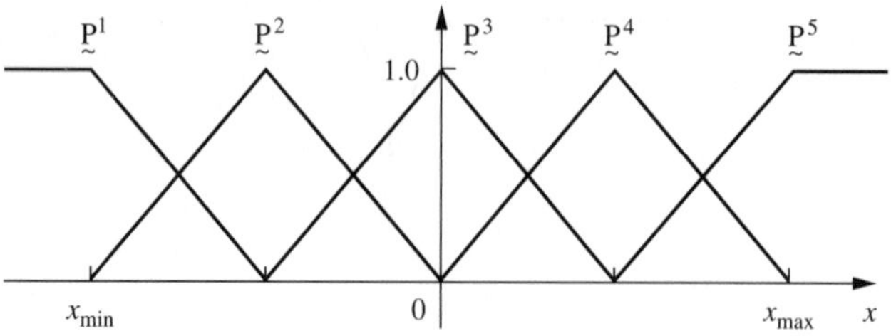

FIGURE 9.4
Fuzzy partitions for equally spaced prototype points [Vadiee, 1993].

the one discussed for case 1 can be used to derive a class of membership functions for each variable. Figure 9.5 illustrates an example of this second situation. The human observer creates partitions that are fuzzy singletons (fuzzy sets with only one element having a nonzero membership) such as partitions $\underset{\sim}{P}_1$, $\underset{\sim}{P}_2$, $\underset{\sim}{P}_4$, and $\underset{\sim}{P}_5$, or crisp sets such as partition $\underset{\sim}{P}_3$ from their knowledge of the system, as shown in Fig. 9.5*a*. The fuzzy partitions $\underset{\sim}{P}_1$, $\underset{\sim}{P}_2$, etc. then take the linguistic values and terms that the human has used for designating their prototypes or have been used by them in their natural language propositions in the form of IF-THEN rules and are fuzzified as shown in Fig. 9.5*b*.

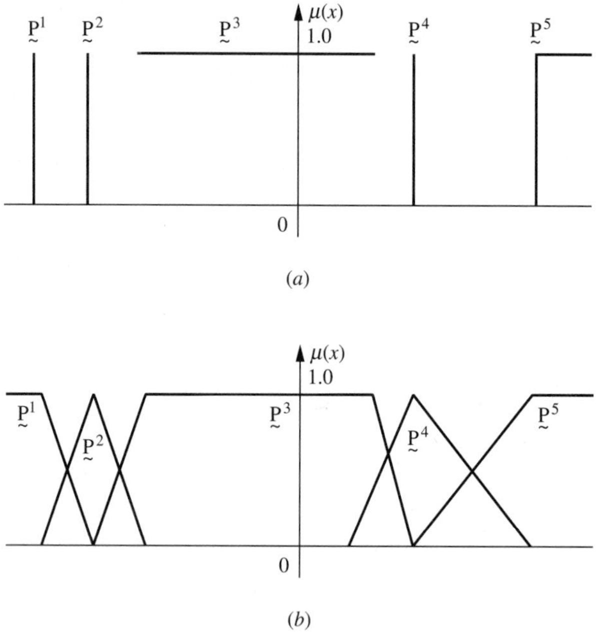

FIGURE 9.5
Intuitive methods for assigning membership functions to linguistic values: (*a*) crisp partition prototype values from a human observer; (*b*) membership functions determined by fuzzifying crisp singletons or crisp sets.

NONLINEAR SIMULATION USING FUZZY RULE-BASED SYSTEMS

A general nonlinear system, such as that in Fig. 9.1, which is comprised of n inputs and m outputs, can be represented by fuzzy relational equations in the form expressed in Table 9.1. Each of the fuzzy relational equations, i.e., $\underset{\sim}{R}^r$, can also be expressed in canonical rule-based form, as shown in Table 9.2.

The rules in Table 9.2 could be connected logically by any of "and," "or," or "else" linguistic connectives; and the variables in Table 9.2, $\underset{\sim}{\mathbf{x}}$ and $\underset{\sim}{\mathbf{y}}$, are the input and output vectors, respectively, of the nonlinear system.

In modeling nonlinear systems, we consider five types of fuzzy rule-based systems. The fuzzy systems discussed are described by a collection of r restrictions in the form of linguistic IF-THEN rules, shown in Table 9.2. The five types are for single-input and single-output systems, but the results can be easily extended to the n-input and m-output systems where input $\mathbf{x}$ is an n-dimensional vector and output $\mathbf{y}$ is an m-dimensional vector. All five types represent special restrictions on the antecedents, $\underset{\sim}{A}^i$, and consequents, $\underset{\sim}{B}^i$, displayed in Table 9.2 [Vadiee, 1993].

1. In the first type of fuzzy models, the input and the output restrictions are given in the form of singletons, i.e., $\underset{\sim}{A}^1 : x = x_1, \underset{\sim}{A}^2 : x = x_2, \ldots$, and $\underset{\sim}{B}^1 : y = y_1, \underset{\sim}{B}^2 : y = y_2, \ldots$. This type is simply a lookup table for the system description. Example 9.1 is of this first type of fuzzy model. The value of the output for a given value of input, e.g., $x = x_1$, is equal to the THEN part value of the ith rule $\underset{\sim}{R}^i$ whose IF part matches exactly the value of input given (see Fig. 9.6), i.e.,

$$\text{IF } \underset{\sim}{A}^i : x = x_i \text{ THEN } \underset{\sim}{B}^i : y = y_i \qquad \text{for } i = 1, 2, \ldots, r \tag{9.12}$$

2. In the second type of fuzzy nonlinear models, the input restrictions are in the form of crisp sets and the output restrictions are given by singletons. This is also a lookup table for the system description. The value of the output for a given value of input is equal to the THEN part value of the ith rule $\underset{\sim}{R}^i$ whose IF part occurs somewhere in the interval, as seen in Fig. 9.7 and given in Eq. (9.13):

$$\text{IF } \underset{\sim}{A}^i : x_{i-1} < x < x_i \text{ THEN } \underset{\sim}{B}^i : y = y_i \qquad \text{for } i = 1, 2, \ldots, r \tag{9.13}$$

This second type is effectively a piecewise-constant approximation of a nonlinear function [Vadiee, 1993]. The restrictions for this second type of fuzzy

TABLE 9.2
Canonical rule-based form of fuzzy relational equations

$\underset{\sim}{R}^1$:	IF x is $\underset{\sim}{A}^1$, THEN y is $\underset{\sim}{B}^1$
$\underset{\sim}{R}^2$:	IF x is $\underset{\sim}{A}^2$, THEN y is $\underset{\sim}{B}^2$
$\vdots$	$\vdots$
$\underset{\sim}{R}^r$:	IF x is $\underset{\sim}{A}^r$, THEN y is $\underset{\sim}{B}^r$

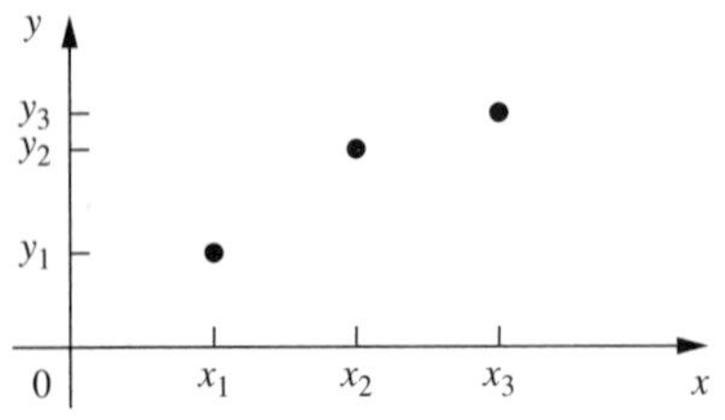

FIGURE 9.6
Input singletons and output singletons.

model might also involve spline functions to represent the output instead of crisp singletons. In this case we simply represent the nonlinear function with functions $f_1(x), f_2(x), \ldots, f_r(x)$, which are nonlinear spline functions (see Fig. 9.8). In this case, Eq. (9.13) becomes

$$\text{IF } \underset{\sim}{A}^i : x_{i-1} < x < x_i \text{ THEN } \underset{\sim}{B}^i : y = f_i(x) \qquad \text{for } i = 1, 2, \ldots, r \qquad (9.14)$$

3. In the third class of fuzzy nonlinear models, the input conditions are crisp sets and the output is expressed as a fuzzy set or described by a fuzzy relation. If the THEN part of the rules is given by fuzzy sets defined on the output universe of discourse, the rules will be in the form

$$\text{IF } \underset{\sim}{A}^i : x_{i-1} < x < x_i \text{ THEN } y = \underset{\sim}{B}^i \qquad \text{for } i = 1, 2, \ldots, r \qquad (9.15)$$

where $\underset{\sim}{B}^i$ for $i = 1, 2, \ldots, r$ are fuzzy sets partitioned on the output space. The output for this type of fuzzy model is a fuzzy set, which can be defuzzified if necessary with any of the defuzzification tools presented in Chapter 5. If the THEN part restrictions are given in the form of fuzzy relations $\underset{\sim}{R}^i$ for $i = 1, 2, \ldots, r$, the fuzzy model will be in the form

$$\text{IF } \underset{\sim}{A}^i : x_{i-1} < x < x_i \text{ THEN } \underset{\sim}{B}^i : y = \underset{\sim}{R}^i \qquad \text{for } i = 1, 2, \ldots, r \qquad (9.16)$$

where the fuzzy relation $\underset{\sim}{R}^i$ is the THEN-consequent restriction of the ith rule whose IF antecedent is contained within the given input crisp set, as seen in Fig. 9.9. In Fig. 9.9 the input space is divided into crisp intervals; the intervals can overlap (not shown in the figure). These inputs map to relations in the output; i.e., $\underset{\sim}{B}^i = \underset{\sim}{R}^i$. The regions, $\underset{\sim}{B}^i$, in the figure represent the output and the mapping. This form of model is slightly more general than that expressed in Eq. (9.15) in that a fuzzy relation can be thought of as a collection of fuzzy sets (e.g., for the discrete case each row of a relation can be thought of as a set).

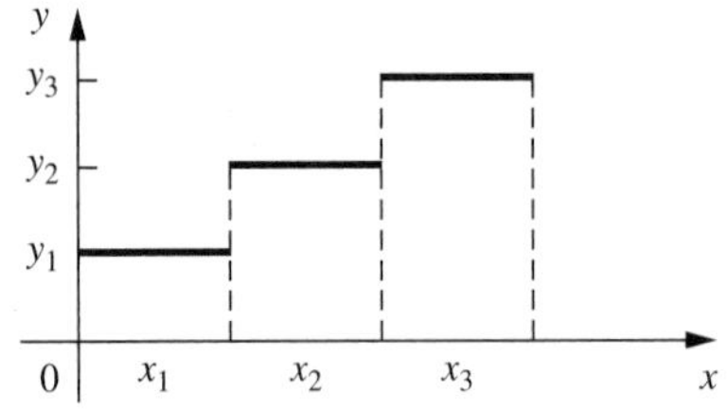

FIGURE 9.7
Input crisp sets (intervals) and output singletons.

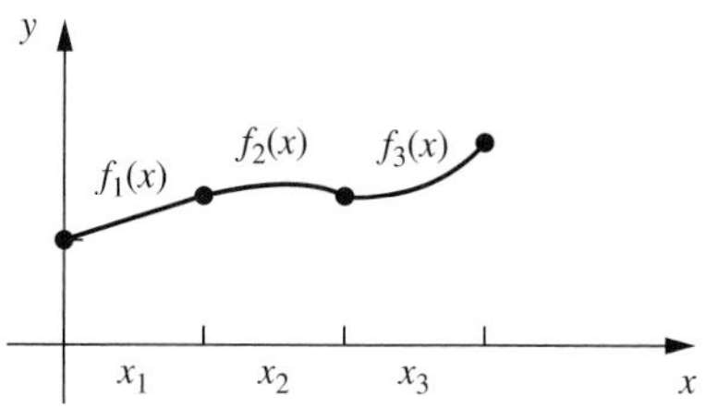

FIGURE 9.8
Input crisp (intervals) sets and output crisp functions.

4. In the fourth type of fuzzy models, the input conditions are given in the form of fuzzy sets, $\underset{\sim}{A}^i$, partitioned on the input universe of discourse, and the outputs are given in the form of singletons or generally nonlinear crisp functions. In the former case, we have

$$\text{IF } x = \underset{\sim}{A}^i \text{ THEN } \underset{\sim}{B}^i : y = y_i \qquad \text{for } i = 1, 2, \ldots, r \tag{9.17}$$

where $\underset{\sim}{A}^i$ for $i = 1, 2, \ldots, r$ are the fuzzy partitions defined on the input space X. In the latter case we can have any of three general nonlinear crisp functions: (*i*) linear functions of the input variables,

$$\text{IF } x = \underset{\sim}{A}^i \text{ THEN } \underset{\sim}{B}^i : y = f_i(x) \qquad \text{for } i = 1, 2, \ldots, r \tag{9.18}$$

where the fuzzy model is called a *quasi-linear fuzzy model* (QLFM); (*ii*) linear polynomial functions of the input variable [a subset of case (*i*)],

$$\text{IF } x = \underset{\sim}{A}^i \text{ THEN } \underset{\sim}{B}^i : y = p_0^i + p_1^i \cdot x_1 + p_2^i \cdot x_2 + \cdots + p_n^i \cdot x_n \qquad \text{for } i = 1, 2, \ldots, r \tag{9.19}$$

where the p_j^i are constants and the fuzzy model is known as static Sugeno's fuzzy model [Sugeno, 1985]; and (*iii*) nonlinear functions of the input variable,

$$\text{IF } x = \underset{\sim}{A}^i \text{ THEN } \underset{\sim}{B}^i : y = f_i(x) \qquad \text{for } i = 1, 2, \ldots, r \tag{9.20}$$

where the fuzzy model is called a *quasi-nonlinear fuzzy model* (QNFM).

By replacing the linear polynomial model in the static Sugeno's model with a difference equation model, the output can be represented by a nonlinear

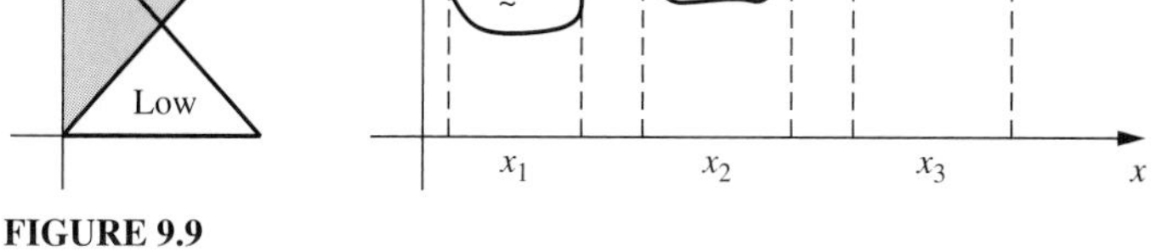

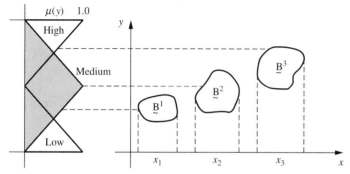

FIGURE 9.9
Crisp set (intervals) inputs and fuzzy set outputs.

function, which represents a quasi-linear dynamic fuzzy system model (Sugeno's dynamic model); more is detailed in Vadiee [1993]. Figure 9.10 shows one form of this fourth type of fuzzy model, where the output is some weighted average of spline functions, Eq. (9.20).

5. The most general fuzzy model for nonlinear simulation is the case in which both the input and output restrictions are described by fuzzy sets; this general form is expressed in Table 9.2,

$$\text{IF } \underset{\sim}{A}^i, \text{ THEN } \underset{\sim}{B}^i \tag{9.21}$$

and shown in Fig. 9.11.

FUZZY ASSOCIATIVE MEMORIES (FAMs)

Consider a fuzzy system with n noninteractive (see Chapter 3) inputs and a single output. Also assume that each input universe of discourse, i.e., $X_1, X_2, \ldots, X_n$, is partitioned into k fuzzy partitions. Based on the canonical fuzzy model given in Table 9.2 for a nonlinear system, the total number of possible rules governing this system is given by

$$l = k^n \tag{9.22a}$$

$$l = (k+1)^n \tag{9.22b}$$

where l is the maximum possible number of canonical rules. Equation (9.22b) is to be used if the partition "anything" is to be used, otherwise Eq. (9.22a) determines the number of possible rules. The actual number of rules, r, necessary to describe a fuzzy system is much less than l, i.e., $r \ll l$, because of the interpolative reason-

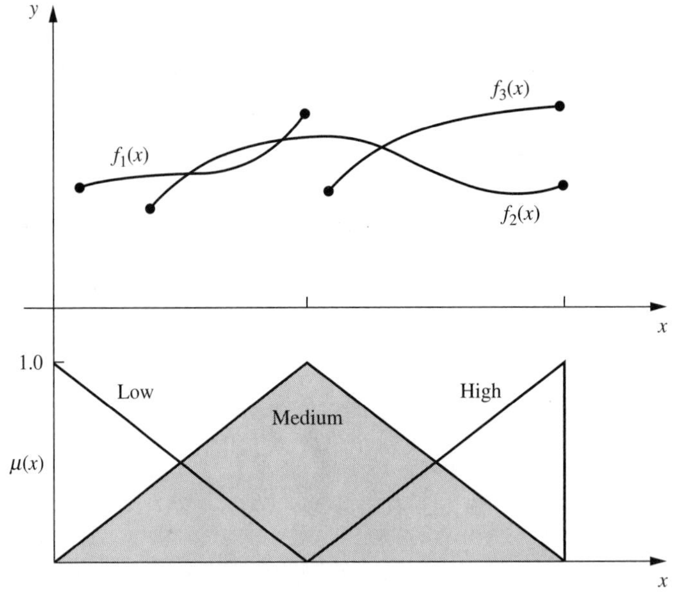

FIGURE 9.10
Fuzzy set inputs and crisp function outputs.

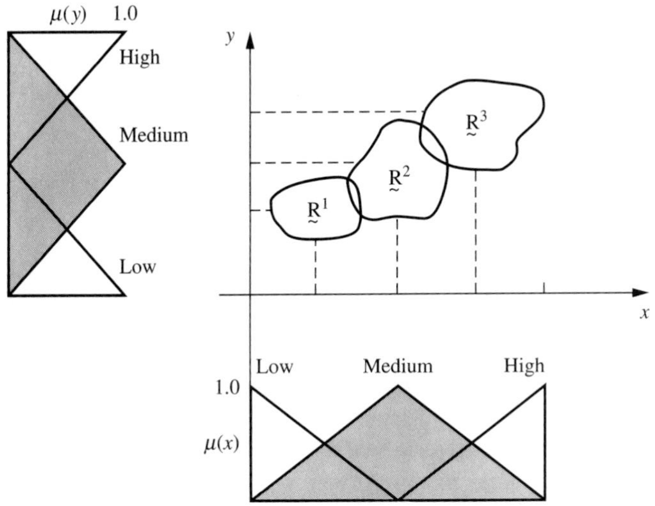

FIGURE 9.11
Fuzzy set inputs and fuzzy set outputs (the most general case).

ing capability of the fuzzy model and because the fuzzy membership functions of the partitions overlap. If each of the n noninteractive inputs is partitioned into a different number of fuzzy partitions, say, X_1 is partitioned into k_1 partitions and X_2 is partitioned into k_2 partitions and so forth, then the maximum number of rules is given by

$$l = k_1 k_2 k_3 \cdots k_n \tag{9.23}$$

For a small number of inputs, e.g., $n = 1$ or $n = 2$, or $n = 3$, there exists a compact form of representing a fuzzy rule-based system. This form is illustrated for $n = 2$ in Fig. 9.12. In the figure there are seven partitions for input A (A_1 to A_7), five partitions for input B (B_1 to B_5), and four partitions for the output variable C (C_1 to C_4). This compact graphical form is called a fuzzy associative memory table, or FAM table. As can be seen from the FAM table, the rule-based system actually represents a general nonlinear mapping from the input space of the fuzzy system to the output space of the fuzzy system. In this mapping, the patches of the input space are being applied to the patches in the output space. Each rule or, equivalently, each fuzzy

Input B \ Input A	A_1	A_2	A_3	A_4	A_5	A_6	A_7
B_1	C_1		C_4	C_4		C_3	C_3
B_2		C_1				C_2	
B_3	C_4		C_1			C_2	C_2
B_4	C_3	C_3		C_1		C_1	C_2
B_5	C_3		C_4	C_4	C_1		C_3

FIGURE 9.12
FAM table for a two-input, single-output fuzzy rule-based system.

relation from input to the output represents a fuzzy point of data that characterizes the nonlinear mapping from the input to the output.

In the FAM table in Fig. 9.12 we see that the maximum number of rules for this situation, using Eq. 9.23, is $l = k_1 k_2 = 7(5) = 35$; but as seen in the figure, the actual number of rules is only $r = 21$.

We will now illustrate the ideas involved in simulation with three examples from various engineering disciplines.

Example 9.2. For the nonlinear function, $y = 10 \sin x_1$, we will develop a fuzzy rule-based system using four simple fuzzy rules to approximate the output y. The universe of discourse for the input variable x_1 will be the interval $[-180, 180]$ in degrees, and the universe of discourse for the output variable y is the interval $[-10, 10]$.

First, we will partition the input space x_1 into five simple partitions on the interval $[-180°, 180°]$, and we will partition the output space y on the interval $[-10, 10]$ into three membership functions, as shown in Fig. 9.13*a* and 9.13*b*, respectively. In these figures the abbreviations NB, NS, Z, PS, and PB refer to the linguistic variables "negative-big," "negative-small," "zero," "positive-small," and "positive-big," respectively.

Second, we develop four simple rules, listed in Table 9.3, that we think emulate the dynamics of the system (in this case the system is the nonlinear equation,

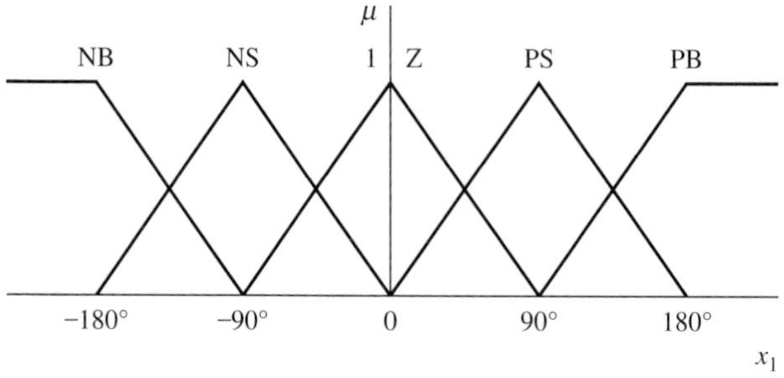

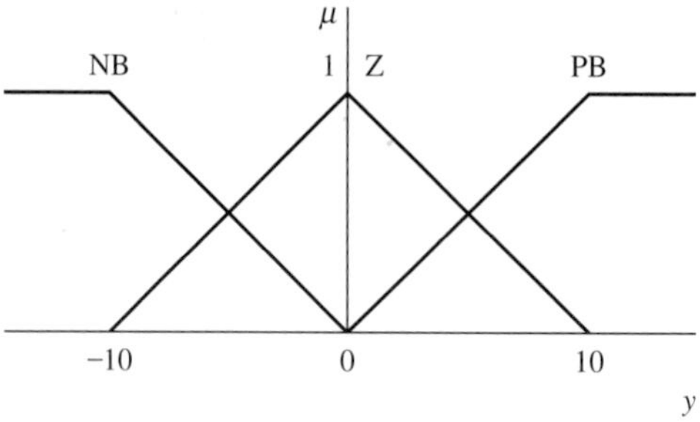

FIGURE 9.13
Fuzzy membership functions for the input and output spaces: (*a*) Five partitions for the input variable, x_1; (*b*) three partitions for the output variable, y.

$y = 10 \sin x_1$ and we are observing the harmonics of this system) and that make use of the linguistic variables in Fig. 9.13. The FAM table for these rules is given in Table 9.4.

The FAM table of Table 9.4 is one-dimensional because there is only one input variable, x_1. As seen in Table 9.4 all rules listed in Table 9.3 are accommodated. Not all the four rules expressed in Table 9.3 are expressed in canonical form (some have disjunctive antecedents), but if they were transformed into canonical form, they would represent the five rules provided in the FAM table in Table 9.4.

In developing an approximate solution for the output y we select a few input points and employ a graphical inference method similar to that illustrated in Chapter 8. We will use the centroid method for defuzzification. Let us choose four crisp singletons as the input:

$$x_1 = \{-135°, -45°, 45°, 135°\}$$

For input $x_1 = -135°$, rules 3 and 4 are fired, as shown in Fig. 9.14*c* and 9.14*d*. For input $x_1 = -45°$, rules 1, 3 and 4 are fired. Figures 9.14*a* and 9.14*b* show the graphical inference for input $x_1 = -45°$ (which fires rule 1), and for $x_1 = 45°$ (which fires rule 2), respectively.

For input $x_1 = -45°$, rules 3 and 4 are also fired, and we get results similar to those shown in Fig. 9.14*c* and 9.14*d* after defuzzification:

$$\text{Rule 3: } y = 0$$
$$\text{Rule 4: } y = -7$$

For $x_1 = 45°$, rules 1, 2, and 3 are fired (see Fig. 9.14*d* for rule 2), and we get the following results for rules 1 and 3 after defuzzification:

$$\text{Rule 1: } y = 0$$
$$\text{Rule 3: } y = 0$$

For $x_1 = 135°$, rules 1 and 2 are fired and we get, after defuzzification, results that are similar to those shown in Fig. 9.14*d*:

$$\text{Rule 1: } y = 0$$
$$\text{Rule 2: } y = 7$$

When we combine the results, we get an aggregated result summarized in Table 9.5 and shown graphically in Fig. 9.15. The y values in each column of Table 9.5 are the defuzzified results from various rules firing for each of the inputs, x_i. When we aggregate the rules using the union operator (disjunctive rules), the effect is to take the maximum value for y in each of the columns in Table 9.5. The plot in Fig. 9.15 represents the maximum y for each of the x_i, and it represents a fairly accurate

TABLE 9.3
Four simple rules for $y = 10 \sin x_1$

1	IF x_1 is Z or PB, THEN y is Z
2	IF x_1 is PS, THEN y is PB
3	IF x_1 is Z or NB, THEN y is Z
4	IF x_1 is NS, THEN y is NB

TABLE 9.4
FAM for the four simple rules in Table 9.3

x_i	NB	NS	Z	PS	PB
y	Z	NB	Z	PB	Z

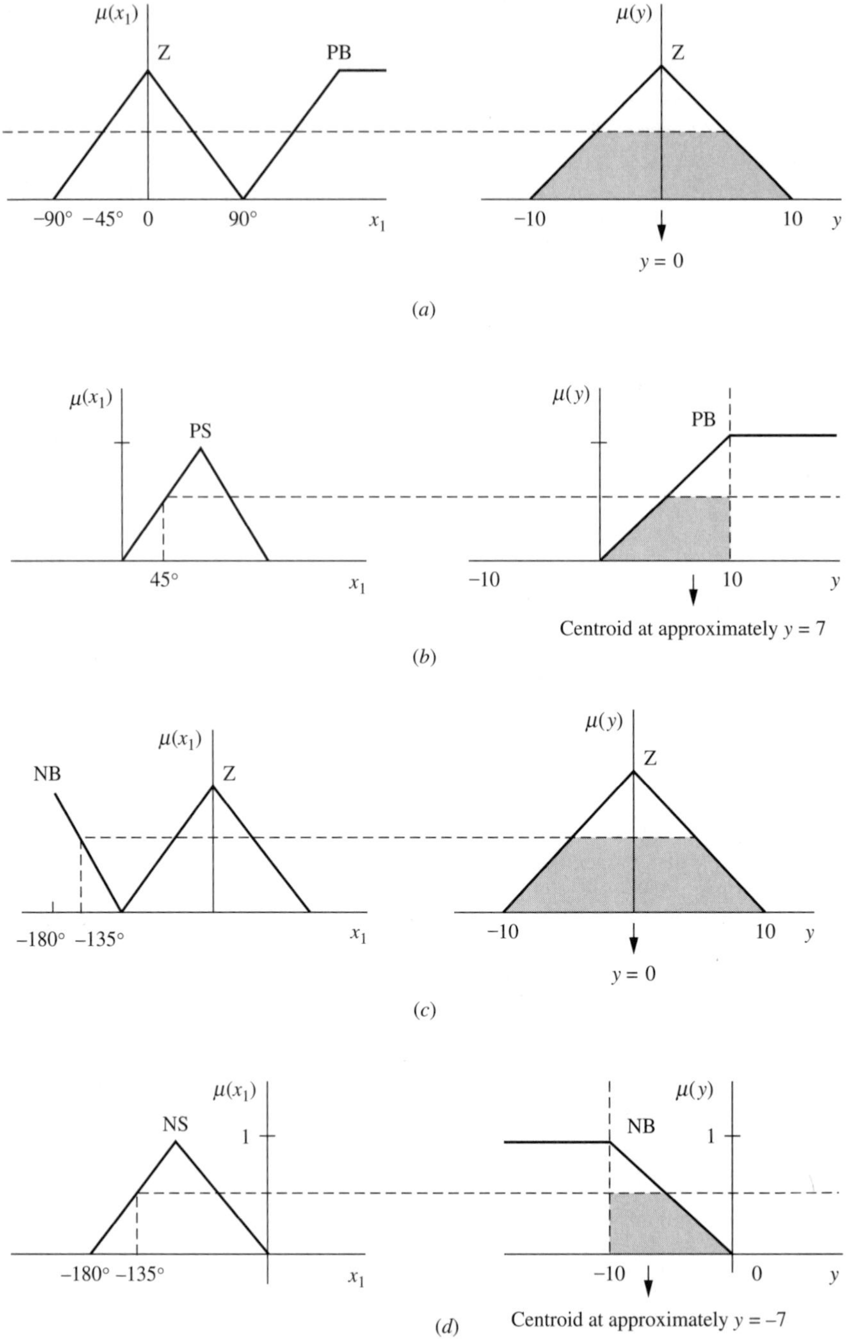

FIGURE 9.14
Graphical inference method showing membership propagation and defuzzification: (*a*) input $x_1 = -45°$ fires rule 1; (*b*) input $x_1 = 45°$ fires rule 2; (*c*) input $x_1 = -135°$ fires rule 3; (*d*) input $x_1 = -135°$ fires rule 4.

TABLE 9.5
Defuzzified results for simulation of $y = 10\sin x_1$

x_1	−135°	−45°	45°	135°
y	0	0	0	0
	−7	0	0	7
		−7	7	

portrayal of the true solution, given only a crude discretization of four inputs and a simple simulation based on four rules. More rules would result in a closer fit to the true sine curve.

We saw in Chapter 6 that it is possible to extend fuzzy input variables through conventional algebraic operations to obtain fuzzy output, using Zadeh's extension principle [Zadeh, 1975]. The nonlinear function, $y = 10\sin x_1$, represents a conventional algebraic operation. It is instructive to compare results of a nonlinear function using the rule-based fuzzy approach just completed with those derived using the extension principle. As before, let x_1 be a variable defined as a fuzzy number on the universe of discourse [−180°, +180°], which is partitioned into five fuzzy partitions as shown in Fig. 9.13*a*.

Let x_1 be defined by the linguistic variables from Table 9.3 and Fig. 9.13 for the four rules of Table 9.4 as

1. x_1 = Z or PB
2. x_1 = PS
3. x_1 = Z or NB
4. x_1 = NS

Further, let $\underset{\sim}{B}$ be a fuzzy set defined on the universe of y consisting of equally spaced discrete points, or elements, $\underset{\sim}{B} = \{-10, -8, -6, -4, -2, 0, 2, 4, 6, 8, 10\}$. To determine

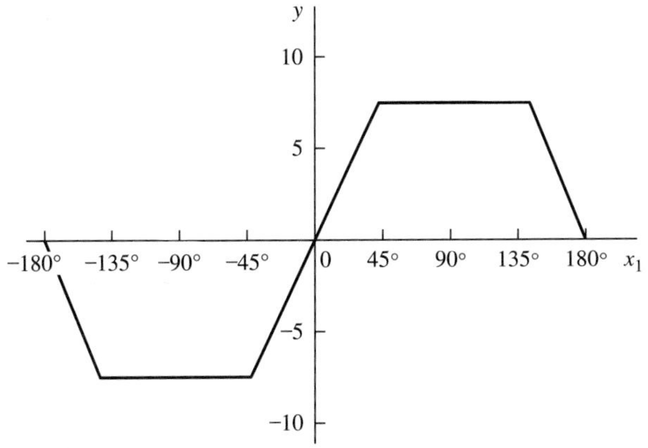

FIGURE 9.15
Simulation of nonlinear system $y = 10\sin x_1$ using a four-rule fuzzy rule-base.

a mapping between x_1 and y we look at the inverse image of $y = f(x_1)$, hence $x_1 = f^{-1}(y)$, which is given in Table 9.6.

For rule 1, x_1 = Z or PB (see Fig. 9.16), we get the following membership values for the mappings displayed in Table 9.6:

$$\begin{aligned}
\mu_y(-10) &= \mu_{A_1}(-90) = 0\\
\mu_y(-8) &= \max\{\mu_{A_1}(-126.9), \mu_{A_1}(-53.1)\} = \max\{0, \mu_{A_1}(-53.1)\} = 0.41\\
\mu_y(-6) &= \max\{0, \mu_{A_1}(-36.9)\} = 0.59\\
\mu_y(-4) &= \max\{0, \mu_{A_1}(-23.6)\} = 0.74\\
\mu_y(-2) &= \max\{0, \mu_{A_1}(-11.5)\} = 0.87\\
\mu_y(0) &= \max\{\mu_{A_1}(0), \mu_{A_1}(180)\} = 1\\
\mu_y(2) &= \max\{\mu_{A_1}(11.5), \mu_{A_1}(168.5)\} = \max\{.87, .87\} = 0.87\\
\mu_y(4) &= \max\{\mu_{A_1}(23.6), \mu_{A_1}(156.4)\} = \max\{.74, .74\} = 0.74\\
\mu_y(6) &= \max\{\mu_{A_1}(36.9), \mu_{A_1}(143.1)\} = 0.59\\
\mu_y(8) &= \max\{0.41, 0.41\} = 0.41\\
\mu_y(10) &= \mu_{A_1}(90) = 0
\end{aligned}$$

These membership values for the various equally spaced points in the y universe are shown in Fig. 9.17.

For rule 2, x_1 = PS (see Fig. 9.18), we get the following membership values for the mappings displayed in Table 9.6:

$$\begin{aligned}
\mu_y(0) &= \max\{\mu_{A_2}(0), \mu_{A_2}(180)\} = \max\{0, 0\} = 0\\
\mu_y(2) &= \max\{\mu_{A_2}(11.5), \mu_{A_2}(168.5)\} = 0.13\\
\mu_y(4) &= \max\{\mu_{A_2}(23.6), \mu_{A_1}(156.4)\} = 0.26\\
\mu_y(6) &= \max\{\mu_{A_1}(36.9), \mu_{A_1}(143.1)\} = 0.41\\
\mu_y(8) &= 0.59\\
\mu_y(10) &= 1
\end{aligned}$$

TABLE 9.6
Inverse image
$x_1 = \sin^{-1}(y/10)$

y	x_1
−10	−90
−8	−126.9, −53.1
−6	−143.1, −36.9
−4	−156.4, −23.6
−2	−168.5, −11.5
0	−180, 0, 180
2	11.5, 168.5
4	23.6, 156.4
6	36.9, 143.1
8	53.1, 126.9
10	90

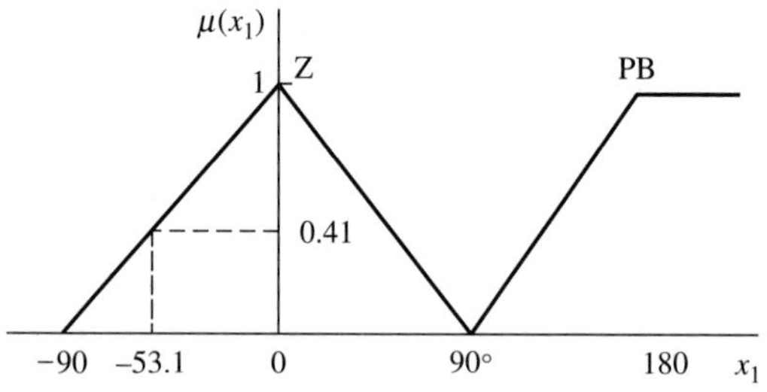

FIGURE 9.16
Graphical representation of x_1 = Z or PB.

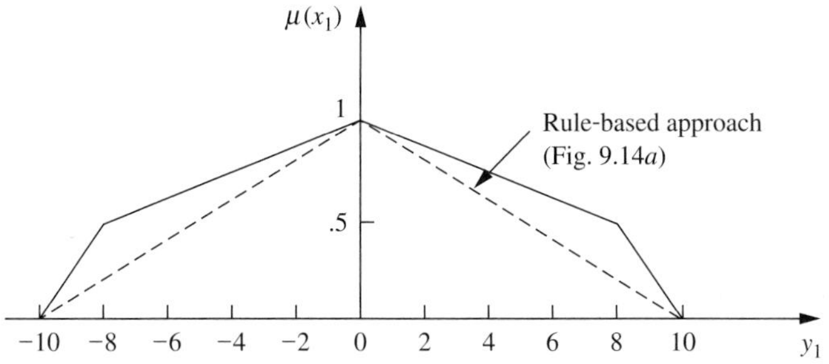

FIGURE 9.17
Membership function for y from rule 1: $x_1 = $ Z or PB.

These membership values for the various equally spaced points in the y universe are shown in Fig. 9.19.

For rule 3, $x_1 = $ Z or NB (see Fig. 9.20), we get the following membership values for the mappings displayed in Table 9.6:

$$\begin{aligned}
\mu_y(-10) &= \mu_{A_3}(-90) = 0 \\
\mu_y(-8) &= \max\{\mu_{A_3}(-126.9), \mu_{A_3}(-53.1)\} = 0.41 \\
\mu_y(-6) &= \max\{\mu_{A_3}(-143.1), \mu_{A_3}(-36.9)\} = 0.59 \\
\mu_y(-4) &= \max\{\mu_{A_3}(-156.4), \mu_{A_3}(-26.3)\} = 0.74 \\
\mu_y(-2) &= \max\{\mu_{A_3}(-168.5), \mu_{A_3}(-11.5)\} = 0.87 \\
\mu_y(0) &= \max\{\mu_{A_3}(0), \mu_{A_3}(-180)\} = 1
\end{aligned}$$

Likewise,

$$\begin{aligned}
\mu_y(2) &= 0.87 \\
\mu_y(4) &= 0.74 \\
\mu_y(6) &= 0.59 \\
\mu_y(8) &= 0.41 \\
\mu_y(10) &= 0
\end{aligned}$$

These membership values for the various equally spaced points in the y universe are shown in Fig. 9.21.

For rule 4, $x_1 = $ NS (see Fig. 9.22), we get the following membership values for the mappings displayed in Table 9.6:

$$\begin{aligned}
\mu_y(0) &= \max\{\mu_{A_4}(0), \mu_{A_4}(-180)\} = 0 \\
\mu_y(-2) &= \max\{\mu_{A_4}(-11.5), \mu_{A_4}(-168.5)\} = 0.13 \\
\mu_y(-4) &= \max\{\mu_{A_4}(-23.6), \mu_{A_4}(-156.4)\} = 0.26 \\
\mu_y(-6) &= \max\{\mu_{A_4}(-36.9), \mu_{A_3}(-143.1)\} = 0.41 \\
\mu_y(-8) &= 0.59 \\
\mu_y(-10) &= 1
\end{aligned}$$

These membership values for the various equally spaced points in the y universe are shown in Fig. 9.23.

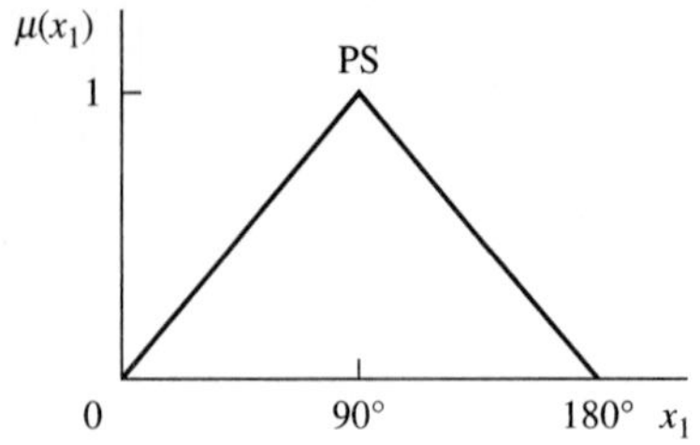

FIGURE 9.18
Graphical representation of x_1 = PS.

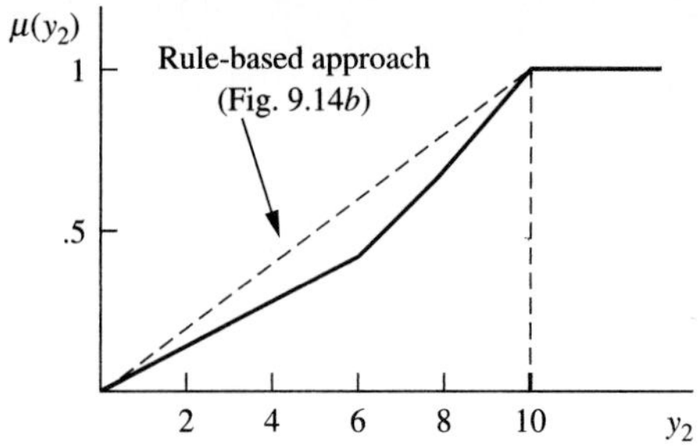

FIGURE 9.19
Membership function for y from rule 2: x_1 = PS.

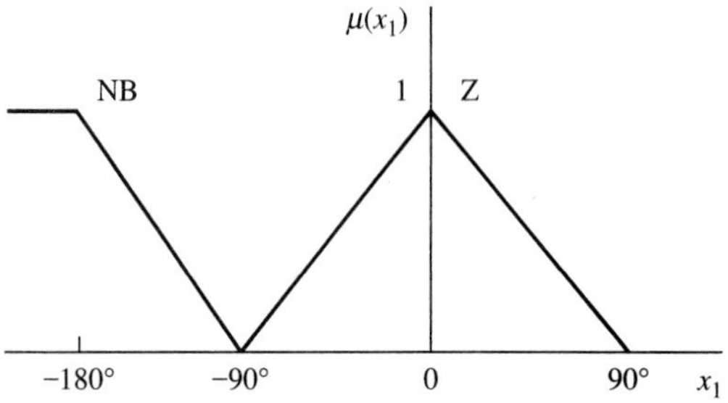

FIGURE 9.20
Graphical representation of x_1 = Z or NB.

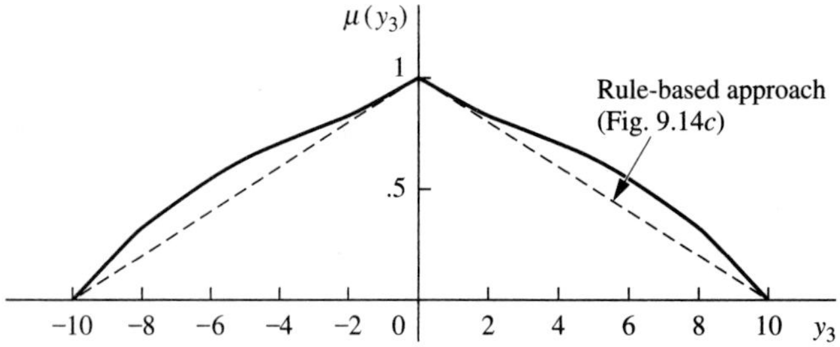

FIGURE 9.21
Membership function for y from rule 3: x_1 = Z or NB.

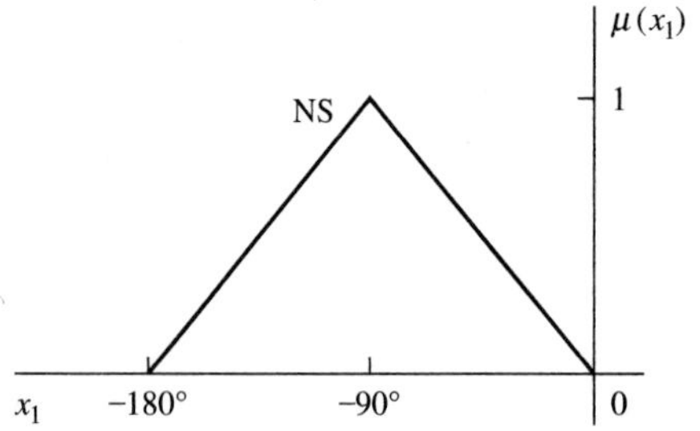

FIGURE 9.22
Graphical representation of x_1 = NS.

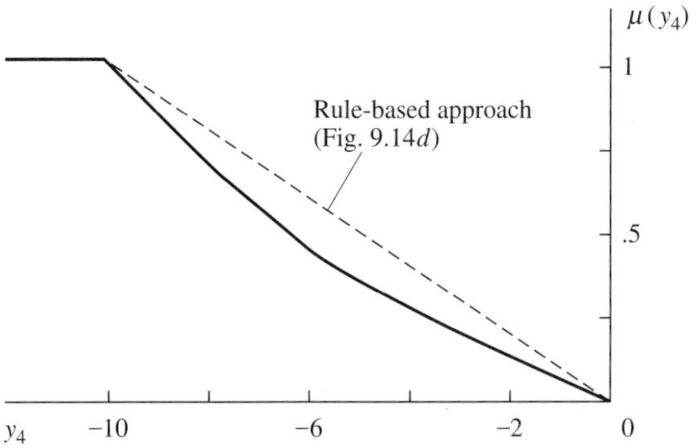

FIGURE 9.23
Membership function for y from rule 4: x_1 = NS.

The reader should note the similarity of Figs. 9.14*a–d* to Figs. 9.17, 9.19, 9.21, and 9.23, respectively, before the former functions are truncated. This simply shows that a graphical approach can give solutions very close to those provided by the extension principle, as seen in the dashed lines in Figs. 9.17, 9.19, 9.21, and 9.23.

Using the membership functions displayed in Figs. 9.16 through 9.23, we can construct fuzzy relations for each of the four rule cases above, i.e., for rules 1–4. The extension principle extends the fuzziness in an input variable through a crisp function or mapping and expresses the fuzziness in the result. Figure 9.24 shows this process graphically. The input, x_1, can be expressed as a fuzzy set, A_i; the "box" is the functional mapping, $f(x_1)$, operating on the input; and the output, y, is also expressed as a fuzzy set, B_i. The functional mapping does, in fact, represent a fuzzy relation, $\underset{\sim}{\text{R}}$. In the following material, to reduce the amount of computation associated with matrix operations, we will work with "reduced" portions of the fuzzy membership functions illustrated in Figs. 9.16 through 9.23.

For rule 1, x_1 = Z or PB, and y = Z, we can simplify the membership function for y by selecting fewer points, i.e., for a reduced universe $y_1 = \{-8, -4, 0, 4, 8\}$ to get (from Fig. 9.17) the reduced fuzzy output set,

$$\underset{\sim}{\text{B}}_1 = \left\{ \frac{.41}{-8} + \frac{.74}{-4} + \frac{1}{0} + \frac{.74}{4} + \frac{.41}{8} \right\}$$

This fuzzy output set corresponds to a reduced set of points in the input universe (rounded off from previous calculations), $x_1 = \{-53, -24, 0, 24, 53, 127, 156, 180\}$, in degrees. Hence, the fuzzy set for the input is given by

$$\underset{\sim}{\text{A}}_1 = \left\{ \frac{.41}{-53} + \frac{.74}{-24} + \frac{1}{0} + \frac{.74}{24} + \frac{.41}{53} + \frac{.41}{127} + \frac{.74}{156} + \frac{1}{180} \right\}$$

We can calculate the Cartesian product of these two fuzzy vectors using a pairwise

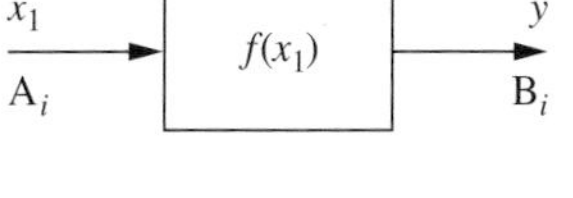

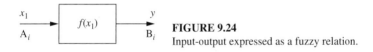

FIGURE 9.24
Input-output expressed as a fuzzy relation.

min-operation, Eq. (3.16),

$$\underset{\sim}{R}_1 = \underset{\sim}{A}_1 \times \underset{\sim}{B}_1 = \begin{bmatrix} .41 \\ .74 \\ 1 \\ .74 \\ .41 \\ .41 \\ .74 \\ 1 \end{bmatrix} \times [.41 \quad .74 \quad 1 \quad .74 \quad .41]$$

and we get a fuzzy relation for rule 1, $\underset{\sim}{R}_1$,

$$\underset{\sim}{R}_1 = \begin{array}{r} \\ -53 \\ -24 \\ 0 \\ 24 \\ 53 \\ 127 \\ 156 \\ 180 \end{array} \begin{array}{c} \begin{array}{ccccc} -8 & -4 & 0 & 4 & 8 \end{array} \\ \begin{bmatrix} .41 & .41 & .41 & .41 & .41 \\ .41 & .74 & .74 & .74 & .41 \\ .41 & .74 & 1 & .74 & .41 \\ .41 & .74 & .74 & .74 & .41 \\ .41 & .41 & .41 & .41 & .41 \\ .41 & .41 & .41 & .41 & .41 \\ .41 & .74 & .74 & .74 & .41 \\ .41 & .74 & 1 & .74 & .41 \end{bmatrix} \end{array}$$

Now, suppose we subject this rule, or relation, to a crisp singleton input $-45°$; this step can be accomplished by translating the crisp value of $-45°$ into a crisp singleton represented as a fuzzy vector, i.e.,

$$\underset{\sim}{A}' = \left\{ \frac{1}{-45°} \right\}$$

The composition of this singleton with $\underset{\sim}{R}_1$ has the effect of interpolating between the values in the relation between the first and second rows since $-53° < -45° < -24°$. One can see that the membership function resulting from this composition might look something like the discretized form in Fig. 9.25. Note that Fig. 9.25 is very close in appearance to the truncated consequent term of rule 1 in Fig. 9.14*a*.

For rule 2, x_1 = PS, and y = PB, we can simplify the membership function for y by selecting fewer points, i.e., for a reduced universe $y_2 = \{2, 6, 10\}$ we get (from

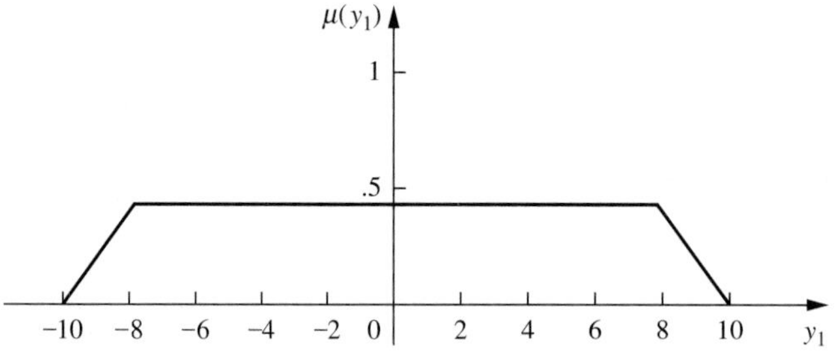

FIGURE 9.25
Truncated membership resulting from composition of $x_1 = -45°$ and relation $\underset{\sim}{R}_1$.

Fig. 9.19) the reduced fuzzy output set,

$$\underset{\sim}{B}_2 = \left\{ \frac{.13}{2} + \frac{.41}{6} + \frac{1}{10} \right\}$$

This fuzzy output set corresponds to a reduced set of points in the input universe (rounded off from previous calculations), $x_2 = \{11, 37, 90, 143, 169\}$, in degrees. Hence, the fuzzy set for the input is given by

$$\underset{\sim}{A}_2 = \left\{ \frac{.13}{11} + \frac{.41}{37} + \frac{1}{90} + \frac{.41}{143} + \frac{.13}{169} \right\}$$

We can calculate the Cartesian product of these two fuzzy vectors using a pairwise min-operation, Eq. (3.16),

$$\underset{\sim}{R}_2 = \begin{bmatrix} .13 \\ .41 \\ 1 \\ .41 \\ .13 \end{bmatrix} \times [.13 \quad .41 \quad 1]$$

and we get a fuzzy relation for rule 2, $\underset{\sim}{R}_2$,

$$\underset{\sim}{R}_2 = \begin{matrix} & \begin{matrix} 2 & 6 & 10 \end{matrix} \\ \begin{matrix} 11 \\ 37 \\ 90 \\ 147 \\ 169 \end{matrix} & \begin{bmatrix} .13 & .13 & .13 \\ .13 & .41 & .41 \\ .13 & .41 & 1 \\ .13 & .41 & .41 \\ .13 & .13 & .13 \end{bmatrix} \end{matrix}$$

Now, suppose we subject this rule, or relation, to a crisp input 45°; this can be accomplished by translating the crisp value of 45° into a crisp singleton represented as a fuzzy vector, i.e.,

$$\underset{\sim}{A}' = \left\{ \frac{1}{45°} \right\}$$

The composition of this singleton with $\underset{\sim}{R}_2$ has the effect of interpolating between the values in the relation between the second and third rows since $37° < 45° < 90°$. One can see that the membership function resulting from this composition might look something like the discretized form of the function in Fig. 9.26. Note that Fig. 9.26 closely resembles the truncated consequent term of rule 2 in Fig. 9.14*b*.

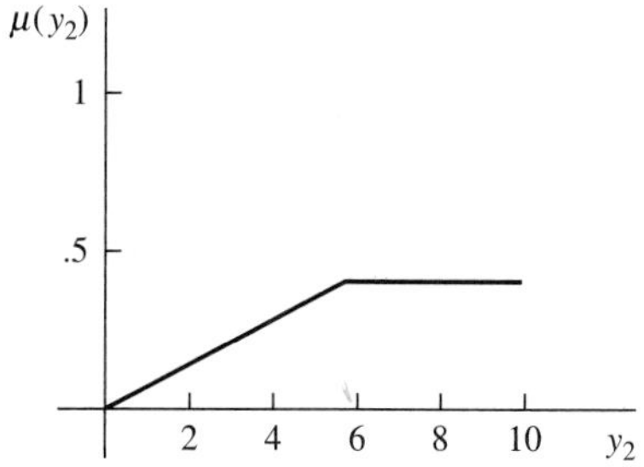

FIGURE 9.26
Truncated membership resulting from composition of $x_1 = 45°$ and relation $\underset{\sim}{R}_2$.

For rule 3, x_1 = Z or NB, and y = Z, we can simplify the membership function for y by selecting fewer points, i.e., for a reduced universe $y_3 = \{-8, -4, 0, 4, 8\}$ to get (from Fig. 9.21) the reduced fuzzy output set,

$$\underset{\sim}{B}_3 = \left\{\frac{.41}{-8} + \frac{.74}{-4} + \frac{1}{0} + \frac{.74}{4} + \frac{.41}{8}\right\}$$

This fuzzy output set corresponds to a reduced set of points in the input universe (rounded off from previous calculations), $x_3 = \{-180, -156, -127, -54, -23, 0, 23, 54\}$, in degrees. Hence, the fuzzy set for the input is given by

$$\underset{\sim}{A}_3 = \left\{\frac{1}{-180} + \frac{.74}{-156} + \frac{.41}{-127} + \frac{.41}{-54} + \frac{.74}{-23} + \frac{1}{0} + \frac{.74}{23} + \frac{.41}{54}\right\}$$

We can calculate the Cartesian product of these two fuzzy vectors using a pairwise min-operation, Eq. (3.16), to get a fuzzy relation for rule 3, $\underset{\sim}{R}_3$:

$$\underset{\sim}{R}_3 = \begin{bmatrix} 1 \\ .74 \\ .41 \\ .41 \\ .74 \\ 1 \\ .74 \\ .41 \end{bmatrix} \times [.41 \quad .74 \quad 1 \quad .74 \quad .41] = \begin{array}{r} \\ -180 \\ -156 \\ -127 \\ -54 \\ -23 \\ 0 \\ 23 \\ 54 \end{array} \begin{array}{c} \begin{array}{ccccc} -8 & -4 & 0 & 4 & 8 \end{array} \\ \begin{bmatrix} .41 & .74 & 1 & .74 & .41 \\ .41 & .74 & .74 & .74 & .41 \\ .41 & .41 & .41 & .41 & .41 \\ .41 & .41 & .41 & .41 & .41 \\ .41 & .74 & .74 & .74 & .41 \\ .41 & .74 & 1 & .74 & .41 \\ .41 & .74 & .74 & .74 & .41 \\ .41 & .41 & .41 & .41 & .41 \end{bmatrix} \end{array}$$

Now, suppose we subject this rule, or relation, to a crisp input $-135°$; this can be accomplished by translating the crisp value of $-135°$ into a crisp singleton represented as a fuzzy vector, i.e.,

$$\underset{\sim}{A}' = \left\{\frac{1}{-135°}\right\}$$

The composition of this singleton with $\underset{\sim}{R}_3$ has the effect of interpolating between the values in the relation between the second and third rows since $-156° < -135° < -127°$. One can see that the membership function resulting from this composition might look something like the discretized form in Fig. 9.27. Note that Fig. 9.27 is very close in appearance to the truncated consequent of rule 3 in Fig. 9.14c.

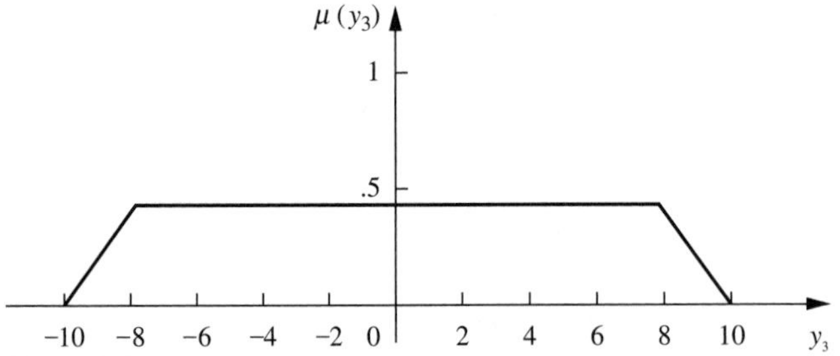

FIGURE 9.27
Truncated membership resulting from composition of $x_1 = -135°$ and relation $\underset{\sim}{R}_3$.

For rule 4, x_1 = NS, and y = NB, we can simplify the membership function for y by selecting fewer points, i.e., for a reduced universe $y_4 = \{-10, -6, -2\}$, to get (from Fig. 9.23) the reduced fuzzy output set,

$$\underset{\sim}{B}_4 = \left\{ \frac{1}{-10} + \frac{.41}{-6} + \frac{.13}{-2} \right\}$$

This fuzzy output set corresponds to a reduced set of points in the input universe (rounded off from previous calculations), $x_4 = \{-169, -144, -90, -38, -11\}$, in degrees. Hence, the fuzzy set for the input is given by

$$\underset{\sim}{A}_4 = \left\{ \frac{.13}{-169} + \frac{.41}{-144} + \frac{1}{-90} + \frac{.41}{-38} + \frac{.13}{-11} \right\}$$

We can calculate the Cartesian product of these two fuzzy vectors using a pairwise min-operation, Eq. (3.16), to get a fuzzy relation for rule 4, $\underset{\sim}{R}_4$:

$$\underset{\sim}{R}_4 = \begin{bmatrix} .13 \\ .41 \\ 1 \\ .41 \\ .13 \end{bmatrix} \times [1 \quad .41 \quad .13] = \begin{array}{r} \\ -169 \\ -144 \\ -90 \\ -38 \\ -11 \end{array} \begin{array}{c} \begin{array}{ccc} -10 & -6 & -2 \end{array} \\ \begin{bmatrix} .13 & .13 & .13 \\ .41 & .41 & .13 \\ 1 & .41 & .13 \\ .41 & .41 & .13 \\ .13 & .13 & .13 \end{bmatrix} \end{array}$$

Now, suppose we subject this rule, or relation, to a crisp input $-135°$; this can be accomplished by translating the crisp value of $-135°$ into a crisp singleton represented as a fuzzy vector, i.e.,

$$\underset{\sim}{A}' = \left\{ \frac{1}{-135°} \right\}$$

The composition of this singleton with $\underset{\sim}{R}_4$ has the effect of interpolating between the values in the relation between the second and third rows since $-144° < -135° < -90°$. One can see that the membership function resulting from this composition might look something like the discretized form in Fig. 9.28. Note that Fig. 9.28 is very close in appearance to the truncated consequent term of rule 4 in Fig. 9.14*d*. As mentioned before, if more rules are used in the simulation the results converge to a sine curve; this solution is left as an exercise for the reader (see Problem 9.16). An additional approach, to improve the simulation, is to use the same number of rules with more values of the input variable; that is, to use a finer partitioning of the input and output space. The

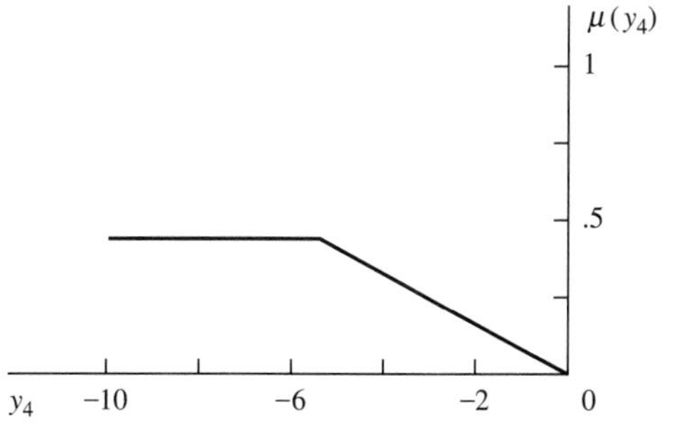

FIGURE 9.28
Truncated membership resulting from composition of $x_1 = -135°$ and relation $\underset{\sim}{R}_3$.

analyst could also adjust the nonlinearity in the simulation by varying the amount of overlap between partitions in the input and output spaces. Eventually more rules would be required to use the finer partitioning effectively, however. For the extension principle this need is manifested in a finer discretization of the input variable, as discussed in Chapter 6.

Example 9.3. Suppose we want to model a serial transmission of a digital signal over a channel using RS232 format. Packets of information transmitted over the channel are ASCII characters composed of start- and stop-bits plus the appropriate binary pattern. If we wanted to know whether a valid bit was sent we could test the magnitude of the signal at the receiver using an absolute value function. For example, suppose we have the voltage (V) versus time trace shown in Fig. 9.29, a typical pattern. In this pattern the ranges for a valid mark and a valid space are as follows:

-12 to -3 V or $+3$ to $+12$ V	A valid mark (denoted by a 1)
-3 to $+3$	A valid space (denoted by a 0)

The absolute value function used to make this distinction is a nonlinear function, as shown in Fig. 9.30. To use this function on the scale of voltages $[-12, +12]$, we will attempt to simulate the nonlinear function, $y = 12|x|$, where the range of x is $[-1, 1]$. First, we partition the input space, $x = [-1, 1]$, into five linguistic partitions as in Fig. 9.31. Next, we partition the output space. This task can usually be accomplished by mapping prototypes of input space to corresponding points in output space, if such information is available. Another approach, as illustrated in Example 9.2, is to use the extension principle. Because we know the functional mapping, the partitioning can be accomplished readily; we will use three equally spaced output partitions as shown in Fig. 9.32.

Since this function is simple and nonlinear, we can propose a few simple rules to simulate its behavior:

1. IF x = zero, THEN y = zero.
2. IF x = NS or PS, THEN y = PS.
3. IF x = NB or PB, THEN y = PB.

We can now conduct a graphical simulation of the nonlinear function expressed by these three rules. Let us assume that we have five input values, the crisp singletons $x = -0.6, -0.3, 0, 0.3$, and 0.6. The input $x = -0.6$ invokes (fires) rules 2 and 3, as shown in Fig. 9.33. The defuzzified output, using the centroid method, for the truncated union of the two consequents is approximately 8. The input $x = -0.3$ invokes (fires) rules 1 and 2, as shown in Fig. 9.34. The defuzzified output for the truncated union of

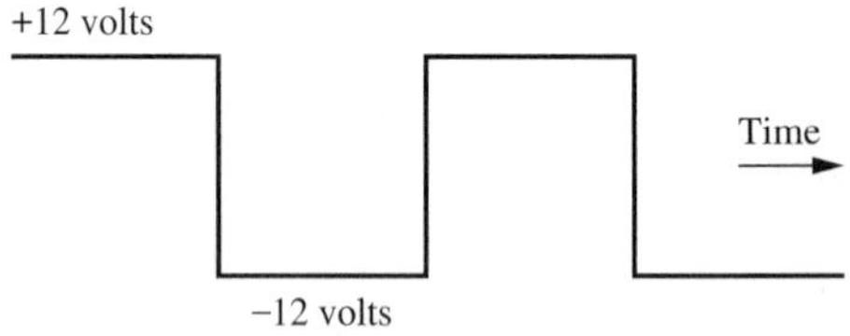

FIGURE 9.29
Typical pattern of voltage vs. time for a valid bit mark.

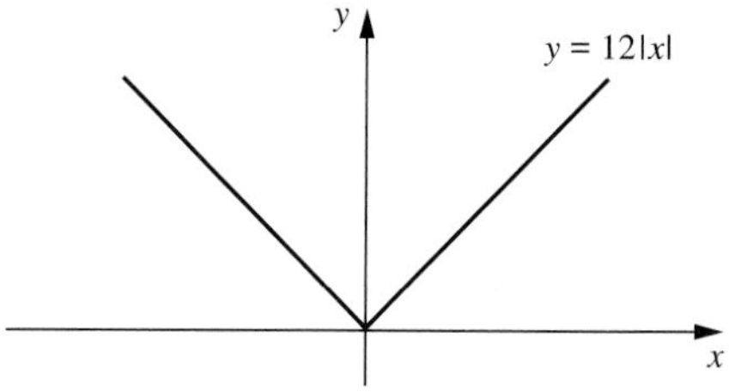

FIGURE 9.30
Nonlinear function $y = 12|x|$.

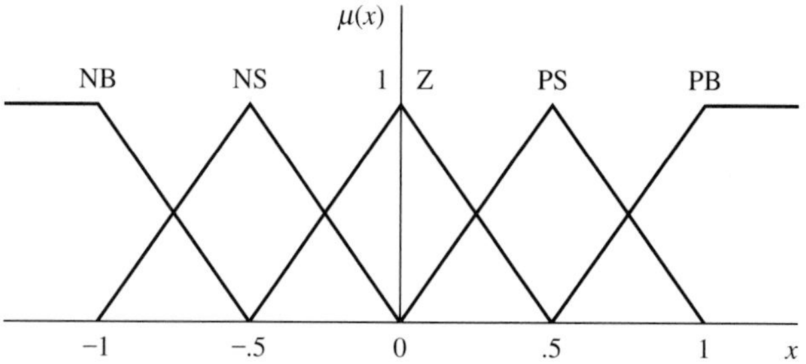

FIGURE 9.31
Partitions on the input space for $x = [-1, 1]$.

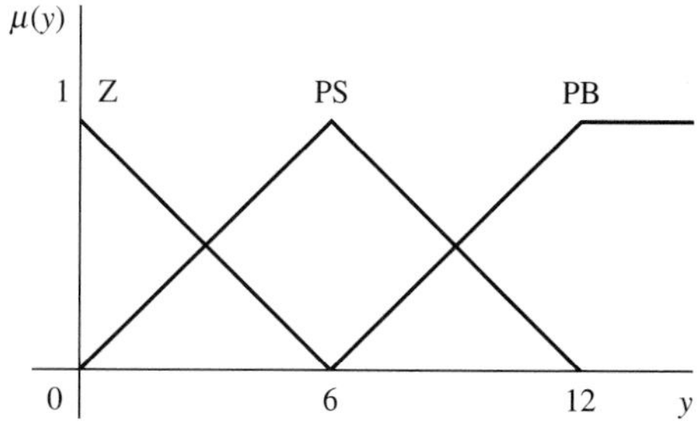

FIGURE 9.32
Output partitions on the range $y = [0, 12]$.

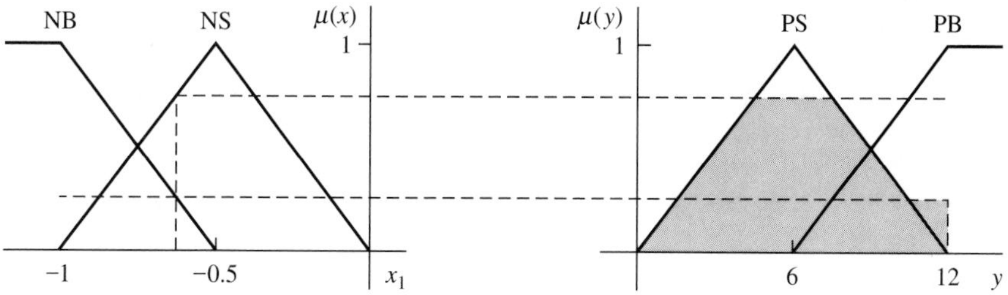

FIGURE 9.33
Graphical simulation for crisp input $x = -0.6$.

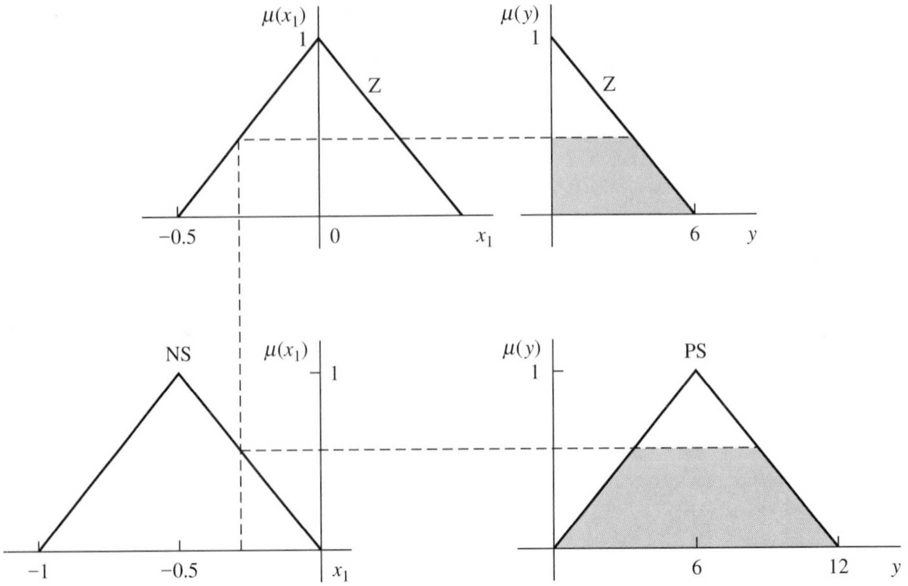

FIGURE 9.34
Graphical simulation for crisp input $x = -0.3$.

the two consequents is approximately 5. The input $x = 0$ invokes (fires) rule 1 only, as shown in Fig. 9.35. The defuzzified output for the truncated consequent ($y = Z$) is a centroidal value of 2. By symmetry it is easy to see that crisp inputs $x = 0.3$ and $x = 0.6$ result in defuzzified values for $y \approx 5$ and $y \approx 8$, respectively.

If we plot these simulated results and compare them to the exact relationship, we get the graph in Fig. 9.36; the simulation, although approximate, is quite good.

Example 9.4. When an aircraft forms a synthetic aperture radar (SAR) image, the pilot needs to calculate the range to a reference point, based on the position of the craft and the output of an inertial navigator, to within some fraction of a wavelength of the transmitted radar pulse. Assume that at position $d = 0$, the aircraft *knows* that the reference point is distance R_0 off the left broadside (angle $= 90°$) of the aircraft, and that the aircraft flies in a straight line; see Fig. 9.37. The question is, what is the range, $r(d)$, to the reference point when the aircraft is at the position d_1? The exact answer is $r(d) = (R_0^2 + d_1^2)^{1/2}$, however, the square root operation is nonlinear, cumbersome, and computationally slow to evaluate. In a typical computation this expression is expanded into a Taylor series. In this example, we wish to use a fuzzy rule-based approach instead.

If we normalize the range, i.e., let $d_1/R_0 = k_1 \cdot x_1$, then $r(x_1) = R_0(1 + k_1^2 x_1^2)^{1/2}$, where now x_1 is a scaled range and k_1 is simply a constant in the scaling process. For example, suppose we are interested in the range $|d_1/R_0| \leq 0.2$; then $k_1 = 0.2$ and $|x_1| \leq 1$. For this particular problem we will let $R_0 = 10{,}000$ meters $= 10$ kilometers (km), then $r(x_1) = 10{,}000(1 + (0.04)x_1^2)^{1/2}$. Table 9.7 shows exact values of $r(x_1)$ for typical values of x_1.

Let $y = r(x_1)$ with x_1 partitioned as shown in Fig. 9.38, and let the output variable, y, be partitioned as shown in Fig. 9.39. In Fig. 9.39, the partitions $\underset{\sim}{S}$ and $\underset{\sim}{L}$ have

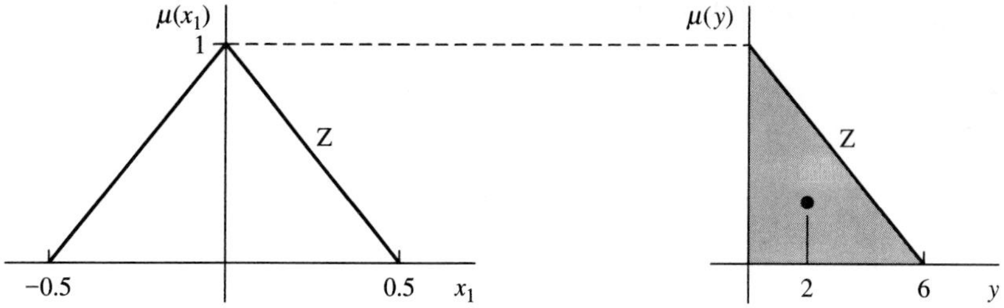

FIGURE 9.35
Graphical simulation for crisp input $x = 0$.

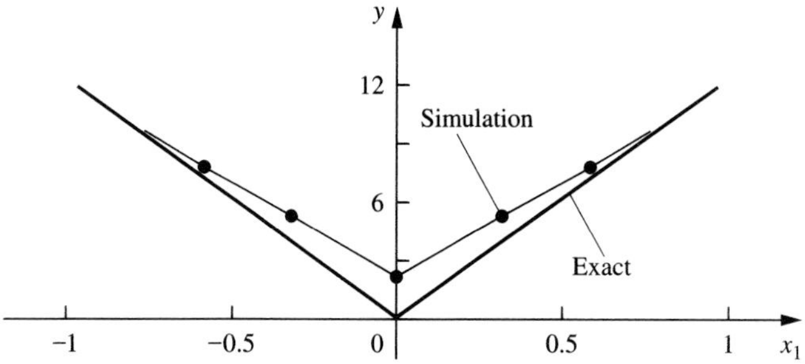

FIGURE 9.36
Simulated versus exact results for Example 9.3.

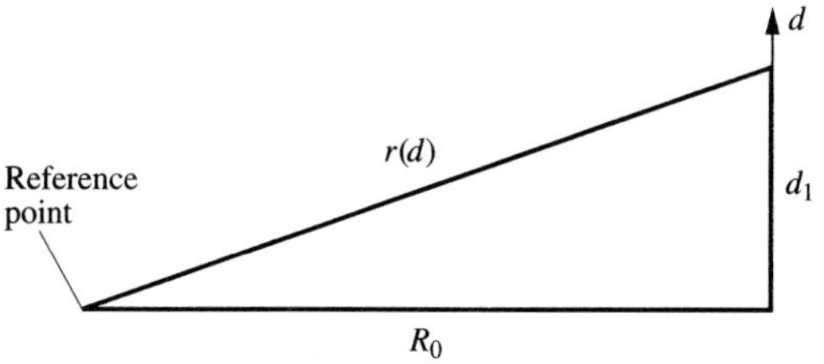

FIGURE 9.37
Schematic of aircraft SAR problem.

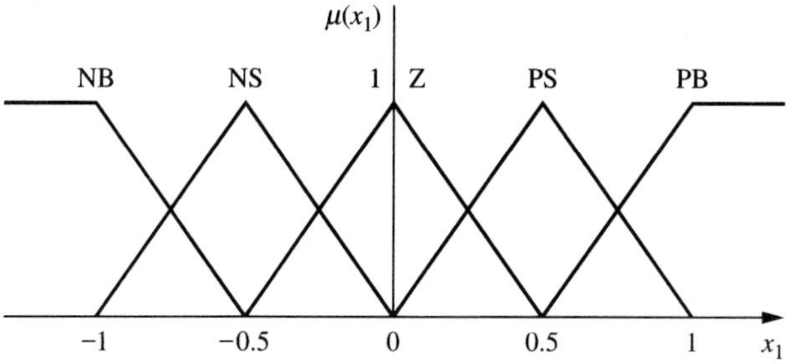

FIGURE 9.38
Partitioning for the input variable, x_1.

TABLE 9.7
Relationships for distance in SAR problem

x_1	$r(x_1)$
−1.0	10,198
−0.5	10,050
0.0	10,000
0.5	10,050
1.0	10,198

symmetrical membership functions. We now pose three simple rules that relate the input and output variables:

Rule 1: IF $x \in \underset{\sim}{Z}$, THEN $y \in \underset{\sim}{S}$.
Rule 2: IF $x \in \underset{\sim}{PS}$ or $\underset{\sim}{NS}$, THEN $y \in \underset{\sim}{M}$.
Rule 3: IF $x \in \underset{\sim}{PB}$ or $\underset{\sim}{NB}$, THEN $y \in \underset{\sim}{L}$.

If we conduct a graphical simulation like that in Example 9.2 we achieve the results shown in Fig. 9.40. In this figure the "o" symbol denotes exact values and the symbol "x" denotes the centroidal value of the fuzzy output as determined in the graphical simulation (in some cases the exact value and the fuzzy value coincide; this is represented by a circle with an x in it). The "approximate" curve follows the exact curve quite well.

SUMMARY

A wide class of complex dynamic processes exists where the knowledge regarding the functional relationship between the input and output variables may be established on numerical or nonnumerical information. The numerical information is usually from a limited number of data points and the nonnumerical information is in the form of vague natural language protocols gathered from interviews with humans familiar with the input-output behavior or the real time control of the system or process. Complexity in the system model arises as a result of many factors such

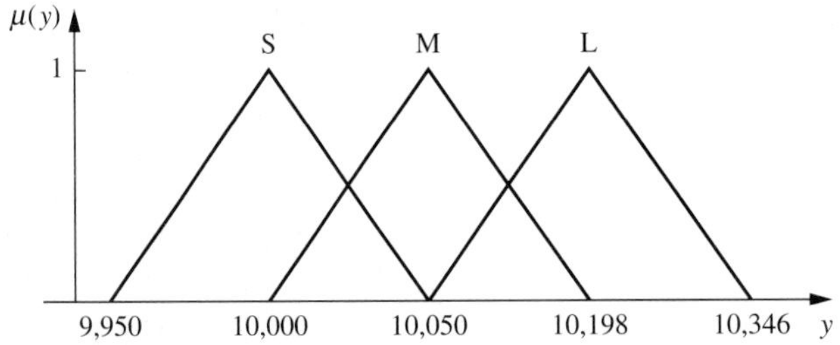

FIGURE 9.39
Partitioning for the output variable, y.

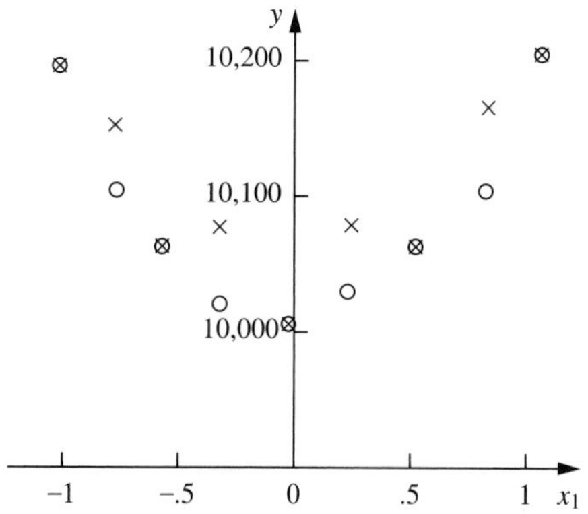

FIGURE 9.40
Exact and fuzzy values compared for SAR problem.

as (1) high dimensionality, (2) too many interacting variables, and (3) unmodeled dynamics such as nonlinearities, time variations, external noise or disturbance, and system perturbations [Vadiee, 1993]. Hence, the information gathered on the system behavior is never complete, sharp, or comprehensive.

Fuzzy mathematics has provided a range of mathematical tools that helps the investigator formalize these ill-defined descriptions about complex systems into the form of linguistic rules and then eventually into mathematical equations, which can then be implemented on digital computers. These rules can be represented by fuzzy associative memories (FAMs). At the expense of relaxing some of the demands on the requirements for precision in some nonlinear systems, a great deal of simplification, ease of computation, speed, and efficiency are gained when using fuzzy models. The ill-defined nonlinear systems can be described with fuzzy relational equations. These relations are expressed in the form of various fuzzy composition operations, which are carried out on classes of membership functions defined on a number of overlapping partitions of the space of possible inputs (antecedents), possible mapping restrictions, and possible output (consequent) responses.

The membership functions used to describe linguistic knowledge are enormously subjective and context-dependent [Vadiee, 1993]. The input variables are assumed to be noninteractive, and the membership functions for them are assigned based on the degree of similarity of a corresponding prototypical element. The membership functions are further assumed to be linearly dependent on the net distance to a prototype input point. In this way the membership functions can be in one of the simple rectangular, triangular, or trapezoidal forms. Appropriate nonlinear transformations or sensory integration and fusion on input and/or output spaces are often used to reduce a complex fuzzy controller to a fuzzy system model. The net effect of this preprocessing on the input data is to decouple and linearize the system dynamics.

This chapter has dealt with the idea of fuzzy nonlinear simulation. The point made in this chapter is not that we can make crude approximations to well-known functions; after all, if we know a function, we certainly don't need fuzzy logic to approximate it. But there are many situations where we can only observe a complicated nonlinear process whose functional relationship we don't know, and whose behavior is known only in the form of linguistic knowledge, such as that expressed for the sine curve example in Table 9.3 or, for more general situations, as that expressed in Table 9.2. Then the power of fuzzy nonlinear simulation is manifested in modeling nonlinear systems whose behavior we can express in the form of input-output data-tuples, or in the form of linguistic rules of knowledge, and whose exact nonlinear specification we don't know. Fuzzy models to address such complex systems are being published in the literature at an accelerating pace; see, for example, Huang and Fan [1993] who address complex hazardous waste problems and Sugeno and Yasukawa [1993] who address problems ranging from a chemical process to a stock price trend model. The ability of fuzzy systems to analyze dynamical systems that are so complex we don't have a mathematical model is the point made in this chapter. As we learn more about a system, the data eventually become robust enough to pose the model in analytic form; at that point we no longer need a fuzzy model.

REFERENCES

Brandt, H. W. (1961). *The fecundity of mathematical methods in economic theory,* trans. Edwin Holstrom, D. Reidel Publishing Company, Dordrecht, Germany.

Huang, Y., and L. Fan (1993). "A fuzzy logic-based approach to building efficient fuzzy rule-based expert systems," *Comput. Chem. Eng.,* vol. 17, no. 2, pp. 188–192.

Sugeno, M. (ed.). (1985). *Industrial applications of fuzzy control,* North-Holland, New York.

Sugeno, M. and T. Yasukawa (1993). "A fuzzy-logic-based approach to qualitative modeling," *IEEE Trans. Fuzzy Syst.,* vol. 1, no. 1, pp. 7–31.

Vadiee, N. (1993). "Fuzzy rule-based expert systems—I and II," in M. Jamshidi, N. Vadiee, and T. Ross, (eds.), *Fuzzy logic and control: software and hardware applications,* Prentice Hall, Englewood Cliffs, N.J., chapters 4 and 5.

Zadeh, L. (1975). "The concept of a linguistic variable and its application to approximate reasoning—I," *Inf. Sci.,* vol. 8, pp. 199–249.

PROBLEMS

9.1. Suppose we are dealing with the simulation of a Junction Field Effect Transistor (JFet). The pertinent variables of the JFet are gate voltage input (V_{GS}), the drain voltage input, V_{DS}, and current through the transistor (I_D). In Fig. P9.1*a* we show the voltage drain (D), and the voltage sources (S). From specifications of the JFet, we can construct a relation, $\underset{\sim}{\mathrm{R}}$, that embodies knowledge about the relationship between input variables V_{GS} and V_{DS} and the output variable, I_D. Assume the following ranges for each of the variables:

V_{DS}: 0 V to +25 V

V_{GS}: −5 V to 0 V

I_D: O amp to 12 amp

(*a*) Divide variable V_{DS} into four partitions: very small positive (VSP), small positive (SP), medium positive (MP), and large positive (LP). Divide variable V_{GS} into four partitions: zero (Z), negative small (NS), negative medium (NM), and negative large (NL). Divide variable I_D into four partitions: very small positive (VSP), small positive (SP), medium positive (MP), and large positive (LP). The partitions for these variables do not have to be of uniform width.

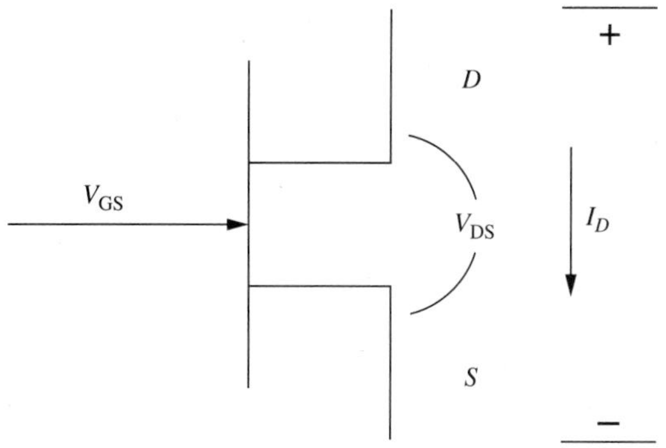

FIGURE P9.1*a*

(*b*) Using Fig. P9.1*b* as a guide, develop rules in the form

IF V_{GS} is NM and V_{DS} is ..., THEN I_D is ...

for the curves in the figure. The antecedents and consequents in the rule, of course, will be the linguistic variables developed in part (*a*). Note in Fig. P9.1*b* that for large values of V_{DS} the value of I_D approaches a constant (crisp) number.

(*c*) Using the rules specified in part (*b*), construct a fuzzy relation matrix, $\underset{\sim}{R}$.

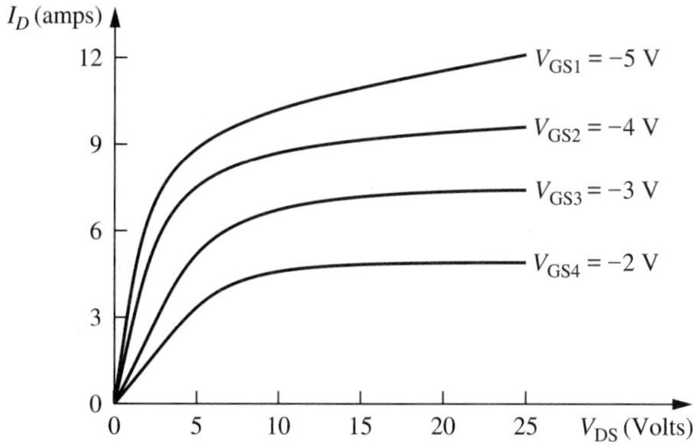

FIGURE P9.1*b*

(*d*) Conduct a simple graphical simulation of this system, using a centroid defuzzification method and at least four of the rules developed in part (*b*), and the following crisp input values:

$$V_{DS} = 10 \text{ V} \qquad \text{and} \qquad V_{GS} = -4 \text{ V}$$

(*e*) Suppose you have new combinations of the gate voltage and drain voltage inputs for the system. Explain how you would determine new values of the transistor current for the new inputs.

9.2. A video monitor's CRT has a nonlinear characteristic between the illuminance output and the voltage input. This nonlinear characteristic is $y = x^{2.2}$, where y is the illumination and x is the voltage. The CCD (charge-coupled device) in a video camera has a linear light-in to voltage-out characteristic. To compensate for the nonlinear characteristic of the monitor, a "gamma correction" circuit is usually employed in a CCD camera. This nonlinear circuit has a transfer function of $y = x^{\text{gamma}}$, where the gamma factor is usually 0.45 (i.e., 1/2.2) to compensate for the 2.2 gamma characteristic of the monitor. The net result should be a linear response between the light incident on the CCD and the light produced by the monitor. Figure P9.2 shows the nonlinear gamma characteristic of a CCD camera (y_{actual}). Both the input, x, and the output, y, have a universe of discourse of [0, 1].

Partition the input variable, x, into three partitions, say small, S, medium, M, and big, B, and partition the output variable, y, into two partitions, say small, SM, and large, L. Using your own few simple rules for the nonlinear function, $y = x^{0.45}$ and the crisp inputs: $x = 0, 0.25, 0.5, 0.75, 1.0$, determine whether your results produce a solution roughly similar to y_{fuzzy} in Fig. P9.2, which was developed with Zadeh's extension principle using three equal partitions for the input variable and three simple rules. Comment on the form of your solution and why it does or does not conform to the actual result.

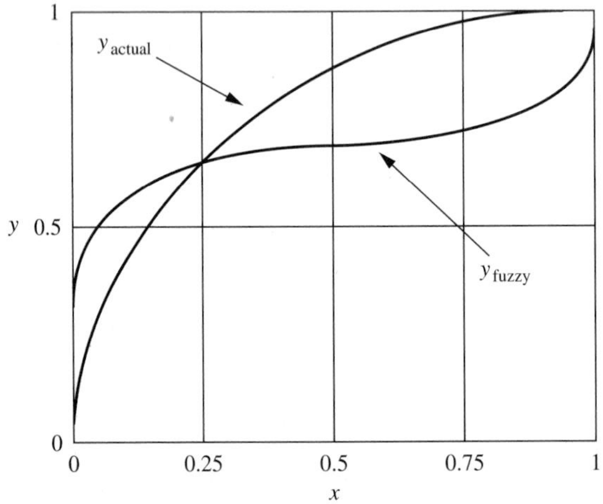

FIGURE P9.2

9.3. A very widely used component in electrical engineering is the diode. The voltage-current relation is extremely nonlinear and is modeled by the expression

$$V_f = V_t \ln(I_f/I_s)$$

where V_f = forward voltage developed across the diode
V_t = terminal voltage ($\sim$ 0.026 V)
I_s = saturation current of a given diode (assume $\sim 10^{-12}$ amps)
I_f = forward current flowing through the diode

The resulting exact voltage-current curve is shown in Fig. P9.3 (rotated 90°). For this highly nonlinear function discuss the following:

(*a*) How would you go about partitioning the input space (I_f) and the output space (V_f)?

(*b*) Propose three to five simple rules to simulate the nonlinearity.

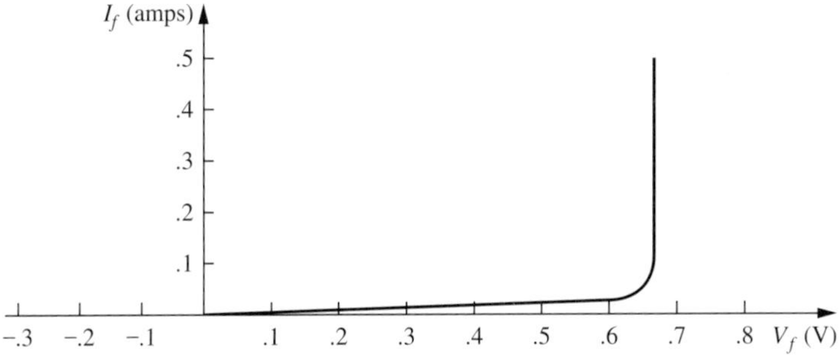

FIGURE P9.3

9.4. A useful nonlinear function used in image processing is the gamma function. It is used to correct for the nonlinearities of image intensity associated with image acquisition (Vidicon) and display (CRT). For this problem, assume a normalized universe of discourse of [0, 1] for the input variable intensity, x, and a normalized universe of discourse of [0, 1] for the output variable intensity, y, since this process can be applied at various locations in a given system (both in analog and digital domains). Use three partitions for the membership functions for x: "dark intensity (D)," "moderate intensity (M)," and "bright intensity (B)." The gamma function is

$$y(x) = \exp\left(\frac{\log(x)}{\gamma}\right)$$

Use a typical value of $\gamma = 2$ for this problem.

(*a*) Use Zadeh's extension principle to derive the membership functions for the output variable, y, corresponding to each of the input membership functions. Label the output membership functions "not nearly so dark," "above moderate intensity," and "bright," corresponding to inputs D, M, and B, respectively.

(*b*) Conduct a numerical simulation of this system using the following three simple rules and calculating the three associated matrix relations:

1. IF x is "dark (D)" THEN y is not nearly so dark
2. IF x is "moderate (M)," THEN y is above moderate intensity
3. IF x is "bright (B)," THEN y is bright

and using three crisp input singletons at $x = 0.2, 0.4$, and 0.8.

(*c*) Plot the results of this simulation versus the exact solution to the gamma function on the intervals $x = [0, 1]$ and $y = [0, 1]$.

9.5. We have a function that is used to relate an input from a particular sensor to the corresponding real value based upon an analog input range of ±4 V. Suppose this nonlinear function is $y = x^3 - x^2 + 5x + 3$. We wish to partition the input space, x, into five membership functions, negative big (NB), negative small (NS), zero (Z), positive small (PS), and positive big (PB), as given here:

$$\text{“NB”} = \left\{\frac{1}{-4} + \frac{.5}{-3} + \frac{0}{-2}\right\}$$

$$\text{“NS”} = \left\{\frac{0}{-4} + \frac{.5}{-3} + \frac{1}{-2} + \frac{.5}{-1} + \frac{0}{0}\right\}$$

$$\text{“Z”} = \left\{ \frac{0}{-2} + \frac{.5}{-1} + \frac{1}{0} + \frac{.5}{1} + \frac{0}{2} \right\}$$

$$\text{“PS”} = \left\{ \frac{0}{0} + \frac{.5}{1} + \frac{1}{2} + \frac{.5}{3} + \frac{0}{4} \right\}$$

$$\text{“PB”} = \left\{ \frac{0}{2} + \frac{.5}{3} + \frac{1}{4} \right\}$$

(*a*) Using the extension principle, find the values of y corresponding to the following four input variables:
  (*i*) y_1: Z or NB
  (*ii*) y_2: Z or PB
  (*iii*) y_3: PS
  (*iv*) y_4: NS

(*b*) Calculate fuzzy relations for the following rules:

1. IF x = Z or NB, THEN $y = y_1$
2. IF x = Z or PB, THEN $y = y_2$
3. IF x = PS, THEN $y = y_3$; ELSE IF x = NS, THEN $y = y_4$

(*c*) For fuzzy triangular membership inputs centered at $x = -3.0, -2, 0, 2,$ and 3.0 and each triangle with a base width of 2.0, conduct a fuzzy nonlinear simulation and plot the results versus the exact results for the function. The fuzzy results should look approximately like those shown in Fig. P9.5.

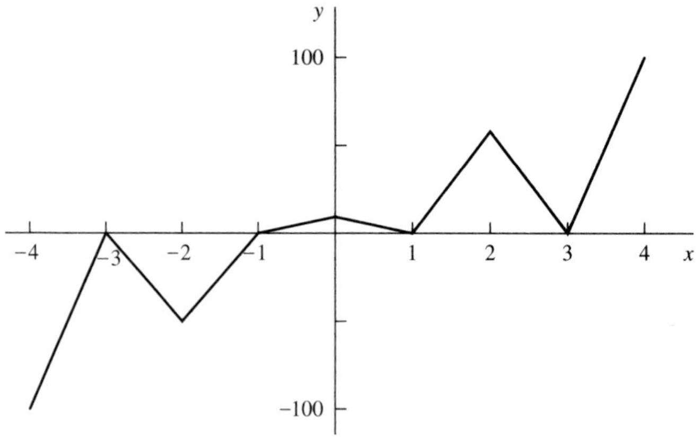

FIGURE P9.5

9.6. A useful nonlinear function is the Gaussian function centered at zero (called the standardized Gaussian), given by $y = a\exp(-x^2)$, where a is a membership value on the interval [0, 1]. This function can be used to "center" a membership function on top of a data point to create a symmetric membership function with infinite tails and a membership value of a at the center point. This function can be used for classification purposes when determining the membership of a datum in the set, and it is the shape of most of the statistical distributions used to model experimental variation.

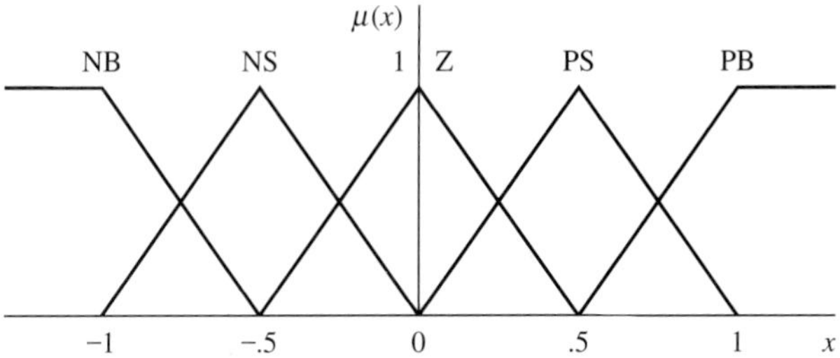

FIGURE P9.6

(*a*) Using the partitioning for the input variable, x, shown in Fig. P9.6, and using the following three simple rules, find membership functions using Zadeh's extension principle:

$$\text{IF } x = \text{Z, THEN } y = \text{MAX}$$
$$\text{IF } x = \text{PS or NS, THEN } y = 76\% \text{ of MAX}$$
$$\text{IF } x = \text{PB or NB, THEN } y = 38\% \text{ of MAX}$$

(*b*) From the union of the output membership function determined in part (*a*), use graphical techniques to determine the defuzzified output for inputs: $x = -1, -0.5, 0, 0.5$, and 1, and plot the results versus the exact solution.

9.7. One of the difficulties with the Gaussian probability distribution is that it has no closed-form integral. Integration of this function must be conducted numerically. Because of this difficulty, approximations to the Gaussian distribution have been developed over the years. One of these approximations is the expression,

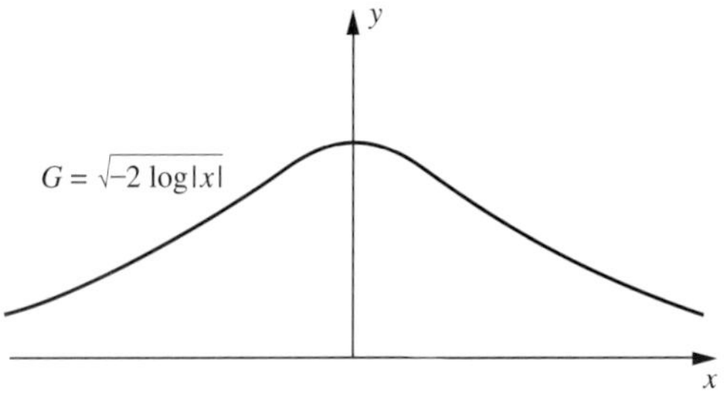

FIGURE P9.7

This expression provides a reasonably good approximation to the Gaussian except for values of x near zero; as can be seen, the function, G, has a singularity at $x = 0$. Table P9.7 shows the exact values for this approximate function, G.

x	**−1**	**−3/4**	**−1/2**	**−1/4**	**0.01**	**1/4**	**1/2**	**3/4**	**1**
G	0	0.5	0.776	1.10	2	1.10	0.776	0.5	0

If one uses the same partitioning for the input variable, x, as given in Problem 9.6, the discrete membership values for each of the quantities x shown in Table P9.7 for the following three fuzzy inputs,

1. x_1 = NB or PB
2. x_2 = Z or PS
3. x_3 = Z or NS

would be

$$x_1 = [1, .5, 0, 0, 0, 0, 0, .5, 1]$$
$$x_2 = [0, 0, 0, .5, 1, .5, 1, .5, 0]$$
$$x_3 = [0, .5, 1, .5, 1, .5, 0, 0, 0]$$

By using Zadeh's extension principle, the membership functions for G for the first five elements in the table (the function is symmetric) corresponding to the three fuzzy inputs would be

$$\mathrm{G}_1 = \left\{\frac{1}{0} + \frac{0.5}{0.5} + \frac{0}{0.776} + \frac{0}{1.10} + \frac{0}{2}\right\} = [1, .5, 0, 0, 0]$$
$$\mathrm{G}_2 = [0, 0, 0, .5, 1]$$
$$\mathrm{G}_3 = [0, .5, 1, .5, 1]$$

(*a*) Develop fuzzy relations (these all will be of size 9 × 5) between the three fuzzy inputs and outputs using a Cartesian product operation.

(*b*) Find the overall fuzzy relation by taking the union of the three relations found in part (*a*).

(*c*) If the matrix relation in part (*b*) is replaced by a continuous surface, composition with crisp singleton inputs for x results in the following table of results for the output G. Verify some of these results.

x	G
−1	0.17
−3/4	0.88
−1/2	1.20
−1/4	1.09
0	1.20
1/4	1.09
1/2	1.20
3/4	0.88
1	0.17

9.8. A constant force, F, acts on a body with mass, m, moving on a smooth surface at velocity, v. The effective power of this force will be EP $= F(v)\cos\theta$ (Fig. P9.8*a*). Using the partitioning for the input variable, θ, as shown in Fig. P9.8*b*, and the partitioning for the output variable, EP, as shown in Fig. P9.8*c*, and the following three simple rules:

1. IF $\underset{\sim}{\mathrm{Z}}$ THEN $\underset{\sim}{\mathrm{ME}}$ (most efficient)
2. IF $\underset{\sim}{\mathrm{NS}}$ or $\underset{\sim}{\mathrm{PS}}$ THEN $\underset{\sim}{\mathrm{NE}}$ (not efficient)
3. IF $\underset{\sim}{\mathrm{PB}}$ or $\underset{\sim}{\mathrm{NB}}$ THEN $\underset{\sim}{\mathrm{NME}}$ (negative most efficient such as braking)

conduct a graphical simulation and plot the results on a graph of EP vs. θ. Show the associated exact solution on this same graph.

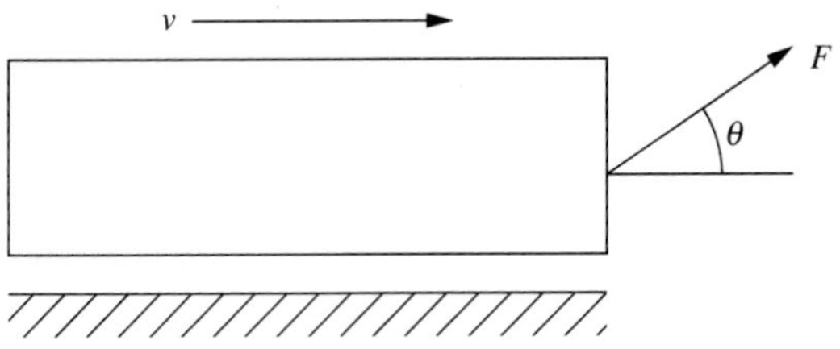

FIGURE P9.8*a*

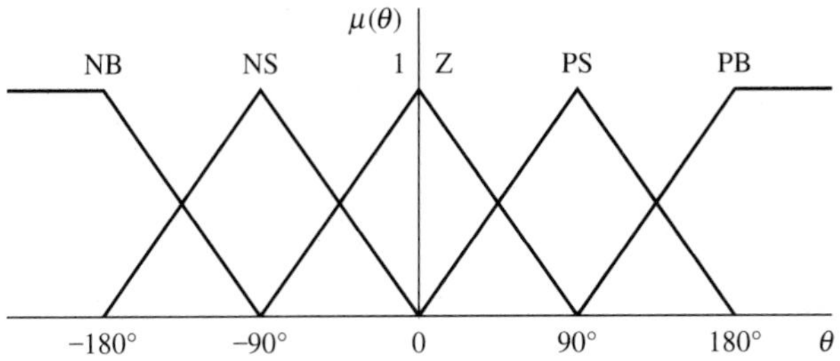

FIGURE P9.8*b*

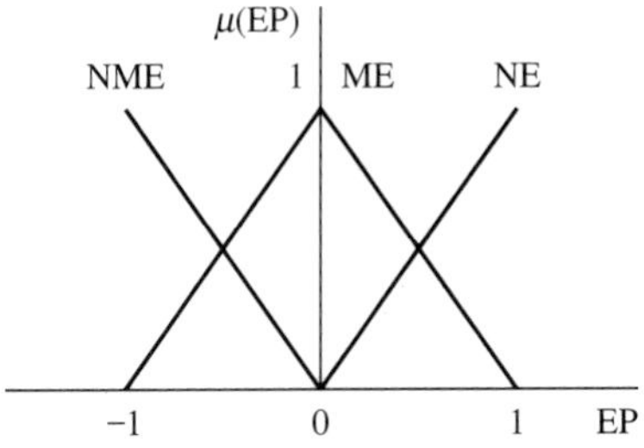

FIGURE P9.8*c*

9.9. Psycho-acoustic research has shown that *white noise* has different effects on people's moods, depending on the average pitch of the tones that make up the noise. Very high and very low pitches make people nervous, whereas midrange noise has a calming effect. The annoyance level of white noise can be approximated by a function of the square of the deviance of the average pitch of the noise from the central pitch of the human hearing range, approximately 10 kHz. As shown in Fig. P9.9*a*, the human annoyance level can be modeled by the nonlinear function, $y = x^2$, where $x =$ deviance (in kHz) from 10 kHz. The range of x is $[-10, 10]$; outside that range pitches are not audible to humans.

If we partition the input variable, x, into five partitions on the range $[-10, 10]$ kHz, as shown in Fig. P9.9*b*, Zadeh's extension principle will yield the partitions for the output space for $y = x^2$ as depicted in Fig. P9.9*c*. Using the following three simple rules,

1. IF $x =$ Z, THEN $y =$ N
2. IF $x =$ NS or PS, THEN $y =$ S
3. IF $x =$ NB or PB, THEN $y =$ V

show how a similar plot of fuzzy results as shown in Fig. 9.9*d* is determined.

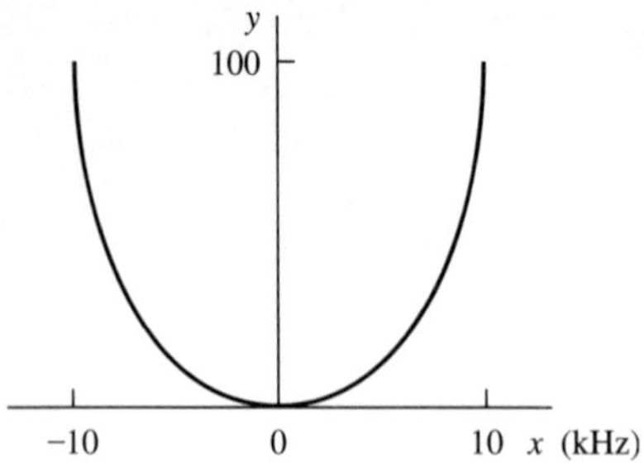

FIGURE P9.9*a*

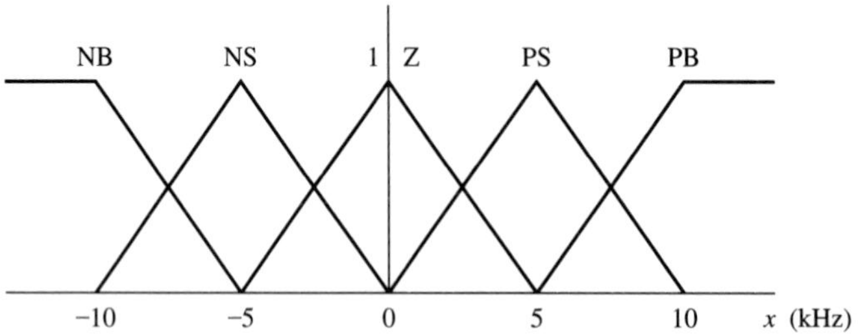

FIGURE P9.9*b*

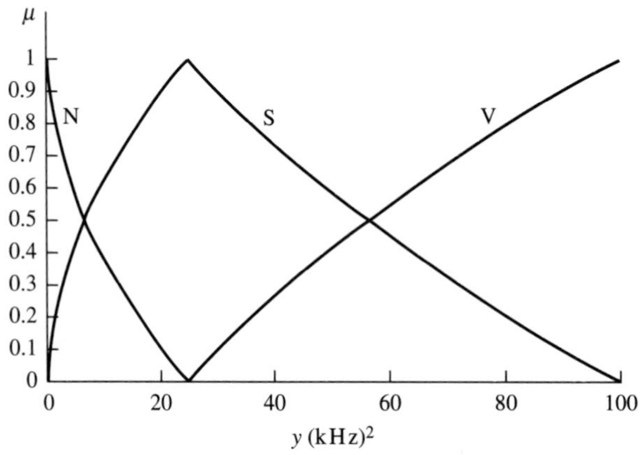

FIGURE P9.9*c*

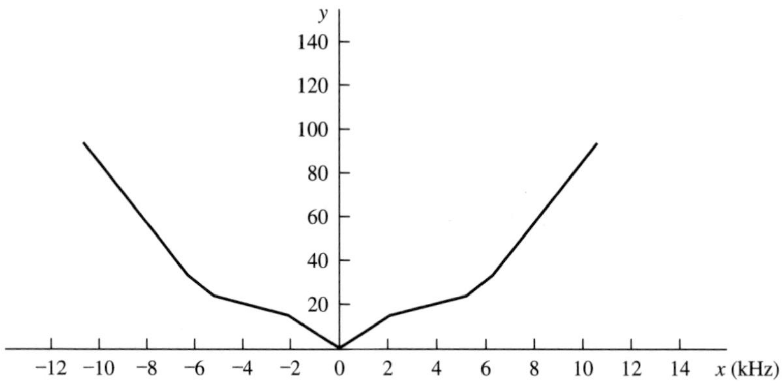

FIGURE P9.9*d*

9.10. The variation of atmospheric pressure with altitude is a nonlinear function that affects airplane and engine performance significantly. This relationship referenced to sea-level standard conditions is

$$\delta = p/p_0 = [1 - (6.857e^{-6})(h)]^{5.2561} \qquad \text{for } h \leq 36{,}089 \text{ feet}$$

Table P9.10 summarizes values of the atmospheric pressure related to altitude and the associated discretized values of the membership functions for three linguistic values of the input variable, airplane altitude, h. Figure P9.10*a* shows the membership function partitions for the input variable.

***h* ft**	***δ***	**Z**	**PS**	**PB**
0	1	1	0	0
9,000	.7148	.5	.5	0
18,000	.4994	0	1	0
27,000	.3398	0	.5	.5
36,000	.2243	0	0	1

(*a*) Using the extension principle, derive membership functions for the output, δ, for each of the input membership functions shown in Fig. P9.10*a*. Label these output functions large, medium, and small, respectively.

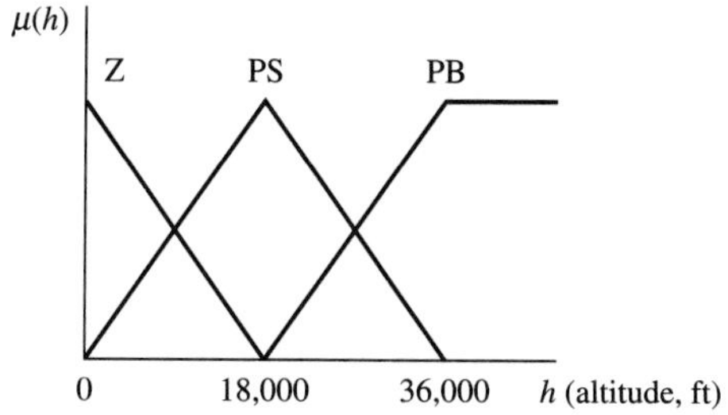

FIGURE P9.10*a*

(*b*) Using the output membership functions from part (*a*), develop fuzzy relations for the following simple rules:

1. IF h is Z, THEN δ is "large"
2. IF h is PS, THEN δ is "medium"
3. IF h is PB, THEN δ is "small"

(*c*) For crisp inputs at h = 0, 9000, 18,000, 27,000, and 36,000 feet conduct a numerical simulation (using compositions on relations or graphical methods) and verify the results shown in Fig. P9.10*b*.

9.11. We can model magnetic field strength as a function of distance from a dipole by approximating the function $y = (10/x^2)$. Dipole strength is usually measured in nanoamperemeters. The universe of discourse for the distance from a dipole is $x \in [0.5, 1]$ in units of decimeters (dm), and the universe of discourse for the output, magnetic field strength, is $y \in [10, 40]$ in units of femtotesla (10^{-15} tesla). Figure P9.11 shows our partitioning of the input space into three linguistic variables, positive small (PS), positive medium (PM), and positive big (PB).

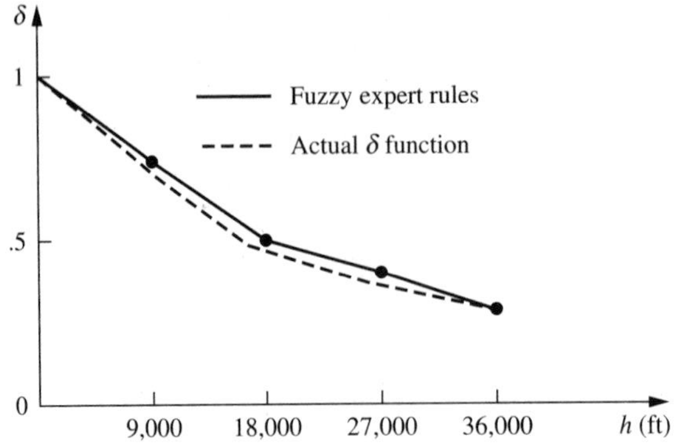

FIGURE P9.10*b*

(*a*) Using the extension principle, find an output membership function for each of the three input partition values and label these membership functions as positive very big (PVB), positive medium (PM), and positive very small (PVS).

(*b*) Develop fuzzy relations, using the following rules, and the associated membership functions for input and output:

IF x is PS, THEN y is PVB
IF x is PM, THEN y is PM
IF x is PB, THEN y is PVS

(*c*) Using disjunctive aggregation of the three fuzzy relations determined in part (*b*), find the combined relation describing the three rules.

(*d*) Using the aggregated relation of part (*c*) and using max-min composition, pose a few fuzzy inputs and find the associated fuzzy outputs to this nonlinear system.

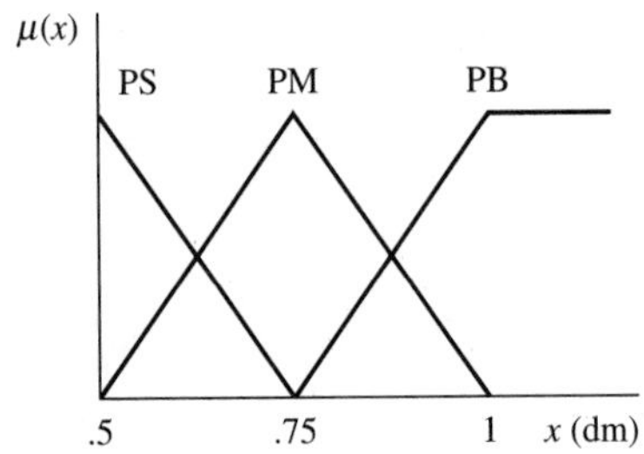

FIGURE P9.11

9.12. Let us consider the case of a series motor under the influence of a load and a constant voltage source, as shown in Fig. P9.12*a*. A series motor should always be operated with some load, otherwise the speed of the motor will become excessively high, resulting in damage to the motor. The speed of the motor, N, in rpm, is inversely related to the armature current, I_a in amps, by the expression $N = k/I_a$, where k is the flux. For this problem, we will estimate the flux parameter based on a motor speed of 1500 rpm at an armature current of 5 amps; hence, $k = 5(1500) = 7500$ rpm-amps. Suppose we consider the armature current to vary in the range $I_a = [-\infty, +\infty]$, and we partition this universe of discourse as shown in Fig. 9.12*b* (note that the extremes at $-\infty$ and $+\infty$

are contained in the partitions NB and PB, respectively). Suppose we also partition the output variable, N, as shown in Fig. P9.12*c*. Using the input and output partitioning provided in Figs. P9.12*b* and P9.12*c* and the following five rules, conduct a graphical

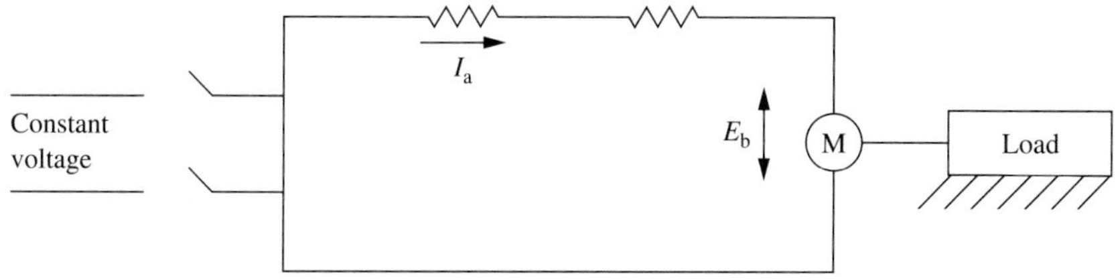

FIGURE P9.12*a*

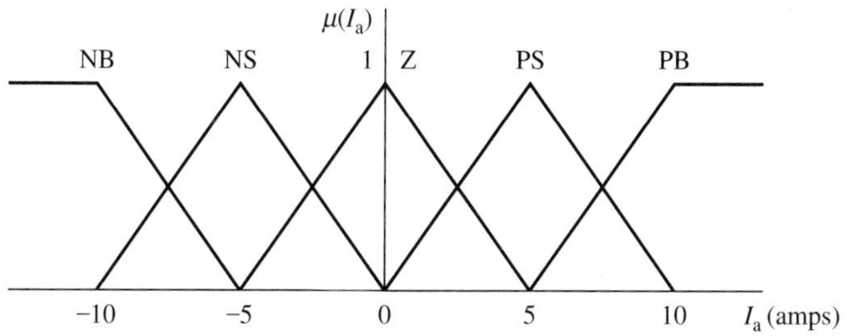

FIGURE P9.12*b*

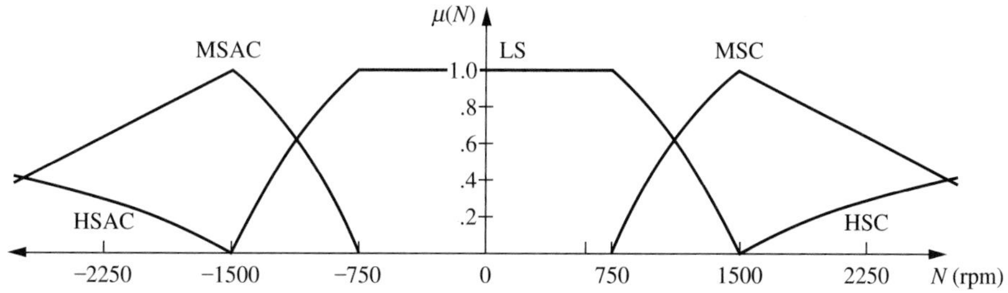

FIGURE P9.12*c*

numerical simulation for the crisp inputs $I_a = -8, -2, 3, 9$ A. Plot this response on a graph of N vs. I_a.

IF I_a is Z, THEN N is HSC or HSAC
IF I_a is PS, THEN N is HSC
IF I_a is NS, THEN N is HSAC
IF I_a is PB, THEN N is MSC
IF I_a is NB, THEN N is MSAC

9.13. The collector current (I_c) in a transistor is a function of the voltage across the base-emitter terminals (V_{be}) and is described by the following equation: $I_c = I_s \exp(V_{be}/V_t)$.

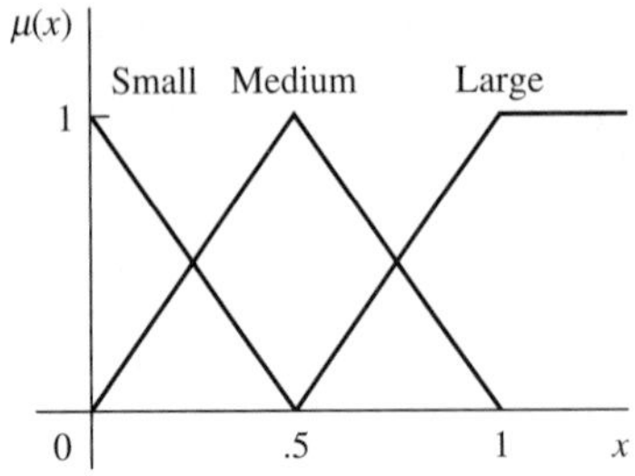

FIGURE P9.13

The voltage V_{be} is the fuzzy input variable on the universe of $V_{be} = [0, 0.7]$, and I_s and V_t are constants. Often it is useful to partition inputs on some normalized universe of discourse, for example, the normalized universe $x = [0, 1]$, as shown in Fig. P9.13. In this way, several different problems can use the same partitioning, with the difference being only a scale factor that accommodates a different universe of discourse and different engineering units. Then the actual input, x', partitioning is given by $x' = k_1(x)$, where k_1 is the scale factor. For this problem involving the input variable V_{be}, we want to use a scale factor of $k_1 = 0.7$; i.e., $V_{be} = k_1(x) = 0.7x$, where x is the standardized input on the interval [0, 1], as shown in Fig. P9.13. Hence, the range for V_{be} is [0, 0.7] in volts.

(*a*) Using $I_s = 1.5$ fA and $V_t = 26$ mV and the extension principle, find membership functions for I_c for the three input partitions shown in Fig. P9.13. Label these output functions, $\underset{\sim}{Y}_1^1$, $\underset{\sim}{Y}_1^2$, and $\underset{\sim}{Y}_1^3$.

(*b*) Construct three fuzzy relations using the derived output membership functions for the following three rules:

$\underset{\sim}{R}_1$: IF V_{be} is small, THEN I_c is $\underset{\sim}{Y}_1^1$
$\underset{\sim}{R}_2$: IF V_{be} is medium, THEN I_c is $\underset{\sim}{Y}_1^2$
$\underset{\sim}{R}_3$: IF V_{be} is large, THEN I_c is $\underset{\sim}{Y}_1^3$

(*c*) For three or four crisp singleton values of the input variable, V_{be}, find defuzzified values for the output, I_c, and compare these to the associated exact values.

9.14. Consider a simple nonlinear problem in mechanics. As shown in Fig. P9.14, a solid piece of any material is under the action of a force F inclined at an angle θ to the longitudinal axis of the material. The problem is to find the axial force component, $F\cos\theta$, acting on the member as θ changes. Since F (force) is not being varied, that is, it is a constant, we need only study the variation of the cosine term, or $y = \cos\theta$. Using methods similar to those posed in Example 9.1 in this chapter for the nonlinear function $y = 10\sin x$, conduct a numerical simulation of the function $y = \cos\theta$.

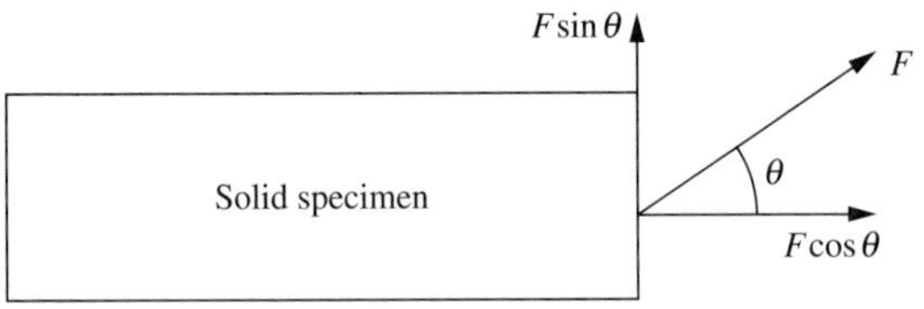

FIGURE P9.14

9.15. In the field of image processing a *limiter* function is used to enhance an image when background lighting is too high. The limiter function is shown in Fig. P9.15*a*.

(*a*) For the input partitioning shown in Fig. P9.15*b* for the input variable, x, verify that the implementation of the extension principle on the limiter function for the five input partitions results in the three output partitions shown in Fig. P9.15*c*.

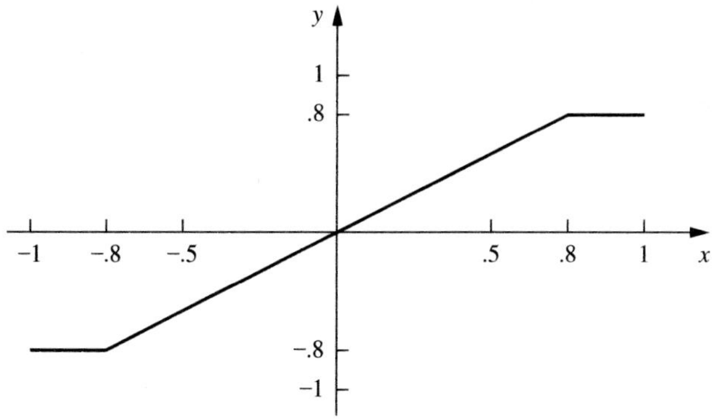

FIGURE P9.15*a*

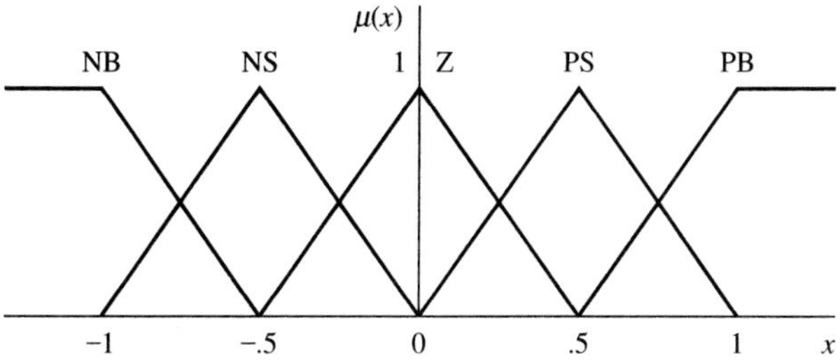

FIGURE P9.15*b*

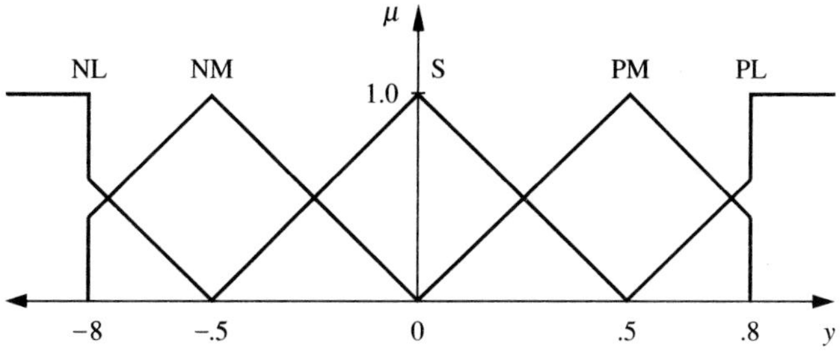

FIGURE P9.15*c*

(*b*) Using the following rules, construct three matrix relations using the input and output partitions:

Rule 1: IF x = Z, THEN y = S
Rule 2: IF x = PB, THEN y = PM
Rule 3: IF x = NB, THEN y = NM

(c) For crisp input values $x = -1, -0.8, -0.6, -0.4, -0.2$, and 0, use graphical techniques or max-min composition and centroidal defuzzification to determine the associated fuzzy outputs. Because of symmetry, values for $0 \leq x \leq 1$ are equal to $|x|$ for $-1 \leq x \leq 0$. Verify that these results follow Fig. P9.15*d*.

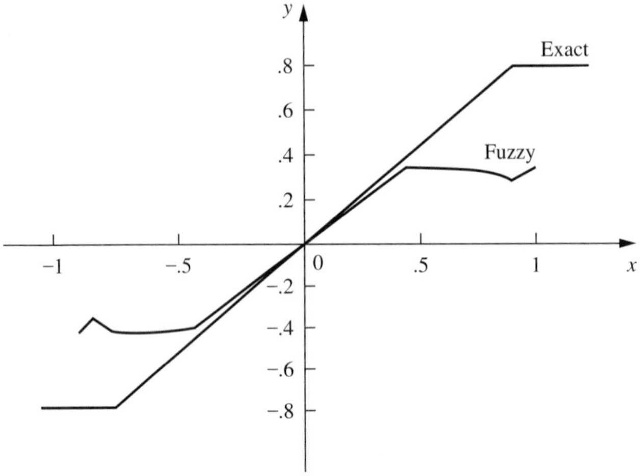

FIGURE P9.15*d*

9.16. Do the example problem on the sine curve, Example 9.2, using (*i*) six rules and (*ii*) eight rules. Does the result look more and more like a sine curve?

CHAPTER 10

FUZZY DECISION MAKING

To be, or not to be: that is the question:
Whether 'tis nobler in the mind to suffer
The slings and arrows of outrageous fortune,
Or to take arms against a sea of troubles,
And by opposing end them.

William Shakespeare
Hamlet, *Act III, Scene I, 1602*

The passage above represents a classic decision situation for humans. It is expressed in natural language—the form of information most used by humans and most ignored in computer-assisted decision making. Consider a more "computer-friendly" expression [Clark, 1992] of the Bard's famous inquiry:

> To be, or not to be—that is the decisional matrix. Whether 'tis preferable within a conceptual framework to suffer the illogic inherent in existence, or to resist the consequences of these limitations, and by opposing them, bring closure to our current existential state.

One can readily see the ludicrous nature of such a venture into the binary world of computers. But, as suggested many other times in this text, this is the nature of the problem engineers face every day: how do we embed natural fuzziness into our otherwise crisp engineering paradigms? Shakespeare would undoubtedly rejoice to learn that his question now has a whole range of possibilities available between the extremes of existence that he originally suggested. Ultimately, the decisions may be binary, as originally posed by Shakespeare in this passage from *Hamlet,* but there certainly should be no restrictions on the usefulness of fuzzy information in the *process* of making a decision or of coming to some consensus.

Decision making is a most important scientific, social, and economic endeavor. To be able to make consistent and correct choices is the essence of any decision process imbued with uncertainty. Most issues in life, as trivial as we might consider them, involve decision processes of one form or another. From the moment we wake in the morning to the time we place our bodies at rest at the day's conclusion we make many, many decisions. What should we wear for the day; should we take an umbrella; what should we eat for breakfast, for lunch, for dinner; should we stop by the gas station on the way to work; what route should we take to work; should we attend that seminar at work; should we write the memorandum to our colleagues before we make the reservations for our next trip out of town; should we go to the store on our way home; should we take the kids to that new museum before, or after, dinner; should we watch the evening news before retiring; and so on and so forth?

Keep in mind when dealing with decision making under uncertainty that *there is a distinct difference between a good decision and a good outcome!* In any decision process we weigh the information about an issue or outcome and choose among two or more alternatives for subsequent action. The information affecting the issue is likely incomplete or uncertain; hence, the outcomes are uncertain, irrespective of the decision made or the alternative chosen. We can make a good decision, and the outcome can be adverse. Alternatively, we can make a bad decision, and the outcome can be advantageous. Such are the vagaries of uncertain events. But in the long run, if we consistently make good decisions, advantageous situations will occur more frequently than bad ones.

To illustrate this notion, consider the choice of whether to take an umbrella on a cloudy, dark morning. As a simple binary matter, the outcomes can be rain or no-rain. We have two alternatives: Take an umbrella, or don't. The information we consider in making this decision could be as unsophisticated as our own feelings about the weather on similar days in the past or as sophisticated as a large-scale meteorological analysis from the National Weather Service. Whatever the source of information, it will be associated with some degree of uncertainty. Suppose we decide to take the umbrella after weighing all the information, and it does not rain. Did we make a bad decision? Perhaps not. Eight times out of 10 in circumstances just like this one, it probably rained. This particular occasion may have been one of the two out of 10 situations when it didn't.

Despite our formal training in this area and despite our common sense about how clear this notion of uncertainty is, we see it violated every day in the business world. A manager makes a good decision, but the outcome is bad and he gets fired. A doctor uses the best established procedures in a medical operation and the patient dies; then she gets sued for malpractice. A child refuses to accept an unsolicited ride home with a distant neighbor on an inclement day, gets soaking wet on the walk home, ruins his shoes, and is reprimanded by his parent for not accepting the ride. A teenager decides to drive on the highway after consuming too many drinks and arrives home safely without incident. In all of these situations the outcomes have nothing to do with the quality of the decisions or with the process itself. The best we can do is to make consistently rational decisions every time we are faced with a choice with the knowledge that in the long run the "goods" will outweigh the "bads."

The problem in making decisions under uncertainty is that the bulk of the information we have about the possible outcomes, about the value of new information, about the way the conditions change with time (dynamic), about the utility of each outcome-action pair, and about our preferences for each action is typically vague, ambiguous, and otherwise fuzzy. If we are fortunate some of the information may be random in character so that we can model it with probability theory. This chapter presents a few paradigms for making decisions within a fuzzy environment. Issues such as personal preferences, multiple objectives, subjective evaluations, and group consensus are presented. The chapter concludes with a rather lengthy development of an area known loosely as fuzzy Bayesian decision making, so named because it involves the introduction of fuzzy information, fuzzy outcomes, and fuzzy actions into the classical probabilistic method of Bayesian decision making. In developing this we are able to compare the value and differences of incorporating both fuzzy and random information into the same representational framework. Acceptance of the new fuzzy approach is therefore eased by its natural accommodation within a classical, and historically popular, decision-making approach.

FUZZY SYNTHETIC EVALUATION

An important application of the fuzzy transform used in developing the extension principle, as discussed in Chapter 6, is synthetic evaluation. The term *synthetic* is used here to connote the process of evaluation whereby several individual elements and components of an evaluation are synthesized into an aggregate form; the whole is a *synthesis* of the parts. The key here is that the various elements can be numeric or non-numeric, and the process of fuzzy synthesis is naturally accommodated using synthetic evaluation. In reality, an evaluation of an object, especially an ill-defined one, is often vague and ambiguous. The evaluation is usually described in natural language terms, since a numerical evaluation is often too complex, too unacceptable, and too ephemeral (transient). For example, when a professor grades a written examination, she might evaluate it from such perspectives as style, grammar, creativity, and so forth. The final grade on the paper might be linguistic instead of numerical, for example, excellent, very good, good, fair, poor, and unsatisfactory. After grading many exams the professor might develop a relation by which she assigns a membership to the relations between the different perspectives, such as style and grammar, and the linguistic grades, such as fair and excellent. A fuzzy relation, $\underset{\sim}{R}$, such as the following one, might result that summarizes the professor's relationship between pairs of grading factors such as *creativity* and grade evaluations such as *very good.*

$$\underset{\sim}{R} = \begin{array}{c} \\ \text{Creative} \\ \text{Grammar} \\ \text{Style} \\ \vdots \end{array} \begin{array}{c} \begin{array}{ccccc} \text{Excellent} & \text{Very good} & \text{Good} & \text{Fair} & \text{Poor} \end{array} \\ \begin{bmatrix} 0.2 & 0.4 & 0.3 & 0.1 & 0 \\ 0 & 0.2 & 0.5 & 0.3 & 0 \\ 0.1 & 0.6 & 0.3 & 0 & 0 \\ \vdots & \vdots & \vdots & \vdots & \vdots \end{bmatrix} \end{array}$$

The professor now wants to assign a grade to each paper. To formalize this approach, let X be a universe of factors and Y be a universe of evaluations, so

$$X = \{x_1, x_2, \ldots, x_n\} \qquad \text{and} \qquad Y = \{y_1, y_2, \ldots, y_m\}$$

Let $\underset{\sim}{R} = [r_{ij}]$ be a fuzzy relation, such as the foregoing grading example, where $i = 1, 2, \ldots, n$ and $j = 1, 2, \ldots, m$. Suppose we introduce a specific paper into the evaluation process on which the professor has given a set of "scores" (w_i) for each of the n grading factors, and we ensure, for convention, that the sum of the scores is unity. Each of these scores is actually a membership value for each of the factors, x_i, and they can be arranged in a fuzzy vector, $\underset{\sim}{w}$. So we have

$$\underset{\sim}{w} = \{w_1, w_2, \ldots, w_n\} \qquad \text{where} \qquad \sum_i w_i = 1 \tag{10.1a}$$

The process of determining a grade for a specific paper is equivalent to the process of determining a membership value for the paper in each of the evaluation categories, y_i. This process is implemented through the composition operation,

$$\underset{\sim}{e} = \underset{\sim}{w} \circ \underset{\sim}{R} \tag{10.1b}$$

where $\underset{\sim}{e}$ is a fuzzy vector containing the membership values for the paper in each of the y_i evaluation categories.

Example 10.1. Suppose we want to measure the value of a microprocessor to a potential client. In conducting this evaluation, the client suggests that certain criteria are important. They can include performance (MIPS), cost ($), availability (AV), and software (SW). Performance is measured by millions of instructions per second, or MIPS; a minimum requirement is 10 MIPS. Cost is the cost of the microprocessor, and a cost requirement of "not to exceed" $500 has been set. Availability relates to how much time after the placement of an order the microprocessor vendor can deliver the part; a maximum of eight weeks has been set. Software represents the availability of operating systems, languages, compilers, and tools to be used with this microprocessor. Suppose further that the client is only able to specify a subjective criterion of having "sufficient" software.

A particular microprocessor (CPU) has been introduced into the market. It is measured against these criteria and given ratings categorized as excellent (e), superior (s), adequate (a), and inferior (i). "Excellent" means that the microprocessor is the best available with respect to the particular criterion. "Superior" means that microprocessor is among the best with respect to this criterion. "Adequate" means that, although not superior, the microprocessor can meet the minimum acceptable requirements for this criterion. "Inferior" means that the microprocessor cannot meet the requirements for the particular criterion. Suppose the microprocessor just introduced has been assigned the following relation based on the consensus of the design team:

$$\mathrm{R} = \begin{array}{r} \\ \text{MIPS} \\ \$ \\ \text{AV} \\ \text{SW} \end{array} \begin{array}{c} \begin{array}{cccc} e & s & a & i \end{array} \\ \begin{bmatrix} .1 & .3 & .4 & .2 \\ 0 & .1 & .8 & .1 \\ .1 & .6 & .2 & .1 \\ .1 & .4 & .3 & .2 \end{bmatrix} \end{array}$$

This relation could have been derived from data using similarity methods such as those discussed in Chapter 3.

If the evaluation team applies a scoring factor of 0.4 for performance, 0.3 for cost, 0.2 for availability, and 0.1 for software, which together form the factor vector, $\underset{\sim}{\mathbf{w}}$, then the composition, $\underset{\sim}{e} = \underset{\sim}{w} \circ \underset{\sim}{\mathrm{R}} = \{0.1, 0.3, 0.4, 0.2\}$ results in an evaluation vector that has its highest membership in the category "adequate."

If we were to perform a crisp analysis for this problem, this CPU would have received two superiors and two adequates from the evaluation team, as shown in the following expression, based on defuzzification to the maximum category in each criteria; that is, the maximum fuzzy value in each row is set to unity, whereas the others are set to zero:

$$\mathrm{R} = \begin{array}{r} \\ \text{MIPS} \\ \$ \\ \text{AV} \\ \text{SW} \end{array} \begin{array}{c} \begin{array}{cccc} e & s & a & i \end{array} \\ \begin{bmatrix} 0 & 0 & 1 & 0 \\ 0 & 0 & 1 & 0 \\ 0 & 1 & 0 & 0 \\ 0 & 1 & 0 & 0 \end{bmatrix} \end{array}$$

The subsequent composition $\underset{\sim}{\mathrm{e}} = \underset{\sim}{\mathrm{w}} \circ \underset{\sim}{\mathrm{R}}$ using the preceding scoring vector would result in an evaluation vector of $\{0, 0.2, 0.4, 0\}$, which would still yield the highest membership for an evaluation of adequate. However, a crisp analysis would not be very useful since it provides little or no assessment of the potential membership in the other categories. Also, the fuzzy approach is more accurate because it reflects the fact that categorizations of the criteria are not crisp. For instance, cost is superior if large numbers of the CPUs are bought—cost is higher and therefore only adequate if the number purchased is small. Similarly, MIPS is somewhat dependent on whatever type of instructions are being executed and, therefore, performance is not crisp but fuzzy.

It is important to point out in concluding this section that the relations expressed in this section are not constrained in that their row sums should equal unity. All of the examples given so far show the row sums equaling unity, a matter of convenience for illustration. However, since the entries in the synthetic evaluation matrix relations are membership values showing the degree of relation between the factors and the evaluations, these values can take on any number between 0 and 1. Hence, row sums could be considerably larger, or smaller, than unity. Some of the problems at the end of this chapter make use of this idea.

FUZZY ORDERING

Decisions are sometimes made on the basis of rank, or ordinal ranking: Which issue is best, which is second best, and so forth. For issues or actions that are deterministic,

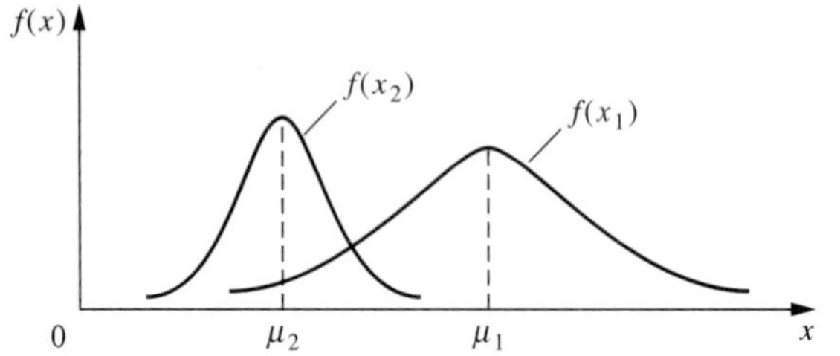

FIGURE 10.1
Density functions for two Gaussian random variables.

such as $y_1 = 5,\ y_2 = 2,\ y_1 \geq y_2$, there is usually no ambiguity in the ranking; we might call this *crisp ordering*. In situations where the issues or actions are associated with uncertainty, either random or fuzzy, rank ordering may be ambiguous. This ambiguity, or uncertainty, can be demonstrated for both random and fuzzy variables. First, let us assume that the uncertainty in rank is random; we can use probability density functions (pdf) to illustrate the random case. Suppose we have one random variable, x_1, whose uncertainty is characterized by a Gaussian pdf with a mean of μ_1 and a standard deviation of σ_1, and another random variable, x_2, also Gaussian with a mean of μ_2 and standard deviation of σ_2. Suppose further that $\sigma_1 > \sigma_2$ and $\mu_1 > \mu_2$. If we plot the pdf's for these two random variables in Fig. 10.1, we see that the question of which variable is greater is not clear.

As an example of this uncertain ranking, suppose x_1 is the height of Italians and x_2 is the height of Swedes. Because this uncertainty is of the random kind, we cannot answer the question, Are Swedes taller than Italians? unless we are dealing with two specific individuals, one each from Sweden and Italy, or we are simply assessing μ_1, average-height Swedes, and μ_2, average-height Italians. But we can ask the question, How frequently are Swedes taller than Italians? We can assess this frequency as the probability that one random variable is greater than another, i.e., $P(x_1 \geq x_2)$, with

$$P(x_1 \geq x_2) = \int_{-\infty}^{\infty} F_{x_2}(x_1)\,dx_1 \tag{10.2a}$$

where F is a cumulative distribution function. Hence, with random variables we can quantify the uncertainty in ordering with a convolution integral, Eq. (10.2*a*).

Second, let us assume that the uncertainty in rank arises because of ambiguity. For example, suppose we are trying to rank people's preferences in colors. In this case the ranking is very subjective and not reducible to the elegant form available for some random variables, such as that given in Eq. (10.2*a*). For fuzzy variables we are also able to quantify the uncertainty in ordering, but in this case we must do so with the notion of membership.

A third type of ranking involves the notion of imprecision [Dubois and Prade, 1980]. To develop this, suppose we have two fuzzy numbers, $\underset{\sim}{I}$ and $\underset{\sim}{J}$. We can use tools provided in Chapter 6 on the extension principle to calculate the truth value of the assertion that fuzzy number $\underset{\sim}{I}$ is greater that fuzzy number $\underset{\sim}{J}$ with the following expression:

$$T(\underset{\sim}{I} \geq \underset{\sim}{J}) = \sup_{x \geq y} \min(\mu_{\underset{\sim}{I}}(x), \mu_{\underset{\sim}{J}}(y)) \tag{10.2b}$$

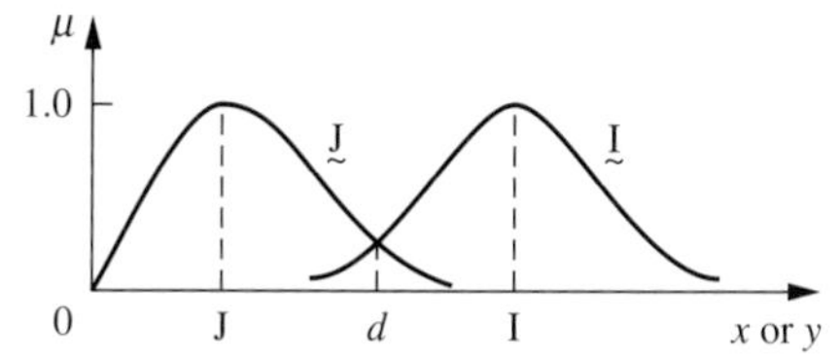

FIGURE 10.2
Two fuzzy numbers as fuzzy sets on the real line.

Figure 10.2 shows the membership functions for two fuzzy numbers $\underset{\sim}{\mathrm{I}}$ and $\underset{\sim}{\mathrm{J}}$. Equation (10.2$b$) is an extension of the inequality $x \geq y$ according to the extension principle. It represents the degree of possibility in the sense that if a specific pair (x, y) exists such that $x \geq y$ and $\mu_{\underset{\sim}{\mathrm{I}}}(x) = \mu_{\underset{\sim}{\mathrm{J}}}(y)$, then $T(\underset{\sim}{\mathrm{I}} \geq \underset{\sim}{\mathrm{J}}) = 1$. Since the fuzzy numbers $\underset{\sim}{\mathrm{I}}$ and $\underset{\sim}{\mathrm{J}}$ are convex, it can be seen from Fig. 10.2 that

$$T(\underset{\sim}{\mathrm{I}} \geq \underset{\sim}{\mathrm{J}}) = 1 \text{ if and only if } \mathrm{I} \geq \mathrm{J} \tag{10.3a}$$

$$T(\underset{\sim}{\mathrm{J}} \geq \underset{\sim}{\mathrm{I}}) = \text{height}(\underset{\sim}{\mathrm{I}} \cap \underset{\sim}{\mathrm{J}}) = \mu_{\underset{\sim}{\mathrm{I}}}(d) = \mu_{\underset{\sim}{\mathrm{J}}}(d) \tag{10.3b}$$

where d is the location of the highest intersection point of the two fuzzy numbers. The operation height $(\underset{\sim}{\mathrm{I}} \cap \underset{\sim}{\mathrm{J}})$ in Equation 10.3b is a good separation metric for two fuzzy numbers; i.e., the closer this metric is to unity, the more difficult it is to distinguish which of the two fuzzy numbers is largest. On the other hand, as this metric approaches zero, the easier is the distinction about rank order (which is largest). Unfortunately, the metric given in Equation 10.3a is not too useful as a ranking metric, because $T(\underset{\sim}{\mathrm{I}} \geq \underset{\sim}{\mathrm{J}}) = 1$ when I is slightly greater and when I is much greater than J. If we know that $\underset{\sim}{\mathrm{I}}$ and $\underset{\sim}{\mathrm{J}}$ are crisp numbers I and J, the truth value becomes $T(\mathrm{I} \geq \mathrm{J}) = 1$ for $\mathrm{I} \geq \mathrm{J}$ and $T(\mathrm{I} \geq \mathrm{J}) = 0$ for $\mathrm{I} < \mathrm{J}$.

The definitions expressed in Eqs. (10.2b) and (10.3) for two fuzzy numbers can be extended to the more general case of many fuzzy sets. Suppose we have k fuzzy sets $\underset{\sim}{\mathrm{I}}_1, \underset{\sim}{\mathrm{I}}_2, \ldots, \underset{\sim}{\mathrm{I}}_k$. Then the truth value of a specified ordinal ranking is given by

$$T(\underset{\sim}{\mathrm{I}} \geq \underset{\sim}{\mathrm{I}}_1, \underset{\sim}{\mathrm{I}}_2, \ldots, \underset{\sim}{\mathrm{I}}_k) = T(\underset{\sim}{\mathrm{I}} \geq \underset{\sim}{\mathrm{I}}_1) \quad \text{and} \quad T(\underset{\sim}{\mathrm{I}} \geq \underset{\sim}{\mathrm{I}}_2) \quad \text{and} \ldots \text{and} \quad T(\underset{\sim}{\mathrm{I}} \geq \underset{\sim}{\mathrm{I}}_k) \tag{10.4}$$

Example 10.2. Suppose we have three fuzzy sets, as described using Zadeh's notation:

$$\underset{\sim}{\mathrm{I}}_1 = \left\{\frac{1}{3} + \frac{0.8}{7}\right\} \quad \text{and} \quad \underset{\sim}{\mathrm{I}}_2 = \left\{\frac{0.7}{4} + \frac{1.0}{6}\right\} \quad \text{and} \quad \underset{\sim}{\mathrm{I}}_3 = \left\{\frac{0.8}{2} + \frac{1}{4} + \frac{0.5}{8}\right\}$$

We can assess the truth value of the inequality, $\underset{\sim}{\mathrm{I}}_1 \geq \underset{\sim}{\mathrm{I}}_2$, as follows:

$$\begin{aligned} T(\underset{\sim}{\mathrm{I}}_1 \geq \underset{\sim}{\mathrm{I}}_2) &= \max_{x_1 \geq x_2} \{\min(\mu_{\underset{\sim}{\mathrm{I}}_1}(x_1), \mu_{\underset{\sim}{\mathrm{I}}_2}(x_2))\} \\ &= \max\{\min(\mu_{\underset{\sim}{\mathrm{I}}_1}(7), \mu_{\underset{\sim}{\mathrm{I}}_2}(4)), \min(\mu_{\underset{\sim}{\mathrm{I}}_1}(7), \mu_{\underset{\sim}{\mathrm{I}}_2}(6))\} \\ &= \max\{\min(0.8, 0.7), \min(0.8, 1.0)\} \\ &= 0.8 \end{aligned}$$

Similarly,

$$\begin{aligned} T(\underset{\sim}{\mathrm{I}}_1 \geq \underset{\sim}{\mathrm{I}}_3) &= 0.8 \qquad & T(\underset{\sim}{\mathrm{I}}_2 \geq \underset{\sim}{\mathrm{I}}_1) &= 1.0 \\ T(\underset{\sim}{\mathrm{I}}_2 \geq \underset{\sim}{\mathrm{I}}_3) &= 1.0 \qquad & T(\underset{\sim}{\mathrm{I}}_3 \geq \underset{\sim}{\mathrm{I}}_1) &= 1.0 \\ T(\underset{\sim}{\mathrm{I}}_3 \geq \underset{\sim}{\mathrm{I}}_2) &= 0.7 \end{aligned}$$

Then
$$T(\underset{\sim}{I}_1 \geq \underset{\sim}{I}_2, \underset{\sim}{I}_3) = 0.8$$
$$T(\underset{\sim}{I}_2 \geq \underset{\sim}{I}_1, \underset{\sim}{I}_3) = 1.0$$
$$T(\underset{\sim}{I}_3 \geq \underset{\sim}{I}_1, \underset{\sim}{I}_2) = 0.7$$

The last three truth values in this example compared one fuzzy set to two others. This calculation is different from pairwise comparisons. To do the latter, one makes use of the minimum function, as prescribed by Eq. (10.4). For example,

$$T(\underset{\sim}{I}_1 \geq \underset{\sim}{I}_2, \underset{\sim}{I}_3) = \min\{(\underset{\sim}{I}_1 \geq \underset{\sim}{I}_2), (\underset{\sim}{I}_1 \geq \underset{\sim}{I}_3)\}$$

Equation (10.4) can be used similarly to obtain the other multiple comparisons. Based on the foregoing ordering, the overall ordering for the three fuzzy sets would be $\underset{\sim}{I}_2$ first, $\underset{\sim}{I}_1$ second, and $\underset{\sim}{I}_3$ last.

When we compare objects that are fuzzy, ambiguous, or vague, we may well encounter a situation where there is a contradiction in the classical notions of ordinal ranking and transitivity in the ranking. For example, suppose we are ordering on the preference of colors. When comparing red to blue, we prefer red; when comparing blue to yellow, we prefer blue; but when comparing red and yellow we might prefer yellow. In this case transitivity of sets representing preference in colors (red $>$ blue and blue $>$ yellow does *not* yield red $>$ yellow) is not maintained.

To accommodate this form of nontransitive ranking (which can be quite normal for noncardinal-type concepts), we introduce a special notion of relativity [Shimura, 1973]. Let x and y be variables defined on universe X, and we define a pairwise function,

$$f_y(x) \text{ as the membership function of } x \text{ with respect to } y$$

and we define another pairwise function

$$f_x(y) \text{ as the membership function of } y \text{ with respect to } x$$

Then the relativity function given by

$$f(x \mid y) = \frac{f_y(x)}{\max[f_y(x), f_x(y)]} \tag{10.5}$$

is a measurement of the membership value of choosing x over y. The relativity function $f(x \mid y)$ can be thought of as the membership of preferring variable x over variable y. Note that the function in Eq. (10.5) uses arithmetic division.

To develop the general case of Eq. (10.5) for many variables, define variables $x_1, x_2, \ldots, x_i, x_{i+1}, \ldots, x_n$ all defined on universe X, and let these variables be collected in a set A; i.e., $A = \{x_1, x_2, \ldots, x_{i-1}, x_i, x_{i+1}, \ldots, x_n\}$. We then define a set identical to set A except this new set will be missing one element, x_i, and this set will be termed A′. The relativity function then becomes

$$\begin{aligned} f(x_i \mid A') &= f(x_i \mid \{x_1, x_2, \ldots, x_{i-1}, x_{i+1}, \ldots, x_n\}) \\ &= \min\{f(x_i \mid x_1), f(x_i \mid x_2), \ldots, f(x_i \mid x_{i-1}), f(x_i \mid x_{i+1}), \ldots, f(x_i \mid x_n)\} \end{aligned} \tag{10.6}$$

which is the fuzzy measurement of choosing x_i over all elements in the set A′. The expression in Eq. (10.6) involves the logical intersection of several variables, hence the minimum function is used. Since the relativity function of one variable with respect to itself is identity, that is,

$$f(x_i \mid x_i) = 1 \tag{10.7}$$

then

$$f(x_i \mid \text{A}') = f(x_i \mid \text{A}) \tag{10.8}$$

We can now form a matrix of relativity values, $f(x_i \mid x_j)$ where $i, j = 1, 2, \ldots, n$, and where x_i and x_j are defined on a universe X. This matrix will be square and of order n, and will be termed the C matrix (C for comparison). The C matrix can be used to rank many different fuzzy sets.

To determine the overall ranking, we need to find the smallest value in each of the rows of the C matrix; that is,

$$\text{C}_i' = \min f(x_i \mid \text{X}), i = 1, 2, \ldots, n \tag{10.9}$$

where C_i' is the membership ranking value for the ith variable. We use the minimum function because this value will have the lowest weight for ranking purposes; further, the maximum function often returns a value of 1 and there would be ambivalence in the ranking process (i.e., ties will result).

Example 10.3. In manufacturing, we often try to compare the capabilities of various microprocessors for their appropriateness to certain applications. For instance, suppose we are trying to select from among four microprocessors the one that is best suited for image-processing applications. Since many factors can affect this decision, including performance, cost, availability, software, and others, coming up with a crisp mathematical model for all these attributes is complicated. Another consideration is that it is much easier to compare these microprocessors subjectively in pairs rather than all four at one time.

Suppose the design team is polled to determine which of the four microprocessors, labeled x_1, x_2, x_3, and x_4, is the most preferred when considered as a group rather than when considered as pairs. First, pairwise membership functions are determined. These represent the subjective measurement of the appropriateness of each microprocessor when compared only to one another. The following pairwise functions are determined:

$$\begin{array}{llll}
f_{x_1}(x_1) = 1 & f_{x_1}(x_2) = .5 & f_{x_1}(x_3) = .3 & f_{x_1}(x_4) = .2 \\
f_{x_2}(x_1) = .7 & f_{x_2}(x_2) = 1 & f_{x_2}(x_3) = .8 & f_{x_2}(x_4) = .9 \\
f_{x_3}(x_1) = .5 & f_{x_3}(x_2) = .3 & f_{x_3}(x_3) = 1 & f_{x_3}(x_4) = .7 \\
f_{x_4}(x_1) = .3 & f_{x_4}(x_2) = .1 & f_{x_4}(x_3) = .3 & f_{x_4}(x_4) = 1
\end{array}$$

For example, microprocessor x_2 has membership 0.5 with respect to microprocessor x_1. Note that if these values were arranged into a matrix, it would not be symmetric. These membership values do not express similarity or relation; they represent membership values of ordering when considered in a particular order. If we now employ Eq. (10.5) to calculate all of the relativity values, the matrix shown below expresses these calculations; this is the so-called comparison, or C, matrix. For example,

$$f(x_2 \mid x_1) = \frac{f_{x_1}(x_2)}{\max[f_{x_1}(x_2), f_{x_2}(x_1)]} = \frac{0.5}{\max[0.5, 0.7]} = 0.71$$

$$\mathrm{C} = \begin{array}{c} \\ x_1 \\ x_2 \\ x_3 \\ x_4 \end{array} \begin{array}{c} \begin{array}{cccc} x_1 & x_2 & x_3 & x_4 \end{array} \\ \left[\begin{array}{cccc} 1 & 1 & 1 & 1 \\ .71 & 1 & .38 & .11 \\ .6 & 1 & 1 & .43 \\ .67 & 1 & 1 & 1 \end{array}\right] \end{array} \begin{array}{c} \min = f(x_i \mid \mathrm{X}) \\ 1 \\ .11 \\ .43 \\ .67 \end{array}$$

The extra column to the right of the foregoing C matrix is the minimum value for each of the rows, i.e., for C'_i, $i = 1, 2, 3, 4$ in Eq. (10.9). For this example problem, the order from best to worst is x_1, x_4, x_3, and x_2. This ranking is much more easily attained with this fuzzy approach than it would have been with some other method where the attributes of each microprocessor are assigned a value measurement and these values are somehow combined. This fuzzy method also contains the subjectiveness inherent in comparing one microprocessor to another. If all four were considered at once, a person's own bias might skew the value assessments to favor a particular microprocessor. By using pairwise comparisons, each microprocessor is compared individually against its peers, which should allow a more fair and less biased comparison.

PREFERENCE AND CONSENSUS

The goal of group decision making typically is to arrive at a consensus concerning a desired action or alternative from among those considered in the decision process. In this context, consensus is usually taken to mean a unanimous agreement by all those in the group concerning their choice. Despite the simplicity in defining consensus, it is another matter altogether to quantify this notion. Most traditional mathematical developments of consensus have used individual preference ranking as their primary feature. In these developments, the individual preferences of those in the decision group are collected to form a group metric whose properties are used to produce a scalar measure of "degree of consensus." However, the underlying axiomatic structure of many of these classical approaches is based on classical set theory. The argument given in this text is that the crisp set approach is too restrictive for the variables relevant to a group decision process. The information in the previous section showed individual preference to be a fuzzy relation.

There can be numerous outcomes of decision groups in developing consensus about a universe, X, of n possible alternatives; i.e., $\mathrm{X} = \{x_1, x_2, \ldots, x_n\}$. To start the development, we define a *reciprocal* relation as a fuzzy relation, $\underset{\sim}{\mathrm{R}}$, of order n, whose individual elements r_{ij} have the following properties [Bezdek et al., 1978]:

$$r_{ii} = 0 \qquad \text{for } 1 \le i \le n \tag{10.10}$$

$$r_{ij} + r_{ji} = 1 \qquad \text{for } i \ne j \tag{10.11}$$

This reciprocal relation, $\underset{\sim}{\mathrm{R}}$, Eqs. (10.10)–(10.11), can be interpreted as follows: r_{ij} is the preference accorded to x_i relative to x_j. Thus, $r_{ij} = 1$ (hence, $r_{ji} = 0$) implies that alternative i is *definitely* preferred to alternative j; this is the crisp case in preference. At the other extreme, we have maximal fuzziness, where $r_{ij} = r_{ji} = 0.5$, and there is equal preference, pairwise. A definite choice of alternative i to all others is manifested in $\underset{\sim}{R}$ as the ith row being all ones (except $r_{ii} = 0$), or the ith column being all zeros.

Two common measures of preference are defined here as *average fuzziness* in $\underset{\sim}{R}$, Eq. (10.12), and *average certainty* in $\underset{\sim}{R}$, Eq. (10.13):

$$F(\underset{\sim}{R}) = \frac{\mathrm{tr}(\underset{\sim}{R}^2)}{n(n-1)/2} \tag{10.12}$$

$$C(\underset{\sim}{R}) = \frac{\mathrm{tr}(\underset{\sim}{R}\underset{\sim}{R}^T)}{n(n-1)/2} \tag{10.13}$$

In Eqs. (10.12)–(10.13), tr ( ) and $(\ )^T$ denote the trace and transpose, respectively, and matrix multiplication is the algebraic kind. Recall that the trace of a matrix is simply the algebraic sum of the diagonal elements, i.e.,

$$\mathrm{tr}(\underset{\sim}{R}) = \sum_{i=1}^{n} r_{ii}$$

The measure, $F(\underset{\sim}{R})$, averages the joint preferences in $\underset{\sim}{R}$ over all distinct pairs in the Cartesian space, X×X. Each term maximizes the measure when $r_{ij} = r_{ji} = 0.5$ and minimizes the measure when $r_{ij} = 1$ and $r_{ji} = 0$; consequently, $F(\underset{\sim}{R})$ is proportional to the fuzziness or uncertainty (also, confusion) about pairwise rankings exhibited by the fuzzy preference relation, $\underset{\sim}{R}$. Conversely, the measure, $C(\underset{\sim}{R})$, averages the individual dominance (assertiveness) of each distinct pair of rankings in the sense that each term maximizes the measure when $r_{ij} = 1$ and $r_{ji} = 0$ and minimizes the measure when $r_{ij} = r_{ji} = 0.5$; hence, $C(\underset{\sim}{R})$ is proportional to the overall certainty in $\underset{\sim}{R}$. The two measures are dependent; they are both on the interval [0, 1]; and it can be shown [Bezdek et al., 1978] that

$$F(\underset{\sim}{R}) + C(\underset{\sim}{R}) = 1 \tag{10.14}$$

It can further be shown that C is a minimum and F is a maximum at $r_{ij} = r_{ji} = 0.5$, and that C is a maximum and F is a minimum at $r_{ij} = 1$, $r_{ji} = 0$. Also, at the state of maximum fuzziness ($r_{ij} = r_{ji} = 0.5$), we get $F(\underset{\sim}{R}) = C(\underset{\sim}{R}) = \frac{1}{2}$; and at the state of no uncertainty ($r_{ij} = 1$, $r_{ji} = 0$) we get $F(\underset{\sim}{R}) = 0$ and $C(\underset{\sim}{R}) = 1$. Moreover, the ranges for these two measures are $0 \le F(\underset{\sim}{R}) \le \frac{1}{2}$ and $\frac{1}{2} \le C(\underset{\sim}{R}) \le 1$.

Measures of preference can be useful in determining consensus. There are different forms of consensus. We have discussed the antithesis of consensus: complete ambivalence, or the maximally fuzzy case where all alternatives are rated equally; call this type of consensus M_1. For M_1 we have a matrix $\underset{\sim}{R}$ where all nondiagonal elements are equal to $\frac{1}{2}$. We have also discussed the converse of M_1, which is the nonfuzzy (crisp) preference where every pair of alternatives is definitely ranked; call this case M_2. In M_2 all nondiagonal elements in $\underset{\sim}{R}$ are equal to one or zero; however, there may not be a clear consensus. Consider following the reciprocity relation, M_2;

$$\mathrm{M}_2 = \begin{bmatrix} 0 & 1 & 0 & 1 \\ 0 & 0 & 1 & 0 \\ 1 & 0 & 0 & 1 \\ 0 & 1 & 0 & 0 \end{bmatrix}$$

Here, the clear pairwise choices are these: alternative 1 over alternative 2, alternative 1 over alternative 4, alternative 2 over alternative 3, and alternative 3 over alternative 4. However, we don't have consensus because alternative 3 is preferred over alternative 1 and alternative 4 is preferred over alternative 2! So for relation M_1 we can't have consensus and for relation M_2 we may not have consensus.

Three types of consensus, however, arise from considerations of the matrix $\underset{\sim}{\mathrm{R}}$. The first type, known as Type I consensus, M_1^*, is a consensus in which there is one clear choice, say alternative i (the ith column is all zeros), and the remaining $(n-1)$ alternatives all have equal secondary preference (i.e., $r_{kj} = \frac{1}{2}$, where $k \neq j$), as shown in the following example, where alternative 2 has a clear consensus:

$$\mathrm{M}_1^* = \begin{bmatrix} 0 & 0 & 0.5 & 0.5 \\ 1 & 0 & 1 & 1 \\ 0.5 & 0 & 0 & 0.5 \\ 0.5 & 0 & 0.5 & 0 \end{bmatrix}$$

In the second type of consensus, called a Type II consensus, M_2^*, there is one clear choice, say alternative i (the ith column is all zeros), but the remaining $(n-1)$ alternatives all have definite secondary preference (i.e., $r_{kj} = 1$, where $k \neq i$), as shown in this example:

$$\mathrm{M}_2^* = \begin{bmatrix} 0 & 0 & 1 & 0 \\ 1 & 0 & 1 & 1 \\ 0 & 0 & 0 & 1 \\ 1 & 0 & 0 & 0 \end{bmatrix}$$

where alternative 2 has a clear consensus, but where there is no clear ordering after the first choice because alternative 1 is preferred to alternative 3, 3 to 4, but alternative 4 is preferred to alternative 1. There can be clear ordering after the first choice in Type II consensus matrices, but it is not a requirement.

Finally, the third type of consensus, called a Type *fuzzy* consensus, M_f^*, occurs where there is a unanimous decision for the most preferred choice, say alternative i again, but the remaining $(n-1)$ alternatives have infinitely many fuzzy secondary preferences. The matrix shown here has a clear choice for alternative 2, but the other secondary preferences are fuzzy to various degrees:

$$\mathrm{M}_f^* = \begin{bmatrix} 0 & 0 & 0.5 & 0.6 \\ 1 & 0 & 1 & 1 \\ 0.5 & 0 & 0 & 0.3 \\ 0.4 & 0 & 0.7 & 0 \end{bmatrix}$$

Mathematically, relations M_1 and M_2 are logical opposites, as are consensus relations M_1^* and M_2^* [Bezdek et al., 1978]. It is interesting to discuss the cardinality of these various preference and consensus relations. In this case, the cardinality of a relation is the number of possible combinations of that matrix type. It is obvious that there is only one possible matrix for M_1. The cardinality of all the preference or consensus relations discussed here is given in Eq. (10.15), where the symbol $|\ |$ denotes cardinality of the particular relation:

$$\begin{aligned} |M_1| &= 1 \\ |M_2| &= 2^{n(n-1)/2} \\ |M_1^*| &= n && \text{(Type I)} \\ |M_2^*| &= (2^{(n^2-3n+2)/2})(n) && \text{(Type II)} \\ |M_f^*| &= \infty && \text{(Type } \textit{fuzzy}\text{)} \end{aligned} \tag{10.15}$$

So, for the examples previously illustrated for $n = 4$ alternatives, there are $64(2^6)$ possible forms of the M_2 preference matrix, there are only 4 Type I (M_1^*) consensus matrices, and there are $32(2^3 \cdot 4)$ possible Type II (M_2^*) consensus matrices.

From the *degree of preference* measures given in Eqs. (10.12)–(10.13), we can construct a *distance to consensus* metric, defined as

$$m(\underset{\sim}{R}) = 1 - (2C(\underset{\sim}{R}) - 1)^{1/2} \tag{10.16}$$

where $m(\underset{\sim}{R}) = 1$ for an M_1 preference relation
$m(\underset{\sim}{R}) = 0$ for an M_2 preference relation
$m(\underset{\sim}{R}) = 1 - (2/n)^{1/2}$ for a Type I (M_1^*) consensus relation
$m(\underset{\sim}{R}) = 0$ for a Type II (M_2^*) consensus relation

We can think of this metric, $m(\underset{\sim}{R})$, as being a distance between the points $M_1(1.0)$ and $M_2(0.0)$ in n-dimensional space. We see that $m(M_1^*) = m(M_2^*)$ for the case where we have only two ($n = 2$) alternatives. For the more general case, where $n > 2$, the distance between Type I and Type II consensus increases with n as it becomes increasingly difficult to develop a consensus choice and simultaneously rank the remaining pairs of alternatives.

Example 10.4. Suppose a reciprocal fuzzy relation, $\underset{\sim}{R}$, is developed by a small group of people in their initial assessments for pairwise preferences for a decision process involving four alternatives, $n = 4$, as shown here:

$$\underset{\sim}{R} = \begin{bmatrix} 0 & 1 & 0.5 & 0.2 \\ 0 & 0 & 0.3 & 0.9 \\ 0.5 & 0.7 & 0 & 0.6 \\ 0.8 & 0.1 & 0.4 & 0 \end{bmatrix}$$

Notice that this matrix carries none of the properties of a consensus type; that is, the group does not reach consensus on their first attempt at ranking the alternatives. However, the group can assess their "degree of consensus" and they can measure how "far" they are from consensus prior to subsequent discussions in the decision process. So, for example, alternative 1 is *definitely* preferred to alternative 2, alternative 1 is rated *equal* to alternative 3, and so forth. For this matrix, $C(\underset{\sim}{R}) = 0.683$ [Eq. (10.13)], $m(\underset{\sim}{R}) = 0.395$, and $m(M_1^*) = 1 - (2/n)^{1/2} = 0.293$ [Eq. (10.16)]. For their first attempt at ranking the four alternatives the group has a degree of consensus of

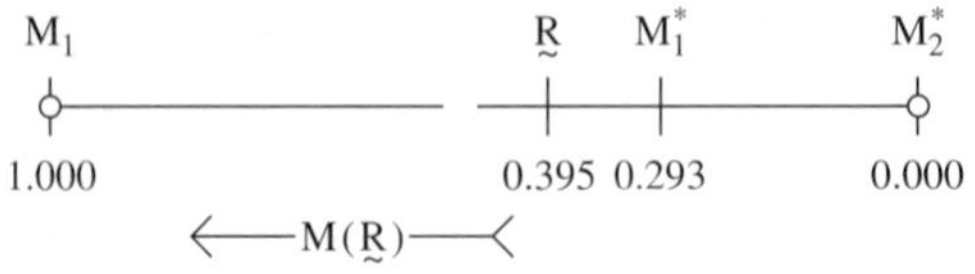

FIGURE 10.3
Illustration of *distance to consensus* [Bezdek et al., 1978].

0.683 (recall a value of 0.5 is completely ambivalent (uncertain) and a value of 1.0 is completely certain). Moreover, the group is $1 - 0.395 = .605$, or 60.5 percent of the way from complete ambivalence (M_1) toward a Type II consensus, or they are $0.605/(1 - 0.293) = 85.5$ percent of the way toward a Type I consensus. These ideas are shown graphically in Fig. 10.3. The value of the *distance to consensus,* $m(\underset{\sim}{R})$, is its use in quantifying the dynamic evolution of a group as the group refines its preferences and moves closer to a Type I or Type II or Type *fuzzy* consensus. It should be noted that the vast majority of group preference situations eventually develop into Type *fuzzy* consensus; Types I and II are typically only useful as boundary conditions.

MULTIOBJECTIVE DECISION MAKING

Many simple decision processes are based on a single objective, such as minimizing cost, maximizing profit, minimizing run time, and so forth. Often, however, decisions must be made in an environment where more than one objective function governs constraints on the problem, and the relative value of each of these objectives is different. For example, suppose we are designing a new computer, and we want simultaneously to minimize cost, maximize CPU, maximize random access memory (RAM), and maximize reliability. Moreover, suppose cost is the most important of our objectives and the other three (CPU, RAM, reliability) carry lesser but equal weight when compared with cost. Two primary issues in multiobjective decision making are to acquire meaningful information regarding the satisfaction of the objectives by the various choices (alternatives) and to rank or weight the relative importance of each of the objectives. The approach illustrated in this section defines a decision calculus that requires only *ordinal* information on the ranking of preferences and importance weights [Yager, 1981].

The typical multiobjective decision problem involves the selection of one alternative, a_i, from a universe of alternatives A given a collection, or set, say {O}, of criteria or objectives that are important to the decision maker. We want to evaluate how well each alternative, or choice, satisfies each objective, and we wish to combine the weighted objectives into an overall decision function in some plausible way. This decision function essentially represents a mapping of the alternatives in A to an ordinal set of ranks. This process naturally requires subjective information from the decision authority concerning the importance of each objective. Ordinal orderings of these importances are usually the easiest to obtain. Numerical values, ratios, and intervals expressing the importance of each objective are difficult to extract and, if attempted and then subsequently altered, can often lead to results inconsistent with the intuition of the decision maker.

To develop this calculus we require some definitions. Define a universe of n alternatives, $A = \{a_1, a_2, \ldots, a_n\}$ and a set of r objectives, $O = \{O_1, O_2, \ldots, O_r\}$.

Let O_i indicate the ith objective. Then the degree of membership of alternative a in O_i, denoted $\mu_{O_i}(a)$ is the degree to which alternative a satisfies the criteria specified for this objective. We seek a decision function that simultaneously satisfies all of the decision objectives; hence, the decision function, D, is given by the intersection of all the objective sets,

$$D = O_1 \cap O_2 \cap \cdots \cap O_r \tag{10.17}$$

Therefore, the grade of membership that the decision function, D, has for each alternative a is given by

$$\mu_D(a) = \min[\mu_{O_1}(a), \mu_{O_2}(a), \ldots, \mu_{O_r}(a)] \tag{10.18}$$

The optimum decision, a^*, will then be the alternative that satisfies

$$\mu_D(a^*) = \max_{a \in A}(\mu_D(a)) \tag{10.19}$$

We now define a set of preferences, {P}, which we will constrain to being linear and ordinal. Elements of this preference set can be linguistic values such as none, low, medium, high, absolute, or perfect; or they could be values on the interval [0, 1]; or they could be values on any other linearly ordered scale, e.g., $[-1, 1]$, $[1, 10]$, etc. These preferences will be attached to each of the objectives to quantify the decision maker's feelings about the influence that each objective should have on the chosen alternative. Let the parameter, b_i, be contained on the set of preferences, {P}, where $i = 1, 2, \ldots, r$. Hence, we have for each objective a measure of how important it is to the decision maker for a given decision.

The decision function, D, now takes on a more general form when each objective is associated with a weight expressing its importance to the decision maker. This function is represented as the intersection of r-tuples, denoted as a decision measure, $M(O_i, b_i)$, involving objectives and preferences,

$$D = M(O_1, b_1) \cap M(O_2, b_2) \cap \cdots \cap M(O_r, b_r) \tag{10.20}$$

A key question is what operation should relate each objective, O_i, and its importance, b_i, that preserves the linear ordering required of the preference set, and at the same time relates the two quantities in a logical way where negation is also accommodated. It turns out that the classical implication operator satisfies all of these requirements. Hence, the decision measure for a particular alternative, a, can be replaced with a classical implication of the form,

$$M(O_i(a), b_i) = b_i \rightarrow O_i(a) = \overline{b_i} \vee O_i(a) \tag{10.21}$$

Justification of the implication as an appropriate measure can be developed using an intuitive argument [Yager, 1981]. The statement "b_i implies O_i" indicates a unique relationship between a preference and its associated objective. Whereas various objectives can have the same preference weighting in a cardinal sense, they will be unique in an ordinal sense even though the equality situation $b_i = b_j$ for $i \neq j$ can exist for some objectives. Ordering will be preserved because $b_i \geq b_j$ will contain the equality case as a subset. Therefore, a reasonable decision model will be the joint intersection of r decision measures,

$$D = \bigcap_{i=1}^{r} (\overline{b_i} \cup O_i) \tag{10.22}$$

and the optimum solution, a^*, is the alternative that maximizes D. If we define

$$C_i = \overline{b_i} \cup O_i \qquad \text{hence} \qquad \mu_{C_i}(a) = \max[\mu_{\overline{b_i}}(a), \mu_{O_i}(a)] \tag{10.23}$$

then the optimum solution, expressed in membership form, is given by

$$\mu_D(a^*) = \max_{a \in A}[\min\{\mu_{C_1}(a), \mu_{C_2}(a), \ldots, \mu_{C_r}(a)\}] \tag{10.24}$$

This model is intuitive in the following manner. As the ith objective becomes more important in the final decision, b_i increases, causing $\overline{b_i}$ to decrease, which in turn causes $C_i(a)$ to decrease, thereby increasing the likelihood that $C_i(a) = O_i(a)$, where now $O_i(a)$ will be the value of the decision function, D, representing alternative a [see Eq. (10.22)]. As we repeat this process for other alternatives, a, Eq. (10.24) reveals that the largest value $O_i(a)$ for other alternatives will eventually result in the choice of the optimum solution, a^*. This is exactly how we would want the process to work.

Yager [1981] gives a good explanation of the value of this approach. For a particular objective, the negation of its importance (preference) acts as a barrier such that all ratings of alternatives below that barrier become equal to the value of that barrier. Here, we disregard all distinctions less than the barrier while keeping distinctions above this barrier. This process is similar to the grading practice of academics who lump all students whose class averages fall below 60 percent into the F category while keeping distinctions of A, B, C, and D for students above this percentile. However, in the decision model developed here this barrier varies, depending upon the preference (importance) of the objective to the decision maker. The more important is the objective, the lower is the barrier, and thus the more levels of distinction there are. As an objective becomes less important the distinction barrier increases, which lessens the penalty to the objective. In the limit, if the objective becomes totally unimportant, then the barrier is raised to its highest level and all alternatives are given the same weight and no distinction is made. Conversely, if the objective becomes the most important, all distinctions remain. In sum, the more important an objective is in the decision process, the more significant its effect on the decision function, D.

A special procedure [Yager, 1981] should be followed in the event of a numerical tie between two or more alternatives. If two alternatives, x and y, are tied, their respective decision values are equal, i.e., $D(x) = D(y) = \max_{a \in A}[D(a)]$, where $a = x = y$. Since $D(a) = \min_i[C_i(a)]$ there exists some alternative k such that $C_k(x) = D(x)$ and some alternative g such that $C_g(y) = D(y)$. Let

$$\hat{D}(x) = \min_{i \neq k}[C_i(x)] \qquad \text{and} \qquad \hat{D}(y) = \min_{i \neq g}[C_i(y)] \tag{10.25}$$

Then, we compare $\hat{\mathrm{D}}(x)$ and $\hat{\mathrm{D}}(y)$ and if, for example, $\hat{\mathrm{D}}(x) > \hat{\mathrm{D}}(y)$ we select x as our optimum alternative. However, if a tie still persists, i.e., if $\hat{\mathrm{D}}(x) = \hat{\mathrm{D}}(y)$, then there exist some other alternatives j and h such that $\hat{\mathrm{D}}(x) = \mathrm{C}_j(x) = \hat{\mathrm{D}}(y) = \mathrm{C}_h(y)$. Then we formulate

$$\hat{\hat{\mathrm{D}}}(x) = \min_{i \neq k, j}[\mathrm{C}_i(x)] \qquad \text{and} \qquad \hat{\hat{\mathrm{D}}}(y) = \min_{i \neq g, h}[\mathrm{C}_i(y)] \tag{10.26}$$

and compare $\hat{\hat{\mathrm{D}}}(x)$ and $\hat{\hat{\mathrm{D}}}(y)$. The tie-breaking procedure continues in this manner until an unambiguous optimum alternative emerges or all of the alternatives have been exhausted. In the latter case where a tie still results some other tie-breaking procedure, such as a refinement in the preference scales, can be used.

Example 10.5. A geotechnical engineer on a construction project must prevent a large mass of soil from sliding into a building site during construction and must retain this mass of soil indefinitely after construction to maintain stability of the area around a new facility to be constructed on the site [Adams, 1994]. She therefore must decide which type of retaining wall design to select for her project. Among the many alternative designs available, the engineer reduces the list of candidate retaining wall designs to three: (*i*) a mechanically stabilized embankment (MSE) wall, (*ii*) a mass concrete spread wall (Conc), and (*iii*) a gabion (Gab) wall. The owner of the facility (the decision maker) has defined four objectives that impact his decision: (*i*) the cost of the wall (Cost), (*ii*) the maintainability (Main) of the wall, (*iii*) whether the design is a standard one (SD), and (*iv*) the environmental (Env) impact of the wall. Moreover, the owner also decides to rank his preferences for these objectives on the unit interval. Hence, the engineer sets up her problem as follows:

$$\mathrm{A} = \{\mathrm{MSE}, \mathrm{Conc}, \mathrm{Gab}\} = \{a_1, a_2, a_3\}$$

$$\mathrm{O} = \{\mathrm{Cost}, \mathrm{Main}, \mathrm{SD}, \mathrm{Env}\} = \{\mathrm{O}_1, \mathrm{O}_2, \mathrm{O}_3, \mathrm{O}_4\}$$

$$\mathrm{P} = \{b_1, b_2, b_3, b_4\} \to [0, 1]$$

From previous experience with various wall designs, the engineer first rates the retaining walls with respect to the objectives, given here. These ratings are fuzzy sets expressed in Zadeh's notation.

$$\underset{\sim}{\mathrm{O}}_1 = \left\{ \frac{0.4}{\mathrm{MSE}} + \frac{1}{\mathrm{Conc}} + \frac{0.1}{\mathrm{Gab}} \right\}$$

$$\underset{\sim}{\mathrm{O}}_2 = \left\{ \frac{0.7}{\mathrm{MSE}} + \frac{0.8}{\mathrm{Conc}} + \frac{0.4}{\mathrm{Gab}} \right\}$$

$$\underset{\sim}{\mathrm{O}}_3 = \left\{ \frac{0.2}{\mathrm{MSE}} + \frac{0.4}{\mathrm{Conc}} + \frac{1}{\mathrm{Gab}} \right\}$$

$$\underset{\sim}{\mathrm{O}}_4 = \left\{ \frac{1}{\mathrm{MSE}} + \frac{0.5}{\mathrm{Conc}} + \frac{0.5}{\mathrm{Gab}} \right\}$$

These membership functions for each of the alternatives are shown graphically in Fig. 10.4.

The engineer wishes to investigate two decision scenarios. Each scenario propagates a different set of preferences from the owner, who wishes to determine the sensitivity of the optimum solutions to his preference ratings. In the first scenario, the owner

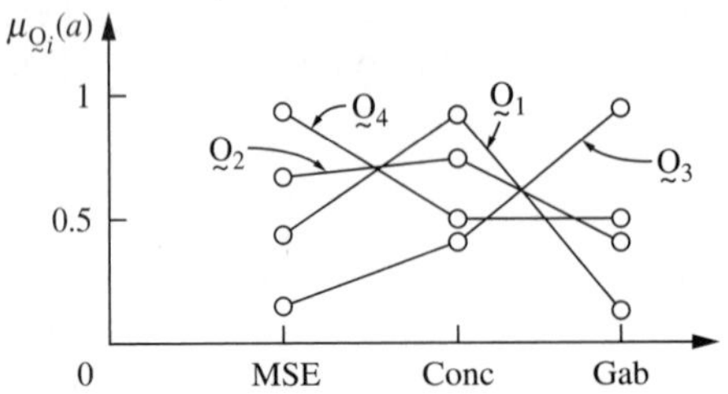

FIGURE 10.4
Membership for each alternative with respect to the objectives.

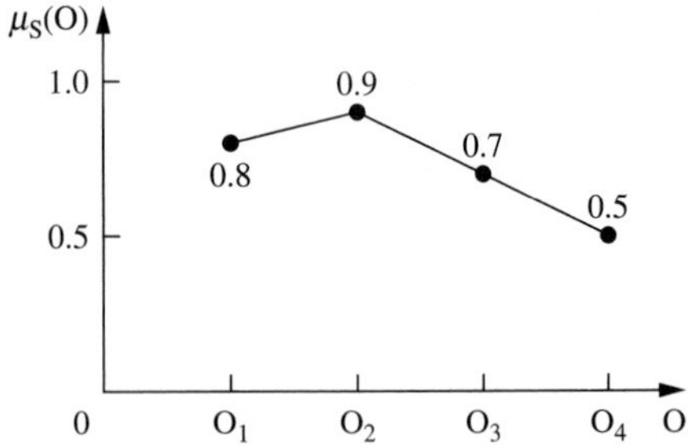

FIGURE 10.5
Preferences in Scenario I.

lists his preferences for each of the four objectives, as shown in Fig. 10.5. From these preference values, the following calculations result:

$$b_1 = 0.8 \qquad b_2 = 0.9 \qquad b_3 = 0.7 \qquad b_4 = 0.5$$
$$\bar{b}_1 = 0.2 \qquad \bar{b}_2 = 0.1 \qquad \bar{b}_3 = 0.3 \qquad \bar{b}_4 = 0.5$$

$$\begin{aligned}
\mathrm{D}(a_1) = \mathrm{D(MSE)} &= (\bar{b}_1 \cup \mathrm{O}_1) \cap (\bar{b}_2 \cup \mathrm{O}_2) \cap (\bar{b}_3 \cup \mathrm{O}_3) \cap (\bar{b}_4 \cup \mathrm{O}_4) \\
&= (0.2 \vee 0.4) \wedge (0.1 \vee 0.7) \wedge (0.3 \vee 0.2) \wedge (0.5 \vee 1) \\
&= 0.4 \wedge 0.7 \wedge 0.3 \wedge 1 = 0.3 \\
\mathrm{D}(a_2) = \mathrm{D(Conc)} &= (0.2 \vee 1) \wedge (0.1 \vee 0.8) \wedge (0.3 \vee 0.4) \wedge (0.5 \vee 0.5) \\
&= 1 \wedge 0.8 \wedge 0.4 \wedge 0.5 = 0.4 \\
\mathrm{D}(a_3) = \mathrm{D(Gab)} &= (0.2 \vee 0.1) \wedge (0.1 \vee 0.4) \wedge (0.3 \vee 1) \wedge (0.5 \vee 0.5) \\
&= 0.2 \wedge 0.4 \wedge 1 \wedge 0.5 = 0.2 \\
\mathrm{D}^* &= \max\{\mathrm{D}(a_1), \mathrm{D}(a_2), \mathrm{D}(a_3)\} = \max\{0.3, 0.4, 0.2\} = 0.4
\end{aligned}$$

Thus, the engineer chose the second alternative, a_2, a concrete (Conc) wall as her retaining design under preference scenario 1.

Now, in the second scenario the engineer was given a different set of preferences by the owner, as shown in Fig. 10.6. From the preference values in Fig. 10.6, the following calculations result:

$$b_1 = 0.5 \qquad b_2 = 0.7 \qquad b_3 = 0.8 \qquad b_4 = 0.7$$
$$\bar{b}_1 = 0.5 \qquad \bar{b}_2 = 0.3 \qquad \bar{b}_3 = 0.2 \qquad \bar{b}_4 = 0.3$$

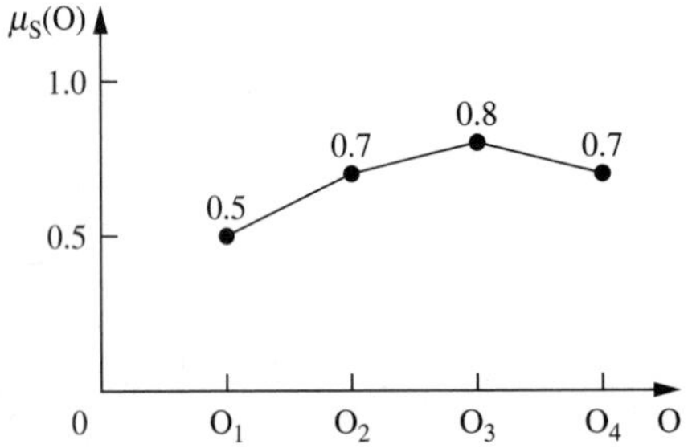

FIGURE 10.6
Preferences in Scenario II.

$$
\begin{aligned}
D(a_1) = D(\text{MSE}) &= (\bar{b}_1 \cup O_1) \cap (\bar{b}_2 \cup O_2) \cap (\bar{b}_3 \cup O_3) \cap (\bar{b}_4 \cup O_4) \\
&= (0.5 \vee 0.4) \wedge (0.3 \vee 0.7) \wedge (0.2 \vee 0.2) \wedge (0.3 \vee 1) \\
&= 0.5 \wedge 0.7 \wedge 0.2 \wedge 1 = 0.2 \\
D(a_2) = D(\text{Conc}) &= (0.5 \vee 1) \wedge (0.3 \vee 0.8) \wedge (0.2 \vee 0.4) \wedge (0.3 \vee 0.5) \\
&= 1 \wedge 0.8 \wedge 0.4 \wedge 0.5 = 0.4 \\
D(a_3) = D(\text{Gab}) &= (0.5 \vee 0.1) \wedge (0.3 \vee 0.4) \wedge (0.2 \vee 1) \wedge (0.3 \vee 0.5) \\
&= 0.5 \wedge 0.4 \wedge 1 \wedge 0.5 = 0.4
\end{aligned}
$$

Therefore, $D^* = \max\{D(a1), D(a2), D(a3)\} = \max\{0.2, 0.4, 0.4\} = 0.4$. But there is a tie between alternative a_2 and a_3. To resolve this tie, the engineer implements Eq. (10.25). She looks closely at $D(a_2)$ and $D(a_3)$ and notes that the decision value of 0.4 for $D(a_2)$ came from the third term [i.e., $C_3(a_2)$; hence $k = 3$ in Eq. (10.25)], and that the decision value of 0.4 for $D(a_3)$ came from the second term [i.e., $C_2(a_3)$; hence $g = 2$ in Eq. (10.25)]. Then the calculations proceed again between the tied choices a_2 and a_3:

$$
\begin{aligned}
\hat{D}(a_2) = \hat{D}(\text{Conc}) &= (0.5 \vee 1) \wedge (0.3 \vee 0.8) \wedge (0.3 \vee 0.5) \\
&= 1 \wedge 0.8 \wedge 0.5 = 0.5 \\
\hat{D}(a_3) = \hat{D}(\text{Gab}) &= (0.5 \vee 0.1) \wedge (0.2 \vee 1) \wedge (0.3 \vee 0.5) \\
&= 0.5 \wedge 1 \wedge 0.5 = 0.5
\end{aligned}
$$

Then $D^* = \max\{\hat{D}(a_2), \hat{D}(a_3)\} = \max\{0.5, 0.5\} = 0.5$, and there is still a tie between alternative a_2 and a_3. To resolve this second tie, the engineer implements Eq. (10.26). She looks closely at $\hat{D}(a_2)$ and $\hat{D}(a_3)$ and notes that the decision value of 0.5 for $\hat{D}(a_2)$ came from the third term [i.e., $C_3(a_2)$; hence $j = 3$ in Eq. (10.26)], and that the decision value of 0.5 for $\hat{D}(a_3)$ came from the first term *and* the second term [i.e., $C_1(a_3) = C_3(a_3)$; hence $h = 1$ and $h = 3$ in Eq. (10.26)]. Then the calculations proceed again between the tied choices a_2 and a_3:

$$
\begin{aligned}
\hat{\hat{D}}(a_2) &= \hat{\hat{D}}(\text{Conc}) = (0.5 \vee 1) \wedge (0.3 \vee 0.8) = 0.8 \\
\hat{\hat{D}}(a_3) &= \hat{\hat{D}}(\text{Gab}) = (0.2 \vee 1) = 1
\end{aligned}
$$

From these results, $D^* = \max\{\hat{D}(a_2), \hat{D}(a_3)\} = 1$, hence the tie is finally broken and the engineer chooses retaining wall a_3, a Gabion wall, for her design under preference scenario 2.

FUZZY BAYESIAN DECISION METHOD

Classical statistical decision making involves the notion that the uncertainty in the future can be characterized probabilistically, as discussed in the introduction to this chapter. When we want to make a decision among various alternatives, our choice is predicated on information about the future, which is normally discretized into various "states of nature." If we knew with certainty the future states of nature, we would not need an analytic method to assess the likelihood of a given outcome. Unfortunately we do not know what the future will entail so we have devised methods to make the best choices given an uncertain environment. Classical Bayesian decision methods presume that future states of nature can be characterized as probability events. For example, consider the condition of "cloudiness" in tomorrow's weather by discretizing the state space into three levels and assessing each level probabilistically: The chance of a very cloudy day is 0.5, a partly cloudy day is 0.2, and a sunny (no clouds) day is 0.3. By convention the probabilities sum to unity. The problem with the Bayesian scheme here is that the events are vague and ambiguous. How many clouds does it take to transition between very cloudy and cloudy? If there is one small cloud in the sky, does this mean it is not sunny? This is the classic sorites paradox discussed in Chapter 7.

The following material first presents Bayesian decision making and then starts to consider ambiguity in the value of new information, in the states of nature, and in the alternatives in the decision process [see Terano et al., 1992]. Examples will illustrate these points.

First we shall consider the formation of probabilistic decision analysis. Let $S = \{s_1, s_2, \ldots, s_n\}$ be a set of possible states of nature; and the probabilities that these states will occur are listed in a vector,

$$\mathbf{P} = \{p(s_1), p(s_2), \ldots, p(s_n)\} \qquad \text{where} \qquad \sum_{i=1}^{n} p(s_i) = 1 \tag{10.27}$$

The probabilities expressed in Eq. (10.27) are called "prior probabilities" in Bayesian jargon because they express prior knowledge about the true states of nature. Assume that the decision maker can choose among m alternatives, $A = \{a_1, a_2, \ldots, a_m\}$, and for a given alternative a_j we assign a utility value, u_{ji}, if the future state of nature turns out to be state s_i. These utility values should be determined by the decision maker since they express value, or cost, for each alternative-state pair, i.e., for each a_j–s_i combination. The utility values are usually arranged in a matrix of the form shown in Table 10.1. The expected utility associated with the jth alternative would be

$$E(u_j) = \sum_{i=1}^{n} u_{ji} p(s_i) \tag{10.28}$$

The most common decision criterion is the *maximum* expected utility among all the alternatives, i.e.,

$$E(u^*) = \max_j E(u_j) \tag{10.29}$$

which leads to the selection of alternative a_k if $u^* = E(u_k)$

Example 10.6. Suppose you are a geological engineer who has been asked by the chief executive officer (CEO) of a large oil firm to help make a decision about whether to

TABLE 10.1
Utility matrix

States s_i Action a_j	s_1	s_2	$\ldots$	s_n
a_1	u_{11}	u_{12}	$\ldots$	u_{1n}
$\vdots$	$\vdots$	$\vdots$		$\vdots$
a_m	u_{m1}	u_{m2}	$\ldots$	u_{mn}

drill for natural gas in a particular geographic region of northwestern New Mexico. You determine for your first attempt at the decision process that there are only two states of nature regarding the existence of natural gas in the region:

$$s_1 = \text{there is natural gas}$$
$$s_2 = \text{there is no natural gas}$$

and you are able to find from previous drilling information that the prior probabilities for each of these states is

$$p(s_1) = 0.5$$
$$p(s_2) = 0.5$$

Note these probabilities sum to unity. You suggest that there are two alternatives in this decision:

$$a_1 = \text{drill for gas}$$
$$a_2 = \overline{a}_1 = \text{do not drill for gas}$$

The decision maker (CEO) helps you assemble a utility matrix to get the process started. He tells you that the best situation for him is to decide to drill for gas, and subsequently find that gas, indeed, was in the geologic formation. He assesses this value (u_{11}) as +5 in nondimensional units; in this case he would have gambled (drilling costs big money) and won. Moreover, he feels that the worst possible situation would be to drill for gas, and subsequently find that there was no gas in the area. Since this would cost him time and money, he determines that the value for this would be $u_{12} = -10$ units; he would have gambled and lost—big. The other two utilities are assessed by the decision maker in nondimensional units as $u_{21} = -2$ and $u_{22} = 4$. Hence, the utility matrix for this situation is given by

$$\mathrm{U} = \begin{bmatrix} 5 & -10 \\ -2 & 4 \end{bmatrix}$$

Figure 10.7 shows the decision tree for this problem, of which the two initial branches correspond to the two possible alternatives, and the second layer of branches corresponds to the two possible states of nature. Superposed on the tree branches are the

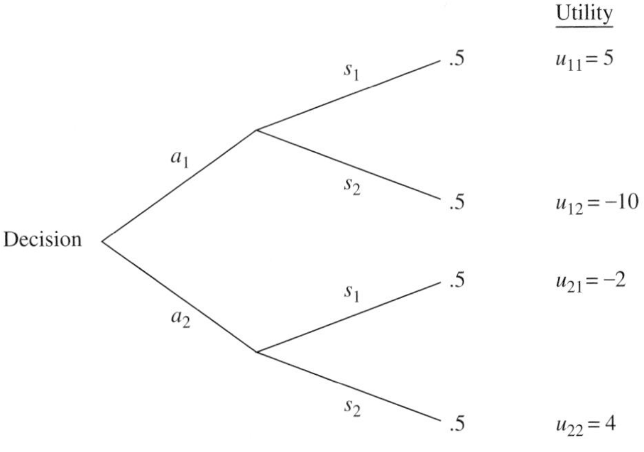

FIGURE 10.7
Decision tree for the two-alternative, two-state problem of Example 10.6.

prior probabilities. The expected utility, in nondimensional units, for each alternative a_1 and a_2 is, from Eq. (10.28),

$$E(u_1) = (0.5)(5) + (0.5)(-10) = -2.5$$
$$E(u_2) = (0.5)(-2) + (0.5)(4) = 1.0$$

and we see that the maximum utility, using Eq. (10.29), is 1.0, which comes from alternative a_2; hence, on the basis of prior information only (prior probabilities) the CEO decides not to drill for natural gas (alternative a_2).

In many decision situations an intermediate issue arises: Should you get more information about the true states of nature prior to deciding? Suppose some new information regarding the true states of nature S is available from r experiments or other observations and is collected in a data vector, $\mathbf{X} = \{x_1, x_2, \ldots, x_r\}$. This information can be used in the Bayesian approach to update the prior probabilities, $p(s_i)$, in the following manner. First, the new information is expressed in the form of conditional probabilities, where the probability of each piece of data, x_k, where $k = 1, 2, \ldots, r$, is assessed according to whether the true state of nature, s_i, is known (not uncertain); these probabilities are presumptions of the future because they are equivalent to the following statement: *Given that we know that the true state of nature is s_i, the probability that the piece of new information x_k confirms that the true state is s_i is $p(x_k \mid s_i)$.* In the literature these conditional probabilities, denoted $p(x_k \mid s_i)$, are also called likelihood values. The likelihood values are then used as weights on the previous information, the prior probabilities $p(s_i)$, to find updated probabilities, known as posterior probabilities, denoted $p(s_i \mid x_k)$. The posterior probabilities are equivalent to this statement: *Given that the piece of new information x_k is true, the probability that the true state of nature is s_i is $p(s_i \mid x_k)$.* These updated probabilities are determined by Bayes rule,

$$p(s_i \mid x_k) = \frac{p(x_k \mid s_i)}{p(x_k)} p(s_i) \tag{10.30}$$

where the term in the denominator of Eq. (10.30), $p(x_k)$, is the marginal probability of the data x_k and is determined using the total probability theorem

$$p(x_k) = \sum_{i=1}^{n} p(x_k \mid s_i) \cdot p(s_i) \tag{10.31}$$

Now the expected utility for the jth alternative, given the data x_k, is determined from the posterior probabilities (instead of the priors),

$$E(u_j \mid x_k) = \sum_{i=1}^{n} u_{ji} p(s_i \mid x_k) \tag{10.32}$$

and the maximum expected utility, given the new data x_k, is now given by

$$E(u^* \mid x_k) = \max_j E(u_j \mid x_k) \tag{10.33}$$

To determine the unconditional maximum expected utility we need to weight each of the r conditional expected utilities given by Eq. (10.33) by the respective marginal probabilities for each datum x_k, i.e., by $p(x_k)$, given in Eq. (10.34) as

$$E(u_x^*) = \sum_{k=1}^{r} E(u^* \mid x_k) \cdot p(x_k) \tag{10.34}$$

We can now introduce a new notion in the decision-making process, called the *value* of information, $V(x)$. In the case we have just introduced where there is some uncertainty about the new information, $X = \{x_1, x_2, \ldots, x_r\}$, we call the information *imperfect* information. The value of this imperfect information, $V(x)$, can be assessed by taking the difference between the maximum expected utility without any new information, Eq. (10.29), and the maximum expected utility with the new information, Eq. (10.34), i.e.,

$$V(x) = E(u_x^*) - E(u^*) \tag{10.35}$$

We now introduce yet another notion in this process, called *perfect* information. This exercise is an attempt to develop a boundary condition for our problem, one that is altruistic in nature, i.e., can never be achieved in reality but nonetheless is quite useful in a mathematical sense to give us some scale on which to assess the value of imperfect information. If information is considered to be perfect (i.e., can predict the future states of nature precisely), we can say that the conditional probabilities are free of dissonance. That is, each new piece of information, or data, predicts one and only one state of nature; hence there is no ambivalence about what state is predicted by the data. However, if there is more than one piece of information the probabilities for a particular state of nature have to be shared by all the data. Mathematically, perfect information is represented by posterior probabilities of 0 or 1, i.e.,

$$p(s_i \mid x_k) = \begin{cases} 1 \\ 0 \end{cases} \tag{10.36}$$

We call this perfect information x_p. For perfect information, the maximum expected utility becomes (see Ex. 10.7)

$$E(u_{x_p}^*) = \sum_{k=1}^{r} E(u_{x_p}^* \mid x_k) p(x_k) \tag{10.37}$$

and the value of perfect information becomes

$$V(x_p) = E(u_{x_p}^*) - E(u^*) \tag{10.38}$$

Example 10.7. We continue with our gas exploration problem, Example 10.6. We had two states of nature—gas, s_1, and no-gas, s_2—and two alternatives—drill, a_1, and no-drill, a_2. The prior probabilities were uniform,

$$p(s_1) = 0.5$$
$$p(s_2) = 0.5$$

TABLE 10.2
Utility matrix for natural gas example

u_{ji}	s_1	s_2
a_1	4	−2
a_2	−1	2

TABLE 10.3
Conditional probabilities for *imperfect* information

	x_1	x_2	x_3	x_4	x_5	x_6	x_7	x_8	
$p(x_k \mid s_1)$	0	.05	.1	.1	.2	.4	.1	.05	$\sum$ row = 1
$p(x_k \mid s_2)$	.05	.1	.4	.2	.1	.1	.05	0	$\sum$ row = 1

TABLE 10.4
Conditional probabilities for *perfect* information

	x_1	x_2	x_3	x_4	x_5	x_6	x_7	x_8	
$p(x_k \mid s_1)$	0	0	0	0	.2	.5	.2	.1	$\sum$ row = 1
$p(x_k \mid s_2)$	.1	.2	.5	.2	0	0	0	0	$\sum$ row = 1

Now, let's suppose the CEO of the natural gas company wants to reconsider his utility values. He provides the utility matrix of Table 10.2 in the same form as Table 10.1. Further, the CEO has asked you to collect new information by taking eight geological boring samples from the region being considered for drilling. You have a natural gas expert examine the results of these eight tests, and she gives you her opinions about the conditional probabilities in the form of a matrix, given in Table 10.3. Moreover, you ask the natural gas expert to give you her assessment about how the conditional probabilities might change if they were perfect tests capable of providing perfect information. She gives you the matrix shown in Table 10.4.

As the engineer assisting the CEO, you now conduct a decision analysis. Since the CEO changed his utility values, you have to recalculate the expected utility of making the decision on the basis of just the prior probabilities, before any new information is acquired. The decision tree for this situation is shown in Figure 10.8.

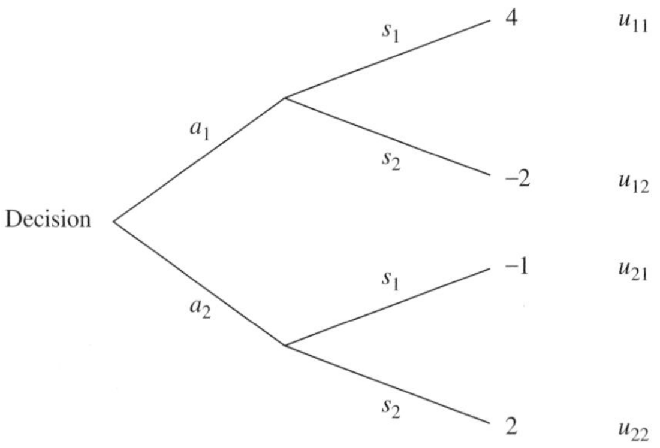

FIGURE 10.8
Decision tree showing utility values.

TABLE 10.5
Posterior probabilities based on imperfect information

	x_1	x_2	x_3	x_4	x_5	x_6	x_7	x_8
$p(s_1 \mid x_k)$	0	$\frac{1}{3}$	$\frac{1}{5}$	$\frac{1}{3}$	$\frac{2}{3}$	$\frac{4}{5}$	$\frac{2}{3}$	1
$p(s_2 \mid x_k)$	1	$\frac{2}{3}$	$\frac{4}{5}$	$\frac{2}{3}$	$\frac{1}{3}$	$\frac{1}{5}$	$\frac{1}{3}$	0
$p(x_k)$	.025	.075	.25	.15	.15	.25	.075	.025
$E(u^* \mid x_k)$	2	1	$\frac{7}{5}$	1	2	$\frac{14}{5}$	2	4
$a_j \mid x_k$	a_2	a_2	a_2	a_2	a_1	a_1	a_1	a_1

The expected utilities and maximum expected utility, based just on prior probabilities, are

$$E(a_1) = (4)(0.5) + (-2)(0.5) = 1.0$$

$$E(a_2) = (-1)(0.5) + (2)(0.5) = 0.5$$

$$E(u^*) = 1 \qquad \text{hence, you choose alternative } a_1\text{, drill for natural gas.}$$

You are now ready to assess the changes in this decision process by considering additional information, both imperfect and perfect. Table 10.5 summarizes your calculations for the new prior probabilities, $p(s_1 \mid x_k)$ and $p(s_2 \mid x_k)$, the marginal probabilities for the new information, $p(x_k)$, the expected conditional utilities, $E(u^* \mid x_k)$, and the expected alternatives, $a_j \mid x_k$.

Typical calculations for the values in Table 10.5 are provided here. For the marginal probabilities for the new *imperfect* information, use Eq (10.31), the conditional probabilites from Table 10.3 and the prior probabilities:

$$p(x_1) = (0)(0.5) + (0.05)(0.5) = 0.025$$

$$p(x_4) = (0.1)(0.5)(0.2)(0.5) = 0.15$$

$$\ldots$$

The posterior probabilities are calculated with the use of Eq. (10.30), using the conditional probabilites from Table 10.3, the prior probabilities, and the marginal probabilities, $p(x_k)$ just determined and summarized in Table 10.5 (fourth row); for example,

$$p(s_1 \mid x_2) = \frac{0.05(0.5)}{0.075} = \frac{1}{3} \qquad p(s_2 \mid x_2) = \frac{0.1(0.5)}{0.075} = \frac{2}{3} \quad \ldots$$

$$p(s_1 \mid x_6) = \frac{0.4(0.5)}{0.25} = \frac{4}{5} \qquad p(s_2 \mid x_6) = \frac{0.1(0.5)}{0.25} = \frac{1}{5} \quad \ldots$$

The conditional expected utilities, $E(u^* \mid x_k)$, are calculated using first Eq. (10.32), then Eq. (10.33); for example,

$$E(u_1 \mid x_3) = (\tfrac{1}{5})(4) + (\tfrac{4}{5})(-2) = -\tfrac{4}{5} \qquad E(u_2 \mid x_3) = (\tfrac{1}{5})(-1) + (\tfrac{4}{5})(2) = \tfrac{7}{5}$$

Hence $E(u^* \mid x_3) = \max(-\frac{4}{5}, \frac{7}{5}) = \frac{7}{5}$ (choose alternative a_2).

$$E(u_1 \mid x_8) = (1)(4) + (0)(-2) = 4 \qquad E(u_2 \mid x_8) = (1)(-1) + (0)(2) = -1$$

Hence $E(u^* \mid x_8) = \max(4, -1) = 4$ (choose alternative a_1).

Now use Eq. (10.34) to calculate the overall unconditional expected utility for *imperfect* information, which is actually the sum of pairwise products of the values in rows 3 and 4 of Table 10.5, e.g.,

$$E(u_x^*) = (0.025)(2) + (0.075)(1) + \cdots + (0.025)(4) = 1.875$$

and the value of the new *imperfect* information, using Eq. (10.35), is

$$V(x) = E(u_x^*) - E(u^*) = 1.875 - 1 = 0.875$$

To decide what alternative to choose, notice in Table 10.5 that the total utility favoring a_1 is 10.8 $(2 + \frac{14}{5} + 2 + 4)$ and the total utility favoring a_2 is 5.4 $(2 + 1 + \frac{7}{5} + 1)$. Hence, the CEO chooses alternative a_1, to drill for gas. In effect, the new information has not changed his mind about drilling.

You begin the process of assessing the changes due to the consideration of the hypothetical *perfect* information. Table 10.6 summarizes your calculations for the new prior probabilities, $p(s_1 \mid x_k)$ and $p(s_2 \mid x_k)$, the marginal probabilities for the perfect information, $p(x_k)$, the expected conditional utilities, $E(u_{x_p}^* \mid x_k)$, and the expected alternatives, $a_j \mid x_k$. These are calculated in the same way as those in Table 10.5, except you make use of the *perfect* conditional probabilities of Table 10.4.

Equation (10.37) is used to calculate the overall unconditional expected utility for *perfect* information, which is actually the sum of pairwise products of the values in rows 3 and 4 of Table 10.6; e.g.,

$$E(u_{x_p}^*) = (0.05)(2) + (0.1)(2) + \cdots + (0.05)(4) = 3.0$$

and the value of the new *perfect* information, using Eq. (10.38), is

$$V(x_p) = E(u_{x_p}^*) - E(u^*) = 3 - 1 = 2.0$$

Alternative a_1 is still the choice here. We note that the hypothetical information has a value of 2 and the imperfect information has a value of less than half of this, 0.875. This difference can be used to assess the value of the imperfect information compared to both no information (1) and perfect information (3).

We now discuss the fact that the new information might be inherently fuzzy [Okuda et al., 1974, 1978]. Suppose the new information, $X = \{x_1, x_2, \ldots, x_r\}$, is a universe of discourse in the units appropriate for the new information. Then we can define fuzzy events, $\underset{\sim}{M}$, on this information, such as "good" information, "mod-

TABLE 10.6
Posterior probabilities based on perfect information

	x_1	x_2	x_3	x_4	x_5	x_6	x_7	x_8
$p(s_1 \mid x_k)$	0	0	0	0	1	1	1	1
$p(s_2 \mid x_k)$	1	1	1	1	0	0	0	0
$p(x_k)$	.05	.1	.25	.1	.1	.25	.1	.05
$E(u^* \mid x_k)$	2	2	2	2	4	4	4	4
$a_j \mid x_k$	a_2	a_2	a_2	a_2	a_1	a_1	a_1	a_1

erate" information, and "poor" information. The fuzzy event will have membership function, $\mu_{\underset{\sim}{M}}(x_k)$, $k = 1, 2, \ldots, r$. We can now define the idea of a "probability of a fuzzy event," i.e., the probability of $\underset{\sim}{M}$, as

$$P(\underset{\sim}{M}) = \sum_{k=1}^{r} \mu_{\underset{\sim}{M}}(x_k) p(x_k) \qquad (10.39)$$

We note in Eq. (10.39) that if the fuzzy event is, in fact, crisp, i.e., $\underset{\sim}{M} = M$, then the probability reduces to

$$P(M) = \sum_{x_k \in M} p(x_k) \qquad \mu_M = \begin{cases} 1 & x_k \in M \\ 0 & \text{otherwise} \end{cases} \qquad (10.40)$$

where Eq. (10.40) describes the probability of a crisp event simply as the sum of the marginal probabilities of those data points, x_k, that are defined to be in the event, M. Based on this, the posterior probability of s_i, given fuzzy information $\underset{\sim}{M}$ is,

$$P(s_i \mid \underset{\sim}{M}) = \frac{\sum_{k=1}^{r} p(x_k \mid s_i) \mu_{\underset{\sim}{M}}(x_k) p(s_i)}{p(\underset{\sim}{M})} = \frac{p(\underset{\sim}{M} \mid s_i) p(s_i)}{p(\underset{\sim}{M})} \qquad (10.41)$$

where

$$p(\underset{\sim}{M} \mid s_i) = \sum_{k=1}^{r} p(x_k \mid s_i) \mu_{\underset{\sim}{M}}(x_k) \qquad (10.42)$$

We now define the collection of all the fuzzy events describing fuzzy information as an *orthogonal* fuzzy information system, $\Phi = \{\underset{\sim}{M}_1, \underset{\sim}{M}_2, \ldots, \underset{\sim}{M}_g\}$ where by *orthogonal* we mean that the sum of the membership values for each fuzzy event, $\underset{\sim}{M}_i$, for every data point in the information universe, x_k, equals unity [Tanaka et al., 1976]. That is,

$$\sum_{t=1}^{g} \mu_{\underset{\sim}{M}_t}(x_k) = 1 \qquad \text{for all } x_k \in X \qquad (10.43)$$

If the fuzzy events on the new information universe are *orthogonal,* we can extend the Bayesian approach to consider fuzzy information. The fuzzy equivalents of Eqs. (10.32), (10.33), and (10.34) become, for a fuzzy event, $\underset{\sim}{M}_t$,

$$E(u_j \mid \underset{\sim}{M}_t) = \sum_{i=1}^{n} u_{ji} \cdot p(s_i \mid \underset{\sim}{M}_t) \qquad (10.44)$$

$$E(u^* \mid \underset{\sim}{M}_t) = \max_j E(u_j \mid \underset{\sim}{M}_t) \qquad (10.45)$$

$$E(u^*_\Phi) = \sum_{t=1}^{g} E(u^* \mid \underset{\sim}{M}_t) \cdot p(\underset{\sim}{M}_t) \qquad (10.46)$$

Now the value of fuzzy information can be determined in an analogous manner as

$$V(\Phi) = E(u^*_\Phi) - E(u^*) \qquad (10.47)$$

TABLE 10.7
Orthogonal membership functions for orthogonal fuzzy events

	x_1	x_2	x_3	x_4	x_5	x_6	x_7	x_8
$\mu_{\underset{\sim}{M}_1}(x_k)$	1	1	.5	0	0	0	0	0
$\mu_{\underset{\sim}{M}_2}(x_k)$	0	0	.5	1	1	.5	0	0
$\mu_{\underset{\sim}{M}_3}(x_k)$	0	0	0	0	0	.5	1	1
$p(x_k)$	.025	.075	.25	.15	.15	.25	.075	.025

Example 10.8 [Continuation of Example 10.7]. We now consider that the test samples you acquired are inherently fuzzy, and define an orthogonal fuzzy information system, Φ,

$$\Phi = \{\underset{\sim}{M}_1, \underset{\sim}{M}_2, \underset{\sim}{M}_3\} = \{\text{poor data, medium data, good data}\}$$

with membership functions in Table 10.7. The fourth row of Table 10.7 repeats the marginal probabilities for each data, x_k, from Table 10.5. As can be seen in Table 10.7 the sum of the membership values in each column (the first three rows) equals unity, as required for *orthogonal* fuzzy sets.

As before, we use Eq. (10.39) to determine the marginal probabilities for each fuzzy event,

$$p(\underset{\sim}{M}_1) = 0.225 \qquad p(\underset{\sim}{M}_2) = 0.55 \qquad p(\underset{\sim}{M}_3) = 0.225$$

and Eq. (10.42) to determine the fuzzy conditional probabilities,

$$p(\underset{\sim}{M}_1 \mid s_1) = 0.1 \qquad p(\underset{\sim}{M}_2 \mid s_1) = 0.55 \qquad p(\underset{\sim}{M}_3 \mid s_1) = 0.35$$
$$p(\underset{\sim}{M}_1 \mid s_2) = 0.35 \qquad p(\underset{\sim}{M}_2 \mid s_2) = 0.55 \qquad p(\underset{\sim}{M}_3 \mid s_2) = 0.1$$

and Eq. (10.41) to determine the fuzzy posterior probabilities,

$$p(s_1 \mid \underset{\sim}{M}_1) = 0.222 \qquad p(s_1 \mid \underset{\sim}{M}_2) = 0.5 \qquad p(s_1 \mid \underset{\sim}{M}_3) = 0.778$$
$$p(s_2 \mid \underset{\sim}{M}_1) = 0.778 \qquad p(s_2 \mid \underset{\sim}{M}_2) = 0.5 \qquad p(s_2 \mid \underset{\sim}{M}_3) = 0.222$$

Now the conditional fuzzy expected utilities can be determined using Eq. (10.44),

$$\underset{\sim}{M}_1: \quad E(u_1 \mid \underset{\sim}{M}_1) = (4)(0.222) + (-2)(0.778) = -0.668$$
$$E(u_2 \mid \underset{\sim}{M}_1) = (-1)(0.222) + (2)(0.778) = 1.334$$

$$\underset{\sim}{M}_2: \quad E(u_1 \mid \underset{\sim}{M}_2) = (4)(0.5) + (-2)(0.5) = 1.0$$
$$E(u_2 \mid \underset{\sim}{M}_2) = (-1)(0.5) + (2)(0.5) = 0.5$$

$$\underset{\sim}{M}_3: \quad E(u_1 \mid \underset{\sim}{M}_3) = (4)(0.778) + (-2)(0.222) = 2.668$$
$$E(u_2 \mid \underset{\sim}{M}_3) = (-1)(0.778) + (2)(0.222) = -0.334$$

and the maximum expected utility from Eq. (10.46), using each of the foregoing three maximum conditional probabilities,

$$E(u_{\Phi}^*) = (0.225)(1.334) + (0.55)(1) + (0.225)(2.668) = 1.45$$

and the value of the fuzzy information from Eq. (10.47),

$$V(\Phi) = 1.45 - 1 = 0.45$$

Here we see that the value of the fuzzy information is less than the value of the perfect information (2.0), and less than the value of the imperfect information (0.875). However, it may turn out that fuzzy information is *far less costly* (remember, precision costs) than either the imperfect or perfect (hypothetical) information. Although not developed in this text, this analysis could be extended to consider *cost* of information.

DECISION MAKING UNDER FUZZY STATES AND FUZZY ACTIONS

The Bayesian method can be further extended to include the possibility that the states of nature are fuzzy and the decision makers' alternatives are also fuzzy [Tanaka et al., 1976]. For example, suppose your company wants to expand and you are considering three fuzzy alternatives in terms of the size of a new facility:

$$\underset{\sim}{A}_1 = \text{small-scale project}$$
$$\underset{\sim}{A}_2 = \text{middle-scale project}$$
$$\underset{\sim}{A}_3 = \text{large-scale project}$$

Just as all fuzzy sets are defined on some universe of discourse, continuous or discrete, the fuzzy alternatives (actions) would also be defined on a universe of discourse, say values of square footage of floor space on a continuous scale of areas in some appropriate units of area. Moreover, suppose further that the economic climate in the future is very fuzzy and you pose the following three possible fuzzy states of nature ($\underset{\sim}{F}_s$, $s = 1, 2, 3$):

$$\underset{\sim}{F}_1 = \text{low rate of economic growth}$$
$$\underset{\sim}{F}_2 = \text{medium rate of economic growth}$$
$$\underset{\sim}{F}_3 = \text{high rate of economic growth}$$

all of which are defined on a universe of numerical rates of economic growth, say S, where $S = \{s_1, s_2, \ldots, s_n\}$ is a discrete universe of economic growth rates (e.g., $-4\%, -3\%, \ldots, 0\%, 1\%, 2\%, \ldots$). The fuzzy states $\underset{\sim}{F}_s$ will be required to be orthogonal fuzzy sets, in order for us to continue to use the Bayesian framework. This orthogonal condition on the fuzzy states will be the same constraint as illustrated in Eq. (10.43), i.e.,

$$\sum_{s=1}^{3} \mu_{\underset{\sim}{F}_s}(s_i) = 1 \qquad i = 1, 2, \ldots, n \tag{10.48}$$

Further, as we need utility values to express the relationship between crisp alternative-state pairs, we still need a utility matrix to express the value of all the fuzzy alternative-state pairings. Such a matrix will have the form shown in Table 10.8.

Proceeding as before, but now with fuzzy states of nature, the expected utility of fuzzy alternative $\underset{\sim}{A}_j$ is

$$E(u_j) = \sum_{s=1}^{3} u_{js} p(\underset{\sim}{F}_s) \tag{10.49}$$

TABLE 10.8
Utility values for fuzzy states and fuzzy alternatives

	$\underset{\sim}{F}_1$	$\underset{\sim}{F}_2$	$\underset{\sim}{F}_3$
$\underset{\sim}{A}_1$	u_{11}	u_{12}	u_{13}
$\underset{\sim}{A}_2$	u_{21}	u_{22}	u_{23}
$\underset{\sim}{A}_3$	u_{31}	u_{32}	u_{33}

where

$$p(\underset{\sim}{F}_s) = \sum_{i=1}^{n} \mu_{\underset{\sim}{F}_s}(s_i)p(s_i) \tag{10.50}$$

and the maximum utility is

$$E(u^*) = \max_j E(u_j) \tag{10.51}$$

We can have crisp or fuzzy information on a universe of information $X = \{x_1, x_2, \ldots, x_r\}$, e.g., rate of increase of gross national product. Our fuzzy information will again reside on a collection of orthogonal fuzzy sets on X, $\Phi = \{\underset{\sim}{M}_1, \underset{\sim}{M}_2, \ldots, \underset{\sim}{M}_g\}$, that are defined on X. We can now derive the posterior probabilities of fuzzy states $\underset{\sim}{F}_s$, given probabilistic information [Eq. (10.52*a*)], x_r, and fuzzy information $\underset{\sim}{M}_t$ [Eq. (10.52*b*)] as follows:

$$p(\underset{\sim}{F}_s \mid x_k) = \frac{\sum_{i=1}^{n} \mu_{\underset{\sim}{F}_s}(s_i)p(x_k \mid s_i)p(s_i)}{p(x_k)} \tag{10.52a}$$

$$p(\underset{\sim}{F}_s \mid \underset{\sim}{M}_t) = \frac{\sum_{i=1}^{n}\sum_{k=1}^{r} \mu_{\underset{\sim}{F}_s}(s_i)\mu_{\underset{\sim}{M}_t}(x_k)p(x_k \mid s_i)p(s_i)}{\sum_{k=1}^{r} \mu_{\underset{\sim}{M}_t}(x_k)p(x_k)} \tag{10.52b}$$

Similarly, the expected utility given the probabilistic [Eq. (10.53*a*)] and fuzzy [Eq. (10.53*b*)] information is then

$$E(u_j \mid x_k) = \sum_{s=1}^{3} u_{js}p(\underset{\sim}{F}_s \mid x_k) \tag{10.53a}$$

$$E(u_j \mid \underset{\sim}{M}_t) = \sum_{s=1}^{3} u_{js}p(\underset{\sim}{F}_s \mid \underset{\sim}{M}_t) \tag{10.53b}$$

where the maximum conditional expected utility for probabilistic [Eq. (10.54*a*)] and fuzzy [Eq. (10.54*b*)] information is

$$E(u^*_{x_k}) = \max_j E(u_j \mid x_k) \tag{10.54a}$$

$$E(u^*_{\underset{\sim}{M}_t}) = \max_j E(u_j \mid \underset{\sim}{M}_t) \tag{10.54b}$$

Finally, the unconditional expected utilities for fuzzy states and probabilistic information, [Eq. (10.55*a*)], or fuzzy information, [Eq. (10.55*b*)], will be

$$E(u^*_x) = \sum_{k=1}^{r} E(u^*_{x_k})p(x_k) \tag{10.55a}$$

$$E(u_{\Phi}^{*}) = \sum_{t=1}^{g} E(u_{\underset{\sim}{M}_t}^{*})p(\underset{\sim}{M}_t) \tag{10.55b}$$

The expected utility given in Eqs. (10.55) now enables us to compute the value of the fuzzy information, within the context of fuzzy states of nature, for probabilistic information [Eq. (10.35)] and fuzzy information [Eq. (10.56)]:

$$V(\Phi) = E(u_{\Phi}^{*}) - E(u^{*}) \tag{10.56}$$

If the new fuzzy information is hypothetically *perfect* (since this would represent the ability to predict a fuzzy state $\underset{\sim}{F}_s$ without dissonance, it is admittedly an untenable boundary condition on the analysis), denoted Φ_p, then we can compute the maximum expected utility of fuzzy *perfect* information using Eq. (10.57). The expected utility of ith alternative $\underset{\sim}{A}_i$ for fuzzy perfect information on state $\underset{\sim}{F}_s$ becomes (from Table 10.8)

$$u(\underset{\sim}{A}_i \mid \underset{\sim}{F}_s) = u(\underset{\sim}{A}_i, \underset{\sim}{F}_s) \tag{10.57}$$

Therefore, the optimum fuzzy action, $\underset{\sim}{A}_{\underset{\sim}{F}_s}^{*}$, is defined by

$$u(\underset{\sim}{A}_{\underset{\sim}{F}_s}^{*} \mid \underset{\sim}{F}_s) = \max_i u(\underset{\sim}{A}_i, \underset{\sim}{F}_s) \tag{10.58}$$

Hence, the total expected utility for fuzzy perfect information is

$$E(u_{\Phi_p}^{*}) = \sum_{j=1}^{3} u(\underset{\sim}{A}_{\underset{\sim}{F}_s}^{*} \mid \underset{\sim}{F}_s)p(\underset{\sim}{F}_s) \tag{10.59}$$

where $p(\underset{\sim}{F}_s)$ are the prior probabilities of the fuzzy states of nature given by Eq. (10.50). The result of Eq. (10.55*b*) for the fuzzy *perfect* case would be denoted $E(u_{\Phi_p}^{*})$, and the value of the fuzzy *perfect* information would be

$$V(\Phi_p) = E(u_{\Phi_p}^{*}) - E(u^{*}) \tag{10.60}$$

Tanaka et al. [1976] have proved that the various values of information conform to the following inequality expression:

$$V(\Phi_p) \geq V(x_p) \geq V(x) \geq V(\Phi) \geq 0 \tag{10.61}$$

The inequalities in Eq. (10.61) are consistent with our intuition. The ordering, $V(x) \geq V(\Phi)$, is due to the fact that information Φ is characterized by fuzziness *and* randomness. The ordering, $V(x_p) \geq V(x)$, is true because x_p is better information than x; it is *perfect*. The ordering, $V(\Phi_p) \geq V(x_p)$, is created by the fact that the uncertainty expressed by the probability $P(\underset{\sim}{F}_i)$ still remains, even if we know the true state, s_i; hence, our interest is not in the crisp states of nature, S, but rather in the fuzzy states, $\underset{\sim}{F}$, which are defined on S.

To illustrate the development of Eqs. (10.48)–(10.61) in expanding a decision problem to consider fuzzy information, fuzzy states, and fuzzy actions in the Bayesian decision framework, the following example in computer engineering is provided.

Example 10.9. One of the decisions your project team faces with each new computer product is what type of printed circuit board (PCB) will be required for the unit.

Depending on the density of tracks (metal interconnect traces on the PCB that act like wire to connect components together), which is related to the density of the components, we may use a single-layer PCB, a double-layer PCB, a four-layer PCB, or a six-layer PCB. A PCB layer is a two-dimensional plane of interconnecting tracks. The number of layers on a PCB is the number of parallel interconnection layers in the PCB. The greater the density of the interconnections in the design, the greater the number of layers required to fit the design onto a PCB of given size. One measure of board track density is the number of nodes required in the design. A node is created at a location in the circuit where two or more lines (wires, tracks) meet. The decision process will comprise the following steps.

1. *Define the fuzzy states of nature.* The density of the PCB is defined as three fuzzy sets on the singleton states $S = (s_1, s_2, s_3, s_4, s_5) = (s_i), i = 1, 2, \ldots, 5$, where i defines the states in terms of a percentage of our most dense (in terms of components and interconnections) PCB. So, your team defines $s_1 = 20\%, s_2 = 40\%, s_3 = 60\%, s_4 = 80\%$, and $s_5 = 100\%$ of the density of the densest PCB; these are singletons on the universe of relative densities. Further, you define the following three fuzzy states, which are defined on the universe of relative density states S:

$$\underset{\sim}{F}_1 = \text{low-density PCB}$$
$$\underset{\sim}{F}_2 = \text{medium-density PCB}$$
$$\underset{\sim}{F}_3 = \text{high-density PCB}$$

2. *Define fuzzy alternatives.* Your decision alternative will represent the type of the PCB we decide to use, as follows (these actions are admittedly not very fuzzy, but in general they can be):

$$\underset{\sim}{A}_1 = \text{use a 2-layer PCB for the new design}$$
$$\underset{\sim}{A}_2 = \text{use a 4-layer PCB for the new design}$$
$$\underset{\sim}{A}_3 = \text{use a 6-layer PCB for the new design}$$

3. *Define new data samples (information).* The universe, $X = (x_1, x_2, \ldots, x_5)$ represents the "measured number of nodes in the PCB schematic;" i.e., the additional information is the measured number of nodes of the schematic that can be calculated by a *schematic capture system.* You propose the following discrete values for number of nodes,

$$x_1 = 100 \text{ nodes}$$
$$x_2 = 200 \text{ nodes}$$
$$x_3 = 300 \text{ nodes}$$
$$x_4 = 400 \text{ nodes}$$
$$x_5 = 500 \text{ nodes}$$

4. *Define orthogonal fuzzy information system.* You determine that the ambiguity in defining the density of nodes can be characterized by three linguistic information sets as $(\underset{\sim}{M}_1, \underset{\sim}{M}_2, \underset{\sim}{M}_3)$, where

$$\underset{\sim}{M}_1 = \text{low number of nodes on PCB [generally} < 300 \text{ nodes]}$$
$$\underset{\sim}{M}_2 = \text{average (medium) number of nodes on PCB [about 300 nodes]}$$
$$\underset{\sim}{M}_3 = \text{high number of nodes on PCB [generally} > 300 \text{ nodes]}$$

5. *Define the prior probabilities.* The prior probabilities of the singleton densities (states) are as follows:

$$p(s_1) = .2$$
$$p(s_2) = .3$$
$$p(s_3) = .3$$
$$p(s_4) = .1$$
$$p(s_5) = .1$$

The preceding numbers indicate that moderately dense boards are the most probable, followed by low-density boards, and high- to very high density boards are the least probable.

6. *Identify the utility values.* You propose the nondimensional utility values shown in Table 10.9 to represent the fuzzy alternative-fuzzy state relationships. The highest utility in Table 10.9 is achieved by the selection of a six-layer PCB for a high-density PCB, since the board layout is achievable. The same high-utility level of 10 is also achieved by selecting the two-layer PCB in conjunction with the low-density PCB, since a two-layer PCB is cheaper than a four- or six-layer PCB. The lowest utility is achieved by the selection of a 2-layer PCB for a high-density PCB; since the layout cannot be done, it will not fit. The second to lowest utility is achieved when a six-layer PCB is chosen, but the design is of low density, so you are wasting money.
7. *Define membership values for each orthogonal fuzzy state.* The fuzzy sets in Table 10.10 satisfy the orthogonality condition, for the sum of each column equals one, $\sum \text{column} = \sum_s \mu_{\underset{\sim}{F}_s}(s_i) = 1$.
8. *Define membership values for each orthogonal fuzzy set on the fuzzy information system.* In Table 10.11, $\sum \text{column} = \sum_t \mu_{\underset{\sim}{M}_t}(x_i) = 1$, hence the fuzzy sets are orthogonal.
9. *Define the conditional probabilities (likelihood values) for the uncertain information.* Table 10.12 shows the conditional probabilities for uncertain (probabilistic) information; note the sum of elements in each row equals unity.
10. *Define the conditional probabilities (likelihood values) for the probabilistic perfect information.* Table 10.13 shows the conditional probabilites for probabilistic perfect information; note that the sum of elements in each row equals unity and that each column only has one entry (i.e., no dissonance).

You are now ready to compute the values of information for this decision process involving fuzzy states, fuzzy alternatives, and fuzzy information.

TABLE 10.9
Utilities for fuzzy states and alternatives

	$\underset{\sim}{F}_1$	$\underset{\sim}{F}_2$	$\underset{\sim}{F}_3$
$\underset{\sim}{A}_1$	10	3	0
$\underset{\sim}{A}_2$	4	9	6
$\underset{\sim}{A}_3$	1	7	10

TABLE 10.10
Orthogonal fuzzy sets for fuzzy states

	s_1	s_2	s_3	s_4	s_5
$\underset{\sim}{F}_1$	1	0.5	0	0	0
$\underset{\sim}{F}_2$	0	0.5	1	0.5	0
$\underset{\sim}{F}_3$	0	0	0	0.5	1

TABLE 10.11
Orthogonal fuzzy sets for fuzzy information

	x_1	x_2	x_3	x_4	x_5
$\underset{\sim}{M}_1$	1	0.4	0	0	0
$\underset{\sim}{M}_2$	0	0.6	1	0.6	0
$\underset{\sim}{M}_3$	0	0	0	0.4	1

TABLE 10.12
Conditional probabilities $p(x_k \mid s_i)$ for uncertain information

	x_1	x_2	x_3	x_4	x_5
$p(x_k \mid s_1)$	0.44	0.35	0.17	0.04	0
$p(x_k \mid s_2)$	0.26	0.32	0.26	0.13	0.03
$p(x_k \mid s_3)$	0.12	0.23	0.30	0.23	0.12
$p(x_k \mid s_4)$	0.03	0.13	0.26	0.32	0.26
$p(x_k \mid s_5)$	0	0.04	0.17	0.35	0.44

Case 1. Crisp states and actions

(i) Utility and optimum decision given no information. Before initiating the no-information case, we must define the nondimensional utility values for this nonfuzzy state situation. Note that the utility values given in Table 10.14 compare the fuzzy alternatives to the singleton states (s_i), as opposed to the fuzzy states for which the utility is defined in Table 10.9. The expected values for this case are determined by using Eq. (10.28), for example,

$$\begin{aligned} E(u_1) &= (10)(0.2) + (8)(0.3) + \cdots + (2)(0.1) \\ &= 6.4 \end{aligned}$$

Similarly, $E(u_2) = 6.3$, and $E(u_3) = 4.4$. Hence, the optimum decision, given no information and with crisp (singleton) states, is alternative one, $\underset{\sim}{A}_1$, i.e., $E(u_1) = 6.4$.

(ii) Utility and optimal decision given uncertain and perfect information.

(*a*) **Probabilistic (uncertain) information.** Table 10.15 summarizes the values of the marginal probability $p(x_k)$, the posterior probabilities, and the maximum expected values for the uncertain case. The values in Table 10.15 have been calculated as in the preceding computation. For example, the marginal probabilities are calculated using Eq. (10.31):

$$\begin{aligned} p(x_1) &= (0.44)(0.2) + (0.26)(0.3) + (0.12)(0.3) + (0.03)(0.1) + (0)(0.1) \\ &= 0.205 \end{aligned}$$

TABLE 10.13
Conditional probabilities $p(x_k \mid s_i)$ for fuzzy perfect information

	x_1	x_2	x_3	x_4	x_5
$p(x_k \mid s_1)$	1.0	0.0	0.0	0.0	0.0
$p(x_k \mid s_2)$	0.0	1.0	0.0	0.0	0.0
$p(x_k \mid s_3)$	0.0	0.0	1.0	0.0	0.0
$p(x_k \mid s_4)$	0.0	0.0	0.0	1.0	0.0
$p(x_k \mid s_5)$	0.0	0.0	0.0	0.0	1.0

TABLE 10.14
Utility values for crisp states

	s_1	s_2	s_3	s_4	s_5
$\underset{\sim}{A}_1$	10	8	6	2	0
$\underset{\sim}{A}_2$	4	6	9	6	4
$\underset{\sim}{A}_3$	1	2	6	8	10

TABLE 10.15
Computed values for uncertain case (nonfuzzy states)

	x_1	x_2	x_3	x_4	x_5
$p(x_k)$	0.205	0.252	0.245	0.183	0.115
$p(s_1 \mid x_k)$	0.429	0.278	0.139	0.044	0.0
$p(s_2 \mid x_k)$	0.380	0.381	0.318	0.213	0.078
$p(s_3 \mid x_k)$	0.176	0.274	0.367	0.377	0.313
$p(s_4 \mid x_k)$	0.015	0.052	0.106	0.175	0.226
$p(s_5 \mid x_k)$	0.0	0.016	0.069	0.191	0.383
$\mathrm{E}(u^* \mid x_k)$	8.42	7.47	6.68	6.66	7.67
$a_j \mid a_k$	1	1	2	2	3

$$p(x_3) = (0.17)(0.2) + (0.26)(0.3) + (0.3)(0.3) + (0.26)(0.1) + (0.17)(0.1)$$
$$= 0.245$$

The posterior probabilities are calculated using Eq. (10.30), the conditional probabilities, and the prior probabilities; for example,

$$p(s_1 \mid x_3) = \frac{(0.17)(0.2)}{(0.245)} = 0.139$$
$$p(s_3 \mid x_2) = \frac{(0.23)(0.3)}{(0.245)} = 0.274$$
$$p(s_5 \mid x_4) = \frac{(0.35)(0.1)}{(0.183)} = 0.191$$

The conditional expected utilities are calculated using Eqs. (10.32)–(10.33); for example for datum x_1,

$$E(u_1 \mid x_1) = (0.429)(10) + (0.380)(8) + (0.176)(6) + (0.015)(2) + (0)(0)$$
$$= 8.42$$
$$E(u_2 \mid x_1) = (0.429)(4) + (0.380)(6) + (0.176)(9) + (0.015)(6) + (0)(4)$$
$$= 5.67$$
$$E(u_3 \mid x_1) = (0.429)(1) + (0.380)(2) + (0.176)(6) + (0.015)(8) + (0)(10)$$
$$= 2.36$$

Therefore, the optimum decision for datum x_1, given uncertain information with crisp states is

$$E(u^* \mid x_1) = \max(8.42, 5.67, 2.36) = 8.42 \qquad \text{(choose action } \underset{\sim}{A}_1\text{)}$$

Now, using Eq. (10.34) to calculate the overall (for all data x_i) unconditional expected utility for the uncertain information, we get

$$E(u_x^*) = (8.42)(0.205) + (7.47)(0.252) + (6.68)(0.245) + \cdots + (7.67)(0.115)$$
$$= 7.37$$

The value of the uncertain information, using Eq. (10.35), is

$$V(x) = 7.37 - 6.4 = 0.97$$

TABLE 10.16
Computed quantities for perfect information and crisp states

	x_1	x_2	x_3	x_4	x_5
$p(x_k)$	0.20	0.30	0.30	0.10	0.10
$p(s_1 \mid x_k)$	1.0	0.0	0.0	0.0	0.0
$p(s_2 \mid x_k)$	0.0	1.0	0.0	0.0	0.0
$p(s_3 \mid x_k)$	0.0	0.0	1.0	0.0	0.0
$p(s_4 \mid x_k)$	0.0	0.0	0.0	1.0	0.0
$p(s_5 \mid x_k)$	0.0	0.0	0.0	0.0	1.0
$\mathrm{E}(u^* \mid x_k)$	10.0	8.0	9.0	8.0	10.0
$a_j \mid a_k$	1	1	2	3	3

(*b*) **Probabilistic perfect information.** Using the same utility values as before, and conditional probabilities as defined in Table 10.12, the marginal probabilities, posterior probabilities, and the expected values are shown in Table 10.16. The unconditional expected utility for probabilistic perfect information is given as

$$E(u^*_{x_p}) = (10)(0.2) + (8)(0.3) + \cdots + (10)(0.1)$$
$$= 8.9$$

and the value of the probabilistic perfect information from Eq. (10.35) is

$$V(x_p) = 8.9 - 6.4 = 2.5$$

Case 2. Fuzzy states and actions

(i) Utility and optimum decision given no information. The utility values for this case are shown in Table 10.9. We calculate the prior probabilities for the fuzzy states using Eq. (10.50). For example,

$$p(\underset{\sim}{F}_1) = (1)(0.2) + (0.5)(0.3) + (0)(0.3) + (0)(0.1) + (0)(0.1)$$
$$= 0.35$$

Similarly, $p(\underset{\sim}{F}_2) = 0.5$, and $p(\underset{\sim}{F}_3) = 0.15$. Therefore, the expected utility is given by Eq. (10.49) as

$$E(u_j) = \begin{bmatrix} 5 \\ 6.8 \\ 5.35 \end{bmatrix}$$

The optimum expected utility of the fuzzy alternatives (actions) for the case of no information using Eq. (10.51) is

$$E(u^*) = 6.8$$

so alternative $\underset{\sim}{A}_2$ is the optimum choice.

(ii) Utility and optimum decision given uncertain and perfect information

(*a*) **Probabilistic (uncertain) information.** Table 10.17 lists the posterior probabilities as determined by Eq. (10.52*a*). For example,

TABLE 10.17
Posterior probabilities for probabilistic information with fuzzy states

	$\underset{\sim}{F}_1$	$\underset{\sim}{F}_2$	$\underset{\sim}{F}_3$
x_1	0.620	0.373	0.007
x_2	0.468	0.49	0.042
x_3	0.298	0.58	0.122
x_4	0.15	0.571	0.279
x_5	0.039	0.465	0.496

TABLE 10.18
Expected utilities for fuzzy alternatives with probabilistic information

	$\underset{\sim}{A}_1$	$\underset{\sim}{A}_2$	$\underset{\sim}{A}_3$
x_1	7.315	5.880	3.305
x_2	6.153	6.534	4.315
x_3	4.718	7.143	5.58
x_4	3.216	7.413	6.934
x_5	1.787	7.317	8.252

$$p(\underset{\sim}{F}_1 \mid x_1) = \frac{(1)(0.44)(0.2) + (0.5)(0.35)(0.3)}{0.205} = 0.620$$

The other values are calculated in a similar manner and are shown in Table 10.17.

The expected utility values for each of the x_k are now calculated using Eq. (10.53a), and these values are given in Table 10.18. The optimum expected utilities for each alternative are found by using Eq. (10.54a),

$$E(u^*_{x_k}) = \max_j E(u_j \mid x_k) = \{7.315, 6.534, 7.143, 7.413, 8.252\}$$

where the optimum choice associated with this value is obviously alternative $\underset{\sim}{A}_3$.

Finally, the expected utility, given by Eq. (10.55), is calculated to be

$$\begin{aligned} E(u^*_{\Phi}) &= \sum_{k=1}^{r} E(u^*_{x_k})p(x_k) \\ &= (7.315)(0.205) + (6.534)(0.252) + (7.143)(0.245) \\ &\quad + (7.413)(0.183) + (8.252)(0.115) = 7.202 \end{aligned}$$

The value of the probabilistic uncertain information for fuzzy states is

$$V(x) = 7.202 - 6.8 = 0.402$$

(b) **Probabilistic perfect information.** Table 10.19 lists the posterior probabilities as determined by Eq. (10.52a). For example,

$$p(\underset{\sim}{F}_1 \mid x_1) = \{(1)(1)(0.2) + (0.5)(0)(0.3)\}/(0.2) = 1.0$$

The other values are calculated in a similar manner and are shown in Table 10.19.

TABLE 10.19
Posterior probabilities for probabilistic *perfect* information with fuzzy states

	$\underset{\sim}{F}_1$	$\underset{\sim}{F}_2$	$\underset{\sim}{F}_3$
x_1	1.0	0.0	0.0
x_2	0.5	0.5	0.0
x_3	0.0	1.0	0.0
x_4	0.0	0.5	0.5
x_5	0.0	0.0	1.0

TABLE 10.20
Expected utilities for fuzzy alternatives with probabilistic *perfect* information

	$\underset{\sim}{A}_1$	$\underset{\sim}{A}_2$	$\underset{\sim}{A}_3$
x_1	10.0	4.0	1.0
x_2	6.5	6.5	4.0
x_3	3.0	9.0	7.0
x_4	1.5	7.5	8.5
x_5	0.0	6.0	10.0

TABLE 10.21
Posterior probabilities for fuzzy information with fuzzy states

	$\underset{\sim}{M}_1$	$\underset{\sim}{M}_2$	$\underset{\sim}{M}_3$
$\underset{\sim}{F}_1$	0.570	0.317	0.082
$\underset{\sim}{F}_2$	0.412	0.551	0.506
$\underset{\sim}{F}_3$	0.019	0.132	0.411

The expected utility values for each of the x_k are now calculated using Eq. (10.53*a*), and these values are given in Table 10.20. The optimum expected utilities for each alternative are found by using Eq. (10.54*a*),

$$E(u^*_{x_k}) = \max_j E(u_j \mid x_k) = \{10.0, 6.5, 9.0, 8.5, 10.0\}$$

where it is not clear which alternative is the optimum choice (i.e., there is a tie between alternatives 1 and 3). Finally, the expected utility, given by Eq. (10.55*a*), is calculated to be

$$\begin{aligned} E(u^*_{x_p}) &= \sum_{k=1}^{r} \mathrm{E}(u^*_{x_k}) p(x_k) \\ &= (10.0)(0.2) + (6.5)(0.3) + (9.0)(0.3) + (8.5)(0.1) + (10.0)(0.1) \\ &= 8.5 \end{aligned}$$

The value of the probabilistic *perfect* information for fuzzy states is

$$\mathrm{V}(x_p) = 8.5 - 6.8 = 1.7$$

(*c*) **Fuzzy information.** For the hypothetical fuzzy information, Table 10.21 summarizes the results of the calculations using Eq. (10.52*b*). An example calculation is shown here:

$$\begin{aligned} p(\underset{\sim}{F}_1 \mid \underset{\sim}{M}_1) = \{&(1)(1)(0.44)(0.2) + (1)(0.4)(0.35)(0.2) + (0.5)(1)(0.26)(0.3) \\ &+ (.5)(0.4)(0.32)(0.3)\} \div \{(1)(0.205) + (0.4)(0.252)\} = 0.57 \end{aligned}$$

Similarly, Table 10.22 summarizes the calculations of the expected utilities using Eq. (10.53*b*).

TABLE 10.22
Posterior probabilities for fuzzy alternatives with fuzzy information

	$\underset{\sim}{M}_1$	$\underset{\sim}{M}_2$	$\underset{\sim}{M}_3$
$\underset{\sim}{A}_1$	6.932	4.821	2.343
$\underset{\sim}{A}_2$	6.096	7.019	7.354
$\underset{\sim}{A}_3$	3.638	5.496	7.740

Now, using Eq. (10.54*b*), we find the optimum expected utility for each of the fuzzy states is

$$E(u^*_{\underset{\sim}{M}_t}) = \max_j E(u_j \mid \underset{\sim}{M}_t) = \{6.932, 7.019, 7.740\}$$

where the optimum choice is again $\underset{\sim}{A}_3$. The marginal probabilities of the fuzzy information sets are calculated using Eq. (10.39); for example, using the marginal probabilities from Table 10.15 and the fuzzy information from Table 10.11, we find

$$p(\underset{\sim}{M}_1) = (1.0)(0.205) + (0.4)(0.252) = 0.306$$

and, along with the other two marginal probabilities, we get

$$p(\underset{\sim}{M}_t) = \begin{bmatrix} 0.306 \\ 0.506 \\ 0.188 \end{bmatrix}$$

The unconditional expected utility using Eq. (10.55*b*) is,

$$E(u^*_{\Phi}) = \sum_{t=1}^{g} E(u^*_{\underset{\sim}{M}_t}) p(\underset{\sim}{M}_t) = 7.128$$

and the value of the perfect information for fuzzy states is $V(\Phi) = 7.128 - 6.8 = 0.328$.

(*d*) **Fuzzy perfect information.** Table 10.23 summarizes the calculations of the expected utilities using Eq. (10.57). Note in Table 10.23 that the expected utilities are the same as the utilities in Table 10.9; this identity arises because the information is presumed perfect, and the conditional probability matrix, $\underset{\sim}{F}_s \mid \underset{\sim}{M}_t$, in Eq. (10.53*b*) is the identity matrix.

Now, using Eq. (10.58), we find the optimum expected utility for each of the fuzzy states is

$$u(A^*_{\underset{\sim}{F}_s} \mid \underset{\sim}{F}_s) = \max_i u(\underset{\sim}{A}_i, \underset{\sim}{F}_s) = \{10.0, 9.0, 10.0\}$$

Finally, using the previously determined prior probabilities of the fuzzy states, $p(\underset{\sim}{F}_s)$ (see section (*i*)), we see the unconditional expected utility using Eq. (10.59) is

$$E(u^*_{\Phi_p}) = \sum_{j=1}^{3} u(A^*_{\underset{\sim}{F}_s} \mid \underset{\sim}{F}_s) p(\underset{\sim}{F}_s) = 10(.35) + 9(.5) + 10(.15) = 9.5$$

where the value of the fuzzy perfect information for fuzzy states is

$$V(\Phi_p) = 9.5 - 6.8 = 2.7$$

TABLE 10.23
Expected utilities for fuzzy alternatives with fuzzy perfect information

	$\underset{\sim}{F}_1$	$\underset{\sim}{F}_2$	$\underset{\sim}{F}_3$
$\underset{\sim}{A}_1$	10.0	3.0	0.0
$\underset{\sim}{A}_2$	4.0	9.0	6.0
$\underset{\sim}{A}_3$	1.0	7.0	10.0

Example summary. A typical decision problem is to decide on a basic policy in a fuzzy environment. This basic policy can be thought of as a fuzzy action. The attributes of such a problem are that there are many states, feasible policy alternatives, and available information. Usually, the utilities for all the states and all the alternatives cannot be formulated because of insufficient data, because of the high cost of obtaining this information, and because of time constraints. On the other hand, a decision maker in top management is generally not concerned with the detail of each element in the decision problem. Mostly, top managers want to decide roughly what alternatives to select as indicators of policy directions. Hence, an approach that can be based on fuzzy states and fuzzy alternatives and that can accommodate fuzzy information is a very powerful tool for making preliminary policy decisions.

The expected utilities and the value of information for the five cases, i.e., for no information, probabilistic (uncertain) information, probabilistic perfect information, fuzzy probabilistic (uncertain) information, and fuzzy perfect information, are summarized in Table 10.24. We can see from this table that the ordering of values of information is in accordance with that described in Eq. (10.61); i.e., $V(\Phi_p) \geq V(x_p) \geq V(x) \geq V(\Phi) \geq 0$. The probabilistic perfect information $[V(x_p) = 1.70]$ has a value much higher than the probabilistic information $[V(x) = 0.40]$. The decision maker needs to ascertain the cost of the hypothetical perfect information when compared with the cost of the uncertain information, the latter being more realistic. On the other hand, there is little difference between the value of fuzzy probabilistic information $[V(\Phi) = 0.33]$ and that of probabilistic information $[V(x) = 0.40]$. This result suggests that the fuzzy probabilistic information is sufficiently valuable compared with the probabilistic information for this problem, because fuzzy information generally costs far less than probabilistic (uncertain) information. Finally, the fact that the fuzzy perfect information $[V(\Phi_p) = 2.70]$ holds more value than the probabilistic perfect information $[V(x_p) = 1.70]$ confirms that our interest is more in the fuzzy states than the crisp states. When utility values, prior probabilities, conditional probabilities, and orthogonal membership values change for any of these scenarios, the elements in Table 10.24 will change and the conclusions derived from them will portray a different situation. The power of this approach is its ability to measure on an ordinal basis the value of the approximate information used in a decision-making problem. When the value of approximate (fuzzy) information approaches that of either probabilistic or perfect information, there is the potential for significant cost savings without reducing the quality of the decision itself.

TABLE 10.24
Summary of expected utility and value of information for fuzzy states and actions for the example

Information	Expected utility	Value of information
No information	6.8	—
Probabilistic information, $V(x)$	7.20	0.40
Perfect information, $V(x_p)$	8.5	1.7
Fuzzy probabilistic information, $V(\Phi)$	7.13	0.33
Fuzzy perfect information, $V(\Phi p)$	9.5	2.7

SUMMARY

The literature is rich with references in the area of fuzzy decision making. This chapter has only presented a few rudimentary ideas in the hope of interesting the readers to continue their learning in this important area. One of the decision metrics in this chapter represents a philosophical approach where an existing crisp theory—Bayesian decision making—is reinterpreted to accept both fuzzy and random uncertainty. New theoretical developments are expanding the field of fuzzy decision making; for example multiobjective situations represent an interesting class of problems that plague optimization in decision making [Sakawa, 1993] as do multiattribute decision problems [Baas and Kwakernaak, 1977]. This philosophical approach has been extended further where fuzzy utilities have been addressed with fuzzy states [Jain, 1976], and where fuzzy utilities are determined in the presence of probabilistic states [Jain, 1978; Watson et al., 1979]. Häage [1978] extended the Bayesian scheme to include possibility distributions (see Chapter 15 for definition of possibility) for the consequences of the decision actions. The other metrics in this chapter extend some specific problems to deal with issues like fuzzy preference relations, fuzzy objective functions, fuzzy ordering, fuzzy consensus, etc. In all of these, there is a compelling need to incorporate fuzziness in human decision making, as originally proposed by Bellman and Zadeh [1970]. In most decision situations the goals, constraints, and consequences of the proposed alternatives are not known with precision. Much of this imprecision is not measureable, and not random. The imprecision can be due to vague, ambiguous, or fuzzy information. Methods to address this form of imprecision are necessary to deal with many of the uncertainties we deal with in humanistic systems.

REFERENCES

Adams, T. (1994). "Retaining structure selection with unequal fuzzy project-level objectives," *J. Intelligent Fuzzy Syst.,* vol. 2, no. 3, pp. 251–266.

Baas, S., and H. Kwakernaak (1977). "Rating and ranking of multiple-aspect alternatives using fuzzy sets," *Automatica,* vol. 13, pp. 47–58.

Bellman, R. and L. Zadeh (1970). "Decision making in a fuzzy environment," *Manage. Sci.,* vol. 17, pp. 141–164.

Bezdek, J., B. Spillman, and R. Spillman (1978). "A fuzzy relation space for group decision theory," *Fuzzy Sets Syst.,* vol. 1, pp. 255–268.

Clark, D. W. (1992). "Computer illogic . . . ," *Mirage Magazine,* Fall, Univ. of New Mexico Alumni Association, pp. 12–13.

Dubois, D., and H. Prade (1980). *Fuzzy sets and systems: Theory and applications,* Academic Press, New York.

Häage, C. (1978). "Possibility and cost in decision analysis," *Fuzzy Sets Syst.,* vol. 1, no. 2, pp. 81–86.

Jain, R. (1976). "Decision-making in the presence of fuzzy variables," *IEEE Trans. Syst., Man, Cybern.,* vol. 6, no. 10, pp. 698–703.

Jain, R. (1978). "Decision-making in the presence of fuzziness and uncertainty," *Proc. IEEE Conf. Decision Control,* New Orleans, pp. 1318–1323.

Okuda, T., H. Tanaka, and K. Asai (1974). "Decision making and information in fuzzy events," *Bull. Univ. Osaka Prefect., Ser. A,* vol. 23, no. 2, pp. 193–202.

Okuda, T., H. Tanaka, and K. Asai (1978). "A formulation of fuzzy decision problems with fuzzy information, using probability measures of fuzzy events," *Inf. Control,* vol. 38, no. 2, pp. 135–147.

Sakawa, M. (1993). *Fuzzy sets and interactive multiobjective optimization,* Plenum Press, New York.
Shimura, M. (1973). "Fuzzy sets concept in rank-ordering objects," *J. Math. Anal. Appl.,* vol. 43, pp. 717–733.
Tanaka, H., T. Okuda, and K. Asai (1976). "A formulation of fuzzy decision problems and its application to an investment problem," *Kybernetes,* vol. 5, pp. 25–30.
Terano, T., K. Asai, and M. Sugeno (1992). *Fuzzy system theory and its applications,* Academic Press, San Diego, CA.
Watson, S., J. Weiss, and M. Donnell (1979). "Fuzzy decision analysis," *IEEE Trans. Syst., Man, Cybern.,* vol. 9, no. 1, pp. 1–9.
Yager, R. (1981). "A new methodology for ordinal multiobjective decisions based on fuzzy sets," *Decision Sci.,* vol. 12, pp. 589–600.

PROBLEMS

Synthetic Evaluation and Rank Ordering

10.1. You wish to use a synthetic evaluation system to evaluate workstation software packages. Such a system could be helpful to a system administrator in selecting new software. You define the following factors and evaluation criteria in this process:

Factors

User-friendliness	x_1
Usefulness of features	x_2
Architectures ported to	x_3
Price—single user	x_4
Price—additional users	x_5

Evaluation criteria

Excellent (e)	y_1
Good (g)	y_2
Fair (f)	y_3
Poor (p)	y_4

You assume that some company provides relational matrices for different types of software, which they keep up-to-date. Categories of software could be databases, spreadsheets, scientific software, etc. You are provided with the following relational matrix for a software package in your category of interest:

$$\underset{\sim}{R} = \begin{array}{c} \begin{matrix} \text{e} & \text{g} & \text{f} & \text{p} \end{matrix} \\ \begin{bmatrix} .2 & .4 & .3 & .1 \\ .1 & .4 & .2 & .3 \\ .2 & .2 & .2 & .4 \\ .2 & .3 & .3 & .2 \\ .1 & .2 & .4 & .3 \end{bmatrix} \end{array} \begin{array}{l} \\ \text{friendly} \\ \text{useful} \\ \text{porting} \\ \text{single user price} \\ \text{additional users price} \end{array}$$

Suppose your main concerns for software packages are that they be useful and inexpensive for a large number of users. You decide to emphasize factors x_2 and x_5, and get a fuzzy weight vector,

$$\underset{\sim}{A} = [.1, .35, .05, .2, .3]$$

Find the corresponding evaluation fuzzy vector. Comment on the value of this method here.

10.2. *Ordering:* Consider a comparison of four chemicals with respect to human toxicity. Assume that c_1 is highly toxic, c_2 is very toxic, c_3 is moderately toxic, and c_4 is slightly toxic. Hence, on a pairwise comparison basis, c_2 resembles c_1 with a fuzzy membership value of 0.75, c_3 resembles c_1 with a fuzzy membership of 0.5, and c_4 resembles c_1 with a fuzzy membership of 0.25. The remainder of the pairwise comparisons follow:

$$
\begin{array}{llll}
f_{c_1}(c_1) = 1 & f_{c_1}(c_2) = .75 & f_{c_1}(c_3) = .5 & f_{c_1}(c_4) = .25 \\
f_{c_2}(c_1) = .75 & f_{c_2}(c_2) = 1 & f_{c_2}(c_3) = .5 & f_{c_2}(c_4) = .25 \\
f_{c_3}(c_1) = .5 & f_{c_3}(c_2) = .25 & f_{c_3}(c_3) = 1 & f_{c_3}(c_4) = .5 \\
f_{c_4}(c_1) = .25 & f_{c_4}(c_2) = .25 & f_{c_4}(c_3) = .5 & f_{c_4}(c_4) = 1
\end{array}
$$

Develop the comparison matrix and determine the overall ranking of toxicity.

10.3. An aircraft control system is a totally *nonlinear system* when the final approach and landing of an aircraft are considered. It involves maneuvering flight in an appropriate course to the airport and then along the optimum glide path trajectory to the runway. We know that this path is usually provided by an instrument landing system, which transmits two radio signals to the aircraft as a navigational aid. These orthogonal radio beams are known as the localizer and the glide slope and are transmitted from the ends of the runway in order to provide the approaching aircraft with the correct trajectory for landing. The pilot executing such a landing must monitor cockpit instruments that display the position of the aircraft relative to the desired flight path and make appropriate corrections to the controls. Presume that four positions are available to the pilot and that four corrections P_1, P_2, P_3, and P_4 from the actual position P are required to put the aircraft on the correct course. Let the subjective estimation be as follows: P_1, P_2, P_3, and P_4 estimate P with fuzzy measurements .3, .5, .6, and .8, respectively. The pairwise comparisons for the four positions are as follows:

$$
\begin{array}{llll}
f_{P_1}(P_1) = 1 & f_{P_1}(P_2) = .5 & f_{P_1}(P_3) = .6 & f_{P_1}(P_4) = .8 \\
f_{P_2}(P_1) = .3 & f_{P_2}(P_2) = 1 & f_{P_2}(P_3) = .4 & f_{P_2}(P_4) = .3 \\
f_{P_3}(P_1) = .6 & f_{P_3}(P_2) = .4 & f_{P_3}(P_3) = 1 & f_{P_3}(P_4) = .6 \\
f_{P_4}(P_1) = 0 & f_{P_4}(P_2) = .3 & f_{P_4}(P_3) = .6 & f_{P_4}(P_4) = 1
\end{array}
$$

Now, from these values, compute the comparison matrix, and determine the overall ranking.

10.4. A piece of property is evaluated on the basis of different factors like price, location, aesthetics of the property, and its floor area. The evaluation criteria may be good, fair and bad. We have a relational matrix as

$$
\underset{\sim}{R} = \begin{array}{r} \\ \text{price} \\ \text{location} \\ \text{aesthetics} \\ \text{floor area} \end{array}
\begin{array}{c} \begin{array}{ccc} \text{g} & \text{f} & \text{b} \end{array} \\ \begin{bmatrix} .2 & .5 & .3 \\ .3 & .5 & .2 \\ .4 & .5 & .1 \\ .45 & .4 & .05 \end{bmatrix} \end{array}
$$

The weighting factors for price, location, aesthetics, and floor area, respectively, are given as $\underset{\sim}{a} = \{.4, .25, .05, .3\}$. Find the corresponding fuzzy vector for the evaluation.

10.5. When evaluating expert system tools, four evaluation criteria are used: (1) excellent, (2) good, (3) fair, and (4) mediocre. There are four aspects : I/O facilities, debugging aids, knowledge base editors, and explanation facilities. The following table shows the relationship matrix:

	Excellent	Good	Fair	Mediocre
I/O	.3	.4	.2	.1
Debug	.2	.5	.3	0
Editors	.5	.2	.2	.1
Explain	.1	.6	.2	.1

Suppose you have weight factor $\underset{\sim}{a} = \{.2, .4, .3, .1\}$. Evaluate the expert system tool.

10.6. Company Z makes chemical additives that are ultimately used for engine oil lubricants. Components such as surfactants, detergents, rust inhibitors, etc. go into the finished engine oil before it is sold to the public. Suppose that company Z makes a product D739.2 that is a detergent additive. You are asked to determine if a particular batch of D739.2 is good enough to be sold to an oil company, which will then make the final product. The detergent is evaluated on the following parameters: actual color of the material, consistency, base number (BN, measure of detergent capacity), and flash point (FP, ignition temperature of the material). After making several hundred batches of the detergent additive D739.2, the following relation matrix is obtained:

$$\underset{\sim}{R} = \begin{array}{c} \\ \text{color} \\ \text{consistency} \\ \text{BN} \\ \text{FP} \end{array} \begin{array}{c} \begin{array}{ccc} \text{excellent} & \text{very good} & \text{fair} \end{array} \\ \begin{bmatrix} .3 & .4 & .3 \\ .1 & .5 & .4 \\ .5 & .4 & .1 \\ .4 & .3 & .3 \end{bmatrix} \end{array}$$

The weight factor for the detergent is $\underset{\sim}{a} = \{.1, .25, .4, .25\}$. Evaluate the quality of the detergent.

10.7. In making a decision to purchase an airplane, airline management will consider the qualities of an airplane's performance with respect to the competitor. The Boeing 737 is the best-selling airplane in aviation history and continues to outsell its more modern competitor, the A320, manufactured by the airbus consortium. The four factors to be considered are these: range, payload, operating costs, and reliability. The criteria will be a comparison of the 737 with respect to the A320: superior (sup.), equivalent (eq.), and deficient (def.).

$$\underset{\sim}{R} = \begin{array}{c} \\ \text{range} \\ \text{payload} \\ \text{cost} \\ \text{reliability} \end{array} \begin{array}{c} \begin{array}{ccc} \text{sup.} & \text{eq.} & \text{def.} \end{array} \\ \begin{bmatrix} 0 & .7 & .3 \\ .1 & .8 & .1 \\ .1 & .5 & .4 \\ .7 & .2 & .1 \end{bmatrix} \end{array}$$

Given a typical airline's weighting factor of the four factors as $\underset{\sim}{a} = \{.15, .15, .3, .4\}$, evaluate the performance of the 737 with respect to the A320.

10.8. A power supply needs to be chosen to go along with an embedded system. Four categories of evaluation criteria are important. The first is the physical size of the power supply. The second is the efficiency of the power supply. The third is the "ripple" voltage of the output of the power supply. This is a measure of how clean the power provided is. The fourth criterion is the peak current provided by the power supply. The following matrix defines the type of power supply required for the embedded system application:

$$\underset{\sim}{R} = \begin{array}{r} \\ \text{physical size} \\ \text{efficiency} \\ \text{ripple voltage} \\ \text{peak current} \end{array} \begin{array}{c} \begin{array}{ccccc} \text{VG} & \text{G} & \text{F} & \text{B} & \text{VB} \end{array} \\ \begin{bmatrix} .2 & .7 & .1 & 0 & 0 \\ .1 & .2 & .4 & .2 & .1 \\ .3 & .4 & .2 & .1 & 0 \\ 0 & .2 & .4 & .3 & .1 \end{bmatrix} \end{array}$$

From this matrix, one can see that for the embedded system in mind, the power supply's physical size is very important as well as its ripple voltage. Of lesser importance is its efficiency, and lesser yet, is its peak current. So a small power supply with clean output voltage is needed. It needs to be somewhat efficient and is not required to provide very much "inrush current" or peak current. Evaluate a power supply with the following characteristics:

$$\underset{\sim}{\text{Power Supply}} = \begin{bmatrix} .5 \\ .1 \\ .2 \\ .2 \end{bmatrix} \begin{array}{l} \text{physical size} \\ \text{efficiency} \\ \text{ripple voltage} \\ \text{peak current} \end{array}$$

10.9. Find out who resembles a father most among his elder son (x_1), his younger son (x_2), and his daughter (x_3). We have subjective estimations from family members to assist us in this problem. For example, the elder son and the younger son resemble their father with fuzzy measurement 0.8 ($f_{x_2}(x_1)$) and 0.5 ($f_{x_1}(x_2)$), respectively, when just the two of them are considered together. The additional pairwise functions are as follows (note: the form of this matrix is transposed from previous examples).

$$\begin{array}{lll} f_{x_1}(x_1) = 1 & f_{x_2}(x_1) = 0.8 & f_{x_3}(x_1) = 0.5 \\ f_{x_1}(x_2) = 0.5 & f_{x_2}(x_2) = 1 & f_{x_3}(x_2) = 0.4 \\ f_{x_1}(x_3) = 0.5 & f_{x_2}(x_3) = 0.7 & f_{x_3}(x_3) = 1 \end{array}$$

10.10. A piece of property is evaluated so that it best suits a client's needs. Different available pieces of properties may have different benefits when compared to each other and to the needs of the client. Assume that four pieces of property are available and the client would like to order them in his order of preference. In the past, the client might just have decided on the basis of total cost alone, even though there might be several other factors (like facilities available) that he should consider. Suppose the client compares four criteria—p_1, p_2, p_3, and p_4—with each other and to his needs. The pairwise functions are as follows:

$$\begin{array}{llll} f_{p_1}(p_1) = 1 & f_{p_1}(p_2) = .5 & f_{p_1}(p_3) = .3 & f_{p_1}(p_4) = .2 \\ f_{p_2}(p_1) = .8 & f_{p_2}(p_2) = 1 & f_{p_2}(p_3) = .4 & f_{p_2}(p_4) = .6 \\ f_{p_3}(p_1) = .5 & f_{p_3}(p_2) = .9 & f_{p_3}(p_3) = 1 & f_{p_3}(p_4) = .95 \\ f_{p_4}(p_1) = .7 & f_{p_4}(p_2) = .4 & f_{p_4}(p_3) = .2 & f_{p_4}(p_4) = 1 \end{array}$$

Develop a comparison matrix based on this information, and determine the overall ranking.

10.11. When designing a radar system for imaging purposes, we frequently need to set priorities in accomplishing certain features. Some features that need to be traded off against each other are these:

1. The ability to penetrate foliage and even the ground to some depth
2. The resolution of the resulting radar image
3. The size of the antenna required for the radar system
4. The amount of power required to operate at a given frequency

It is useful to determine the order of importance of these features in selecting an operating frequency for the radar. Let x_1 represent penetration; x_2, resolution; x_3, antenna size; and x_4, power. A crisp ordering will have trouble resolving the importance of penetration compared to resolution, resolution compared to antenna size, and antenna size compared to penetration. These are entities that can only be compared in a very subjective manner, ideal for fuzzy techniques and difficult for crisp techniques.

Let $f_{x_i}(x_j)$ be the relative importance of feature x_j with respect to x_i. The comparisons $f_{x_i}(x_j)$ are subjectively assigned as follows:

		x_j			
		x_1	x_2	x_3	x_4
x_i	x_1	1	.6	.5	.9
	x_2	.5	1	.7	.8
	x_3	.9	.8	1	.5
	x_4	.3	.2	.3	1

Develop a comparison matrix and determine the overall ranking of the importance of each feature.

10.12. As a manufacturer of audio speakers, you would like a way of comparing the fidelity of various speaker designs. The sound of musical instruments reproduced through the speakers is compared to the sound of the live musical instruments. Judgment of fidelity is very subjective. A way is needed to order the speaker designs in terms of fidelity. The four speaker designs to compare are $X = \{x_1, x_2, x_3, x_4\}$.

$$f = \begin{bmatrix} 1 & .2 & .6 & .5 \\ .4 & 1 & 1 & .8 \\ .1 & .3 & 1 & .9 \\ .5 & 1 & .1 & 1 \end{bmatrix}$$

Compute the comparison matrix using the pairwise comparisons in the matrix above.

10.13. Formalisms for reasoning with uncertainty and making inferences with uncertain information have been useful in some applications as desiderata. Suppose there are five formalisms to choose from for a particular application, and we want to rank order them by using the desiderata as a comparison metric. The five formalisms are as follows:

x_1: Modified Bayesian

x_2: Evidential reasoning

x_3: Fuzzy necessity and possibility

x_4: Reasoning with uncertainty module (RUM)

x_5: Reasoned assumptions

To rank order the five formalisms, we use the pairwise function, $f_y(x)$ as the membership function of x with respect to y. The following pairwise functions give fuzzy measures of how close the performance of the pair measures up to the ideal formalism when only the pair is considered alone.

$$
\begin{array}{lllll}
f_{x_1}(x_1) = 1 & f_{x_2}(x_1) = .25 & f_{x_3}(x_1) = .3 & f_{x_4}(x_1) = .15 & f_{x_5}(x_1) = .4 \\
f_{x_1}(x_2) = .66 & f_{x_2}(x_2) = 1 & f_{x_3}(x_2) = .3 & f_{x_4}(x_2) = .4 & f_{x_5}(x_2) = .7 \\
f_{x_1}(x_3) = .9 & f_{x_2}(x_3) = .4 & f_{x_3}(x_3) = 1 & f_{x_4}(x_3) = .8 & f_{x_5}(x_3) = .8 \\
f_{x_1}(x_4) = .5 & f_{x_2}(x_4) = .6 & f_{x_3}(x_4) = .9 & f_{x_4}(x_4) = 1 & f_{x_5}(x_4) = .7 \\
f_{x_1}(x_5) = .8 & f_{x_2}(x_5) = .6 & f_{x_3}(x_5) = .5 & f_{x_4}(x_5) = .4 & f_{x_5}(x_5) = 1
\end{array}
$$

Form a comparison matrix C for these data.

10.14. In tracking soil particles, a tracked particle can be occluded by other objects. To find out which one is the tracked particle, one can choose or pick a particle that is a certain distance from the tracked particle. Suppose there are four particles in the region of interest where the tracked particle is. Furthermore, let x_1, x_2, x_3, and x_4 resemble the tracked particle with fuzzy measurement .3, .4, .6, .7, respectively, when they alone are considered. Note $f_{x_j}(x_i)$ means how close x_i is to the tracked particle with respect to x_j.

$$
\begin{array}{llll}
f_{x_1}(x_1) = 1 & f_{x_1}(x_2) = .6 & f_{x_1}(x_3) = .4 & f_{x_1}(x_4) = .3 \\
f_{x_2}(x_1) = .7 & f_{x_2}(x_2) = 1 & f_{x_2}(x_3) = .1 & f_{x_2}(x_4) = .4 \\
f_{x_3}(x_1) = .2 & f_{x_3}(x_2) = .4 & f_{x_3}(x_3) = 1 & f_{x_3}(x_4) = .3 \\
f_{x_4}(x_1) = .5 & f_{x_4}(x_2) = .3 & f_{x_4}(x_3) = .4 & f_{x_4}(x_4) = 1
\end{array}
$$

Develop a comparison matrix, and determine which particle is closest to the tracked particle.

10.15. Suppose a wine manufacturer was interested in introducing a new wine to the market. A very good but somewhat expensive Chenin Blanc was already available to consumers and was very profitable to the company. To enter the lower-priced market of wine consumers, the wine manufacturer decided to make a less expensive wine that *tasted* similar to the very profitable Chenin Blanc already sold. After much market research and production knowledge, the manufacturer settled on four possible wines to introduce into the market. The fuzzy criteria of evaluation is *taste*, and we would like to know which wine tastes the most like the expensive Chenin Blanc. Define the following subjective estimations: Universe X $= \{x_1, x_2, x_3, x_4\}$. A panel of wine tasters tasted each of the wines x_1, x_2, x_3, and x_4 and made the following estimations:

$$
\begin{array}{llll}
f_{x_1}(x_1) = 1 & f_{x_1}(x_2) = .4 & f_{x_1}(x_3) = .8 & f_{x_1}(x_4) = .5 \\
f_{x_2}(x_1) = .2 & f_{x_2}(x_2) = 1 & f_{x_2}(x_3) = .7 & f_{x_2}(x_4) = .4 \\
f_{x_3}(x_1) = .3 & f_{x_3}(x_2) = .2 & f_{x_3}(x_3) = 1 & f_{x_3}(x_4) = .5 \\
f_{x_4}(x_1) = .7 & f_{x_4}(x_2) = .5 & f_{x_4}(x_3) = .8 & f_{x_4}(x_4) = 1
\end{array}
$$

Develop a comparison matrix, and determine which of the four wines tastes most like the expensive Chenin Blanc.

Multiobjective Decision Making

10.16. A carcinogen, trichloroethylene (TCE), has been detected in soil and groundwater at levels higher than the EPA maximum contaminant levels (MCLs). There is an immediate need to remediate soil and groundwater. Three remediation alternatives—(*i*) pump and treat with air stripping (PTA), (*ii*) pump and treat with photooxidation (PTP), and (*iii*) bioremediation of soil with pump and treat and air stripping (BPTA)—are investigated.

The objectives are these: cost (O_1), effectiveness (O_2, capacity to reduce the contaminant concentration), duration (O_3), and speed of implementation (O_4). The ranking of the alternatives on each objective are given as follows:

$$\underset{\sim}{O}_1 = \left\{ \frac{0.7}{\text{PTA}} + \frac{0.9}{\text{PTP}} + \frac{0.3}{\text{BPTA}} \right\}$$

$$\underset{\sim}{O}_2 = \left\{ \frac{0.4}{\text{PTA}} + \frac{0.6}{\text{PTP}} + \frac{0.8}{\text{BPTA}} \right\}$$

$$\underset{\sim}{O}_3 = \left\{ \frac{0.7}{\text{PTA}} + \frac{0.3}{\text{PTP}} + \frac{0.6}{\text{BPTA}} \right\}$$

$$\underset{\sim}{O}_4 = \left\{ \frac{0.8}{\text{PTA}} + \frac{0.5}{\text{PTP}} + \frac{0.5}{\text{BPTA}} \right\}$$

The preferences for each objective are P $= \{0.6, 0.8, 0.7, 0.5\}$. Determine the optimum choice of a remediation alternative.

10.17. An aluminum company needs to improve the performance of its cold rolling mill in order to meet new customer specifications. One way to improve the mill performance is to keep the mill equipment and upgrade the motor drives, cylinder, and control system that controls the final thickness of the metal off the mill (the customer specification that needs to be improved). The alternatives being considered for upgrading the control system are these: (*i*) developing and installing an in-house proprietary system (IN); (*ii*) developing a specification and getting an outside vendor to develop and install the system (OUT); and (*iii*) just keep the current system and make minor modifications (OLD).

The company has several objectives that it would like to meet with respect to the project. The first is the COST—it wants to minimize this. The second is the FLEXIBILITY—it wants the ability to make future changes and respond quickly to market demands. The third is the DOWNTIME—the company wants to minimize the downtime associated with upgrading to reduce loss production time. The fourth is the "PERFORMANCE against new market tolerances"—it needs to meet new market specifications. The project team, which includes a control engineer, production manager, and mechanical engineer, rate the alternatives against the objectives as:

COST $\quad \underset{\sim}{O}_1 = \left\{ \frac{0.4}{\text{IN}} + \frac{0.6}{\text{OUT}} + \frac{1}{\text{OLD}} \right\}$

FLEXIBILITY $\quad \underset{\sim}{O}_2 = \left\{ \frac{0.9}{\text{IN}} + \frac{0.7}{\text{OUT}} + \frac{0.2}{\text{OLD}} \right\}$

DOWNTIME $\quad \underset{\sim}{O}_3 = \left\{ \frac{0.5}{\text{IN}} + \frac{0.5}{\text{OUT}} + \frac{1}{\text{OLD}} \right\}$

PERFORMANCE $\quad \underset{\sim}{O}_4 = \left\{ \frac{1}{\text{IN}} + \frac{0.8}{\text{OUT}} + \frac{0.2}{\text{OLD}} \right\}$

Preferences defined by the company are COST, 0.6; FLEXIBILITY, 0.7; DOWNTIME, 0.8; and PERFORMANCE, 0.9. Determine the best alternative for upgrading the control system.

10.18. You are responsible for selecting an electronic CAE package to do schematic capture, printed circuit board design, and design simulation. The preliminary selection comes up with the high-performance packages Newton Graphics (NEWT), Cadence (CAD), and Racal Redac (RAC). The objectives that affect the final decision are (*i*) initial cost of the packages, (*ii*) maintenance costs, (*iii*) hardware and operating systems support, and (*iv*) speed of processing a milestone model. Based on prior experience on maintenance costs and performance, the objectives are ranked as follows:

$$\underset{\sim}{O}_{\text{cost}} = \left\{ \frac{0.4}{\text{NEWT}} + \frac{1}{\text{CAD}} + \frac{0.6}{\text{RAC}} \right\}$$

$$\underset{\sim}{O}_{\text{maint}} = \left\{ \frac{0.5}{\text{NEWT}} + \frac{0.7}{\text{CAD}} + \frac{0.6}{\text{RAC}} \right\}$$

$$\underset{\sim}{O}_{\text{hardware}} = \left\{ \frac{1}{\text{NEWT}} + \frac{0.6}{\text{CAD}} + \frac{0.8}{\text{RAC}} \right\}$$

$$\underset{\sim}{O}_{\text{speed}} = \left\{ \frac{0.9}{\text{NEWT}} + \frac{0.7}{\text{CAD}} + \frac{0.6}{\text{RAC}} \right\}$$

The preferences assigned to each objective are

$$b_1 = 0.9, \qquad b_2 = 0.7, \qquad b_3 = 0.5, \qquad b_4 = 0.5$$

Determine the CAE package that is suitable for the project.

10.19. Evaluate three different approaches to controlling conditions of an aluminum smelting cell (with respect to voltage across the cell and alumina concentration in the bath). The control approaches are

a_1 = AGT: aggressive control tuning (very reactive)

a_2 = MOD: moderate control tuning (mildly reactive)

a_3 = MAN: essentially manual operation (very little computer control)

There are several objectives to consider:

$\underset{\sim}{O}_1$: minimum power consumption (power/lb of aluminum produced)

$\underset{\sim}{O}_2$: overall operating stability

$\underset{\sim}{O}_3$: minimum environmental impact

The control approaches are rated as follows:

$$\underset{\sim}{O}_1 = \left\{ \frac{0.7}{\text{AGT}} + \frac{0.6}{\text{MOD}} + \frac{0.3}{\text{MAN}} \right\}$$

$$\underset{\sim}{O}_2 = \left\{ \frac{0.45}{\text{AGT}} + \frac{0.8}{\text{MOD}} + \frac{0.6}{\text{MAN}} \right\}$$

$$\underset{\sim}{O}_3 = \left\{ \frac{0.5}{\text{AGT}} + \frac{0.62}{\text{MOD}} + \frac{0.4}{\text{MAN}} \right\}$$

The preferences are given by $b_1 = 0.8$, $b_2 = 0.5$, and $b_3 = 0.6$. What is the best choice of control ?

10.20. One of the most important activities of a ship owner is the preservation of its ships. One of the factors in preserving a ship in a serviceable condition is painting since sea water is highly corrosive. The criteria used by ship owners in selecting marine paints are (*i*) corrosion resistance (COR), (*ii*) durability (DUR), (*iii*) availability (AV), (*iv*) toxicity (TOX), and (*v*) cost. A shipping company is presented with five choices of paint (B_i, $i = 1, 5$). The company needs to select one of the brands so that the purchasing department can negotiate a price with the paint manufacturers. The engineering department evaluated the five brands of paint with respect to the objectives as follows:

$$\underset{\sim}{O}_{COR} = \left\{ \frac{0.4}{B_1} + \frac{0.7}{B_2} + \frac{0.8}{B_3} + \frac{0.6}{B_4} + \frac{0.3}{B_5} \right\}$$

$$\underset{\sim}{O}_{DUR} = \left\{ \frac{0.5}{B_1} + \frac{0.9}{B_2} + \frac{0.5}{B_3} + \frac{0.8}{B_4} + \frac{0.2}{B_5} \right\}$$

$$\underset{\sim}{O}_{AV} = \left\{ \frac{0.9}{B_1} + \frac{0.7}{B_2} + \frac{0.4}{B_3} + \frac{0.3}{B_4} + \frac{0.7}{B_5} \right\}$$

$$\underset{\sim}{O}_{TOX} = \left\{ \frac{0.6}{B_1} + \frac{0.9}{B_2} + \frac{0.8}{B_3} + \frac{0.3}{B_4} + \frac{0.9}{B_5} \right\}$$

$$\underset{\sim}{O}_{COST} = \left\{ \frac{0.3}{B_1} + \frac{0.4}{B_2} + \frac{0.2}{B_3} + \frac{0.5}{B_4} + \frac{0.4}{B_5} \right\}$$

The ship owner ranked the purchasing objectives as corrosion = 0.9, durability = 1.0, availability = 0.6, toxicity = 0.7, and cost = 0.3. Find which brand of paint the shipping company should select for its ship preservation program.

Bayesian Decision Making

10.21. A company produces printed circuit boards as a subcomponent for a system that is integrated (with other subcomponents) by another company. The system integration company cannot give precise information on how many PC boards it needs other than "approximately 10,000." They may require more or less than this number. The PC board manufacturer has three courses of action from which to choose: (*i*) build somewhat less than 10,000 PC boards, $\underset{\sim}{A}_1$; (*ii*) build approximately 10,000 PC boards, $\underset{\sim}{A}_2$; and (*iii*) build somewhat more than 10,000 PC boards, $\underset{\sim}{A}_3$.

The systems integration company will need the PC boards to meet the demand for their final product. The following are the three fuzzy states of nature:

1. Low demand, $\underset{\sim}{D}_1$
2. Medium demand, $\underset{\sim}{D}_2$
3. High demand, $\underset{\sim}{D}_3$

The utility function is given in this table:

	$\underset{\sim}{D}_1$	$\underset{\sim}{D}_2$	$\underset{\sim}{D}_3$
$\underset{\sim}{A}_1$	3	2	−1
$\underset{\sim}{A}_2$	−1	4	2
$\underset{\sim}{A}_3$	−5	2	5

There are six discrete states of nature, s_1 through s_6, on which the fuzzy states are defined. The membership functions for the fuzzy states and the prior probabilities $p(s_i)$ of the discrete states are shown in the following table:

	s_1	s_2	s_3	s_4	s_5	s_6
$\mu_{\underset{\sim}{D}_1}$	1.0	0.7	0.1	0.0	0.0	0.0
$\mu_{\underset{\sim}{D}_2}$	0.0	0.3	0.9	0.9	0.3	0.0
$\mu_{\underset{\sim}{D}_3}$	0.0	0	0.0	0.1	0.7	1.0
$p(s_i)$	0.2	0.1	0.4	0.1	0.1	0.1

The demand for the system integrator's product is related to the growth of refineries, as the final product is used in refineries. The new samples of refinery growth information are x; and $\underset{\sim}{M}_i$ are the fuzzy sets on this information, defined as

1. Low growth, $\underset{\sim}{M}_1$
2. Medium growth, $\underset{\sim}{M}_2$
3. High growth, $\underset{\sim}{M}_3$

	x_1	x_2	x_3	x_4	x_5	x_6
$\mu_{\underset{\sim}{M}_1}$	1.0	0.7	0.2	0.0	0.0	0.0
$\mu_{\underset{\sim}{M}_2}$	0.0	0.3	0.8	0.8	0.3	0.0
$\mu_{\underset{\sim}{M}_3}$	0.0	0.0	0.0	0.2	0.7	1.0

The likelihood values for the probabilistic uncertain information for the data samples are shown here:

	x_1	x_2	x_3	x_4	x_5	x_6
s_1	0.1	0.1	0.5	0.1	0.1	0.1
s_2	0.0	0.0	0.1	0.4	0.4	0.1
s_3	0.1	0.2	0.4	0.2	0.1	0.0
s_4	0.5	0.1	0.0	0.0	0.2	0.2
s_5	0.0	0.0	0.0	0.1	0.3	0.6
s_6	0.1	0.7	0.2	0.0	0.0	0.0

The likelihood values for the probabilistic perfect information for the data samples are shown next:

	x_1	x_2	x_3	x_4	x_5	x_6
s_1	0.0	0.0	1.0	0.0	0.0	0.0
s_2	0.0	0.0	0.0	1.0	0.0	0.0
s_3	0.0	0.0	0.0	0.0	1.0	0.0
s_4	1.0	0.0	0.0	0.0	0.0	0.0
s_5	0.0	0.0	0.0	0.0	0.0	1.0
s_6	0.0	1.0	0.0	0.0	0.0	0.0

For the information just presented, compare the following for perfect and imperfect information:

(*a*) Posterior probabilities of fuzzy state 2 ($\underset{\sim}{D}_2$) given the fuzzy information 3 ($\underset{\sim}{M}_3$)

(*b*) Conditional expected utility for action 1 ($\underset{\sim}{A}_1$) and fuzzy information 2 ($\underset{\sim}{M}_2$)

10.22. In a particular region a water authority must decide whether to build dikes to prevent flooding in case of excess rainfall. Three fuzzy courses of action may be considered:

1. Build a permanent dike ($\underset{\sim}{A}_1$)
2. Build a temporary dike ($\underset{\sim}{A}_2$)
3. Do not build a dike ($\underset{\sim}{A}_3$)

The sets $\underset{\sim}{A}_1$, $\underset{\sim}{A}_2$, and $\underset{\sim}{A}_3$ are fuzzy sets depending on the type and size of the dike to be built. The utility from each of these investments depends on the rainfall in the region. The crisp states of nature, $S = \{s_1, s_2, s_3, s_4, s_5\}$ are the amount of total rainfall in millimeters in the region. The utility for each of the alternatives has been developed for three levels of rainfall: (1) low ($\underset{\sim}{F}_1$), (2) medium ($\underset{\sim}{F}_2$), and (3) heavy ($\underset{\sim}{F}_3$), which are defined by fuzzy sets on S. The utility matrix may be given as follows:

u_{ij}	$\underset{\sim}{F}_1$	$\underset{\sim}{F}_2$	$\underset{\sim}{F}_3$
$\underset{\sim}{A}_1$	−2	4	10
$\underset{\sim}{A}_2$	1	8	−10
$\underset{\sim}{A}_3$	10	−5	−20

The membership functions of $\underset{\sim}{F}_1$, $\underset{\sim}{F}_2$, $\underset{\sim}{F}_3$, and the prior probabilities are given here:

	s_1	s_2	s_3	s_4	s_5
$\mu_{\underset{\sim}{F}_1}(s_i)$	1	.4	.05	0	0
$\mu_{\underset{\sim}{F}_2}(s_i)$	0	.6	.85	.15	0
$\mu_{\underset{\sim}{F}_3}(s_i)$	0	0	.1	.85	1
$p(s_i)$	0.1	0.2	0.2	0.35	0.15

Let $X = \{x_1, x_2, x_3, x_4\}$ be the set of amount of rainfall in the next year. This represents the new information. The conditional probabilities $p(x_j \mid s_i)$ for probabilistic uncertain information are as given below.

	x_1	x_2	x_3	x_4
s_1	0.7	0.2	0.1	0.0
s_2	0.1	0.7	0.2	0.0
s_3	0.1	0.2	0.7	0.0
s_4	0.0	0.1	0.2	0.7
s_5	0.0	0.0	0.3	0.7

Consider a fuzzy information system,

$$\underset{\sim}{M} = \{\underset{\sim}{M}_1, \underset{\sim}{M}_2, \underset{\sim}{M}_3\}$$

where $\underset{\sim}{M}_1$ = rainfall is less than approximately 35 mm

$\underset{\sim}{M}_2$ = rainfall is equal to approximately 35 mm

$\underset{\sim}{M}_3$ = rainfall is greater than approximately 35 mm

The membership functions for the new fuzzy information that satisfy the orthogonality condition are given here:

	x_1	x_2	x_3	x_4
$\mu_{\underset{\sim}{M}_1}(x_i)$	1.0	0.3	0.1	0.0
$\mu_{\underset{\sim}{M}_2}(x_i)$	0.0	0.7	0.8	0.1
$\mu_{\underset{\sim}{M}_3}(x_i)$	0.0	0.0	0.1	0.9

Determine the following:

(a) Posterior probabilities for fuzzy state $\underset{\sim}{F}_2$ and fuzzy information $\underset{\sim}{M}_1$, and for fuzzy state $\underset{\sim}{F}_3$ and fuzzy information $\underset{\sim}{M}_3$

(b) Conditional expected utility of building a permanent dike ($\underset{\sim}{A}_1$) when fuzzy information $\underset{\sim}{M}_3$ is given

10.23. The merging of new hardware and new software systems can be a daunting process. If there are a large number of problems, completing the task can take a long time, costing market share, revenue, and possibly the company's success. If more hardware systems and machines are available, then more problems can be fixed and the computer can be delivered to market sooner. However, machines are expensive.

Define three fuzzy states of nature representing the number of problems to be found in a system: $\underset{\sim}{F}_1$ = low, $\underset{\sim}{F}_2$ = medium, and $\underset{\sim}{F}_3$ = high. Define three fuzzy actions representing the number of machines available for integration: $\underset{\sim}{A}_1$ = low, $\underset{\sim}{A}_2$ = medium, and $\underset{\sim}{A}_3$ = high. Management has assigned this utility matrix:

	$\underset{\sim}{F}_1$	$\underset{\sim}{F}_2$	$\underset{\sim}{F}_3$
$u(\underset{\sim}{A}_1, \underset{\sim}{F}_i)$	10	−10	−30
$u(\underset{\sim}{A}_2, \underset{\sim}{F}_i)$	0	15	−15
$u(\underset{\sim}{A}_3, \underset{\sim}{F}_i)$	5	0	20

Define the complexity of the system as set X = $\{x_1, x_2, x_3, x_4, x_5\}$ and the number of problems as set S = $\{s_1, s_2, s_3, s_4, s_5\}$. The membership and prior probability of s_i are as follows:

	s_1	s_2	s_3	s_4	s_5
$\mu_{\underset{\sim}{F}_1}(s_i)$	1.0	0.7	0.2	0.0	0.0
$\mu_{\underset{\sim}{F}_2}(s_i)$	0.0	0.3	0.7	0.3	0.0
$\mu_{\underset{\sim}{F}_3}(s_i)$	0.0	0.0	0.1	0.7	1.0
$p(s_i)$	0.05	0.05	0.2	0.3	0.4

Consider the fuzzy information system $\underset{\sim}{M}_1 = \{\underset{\sim}{M}_1, \underset{\sim}{M}_2, \underset{\sim}{M}_3\}$, where

	x_1	x_2	x_3	x_4	x_5
$\mu_{\underset{\sim}{M}_1}(x_i)$	1.0	0.5	0.1	0.0	0.0
$\mu_{\underset{\sim}{M}_2}(x_i)$	0.0	0.5	0.7	0.2	0.0
$\mu_{\underset{\sim}{M}_3}(x_i)$	0.0	0.0	0.2	0.8	1.0

(*a*) Determine the actions to take for each fuzzy information, $\underset{\sim}{M}_i$, $i = 1, 2, 3$, for probabilistic uncertain information $p(x_j \mid s_i)$ as given in the following table:

	x_1	x_2	x_3	x_4	x_5
s_1	0.7	0.3	0.0	0.0	0.0
s_2	0.2	0.5	0.3	0.0	0.0
s_3	0.0	0.1	0.8	0.1	0.0
s_4	0.0	0.0	0.2	0.7	0.1
s_5	0.0	0.0	0.0	0.1	0.9

(*b*) Determine the actions to take for each fuzzy information, $\underset{\sim}{M}_i$, $i = 1, 2, 3$, for perfect information.

	x_1	x_2	x_3	x_4	x_5
s_1	1	0	0	0	0
s_2	0	1	0	0	0
s_3	0	0	1	0	0
s_4	0	0	0	1	0
s_5	0	0	0	0	1

10.24. One oil company planned to order a new oil tanker to meet the needs of plants in the booming southern coastal area. They asked you to help them investigate the design of that new ship. The displacement (capacity) of the ship is related to the cost and the operating efficiency, which largely depend on the demand of the oil in the industrial area. You can select from three possible actions, i.e., $\underset{\sim}{A}_1$ (small ship), $\underset{\sim}{A}_2$ (medium ship), $\underset{\sim}{A}_3$ (large ship); and the oil demand is classified as $\underset{\sim}{D}_1$ (low demand), $\underset{\sim}{D}_2$ (medium demand), $\underset{\sim}{D}_3$ (high demand). The utility function $U(\underset{\sim}{A}_i, \underset{\sim}{D}_j)$ is written as

	$\underset{\sim}{D}_1$	$\underset{\sim}{D}_2$	$\underset{\sim}{D}_3$
$\underset{\sim}{A}_1$	100	120	130
$\underset{\sim}{A}_2$	0	200	250
$\underset{\sim}{A}_3$	−50	100	300

The demand for oil is based on the orders the company receives, which can be divided into several categories, say five, d_1 to d_5. Suppose there is some historical record to predict the probability of possible situations, such as prior probabilities. Fuzzy numbers $\underset{\sim}{D}_1$ to $\underset{\sim}{D}_3$ are assigned membership function values on that.

The membership functions of $\underset{\sim}{D}_1$, $\underset{\sim}{D}_2$, and $\underset{\sim}{D}_3$ and prior probabilities $p(d_i)$ are given in the following table:

	d_1	d_2	d_3	d_4	d_5
$\mu_{\underset{\sim}{D}_1}(d_i)$	1.0	0.5	0.0	0.0	0.0
$\mu_{\underset{\sim}{D}_2}(d_i)$	0.0	0.5	1.0	0.5	0.0
$\mu_{\underset{\sim}{D}_3}(d_i)$	0.0	0.0	0.0	0.5	1.0
$p(d_i)$	0.2	0.4	0.2	0.1	0.1

The probability is based on a stock index, which can be regarded as a symptom of the economic factors that influences the oil demand, say $X = \{x_1, \ldots, x_7\}$. This probability is obtained from worldwide economic indices:

	x_1	x_2	x_3	x_4	x_5	x_6	x_7
d_1	0.6	0.3	0.1	0.0	0.0	0.0	0.0
d_2	0.1	0.7	0.2	0.0	0.0	0.0	0.0
d_3	0.0	0.15	0.4	0.4	0.05	0.0	0.0
d_4	0.0	0.0	0.05	0.05	0.1	0.5	0.3
d_5	0.0	0.0	0.0	0.05	0.05	0.2	0.7

We also define three fuzzy numbers based on $X = \{x_1, \ldots, x_7\}$: $\underset{\sim}{M}_1$—low; $\underset{\sim}{M}_2$—medium; $\underset{\sim}{M}_3$—high:

	x_1	x_2	x_3	x_4	x_5	x_6	x_7
$\mu\underset{\sim}{M}_1(x_j)$	1.0	0.8	0.2	0.0	0.0	0.0	0.0
$\mu\underset{\sim}{M}_2(x_j)$	0.0	0.2	0.8	1.0	0.8	0.2	0.0
$\mu\underset{\sim}{M}_3(x_j)$	0.0	0.0	0.0	0.0	0.2	0.8	1.0

If the future situation can be predicted precisely, "perfect" information can be used here instead of probabilistic information. The following table gives the $p(x_j \mid d_i)$:

	x_1	x_2	x_3	x_4	x_5	x_6	x_7
d_1	0.7	0.3	0.0	0.0	0.0	0.0	0.0
d_2	0.0	0.0	0.6	0.4	0.0	0.0	0.0
d_3	0.0	0.0	0.0	0.0	1.0	0.0	0.0
d_4	0.0	0.0	0.0	0.0	0.0	1.0	0.0
d_5	0.0	0.0	0.0	0.0	0.0	0.0	1.0

Compute the conditional expected utility for action $\underset{\sim}{A}_1$ and fuzzy information $\underset{\sim}{M}_2$ for uncertain and perfect information.

10.25. Your design team needs to determine what level of technology to incorporate in a new product. As is usually the case, current technology is least expensive whereas the most advanced or leading edge technology is the most expensive. A given technology usually comes down in price with time. The decision cycle of your project is several years. The team must decide what level of technology to incorporate in the product based on the future expected cost. If the technology is still expensive by the time the product goes to the market, the product will not sell. If you do not incorporate the latest affordable technology, your product may not be so advanced as that of the competition and therefore sales may be poor. Consider the following:

Actual discrete states of nature:

s_1: Cost is low

s_2: Cost is moderate

s_3: Cost is high

Fuzzy actions:

$\underset{\sim}{A}_1$: Use current /well-established technology

$\underset{\sim}{A}_2$: Use newer / leading edge / advanced technology

Fuzzy states on fuzzy information system, μ:

$\underset{\sim}{M}_1$: Cost is approximately the cost of implementing with current technology

$\underset{\sim}{M}_2$: Cost is approximately 2 times the cost of the current technology

$\underset{\sim}{M}_3$: Cost is approximately 10 times the cost of current technology

Let $X = \{x_1, x_2, x_3, x_4, x_5\}$ be the set of rates of increase in usage of advanced technology in the next term. Then we have the following:

Fuzzy states of nature:

$\underset{\sim}{F}_1$: Low cost

$\underset{\sim}{F}_2$: Medium cost

$\underset{\sim}{F}_3$: High cost

Prior probabilities:

$$p(s_i) = \begin{bmatrix} 0.25 \\ 0.5 \\ 0.25 \end{bmatrix} \begin{matrix} s_1 \\ s_2 \\ s_3 \end{matrix}$$

Utility matrix:

$$u = \begin{matrix} & \begin{matrix} s_1 & s_2 & s_3 \end{matrix} \\ & \begin{bmatrix} -8 & -5 & 0 \\ 10 & -5 & -10 \end{bmatrix} \begin{matrix} \underset{\sim}{A}_1 \\ \underset{\sim}{A}_2 \end{matrix} \end{matrix}$$

Membership values for each orthogonal fuzzy state on the actual state system:

$$\mu_{\underset{\sim}{F}} = \begin{matrix} & \begin{matrix} s_1 & s_2 & s_3 \end{matrix} \\ & \begin{bmatrix} .8 & .1 & 0 \\ .2 & .8 & .2 \\ 0 & .1 & .8 \end{bmatrix} \begin{matrix} \underset{\sim}{F}_1 \\ \underset{\sim}{F}_2 \\ \underset{\sim}{F}_3 \end{matrix} \end{matrix}$$

Membership values for each orthogonal fuzzy set on the fuzzy information system:

$$
\mu_M = \begin{array}{c} \\ \\ \\ \end{array}\begin{array}{ccccc} x_1 & x_2 & x_3 & x_4 & x_5 \\ \end{array} \\
\left[\begin{array}{ccccc} 1 & .5 & 0 & 0 & 0 \\ 0 & .5 & 1 & .5 & 0 \\ 0 & 0 & 0 & .5 & 1 \end{array}\right]\begin{array}{c} M_1 \\ M_2 \\ M_3 \end{array}
$$

Utility matrix for fuzzy information:

$$
u = \begin{array}{ccc} F_1 & F_2 & F_3 \end{array} \\
\left[\begin{array}{ccc} -5 & 0 & 5 \\ 10 & 2 & -10 \end{array}\right]\begin{array}{c} A_1 \\ A_2 \end{array}
$$

Likelihood values for probabilistic (uncertain) information for the data samples:

$$
p(x_i \mid s_k) = \begin{array}{ccccc} x_1 & x_2 & x_3 & x_4 & x_5 \end{array} \\
\left[\begin{array}{ccccc} .1 & .25 & .15 & .35 & .15 \\ .3 & .05 & .1 & .1 & .45 \\ .2 & .4 & .35 & 0 & .05 \end{array}\right]\begin{array}{c} s_1 \\ s_2 \\ s_3 \end{array}
$$

Likelihood values for probabilistic perfect information for the data samples:

$$
p(x_i \mid s_k) = \begin{array}{ccccc} x_1 & x_2 & x_3 & x_4 & x_5 \end{array} \\
\left[\begin{array}{ccccc} 0 & 0 & 0 & 1 & 0 \\ .4 & 0 & 0 & 0 & .6 \\ 0 & .55 & .45 & 0 & 0 \end{array}\right]\begin{array}{c} s_1 \\ s_2 \\ s_3 \end{array}
$$

(a) Determine the value of information for the fuzzy states and fuzzy actions for uncertain probabilistic information.

(b) Determine the value of information for the fuzzy states and fuzzy actions for perfect probabilistic information.

Fuzzy Preference and Consensus

10.26. An electronics manufacturer of audio speakers is assessing the fidelity of four speakers so that the best speaker can be manufactured on a large scale. The pairwise preferences for the speakers are given as

$$
R = \left[\begin{array}{cccc} 0 & 0.4 & 0.7 & 0.5 \\ 0.6 & 0 & 0.9 & 0.8 \\ 0.3 & 0.1 & 0 & 0.9 \\ 0.5 & 0.2 & 0.1 & 0 \end{array}\right]
$$

Determine the average fuzziness, and average certainty in R.

10.27. Four professional photographers are asked to suggest the best camera to purchase among the four latest models of A, B, C, and D. The pairwise relationship matrix formed by the photographers is shown here:

$$
R = \left[\begin{array}{cccc} 0 & 0.8 & 0.5 & 0.4 \\ 0.2 & 0 & 0.4 & 0.2 \\ 0.5 & 0.6 & 0 & 0.6 \\ 0.6 & 0.8 & 0.4 & 0 \end{array}\right]
$$

For the preceding relationship determine the following:

(*a*) Average fuzziness in the relationship

(*b*) The distance to type I consensus

10.28. The Environmental Protection Agency (EPA) is faced with the challenge of cleaning up contaminated groundwaters at many sites around the country. In order to ensure an efficient cleanup process, it is crucial to select a firm that offers the best remediation technology at a reasonable cost. The EPA is deciding among four environmental firms. The professional engineers at the EPA compared the four firms and created a relationship matrix, shown here:

$$\underset{\sim}{R} = \begin{bmatrix} 0 & 0.5 & 0.7 & 0.4 \\ 0.5 & 0 & 0.9 & 0.2 \\ 0.3 & 0.1 & 0 & 0.3 \\ 0.6 & 0.8 & 0.7 & 0 \end{bmatrix}$$

Compute the distance to Type *fuzzy* consensus.

10.29. An international construction company is considering updating its radio system. They are evaluating the capabilities and costs of five systems presented by five different electronics firms. They develop a reciprocal fuzzy relation for the pairwise preference based on the subjective evaluations of the five systems, as shown next:

$$\underset{\sim}{R} = \begin{bmatrix} 0 & 1.0 & 0.5 & 0.3 & 0.6 \\ 0 & 0 & 0.4 & 0.3 & 0.7 \\ 0.5 & 0.6 & 0 & 0.4 & 0.7 \\ 0.7 & 0.7 & 0.6 & 0 & 0.9 \\ 0.4 & 0.3 & 0.3 & 0.1 & 0 \end{bmatrix}$$

Calculate the average fuzziness in $\underset{\sim}{R}$, and the distance to Type I and to Type II consensus.

10.30. Britz manufacturing company is planning to purchase a lathe, and they are assessing the proposals from four lathe manufacturers. They have developed a reciprocal relation for the four manufacturers based on the speed of delivery of the lathes and the cost. The relation is

$$\underset{\sim}{R} = \begin{bmatrix} 0 & 0.1 & 0.7 & 0.2 \\ 0.9 & 0 & 0.6 & 1 \\ 0.3 & 0.4 & 0 & 0.5 \\ 0.8 & 0 & 0.5 & 0 \end{bmatrix}$$

Calculate the degree of preference measures, and the distance to Type I, Type II, and Type *fuzzy* consensus. Explain the differences between the distances to the three consensuses.

CHAPTER 11

FUZZY CLASSIFICATION

From causes which appear similar, we expect similar effects. This is the sum total of all our experimental conclusions.

David Hume
Scottish philosopher
Enquiry Concerning Human Understanding, *1748*

There is structure in nature. Much of this structure is known to us and is quite beautiful. Consider the natural sphericity of rain drops and bubbles; why do balloons take this shape? How about the elegant beauty of crystals, rhombic solids with rectangular, pentagonal, or hexagonal cross sections? Why do these naturally beautiful, geometric shapes exist? What causes the natural repetition of the mounds of sand dunes? Some phenomena we cannot see directly, for example, the elliptical shape of the magnetic field around the Earth; or we can see only when certain atmospheric conditions exist such as the beautiful and circular appearance of a rainbow or the repetitive patterns of the aurora borealis in the night sky near the North Pole. Some patterns, such as the helical appearance of DNA or the cylindrical shape of some bacteria, have only appeared to us since the advent of extremely powerful electron microscopes. Consider the geometry and colorful patterns of a butterfly's wings; why do these patterns exist in our physical world? The answers to some of these questions are still unknown; many others have been discovered through increased understanding of physics, chemistry, and biology.

Just as there is structure in nature, we believe there is an underlying structure in most of the phenomena we wish to understand. Examples abound in image recognition, molecular biology applications such as protein folding and 3D molecular structure, oil exploration, cancer detection, and many others. For fields dealing with diagnosis we often seek to find structure in the data obtained from observation. Our observations can be visual, audio, or any of a variety of sensor-based electronic or optical signals. Finding the structure in data is the essence of classification. As the quotation at the beginning of this chapter suggests, our experimental observations lead us to develop relationships between the inputs and outputs of an experiment. As we are able to conduct more experiments we see the relationships forming some recognizable, or classifiable, structure. By finding structure, we are classifying the data according to similar patterns, attributes, features, and other characteristics. The general area is known as *classification*.

In classification, also termed *clustering,* the most important issue is deciding what criteria to classify against. For example, suppose we want to classify people. In describing people we will look at their height, weight, gender, religion, education, appearance, and so on. Many of these features are numerical quantities such as height and weight; other features are simply linguistic descriptors and these can be quite nonnumeric. We can easily classify people according to sex; we would only need one of the many features available to describe people—gender. For this classification the criterion is simple: female or male. We might want to classify people into three size categories—small, medium, and large. For this classification we might only need two of the features describing people: height and weight. Here, the classification criterion might be some algebraic combination of height and weight. Suppose we want to classify people according to whether we would want them as *neighbors*. Here the number of features to be used in the classification is not at all clear, and we might also have trouble developing a criterion for this classification. Nevertheless, a criterion for classification must be prepared before we can segregate the data into definable classes. As is often the case in classification studies, the number and kind of features and the type of classification criteria are choices that are continually changed as the data are manipulated; and this iteration continues until we think we have a grouping of the data that seems plausible from a structural and physical perspective.

This chapter summarizes only two popular methods of classification. The first is classification using equivalent relations [Zadeh, 1971; Bezdek and Harris, 1978]. This approach makes use of certain special properties of equivalent relations and the concept of defuzzification known as lambda-cuts on the relations. The second method of classification is a very popular method known as *fuzzy c-means,* so named because of its close analog in the crisp world, *hard c-means* [Bezdek, 1981]. This method uses concepts in n-dimensional Euclidean space to determine the geometric *closeness* of data points by assigning them to various clusters or classes and then determining the distance between the clusters.

CLASSIFICATION BY EQUIVALENCE RELATIONS

Crisp Relations

Define a set, $[x_i] = \{x_j \mid (x_i, x_j) \in \text{R}\}$, as the equivalent class of x_i on a universe of data points, X. This class is contained in a special relation, R, known as an equivalent relation (see Chapter 3). This class is a set of all elements related to x_i that have the following properties [Bezdek, 1974]:

1. $x_i \in [x_i] \qquad \therefore (x_i, x_i) \in \mathrm{R}$
2. $[x_i] \neq [x_j] \Rightarrow [x_i] \cap [x_j] = \varnothing$
3. $\bigcup_{x \in \mathrm{X}} [x] = \mathrm{X}$

The first property is that of reflexivity (see Chapter 3), the second property indicates that equivalent classes do not overlap, and the third property simply expresses that the union of all equivalent classes exhausts the universe. Hence, the equivalence relation R can divide the universe X into mutually exclusive equivalent classes, i.e.,

$$\mathrm{X} \mid \mathrm{R} = \{[x] \mid x \in \mathrm{X}\} \tag{11.1}$$

where X|R is called the quotient set. The quotient set of X relative to R, denoted X|R, is the set whose elements are the equivalence classes of X under the equivalence relation R. The cardinality of X|R (i.e., the number of distinct equivalence classes of X under R) is called the rank of the matrix R.

Example 11.1 [Dong, 1987]. Define a universe of integers, $\mathrm{X} = \{1, 2, 3, \ldots, 10\}$ and define R as the crisp relation for "the identical remainder after dividing each element of the universe by 3." We have

$$\mathrm{R} = \begin{array}{c} \\ 1 \\ 2 \\ 3 \\ 4 \\ 5 \\ 6 \\ 7 \\ 8 \\ 9 \\ 10 \end{array} \begin{array}{c} \begin{array}{cccccccccc} 1 & 2 & 3 & 4 & 5 & 6 & 7 & 8 & 9 & 10 \end{array} \\ \begin{bmatrix} 1 & 0 & 0 & 1 & 0 & 0 & 1 & 0 & 0 & 1 \\ 0 & 1 & 0 & 0 & 1 & 0 & 0 & 1 & 0 & 0 \\ 0 & 0 & 1 & 0 & 0 & 1 & 0 & 0 & 1 & 0 \\ 1 & 0 & 0 & 1 & 0 & 0 & 1 & 0 & 0 & 1 \\ 0 & 1 & 0 & 0 & 1 & 0 & 0 & 1 & 0 & 0 \\ 0 & 0 & 1 & 0 & 0 & 1 & 0 & 0 & 1 & 0 \\ 1 & 0 & 0 & 1 & 0 & 0 & 1 & 0 & 0 & 1 \\ 0 & 1 & 0 & 0 & 1 & 0 & 0 & 1 & 0 & 0 \\ 0 & 0 & 1 & 0 & 0 & 1 & 0 & 0 & 1 & 0 \\ 1 & 0 & 0 & 1 & 0 & 0 & 1 & 0 & 0 & 1 \end{bmatrix} \end{array}$$

We note that this relation is reflexive, it is symmetric, and, as can be determined by inspection (see Chapter 3), it is also transitive; hence the matrix is an equivalence relation. We can group the elements of the universe into classes as follows:

$$\begin{aligned} [1] = [4] = [7] = [10] &= \{1, 4, 7, 10\} \qquad \text{with remainder} = 1 \\ [2] = [5] = [8] &= \{2, 5, 8\} \qquad \text{with remainder} = 2 \\ [3] = [6] = [9] &= \{3, 6, 9\} \qquad \text{with remainder} = 0 \end{aligned}$$

Then we can show that the classes do not overlap, i.e., they are mutually exclusive:

$$[1] \cap [2] = \varnothing \text{ and } [2] \cap [3] = \varnothing$$

and that the union of all the classes exhausts (comprises) the universe.

$$\bigcup [x] = \mathrm{X}$$

The quotient set is then determined to have three classes,

$$\mathrm{X} \mid \mathrm{R} = \{(1, 4, 7, 10), (2, 5, 8), (3, 6, 9)\}$$

Not all relations are equivalent, but if a relation is at least a tolerance relation (i.e., it exhibits properties of reflexivity and symmetry) then it can be converted to an equivalent relation through max-min compositions with itself.

Example 11.2. Suppose you have a collection (universe) of five data points,

$$\mathrm{X} = \{x_1, x_2, x_3, x_4, x_5\}$$

and these data points show similarity to one another according to the following tolerance relation, which is reflexive and symmetric:

$$\mathrm{R}_1 = \begin{bmatrix} 1 & 1 & 0 & 0 & 0 \\ 1 & 1 & 0 & 0 & 1 \\ 0 & 0 & 1 & 0 & 0 \\ 0 & 0 & 0 & 1 & 0 \\ 0 & 1 & 0 & 0 & 1 \end{bmatrix}$$

We see that this tolerance relation is not transitive from the expression

$$(x_1, x_2) \in \mathrm{R}_1, \qquad (x_2, x_5) \in \mathrm{R}_1 \qquad \text{but} \qquad (x_1, x_5) \notin \mathrm{R}_1$$

As indicated in Chapter 3, any tolerance relation can be reformed into an equivalence relation through at most $n - 1$ compositions with itself. In this case one composition of R_1 with itself results in an equivalence relation,

$$\mathrm{R}_1 \circ \mathrm{R}_1 = \begin{bmatrix} 1 & 1 & 0 & 0 & 1 \\ 1 & 1 & 0 & 0 & 1 \\ 0 & 0 & 1 & 0 & 0 \\ 0 & 0 & 0 & 1 & 0 \\ 1 & 1 & 0 & 0 & 1 \end{bmatrix} = \mathrm{R}$$

As one can see in the relation, R, there are three classes. Columns 1, 2, and 5 are identical and columns 4 and 5 are each unique. The data points can then be classified into three groups or classes, as delineated below:

$$[x_1] = [x_2] = [x_5] = \{x_1, x_2, x_5\} \qquad [x_3] = \{x_3\} \qquad [x_4] = \{x_4\}$$

Fuzzy Relations

As already illustrated, crisp equivalent relations can be used to divide the universe X into mutually exclusive classes. In the case of fuzzy relations, for all fuzzy equivalent relations, their λ-cuts are equivalent ordinary relations. Hence, to classify data points in the universe using fuzzy relations, we need to find the associated fuzzy equivalent relation.

Example 11.3. Example 3.11 had a tolerance relation, say R_t, describing five data points, that was formed into a fuzzy equivalence relation, R, by composition; this process is repeated here for this classification example.

$$\mathrm{R}_t = \begin{bmatrix} 1 & 0.8 & 0 & 0.1 & 0.2 \\ 0.8 & 1 & 0.4 & 0 & 0.9 \\ 0 & 0.4 & 1 & 0 & 0 \\ 0.1 & 0 & 0 & 1 & 0.5 \\ 0.2 & 0.9 & 0 & 0.5 & 1 \end{bmatrix} \rightarrow \mathrm{R} = \begin{bmatrix} 1 & 0.8 & 0.4 & 0.5 & 0.8 \\ 0.8 & 1 & 0.4 & 0.5 & 0.9 \\ 0.4 & 0.4 & 1 & 0.4 & 0.4 \\ 0.5 & 0.5 & 0.4 & 1 & 0.5 \\ 0.8 & 0.9 & 0.4 & 0.5 & 1 \end{bmatrix}$$

By taking λ-cuts of fuzzy equivalent relation R at values of $\lambda = 1, 0.9, 0.8, 0.5$, and 0.4, we get the following:

$$R_1 = \begin{bmatrix} 1 & & & & 0 \\ & 1 & & & \\ & & 1 & & \\ & & & 1 & \\ 0 & & & & 1 \end{bmatrix} \quad R_{0.9} = \begin{bmatrix} 1 & 0 & 0 & 0 & 0 \\ 0 & 1 & 0 & 0 & 1 \\ 0 & 0 & 1 & 0 & 0 \\ 0 & 0 & 0 & 1 & 0 \\ 0 & 1 & 0 & 0 & 1 \end{bmatrix} \quad R_{0.8} = \begin{bmatrix} 1 & 1 & 0 & 0 & 1 \\ 1 & 1 & 0 & 0 & 1 \\ 0 & 0 & 1 & 0 & 0 \\ 0 & 0 & 0 & 1 & 0 \\ 1 & 1 & 0 & 0 & 1 \end{bmatrix}$$

$$R_{0.5} = \begin{bmatrix} 1 & 1 & 0 & 1 & 1 \\ 1 & 1 & 0 & 1 & 1 \\ 0 & 0 & 1 & 0 & 0 \\ 1 & 1 & 0 & 1 & 1 \\ 1 & 1 & 0 & 1 & 1 \end{bmatrix} \quad R_{0.4} = \begin{bmatrix} 1 & 1 & 1 & 1 & 1 \\ 1 & 1 & 1 & 1 & 1 \\ 1 & 1 & 1 & 1 & 1 \\ 1 & 1 & 1 & 1 & 1 \\ 1 & 1 & 1 & 1 & 1 \end{bmatrix}$$

where we can see that the clustering of the five data points according to the λ-cut level is as shown in Table 11.1.

We can express the classification scenario described in Table 11.1 with a systematic classification diagram, as shown in Fig. 11.1. In the figure you can see that the higher the value of λ, the finer is the classification. That is, as λ gets larger the tendency of classification tends to approach the trivial case where each data point is assigned to its own class.

Another example in fuzzy classification considers grouping photographs of family members together according to visual similarity in attempting to determine genetics of the family tree when considering only facial image.

Example 11.4 [Tamura et al., 1971]. Three families exist that have a total of 16 people, all of whom are related by blood. Each person has his or her photo taken, and the 16 photos are mixed. A person not familiar with the members of the three families is asked to view the photographs to grade their resemblance to one another. In conducting this study the person assigns the similarity relation matrix, r_{ij} as shown in Table 11.2. The matrix developed by the person is a tolerance fuzzy relation, but it does not have properties of equivalence, i.e.,

$$r_{ij} = 1 \quad \text{for } i = j$$

$$r_{ij} = r_{ji}$$

$$r_{ij} \geq \lambda_1 \quad \text{and} \quad r_{jk} \geq \lambda_2 \quad \text{but} \quad r_{ik} < \min(\lambda_1, \lambda_2); \quad \text{i.e., transitivity does not hold.}$$

TABLE 11.1
Classification of five data points according to λ-cut level

λ-cut level	Classification
1.0	$[x_1][x_2][x_3][x_4][x_5]$
0.9	$[x_1]\{x_2, x_5\}[x_3][x_4]$
0.8	$\{x_1, x_2, x_5\}[x_3][x_4]$
0.5	$\{x_1, x_2, x_4, x_5\}[x_3]$
0.4	$\{x_1, x_2, x_3, x_4, x_5\}$

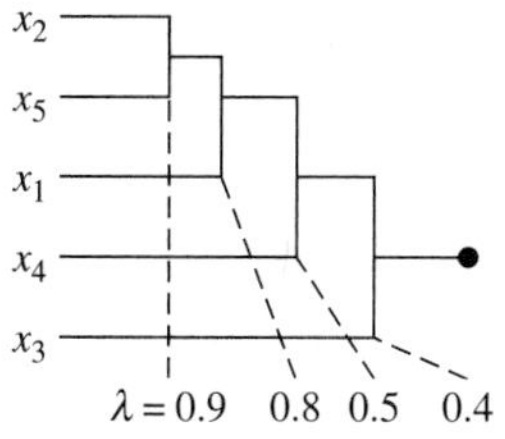

FIGURE 11.1
Classification diagram for Example 11.3.

TABLE 11.2
Similarity relation matrix, r_{ij}

	1	2	3	4	5	6	7	8	9	10	11	12	13	14	15	16
1	1.0															
2	0.0	1.0														
3	0.0	0.0	1.0													
4	0.0	0.0	0.4	1.0												
5	0.0	0.8	0.0	0.0	1.0											
6	0.5	0.0	0.2	0.2	0.0	1.0										
7	0.0	0.8	0.0	0.0	0.4	0.0	1.0									
8	0.4	0.2	0.2	0.5	0.0	0.8	0.0	1.0								
9	0.0	0.4	0.0	0.8	0.4	0.2	0.4	0.0	1.0							
10	0.0	0.0	0.2	0.2	0.0	0.0	0.2	0.0	0.2	1.0						
11	0.0	0.5	0.2	0.2	0.0	0.0	0.8	0.0	0.4	0.2	1.0					
12	0.0	0.0	0.2	0.8	0.0	0.0	0.0	0.0	0.4	0.8	0.0	1.0				
13	0.8	0.0	0.2	0.4	0.0	0.4	0.0	0.4	0.0	0.0	0.0	0.0	1.0			
14	0.0	0.8	0.0	0.2	0.4	0.0	0.8	0.0	0.2	0.2	0.6	0.0	0.0	1.0		
15	0.0	0.0	0.4	0.8	0.0	0.2	0.0	0.0	0.2	0.0	0.0	0.2	0.2	0.0	1.0	
16	0.6	0.0	0.0	0.2	0.2	0.8	0.0	0.4	0.0	0.0	0.0	0.0	0.4	0.2	0.4	1.0

For example,

$$r_{16} = 0.5, \qquad r_{68} = 0.8, \qquad \text{but} \qquad r_{18} = 0.4 < 0.5$$

By composition the equivalence relation shown in Table 11.3 is obtained.

When we take a λ-cut of this fuzzy equivalent relation at $\lambda = 0.6$, we get the defuzzified relation shown in Table 11.4.

TABLE 11.3
Equivalence relation

	1	2	3	4	5	6	7	8	9	10	11	12	13	14	15	16
1	1.0															
2	0.4	1.0														
3	0.4	0.4	1.0													
4	0.5	0.4	0.4	1.0												
5	0.4	0.8	0.4	0.4	1.0											
6	0.6	0.4	0.4	0.5	0.4	1.0										
7	0.4	0.8	0.4	0.4	0.8	0.4	1.0									
8	0.6	0.4	0.4	0.5	0.4	0.8	0.4	1.0								
9	0.5	0.4	0.4	0.8	0.4	0.5	0.4	0.5	1.0							
10	0.5	0.4	0.4	0.8	0.4	0.5	0.4	0.5	0.8	1.0						
11	0.4	0.8	0.4	0.4	0.8	0.4	0.8	0.4	0.4	0.4	1.0					
12	0.5	0.4	0.4	0.8	0.4	0.5	0.4	0.5	0.8	0.8	0.4	1.0				
13	0.8	0.4	0.4	0.5	0.4	0.6	0.4	0.6	0.5	0.5	0.4	0.5	1.0			
14	0.4	0.8	0.4	0.4	0.8	0.4	0.8	0.4	0.4	0.4	0.8	0.4	0.4	1.0		
15	0.5	0.4	0.4	0.8	0.4	0.5	0.4	0.5	0.8	0.8	0.4	0.8	0.5	0.4	1.0	
16	0.6	0.4	0.4	0.5	0.4	0.8	0.4	0.8	0.5	0.5	0.4	0.5	0.6	0.4	0.5	1.0

TABLE 11.4
Defuzzified relation

	1	2	3	4	5	6	7	8	9	10	11	12	13	14	15	16
1	1															
2	0	1														
3	0	0	1													
4	0	0	0	1												
5	0	1	0	0	1											
6	1	0	0	0	0	1										
7	0	0	0	0	1	0	1									
8	1	0	0	0	0	1	0	1								
9	0	0	0	1	0	0	0	0	1							
10	0	0	0	1	0	0	0	0	1	1						
11	0	1	0	0	1	0	1	0	0	0	1					
12	0	0	0	1	0	0	0	0	1	1	0	1				
13	1	0	0	0	0	1	0	1	0	0	0	0	1			
14	0	1	0	0	1	0	1	0	0	0	1	0	0	1		
15	0	0	0	1	0	0	0	0	1	1	0	1	0	0	1	
16	1	0	0	0	0	1	0	1	0	0	0	0	1	0	0	1

Four distinct classes are identified:

$$\{1, 6, 8, 13, 16\}, \qquad \{2, 5, 7, 11, 14\}, \qquad \{3\}, \qquad \{4, 9, 10, 12, 15\}$$

From this clustering it seems that only photograph number 3 cannot be identified with any of the three families. Perhaps a lower value of λ might assign photograph 3 to one of the other three classes. The other three clusters are all correct in that the members identified in each class are, in fact, the members of the correct families as described in Tamura et al (1971).

Classification using equivalence relations can also be employed to segregate data that are originally developed as a similarity relation using some of the similarity methods developed at the end of Chapter 3. The following problem is an example of this, involving earthquake damage assessment. It was first introduced in Chapter 3 as Example 3.12.

Example 11.5. Five regions have suffered damage from a recent earthquake (see Example 3.12). The buildings in each region are characterized according to three damage levels: no damage, medium damage, and serious damage. The percentage of buildings for a given region in each of the damage levels is given in Table 11.5.

TABLE 11.5
Proportion of buildings damaged, in three levels by region

	Regions				
	x_1	x_2	x_3	x_4	x_5
x_{i1}—Ratio with no damage	0.3	0.2	0.1	0.7	0.4
x_{i2}—Ratio with medium damage	0.6	0.4	0.6	0.2	0.6
x_{i3}—Ratio with serious damage	0.1	0.4	0.3	0.1	0

Using the cosine amplitude approach, described in Chapter 3, we obtain the following tolerance relation, $\underset{\sim}{R}_1$:

$$\underset{\sim}{R}_1 = \begin{bmatrix} 1 & & & & \\ 0.836 & 1 & & \text{sym} & \\ 0.914 & 0.934 & 1 & & \\ 0.682 & 0.6 & 0.441 & 1 & \\ 0.982 & 0.74 & 0.818 & 0.774 & 1 \end{bmatrix}$$

Three max-min compositions produce a fuzzy equivalence relation,

$$\underset{\sim}{R} = \underset{\sim}{R}_1^3 = \begin{bmatrix} 1 & & & & \\ 0.914 & 1 & & \text{sym} & \\ 0.914 & 0.934 & 1 & & \\ 0.774 & 0.774 & 0.774 & 1 & \\ 0.982 & 0.914 & 0.914 & 0.774 & 1 \end{bmatrix}$$

Now, if we take λ-cuts at two different values of λ, say $\lambda = 0.914$ and $\lambda = 0.934$, the following defuzzified crisp equivalence relations and their associated classes are derived:

$$\lambda = 0.914: \quad R_\lambda = \begin{bmatrix} 1 & 1 & 1 & 0 & 1 \\ 1 & 1 & 1 & 0 & 1 \\ 1 & 1 & 1 & 0 & 1 \\ 0 & 0 & 0 & 1 & 0 \\ 1 & 1 & 1 & 0 & 1 \end{bmatrix}$$
$$\{x_1, \ x_2, \ x_3, \ x_5\}, \ \{x_4\}$$

$$\lambda = 0.934: \quad R_\lambda = \begin{bmatrix} 1 & 0 & 0 & 0 & 1 \\ 0 & 1 & 1 & 0 & 0 \\ 0 & 1 & 1 & 0 & 0 \\ 0 & 0 & 0 & 1 & 0 \\ 1 & 0 & 0 & 0 & 1 \end{bmatrix}$$
$$\{x_1, \ x_5\}, \ \{x_2, \ x_3\}, \ \{x_4\}$$

Hence, if we wanted to classify the earthquake damage for purposes of insurance payout into, say, two intensities on the modified Mercalli scale (the Mercalli scale is a measure of an earthquake's strength in terms of average damage the earthquake causes in structures in a given region), then regions 1, 2, 3, and 5 belong to a larger Mercalli intensity and region 4 belongs to a smaller Mercalli intensity (see $\lambda = 0.914$). But if we wanted to have a finer division for, say, three Mercalli scales, we could assign the regions shown in Table 11.6.

TABLE 11.6
Classification of earthquake damage by region for $\lambda = 0.934$

Regions	Mercalli intensity
$\{\mathbf{x}_4\}$	VII
$\{\mathbf{x}_1, \mathbf{x}_5\}$	VIII
$\{\mathbf{x}_2, \mathbf{x}_3\}$	IX

CLUSTER ANALYSIS

Clustering refers to identifying the number of subclasses of c clusters in a data universe X comprised of n data samples, and partitioning X into c clusters ($2 \leq c < n$). Note that $c = 1$ denotes rejection of the hypothesis that there are clusters in the data, whereas $c = n$ constitutes the trivial case where each sample is in a "cluster" by itself. There are two kinds of c-partitions of data: hard (or crisp) and soft (or fuzzy). For numerical data one assumes that the members of each cluster bear more mathematical similarity to each other than to members of other clusters. Two important issues to consider in this regard are how to measure the similarity between pairs of observations and how to evaluate the partitions once they are formed.

One of the simplest similarity measures is distance between pairs of feature vectors in the feature space. If one can determine a suitable distance measure and compute the distance between all pairs of observations, then one may expect that the distance between points in the same cluster will be considerably less than the distance between points in different clusters. Several circumstances, however, mitigate the general utility of this approach, such as the combination of values of incompatible features, as would be the case, for example, when different features have significantly different scales. The clustering method described in this chapter defines "optimum" partitions through a global criterion function that measures the extent to which candidate partitions optimize a weighted sum of squared errors between data points and *cluster centers* in feature space. Many other clustering algorithms have been proposed for distinguishing substructure in high-dimensional data [Bezdek et al., 1986]. It is emphasized here that the method of clustering must be closely matched with the particular data under study for successful interpretation of substructure in the data.

CLUSTER VALIDITY

In many cases, the number c of clusters in the data is known. In other cases, however, it may be reasonable to expect cluster substructure at more than one value of c. In this situation it is necessary to identify the value of c that gives the most plausible number of clusters in the data for the analysis at hand. This problem is known as cluster validity [see Duda and Hart, 1973, or Bezdek, 1981]. If the data used are labeled, there is a unique and absolute measure of cluster validity—viz., the c that is given. For unlabeled data, no absolute measure of clustering validity exists. Although the importance of these differences is not known, it is clear that the features nominated should be sensitive to the phenomena of interest and not to other variations that might not matter to the applications at hand.

c-MEANS CLUSTERING

Bezdek [1981] developed an extremely powerful classification method to accommodate fuzzy data. It is an extension of a method known as c-means, or hard c-means, when employed in a crisp classification sense. To introduce this method, we define a sample set of n data samples that we wish to classify:

$$\text{X} = \{\mathbf{x}_1, \mathbf{x}_2, \mathbf{x}_3, \ldots, \mathbf{x}_n\} \tag{11.2}$$

Each data sample, $\mathbf{x}_i$, is defined by m features, i.e.,

$$\mathbf{x}_i = \{x_{i1}, x_{i2}, x_{i3}, \ldots x_{im}\} \tag{11.3}$$

where each $\mathbf{x}_i$ in the universe X is an m-dimensional vector of m elements or m features. Since the m features all can have different units, in general, we have to normalize each of the features to a unified scale before classification. In a geometric sense, each $\mathbf{x}_i$ is a *point* in m-dimensional feature space, and the universe of the data sample, X, is a *point set* with n elements in the sample space.

Bezdek [1981] suggested using an objective function approach for clustering the data into hyperspherical clusters. This idea for hard clustering is shown in three-dimensional feature space in Fig. 11.2. In this figure, each cluster of data is shown as a hyperspherical shape with a hypothetical geometric cluster center. The objective function is developed so as to do two things simultaneously: first, minimize the Euclidean distance between each data point in a cluster and its cluster center (a calculated point), and second, maximize the Euclidean distance between cluster centers.

Hard c-Means (HCM)

HCM is used to classify data in a crisp sense. By this we mean that each data point will be assigned to one, and only one, data cluster. In this sense these clusters are also called *partitions*—that is, partitions of the data. Define a family of sets $\{A_i, i = 1, 2, \ldots, c\}$ as a hard c-partition of X, where the following set-theoretic forms apply to these partitions:

$$\bigcup_{i=1}^{c} A_i = X \tag{11.4}$$

$$A_i \cap A_j = \varnothing \qquad \text{all } i \neq j \tag{11.5}$$

$$\varnothing \subset A_i \subset X \qquad \text{all } i \tag{11.6}$$

again, where $X = \{\mathbf{x}_1, \mathbf{x}_2, \mathbf{x}_3, \ldots, \mathbf{x}_n\}$ is a finite set space comprised of the universe of data samples, and c is the number of classes, or partitions, or clusters, into which we want to classify the data. We note the obvious,

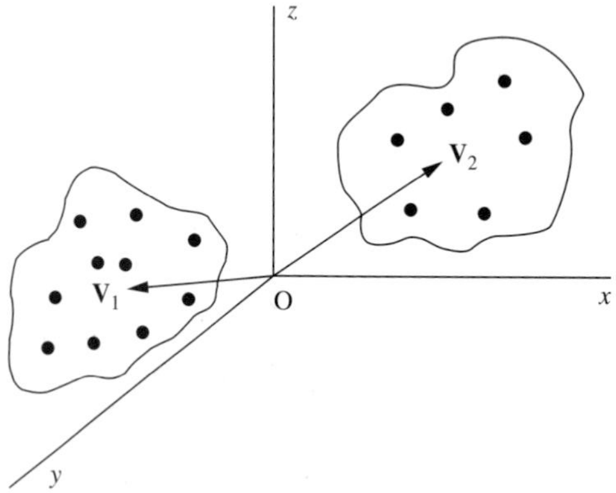

FIGURE 11.2
Cluster idea with hard c-means.

$$2 \leq c < n \tag{11.7}$$

where $c = n$ classes just places each data sample into its own class, and $c = 1$ places all data samples into the same class; neither case requires any effort in classification, and both are intrinsically uninteresting. Equation (11.4) expresses the fact that the set of all classes exhausts the universe of data samples. Equation (11.5) indicates that none of the classes overlap in the sense that a data sample can belong to more than one class. Equation (11.6) simply expresses that a class cannot be empty and it cannot contain all the data samples.

Suppose we have the case where $c = 2$. Equations (11.4) and (11.5) are then manifested in the following set expressions:

$$\mathrm{A}_2 = \overline{\mathrm{A}}_1 \qquad \mathrm{A}_1 \cup \overline{\mathrm{A}}_1 = \mathrm{X} \qquad \text{and} \qquad \mathrm{A}_1 \cap \overline{\mathrm{A}}_1 = \varnothing$$

These set expressions are equivalent to the excluded middle laws [Eqs. 2.12].

The function-theoretic expressions associated with Eqs. (11.4), (11.5), and (11.6) are these

$$\bigvee_{i=1}^{c} \chi_{\mathrm{A}_i}(\mathbf{x}_k) = 1 \qquad \text{for all } k \tag{11.8}$$

$$\chi_{\mathrm{A}_i}(x_k) \wedge \chi_{\mathrm{A}_j}(\mathbf{x}_k) = 0 \qquad \text{for all } k \tag{11.9}$$

$$0 < \sum_{k=1}^{n} \chi_{\mathrm{A}_i}(\mathbf{x}_k) < n \qquad \text{for all } i \tag{11.10}$$

where the characteristic function $\chi_{\mathrm{A}_i}(\mathbf{x}_k)$ is defined once again as

$$\chi_{\mathrm{A}_i}(x_k) = \begin{Bmatrix} 1 & \mathbf{x}_k \in \mathrm{A}_i \\ 0 & \mathbf{x}_k \notin \mathrm{A}_i \end{Bmatrix} \tag{11.11}$$

Equations (11.8) and (11.9) explain that any sample $\mathbf{x}_k$ can only and definitely belong to one of the c classes. Equation (11.10) implies that no class is empty and no class is the whole set X (that is, the universe).

For simplicity in notation, our membership assignment of the jth data point in the ith cluster, or class, is defined to be $\chi_{ij} \equiv \chi_{\mathrm{A}_i}(\mathbf{x}_j)$. Now define a matrix U comprised of elements $\chi_{ij}(i = 1, 2, \ldots, c;\ j = 1, 2, \ldots, n)$; hence, U is a matrix with c rows and n columns. Then we define a hard c-partition space for X as the following matrix set:

$$\mathrm{M}_c = \left\{ \mathrm{U} \mid \chi_{ij} \in \{0, 1\}, \sum_{i=1}^{c} \chi_{ik} = 1,\ 0 < \sum_{k=1}^{n} \chi_{ik} < n \right\} \tag{11.12}$$

Any matrix $\mathrm{U} \in \mathrm{M}_c$ is a hard c-partition. The cardinality of any hard c-partition, M_c, is

$$\eta_{\mathrm{M}_c} = \left(\frac{1}{c!}\right) \left[\sum_{i=1}^{c} \binom{c}{i} (-1)^{c-i} \cdot i^n \right] \tag{11.13}$$

where the expression $\binom{c}{i}$ is the binomial coefficient of c things taken i at a time.

Example 11.6. Suppose we have five data points in a universe, $X = \{x_1, x_2, x_3, x_4, x_5\}$. Also, suppose we want to cluster these five points into two classes. For this case we have $n = 5$ and $c = 2$. The cardinality, using Eq. (11.13), of this hard c-partition is given by

$$\eta_{M_c} = \tfrac{1}{2}[2(-1) + 2^5] = 15$$

Some of the 15 possible hard 2-partitions are listed here:

$$\begin{bmatrix} 1 & 1 & 1 & 1 & 0 \\ 0 & 0 & 0 & 0 & 1 \end{bmatrix} \quad \begin{bmatrix} 1 & 1 & 1 & 0 & 0 \\ 0 & 0 & 0 & 1 & 1 \end{bmatrix} \quad \begin{bmatrix} 1 & 1 & 0 & 0 & 0 \\ 0 & 0 & 1 & 1 & 1 \end{bmatrix} \quad \begin{bmatrix} 1 & 0 & 0 & 0 & 0 \\ 0 & 1 & 1 & 1 & 1 \end{bmatrix}$$

$$\begin{bmatrix} 1 & 0 & 1 & 0 & 0 \\ 0 & 1 & 0 & 1 & 1 \end{bmatrix} \quad \begin{bmatrix} 1 & 0 & 0 & 1 & 0 \\ 0 & 1 & 1 & 0 & 1 \end{bmatrix} \quad \begin{bmatrix} 1 & 0 & 0 & 0 & 1 \\ 0 & 1 & 1 & 1 & 0 \end{bmatrix}$$

and so on.

Notice that these two matrices,

$$\begin{bmatrix} 1 & 1 & 1 & 1 & 0 \\ 0 & 0 & 0 & 0 & 1 \end{bmatrix} \quad \text{and} \quad \begin{bmatrix} 0 & 0 & 0 & 0 & 1 \\ 1 & 1 & 1 & 1 & 0 \end{bmatrix}$$

are not different-clustering 2-partitions. In fact, they are the same 2-partition irrespective of an arbitrary row-swap. If we label the first row of the first U-matrix class c_1 and we label the second row class c_2, we would get the same classification for the second U-matrix by simply relabeling each row: The first row is c_2 and the second row is c_1. The cardinality measure given in Eq. (11.13) gives the number of *unique* c-partitions for n data points.

An interesting question now arises: Of all the possible c-partitions for n data samples, how can we select the most reasonable c-partition for the partition space M_c? For instance, in the example just provided, which of the 15 possible hard 2-partitions for 5 data points and 2 classes is the best? The answer to this question is provided by the objective function (or classification criteria) to be used to classify or cluster the data. The one proposed for the hard c-means the algorithm is known as a within-class sum of squared errors approach using a Euclidean norm to characterize distance. This algorithm is denoted $J(U, \mathbf{v})$, where U is the partition matrix, and the parameter, $\mathbf{v}$, is a vector of cluster centers. This objective function is given by

$$J(U,\mathbf{v}) = \sum_{k=1}^{n} \sum_{i=1}^{c} \chi_{ik}(d_{ik})^2 \tag{11.14}$$

where d_{ik} is a Euclidean distance measure (in m-dimensional feature space, R^m) between the kth data sample x_k and ith cluster center $\mathbf{v}_i$, given by

$$d_{ik} = d(x_k - v_i) = \|x_k - \mathbf{v}_i\| = \left[\sum_{j=1}^{m} (x_{kj} - v_{ij})^2\right]^{1/2} \tag{11.15}$$

Since each data sample requires m coordinates to describe its location in R^m-space, each cluster center also requires m coordinates to describe its location in this same space. Therefore, the ith cluster center is a vector of length m,

$$\mathbf{v}_i = \{v_{i1}, v_{i2}, \ldots, v_{im}\}$$

where the *j*th coordinate is calculated by

$$v_{ij} = \frac{\sum_{k=1}^{n} \chi_{ik} \cdot x_{kj}}{\sum_{k=1}^{n} \chi_{ik}} \tag{11.16}$$

We seek the optimum partition, U^*, to be the partition that produces the minimum value for the function, J. That is,

$$\mathrm{J}(\mathrm{U}^*, \mathbf{v}^*) = \min_{\mathrm{U} \in \mathrm{M}_c} \mathrm{J}(\mathrm{U}, \mathbf{v}) \tag{11.17}$$

Finding the optimum partition matrix, U^*, is exceedingly difficult for practical problems because $\mathrm{M}_c \to \infty$ for even modest-sized problems. For example, for the case where $n = 25$ and $c = 10$, the cardinality approaches an extremely large number, i.e., $\mathrm{M}_c \to 10^{18}$! Obviously, a search for optimality by exhaustion is *not* computationally feasible for problems of reasonable interest. Fortunately, very useful and effective alternative search algorithms have been devised [Bezdek, 1981].

One such search algorithm is known as *iterative optimization*. Basically, this method is like many other iterative methods in that we start with an initial guess at the U matrix. From this assumed matrix, input values for the number of classes, and iteration tolerance (the accuracy we demand in the solution), we calculate the centers of the clusters (classes). From these cluster, or class, centers we recalculate the membership values that each data point has in the cluster. We compare these values with the assumed values and continue this process until the changes from cycle to cycle are within our prescribed tolerance level.

The step-by-step procedures in this iterative optimization method are provided here [Bezdek, 1981]:

1. Fix $c (2 \leq c < n)$ and initialize the U matrix:

$$\mathrm{U}^{(0)} \in \mathrm{M}_c$$

Then do $r = 0, 1, 2, \ldots$

2. Calculate the c center vectors:

$$\{\mathbf{v}_i^{(r)} \text{ with } \mathrm{U}^{(r)}\}$$

3. Update $\mathrm{U}^{(r)}$; calculate the updated characteristic functions (for all i, k):

$$\chi_{ik}^{(r+1)} = \begin{cases} 1 & d_{ik}^{(r)} = \min\{d_{jk}^{(r)}\} \text{ for all } j \in c \\ 0 & \text{otherwise} \end{cases} \tag{11.18}$$

4. If

$$\|\mathrm{U}^{(r+1)} - \mathrm{U}^{(r)}\| \leq \varepsilon \text{ (tolerance level)} \tag{11.19}$$

stop; otherwise set $r = r + 1$ and return to step 2.

In step 4, the notation $\| \|$ is any matrix norm such as the Euclidean norm.

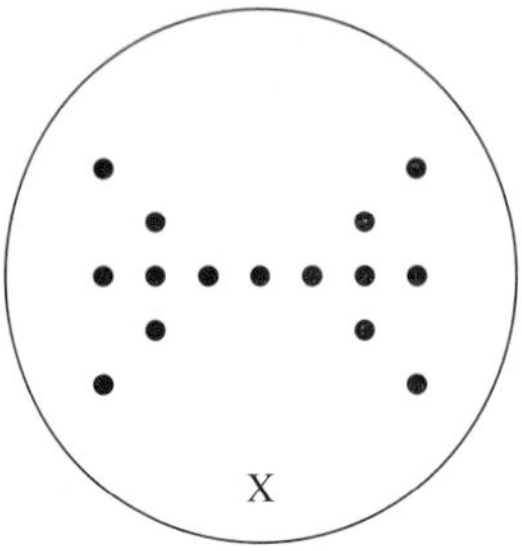

FIGURE 11.3
Butterfly classification problem [Bezdek, 1981].

Example 11.7 [Bezdek, 1981]. A good illustration of the iterative optimization method is provided with the "butterfly problem" shown in Fig. 11.3. In this problem we have 15 data points and one of them is on a vertical line of symmetry (the point in the middle of the data cluster). Suppose we want to cluster our data into two classes. We can see that the points to the left of the line of symmetry should be in one class and the points to the right of the line of symmetry should be in the other class. The problem lies in assigning the point on the line of symmetry to a class. To which class should this point belong? Whichever class the algorithm assigns this point to, there will be a good argument that it should be a member of the other class. Alternatively, the argument may revolve around the fact that the choice of two classes is a poor one for this problem. Three classes might be the best choice, but the physics underlying the data might be binary and two classes may be the only option.

In conducting the iterative optimization approach we have to assume an initial U matrix. This matrix will have two rows (two classes, $c = 2$) and 15 columns (15 data points, $n = 15$). It is important to understand that the classes may be unlabeled in this process. That is, we can look at the structure of the data without the need for the assignment of labels to the classes. This is often the case when one is first looking at a group of data. After several iterations with the data, and as we become more and more knowledgeable about the data, we can then assign labels to the classes. We start the solution with the assumption that the point in the middle (i.e., the eighth column) is assigned to the class represented by the bottom row of the initial U matrix, $U^{(0)}$:

$$U^{(0)} = \begin{bmatrix} 1 & 1 & 1 & 1 & 1 & 1 & 0 & 0 & 0 & 0 & 0 & 0 & 0 & 0 \\ 0 & 0 & 0 & 0 & 0 & 0 & 1 & 1 & 1 & 1 & 1 & 1 & 1 & 1 \end{bmatrix}$$

After four iterations [Bezdek, 1981] this method converges to within a tolerance level of $\varepsilon = 0.01$, as

$$U^{(4)} = \begin{bmatrix} 1 & 1 & 1 & 1 & 1 & 1 & 1 & 0 & 0 & 0 & 0 & 0 & 0 & 0 & 0 \\ 0 & 0 & 0 & 0 & 0 & 0 & 0 & 1 & 1 & 1 & 1 & 1 & 1 & 1 & 1 \end{bmatrix}$$

We note that the point on the line of symmetry (i.e., the eighth column) is still assigned to the class represented by the second row of the U matrix. The elements in the U matrix indicate membership of that data point in the first or second class. For example, the point on the line of symmetry has full membership in the second class and no membership in the first class; yet it is plain to see from Fig. 11.3 that physically it should probably share membership with each class. This is not possible with crisp classification; membership is binary—a point is either a member of a class or it is not.

The following example illustrates again the crisp classification method. The process will be instructive because of its similarity to the subsequent algorithm to be developed for the fuzzy classification method.

Example 11.8. In a chemical engineering process involving an automobile's catalytic converter (which converts carbon monoxide to carbon dioxide) we have a relationship between the conversion efficiency of the catalytic converter and the *inverse of the temperature* of the catalyst. Two classes of data are known from the reaction efficiency. Points of high conversion efficiency and high temperature are indicators of a nonpolluting system (class c_1), and points of low conversion efficiency and low temperature are indicative of a polluting system (class c_2). Suppose you measure the conversion efficiency and temperature (T) of four different catalytic converters and attempt to characterize them as polluting or nonpolluting. The four data points ($n = 4$) are shown in Fig. 11.4, where the y axis is conversion efficiency and the x axis is the inverse of the temperature [in a conversion process like this the exact solution takes the form of ln $(1/T)$]. The data are described by two features ($m = 2$), and have the following coordinates in 2D space:

$$\mathbf{x}_1 = \{1, 3\}$$
$$\mathbf{x}_2 = \{1.5, 3.2\}$$
$$\mathbf{x}_3 = \{1.3, 2.8\}$$
$$\mathbf{x}_4 = \{3, 1\}$$

We desire to classify these data points into two classes ($c = 2$). It is sometimes useful to calculate the cardinality of the possible number of crisp partitions for this system, i.e., to find η_{M_c} using Eq. (11.13); thus,

$$\eta_{\mathrm{M}_c} = \left(\frac{1}{c!}\right)\left[\sum \binom{c}{i}(-1)^{c-i} i^n\right] = \frac{1}{2!}\left[\binom{2}{1}(-1)^1(1)^4 + \binom{2}{2}(-1)^0(2)^4\right]$$
$$= \frac{1}{2}[-2 + 16] = 7$$

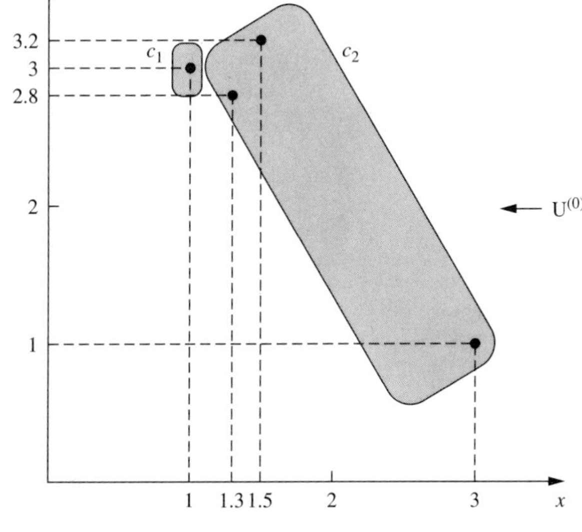

FIGURE 11.4
Four data points in two-dimensional feature space.

which says that there are seven unique ways (irrespective of row swaps) to classify the four points into two clusters. Let's begin the iterative optimization algorithm with an initial guess of the crisp partition, U, by assuming $\mathbf{x}_1$ to be in class 1 and $\mathbf{x}_2, \mathbf{x}_3, \mathbf{x}_4$ to be in class 2, as shown in Fig. 11.4, i.e.,

$$\mathrm{U}^{(0)} = \begin{bmatrix} 1 & 0 & 0 & 0 \\ 0 & 1 & 1 & 1 \end{bmatrix}$$

Now, from the initial $\mathrm{U}^{(0)}$ (which is one of the seven possible crisp partitions) we seek the optimum partition U^*; that is,

$$\mathrm{U}^{(0)} \to \mathrm{U}^{(1)} \to \mathrm{U}^{(2)} \to \cdots \to \mathrm{U}^*$$

Of course, optimality is defined in terms of the desired tolerance or convergence level, ε.

In general, for class 1 we calculate the coordinates of the cluster center,

$$v_{1j} = \frac{\chi_1 x_{1j} + \chi_2 x_{2j} + \chi_3 x_{3j} + \chi_4 x_{4j}}{\chi_1 + \chi_2 + \chi_3 + \chi_4} = \frac{(1)x_{1j} + (0)x_{2j} + (0)x_{3j} + (0)x_{4j}}{1+0+0+0}$$

and $\quad \mathbf{v}_i = \{v_{i1}, v_{i2}, \ldots, v_{im}\}$

In this case $m = 2$, which means we deal with two coordinates for each data point. Therefore,

$$\mathbf{v}_i = \{v_{i1}, v_{i2}\}$$

where for $c = 1$ (which is class 1), $\mathbf{v}_1 = \{v_{11}, v_{12}\}$
for $c = 2$ (which is class 2), $\mathbf{v}_2 = \{v_{21}, v_{22}\}$

Therefore, using the expression for v_{ij} for $c = 1$, and $j = 1$ and 2, respectively,

$$\left.\begin{aligned} v_{11} &= \frac{1(1)}{1} = 1 && \to x \text{ coordinate} \\ v_{12} &= \frac{1(3)}{1} = 3 && \to y \text{ coordinate} \end{aligned}\right\} \Rightarrow \mathbf{v}_1 = \{1, 3\}$$

which just happens to be the coordinates of point x_1, since this is the only point in the class for the assumed initial partition, $\mathrm{U}^{(0)}$. For $c = 2$ or class 2, we get cluster center coordinates

$$v_{2j} = \frac{(0)x_{1j} + (1)x_{2j} + (1)x_{3j} + (1)x_{4j}}{0+1+1+1} = \frac{x_{2j} + x_{3j} + x_{4j}}{3}$$

Hence, for $c = 2$ and $j = 1$ and 2, respectively,

$$\left.\begin{aligned} v_{21} &= \frac{1(1.5) + 1(1.3) + 1(3)}{3} = 1.93 && \to x \text{ coordinate} \\ v_{22} &= \frac{1(3.2) + 1(2.8) + 1(1)}{3} = 2.33 && \to y \text{ coordinate} \end{aligned}\right\} \Rightarrow \mathbf{v}_2 = \{1.93, 2.33\}$$

Now, we compute the values for d_{ik}, or the distances from the sample $\mathbf{x}_k$ (a data set) to the center, $\mathbf{v}_i$, of the ith class. Using Eq. (11.15),

$$d_{ik} = \left\{ \sum_{j=1}^{m} (x_{kj} - v_{ij})^2 \right\}^{1/2} \tag{11.15}$$

we get, for example, for $c = 1$: $d_{1k} = [(x_{k1} - v_{11})^2 + (x_{k2} - v_{12})^2]^{1/2}$. Therefore, for each data set $k = 1$ to 4, we compute the values of d_{ik} as follows, for cluster 1:

$$\begin{aligned} d_{11} &= \sqrt{(1-1)^2 + (3-3)^2} = 0.0 \\ d_{12} &= \sqrt{(1.5-1)^2 + (3.2-3)^2} = 0.54 \\ d_{13} &= \sqrt{(1.3-1)^2 + (2.8-3)^2} = 0.36 \\ d_{14} &= \sqrt{(3-1)^2 + (1-3)^2} = 2.83 \end{aligned}$$

and for cluster 2:

$$\begin{aligned} d_{21} &= \sqrt{(1-1.93)^2 + (3-2.33)^2} = 1.14 \\ d_{22} &= \sqrt{(1.5-1.93)^2 + (3.2-2.33)^2} = 0.97 \\ d_{23} &= \sqrt{(1.3-1.93)^2 + (2.8-2.33)^2} = 0.78 \\ d_{24} &= \sqrt{(3-1.93)^2 + (1-2.33)^2} = 1.70 \end{aligned}$$

Now, we update the partition to $U^{(1)}$ for each data point (for $(c-1)$ clusters) using Eq. (11.18). Hence, for class 1 we compare d_{ik} against the minimum of $\{d_{ik}, d_{2k}\}$:

For $k = 1$,

$$d_{11} = 0.0, \quad \min(d_{11}, d_{21}) = \min(0, 1.14) = 0.0; \quad \text{thus } \chi_{11} = 1$$

For $k = 2$,

$$d_{12} = 0.54, \quad \min(d_{12}, d_{22}) = \min(0.54, 0.97) = 0.54; \quad \text{thus } \chi_{12} = 1$$

For $k = 3$,

$$d_{13} = 0.36, \quad \min(d_{13}, d_{23}) = \min(0.36, 0.78) = 0.36; \quad \text{thus } \chi_{13} = 1$$

For $k = 4$,

$$d_{14} = 2.82, \quad \min(d_{14}, d_{24}) = \min(2.83, 1.70) = 1.70; \quad \text{thus } \chi_{14} = 0$$

Therefore, the updated partition is

$$U^{(10)} = \begin{bmatrix} 1 & 1 & 1 & 0 \\ 0 & 0 & 0 & 1 \end{bmatrix}$$

Since the partitions $U^{(0)}$ and $U^{(1)}$ are different, we repeat the same procedure based on the new setup of two classes. For $c = 1$, the center coordinates are

$$\mathbf{v}_{1j} \quad \text{or} \quad \mathbf{v}_j = \frac{x_{1j} + x_{2j} + x_{3j}}{1+1+1+0}, \quad \text{since } x_{14} = 0$$

$$\left.\begin{aligned} \mathbf{v}_{11} &= \frac{x_{11} + x_{21} + x_{31}}{3} = \frac{1 + 1.5 + 1.3}{3} = 1.26 \\ \mathbf{v}_{12} &= \frac{x_{12} + x_{22} + x_{32}}{3} = \frac{3 + 3.2 + 2.8}{3} = 3.0 \end{aligned}\right\} \quad \mathbf{v}_1 = \{1.26, 3.0\}$$

and for $c = 2$, the center coordinates are

$$\mathbf{v}_{2j} \quad \text{or} \quad \mathbf{v}_j = \frac{x_{4j}}{0+0+0+1}, \quad \text{since } x_{21} = x_{22} = x_{23} = 0$$

$$\left.\begin{aligned} \mathbf{v}_{21} &= \tfrac{3}{1} = 3 \\ \mathbf{v}_{22} &= \tfrac{1}{1} = 1 \end{aligned}\right\} \quad \mathbf{v}_2 = \{3, 1\}$$

Now, we calculate distance measures again:

$$d_{11} = \sqrt{(1-1.26)^2 + (3-3)^2} = 0.26 \qquad d_{21} = \sqrt{(1-3)^2 + (3-1)^2} = 2.83$$

$$d_{12} = \sqrt{(1.5-1.26)^2 + (3.2-3)^2} = 0.31 \qquad d_{22} = \sqrt{(1.5-3)^2 + (3.2-1)^2} = 2.66$$

$$d_{13} = \sqrt{(1.3-1.26)^2 + (2.8-3)^2} = 0.20 \qquad d_{23} = \sqrt{(1.3-3)^2 + (2.8-1)^2} = 2.47$$

$$d_{14} = \sqrt{(3-1.26)^2 + (1-3)^2} = 2.65 \qquad d_{24} = \sqrt{(3-3)^2 + (1-1)^2} = 0.0$$

and again update the partition $U^{(1)}$ to $U^{(2)}$:

For $k = 1$,

$$d_{11} = 0.26, \quad \min(d_{11}, d_{21}) = \min(0.26, 2.83) = 0.26; \quad \text{thus } \chi_{11} = 1$$

For $k = 2$,

$$d_{12} = 0.31, \quad \min(d_{12}, d_{22}) = \min(0.31, 2.66) = 0.31; \quad \text{thus } \chi_{12} = 1$$

For $k = 3$,

$$d_{13} = 0.20, \quad \min(d_{13}, d_{23}) = \min(0.20, 2.47) = 0.20; \quad \text{thus } \chi_{13} = 1$$

For $k = 4$,

$$d_{14} = 2.65, \quad \min(d_{14}, d_{24}) = \min(2.65, 0.0) = 0.0; \quad \text{thus } \chi_{14} = 0$$

Because the partitions $U^{(1)}$ and $U^{(2)}$ are identical, we could say the iterative process has converged; therefore, the optimum hard partition (crisp) is

$$U^{(*)} = \begin{bmatrix} 1 & 1 & 1 & 0 \\ 0 & 0 & 0 & 1 \end{bmatrix}$$

This optimum partition tells us that for this catalytic converter example, the data points $\mathbf{x}_1$, $\mathbf{x}_2$, and $\mathbf{x}_3$ are more similar in the 2D feature space, and different from data point $\mathbf{x}_4$. We could say that points $\mathbf{x}_1$, $\mathbf{x}_2$, and $\mathbf{x}_3$ are more indicative of a nonpolluting converter than is data point $\mathbf{x}_4$.

Fuzzy *c*-Means (FCM)

Let us consider whether the butterfly example in Fig. 11.3 could be improved with the use of fuzzy set methods. To develop these methods in classification, we define a family of fuzzy sets $\{\underset{\sim}{A}_i, i = 1, 2, \ldots, c\}$ as a fuzzy c-partition on a universe of data points, X. Because fuzzy sets allow for degrees of membership we can extend the crisp classification idea into a fuzzy classification notion. Then we can assign membership to the various data points in each fuzzy set (fuzzy class, fuzzy cluster). Hence, a single point can have partial membership in more than one class. It will be useful to describe the membership value that the kth data point has in the ith class with the following notation:

$$\mu_{ik} = \mu_{\underset{\sim}{A}_i}(x_k) \in [0, 1]$$

with the restriction (as with crisp classification) that the sum of all membership values for a single data point in all of the classes has to be unity:

$$\sum_{i=1}^{c} \mu_{ik} = 1 \qquad \text{for all } k = 1, 2, \ldots, n \tag{11.20}$$

As before in crisp classification, there can be no empty classes and there can be no class that contains all the data points. This qualification is manifested in the following expression:

$$0 < \sum_{k=1}^{n} \mu_{ik} < n \tag{11.21}$$

Because each data point can have partial membership in more than one class, the restriction of Eq. (11.9) is not present in the fuzzy classification case, i.e.,

$$\mu_{ik} \wedge \mu_{jk} \neq 0 \tag{11.22}$$

The provisions of Eqs. (11.8) and (11.10) still hold for the fuzzy case, however,

$$\bigvee_{i=1}^{c} \mu_{\mathrm{A}_i}(x_k) = 1 \qquad \text{for all } k \tag{11.23}$$

$$0 < \sum_{k=1}^{n} \mu_{\mathrm{A}_i}(x_k) < n \qquad \text{for all } i \tag{11.24}$$

Before, in the case of $c = 2$, the classification problem reduced to that of the excluded middle laws for crisp classification. Since we now allow partial membership, the case of $c = 2$ does not follow the restrictions of the excluded middle laws; i.e., for two classes $\underset{\sim}{\mathrm{A}}_i$ and $\underset{\sim}{\mathrm{A}}_j$,

$$\underset{\sim}{\mathrm{A}}_i \cap \mathrm{A}_j \neq \varnothing \tag{11.25}$$

$$\varnothing \subset \underset{\sim}{\mathrm{A}}_i \subset \mathrm{X} \tag{11.26}$$

We can now define a family of fuzzy partition matrices, M_{fc}, for the classification involving c classes and n data points,

$$\mathrm{M}_{fc} = \left\{ \underset{\sim}{\mathrm{U}} \mid \mu_{ik} \in [0, 1]; \sum_{i=1}^{c} \mu_{ik} = 1; 0 < \sum_{k=1}^{n} \mu_{ik} < n \right\} \tag{11.27}$$

where $i = 1, 2, \ldots, c$ and $k = 1, 2, \ldots, n$.

Any $\underset{\sim}{\mathrm{U}} \in \mathrm{M}_{fc}$ is a fuzzy c-partition, and it follows from the overlapping character of the classes and the infinite number of membership values possible for describing class membership that the cardinality of M_{fc} is also infinity, i.e., $\eta_{\mathrm{M}_{fc}} = \infty$.

Example 11.9 [Similar to Bezdek, 1981]. Suppose you are a fruit geneticist interested in genetic relationships among fruits. In particular, you know that a tangelo is a cross between a grapefruit and a tangerine. You describe the fruit with such features as color, weight, sphericity, sugar content, skin texture, and so on. Hence, your feature space could be highly dimensional. Suppose you have three fruits (three data points):

$$\mathrm{X} = \{x_1 = \text{grapefruit}, x_2 = \text{tangelo}, x_3 = \text{tangerine}\}$$

These data points are described by m features, as discussed. You want to class the three fruits into two classes to determine the genetic assignment for the three fruits. In the crisp case, the classification matrix can take one of three forms; i.e., the cardinality for this case where $n = 3$ and $c = 2$ is $\eta_{M_c} = 3$ [see Eq. (11.13)]. Suppose you arrange your $\underset{\sim}{U}$ matrix as follows:

$$\underset{\sim}{U} = \begin{matrix} & x_1 & x_2 & x_3 \\ c_1 \\ c_2 \end{matrix} \begin{bmatrix} 1 & 0 & 0 \\ 0 & 1 & 1 \end{bmatrix}$$

The three possible partitions of the matrix are

$$\begin{bmatrix} 1 & 0 & 0 \\ 0 & 1 & 1 \end{bmatrix} \quad \begin{bmatrix} 1 & 1 & 0 \\ 0 & 0 & 1 \end{bmatrix} \quad \begin{bmatrix} 1 & 0 & 1 \\ 0 & 1 & 0 \end{bmatrix}$$

Notice in the first partition that we are left with the uncomfortable segregation of the grapefruit in one class and the tangelo and the tangerine in the other; the tangelo shares nothing in common with the grapefruit! In the second partition, the grapefruit and the tangelo are in a class, suggesting that they share nothing in common with the tangerine! Finally, the third partition is the most genetically discomforting of all, because here the tangelo is in a class by itself, sharing nothing in common with its progenitors! One of these three partitions will be the final partition when any algorithm is used. The question is, which one is best? Intuitively the answer is none, but in crisp classification we have to use one of these.

In the fuzzy case this segregation and genetic absurdity are not a problem. We can have the most intuitive situation where the tangelo shares membership with both classes with the parents. For example, the following partition might be a typical outcome for the fruit genetics problem:

$$\underset{\sim}{U} = \begin{matrix} & x_1 & x_2 & x_3 \\ 1 \\ 2 \end{matrix} \begin{bmatrix} .91 & .58 & .13 \\ .09 & .42 & .87 \end{bmatrix}$$

In this case, Eq. (11.24) shows that the sum of each row is a number between 0 and n, or

$$0 < \sum_k \mu_{1k} = 1.62 < 3$$

$$0 < \sum_k \mu_{2k} = 1.38 < 3$$

and for Eq. (11.22) there is overlap among the classes for each data point,

$$\mu_{11} \wedge \mu_{21} = \min(.91, .09) = .09 \neq 0$$
$$\mu_{12} \wedge \mu_{22} = \min(.58, .42) = .42 \neq 0$$
$$\mu_{13} \wedge \mu_{23} = \min(.13, .87) = .13 \neq 0$$

FUZZY *c*-MEANS ALGORITHM. To describe a method to determine the fuzzy c-partition matrix $\underset{\sim}{U}$ for grouping a collection of n data sets into c classes, we define an objective function J_m for a fuzzy c-partition,

$$J_m(\underset{\sim}{U}, \mathbf{v}) = \sum_{k=1}^{n} \sum_{i=1}^{c} (\mu_{ik})^{m'} (d_{ik})^2 \tag{11.28}$$

where

$$d_{ik} = d(\mathbf{x}_k - \mathbf{v}_i) = \left[\sum_{j=1}^{m}(x_{kj} - v_{ij})^2\right]^{1/2} \tag{11.29}$$

and where μ_{ik} is the membership of the kth data point in the ith class.

As with crisp classification, the function J_m can have a large number of values, the smallest one associated with the *best* clustering. Because of the large number of possible values, now infinite because of the infinite cardinality of fuzzy sets, we seek to find the best possible, or optimum, solution without resorting to an exhaustive, or expensive, search. The distance measure, d_{ik} in Eq. (11.29), is again a Euclidean distance between the ith cluster center and the kth data set (data point in m-space). A new parameter is introduced in Eq. (11.28) called a weighting parameter, m' [Bezdek, 1981]. This value has a range $m' \in [1, \infty)$. This parameter controls the amount of fuzziness in the classification process and is discussed shortly. Also, as before, $\mathbf{v}_i$ is the ith cluster center, which is described by m features (m coordinates) and can be arranged in vector form as before, $\mathbf{v}_i = \{v_{i1}, v_{i2}, \ldots, v_{im}\}$.

Each of the cluster coordinates for each class can be calculated in a manner similar to the calculation in the crisp case [see Eq. (11.16)],

$$v_{ij} = \frac{\sum_{k=1}^{n} \mu_{ik}^{m'} \cdot x_{kj}}{\sum_{k=1}^{n} \mu_{ik}^{m'}} \tag{11.30}$$

where j is a variable on the feature space, i.e., $j = 1, 2, \ldots, m$.

As in the crisp case the optimum fuzzy c-partition will be the smallest of the partitions described in Eq. (11.28); that is,

$$J_m^*(\underset{\sim}{\mathrm{U}}^*, \mathbf{v}^*) = \min_{\mathrm{M_{fc}}} J(\underset{\sim}{\mathrm{U}}, \mathbf{v}) \tag{11.31}$$

As with many optimization processes (see Chapter 14), the solution to Eq. (11.31) cannot be guaranteed to be a global optimum, i.e., the best of the best. What we seek is the best solution available within a pre-specified level of accuracy. A recent, effective algorithm for fuzzy classification, called iterative optimization, was proposed by Bezdek [1981]. The steps in this algorithm are as follows.

Algorithm.

1. Fix c ($2 \le c < n$) and select a value for parameter m'. Initialize the partition matrix, $\underset{\sim}{\mathrm{U}}^{(0)}$. Each step in this algorithm will be labeled r, where $r = 0, 1, 2, \ldots$.
2. Calculate the c centers $\{\mathbf{v}_i^{(r)}\}$ for each step.
3. Update the partition matrix for the rth step, $\underset{\sim}{\mathrm{U}}^{(r)}$ as follows:

$$\mu_{ik}^{(r+1)} = \left[\sum_{j=1}^{c}\left(\frac{d_{ik}^{(r)}}{d_{jk}^{(r)}}\right)^{2/(m'-1)}\right]^{-1} \quad \text{for } \mathrm{I}_k = \varnothing \tag{11.32a}$$

or

$$\mu_{ik}^{(r+1)} = 0 \qquad \text{for all classes } i \text{ where } i \in \tilde{\mathrm{I}}_k \tag{11.32b}$$

where

$$\mathrm{I}_k = \{i \mid 2 \leq c < n;\ d_{ik}^{(r)} = 0\} \tag{11.33}$$

and

$$\tilde{\mathrm{I}}_k = \{1, 2, \ldots, c\} - \mathrm{I}_k \tag{11.34}$$

and

$$\sum_{i \in \mathrm{I}_k} \mu_{ik}^{(r+1)} = 1 \tag{11.35}$$

4. If $\|\underset{\sim}{\mathrm{U}}^{(r+1)} - \underset{\sim}{\mathrm{U}}^{(r)}\| \leq \varepsilon_L$, stop; otherwise set $r = r + 1$ and return to step 2.

In step 4 we compare a matrix norm $\|\ \|$ of two successive fuzzy partitions to a prescribed level of accuracy, ε_L, to determine whether the solution is good enough. In step 3 there is a considerable amount of logic involved in Eqs. (11.32) to (11.35). Equation (11.32*a*) is straightforward enough, except when the variable d_{jk} is zero. Since this variable is in the denominator of a fraction, the operation is undefined mathematically, and computer calculations are abruptly halted. So the parameters I_k and $\tilde{\mathrm{I}}_k$ comprise a bookkeeping system to handle situations when some of the distance measures, d_{ij}, are zero, or extremely small in a computational sense. If a zero value is detected, Eq. (11.32*b*) sets the membership for that partition value to be zero. Equations (11.33) and (11.34) describe the bookkeeping parameters I_k and $\tilde{\mathrm{I}}_k$, respectively, for each of the classes. Equation (11.35) simply says that all the nonzero partition elements in each column of the fuzzy classification partition, $\underset{\sim}{\mathrm{U}}$, sum to unity. The following example serves to illustrate Eqs. (11.32)–(11.35).

Example 11.10. Suppose we have calculated the following distance measures for one step in our iterative algorithm for a classification problem involving three classes and five data points. The values in Table 11.7 are simple numbers for ease in illustration. The bookkeeping parameters I_k and $\tilde{\mathrm{I}}_k$, where in this example $k = 1, 2, 3, 4, 5$, are given next, as illustration of the use of Eqs. (11.33) and (11.34).

TABLE 11.7
Distance measures for hypothetical example ($c = 3, n = 5$)

$d_{11} = 1$	$d_{21} = 2$	$d_{31} = 3$
$d_{12} = 0$	$d_{22} = 0.5$	$d_{32} = 1$
$d_{13} = 1$	$d_{23} = 0$	$d_{33} = 0$
$d_{14} = 3$	$d_{24} = 1$	$d_{34} = 1$
$d_{15} = 0$	$d_{25} = 4$	$d_{35} = 0$

$$
\begin{aligned}
I_1 &= \varnothing & \tilde{I}_1 &= \{1, 2, 3\} - \varnothing = \{1, 2, 3\} \\
I_2 &= \{1\} & \tilde{I}_2 &= \{1, 2, 3\} - \{1\} = \{2, 3\} \\
I_3 &= \{2, 3\} & \tilde{I}_3 &= \{1, 2, 3\} - \{2, 3\} = \{1\} \\
I_4 &= \varnothing & \tilde{I}_4 &= \{1, 2, 3\} - \varnothing = \{1, 2, 3\} \\
I_5 &= \{1, 3\} & \tilde{I}_5 &= \{1, 2, 3\} - \{1, 3\} = \{2\}
\end{aligned}
$$

Now, Eqs. (11.32) and (11.35) are illustrated:

For data point 1: $\mu_{11}, \mu_{21}, \mu_{31} \neq 0$ and $\mu_{11} + \mu_{21} + \mu_{31} = 1$

For data point 2: $\mu_{12} = 0$ and $\mu_{22}, \mu_{32} \neq 0$ and $\mu_{22} + \mu_{23} = 1$

For data point 3: $\mu_{13} = 1$ and $\mu_{23} = \mu_{33} = 0$

For data point 4: $\mu_{14}, \mu_{24}, \mu_{34} \neq 0$ and $\mu_{14} + \mu_{24} + \mu_{34} = 1$

For data point 5: $\mu_{25} = 1$ and $\mu_{15} = \mu_{35} = 0$

The algorithm given in Eq. (11.28) is a *least squares* function, where the parameter n is the number of data sets and c is the number of classes (partitions) into which one is trying to classify the data sets. The squared distance, d_{ik}^2, is then weighted by a measure, $(u_{ik})^{m'}$, of the membership of x_k in the ith cluster. The value of J_m is then a measure of the sum of all the *weighted* squared errors; this value is then minimized with respect to two constraint functions. First, J_m is minimized with respect to the squared errors within each cluster, i.e., for each specific value of c. Simultaneously, the distance between cluster centers is maximized, i.e., $\max\|\mathbf{v}_i - \mathbf{v}_j\|, i \neq j$.

As indicated, the range for the membership exponent is $m' \in [1, \infty)$. For the case $m' = 1$, the distance norm is Euclidean and the FCM algorithm approaches a hard c-means algorithm; i.e., only 0's and 1's come out of the clustering. Conversely, as $m' \rightarrow \infty$, the value of the function $J_m \rightarrow 0$. This result seems intuitive, because the membership values are numbers less than or equal to 1, and large powers of fractions less than one approach zero. In general, the larger m' is, the fuzzier are the membership assignments of the clustering; conversely, as $m' \rightarrow 1$, the clustering values become hard, i.e., 0 or 1. The exponent m' thus controls the extent of membership sharing between fuzzy clusters. If all other algorithmic parameters are fixed, then increasing m' will result in decreasing J_m. No theoretical optimum choice of m' has emerged in the literature. However, the bulk of the literature seems to report values in the range 1.25 to 2. Convergence of the algorithm tends to be slower as the value of m' increases.

The algorithm described here can be remarkably accurate and robust in the sense that poor guesses for the initial partition matrix, $\underset{\sim}{U}^{(0)}$, can be overcome quickly, as illustrated in the next example.

Example 11.11. Continuing with the chemical engineering example on a catalytic converter shown in Fig. 11.4, we can see that a visual display of these points in two-dimensional feature space ($m = 2$) makes it easy for the human to cluster the data into two convenient classes based on the proximity of the points to one another. The fuzzy classification method generally converges quite rapidly, even when the initial guess for the fuzzy partition is quite poor, in a classification sense. The fuzzy iterative optimization method for this case would proceed as follows.

Using U* from the previous example as the initial fuzzy partition, $\underset{\sim}{U}^{(0)}$, and assuming a weighting factor of $m' = 2$ and a criterion for convergence of $\varepsilon_L = 0.01$, i.e.,

$$\max_{i,k} \mid \mu_{ik}^{(r+1)} - \mu_{ik}^{(r)} \mid \leq 0.01$$

we want to determine the optimum fuzzy 2-partition $\underset{\sim}{U}^*$. To begin, the initial fuzzy partition is

$$\underset{\sim}{U}^{(0)} = \begin{bmatrix} 1 & 1 & 1 & 0 \\ 0 & 0 & 0 & 1 \end{bmatrix}$$

Next is the calculation of the initial cluster centers using Eq. (11.30), where $m' = 2$;

$$v_{ij} = \frac{\sum_{k=1}^{n} (\mu_{ik})^2 \cdot x_{kj}}{\sum_{k=1}^{n} (\mu_{ik})^2}$$

where for $c = 1$,

$$\begin{aligned} v_{1j} &= \frac{\mu_1^2 x_{1j} + \mu_2^2 x_{2j} + \mu_3^2 x_{3j} + \mu_4^2 x_{4j}}{\mu_1^2 + \mu_2^2 + \mu_3^2 + \mu_4^2} \\ &= \frac{(1)x_{1j} + (1)x_{2j} + (1)x_{3j} + (0)x_{4j}}{1 + 1 + 1 + 0} = \frac{x_{1j} + x_{2j} + x_{3j}}{1^2 + 1^2 + 1^2 + 0} \end{aligned}$$

$$\left.\begin{aligned} v_{11} &= \frac{1 + 1.5 + 1.3}{3} = 1.26 \\ v_{12} &= \frac{3 + 3.2 + 2.8}{3} = 3.0 \end{aligned}\right\} \quad \mathbf{v}_1 = \{1.26, 3.0\}$$

and for $c = 2$,

$$v_{2j} \quad \text{or} \quad \mathbf{v}_j = \frac{x_{4j}}{0 + 0 + 0 + 1^2}$$

$$\left.\begin{aligned} v_{21} &= \tfrac{3}{1} = 3 \\ v_{22} &= \tfrac{1}{1} = 1 \end{aligned}\right\} \quad \mathbf{v}_2 = \{3, 1\}$$

Now the distance measures (distances of each data point from each cluster center) are found using Eq. (11.29):

$$\begin{aligned} d_{11} &= \sqrt{(1 - 1.26)^2 + (3 - 3)^2} = 0.26 & d_{21} &= \sqrt{(1 - 3)^2 + (3 - 1)^2} = 2.82 \\ d_{12} &= \sqrt{(1.5 - 1.26)^2 + (3.2 - 3)^2} = 0.31 & d_{22} &= \sqrt{(1.5 - 3)^2 + (3.2 - 1)^2} = 2.66 \\ d_{13} &= \sqrt{(1.3 - 1.26)^2 + (2.8 - 3)^2} = 0.20 & d_{23} &= \sqrt{(1.3 - 3)^2 + (2.8 - 1)^2} = 2.47 \\ d_{14} &= \sqrt{(3 - 1.26)^2 + (1 - 3)^2} = 2.65 & d_{24} &= \sqrt{(3 - 3)^2 + (1 - 1)^2} = 0.0 \end{aligned}$$

With the distance measures, we can now update $\underset{\sim}{U}$ using Eqs. (11.33) to (11.35) (for $m' = 2$), i.e.,

$$\mu_{ik}^{(r+1)} = \left[\sum_{j=1}^{c} \left(\frac{d_{ik}^{(r)}}{d_{jk}^{(r)}} \right)^2 \right]^{-1}$$

and we get,

$$\mu_{11} = \left[\sum_{j=1}^{c}\left(\frac{d_{11}}{d_{j1}}\right)^2\right]^{-1} = \left[\left(\frac{d_{11}}{d_{11}}\right)^2 + \left(\frac{d_{11}}{d_{21}}\right)^2\right]^{-1} = \left[\left(\frac{0.26}{0.26}\right)^2 + \left(\frac{0.26}{2.82}\right)^2\right]^{-1} = 0.991$$

$$\mu_{12} = \left[\left(\frac{d_{12}}{d_{12}}\right)^2 + \left(\frac{d_{12}}{d_{22}}\right)^2\right]^{-1} = \left[1 + \left(\frac{0.31}{2.66}\right)^2\right]^{-1} = 0.986$$

$$\mu_{13} = \left[\left(\frac{d_{13}}{d_{13}}\right)^2 + \left(\frac{d_{13}}{d_{23}}\right)^2\right]^{-1} = \left[1 + \left(\frac{0.20}{2.47}\right)^2\right]^{-1} = 0.993$$

$$\mu_{14} = \left[\left(\frac{d_{14}}{d_{14}}\right)^2 + \left(\frac{d_{14}}{d_{24}}\right)^2\right]^{-1} = \left[1 + \left(\frac{2.65}{0}\right)^2\right]^{-1} \rightarrow 0.0, \qquad \text{for } I_4 \neq \emptyset$$

Using Eq. (11.20) for the other partition values, μ_{2j}, for $j = 1, 2, 3, 4$, the new membership functions form an updated fuzzy partition given by

$$\underset{\sim}{U}^{(1)} = \begin{bmatrix} 0.991 & 0.986 & 0.993 & 0 \\ 0.009 & 0.014 & 0.007 & 1 \end{bmatrix}$$

To determine whether we have achieved convergence, we choose a matrix norm $\| \|$ such as the maximum absolute value of pairwise comparisons of each of the values in $\underset{\sim}{U}^{(0)}$ and $\underset{\sim}{U}^{(1)}$; e.g.,

$$\max_{i,k} \mid \mu_{ik}^{(1)} - \mu_{ik}^{(0)} \mid = 0.0134 > 0.01$$

This result suggests our convergence criteria have not yet been satisfied, so we need another iteration of the method.

For the next iteration we proceed by again calculating cluster centers, but now from values from the latest fuzzy partition, $\underset{\sim}{U}^{(1)}$; for $c = 1$,

$$v_{1j} = \frac{(0.991)^2 x_{1j} + (0.986)^2 x_{2j} + (0.993)^2 x_{3j} + (0) x_{4j}}{0.991^2 + 0.986^2 + 0.993^2 + 0}$$

$$\left.\begin{aligned} v_{11} &= \frac{0.98(1) + 0.97(1.5) + 0.99(1.3)}{2.94} = \frac{3.719}{2.94} \approx 1.26 \\ v_{12} &= \frac{0.98(3) + 0.97(3.2) + 0.99(2.8)}{2.94} = \frac{8.816}{2.94} \approx 3.0 \end{aligned}\right\} \quad \mathbf{v}_1 = \{1.26, 3.0\}$$

and for $c = 2$,

$$v_{2j} = \frac{(0.009)^2 x_{1j} + (0.014)^2 x_{2j} + (0.007)^2 x_{3j} + (1)^2 x_{4j}}{0.009^2 + 0.014^2 + 0.007^2 + 1^2}$$

$$\left.\begin{aligned} v_{21} &= \frac{0.009^2(1) + 0.014^2(1.5) + 0.007^2(1.3) + 1(3)}{1.000} \approx 3.0 \\ v_{22} &= \frac{0.009^2(3) + 0.014^2(3.2) + 0.007^2(2.8) + 1(1)}{1.000} \approx 1.0 \end{aligned}\right\} \quad \mathbf{v}_2 = \{3.0, 1.0\}$$

By continuing in this manner, next with calculation of new distance measures, then with new partition values, it will be found that the slight difference in the partition values, $\mu^{(r+1)}$, for the next iteration will result in convergence to a value less than that specified (0.01) and the final partition matrix will be unchanged, to an accuracy of two

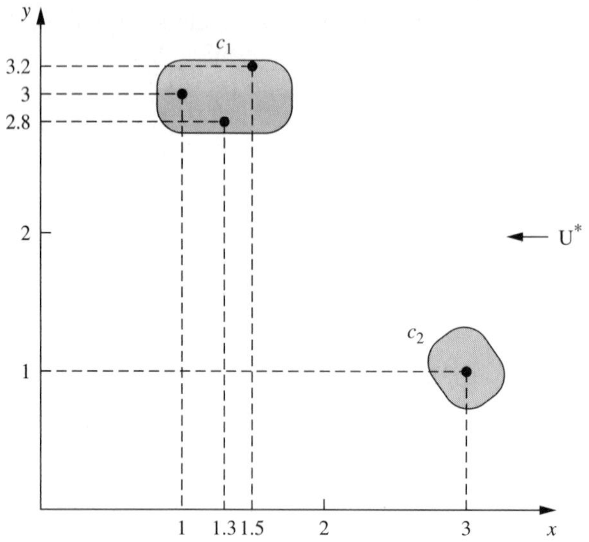

FIGURE 11.5
Converged fuzzy partition for catalytic converter example.

digits, from that obtained in the previous iteration. As suggested earlier, convergence is rapid, at least for this example. The final partition, $\underset{\sim}{U}^{(2)}$, will result in a classification shown in Fig. 11.5.

CLASSIFICATION METRIC

In most studies involving fuzzy pattern classification, the data used in the classification process typically come from electrically active transducer readings [Bezdek et al., 1986]. When a fuzzy clustering is accomplished, a question remains concerning the uncertainty of the clustering in terms of the features used. That is, our interest should lie with the extent to which pairs of fuzzy classes of $\underset{\sim}{U}$ overlap; a true classification with no uncertainty would contain classes with no overlap. The question then is, "How fuzzy is a fuzzy c-partition?" Suppose we compare two memberships for a given data set, x_k, pairwise, using the minimum function, i.e.,

$$\min\{u_i(x_k), u_j(x_k)\} > 0 \tag{11.36}$$

This comparison would indicate that membership of x_k is *shared* by u_i and u_j, whereas the minimum of these two values reflects the minimum amount of *unshared* membership x_k can claim in either u_i or u_j. Hence, a fuzziness measure based on functions $\min\{u_i(x_k), u_j(x_k)\}$ would constitute a point-by-point assessment—not of overlap, but of "anti-overlap" [Bezdek, 1974]. A more useful measure of fuzziness in this context, which has values directly dependent on the relative overlap between nonempty fuzzy class intersections, can be found with the algebraic product of u_i and u_j, or the form $u_i u_j(x_k)$. An interesting interpretation of the fuzzy clustering results is to compute the fuzzy partition coefficient,

$$F_c(\underset{\sim}{U}) = \frac{\operatorname{tr}(\underset{\sim}{U}^* \underset{\sim}{U}^{\mathrm{T}})}{n} \tag{11.37}$$

where $\underset{\sim}{U}$ is the fuzzy partition matrix being segregated into c classes (partitions), n is the number of data sets, and the operation "$*$" is standard matrix multiplication. The product $\underset{\sim}{U}^*\underset{\sim}{U}^T$ is a matrix of size $c \times c$. The partition coefficient, $F_c(\underset{\sim}{U})$, has some special properties [Bezdek, 1974]: $F_c(\underset{\sim}{U}) = 1$ if the partitioning in $\underset{\sim}{U}$ is crisp (comprised of zeros and ones); $F_c(\underset{\sim}{U}) = 1/c$ if all the values $u_i = 1/c$ (complete ambiguity); and in general $1/c \leq F_c(\underset{\sim}{U}) \leq 1$. The diagonal entries of $\underset{\sim}{U}^*\underset{\sim}{U}^T$ are proportional to the amount of *unshared* membership of the data sets in the fuzzy clusters, whereas the off-diagonal elements of $\underset{\sim}{U}^*\underset{\sim}{U}^T$ represent the amount of membership *shared* between pairs of fuzzy clusters of $\underset{\sim}{U}$. If the off-diagonal elements of $\underset{\sim}{U}^*\underset{\sim}{U}^T$ are zero, then the partitioning (clustering) is crisp. As the partition coefficient approaches a value of unity, the fuzziness in overlap in classes is minimized. Hence, as $F_c(\underset{\sim}{U})$ increases, the decomposition of the data sets into the classes chosen is more successful.

Example 11.12 [Ross et al., 1993]. Forced response dynamics of a simple mechanical two-degrees-of-freedom (2-DOF) oscillating system, as shown in Fig. 11.6, are conducted. In the figure the parameters m_1, k_1, and c_1, and m_2, k_2, and c_2 are the mass, stiffness, and damping coefficients for the two masses, respectively, and the base of the system is assumed fixed to ground. The associated frequency and damping ratios for the two modes of free vibration for this system are summarized in Table 11.8. The 2-DOF system is excited with a force actuator on the larger of the two masses (see Fig. 11.6), and a displacement sensor on this same mass collects data on displacement vs. time.

The response is computed for the displacement sensor in the form of frequency response functions (FRF). The derivatives of these FRF were computed for a specific exciting frequency of 1.0 rad/s with respect to the six modal mass and stiffness matrix elements (denoted in Table 11.9 as $x_1, x_2, \ldots, x_6$). Amplitude derivatives and the covariance matrix entries of these parameters are given in Table 11.9.

A simple FCM classification approach was conducted on the feature data ($m = 2$) given in Table 11.9 and the fuzzy partitions shown in Table 11.10 resulted for a 2-class case ($c = 2$) and a 3-class case ($c = 3$).

The resulting values of $F_c(\underset{\sim}{U})$ from Eq. (11.37) for the two clustering cases are listed in Table 11.11. Of course, the result for $c = 3$ is intuitively obvious from inspection of Table 11.10, which is crisp. However, such obviousness is quickly lost when one deals with problems characterized by a large data base.

TABLE 11.8
Modal response parameters for 2-DOF example

Mode	Frequency (rad/sec)	Damping ratio
1	0.98617	0.01
2	1.17214	0.01

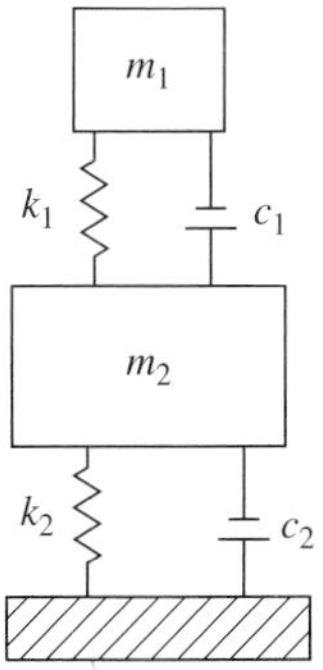

FIGURE 11.6
Mechanical system with two degrees of freedom.

TABLE 11.9
FRF sensitivity-parameter uncertainty data sets

Data set	FRF derivative	Variance
x_1	3.5951	0.2370
x_2	1.7842	0.2906
x_3	0.1018	0.3187
x_4	3.3964	0.2763
x_5	1.7620	0.2985
x_6	0.1021	0.4142

TABLE 11.10
Clustering results for a simple 2-DOF problem

	Data pairs ($c = 2$)					
	x_1	x_2	x_3	x_4	x_5	x_6
Class 1	0.000	0.973	0.998	0.000	0.976	0.998
Class 2	1.000	0.027	0.002	1.000	0.024	0.002
	Data pairs ($c = 3$)					
Class 1	0	0	1	0	0	1
Class 2	0	1	0	0	1	0
Class 3	1	0	0	1	0	0

HARDENING THE FUZZY *c*-PARTITION

There are two popular methods, among many others, to defuzzify fuzzy partitions, $\underset{\sim}{U}$, i.e., for hardening the fuzzy classification matrix. This defuzzification may be required in the ultimate assignment of data to a particular class. These two methods are called the *maximum-membership method* and the *nearest-center classifier*.

In the max-membership method, the largest element in each column of the $\underset{\sim}{U}$ matrix is assigned a membership of unity and all other elements in each column are assigned a membership value of zero. In mathematical terms if the largest membership in the kth column is μ_{ik}, then x_k belongs to class i, i.e., if

$$\mu_{ik} = \max_{j \in c}\{\mu_{ik}\} \text{ then } \mu_{ik} = 1; \qquad \mu_{jk} = 0 \text{ for all } j \neq i \tag{11.38}$$

for $i = 2, \ldots, c$ and $k = 1, 2, \ldots, n$.

In the nearest-center classifier, each of the data points is assigned to the class that it is closest to; i.e., the minimum Euclidean distance from a given data point and the c cluster centers dictates the class assignment of that point. In mathematical terms, if

$$d_{ik} = \min_{j \in c}\{d_{jk}\} = \min_{j \in c} \|\mathbf{x}_k - \mathbf{v}_j\|$$

then

$$\begin{aligned} \mu_{ik} &= 1 \\ \mu_{jk} &= 0 \qquad \text{for all } j \neq i \end{aligned} \tag{11.39}$$

TABLE 11.11
Partitioning coefficient for two different classes

c	$F_c(\underset{\sim}{U})$
2	0.982
3	1.000

Example 11.13. If we take the partition matrix, $\underset{\sim}{U}$, developed on the catalytic converter in Example 11.8 as shown in Fig. 11.5, and harden it using the methods in Eqs. (11.38) and (11.39), we get the following:

$$\underset{\sim}{U} = \begin{bmatrix} 0.991 & 0.986 & 0.993 & 0 \\ 0.009 & 0.014 & 0.007 & 1 \end{bmatrix}$$

Max-Membership Method.

$$U^{\text{Hard}} = \begin{bmatrix} 1 & 1 & 1 & 0 \\ 0 & 0 & 0 & 1 \end{bmatrix}$$

Nearest-Center Classifier. If we take the distance measures from the catalytic converter problem, i.e.,

$$\begin{aligned} d_{12} &= 0.26 \qquad & d_{21} &= 2.82 \\ d_{12} &= 0.31 & d_{21} &= 2.66 \\ d_{12} &= 0.20 & d_{21} &= 2.47 \\ d_{12} &= 2.65 & d_{21} &= 0 \end{aligned}$$

and arrange these values in a 2×4 matrix, such as

$$d_{ij} = \begin{bmatrix} 0.26 & 0.31 & 0.20 & 2.65 \\ 2.82 & 2.66 & 2.47 & 0 \end{bmatrix}$$

then the minimum value (distance) in each column is set to unity, and all other values (distances) in that column are set to zero. This process results in the following hard c-partition:

$$U^{\text{Hard}} = \begin{bmatrix} 1 & 1 & 1 & 0 \\ 0 & 0 & 0 & 1 \end{bmatrix}$$

which, for this example, happens to be the same partition that is derived using the max-membership hardening method.

SIMILARITY RELATIONS FROM CLUSTERING

The classification idea can be recast in the form of a similarity relation that is also a tolerance relation. This idea represents another way to look at the structure in data, by comparing the data points to one another, pairwise, in a similarity analysis. In classification we seek to segregate data into clusters where points in each cluster are as "similar" to one another as possible and where clusters are dissimilar to one another. This notion of similarity, then, is central to classification. The use of a fuzzy similarity relation can be useful in the classification process [Bezdek and Harris, 1978].

A fuzzy relation $\underset{\sim}{R}$ can be constructed from the fuzzy partition $\underset{\sim}{U}$ as follows:

$$\underset{\sim}{R} = \left(\underset{\sim}{U}^{\mathrm{T}}\left(\sum\wedge\right)\underset{\sim}{U}\right) = [r_{kj}] \tag{11.40}$$

$$r_{kj} = \sum_{i=1}^{c} \mu_{ik} \wedge \mu_{ij} \tag{11.41}$$

where the symbol $(\sum \wedge)$ denotes "sum of mins."

Example 11.14. We take the fuzzy partition $\underset{\sim}{U}$ from the fruit genetics example (Example 11.9) and perform the mixed algebraic and set operations as provided in Eqs. (11.40) and (11.41). So for

$$\underset{\sim}{U}^{T} = \begin{bmatrix} .91 & .09 \\ .58 & .42 \\ .13 & .87 \end{bmatrix} \quad \text{and} \quad \underset{\sim}{U} = \begin{bmatrix} .91 & .58 & .13 \\ .09 & .42 & .87 \end{bmatrix}$$

we get

$$\begin{aligned} r_{11} &= \min(.91, .91) + \min(.09, .09) = 1 \\ r_{12} &= \min(.91, .58) + \min(.09, .42) = .67 \\ r_{13} &= \min(.91, .13) + \min(.09, .87) = .22 \\ r_{23} &= \min(.58, .13) + \min(.42, .87) = .55 \end{aligned}$$

and so forth, and the following fuzzy similarity relation results:

$$\underset{\sim}{R} = \begin{bmatrix} 1 & .67 & .22 \\ .67 & 1 & .55 \\ .22 & .55 & 1 \end{bmatrix}$$

The fuzzy similarity relation $\underset{\sim}{R}$ provides similar information about clustering as does the original fuzzy partition, $\underset{\sim}{U}$. The fuzzy classification partition groups the data according to class type; the fuzzy relation shows the pairwise similarity of the data without regard to class type. Data that have strong similarity, or high membership values in $\underset{\sim}{R}$, should tend to have high membership in the same class in $\underset{\sim}{U}$. Although the two measures are based on the same data (i.e., the features describing each data point), their information content is slightly different.

SUMMARY

The concept of a fuzzy set first arose in the study of problems related to pattern classification [Bellman et al., 1966]. Since the recognition and classification of patterns is integral to human perception, and since these perceptions are fuzzy, this study seems a likely beginning. This chapter has presented a simple idea in the area of classification involving equivalence relations and has dealt in depth with a particular form of classification using a popular clustering method: *fuzzy c-means*. The objective in clustering is to partition a given data set into homogeneous clusters; by homogeneous we mean that all points in the same cluster share *similar* attributes and they do not share similar attributes with points in other clusters. However, the separation of clusters and the meaning of *similar* are fuzzy notions and can be described as such. One of the first introductions to the clustering of data was in the area of fuzzy partitions [Ruspini, 1969, 1970, 1973*a*], where similarity was measured using membership values. In this case, the classification metric was a function involving a distance measure that was minimized. Ruspini [1973*b*] points out that a definite benefit of fuzzy clustering is that stray points (outliers) or points isolated between clusters (see Fig. 11.2) may be classified this way; they will have low membership values in the clusters from which they are isolated. In crisp classification methods these stray points need to belong to at least one of the clusters, and their membership in the cluster to which they are assigned is unity; their distance, or the extent of their isolation, cannot be measured by their membership. These notions of fuzzy classification described in this chapter provide for a point of departure in the recognition of known patterns—the subject of the next chapter.

REFERENCES

Bellman, R., R. Kalaba, and L. Zadeh (1966). "Abstraction and pattern classification," *J. Math. Anal. Appl.,* vol. 13, pp. 1–7.

Bezdek, J. (1974). "Numerical taxonomy with fuzzy sets," *J. Math. Biol.,* vol. 1, pp. 57–71.

Bezdek, J. (1981). *Pattern recognition with fuzzy objective function algorithms,* Plenum, New York.

Bezdek, J., and J. Harris (1978). "Fuzzy partitions and relations: an axiomatic basis for clustering," *Fuzzy Sets Syst.,* vol. 1, pp. 111–127.

Bezdek, J., N. Grimball, J. Carson, and T. Ross (1986). "Structural failure determination with fuzzy sets," *Civ. Eng. Syst.,* vol. 3, pp. 82–92.

Dong, W. (1987). Personal notes.

Duda, R., and P. Hart (1973). *Pattern classification and scene analysis,* John Wiley & Sons, New York.

Ross, T., T. Hasselman, and J. Chrostowski (1993). "Fuzzy set methods in assessing uncertainty in the modeling and control of space structures," *J. Intelligent Fuzzy Syst.,* vol. 1, no. 2, pp. 135–155.

Ruspini, E. (1969). "A new approach to clustering," *Inf. Control,* vol. 15, pp. 22–32.

Ruspini, E. (1970). "Numerical methods for fuzzy clustering," *Inf. Sci.,* vol. 2, pp. 319–350.

Ruspini, E. (1973*a*). "New experimental results in fuzzy clustering," *Inf. Sci.,* vol. 6, pp. 273–284.

Ruspini, E. (1973*b*). "A fast method for probabilistic and fuzzy cluster analysis using association measures," *Proc. 6th Int. Conf. Syst. Sci.,* Hawaii, pp. 56–58.

Tamura, S., S. Higuchi, and K. Tanaka (1971). "Pattern classification based on fuzzy relations," *IEEE Trans. Syst. Man Cybern.,* vol. 1, pp. 61–66.

Zadeh, L. (1971). "Similarity relations and fuzzy orderings," *Inf. Sci.,* vol. 3, pp. 177–200.

PROBLEMS

Exercises for Equivalence Classification

11.1. A fuzzy tolerance relation, $\underset{\sim}{R}$, is reflexive and symmetric. Find the equivalence relation $\underset{\sim}{R}_e$ and then classify it according to λ-cut levels $= \{0.9, 0.8, 0.5\}$.

$$\underset{\sim}{R} = \begin{Bmatrix} 1 & 0.8 & 0 & 0.2 & 0.1 \\ 0.8 & 1 & 0.9 & 0 & 0.4 \\ 0 & 0.9 & 1 & 0 & 0.3 \\ 0.2 & 0 & 0 & 1 & 0.5 \\ 0.1 & 0.4 & 0.3 & 0.5 & 1 \end{Bmatrix}$$

11.2. There are four groups of graduate students in a first postgraduate year. A fuzzy tolerance relation $\underset{\sim}{R}$ is produced according to the level of academic preparation of the students. If λ-cut levels are $\{0.9, 0.8\}$, find classes into which these four groups can be classified.

$$\underset{\sim}{R} = \begin{bmatrix} 1 & 0.8 & 0.3 & 0.7 \\ 0.8 & 1 & 0.9 & 0.1 \\ 0.3 & 0.9 & 1 & 0.6 \\ 0.7 & 0.1 & 0.6 & 1 \end{bmatrix}$$

11.3. A customer evaluates five banks by their mortgage policy, loan interest, and dividend benefits. She has developed a fuzzy tolerance relation $\underset{\sim}{R}$ according to these three criteria for banks. If she uses a λ-cut level of 0.8, then how many classes can she make from these banks?

$$\underset{\sim}{R} = \begin{bmatrix} 1 & 0.2 & 0.5 & 0.9 & 0.7 \\ 0.2 & 1 & 0.6 & 0.4 & 0.8 \\ 0.5 & 0.6 & 1 & 0.4 & 0.3 \\ 0.9 & 0.4 & 0.4 & 1 & 0.5 \\ 0.7 & 0.8 & 0.3 & 0.5 & 1 \end{bmatrix}$$

11.4. In a pattern recognition test, four unknown patterns need to be classified according to three known patterns (primitives) *a*, *b*, and *c*. The relationship between primitives and unknown patterns is in the following table:

	x_1	x_2	x_3	x_4
a	0.6	0.2	0.1	0.8
b	0.3	0.2	0.7	0.1
c	0.1	0.6	0.2	0.1

If a λ-cut level is 0.5, then into how many classes can these patterns be divided? [*Hint:* Use a max-min method (see Chapter 3) to generate a fuzzy similarity relation $\underset{\sim}{R}$.]

11.5. As a first step in automatic segmentation of magnetic resonance imaging (MRI) data regarding the head, it is necessary to determine the orientation of a data set to be segmented. The standard radiological orientations are sagittal, coronal, and horizontal. One way to classify the orientation of the new data would be to compare a slice of the new data to slices of known orientation. To do the classification we will use a simple metric obtained by overlaying slice images and obtaining an area of intersection, then normalizing these, based on the largest area of intersection. This metric will be our "degree of resemblance" for the equivalence relation. From data you have the following fuzzy relation:

	S	C	H	N
Sagittal	1	.6	.4	.7
Coronal	.6	1	.5	.7
Horizontal	.4	.5	1	.5
New slice	.7	.7	.5	1

(*a*) What kind of relation is this?

(*b*) Determine the equivalence relation and conduct a classification at λ-cut levels of 0.4, 0.6, and 0.7.

Problems for Fuzzy *c*-Means

11.6. The aerodynamic model for a pilot training simulator is composed of numerous lookup data tables. The data in these tables are representative of characteristics of the actual airplanes. These data tables are developed using data recorded over time during flights of an instrumented airplane. The flight test data generally cluster around the breakpoints used in the data tables. The flight test data whose fit is "best" at a given table

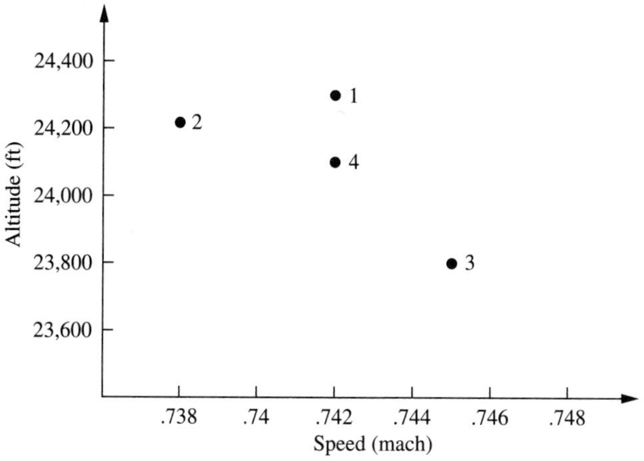

FIGURE P11.6

breakpoint can be identified using fuzzy classification methods. Given the four data points shown in Fig. P11.6 and summarized as

$$\begin{aligned} FT_1 &= \{.742 \text{ mach}, 24{,}300 \text{ ft altitude}\} \\ FT_2 &= \{.738 \text{ mach}, 24{,}200 \text{ ft altitude}\} \\ FT_3 &= \{.745 \text{ mach}, 23{,}800 \text{ ft altitude}\} \\ FT_4 &= \{.742 \text{ mach}, 24{,}100 \text{ ft altitude}\} \end{aligned}$$

with a fuzzification value of $m' = 1.5$, a convergence level of 0.1, and the following initial fuzzy partition:

$$\underset{\sim}{U}^{(0)} = \begin{bmatrix} 1 & 1 & 0 & 1 \\ 0 & 0 & 1 & 0 \end{bmatrix}$$

find the converged classification partition.

11.7. (*Note:* This problem will require a computerized form of the c-means algorithm.) Suppose we conduct a tensile strength test of four kinds of unidentified material. We know from other sources that the materials are from two different categories. From the yield stress, σ_y, and yield strain, Δ_y, data determine which materials are from the two different categories (see Fig. P11.7).

	m_1	m_2	m_3	m_4
σ_y	8.9	8.1	7.3	8.3
Δ_y	1.4	1.6	1.8	1.9

Determine which values for m' and ε_L would give the following results after twenty-five iterations, i.e.,

$$\underset{\sim}{U}^{(25)} = \begin{bmatrix} .911 & .824 & .002 & .906 \\ .089 & .176 & .998 & .094 \end{bmatrix}$$

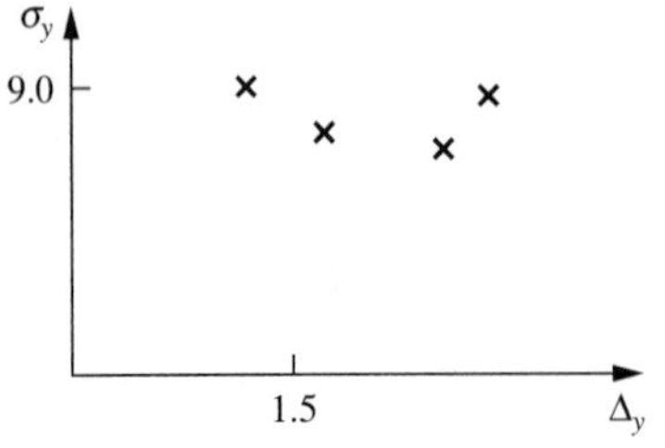

FIGURE P11.7

and final cluster centers of

$$\mathbf{v}_1 = \{8.458\ 1.634\} \qquad \mathbf{v}_2 = \{7.346\ 1.792\}$$

11.8. Suppose we have data points in two dimensions that represent computer processing characteristics. These characteristics are the number of floating point operations per second and memory size [a point is denoted by the pair (megaflops, megabytes)]. Say we have a set of four data points $\{\mathbf{x}_1, \mathbf{x}_2, \mathbf{x}_3, \mathbf{x}_4\}$ and we want to classify each data point as a member of one of the two classes c_1 = supercomputer or c_2 = PC.

Further, let us establish the following:

$$\begin{aligned} \mathbf{x}_1 &= (.001, .064) \\ \mathbf{x}_2 &= (.5, 8) \\ \mathbf{x}_3 &= (100, 32) \\ \mathbf{x}_4 &= (1000, 128) \\ m' &= 1.75 \\ \varepsilon_L &\leq 0.001 \end{aligned}$$

Then, let us make an initial guess

$$\underset{\sim}{\mathrm{U}}^{(0)} = \begin{bmatrix} 0 & 0 & 0 & 1 \\ 1 & 1 & 1 & 0 \end{bmatrix}$$

Find the final fuzzy partition.

11.9. This problem deals with categorizing television picture tubes (CRTs). Red (R), green (G), and blue (B) brightness measurements have been taken on four CRTs. We would like to determine how the data are clustered if the data involved are divided into two classes. You are to start by finding the hard c-partition and then using the resulting hard 2-partition to seed the process of finding the fuzzy 2-partition. Use the following parameters:

$n = 4,$	number of data points	$k = 1, 2, \ldots, n$
$c = 2,$	number of classes	$i = 1, 2, \ldots, c$
$m = 3,$	number of coordinates	$j = 1, 2, \ldots, m$

Data points:

$$x = \begin{array}{c} \begin{array}{ccc} \mathrm{R} & \mathrm{G} & \mathrm{B} \end{array} \\ \begin{bmatrix} .30 & .59 & .11 \\ .38 & .61 & .06 \\ .20 & .58 & .20 \\ .41 & .56 & .05 \end{bmatrix} \end{array} \begin{array}{l} = \mathbf{x}_1 \\ = \mathbf{x}_2 \\ = \mathbf{x}_3 \\ = \mathbf{x}_4 \end{array}$$

Use an initial guess for $\underset{\sim}{U}^{(0)}$ of

$$\underset{\sim}{U}^{(0)} = \begin{bmatrix} 1 & 0 & 1 & 0 \\ 0 & 1 & 0 & 1 \end{bmatrix}$$

and use a fuzzy parameter value of $m' = 1.5$ and a convergence level of $\varepsilon_L \leq 0.01$.

11.10. For a camera with a fill-in flash, we are interested in the distribution of combinations of ambient light level and subject distances found in common picture-taking situations. Clustering of these points would give an idea of how the fill-in flash would be used. Suppose that a test produced 4 pictures taken at these ambient light/distance combinations:

$\mathbf{x}_1$	600 foot-candles, 10.7 ft
$\mathbf{x}_2$	40 foot-candles, 1.7 ft
$\mathbf{x}_3$	115 foot-candles, 3.5 ft
$\mathbf{x}_4$	325 foot-candles, 4 ft

Using the initial partition illustrated in Fig. P11.10, $m' = 1.25$, and a convergence level of 0.1, find the optimum fuzzy 2-partition.

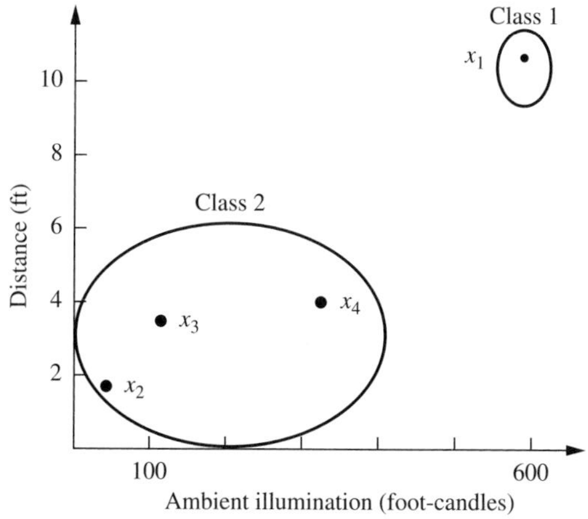

FIGURE P11.10

For the following problems, use $m' = 2.0$ *and* $\varepsilon_L \leq 0.01$.

11.11. A problem in construction management is to allocate four different job sites to two different construction teams such that the time wasted in shuttling between the sites is minimized. Let the job sites be designated as x_i and combined to give a universe, $X = \{\mathbf{x}_1, \mathbf{x}_2, \mathbf{x}_3, \mathbf{x}_4\}$. If the head office, where the construction teams start every day, has coordinates $\{0, 0\}$, the following vectors give the locations of the four job sites:

$$\mathbf{x}_1 = \{4, 5\}$$
$$\mathbf{x}_2 = \{3, 4\}$$
$$\mathbf{x}_3 = \{8, 10\}$$
$$\mathbf{x}_4 = \{9, 12\}$$

Conduct a fuzzy c-means calculation to determine the optimum partition, $\underset{\sim}{U}^*$. Start with the following initial 2-partition:

$$\underset{\sim}{U}^{(0)} = \begin{Bmatrix} 1 & 1 & 0 & 0 \\ 0 & 0 & 1 & 1 \end{Bmatrix}$$

11.12. A radar image of a vehicle is a mapping of the bright (most reflective) parts of it. Suppose we have a radar image that we know contains two vehicles parked close together. The threshold on the instrument has been set such that the image contains seven bright dots. We wish to classify the dots as belonging to one or the other vehicle with a fuzzy membership before we conduct a recognition of the vehicle type. The seven bright dots are arranged in a matrix X, and we seek to find an optimum membership matrix $\underset{\sim}{U}^*$. The features defining each of the seven dots are given here:

2	9	9	5	8	5	6
7	3	5	6	8	10	4

Start the calculation with the following initial 2-partition:

$$\underset{\sim}{U}^0 = \begin{bmatrix} 0 & 0 & 0 & 0 & 0 & 0 & 1 \\ 1 & 1 & 1 & 1 & 1 & 1 & 0 \end{bmatrix}$$

Find the converged optimal 2-partition.

11.13. As a company issuing bank cards, we want to separate individuals holding personal credit cards into two groups: profitable and unprofitable. Profitable card holders usually let a nonzero balance on the card go from month to month, thereby accruing interest; and they eventually always pay the total balance. Unprofitable card holders often do not pay a balance, and card privileges usually have to be canceled. Suppose the data for the jth card holder consist of three features:

$$\mathbf{x}_j = (x_{11}, x_{21}, x_{31})$$

where x_{1j} = account balance
x_{2j} = amount paid
x_{3j} = amount purchased

Suppose we want to classify four individuals, each characterized by the following normalized data:

$$\mathbf{x}_1 = (1, .75, 1)$$
$$\mathbf{x}_2 = (0, 0, -.5)$$
$$\mathbf{x}_3 = (.5, .5, .75)$$
$$\mathbf{x}_4 = (1, -.5, -.5)$$

Note that the values (features) in each vector are normalized Gaussian variables such that $\mu = 0$ and $\sigma = 1$, where

$$\text{value} = \frac{X - \mu_c}{\sigma_c}$$

Hence, values at zero are at the mean of a class (c), positive values indicate a variable greater than the mean, and negative values indicate a variable less than the mean.

Use the following 2-partition as the initial guess, and find the optimum 2-partition.

$$\underset{\sim}{U}^{(0)} = \begin{Bmatrix} 1 & 1 & 0 & 0 \\ 0 & 0 & 1 & 1 \end{Bmatrix}$$

11.14. We are classifying imperfections on prints of film. Features on the film are m_1 = film density; m_2 = length of spot. We wish to classify three simple data points (in non-dimensional units) into two classes and conduct a fuzzy 2-partition using the following data:

$$\mathbf{x}_1 = \{5, 3\}$$
$$\mathbf{x}_2 = \{4.8, 3.5\}$$
$$\mathbf{x}_3 = \{5.2, .5\}$$

and the following initial 2-partition:

$$\underset{\sim}{U}^{(0)} = \begin{bmatrix} 1 & 1 & 0 \\ 0 & 0 & 1 \end{bmatrix}$$

Find the optimal 2-partition.

11.15. In a magnetoencephalography (MEG) experiment, we attempt to partition the space of dipole model order versus reduced chi square value for the dipole fit. This could be useful to an MEG researcher in determining any trends in his or her data-fitting procedures. Typical ranges for these parameters would be as follows:

$$\text{Dipole model order} = (1, 2, \ldots, 6) = x_{1i}$$
$$\text{Reduced } \chi^2 \in (1, 3) = x_{2i}$$

Suppose we have three MEG data points, $\mathbf{x}_i = (x_{1i}, x_{2i})$, $i = 1, 2, 3$, to classify into two classes. The data are

$$\mathbf{x}_1 = (2, 1.5) \qquad \mathbf{x}_2 = (3, 2.2) \qquad \mathbf{x}_3 = (4, 2)$$

Find the optimum 2-partition using the following initial partition:

$$\underset{\sim}{U}^{(0)} = \begin{bmatrix} 1 & 0 & 0 \\ 0 & 1 & 1 \end{bmatrix}$$

11.16. One way to express the *readability* of a document is to determine the ratio of the number of paragraphs to the number of words; the higher this ratio, the more readable the document. Given the following three documents, and the number of paragraphs and words in each, classify them into two categories:

Document	Number of paragraphs (feature 1)	Number of words (feature 2)
x_1	10	500
x_2	8	2600
x_3	8	460

Use an initial partition $\underset{\sim}{U}^{(0)}$, and find the optimum 2-partition.

$$\underset{\sim}{U}^{(0)} = \begin{matrix} \\ c_1 \\ c_2 \end{matrix} \begin{matrix} \mathbf{x}_1 & \mathbf{x}_2 & \mathbf{x}_3 \\ \left[\begin{matrix} 1 \\ 0 \end{matrix}\right. & \begin{matrix} 0 \\ 1 \end{matrix} & \left.\begin{matrix} 1 \\ 0 \end{matrix}\right] \end{matrix}$$

11.17. A paleontologist found several ancient reptile remains in an excavation. He knows that the pieces are characteristic of only two periods in the Mesozoic era: Jurassic and Triassic. He has to classify the remains according to their periods. He takes three samples from among his findings to conduct an initial examination, and describes them with two paleontological features, both dimensionless. The three data sets are $\mathbf{x}_1 = \{0.1, 1.1\}$, $\mathbf{x}_2 = \{0.15, 0.9\}$, and $\mathbf{x}_3 = \{0.5, 0.95\}$. These data points are plotted as shown in Fig. P11.17. Although visually the classification seems clear, find the optimum fuzzy 2-partition with the initial guess

$$\underset{\sim}{U}^{(0)} = \begin{matrix} c_1 \\ c_2 \end{matrix} \begin{bmatrix} 1 & 1 & 0 \\ 0 & 0 & 1 \end{bmatrix}$$

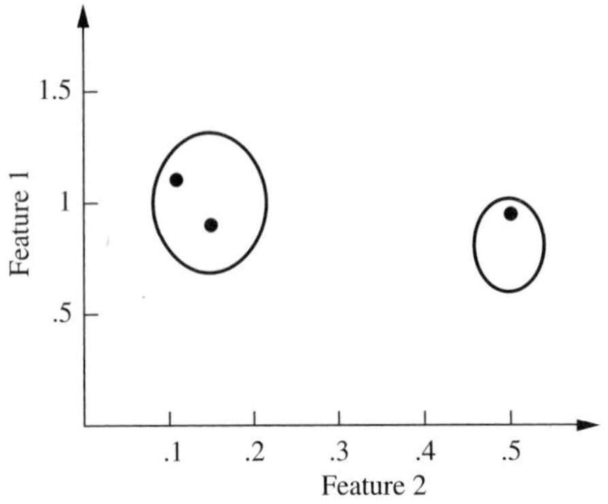

FIGURE P11.17

11.18. Suppose we want to sample a complex signal from a demodulator circuit and classify it into two sets, $A_\varnothing$ or A_1. The sample points are $\mathbf{x}_1 = (-3, 1)$, $\mathbf{x}_2 = (-2, 2)$, $\mathbf{x}_3 = (-1, 1.5)$, and $\mathbf{x}_4 = (1, 2)$ as shown in Fig. P11.18. If the first row of your initial 2-partition is [1 0 0 0], find the fuzzy 2-partition after three iterations; i.e., find $\underset{\sim}{U}^{(3)}$.

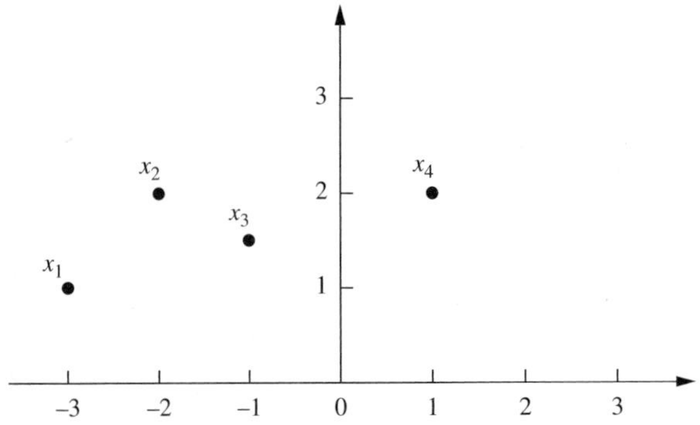

FIGURE P11.18

11.19. A small number of sequential operations can effectively limit the speedup of a parallel algorithm. Let f be the fraction of operations in a computation that must be performed sequentially, where $0 \le f \le 1$. According to Amdahl's law, the maximum speedup s achievable by a parallel computer with p processors is

$$s \le \frac{1}{f + (1 - f)/p}$$

Suppose we have three data points, each described by two features: the fraction of sequential operations (f), and the maximum efficiency (s). These data points and their features are given in the following table:

	x_1	x_2	x_3
f	0.3	0.2	0.1
s	0.5	0.4	0.8

We want to classify these three data points into two classes ($c = 2$) according to the curves of Amdahl's law. Of the three possible hard partitions ($\eta_{M_c} = 3$), the one that seems graphically plausible is

$$\begin{bmatrix} 1 & 1 & 0 \\ 0 & 0 & 1 \end{bmatrix}$$

Using this partition as the initial guess, find the fuzzy 2-partition after two cycles, $\underset{\sim}{U}^{(2)}$.

11.20. In the field of electrical engineering we want to classify a system's complexity by the number of inputs and by the order of the system's response to a unit step input. We want to classify four systems (four data points), whose coordinates in features space are

$$\mathbf{x}_1 = (1, 2)$$
$$\mathbf{x}_2 = (2, 3)$$
$$\mathbf{x}_3 = (3, 3)$$
$$\mathbf{x}_4 = (4, 3)$$

as seen in Fig. P11.20. If the first row of the initial partition is [0, 0.6, 0.7, 0.9], find the optimum fuzzy 2-partition.

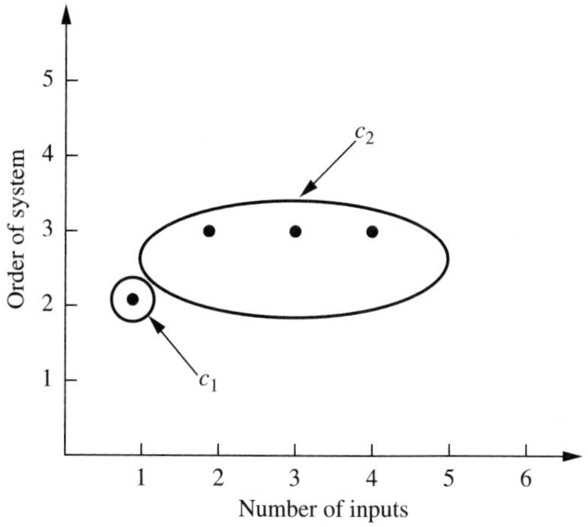

FIGURE P11.20

11.21. We want to classify the performance of three computer systems based on throughput (in mips) and response time (in seconds). The data points in our sample, $X = \{\mathbf{x}_1, \mathbf{x}_2, \mathbf{x}_3\}$, are $\mathbf{x}_1 = (50, 10)$, $\mathbf{x}_2 = (40, 12)$, and $\mathbf{x}_3 = (20, 5)$. Using the initial 2-partition,

$$\underset{\sim}{U}^{(0)} = \begin{bmatrix} 1 & 1 & 0 \\ 0 & 0 & 1 \end{bmatrix}$$

verify that the optimum fuzzy 2-partition after two cycles is

$$\underset{\sim}{U}^{(2)} = \begin{bmatrix} .974 & .9418 & 0 \\ .026 & .0582 & 1 \end{bmatrix}$$

11.22. We are conducting some market studies of four new video camera designs that our company wants to manufacture, then sell. Each of these four data points (cameras) has the following three features: horizontal resolution measured in TV lines, signal-to-noise ratio (SNR) measured in decibels (db), and sensitivity in millivolts (mV). We want to conduct a fuzzy partitioning to classify these four cameras into two fuzzy classes; the classes can be labeled "best design" and "adequate design." This classification will serve to reduce the number of camera designs we consider beyond our preliminary design stage. You are given a set of sample data $X = \{\mathbf{x}_1, \mathbf{x}_2, \mathbf{x}_3, \mathbf{x}_4\}$ with these element vectors (features):

$$\mathbf{x}_1 = \begin{pmatrix} 330 \\ 50 \\ 200 \end{pmatrix} \quad \mathbf{x}_2 = \begin{pmatrix} 580 \\ 40 \\ 350 \end{pmatrix} \quad \mathbf{x}_3 = \begin{pmatrix} 560 \\ 45 \\ 150 \end{pmatrix} \quad \mathbf{x}_4 = \begin{pmatrix} 320 \\ 47 \\ 135 \end{pmatrix}$$

(*a*) Find the optimum *hard* 2-partition, U^*, with initial guess

$$U^{(0)} = \begin{bmatrix} 1 & 0 & 0 & 1 \\ 0 & 1 & 1 & 0 \end{bmatrix}$$

(*b*) Using the final result from part (*a*) as the initial guess, verify that the optimum *fuzzy* 2-partition after four cycles is

$$\underset{\sim}{U}^{(4)} = \begin{bmatrix} .983 & .074 & .194 & .985 \\ .017 & .926 & .806 & .015 \end{bmatrix}$$

and show that the cluster centers are given by

$$\mathbf{v}^{(4)} = \begin{bmatrix} 330 & 571 \\ 48 & 42 \\ 168 & 263 \end{bmatrix}$$

(*c*) Comment on how the fuzzy algorithm clusters the cameras according to resolution, SNR, and sensitivity values.

11.23. [Dong, 1987]. You are given a set of sample data $X = \{\mathbf{x}_1, \mathbf{x}_2, \mathbf{x}_3, \mathbf{x}_4\}$ with these element vectors (features):

$$\mathbf{x}_1 = \{1, 3\}$$
$$\mathbf{x}_2 = \{1.5, 3.2\}$$
$$\mathbf{x}_3 = \{1.3, 2.8\}$$
$$\mathbf{x}_4 = \{3, 1\}$$

For $c = 2$,

(*a*) Calculate the cardinal number, η_{M_c} of the *hard* c-partition space, M_c.

(*b*) List all possible hard c-partitions ($c = 2$).

(*c*) Determine the optimum hard 2-partition, U^*.

11.24. Using U^* from Problem 11.23 as $\underset{\sim}{U}^{(0)}$, determine the optimum fuzzy 2-partition, $\underset{\sim}{U}^*$.

CHAPTER
12

FUZZY PATTERN RECOGNITION

Precision is not truth.

Henri E. B. Matisse, 1869–1954
Impressionist painter

Pattern recognition can be defined as a process of identifying structure in data by comparisons to known structure; the known structure is developed through methods of classification [Bezdek, 1981] as illustrated in Chapter 11. In the statistical approach to numerical pattern recognition, which is treated thoroughly by Fukunaga [1972], each input observation is represented as a multi-dimensional data vector (feature vector) where each component is called a feature. The purpose of the pattern recognition system is to assign each input to one of c possible pattern classes (or data clusters). Presumably, different input observations should be assigned to the same class if they have similar features and to different classes if they have dissimilar features. Statistical pattern recognition systems rest on mathematical models; it is crucial that the measure of mathematical similarity used to match feature vectors with classes assess a property shared by physically similar components of the process generating the data. The components of a typical pattern recognition system are illustrated in Fig. 12.1.

The data used to design a pattern recognition system are usually divided into two categories: design (or training) data and test data, much like the categorization used in neural networks. Design data are used to establish the algorithmic parameters of the pattern recognition system. The design samples may be labeled (the class to

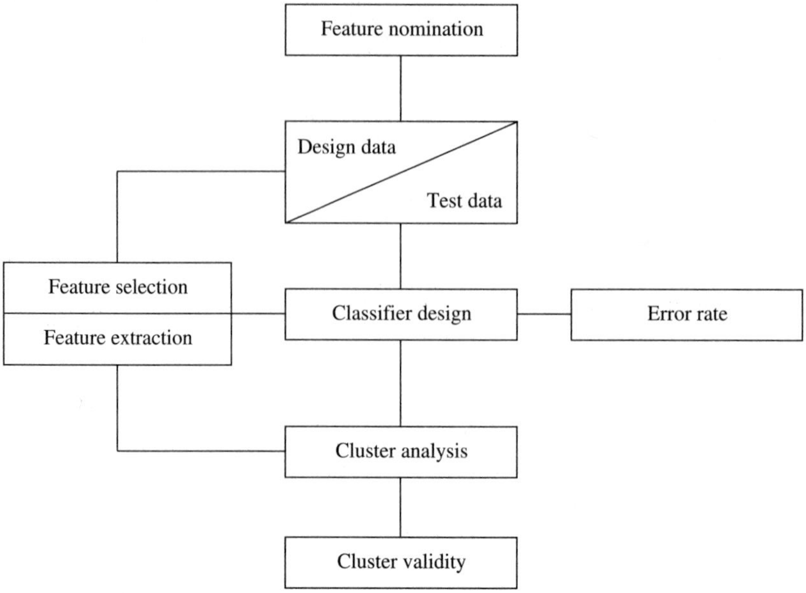

FIGURE 12.1
Pattern recognition systems [Bezdek et al., 1986].

which each observation belongs is known) or unlabeled (the class to which each data sample belongs is unknown). Test data are labeled samples used to test the overall performance of the pattern recognition system.

In the descriptions that follow, the following notation is used:

$$X = \{x_1, x_2, \ldots, x_n\} = \text{the universe of data samples}$$
$$n = \text{number of data samples in universe}$$
$$p = \text{number of original (nominated) features}$$
$$x_k \in R^p;\ k\text{th data sample in X, in } p\text{-dimensional feature space}$$
$$x_{kj} \in R;\ j\text{th measured feature of } x_k$$
$$s = \text{number of selected or extracted features}$$
$$c = \text{number of clusters or classes}$$

There are many similarities between classification and pattern recognition. The information provided in Fig. 12.2 summarizes the distinction between the two made in this textbook. Basically, classification establishes (or seeks to determine) the structure in data, whereas pattern recognition attempts to take new data and assign them to one of the classes defined in the classification process. Simply stated, classification *defines* the patterns and pattern recognition *assigns* data to a class; hence, the processes of define and assign are a coupled pair in the process described in Fig. 12.2. In both the classification process and the pattern recognition process there are necessary feedback loops; the first loop in classification is required when one is seeking a better segmentation of the data (i.e., better class distinctions), and the second loop is required when pattern matching fails (i.e., no useful assignment can be made).

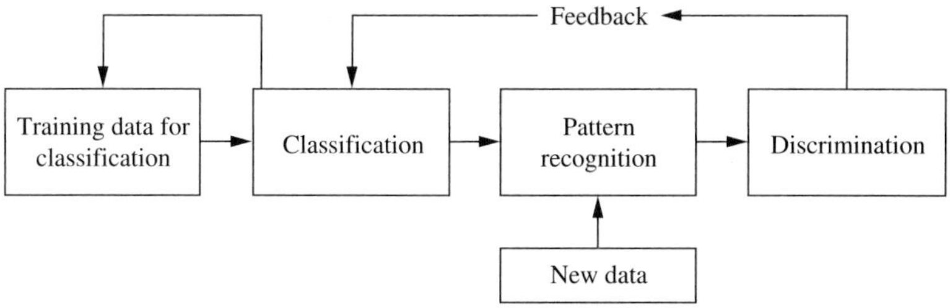

FIGURE 12.2
Difference between classification and pattern recognition (Dong, 1987).

FEATURE ANALYSIS

Feature analysis refers to methods for conditioning the raw data so that the information that is most relevant for classification and interpretation (recognition) is enhanced and represented by a minimal number of features. Feature analysis consists of three components: nomination, selection, and extraction. Feature nomination (FN) refers to the process of proposing the original p features; it is usually done by workers close to the physical process and may be heavily influenced by physical constraints, e.g., what can be measured by a particular sensor. For example, the nominated features can correspond to simple characteristics of various sensors that are represented by digitization of the sensor records. Feature selection (FS) refers to choosing the "best" subset of s features ($s < p$) from the original p features. Feature extraction (FE) describes the process of transforming the original p-dimensional feature space into an s-dimensional space in some manner that "best" preserves or enhances the information available in the original p-space. This is usually accomplished mathematically by means of some linear combination of the initial measurements.

Another method of feature extraction that lies closer to the expertise of the engineer is heuristic nomination and/or extraction. In other words, the process being examined may suggest choices for analytic features, e.g., in sensor records of measured pressures, the slopes (of rise and decay), areas (impulse or energy) during rise and decay, or even transformations (Fourier or Laplace) of the pressure waveform. Implicit in both FS and FE is a means for evaluating feature sets chosen by a particular procedure. The usual benchmark of feature quality is the empirical error rate achieved by a classifier on labeled test data. A second method of assessing feature quality is to refer algorithmic interpretations of the data to domain experts: do the computed results make sense? This latter test is less esoteric than the mathematical criteria, but very important.

PARTITIONS OF THE FEATURE SPACE

Partitioning the feature space into c regions, one for each subclass in the data, is usually in the domain of classifier design. More specifically, crisp classifiers partition R^p (or R^s) into disjoint subsets, whereas fuzzy classifiers assign fuzzy label vectors to each vector in feature space. The ideal classifier never errs; since this is usually impossible, one seeks designs that minimize the expected probability of error or, if some mistakes are more costly than others, minimize the average cost of errors, or both. An in-depth analysis of classifier design is available in Duda and Hart [1973]. Because many sources of data, such as that in image processing, do not lend themselves readily to classifier design, the discussion presented here does not consider this aspect of pattern classification. Chapter 11 has discussed the nature and importance of cluster analysis and cluster validity in terms of feature selection.

SINGLE SAMPLE IDENTIFICATION

A typical problem in pattern recognition is to collect data from a physical process and classify them into known patterns. The known patterns typically are represented as class structures, where each class structure is described by a number of features. For simplicity in presentation, the material that follows represents classes or patterns characterized by one feature; hence, the representation can be considered one-dimensional.

Suppose we have several typical patterns stored in our knowledge base (i.e., the computer), and we are given a new data sample that has not yet been classified. We want to determine which pattern the sample most closely resembles. Express the typical patterns as fuzzy sets $\underset{\sim}{A}_1, \underset{\sim}{A}_2, \ldots, \underset{\sim}{A}_m$. Now suppose we are given a new data sample, which is characterized by the crisp singleton, x_0. Using the simple criterion of maximum membership, the typical pattern that the data sample most closely resembles is found by the following expression:

$$\mu_{\underset{\sim}{A}_i}(x_0) = \max\{\mu_{\underset{\sim}{A}_1}(x_0), \mu_{\underset{\sim}{A}_2}(x_0), \ldots, \mu_{\underset{\sim}{A}_m}(x_0)\} \tag{12.1}$$

where x_0 belongs to the fuzzy set $\underset{\sim}{A}_i$, which is the set indication for the set with the highest membership at point x_0. Figure 12.3 shows the idea expressed by Eq. (12.1), where clearly the new data sample defined by the singleton expressed by x_0 most closely resembles the pattern described by fuzzy set $\underset{\sim}{A}_2$.

Example 12.1 [Dong, 1987]. We can illustrate the single data sample example using the problem of identifying a triangle, as described in Chapter 4. Suppose the single data sample is described by a data triplet, where the three coordinates are the angles of a specific triangle, for example, the triangle as shown in Fig. 4.6, $x_0 = \{A = 85°, B = 50°, C = 45°\}$. Recall from Chapter 4 that there were five known patterns stored: isosceles,

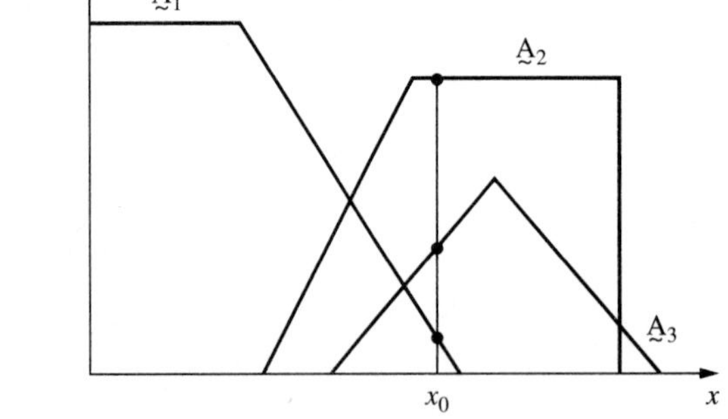

FIGURE 12.3
Single data sample using max-membership criteria.

right, right and isosceles, equilateral, and all other triangles. If we take this single triangle and determine its membership in each of the known patterns, we get the following results (as we did before in Chapter 4):

$$\begin{aligned}\mu_I(85, 50, 45) &= 1 - \tfrac{5}{60} = 0.916\\ \mu_R(85, 50, 45) &= 1 - \tfrac{5}{90} = 0.94\\ \mu_{IR}(85, 50, 45) &= 1 - \max\left[\tfrac{5}{60}, \tfrac{5}{90}\right] = 0.916\\ \mu_E(85, 50, 45) &= 1 - \tfrac{1}{180}(40) = 0.78\\ \mu_T(85, 50, 45) &= \tfrac{1}{180}\min[3.35, 3(5), \tfrac{2}{5}, 40] = 0.05\end{aligned}$$

Using the criterion of maximum membership, we see from these values that x_0 most closely resembles the right triangle pattern, $\underset{\sim}{\mathbf{R}}$.

Now let us extend the paradigm to consider the case where the new data sample is not crisp, but rather a fuzzy set itself. Suppose we have m typical patterns represented as fuzzy sets $\underset{\sim}{A}_i$ on X($i = 1, 2, \ldots, m$) and a new piece of data, perhaps consisting of a group of observations, is represented by a fuzzy set $\underset{\sim}{B}$ on X. The task now is to find which $\underset{\sim}{A}_i$ the sample $\underset{\sim}{B}$ most closely matches. To address this issue, we refer to the section that discussed fuzzy vectors in Chapter 6.

If we define two fuzzy vectors, say $\underset{\sim}{\mathbf{a}}$ and $\underset{\sim}{\mathbf{b}}$, then from Chapter 6 recall that when the two vectors are identical (i.e., they are the same length and they contain the same elements in an ordered sense), then the inner product, $\underset{\sim}{\mathbf{a}} \bullet \underset{\sim}{\mathbf{b}}^{\mathrm{T}}$ [Eq. (6.36)], reaches a maximum value as the outer product $\underset{\sim}{\mathbf{a}} \oplus \underset{\sim}{\mathbf{b}}^{\mathrm{T}}$ [Eq. (6.37)] reaches a minimum. These two norms, the inner product and the outer product, can be used simultaneously in pattern recognition studies because they measure *closeness* or *similarity*.

As suggested in Chapter 6, we can extend the fuzzy vectors to the case of fuzzy sets. Whereas vectors are defined on a finite countable universe, sets can be used to address infinite-valued universes (see example below using Gaussian membership functions). Let $\mathrm{P}^*(\mathrm{X})$ be a group of fuzzy sets with $\underset{\sim}{A}_i \neq \emptyset$, and $\underset{\sim}{A}_i \neq \mathrm{X}$. Now we define two fuzzy sets from this family of sets, i.e., $\underset{\sim}{A}, \underset{\sim}{B} \in \mathrm{P}^*(\mathrm{X})$; then either of the expressions [Eqs. (12.2) and (12.3)]

$$(\underset{\sim}{A}, \underset{\sim}{B})_1 = (\underset{\sim}{A} \bullet \underset{\sim}{B}) \wedge (\overline{\underset{\sim}{A} \oplus \underset{\sim}{B}}) \tag{12.2}$$

$$(\underset{\sim}{A}, \underset{\sim}{B})_2 = \tfrac{1}{2}[(\underset{\sim}{A} \bullet \underset{\sim}{B}) + (\overline{\underset{\sim}{A} \oplus \underset{\sim}{B}})] \tag{12.3}$$

describe two metrics to assess the degree of similarity of the two sets $\underset{\sim}{A}$ and $\underset{\sim}{B}$:

$$(\underset{\sim}{A}, \underset{\sim}{B}) = (\underset{\sim}{A}, \underset{\sim}{B})_1, \qquad \text{or} \qquad (\underset{\sim}{A}, \underset{\sim}{B}) = (\underset{\sim}{A}, \underset{\sim}{B})_2 \tag{12.4}$$

In particular, when either of the values of $(\underset{\sim}{A}, \underset{\sim}{B})$ from Eq. (12.4) approaches 1, then the two fuzzy sets $\underset{\sim}{A}$ and $\underset{\sim}{B}$ are "more closely similar;" when either of the values $(\underset{\sim}{A}, \underset{\sim}{B})$ from Eq. (12.4) approaches a value of 0, the two fuzzy sets are "more far apart" (dissimilar). The metric in Eq. (12.2) uses a minimum property to describe similarity, and the expression in Eq. (12.3) uses an arithmetic metric to describe similarity. It can be shown (see Problem 12.3) that the first metric [Eq. (12.2)] always gives a value that is less than the value obtained from the second metric [Eq. (12.3)].

Both of these metrics represent a concept that has been called the *approaching degree* [Wang, 1983].

Example 12.2. Suppose we have a universe of five discrete elements, X = $\{x_1, x_2, x_3, x_4, x_5\}$, and we define two fuzzy sets, $\underset{\sim}{A}$ and $\underset{\sim}{B}$, on this universe. Note that the two fuzzy sets are special: They are actually crisp sets and both are complements of one another.

$$\underset{\sim}{A} = \left\{\frac{1}{x_1} + \frac{1}{x_2} + \frac{0}{x_3} + \frac{0}{x_4} + \frac{0}{x_5}\right\}$$

$$\underset{\sim}{B} = \overline{\underset{\sim}{A}} = \left\{\frac{0}{x_1} + \frac{0}{x_2} + \frac{1}{x_3} + \frac{1}{x_4} + \frac{1}{x_5}\right\}$$

If we calculate the quantities expressed by Eqs. (12.2)–(12.4), we obtain the following values:

$$\underset{\sim}{A} \bullet \underset{\sim}{B} = 0$$

$$\underset{\sim}{A} \oplus \underset{\sim}{B} = 1$$

$$(\underset{\sim}{A}, \underset{\sim}{B})_1 = (\underset{\sim}{A}, \underset{\sim}{B})_2 = 0$$

The conclusion is that a crisp set and its complement are completely dissimilar.

The value of the approaching degree in the previous example should be intuitive. Since each set is the crisp complement of the other, they should be considered distinctly different patterns, i.e., there is no overlap. The inner product being zero and the outer product being unity confirms this mathematically. Conversely, if we assume fuzzy set $\underset{\sim}{B}$ to be identical to $\underset{\sim}{A}$, i.e., $\underset{\sim}{B} = \underset{\sim}{A}$, then we would find that the inner product equals one and the outer product equals zero and the approaching degree [Eqs. (12.4)] would equal unity. The reader is asked to confirm this (see Problem 12.2). This proof simply reinforces the notion that a set is most similar to itself.

Example 12.3 [Dong, 1987]. Suppose we have a one-dimensional universe on the real line, X = $[-\infty, \infty]$; and we define two fuzzy sets having normal, Gaussian membership functions, $\underset{\sim}{A}$, $\underset{\sim}{B}$, which are defined mathematically as

$$\mu_{\underset{\sim}{A}}(x) = \exp\left[\frac{-(x-a)^2}{\sigma_a^2}\right] \qquad \mu_{\underset{\sim}{B}}(x) = \exp\left[\frac{-(x-b)^2}{\sigma_b^2}\right]$$

and shown graphically in Fig. 12.4. It can be shown that the inner product of the two fuzzy sets is equal to

$$\underset{\sim}{A} \bullet \underset{\sim}{B} = \exp\left[\frac{-(a-b)^2}{(\sigma_a + \sigma_b)^2}\right] = \mu_{\underset{\sim}{A}}(x_0) = \mu_{\underset{\sim}{B}}(x_0)$$

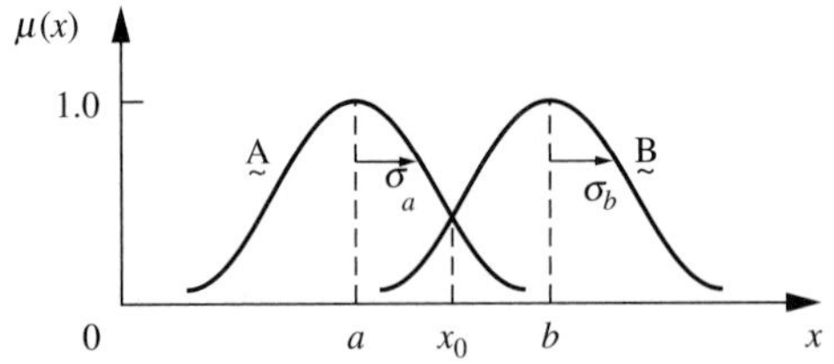

FIGURE 12.4
Two Gaussian membership functions.

where $x_0 = \dfrac{\sigma_a \cdot b + \sigma_b \cdot a}{\sigma_a + \sigma_b}$

and that the outer product is calculated to be $\underset{\sim}{A} \oplus \underset{\sim}{B} = 0$. Hence, the values of Eqs. (12.2) and (12.3) are

$$(\underset{\sim}{A}, \underset{\sim}{B})_1 = \exp\left[\frac{-(a-b)^2}{(\sigma_a + \sigma_b)^2}\right] \wedge 1 \qquad \text{and} \qquad (\underset{\sim}{A}, \underset{\sim}{B})_2 = \frac{1}{2}\left[\exp\left[\frac{-(a-b)^2}{(\sigma_a + \sigma_b)^2}\right] + 1\right]$$

The preceding material has presented some examples in which a new data sample is compared to a single known pattern. In the usual pattern recognition problem we are interested in comparing a data sample to a number of known patterns. Suppose we have a collection of m patterns, each represented by a fuzzy set, $\underset{\sim}{A}_i$, where $i = 1, 2, \ldots, m$, and a sample pattern $\underset{\sim}{B}$, all defined on universe X. Then the question is, Which known pattern $\underset{\sim}{A}_i$ does data sample $\underset{\sim}{B}$ most closely resemble? A useful metric that has appeared in the literature is to compare the data sample to each of the known patterns in a pairwise fashion, determine the approaching degree value for each of these pairwise comparisons, then select the pair with the largest approaching degree value as the one governing the pattern recognition process. The known pattern that is involved in the maximum approaching degree value is then the pattern the data sample most closely resembles in a maximal sense. This concept has been termed the *maximum approaching degree* [Wang, 1983]. Equation (12.5) shows this concept for m known patterns:

$$(\underset{\sim}{B}, \underset{\sim}{A}_i) = \max\{(\underset{\sim}{B}, \underset{\sim}{A}_1), (\underset{\sim}{B}, \underset{\sim}{A}_2), \ldots, (\underset{\sim}{B}, \underset{\sim}{A}_m)\} \qquad (12.5)$$

Example 12.4 [Dong, 1987]. Suppose you are an earthquake engineering consultant hired by the state of California to assess earthquake damage in a region just hit by a large earthquake. Your assessment of damage will be very important to residents of the area because insurance companies will base their claim payouts on your assessment. You must be as impartial as possible. From previous historical records you determine that the six categories of the modified Mercalli intensity (I) scale (VI) to (XI) are most appropriate for the range of damage to the buildings in this region. These damage patterns can all be represented by Gaussian membership functions, $\underset{\sim}{A}_i$, $i = 1, 2, \ldots, 6$, of the following form:

$$\mu_{\underset{\sim}{A}_i}(x) = \exp\left(\frac{-(x - a_i)^2}{\sigma_a^2}\right)$$

where parameters a_i and σ_i define the shape of each membership function. Your historical database provides the information shown in Table 12.1 for the parameters for the six regions.

TABLE 12.1
Parameters for Gaussian membership functions

	I					
	$\underset{\sim}{A}_1$, VI	$\underset{\sim}{A}_2$, VII	$\underset{\sim}{A}_3$, VIII	$\underset{\sim}{A}_4$, IX	$\underset{\sim}{A}_5$, X	$\underset{\sim}{A}_6$, XI
a_i	5	20	35	49	71	92
σ_{a_i}	3	10	13	26	18	4

You determine via inspection that the pattern of damage to buildings in a given location is represented by a fuzzy set $\underset{\sim}{B}$, with the following characteristics:

$$\mu_{\underset{\sim}{B}}(x) = \exp\left(\frac{-(x-b)^2}{\sigma_b^2}\right); \qquad b = 41; \qquad \sigma_b = 10$$

The system you now have is shown graphically in Fig. 12.5. You then conduct the following calculations, using the similarity metric from Eq. (12.3) to determine the maximum approaching degree:

$$\begin{aligned} (\underset{\sim}{B}, \underset{\sim}{A}_1) &= \tfrac{1}{2}(.0004 + 1) \approx 0.5 & \quad (\underset{\sim}{B}, \underset{\sim}{A}_4) &= 0.98 \\ (\underset{\sim}{B}, \underset{\sim}{A}_2) &= 0.67 & \quad (\underset{\sim}{B}, \underset{\sim}{A}_5) &= 0.65 \\ (\underset{\sim}{B}, \underset{\sim}{A}_3) &= 0.97 & \quad (\underset{\sim}{B}, \underset{\sim}{A}_6) &= 0.5 \end{aligned}$$

From this list we see that Mercalli intensity IX ($\underset{\sim}{A}_4$) most closely resembles the damaged area because of the maximum membership value of 0.98.

Suppose you assume the membership function of the damaged region to be a simple singleton with the following characteristics:

$$\mu_{\underset{\sim}{B}}(41) = 1 \qquad \text{and} \qquad \mu_{\underset{\sim}{B}}(x \neq 41) = 0 \qquad x_0 = 41$$

as shown in Fig. 12.6. This example reduces to the single data sample problem posed earlier, i.e.,

$$(\underset{\sim}{B}, \underset{\sim}{A}_i) = \mu_{\underset{\sim}{A}_i}(x_0) \wedge 1 = \mu_{\underset{\sim}{A}_i}(x_0)$$

Your calculations, again using Eq. (12.3), produce the following results:

$$\begin{aligned} \mu_{\underset{\sim}{A}_1}(41) &\approx 0 & \quad \mu_{\underset{\sim}{A}_4}(41) &= .91 \\ \mu_{\underset{\sim}{A}_2}(41) &= .01 & \quad \mu_{\underset{\sim}{A}_5}(41) &= .06 \\ \mu_{\underset{\sim}{A}_3}(41) &= .81 & \quad \mu_{\underset{\sim}{A}_6}(41) &\approx 0 \end{aligned}$$

Again, Mercalli scale IX ($\underset{\sim}{A}_4$) would be chosen on the basis of maximum membership (0.91). If we were to make the selection without regard to the shapes of the membership values, as shown in Fig. 12.6, but instead only considered the mean value of each region, we would be inclined erroneously to select region VIII because its mean value of 35 is closer to the singleton at 41 than it is to the mean value of region IX, i.e., to 49.

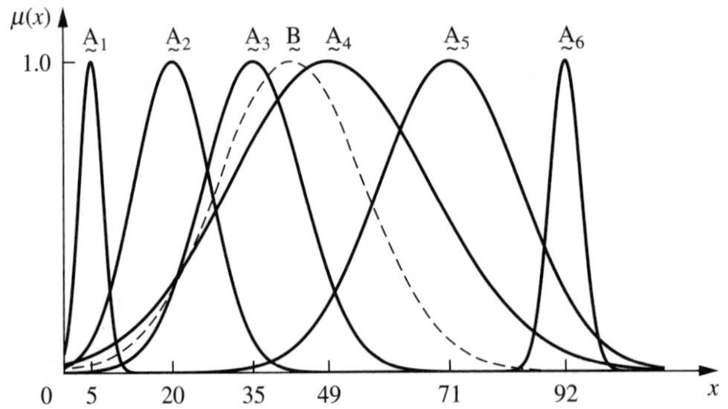

FIGURE 12.5
Six known patterns and a new fuzzy set data sample.

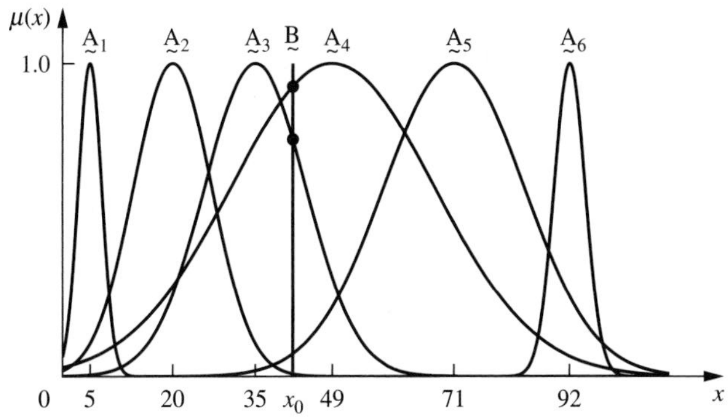

FIGURE 12.6
Six known patterns and a new singleton data sample.

MULTIFEATURE PATTERN RECOGNITION

In the material covered so far in this chapter we have considered only one-dimensional pattern recognition; that is, the patterns here have been constructed based only on a single feature, such as Mercalli earthquake intensity. Suppose the preceding example on earthquake damage also considered, in addition to earthquake intensity, importance of the particular building (schools versus industrial plants), seismicity of the region, previous history of damaging quakes, and so forth. How could we address the consideration of many features in the pattern recognition process? The literature develops many answers to this question, but this text summarizes three popular and easy approaches: (*i*) nearest neighbor classifier, (*ii*) nearest center classifier, and (*iii*) weighted approaching degree. The first two methods are restricted to the recognition of crisp singleton data samples.

In the *nearest neighbor classifier* we can consider m features for each data sample. So each sample (x_i) is a vector of features,

$$\mathrm{x}_i = \{x_{i1}, x_{i2}, x_{i3}, \ldots, x_{im}\} \tag{12.6}$$

Now suppose we have n data samples in a universe, or $\mathrm{X} = \{\mathbf{x}_1, \mathbf{x}_2, \mathbf{x}_3, \ldots, \mathbf{x}_n\}$. Using a conventional fuzzy classification approach, we can cluster the samples into c-fuzzy partitions, then get c-hard partitions from these by using the equivalent relations idea or by "hardening" the soft partition $\underset{\sim}{\mathrm{U}}$, both of which were described in Chapter 11. This would result in hard classes with the following properties:

$$\mathrm{X} = \bigcup_{i=1}^{c} \mathrm{A}_i \qquad \mathrm{A}_i \cap \mathrm{A}_j = \varnothing \qquad i \neq j$$

Now if we have a new singleton data sample, say $\mathbf{x}$, then the nearest neighbor classifier is given by the following distance measure, d:

$$d(\mathbf{x}, \mathbf{x}_i) = \min_{1 \leq k \leq n} \{d(\mathbf{x}, \mathbf{x}_k)\} \tag{12.7}$$

for each of the n data samples where $\mathbf{x}_i \in \mathrm{A}_j$. That is, points $\mathbf{x}$ and $\mathbf{x}_i$ are nearest neighbors, hence both would belong to the same class.

In another method for singleton recognition, the nearest center classifier method works as follows. We again start with n known data samples, $\mathbf{X} = \{\mathbf{x}_1, \mathbf{x}_2, \mathbf{x}_3, \ldots, \mathbf{x}_n\}$ and each data sample is m-dimensional (characterized by m features). We then cluster these samples into c classes using a fuzzy classification method such as the fuzzy c-means approach described in Chapter 11. These fuzzy classes each have a class center, so

$$\mathbf{V} = \{\mathbf{v}_1, \mathbf{v}_2, \mathbf{v}_3, \ldots, \mathbf{v}_c\}$$

is a vector of the c class centers. If we have a new singleton data sample, say x, the nearest center classifier is then given by

$$d(\mathbf{x}, \mathbf{v}_i) = \min_{1 \le k \le c} \{d(\mathbf{x}, \mathbf{v}_k)\} \tag{12.8}$$

and now the data singleton, $\mathbf{x}$, is classified as belonging to fuzzy partition, $\underset{\sim}{A}_i$.

In the third method for addressing multifeature pattern recognition for a sample with several (m) fuzzy features, we will use the approaching degree concept again to compare the new data pattern with some known data patterns. Define a new data sample characterized by m features as a collection of noninteractive fuzzy sets, $\underset{\sim}{B} = \{\underset{\sim}{B}_1, \underset{\sim}{B}_2, \ldots, \underset{\sim}{B}_m\}$. Because the new data sample is characterized by m features, each of the known patterns, $\underset{\sim}{A}_i$, is also described by m features. Hence, each known pattern in m-dimensional space is a fuzzy class (pattern) given by, $\underset{\sim}{A}_i = \{\underset{\sim}{A}_{i1}, \underset{\sim}{A}_{i2}, \ldots, \underset{\sim}{A}_{im}\}$ where $i = 1, 2, \ldots, c$ describes c-classes (c-patterns). Since some of the features may be more important than others in the pattern recognition process, we introduce normalized weighting factors w_j, where

$$\sum_{j=1}^{m} w_j = 1 \tag{12.9}$$

Then either Eq. (12.2) or (12.3) in the approaching degree concept is modified for each of the known c-patterns ($i = 1, 2, \ldots, c$) by

$$(\underset{\sim}{B}, \underset{\sim}{A}_i) = \sum_{j=1}^{m} w_j \left(\underset{\sim}{B}_j, \underset{\sim}{A}_{ij}\right) \tag{12.10}$$

As before in the maximum approaching degree, sample $\underset{\sim}{B}$ is closest to pattern $\underset{\sim}{A}_j$ when

$$(\underset{\sim}{B}, \underset{\sim}{A}_j) = \max_{1 \le i \le c} \{(\underset{\sim}{B}, \underset{\sim}{A}_i)\} \tag{12.11}$$

Note that when the collection of fuzzy sets $\underset{\sim}{B} = \{\underset{\sim}{B}_1, \underset{\sim}{B}_2, \ldots, \underset{\sim}{B}_m\}$ reduces to a collection of crisp singletons, i.e., $\underset{\sim}{B} = \{\mathbf{x}_1, \mathbf{x}_2, \ldots, \mathbf{x}_m\}$, then Eq. (12.10) reduces to

$$\mu_{\underset{\sim}{A}_i}(x) = \sum_{j=1}^{m} w_j \cdot \mu_{\underset{\sim}{A}_{ij}}(\mathbf{x}_j) \tag{12.12}$$

As before in the maximum approaching degree, sample singleton, $\mathbf{x}$, is closest to pattern $\underset{\sim}{A}_j$ when Eq. (12.11) reduces to

$$\mu_{\underset{\sim}{A}}(\mathbf{x}) = \max_{1 \le i \le c} \{\mu_{\underset{\sim}{A}_i}(\mathbf{x})\} \tag{12.13}$$

Example 12.5. An example of multifeature pattern recognition is given where $m = 2$; the patterns can be illustrated in 3D images. Suppose we have a new pattern, $\underset{\sim}{B}$, that we wish to recognize by comparing it to other known patterns. This new pattern is characterized by two features; hence, it can be represented by a vector of its two noninteractive projections, $\underset{\sim}{B}_1$ and $\underset{\sim}{B}_2$. That is,

$$\underset{\sim}{B} = \{\underset{\sim}{B}_1, \underset{\sim}{B}_2\}$$

where the noninteractive patterns $\underset{\sim}{B}_1$ and $\underset{\sim}{B}_2$ are defined on their respective universes of discourse, X_1 and X_2. The two projections together [using Eq. (3.19)] produce a three-dimensional pattern in the shape of a pyramid, as shown in Fig. 12.7.

Further suppose that we have two ($c = 2$) patterns to which we wish to compare our new pattern; call them patterns $\underset{\sim}{A}_1$ and $\underset{\sim}{A}_2$. Each of these two known patterns could also be represented by their respective noninteractive projections, as shown in Figs. 12.8*a* and 12.8*b*, where the projections of each known pattern are also defined on X_1 and X_2.

The last step in this process is to assign weights to the various known patterns. Let us assume that $w_1 = 0.3$ and $w_2 = 0.7$, since $0.3 + 0.7 = 1$ by Eq. (12.9). We compare the new pattern with the two known patterns using Eq. (12.10),

$$\left.\begin{aligned} (\underset{\sim}{B}, \underset{\sim}{A}_1) &= w_1\,(\underset{\sim}{B}_1, \underset{\sim}{A}_{11}) + w_2\,(\underset{\sim}{B}_2, \underset{\sim}{A}_{12}) \\ (\underset{\sim}{B}, \underset{\sim}{A}_2) &= w_1\,(\underset{\sim}{B}_1, \underset{\sim}{A}_{21}) + w_2\,(\underset{\sim}{B}_2, \underset{\sim}{A}_{22}) \end{aligned}\right\}$$

where each of the operations in the preceding expressions is determined using the method of the approaching degree as described in Eqs. (12.2) or (12.3) for the *i*th pattern, i.e.,

$$(\underset{\sim}{B}, \underset{\sim}{A}_z) = \left[(\underset{\sim}{B} \bullet \underset{\sim}{A}_z) \wedge (\overline{\underset{\sim}{B} \oplus \underset{\sim}{A}_z})\right] \qquad \text{or} \qquad (\underset{\sim}{B}, \underset{\sim}{A}_z) = \tfrac{1}{2}\,[(\underset{\sim}{B} \bullet \underset{\sim}{A}_z) + (\overline{\underset{\sim}{B} \oplus \underset{\sim}{A}_z})]$$

Then we assign the new pattern to the known pattern most closely resembling the new pattern using Eq. (12.11), i.e.,

$$(\underset{\sim}{B}, \underset{\sim}{A}_i) = \max\{(\underset{\sim}{B}, \underset{\sim}{A}_1), (\underset{\sim}{B}, \underset{\sim}{A}_2)\}$$

The remainder of this example is left as an exercise for the reader.

Although it is not possible to sketch the membership functions for problems dealing with three or more features, the procedures outlined for multifeature pattern recognition work just as they did with the previous example. The following example in chemical engineering illustrates the multidimensional issues of Eqs. (12.9) to (12.13).

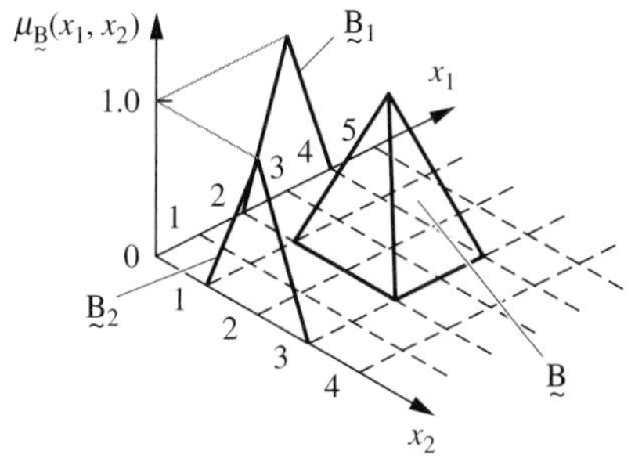

FIGURE 12.7
New pattern B and its noninteractive projections, $\underset{\sim}{B}_1$ and $\underset{\sim}{B}_2$.

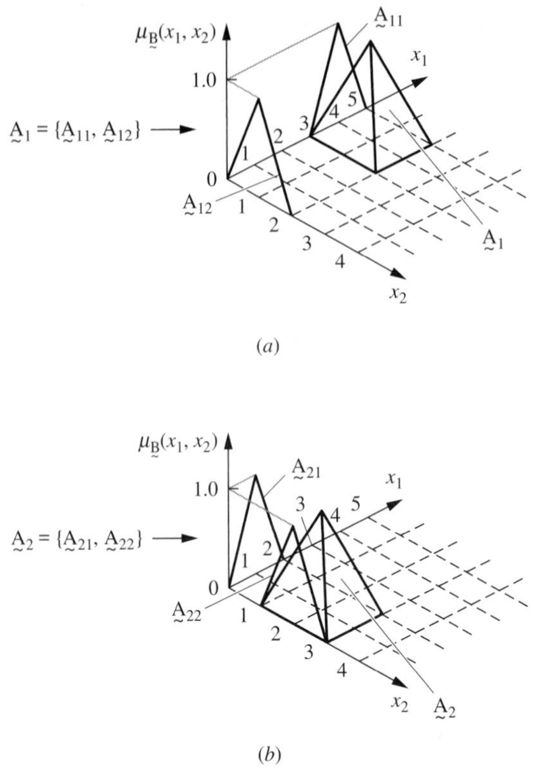

FIGURE 12.8
Multifeature pattern recognition: (*a*) known pattern $\underset{\sim}{A}_1$ and its noninteractive projections, $\underset{\sim}{A}_{11}$ and $\underset{\sim}{A}_{12}$; (*b*) known pattern $\underset{\sim}{A}_2$ and its noninteractive projections, $\underset{\sim}{A}_{21}$ and $\underset{\sim}{A}_{22}$.

Example 12.6. A certain industrial production process can be characterized by three features: (*i*) pressure, (*ii*) temperature, and (*iii*) flow rate. Combinations of these features are used to indicate the current mode (pattern) of operation of the production process. Typical linguistic values for each feature for each mode of operation are defined by the fuzzy sets given in Table 12.2. The pattern recognition task is described as follows: The system *reads* sensor indicators of each feature (pressure, temperature, flow rate), manifested as crisp read-out values; and it then determines the current mode of operation (i.e., it attempts to recognize a pattern of operation), and then the results are logged.

TABLE 12.2
Relationships between operation mode and feature values

Mode (pattern)	Pressure	Temperature	Flow rate
Autoclaving	High	High	Zero
Annealing	High	Low	Zero
Sintering	Low	Zero	Low
Transport	Zero	Zero	High

The four modes (patterns) of operation, and their associated linguistic values for each feature, are as follows:

1. *Autoclaving:* Here the pressure is high, temperature is high, and the flow rate is zero.
2. *Annealing:* Here the pressure is high, temperature is low, and the flow rate is zero.
3. *Sintering:* Here the pressure is low, temperature is zero, and the flow rate is low.
4. *Transport:* Here the pressure is zero, temperature is zero, and the flow rate is high.

This linguistic information is summarized in Table 12.2.

The features of pressure, temperature, and flow rate are expressed in the engineering units of kPa (kilopascals), °C (degrees Celsius), and gph (gallons per hour), respectively. Membership functions for these three features are shown in Figs. 12.9 to 12.11.

Now, suppose the system reads from a group of sensors a set of crisp readings (pressure = 5 kPa, temperature = 150°C, flow = 5 gph). We want to assign (recognize) this group of sensor readings to one of our four patterns (modes of operation). To begin, we need to assign weights to each of the features using Eq. (12.9). Since there is an explosion hazard associated with the production pressure value (5 kPa), we will weight it more heavily than the other two features:

$$w_{\text{pressure}} = 0.5$$
$$w_{\text{temperature}} = 0.25$$
$$w_{\text{flow}} = 0.25$$

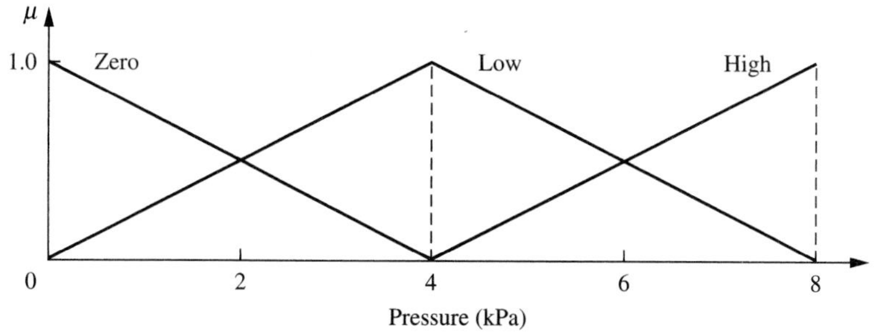

FIGURE 12.9
Membership functions for pressure.

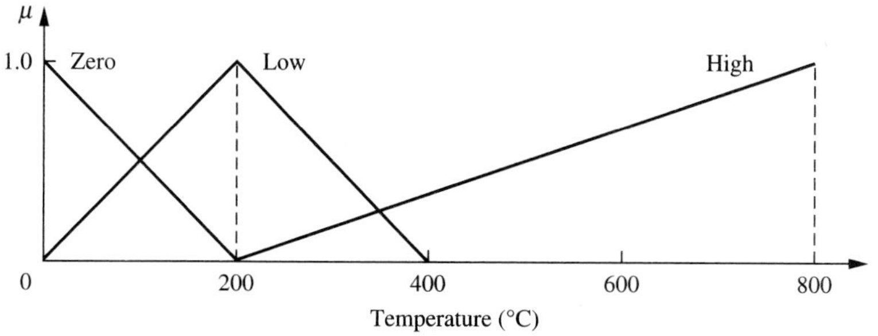

FIGURE 12.10
Membership functions for temperature.

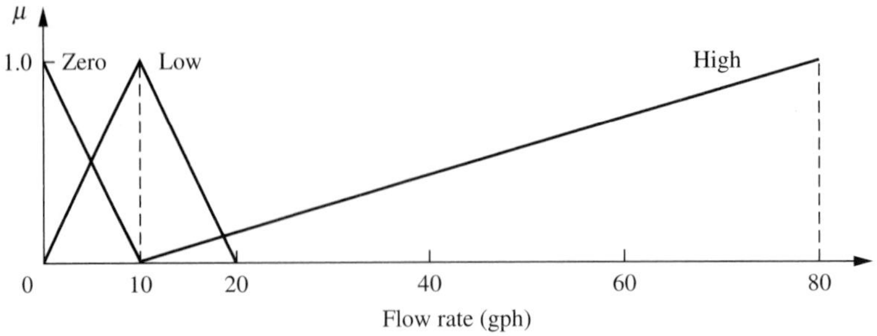

FIGURE 12.11
Membership functions for flow rate.

Now we will use Eqs. (12.12) and (12.13) to employ the approaching degree to find which mode of operation is indicated by the above crisp values (5 kPa, 150°C, 5 gph). Using Eq. (12.12) and these two expressions,

$$X = \{5 \text{ kPa}, 150°\text{C}, 5 \text{ gph}\}$$
$$W = \{0.5, 0.25, 0.25\}$$

we find that

$$\mu_{\text{autoclaving}}(x) = (0.5)\cdot(0.25) + (0.25)\cdot(0) + (0.25)\cdot(0.5) = 0.25$$
$$\mu_{\text{annealing}}(x) = (0.5)\cdot(0.25) + (0.25)\cdot(0.75) + (0.25)\cdot(0.5) = 0.4375$$
$$\mu_{\text{sintering}}(x) = (0.5)\cdot(0.75) + (0.25)\cdot(0.25) + (0.25)\cdot(0.5) = 0.5625$$
$$\mu_{\text{transport}}(x) = (0.5)\cdot(0) + (0.25)\cdot(0.25) + (0.25)\cdot(0) = 0.0625$$

The crisp set, X = {5 kPa, 150°C, 5 gph}, most closely matches the values of pressure, temperature, and flow associated with *sintering*. Therefore, we write the production mode "sintering" in the logbook as the current production mode indicated by crisp readings from our three sensors.

Now suppose for this industrial process we use the same patterns (autoclaving, annealing, etc.). Suppose now that the readings or information on pressure, temperature, and flow are fuzzy sets rather than crisp singletons, i.e., $\underset{\sim}{B} = \{\underset{\sim}{B}_{\text{pressure}}, \underset{\sim}{B}_{\text{temperature}}, \underset{\sim}{B}_{\text{flow}}\}$.

These fuzzy sets are defined in Figs. 12.12–12.14. Given these fuzzy definitions for our new pattern $\underset{\sim}{B}$, we use Eq. (12.11) to find which pattern is best matched by the new values $\underset{\sim}{B}$.

For the approaching degree between our new pattern's pressure feature and the stored autoclaving pattern's pressure feature we get

$$\underset{\sim}{B}_{\text{pressure}} \bullet \text{autoclaving pressure} = 0 \text{ (min of maxes)}$$
$$\underset{\sim}{B}_{\text{pressure}} \oplus \text{autoclaving pressure} = 0 \text{ (max of mins)}$$

as summarized in Fig. 12.15.

For the approaching degree between our new pattern's temperature feature and the stored autoclaving pattern's temperature feature we get

$$\underset{\sim}{B}_{\text{temperature}} \bullet \text{autoclaving temperature}$$
$$= \max[(0 \wedge 0), (0 \wedge 0.5), (0.166 \wedge 1), (0.33 \wedge 0.5), (0 \wedge 0)] = 0.33$$

$$\underset{\sim}{B}_{\text{temperature}} \oplus \text{autoclaving temperature}$$
$$= \min[(0 \vee 0), (0 \vee 0.5), (0.166 \vee 1), (0.33 \vee 0.5), (0 \vee 0)] = 0$$

as summarized in Fig. 12.16.

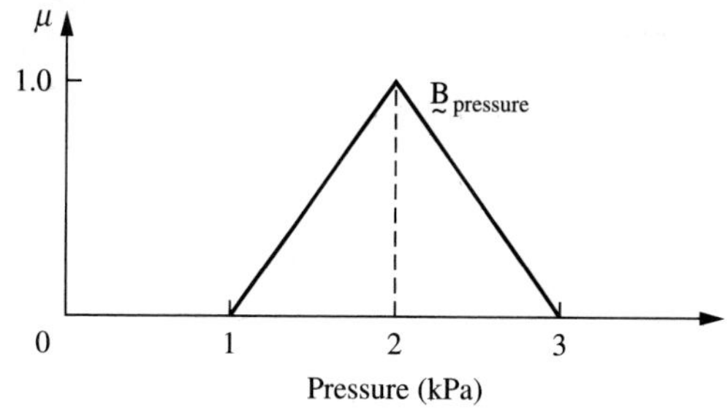

FIGURE 12.12
Fuzzy sensor reading for pressure.

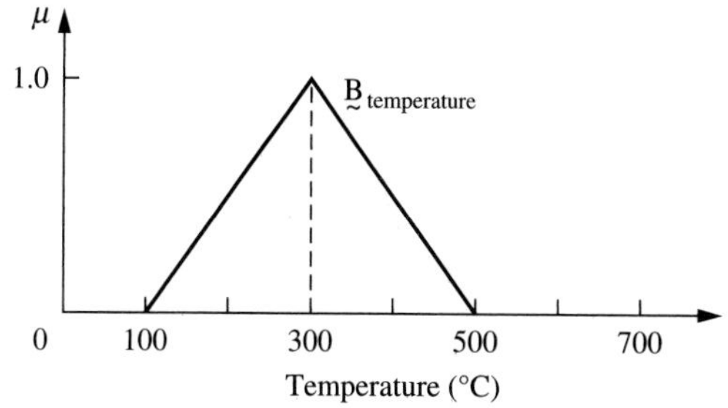

FIGURE 12.13
Fuzzy sensor reading for temperature.

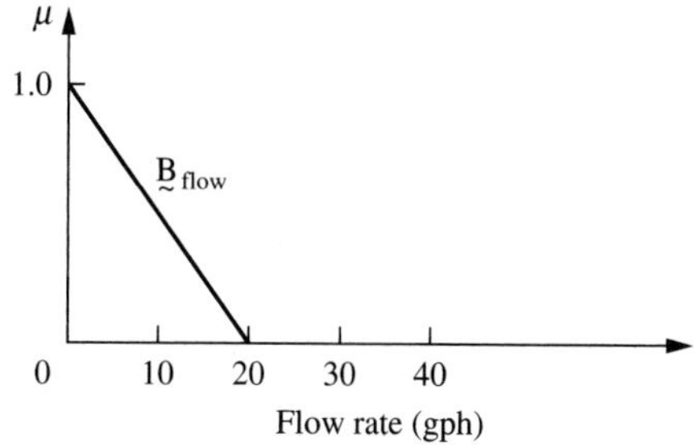

FIGURE 12.14
Fuzzy sensor reading for flow rate.

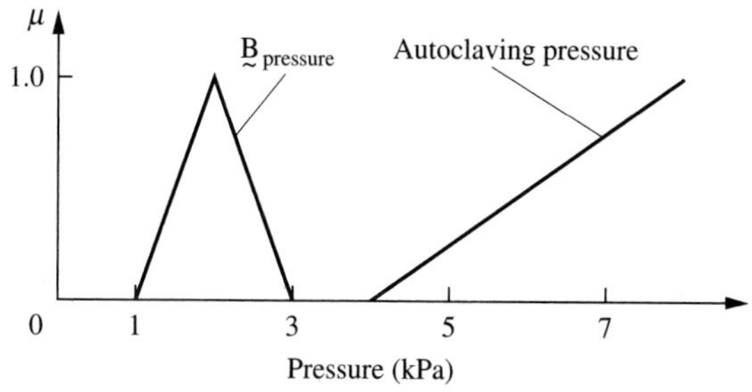

FIGURE 12.15
Pressure comparisons for autoclaving pattern.

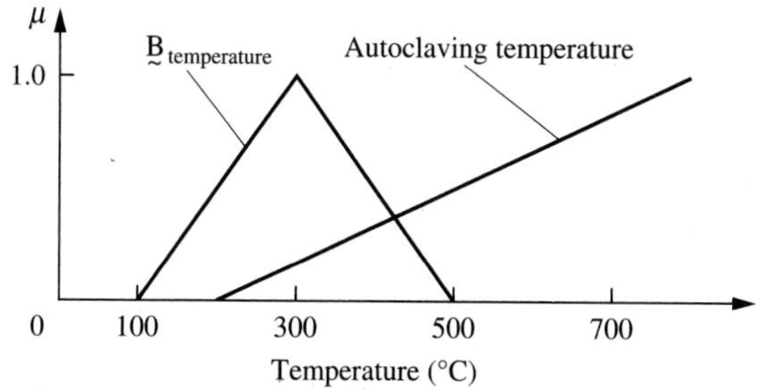

FIGURE 12.16
Temperature comparisons for autoclaving pattern.

For the approaching degree between our new pattern's flow rate feature and the stored autoclaving pattern's flow rate feature we get

$$\underset{\sim}{B}_{\text{flow}} \bullet \text{autoclaving flow} = \max[(1 \wedge 1), (0 \wedge 0.5), (0 \wedge 0)] = 1.0$$
$$\underset{\sim}{B}_{\text{flow}} \oplus \text{autoclaving flow} = \min[(1 \vee 1), (0 \vee 0.5), (0 \vee 0)] = 0$$

as summarized in Fig. 12.17.

Now the use of Eqs. (12.3) and (12.10) enables us to calculate the approaching degree value between the new sensor pattern and the autoclaving pattern:

$$\begin{aligned}(\underset{\sim}{B}, \text{autoclaving}) &= 0.5\left(\tfrac{1}{2}(0 + 1)\right) + 0.25\left(\tfrac{1}{2}(0.33 + 1.0)\right) + 0.25\left(\tfrac{1}{2}(1 + 1)\right) \\ &= (0.5)(0.5) + (0.25)(0.66) + (0.25)(1) = 0.665\end{aligned}$$

For the next possible pattern, annealing, we again use Eq. (12.11) to determine the approaching degree between the new sensor pressure and the annealing pressure. Because they are disjoint,

$$\underset{\sim}{B}_{\text{pressure}} \bullet \text{annealing pressure} = 0$$
$$\underset{\sim}{B}_{\text{pressure}} \oplus \text{annealing pressure} = 0$$

as summarized in Fig. 12.18.

The approaching degree between the new sensor temperature and the annealing temperature is given by

$$\underset{\sim}{B}_{\text{temperature}} \bullet \text{annealing temperature} = 0.75$$

by inspection of max of mins, and

$$\begin{aligned}&\underset{\sim}{B}_{\text{temperature}} \oplus \text{annealing temperature} \\ &\quad = \min[(0 \vee 0), (0 \vee 0.5), (0.5 \vee 1), (1 \vee 0.5), (0.5 \vee 0), (0 \vee 0)] = 0\end{aligned}$$

as summarized in Fig. 12.19.

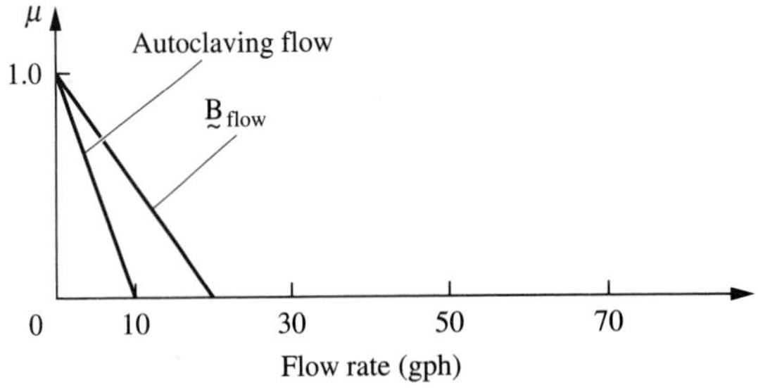

FIGURE 12.17
Flow rate comparisons for autoclaving pattern.

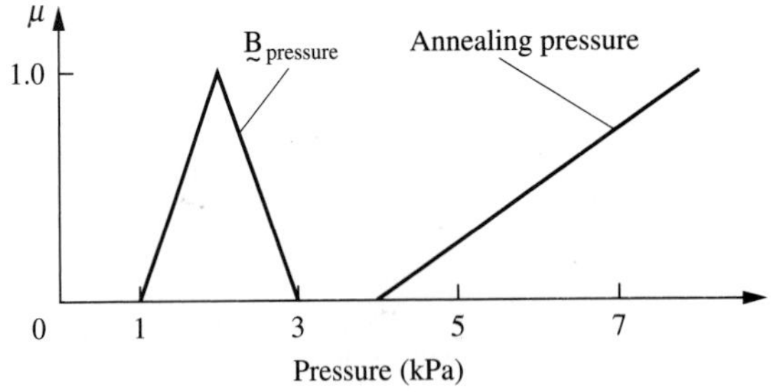

FIGURE 12.18
Pressure comparisons for annealing pattern.

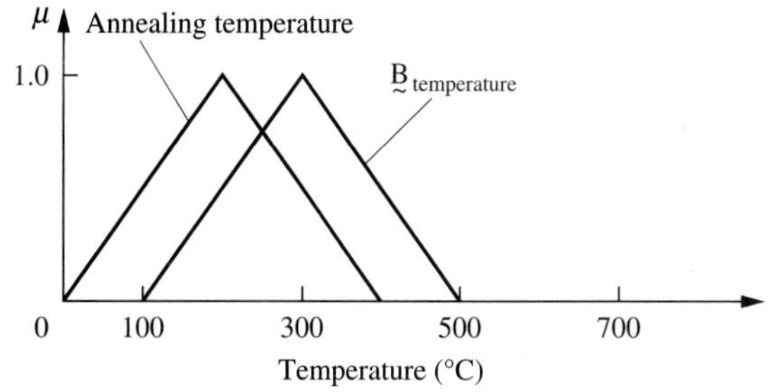

FIGURE 12.19
Temperature comparisons for annealing pattern.

The approaching degree between the new sensor flow rate and the annealing flow rate is given by

$$\underset{\sim}{B}_{\text{flow}} \bullet \text{annealing flow} = 1$$
$$\underset{\sim}{B}_{\text{flow}} \oplus \text{annealing flow} = 0$$

(identical to $\underset{\sim}{B}_{\text{flow}}$ and autoclaving flow).

Again using Eqs. (12.3) and (12.10), we get the approaching degree value for the annealing pattern,

$$\begin{aligned}(\underset{\sim}{B},\ \text{annealing}) &= 0.5\left(\tfrac{1}{2}(0+1)\right) + 0.25\left(\tfrac{1}{2}(0.75+1)\right) + 0.25\left(\tfrac{1}{2}(1+1)\right) \\ &= (0.5)(0.5) + (0.25)(0.87) + (0.25)(1) = 0.7175\end{aligned}$$

Now moving to the next pattern, sintering, we again use Eq. (12.11) for each of the features. The first is pressure:

$$\underset{\sim}{B}_{\text{pressure}} \bullet \text{sintering pressure} \approx 0.6$$

by inspection of max (mins), and

$$\begin{aligned}&\underset{\sim}{B}_{\text{pressure}} \oplus \text{sintering pressure} \\ &\qquad = \min[(0 \vee 0), (0 \vee 0.25), (1 \vee 0.5), (0 \vee 0.75), (0 \vee 1), \ldots] = 0\end{aligned}$$

as summarized in Fig. 12.20.

Next is temperature:

$$\underset{\sim}{B}_{\text{temperature}} \bullet \text{sintering temperature} = 0.25$$

by inspection of max (mins), and

$$\begin{aligned}&\underset{\sim}{B}_{\text{temperature}} \oplus \text{sintering temperature} \\ &\qquad = \min[(0 \vee 1), (0 \vee 0.5), (0.5 \vee 0), (1 \vee 0), (0.5 \vee 0), (0 \vee 0)] = 0\end{aligned}$$

as summarized in Fig. 12.21.

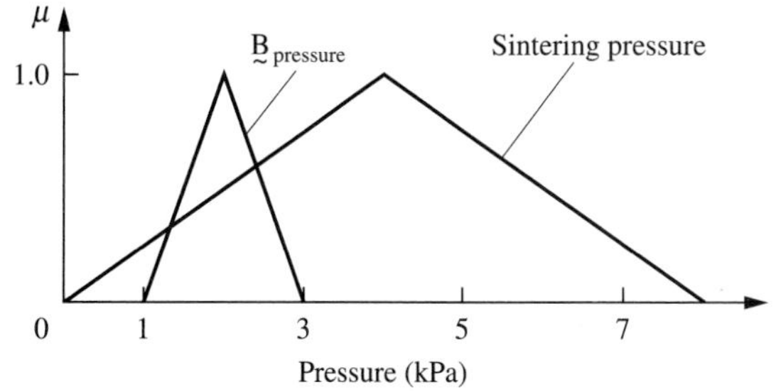

FIGURE 12.20
Pressure comparisons for sintering pattern.

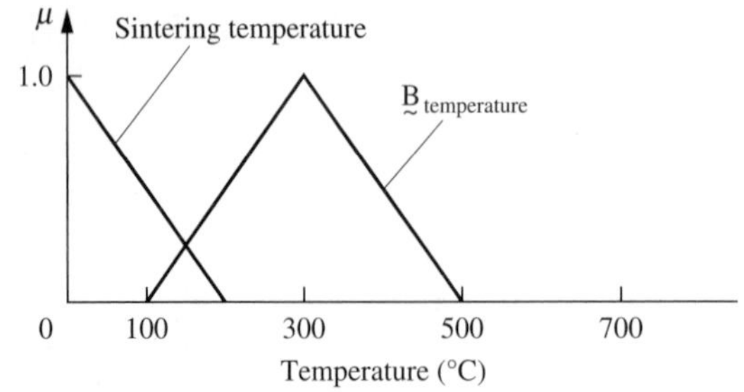

FIGURE 12.21
Temperature comparisons for sintering pattern.

Finally we consider the flow rate:

$$\underset{\sim}{B}_{\text{flow}} \bullet \text{sintering flow} = 0.7$$

by inspection of max (mins), and

$$\underset{\sim}{B}_{\text{flow}} \oplus \text{sintering flow} = \min[(0 \vee 1), (1 \vee 0), (0 \vee 0)] = 0$$

as summarized in Fig. 12.22.

Using Eqs. (12.3) and (12.10) with the metric from Eq. (12.3), we get

$$\begin{aligned}(\underset{\sim}{B}, \text{ sintering}) &= 0.5\left(\tfrac{1}{2}(0.6 + 1)\right) + 0.25\left(\tfrac{1}{2}(0.25 + 1)\right) + 0.25\left(\tfrac{1}{2}(0.7 + 1)\right) \\ &= (0.5)(0.8) + (0.25)(0.625) + (0.25)(0.85) = 0.7687\end{aligned}$$

Finally we consider the last pattern, the transport mode of operation. Using Eq. (12.11) for each feature, we begin first with pressure:

$$\underset{\sim}{B}_{\text{pressure}} \bullet \text{transport pressure} \approx 0.7$$

by inspection of max (mins), and

$$\begin{aligned}&\underset{\sim}{B}_{\text{pressure}} \oplus \text{transport pressure} \\ &\qquad = \min[(0 \vee 1), (0 \vee 0.75), (1 \vee 0.5), (0.25 \vee 0), (0 \vee 0)] = 0\end{aligned}$$

as summarized in Fig. 12.23.

Then, moving to temperature:

$$\begin{aligned}\underset{\sim}{B}_{\text{temperature}} \bullet \text{transport temperature} &= 0.25 \\ \underset{\sim}{B}_{\text{temperature}} \oplus \text{transport temperature} &= 0\end{aligned}$$

as summarized in Fig. 12.24.

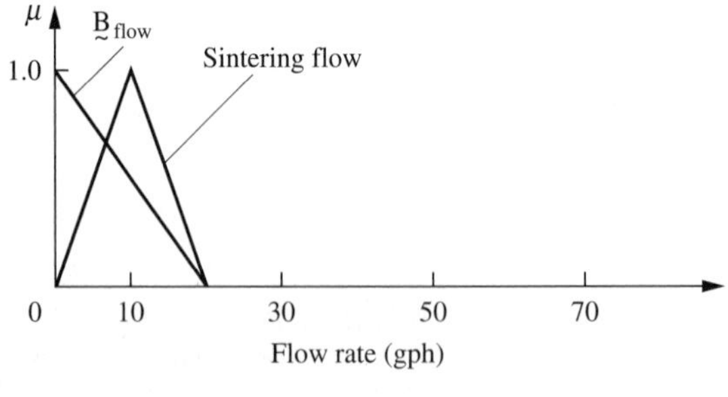

FIGURE 12.22
Flow rate comparisons for sintering pattern.

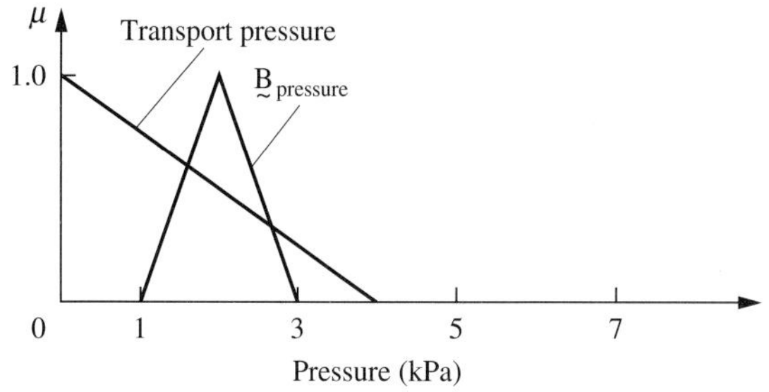

FIGURE 12.23
Pressure comparisons for transport pattern.

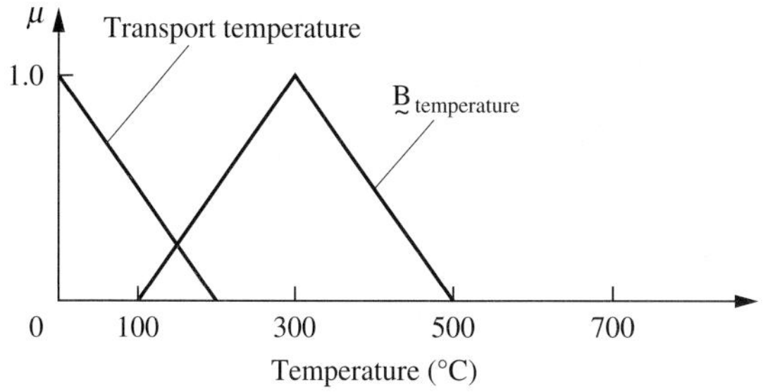

FIGURE 12.24
Temperature comparisons for transport pattern.

And last, moving to flow rate:

$$\underset{\sim}{B}_{\text{flow}} \bullet \text{transport flow} = 0.1$$
$$\underset{\sim}{B}_{\text{flow}} \oplus \text{transport flow} = 0$$

as summarized in Fig. 12.25.

To conclude the calculations using the approaching degree on the last pattern of transport, Eqs. (12.3) and (12.10) are used to determine

$$\begin{aligned}(\underset{\sim}{B}, \text{transport}) &= 0.5\left(\tfrac{1}{2}(0.7+1)\right) + 0.25\left(\tfrac{1}{2}(0.25+1)\right) + 0.25\left(\tfrac{1}{2}(0.1+1)\right) \\ &= (0.5)(0.85) + (0.25)(0.625) + (0.25)(0.55) = 0.7188\end{aligned}$$

Summarizing the results for the four possible patterns, we have

$$\begin{aligned}(\underset{\sim}{B}, \text{autoclaving}) &= 0.665 \\ (\underset{\sim}{B}, \text{annealing}) &= 0.7175 \\ (\underset{\sim}{B}, \text{sintering}) &= 0.7687 \qquad \text{(max is here)} \\ (\underset{\sim}{B}, \text{transport}) &= 0.7188\end{aligned}$$

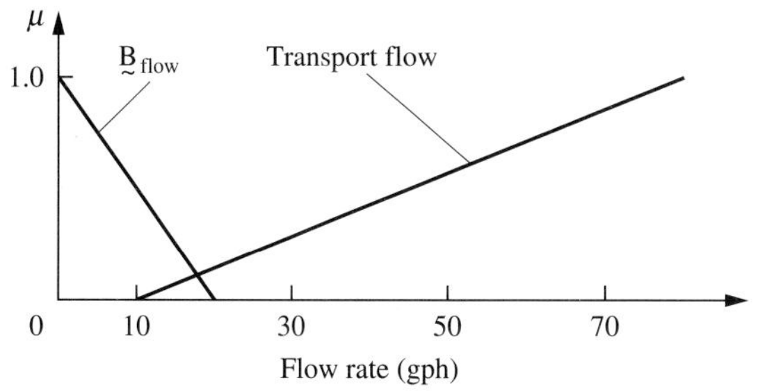

FIGURE 12.25
Flow rate comparisons for transport pattern.

The fuzzy readings of pressure, temperature, and flow collectively match most closely, in an approaching degree sense, the *sintering* pattern. We therefore write the production mode "sintering" in our log.

IMAGE PROCESSING

An image (having various shades of gray) is represented mathematically by a spatial brightness function $f(m, n)$ where (m, n) denotes the spatial coordinate of a point in the (flat) image. The value of $f(m, n)$, $0 < f(m, n) < \infty$, is proportional to the brightness value or gray level of the image at the point (m, n). For computer processing, the continuous function $f(m, n)$ has been discretized both in spatial coordinates and in brightness. Such an approximated image X(digitized) can be considered as an $M \times N$ array,

$$\mathrm{X} = f(m,n) = \begin{bmatrix} x_{11} & x_{12} & \cdots & x_{1n} & \cdots & x_{1N} \\ x_{21} & x_{22} & \cdots & x_{2n} & \cdots & x_{2N} \\ x_{31} & x_{32} & \cdots & x_{3n} & \cdots & x_{3N} \\ \vdots & \vdots & \cdots & \vdots & \cdots & \vdots \\ x_{M1} & x_{M2} & \cdots & x_{Mn} & \cdots & x_{MN} \end{bmatrix} \tag{12.14}$$

whose row and column indices identify a point (m, n) in the image, and the corresponding matrix element value $x_{mn}[\sim f(m, n)]$ denotes the gray level at that point.

The right side of Eq. (12.14) represents what is called a digital image. Each element of the matrix, which is a discrete quantity, is referred to as an image element, picture element, pixel, or pel, with the last two names commonly used as abbreviations of picture element. From now on the terms *image* and *pixels* will be used to denote a digital image and its elements, respectively. For the purpose of processing, this image along with the coordinates of its pixels is stored in the computer in the form of an $M \times N$ array of numbers.

The methods so far developed for image processing may be categorized into two broad classes, namely, frequency domain methods and spatial domain methods. The techniques in the first category depend on modifying the Fourier transform of an image by transforming pixel intensity to pixel frequency, whereas in spatial domain methods the direct manipulation of the pixel is adopted. Some fairly simple and yet powerful processing approaches are formulated in the spatial domain.

In frequency domain methods, processing is done with various kinds of frequency filters. For example, low frequencies are associated with uniformly gray areas, and high frequencies are associated with regions where there are abrupt changes in pixel brightness. In the spatial domain methods pixel intensities can be modified independently of pixel location, or they can be modified according to their neighboring pixel intensities. Examples of these methods include (1) contrast stretching, where the range of pixel intensities is enlarged to accommodate a larger range of values, (2) image smoothing, where "salt and pepper" noise is removed from the image, and (3) image sharpening, which involves edge or contour detection and extraction.

Although many of the crisp methods for image processing have a good physical and theoretical basis, they often correlate poorly with the recognition of an image judged by a human because the human visual system does not process the image in a point-by-point fashion. When pattern recognition becomes indeterminate because the underlying variability is vague and imprecise, fuzzy methods can be very useful. A good example of this is the recognition of human speech. Speech carries information regarding the message and the speaker's sex, age, health, and mind; hence it is to a large extent fuzzy in nature. Similarly, an image carries significant fuzzy information. With this in mind we could consider an image as an array of fuzzy singletons, each with a value of membership function denoting the degree of brightness, or "grayness."

Before one is able to conduct meaningful pattern recognition exercises with images, one may need to preprocess the image to achieve the best image possible for the recognition process. The original image might be polluted with considerable noise, which would make the recognition process difficult. Processing, reducing, or eliminating this noise will be a useful step in the process. An image can be thought of as an ordered array of pixels, each characterized by gray tone. These levels might vary from a state of no brightness, or completely black, to a state of complete brightness, or totally white. Gray tone levels in between these two extremes would get increasingly lighter as we go from black to white. Various preprocessing techniques such as contrast enhancement, filtering, edge detection, ambiguity measure, segmentation, and others are described in the literature [Pal and Majumder, 1986]. For this chapter we will introduce only contrast enhancement using fuzzy procedures.

An image X of $M \times N$ dimensions can be considered as an array of fuzzy singletons, each with a value of membership denoting the degree of brightness level p, $p = 0, 1, 2, \ldots, P-1$ (e.g., a range of densities from $p = 0$ to $p = 255$), or some relative pixel density. Using the notation of fuzzy sets, we can then write Eq. (12.14) as

$$\mathrm{X} = \begin{bmatrix} \mu_{11}/x_{11} & \mu_{12}/x_{12} & \cdots & \mu_{1n}/x_{1n} & \cdots & \mu_{1N}/x_{1N} \\ \mu_{21}/x_{21} & \mu_{22}/x_{22} & \cdots & \mu_{2n}/x_{2n} & \cdots & \mu_{2N}/x_{2N} \\ \mu_{31}/x_{31} & \mu_{32}/x_{32} & \cdots & \mu_{3n}/x_{3n} & \cdots & \mu_{3N}/x_{3N} \\ \vdots & \vdots & \cdots & \vdots & \cdots & \vdots \\ \mu_{M1}/x_{M1} & \mu_{M2}/x_{M2} & \cdots & \mu_{Mn}/x_{Mn} & \cdots & \mu_{MN}/x_{MN} \end{bmatrix} \tag{12.15}$$

where $m = 1, 2, \ldots, M$ and $n = 1, 2, \ldots, N$, and where $\mu_{mn}/x_{mn} (0 \leq \mu_{mn} \leq 1)$ represents the grade of possessing some property μ_{mn} by the (m, n)th pixel x_{mn}. This fuzzy property μ_{mn} may be defined in a number of ways with respect to any brightness level (pixel density) depending on the problem.

Contrast within an image is the measure of difference between the gray levels in an image. The greater the contrast, the greater is the distinction between gray levels in the image. Images of high contrast have either all black or all white regions; there is very little gray in the image. Low-contrast images have lots of similar gray levels in the image, and very few black or white regions. High-contrast images can be thought of as crisp, and low-contrast ones as completely fuzzy. Images with good gradation of grays between black and white are usually the best images for purposes of recognition by humans. Heretofore, computers have worked best with images that have had high contrast, although algorithms based on fuzzy sets have been successful with both.

The object of contrast enhancement is to process a given image so that the result is more suitable than the original for a specific application in pattern recognition. As with all image-processing techniques we have to be especially careful that the processed image is not distinctly different from the original image, making the identification process worthless. The technique used here makes use of modifications to the brightness membership value in stretching or contracting the contrast of an image. Many contrast enhancement methods work as shown in Fig. 12.26, where the procedure involves a primary enhancement of an image, denoted with an E_1 in the figure, followed by a smoothing algorithm, denoted by an S, and a subsequent final enhancement, step E_2. The fuzzy operator defined in Eq. (8.8), called intensification, is often used as a tool to accomplish the primary and final enhancement phases shown in Fig. 12.26.

The function of the smoothing portion of this method (the S block in Fig. 12.26) is to blur (make more fuzzy) the image, and this increased blurriness then requires the use of the final enhancement step, E_2. Smoothing is based on the property that adjacent image points (points that are close spatially) tend to possess nearly equal gray levels. Generally, smoothing algorithms distribute a portion of the intensity of one pixel in the image to adjacent pixels. This distribution is greatest for pixels nearest to the pixel being smoothed, and it decreases for pixels farther from the pixel being smoothed.

The contrast intensification operator, Eq. (8.8), on a fuzzy set $\underset{\sim}{A}$ generates another fuzzy set, $\underset{\sim}{A}' = \text{INT}(\underset{\sim}{A})$ in which the fuzziness is reduced by increasing the values of $\mu_{\underset{\sim}{A}}(x)$ that are greater than 0.5 and by decreasing the values of $\mu_{\underset{\sim}{A}}(x)$ that are less than 0.5 [Pal and King, 1980]. If we define this transformation T_1, we can define T_1 for the membership values of brightness for an image as

$$\begin{aligned} T_1(\mu_{mn}) &= T_1'(\mu_{mn}) = 2\mu_{mn}^2 \qquad 0 \le \mu_{mn} \le 0.5 \\ &= T_1''(\mu_{mn}) = 1 - 2(1 - \mu_{mn})^2 \qquad 0.5 \le \mu_{mn} \le 1 \end{aligned} \tag{12.16}$$

In general, each μ_{mn} in X may be modified to μ'_{mn} to enhance the image X in the property domain by a transformation function, T_r, where

$$\begin{aligned} \mu'_{mn} &= T_r(\mu_{mn}) = T_r'(\mu_{mn}) \qquad 0 \le \mu_{mn} \le 0.5 \\ &= T_r''(\mu_{mn}) \qquad 0.5 \le \mu_{mn} \le 1 \end{aligned} \tag{12.17}$$

and $T_1(\mu_{mn})$ represents the operator INT as defined in Eq. (8.8). The transformation T_r is defined as successive applications of T_1 by the recursive relation,

$$T_r(\mu_{mn}) = T_1\{T_{r-1}(\mu_{mn})\} \qquad r = 1, 2, \ldots \tag{12.18}$$

The graphical effect of this recursive transformation for a typical membership function is shown in Fig. 12.27. As r (i.e., the number of successive applications of the INT function) increases, the slope of the curve gets steeper. As r approaches infinity,

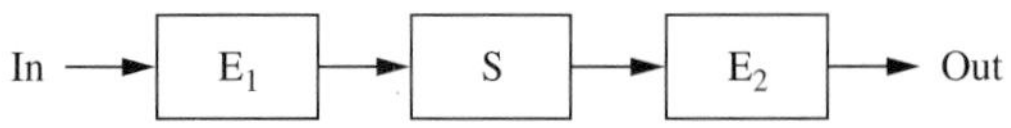

FIGURE 12.26
Diagram of the enhancement model [adapted from Pal and Majumder, 1986].

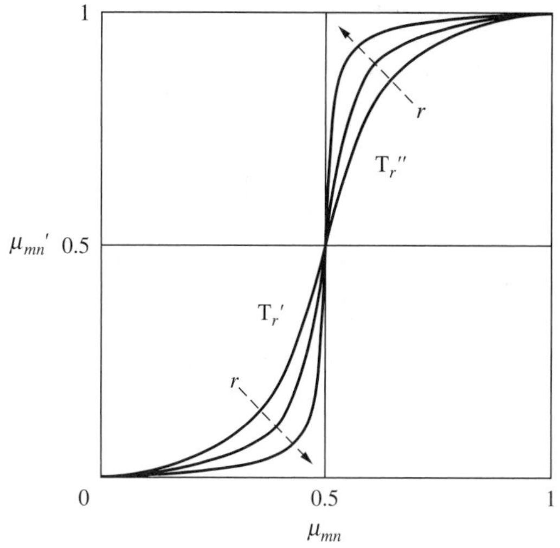

FIGURE 12.27
INT transformation function for contrast enhancement [Pal and Majumder, 1986].

the shape approaches a crisp (binary) function. The parameter r allows the user to use an appropriate level of enhancement for domain-specific situations.

Example 12.7. We will demonstrate enhancement of the image shown in Figure 12.28*a*. The dark square image of Fig. 12.28*a* has a lighter square box in it that is not very apparent because the shade of the background is very nearly the same as that of the lighter box itself. Table 12.3 shows the 256 gray-scale intensity values of pixels of the 10×10 pixel array of the image shown in Fig. 12.28*a*. If we take the intensity values from Table 12.4 and scale them on the interval [0, 255], we get membership values in the density set *white* (low values are close to black, high values close to white). These values, of course, will be between 0 and 1 as membership values.

Using Eq. (12.16), we modify the pixel values to obtain the matrix shown in Table 12.5. The reader should notice that the intensity values above and below 0.5 have

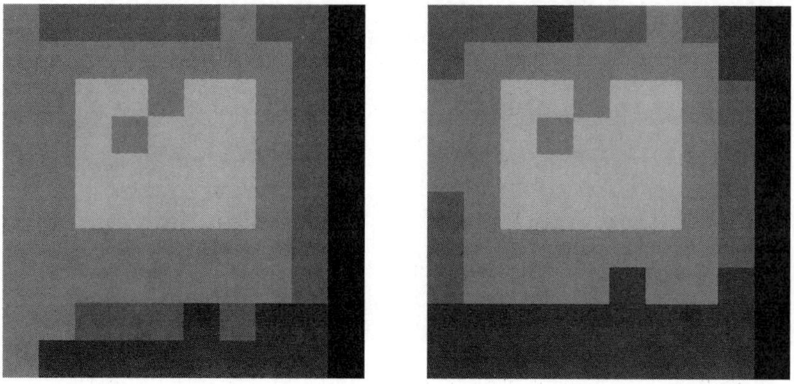

(*a*) (*b*)

FIGURE 12.28
Lighter square inside smaller square: (*a*) original image; (*b*) image after one application of INT operator [Eq. (12.16)].

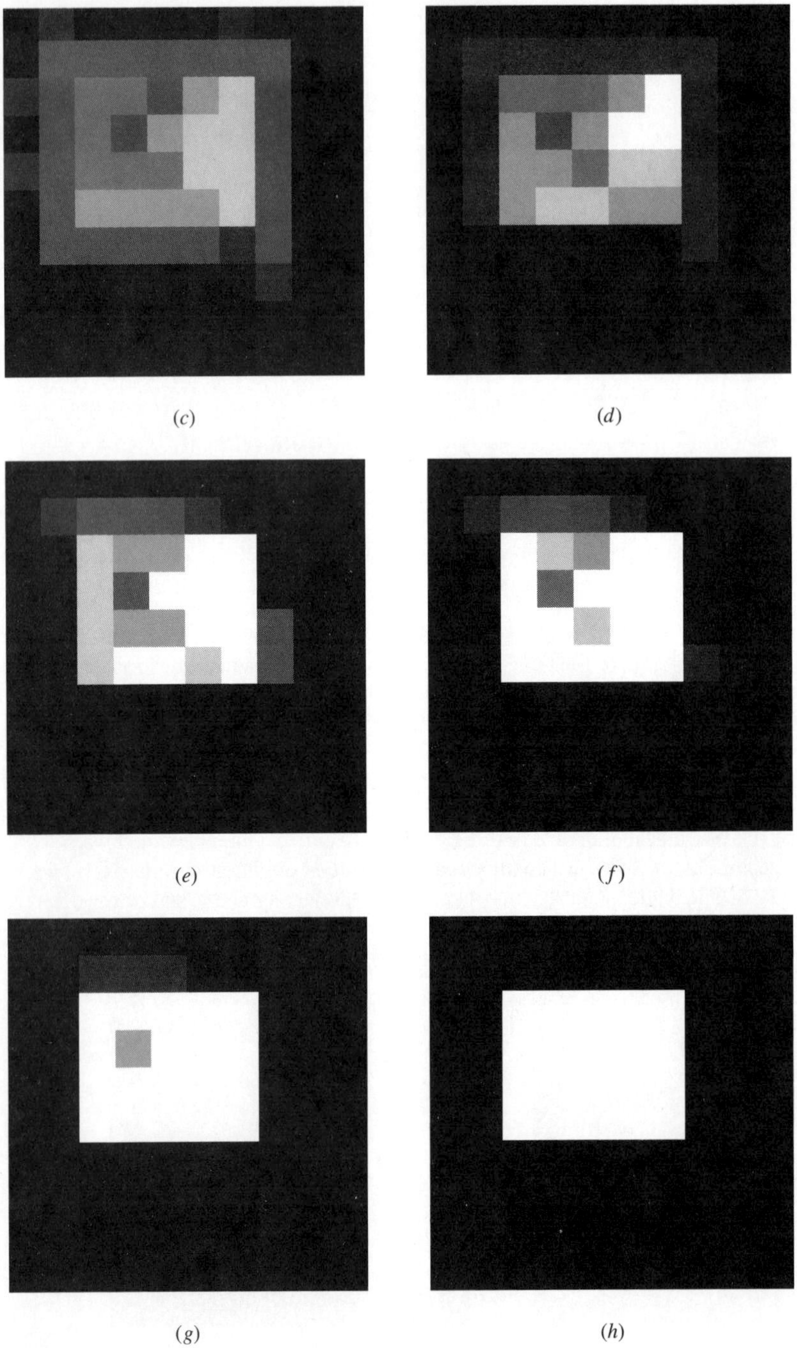

FIGURE 12.28 (*Continued*)
Lighter square inside smaller square: (*c*)–(*h*) successive applications of INT operator on original image (*a*).

TABLE 12.3
Gray-scale intensity values of pixels in a 10 × 10 pixel array of the image shown in Fig. 12.28*a*

77	89	77	64	77	71	99	56	51	38
77	122	125	125	125	122	117	115	51	26
97	115	140	135	133	153	166	112	56	31
82	112	145	130	150	166	166	107	74	23
84	107	140	138	135	158	158	120	71	18
77	110	143	148	153	145	148	122	77	13
79	102	99	102	97	94	92	115	77	18
71	77	74	77	71	64	77	89	51	20
64	64	48	51	51	38	51	31	26	18
51	38	26	26	31	13	26	26	26	13

TABLE 12.4
Scaled matrix of the intensity values in Table 12.3

0.30	0.35	0.30	0.25	0.30	0.28	0.39	0.22	0.20	0.15
0.30	0.48	0.49	0.49	0.49	0.48	0.46	0.45	0.20	0.10
0.38	0.45	0.55	0.53	0.52	0.60	0.65	0.44	0.22	0.12
0.32	0.44	0.57	0.51	0.59	0.65	0.65	0.42	0.29	0.09
0.33	0.42	0.55	0.54	0.53	0.62	0.62	0.47	0.28	0.07
0.30	0.43	0.56	0.58	0.60	0.57	0.58	0.48	0.30	0.05
0.31	0.40	0.39	0.40	0.38	0.37	0.36	0.45	0.30	0.07
0.28	0.30	0.29	0.30	0.28	0.25	0.30	0.35	0.20	0.08
0.25	0.25	0.19	0.20	0.20	0.15	0.20	0.12	0.10	0.07
0.20	0.15	0.10	0.10	0.12	0.05	0.10	0.10	0.10	0.05

TABLE 12.5
Intensity matrix after applying the enhancement algorithm

0.18	0.24	0.18	0.12	0.18	0.16	0.30	0.10	0.08	0.05
0.18	0.46	0.48	0.48	0.48	0.46	0.42	0.40	0.08	0.02
0.29	0.40	0.60	0.56	0.54	0.68	0.75	0.39	0.10	0.03
0.20	0.39	0.63	0.52	0.66	0.75	0.75	0.35	0.17	0.02
0.22	0.35	0.60	0.58	0.56	0.71	0.71	0.44	0.16	0.01
0.18	0.37	0.61	0.65	0.68	0.63	0.65	0.46	0.18	0.01
0.19	0.32	0.30	0.32	0.29	0.27	0.26	0.40	0.18	0.01
0.16	0.18	0.17	0.18	0.16	0.12	0.18	0.24	0.08	0.01
0.12	0.12	0.07	0.08	0.08	0.05	0.08	0.03	0.02	0.01
0.01	0.01	0.01	0.01	0.01	0.01	0.01	0.01	0.01	0.01

been suitably modified to increase the contrast between the intensities. The enhanced image is shown in Fig. 12.28*b*. Results of successive enhancements of the image by using Eq. (12.16) repeatedly are shown in Figs. 12.28*c* through 12.28*h*.

A useful smoothing algorithm is called defocusing. The (m, n)th smoothed pixel intensity is found [Pal and King, 1981] from

$$\mu'_{mn} = a_0\mu_{mn} + a_1\sum_{Q_1}\mu_{ij} + a_2\sum_{Q_2}\mu_{ij} + \cdots + a_s\sum_{Q_s}\mu_{ij} \qquad (12.19)$$

where
$$a_0 + N_1a_1 + N_2a_2 + \cdots + N_sa_s = 1$$
$$1 > a_1 > a_2 \cdots a_s > 0$$
$$(i, j) \neq (m, n)$$

In Eq. (12.19), μ_{mn} represents the (m, n)th pixel intensity, expressed as a membership value; Q_1 denotes a set of N_1 coordinates (i, j) that are on or within a circle of radius R_1 centered at the point (m, n); Q_s denotes a set of N_s coordinates (i, j) that are on or within a circle of radius R_s centered at the (m, n)th point but that do not fall into Q_{s-1}; and so on. For example, $Q = \{(m, n+1), (m, n-1), (m+1, n), (m-1, n)\}$ is a set of coordinates that are on or within a circle of unit radius from a point (m, n). Hence, in this smoothing algorithm, a part of the intensity of the (m, n)th pixel is being distributed to its neighbors. The amount of energy distributed to a neighboring point decreases as its distance from the (m, n)th pixel increases. The parameter a_0 represents the fraction retained by a pixel after distribution of part of its energy (intensity) to its neighbors. The set of coefficients a_i is important in the algorithm, and specific values are problem-dependent.

Example 12.8. The final enhanced image in Example 12.7 is used here with some random "salt and pepper" noise introduced into it. "Salt and pepper" noise is the occurrence of black and white pixels scattered randomly throughout the image. In Fig. 12.29, five pixels are shown to have intensity values different from what they should be (i.e., compared with the image shown in Fig. 12.28*h*, for example).

We use the image-smoothing algorithm presented in Eq. (12.19) to reduce the "salt and pepper" noise. Using $a_0 = a_2 = a_3 = a_4 = \cdots = 0$ as the values for the coefficients in Eq. (12.19) gives us the expression for the intensity (membership value) for a pixel as

$$\mu_{00} = \tfrac{1}{4}(\mu_{-10} + \mu_{10} + \mu_{01} + \mu_{0-1})$$

as shown in Fig. 12.30. The expression for μ_{00} does limit the pixels that can be smoothed. The pixels along the edges of the image cannot be smoothed, because all the intensity values around the pixel of interest would not be available. The user should thus be careful when programming for this algorithm.

To start the algorithm, we begin with the initial pixel values describing the image in Fig. 12.29. These initial values are presented in a normalized fashion as member-

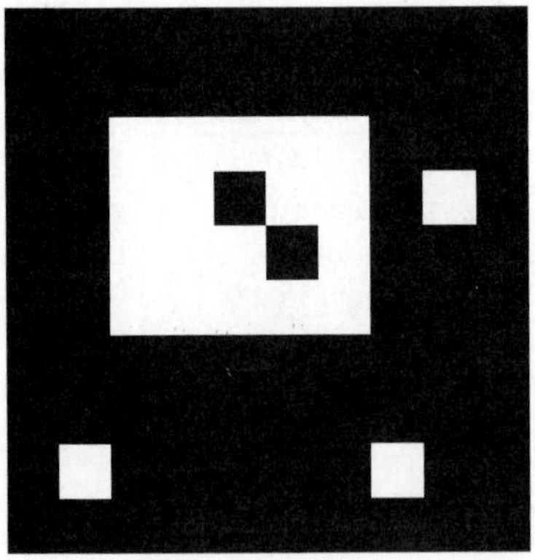

FIGURE 12.29
Image with five "salt and pepper" noise points.

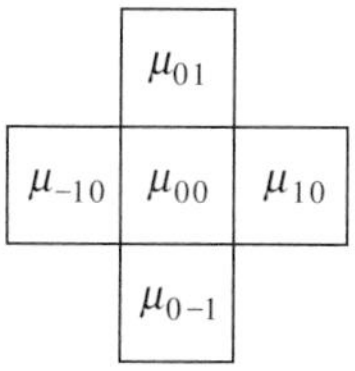

FIGURE 12.30
The pixels required around the center pixel to use the smoothing algorithm for reducing "salt and pepper" noise.

ship values in the set *white* ($\mu = 0$ means the pixel has no membership in *white,* and complete membership in the complement of white, or *black*) as seen in Table 12.6.

After one application of the smoothing algorithm, Eq. (12.19), the intensity matrix of Table 12.6 is modified as shown in Table 12.7, and the associated smoothed image is shown in Fig. 12.31.

SYNTACTIC RECOGNITION

In many recognition problems structural information plays an important role in describing the patterns. Some examples include image recognition, fingerprint recognition, chromosome analysis, character recognition, scene analysis, etc. In such cases, when the patterns are complex and the number of possible descriptions is very large, it is impractical to regard each description as defining a class; rather, description of

TABLE 12.6
Scaled intensity values (black=0, white=1) for the image shown in Fig. 12.29

0.00	0.00	0.00	0.00	0.00	0.00	0.00	0.00	0.00	0.00
0.00	0.00	0.00	0.00	0.00	0.00	0.00	0.00	0.00	0.00
0.00	0.00	1.00	1.00	1.00	1.00	1.00	0.00	0.00	0.00
0.00	0.00	1.00	1.00	0.00	1.00	1.00	0.00	1.00	0.00
0.00	0.00	1.00	1.00	1.00	0.00	1.00	0.00	0.00	0.00
0.00	0.00	1.00	1.00	1.00	1.00	1.00	0.00	0.00	0.00
0.00	0.00	0.00	0.00	0.00	0.00	0.00	0.00	0.00	0.00
0.00	0.00	0.00	0.00	0.00	0.00	0.00	0.00	0.00	0.00
0.00	1.00	0.00	0.00	0.00	0.00	0.00	1.00	0.00	0.00
0.00	0.00	0.00	0.00	0.00	0.00	0.00	0.00	0.00	0.00

TABLE 12.7
Intensity matrix after smoothing the image once

0.00	0.00	0.00	0.00	0.00	0.00	0.00	0.00	0.00	0.00
0.00	0.00	0.25	0.31	0.33	0.33	0.33	0.08	0.02	0.00
0.00	0.25	0.62	0.73	0.52	0.71	0.51	0.15	0.29	0.00
0.00	0.31	0.73	0.62	0.78	0.62	0.53	0.42	0.18	0.00
0.00	0.33	0.77	0.85	0.66	0.82	0.59	0.25	0.11	0.00
0.00	0.33	0.52	0.59	0.56	0.60	0.30	0.14	0.06	0.00
0.00	0.08	0.15	0.19	0.19	0.20	0.12	0.07	0.03	0.00
0.00	0.27	0.11	0.07	0.07	0.07	0.05	0.28	0.08	0.00
0.00	0.07	0.04	0.03	0.02	0.02	0.27	0.14	0.05	0.00
0.00	0.00	0.00	0.00	0.00	0.00	0.00	0.00	0.00	0.00

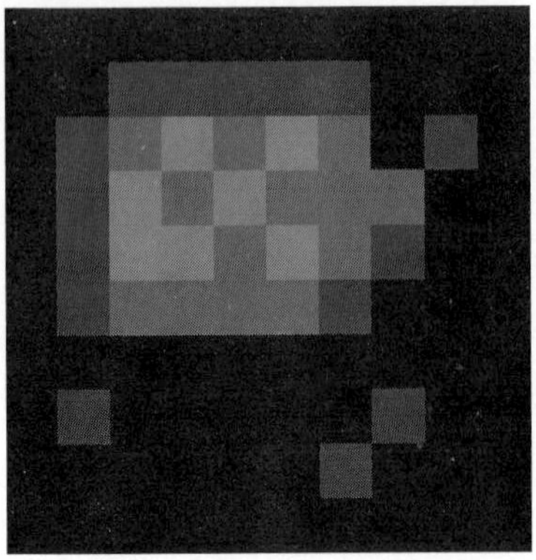

FIGURE 12.31
Image from Fig. 12.29 after one application of smoothing algorithm.

the patterns in terms of small sets of simple subpatterns of primitives and grammatical rules derived from formal language theory becomes necessary [Fu, 1982].

The application of the concepts of formal language theory to structural pattern recognition problems can be illustrated as follows. Let us consider the simplest case in which there are two pattern classes, C_1 and C_2. Let us consider a set of the simplest subpatterns, in terms of which the patterns of both classes can be described completely. We call these the primitives. They can be identified with the terminals of formal language theory. Accordingly, we denote the set of primitives by V_T. Each pattern may then be looked upon as a string or sentence. Let us suppose we can find grammars G_1 and G_2 such that the sets of strings generated by them, $L(G_1)$ and $L(G_2)$, are exactly the same as those corresponding to pattern classes C_1 and C_2, respectively. Clearly, then, if a string corresponding to an unknown pattern is seen to be a member of $L(G_i)$, $i = 1, 2$, we can classify the pattern into C_i. Of course, if it is certain that the unknown string can only come from either of the two classes, then it is sufficient to have just one grammar corresponding to any one of the two classes, say, C_1. In this case, if a string is not from $L(G_1)$, it is automatically assumed to be from C_2. The procedure by which one determines whether a given string is syntactically correct with respect to a given grammar is called syntax analysis or parsing [Fu, 1982].

Example 12.9. [Pal and Majumder, 1986]. Let us consider a very simple problem in which we wish to distinguish "squares" from "rectangles." Obvious primitives for this problem are horizontal and vertical line segments of unit length, which we denote by a and b, respectively. Let us suppose for the sake of simplicity that the dimensions of the figures under consideration are integral multiples of the unit used. Then the two classes can be described as

$$C_{\text{squares}} = \{a^n b^n a^n b^n \mid n \geq 1\}$$

$$C_{\text{rectangles}} = \{a^m b^n a^m b^n \mid m, n \geq 1, m \neq n\}$$

One can easily see that C_{squares} is the same as L(G), the language generated by a grammar (see Eq. 12.21)

$$G = (Y_N, V_T, P, S)$$

where $V_N = \{S, A, B, C\}$

$V_T = \{a, b\}$

$P = \{Sa, aAb, Sabab, aAbaaAbb, aAbaaBbb, BbCa, bCabbCaa, Cba\}$

where P is a collection of various concatenations making up squares and rectangles. Therefore, any pattern for which the corresponding string can be parsed by G is classified as a square. Otherwise, it is a rectangle.

The foregoing approach has a straightforward generalization for an m-class pattern recognition problem. Depending on whether or not the m classes exhaust the pattern space, we choose $m-1$ or m grammars. In the first case, we classify a pattern into C_i if the corresponding string belongs to $L(G_i)$, $i = 1, 2, \ldots, m-1$. Otherwise, the pattern is classified into C_m. In the second case, a pattern is identified as coming from C_i, $i = 1, 2, \ldots, m$ if the string corresponding to it can be parsed by G_i. If not, the pattern is reckoned to be noisy and is rejected.

The syntactic (structural) approach, which draws an analogy between the hierarchical structure of patterns and the syntax of languages, is a powerful method. After identifying each of the primitives within the pattern, the recognition process is accomplished by performing a syntax analysis of the "sentence" describing the given pattern. In the syntactic method, the ability to select and classify the simple pattern primitives and their relationships represented by the composition operations is the vital means by which a system is effective. Since the techniques of composition of primitives into patterns are usually governed by the formal language theory, the approach is often referred to as a *linguistic approach*. This learned description is then used to analyze and to produce syntactic descriptions of unknown patterns.

For the purposes of recognition by computer, the patterns are usually digitized (in time or in space) to carry out the previously mentioned preprocessing tasks. In the syntax analyzer the decision is made whether the pattern representation is syntactically correct. If the pattern is not syntactically correct, it is either rejected or analyzed on the basis of other grammars, which presumably represent other possible classes of patterns under consideration. For the purpose of recognition, the string of primitives of an input pattern is matched with those of the prototypes representing the classes. The class of the reference pattern having the best match is decided to be an appropriate category.

In practical situations, most patterns encountered are noisy or distorted. That is, the string corresponding to a noisy pattern may not be recognized by any of the pattern grammars, or ambiguity may occur in the sense that patterns belonging to the different classes may appear to be the same. In light of these observations, the foregoing approach may seem to have little practical importance. However, efforts have been made to incorporate features that can help in dealing with noisy patterns. The more noteworthy of them are the following [Fu, 1982]:

(*a*) The use of approximation
(*b*) The use of transformational grammars
(*c*) The use of similarity and error-correcting parsing
(*d*) The use of stochastic grammars
(*e*) The use of fuzzy grammars

The first approach proposes to reduce the effect of noise and distortion by approximation at the preprocessing and primitive extraction stage. The second approach attempts to define the relation between noisy patterns and their corresponding noise-free patterns by a transformational grammar. If it is possible to determine such a transformational grammar, the problem of recognizing noisy or distorted patterns can be transformed into one of recognizing noise-free patterns. The third approach defines distance measures between two strings and extends these to define distance measures between a string and a language.

In the stochastic approach, when ambiguity occurs, that is, when two or more patterns have the same structural description or when a single (noisy) string is accepted by more than one pattern grammar, it means that languages describing different classes overlap. The incorporation of the element of probability into the pattern grammars gives a more realistic model of such situations and gives rise to the concept of stochastic languages.

One natural way of generating stochastic language from ordinary formal language is to randomize the productions of the corresponding grammars. This leads to the concept of stochastic phrase-structure grammars, which are the same as ordinary phrase-structure grammars except that every production rule has a probability associated with it. Also, if a pattern grammar is being heuristically constructed, then we can tackle the problem of "unwanted" strings (strings that do not represent patterns in the class) by assigning very small probabilities to such strings. Reviews of the syntactic methods and their applications are available in Fu [1982]. The last approach is addressed next.

Formal Grammar

The concept of a grammar was formalized by linguists with a view to finding a means of obtaining structural descriptions of sentences of a language that could not only recognize but also generate the language [Fu, 1982; Hopcroft and Ullman, 1969]. Although satisfactory formal grammars have not been obtained to date for describing the English language, the concept of a formal grammar can easily be explained with certain ideas borrowed from English grammar.

An *alphabet* or vocabulary is any finite set of symbols. A *sentence* (or string or word) over an alphabet is any *string* of finite length composed of symbols from the alphabet. If V is an alphabet, then V^* denotes the set of all sentences composed of symbols of V, including the empty sentence Λ. A *language* is any set of sentences over an alphabet. For example,

$$\text{If } V = \{a, b\}, \text{ then } V^* = \{\Lambda, a, b, ab, ba, \ldots\}$$

and a few examples of languages over V are

$$L_1 = \{a^n, n = 1, 2, \ldots \text{ (denoted as written, or repeated, } n \text{ times}\}$$
$$L_2 = \{a^m b^n, m \neq n + 1, m, n = 1, 2, \ldots\}$$
$$L_3 = \{a, b, ab\}$$

and so forth [Pal and Majumder, 1986]. The formal prescription of a language theory is useful in syntactic recognition from the axiom, "If there exists a procedure (or algorithm) for recognizing a language, then there also exists a procedure for generating it."

Suppose we want to parse a simple English sentence, "The athlete jumped high." The rules that one applies to parsing can easily be described as follows:

$$\begin{aligned} &\langle\text{sentence}\rangle \rightarrow \langle\text{noun phrase}\rangle\langle\text{verb phrase}\rangle \\ &\langle\text{noun phrase}\rangle \rightarrow \langle\text{article}\rangle\langle\text{noun}\rangle \\ &\langle\text{verb phrase}\rangle \rightarrow \langle\text{verb}\rangle\langle\text{adverb}\rangle \\ &\langle\text{article}\rangle \rightarrow \text{The} \\ &\langle\text{noun}\rangle \rightarrow \text{athlete} \\ &\langle\text{verb}\rangle \rightarrow \text{jumped} \\ &\langle\text{adverb}\rangle \rightarrow \text{high} \end{aligned} \tag{12.20}$$

where the symbol $\rightarrow$ denotes "can be written as." We can now define a formal phrase-grammar, G, as a four-tuple [Pal and Majumder, 1986],

$$G = (V_N, V_T, P, S) \tag{12.21}$$

where V_N and V_T are the nonterminal and terminal vocabularies of G. Essentially, the nonterminal vocabularies are the phrases just illustrated and the terminal vocabularies are the alphabet of the language; in a sense V_T is the collection of singletons of the language (smallest elements) and V_N is the collections, or sets, containing the singletons. The symbols P and S denote the finite set of production rules of the type $\alpha \rightarrow \beta$ where α and β are strings over $V = V_N \cup V_T$, with α having at least one symbol of V_N and $S \in V_N$ is a starting symbol or a sentence (sentence or object to be recognized). In the preceding example,

$$\begin{aligned} V_N &= \{\langle\text{sentence}\rangle, \langle\text{noun phrase}\rangle, \langle\text{verb phrase}\rangle, \langle\text{article}\rangle, \langle\text{noun}\rangle, \langle\text{verb}\rangle, \langle\text{adverb}\rangle\} \\ V_T &= \{\text{the, athlete, jumped, high}\} \\ P &= \text{the set of rules, Eqs. (12.20)} \\ S &= \langle\text{sentence}\rangle \end{aligned}$$

Fuzzy Grammar and Syntactic Recognition

The concept of a formal grammar is often found to be too rigid to handle real patterns, which are usually distorted or noisy yet still retain underlying structure. When the indeterminacy of the patterns is due to inherent vagueness rather than randomness, fuzzy language can be a better tool for describing the ill-defined structural information [see Pal and Majumder, 1986]. In this case, the generative power of a grammar is increased by introducing fuzziness either in the definition of primitives (labels of the fuzzy sets) or in the physical relations among primitives (fuzzified production rules), or in both of these. A fuzzy grammar produces a language that is a fuzzy set of strings with the membership value of each string denoting the degree of belonging of the string in that language. The grade of membership of an unknown pattern in a class described by the grammar is obtained using a max-min composition rule.

Let V_T^* denote the set of finite strings of alphabet V_T, including the null string, Λ. Then, a fuzzy language (FL) on V_T is defined as a fuzzy subset of V_T^*. Thus, FL is the fuzzy set,

$$\text{FL} = \sum_{x \in V_T^*} \frac{\mu_{\text{FL}}(x)}{x} \tag{12.22}$$

where $\mu_{\text{FL}}(x)$ is the grade of membership of the string x in FL. It is further assumed that all other strings in V_T^* have 0 membership in FL.

For two fuzzy languages FL_1 and FL_2 the operations of containment, equivalence, union, intersection, and complement follow the same definitions for the resulting membership as those delineated in Chapters 2 and 3.

Informally, a fuzzy grammar may be viewed as a set of rules for generating a fuzzy subset of V_T^*. A fuzzy grammar FG is a 6-tuple given by

$$\text{FG} = (V_N, V_T, P, S, J, \mu) \tag{12.23}$$

where, in addition to the definitions given for Eq. (12.20), we have J: $\{r_i \mid i = 1, 2, \ldots n$, and $n =$ cardinality of P$\}$, i.e., the number of production rules, and μ is a mapping $\mu : J \to [0, 1]$, such that $\mu(r_i)$ denotes the membership in P of the rule labeled r_i. A fuzzy grammar generates a fuzzy language L(FG) as follows.

A string $x \in V_T^*$ is said to be in L(FG) if and only if it is derivable from S and its grade of membership $\mu_{\text{L(FG)}}(x)$ in L(FG) is > 0, where

$$\mu_{\text{L(FG)}}(x) = \max_{1 \le k \le m} \left[\min_{1 \le k \le l_k} \mu(r_i^k) \right] \tag{12.24}$$

where m is the number of derivations that x has in FG; l_k is the length of the kth derivation chain, $k = 1(1)m$; and r_i^k is the label of the ith production used in the kth derivation chain, $i = 1, 2, \ldots, l_k$. Clearly if the production rule $\alpha \to \beta$ is visualized as a chain link of strength $\mu(r)$, r being the label of the rule, then the strength of a derivation chain is the strength of its weakest link, and hence,

$$\mu_{\text{L(FG)}}(x) = \text{strength of the strongest derivation chain for S to } x \text{, for all } x \in V_T^*$$

Example 12.10 [Pal and Majumder, 1986]. Suppose $\text{FG}_1 = (\{A, B, S\}, \{a, b\}, P, S, \{1, 2, 3, 4\}, \mu)$, where J, P, and μ are as follows:

1. $S \to AB$	with	$\mu(1) = 0.8$
2. $S \to aSb$	with	$\mu(2) = 0.2$
3. $A \to a$	with	$\mu(3) = 1$
4. $B \to b$	with	$\mu(4) = 1$

Then the fuzzy language must be $\text{FL}_1 = \{x \mid x = a^n b^n, n = 1, 2, \ldots\}$ with

$$\mu_{\text{FL}_1}(ab) = \begin{cases} 0.8 & \text{if } n = 1 \\ 0.2 & \text{if } n \ge 2 \end{cases}$$

Careful inspection of rules 1 and 2 shows that rule 2 can be repeated over and over again, generating first, *ab* (rule 1), second, *aabb* (rule 2 using the new value for S), third, *aaabbb*, and so on, recursively, with increasing n.

These grammars, of course, can be used in a wide array of pattern recognition problems where the alphabets become line segments and the words, or vocabulary, become geometric shapes as illustrated in the following three examples. The following example from Pal and Majumder [1986], for a right triangle, illustrates this idea.

Example 12.11 [Pal and Majumder, 1986]. Suppose a fuzzy grammar is given by

$$\mathrm{FG}_2 = (\{\mathrm{S, A, B}\}, (a, b, c), \mathrm{P, S, J}, \mu)$$

where J, P, and *m* are as follows:

$$\begin{aligned} r_1 &: \mathrm{S} \rightarrow a\mathrm{A} \quad \text{with} \quad \mu(r_1) = \mu_{\mathrm{H}}(a) \\ r_2 &: \mathrm{A} \rightarrow b\mathrm{B} \quad \text{with} \quad \mu(r_2) = \mu_{\mathrm{V}}(a) \\ r_3 &: \mathrm{B} \rightarrow c \quad \text{with} \quad \mu(r_3) = \mu_{\mathrm{ob}}(a) \end{aligned}$$

with the primitives *a*, *b*, and *c* being *horizontal, vertical*, and *oblique* directed line segments, respectively, as seen in Fig. 12.32. The membership functions for these line segments are given here with reference to Fig. 12.33.

$$\mu_{\mathrm{H}}(\theta) = 1 - |\tan\theta| \quad \text{and} \quad \mu_{\mathrm{V}}(\theta) = 1 - \left|\frac{1}{\tan\theta}\right|$$

$$\text{and} \quad \mu_{\mathrm{ob}}(\theta) = 1 - \left|\frac{\theta - 45^\circ}{45^\circ}\right|$$

From the three rules, the only string generated is $x = abc$. This string is, of course, a right triangle, as seen in Fig. 12.33, and is formed from the specified sequence in the

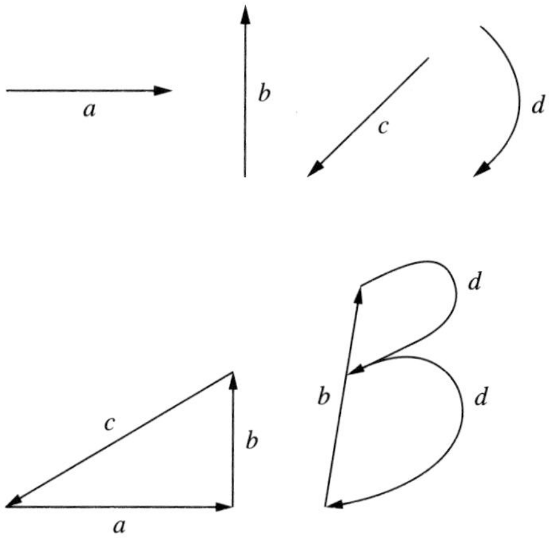

FIGURE 12.32
Primitive line segments and production of a triangle [Pal and Majumder, 1986].

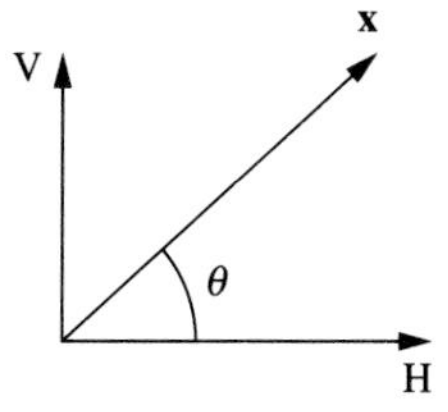

FIGURE 12.33
Membership functions for horizontal, vertical, and oblique lines.

string *abc*. In this syntactic recognition, the primitives (line segments) will be concatenated in the "head-to-tail" style.

Another example for a geometric shape is given in Example 12.12 for a trapezoid.

Example 12.12. Consider a fuzzy grammar

$$\text{FG} = (\{\text{S, A, B, C}\}, \{a, b, c, d\}, \text{P, S, J}, \mu)$$

with

$$\begin{aligned} \text{P}: r_1 &: \text{S} \rightarrow a + \text{A} & \mu(r_1) &= \mu_{\text{PO}}(a) \\ r_2 &: \text{A} \rightarrow b + \text{B} & \mu(r_2) &= \mu_{\text{FH}}(b) \\ r_3 &: \text{B} \rightarrow c + \text{C} & \mu(r_3) &= \mu_{\text{NO}}(c) \\ r_4 &: \text{C} \rightarrow {\sim}b & \mu(r_4) &= \mu_{\text{IH}}({\sim}b) \end{aligned}$$

The primitives a, b, c, and $\sim b$ represent *positive oblique, forward horizontal, negative oblique,* and *inverse horizontal* directed line segments, respectively, as in Fig. 12.34.

The fuzzy membership functions for positive oblique (PO), forward horizontal (FH), negative oblique (NO), and inverse horizontal (IH) are as follows:

$$\mu_{\text{PO}} = \begin{cases} 1 & 0 \leq \theta \leq 90^\circ \\ 0 & \text{elsewhere} \end{cases}$$

$$\mu_{\text{NO}} = \begin{cases} 1 & 90^\circ \leq \theta \leq 180^\circ \\ 0 & \text{elsewhere} \end{cases}$$

$$\mu_{\text{FH}} = \cos\theta$$

$$\mu_{\text{IH}} = -\cos(\theta + 90^\circ)$$

where the angle θ is as defined in Fig. 12.35.

Making use of a head-tail type of concatenation, the foregoing fuzzy grammar FG produces a structure of a trapezoid with the membership function

$$\mu_{\text{L(FG)}}(x) = \min(\mu_{\text{PO}}(a), \mu_{\text{FH}}(b), \mu_{\text{NO}}(c), \mu_{\text{IH}}({\sim}b))$$

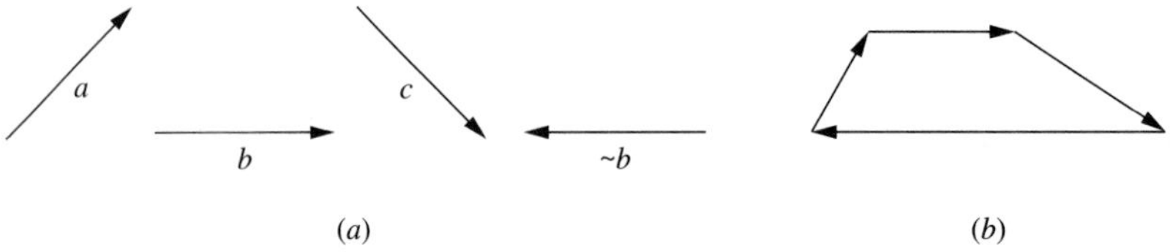

FIGURE 12.34
Directed line segments for Example 12.12: (*a*) primitives; (*b*) production of trapezoid.

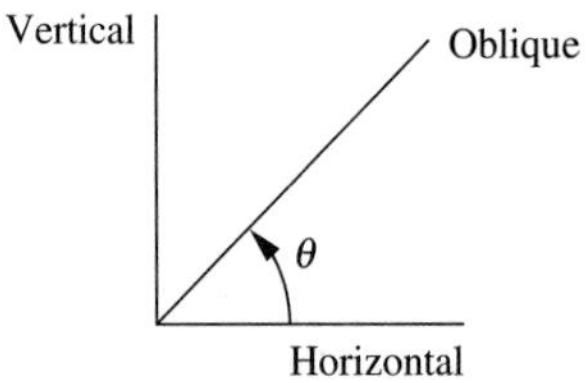

FIGURE 12.35
Geometric orientation for membership functions.

where $x = a + b + c + (\sim b)$. The structure is shown in Fig. 12.34b. If membership values for $\mu_{\text{PO}}(a) = 1$, $\mu_{\text{FH}}(b) = 1$, $\mu_{\text{NO}}(c) = 1$, and $\mu_{\text{IH}}(\sim b) = 0.9$, then the membership value of the trapezoid is

$$\min(1, 1, 1, 0.9) = 0.9$$

This geometric idea in syntactic recognition can be extended to include other symbols, such as illustrated in the following example.

Example 12.13. For the syntactic pattern recognition of electrical symbols in an electrical design diagram such as a resistor, an inductor, and a capacitor, etc., several fuzzy grammars can be used for each different symbol. An inductor and a resistor can be assigned to the same fuzzy language. If the fuzzy grammar for this language is called FG_1, then FG_1 is defined as

$$\text{FG}_1 = (\{\text{A}, \text{S}\}, \{a_i\}, \text{P}, \text{S}, \{1, 2, 3\}, \mu)$$

where J, P, and μ are as follows:

1. $\text{S} \rightarrow \text{A}$ with $\mu(1) = 0.15$
2. $\text{S} \rightarrow a_i\text{S}$ with $\mu(2) = 0.85$
3. $\text{A} \rightarrow a_i$ with $\mu(3) = 1$

and a_i is a primitive. Suppose a_1 ($i = 1$) represents the primitive (inductor symbol) 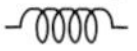 and a_2 ($i = 2$) represents the primitive (resistor symbol) 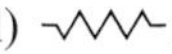. The preceding fuzzy grammar generates the fuzzy language as

$$\text{L}(\text{FG}_1) = \{x \mid x = a_i^n,\ n = 1, 2, 3, \ldots\}$$

Further, if the concatenation of a head-tail type is used for the generation of string x, then $x = a_i^n$ infers the pattern of an inductor when $i = 1$ or a resistor when $i = 2$. These ideas are shown in Fig. 12.36. The membership values for these two patterns can be expressed as

1. $i = 1$, which infers a pattern of an inductor

$$\mu_{\text{L}(\text{FG}_1)}(a_1^n) = \begin{cases} 0.15 & n = 1 \\ 0.85 & n \geq 2 \end{cases}$$

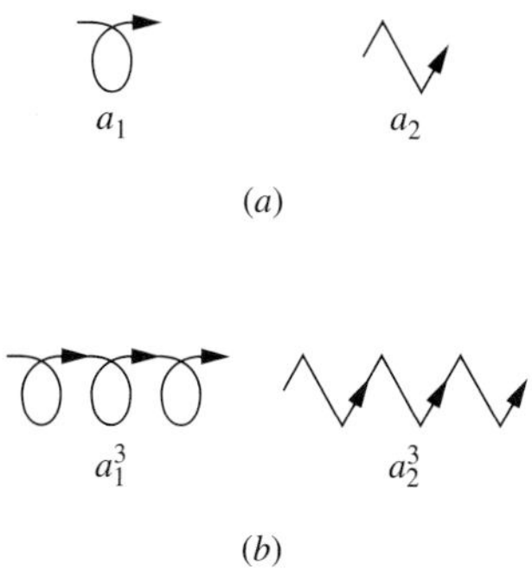

FIGURE 12.36
Directed line segments for Example 12.13: (a) primitives and (b) patterns produced by the primitives.

2. $i = 2$, which infers a pattern of a resistor

$$\mu_{\mathrm{L(FG_1)}}(a_2^n) = \begin{cases} 0.15 & n = 1 \\ 0.85 & n \geq 2 \end{cases}$$

and these patterns and associated membership values are summarized in Table 12.8.

Another fuzzy grammar $\mathrm{FG_2}$ can be developed to recognize the pattern of a capacitor. If $\mathrm{FG_2}$ is expressed in a general form,

$$\mathrm{FG_2} = (\mathrm{V_N}, \mathrm{V_T}, \mathrm{P}, \mathrm{S}, \mathrm{J}, \mu)$$

then

$$\begin{aligned} \mathrm{V_N} &= \{\mathrm{S}, \mathrm{R}\} \\ \mathrm{V_T} &= \{a_3, a_4\} \qquad \text{or} \qquad \mathrm{V_T} = \{a_4, a_5\} \\ \mathrm{P}: \mathrm{S} &\rightarrow \mathrm{L}(a_3, a_4) \qquad \text{or} \qquad \mathrm{S} \rightarrow \mathrm{L}(a_4, a_5) \end{aligned}$$

where $a_i (i = 3, 4, 5)$ is a primitive in which a_3 represents the symbol ")", a_4 represents the symbol "|", and a_5 represents the symbol "(". L(x, y) means "x is to the right of y." Therefore, the fuzzy language decided by $\mathrm{FG_2}$ represents a capacitor in reality. A pattern S that meets $\mathrm{FG_2}$ can be considered a capacitor. Besides, $\mathrm{FG_2}$ belongs to a context-free grammar.

If an unknown pattern S has the primitives of a_3 and a_4, and the membership values for a_3 and a_4 are given by

$$\mu_{\mathrm{L(FG_2)}}(a_3) = 0.8 \qquad \mu_{\mathrm{L(FG_2)}}(a_4) = 1$$

then the membership of S representing a capacitor is

$$\mu_c(\mathrm{S}) = \min(\mu_{\mathrm{L(FG_2)}}(a_3), \mu_{\mathrm{L(FG_2)}}(a_4)) = \min(0.8, 1) = 0.8$$

Similarly, if an unknown pattern S has the primitives of a_4 and a_5, and the membership values for a_4 and a_5 are

$$\mu_{\mathrm{L(FG_2)}}(a_4) = 0.9 \qquad \mu_{\mathrm{L(FG_2)}}(a_5) = 0.8$$

then the membership for S as a capacitor is

$$\mu_c(\mathrm{S}) = \min(\mu_{\mathrm{L(FG_2)}}(a_4), \mu_{\mathrm{L(FG_2)}}(a_5)) = \min(0.9, 0.8) = 0.8$$

By modification of the fuzzy grammar used for capacitors, $\mathrm{FG_2}$ can be used to develop a fuzzy language for the electrical source (AC, DC) patterns of  and . In those cases,

TABLE 12.8
The recognition of electrical elements of inductors and resistors

	Membership of primitives		Inference of the element
Number	a_1	a_2	
$n = 1$	0.15	0	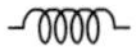
$n \geq 2$	0.85	0	
$n = 1$	0	0.15	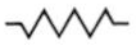
$n \geq 2$	0	0.85	

V_T and P are changed to meet requirements of different patterns. In general, for the pattern recognition of an electrical element, the unknown pattern is first classified into a certain grammar, such as FG_1 or FG_2, then the recognition is carried out according to different primitives, production rules, and membership functions.

SUMMARY

This chapter has introduced only the most elementary forms of fuzzy pattern recognition. In the first section of this chapter, a simple similarity metric called the *approaching degree* (the name is arbitrary; other pseudonyms are possible) is used to assess "closeness" between a known one-dimensional element and an unrecognized one-dimensional element. The idea involved in the approaching degree can be extended to higher-dimensional problems, as illustrated in this chapter, with the use of noninteractive membership functions. The areas of image processing and syntactic recognition are also just briefly introduced to stimulate the reader into some exploratory thinking about the wealth of other possibilities in both these fields. The references to this chapter can be explored further to enrich the reader's background in these, and other pattern recognition, applications.

REFERENCES

Bezdek, J. (1981). *Pattern recognition with fuzzy objective function algorithms,* Plenum Press, New York.

Bezdek, J., N. Grimball, J. Carson, and T. Ross (1986). "Structural failure determination with fuzzy sets," *Civ. Eng. Syst.,* vol. 3, pp. 82–92.

Dong, W. (1987). Personal notes.

Duda, R., and R. Hart. (1973). *Pattern classification and scene analysis,* John Wiley & Sons, New York.

Fu, K. S. (1982). *Syntactic pattern recognition and applications,* Prentice Hall, Englewood Cliffs, NJ.

Fukunaga, K. (1972). *Introduction to statistical pattern recognition,* Academic, New York.

Hopcroft, J., and J. Ullman. (1969). *Formal languages and their relation to automata,* Addison-Wesley, Reading, MA.

Pal, S., and R. King. (1980). "Image enhancement using fuzzy sets," *Electron. Lett.,* vol. 16, pp. 376–378.

Pal, S., and R. King. (1981). "Image enhancement using smoothing with fuzzy sets," *IEEE Trans. Syst. Man Cybern.,* vol. SMC-11, pp. 494–501.

Pal, S., and D. Majumder. (1986). in *Fuzzy mathematical approach to pattern recognition,* John Wiley & Sons, New York.

Wang, P. (1983). in "Approaching degree method," in *Fuzzy sets theory and its applications,* Science and Technology Press, Shanghai, P.R.C. (in Chinese).

PROBLEMS

12.1. For two fuzzy vectors **a** and **b**, prove the following expressions (transpose on the second vector in each operation is presumed):

(*a*) $\overline{\mathbf{a} \bullet \mathbf{b}} = \overline{\mathbf{a}} \oplus \overline{\mathbf{b}}$

(*b*) $\mathbf{a} \bullet \mathbf{a} \leq 0.5$

12.2. Prove the following:

(*a*) For any $\underset{\sim}{A} \in P^*(X)$, prove that $(\underset{\sim}{A}, \underset{\sim}{A})_{1 \text{ or } 2} = 1$

(*b*) For any $\underset{\sim}{A}$ on X, that

$$(\underset{\sim}{A}, \overline{\underset{\sim}{A}})_1 \leq \tfrac{1}{2}$$
$$(\underset{\sim}{A}, \overline{\underset{\sim}{A}})_2 \leq \tfrac{1}{2}$$

12.3. Show that the metric in Eq. (12.2) always gives a value less than or equal to the metric in Eq. (12.3) for any pair of fuzzy sets.

12.4. In signal processing the properties of an electrical signal can be important. The most sought-after properties of continuous time signals are its magnitude, phase, and frequency exponents. Three of these properties together determine one sinusoidal component of a signal where a sinusoid can be represented by the following voltage:

$$V(t) = A\sin(f_0 t - \theta)$$

where A = magnitude (or amplitude) of the sinusoidal component
f_0 = fundamental frequency of the sinusoidal component
θ = phase of the sinusoidal component

With each of these properties representing a "feature" of the electrical signal, it is possible to model a fuzzy pattern recognition system to detect what type of sinusoidal components are present. Let us define the prototypical values for patterns of magnitude, frequency, and phase that we are interested in:

Components	**Prototypical values**
$0 \leq A \leq 12$ V	3 V, 6 V, 9 V, 12 V
$0 \leq f_0 \leq 80$ Hz	20 Hz, 40 Hz, 60 Hz, 80 Hz
$0 \leq \theta \leq 180°$	45°, 90°, 135°, 180°

Draw the resulting three-feature membership graphs.

Now let the input sinusoidal signal vector **B** comprise three crisp singletons, i.e., **B** = {5 V, 45 Hz, 45°}, with weights of 0.6, 0.2, and 0.2 assigned to each of the corresponding features. Determine which pattern vector **B** most closely resembles.

12.5. Using the same patterns as in Problem 12.4, but with a new input fuzzy pattern $\underset{\sim}{B}$ and features given by

$$\underset{\sim}{B}_{\text{voltage}} = \left\{\frac{0}{1} + \frac{0.2}{2} + \frac{0.7}{3} + \frac{1.0}{5} + \frac{0}{6}\right\}$$

$$\underset{\sim}{B}_{\text{frequency}} = \left\{\frac{0}{20} + \frac{0.5}{30} + \frac{1.0}{40} + \frac{0.4}{50} + \frac{0}{60}\right\}$$

$$\underset{\sim}{B}_{\text{phase}} = \left\{\frac{0}{50} + \frac{0.3}{70} + \frac{0.7}{90} + \frac{1}{110} + \frac{0.7}{120} + \frac{0}{130}\right\}$$

determine the pattern that most closely matches the input pattern.

12.6. A robot is being trained to recognize three distinct fruits: apple, banana, and orange. All other fruits are in the category "others." In the robot's memory are four patterns with three features each: *Apple, Banana, Orange*, and *Other*. The features associated with these patterns are shape, color, and weight. The membership functions for the

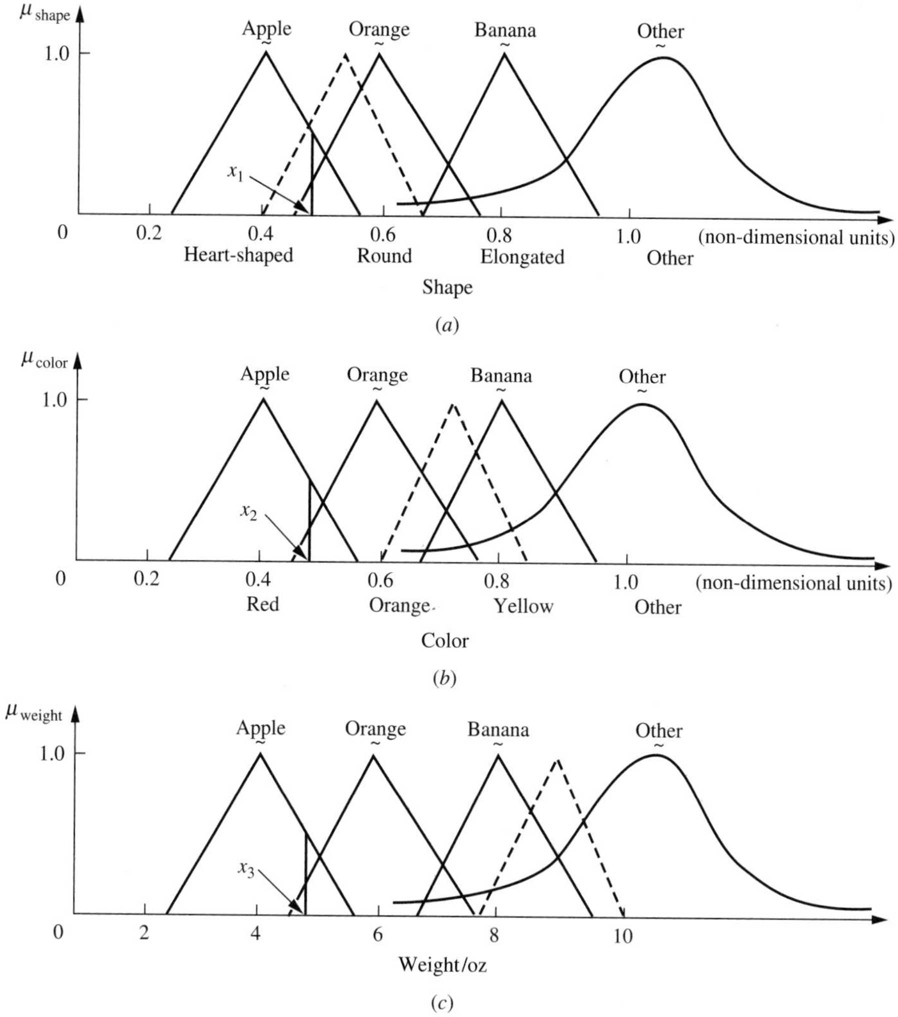

FIGURE P12.6

features are given in Fig. P12.6. The weighting factors for each of the features are these:

Feature 1: Shape	$w_1 = 0.40$
Feature 2: Color	$w_2 = 0.35$
Feature 3: Weight	$w_3 = 0.25$

Determine the fruit pattern that most closely matches the new pattern under these conditions:

(*a*) The new pattern is a crisp singleton, $\underset{\sim}{B}$ = { "shape = roundish with flat top," "color = red with tint of orange," "weight ≈ 4.4 oz"}, indicated as x_1, x_2, x_3 on the figure.

(*b*) The new pattern is a fuzzy pattern shown as a dotted triangle in Figs. P12.6.

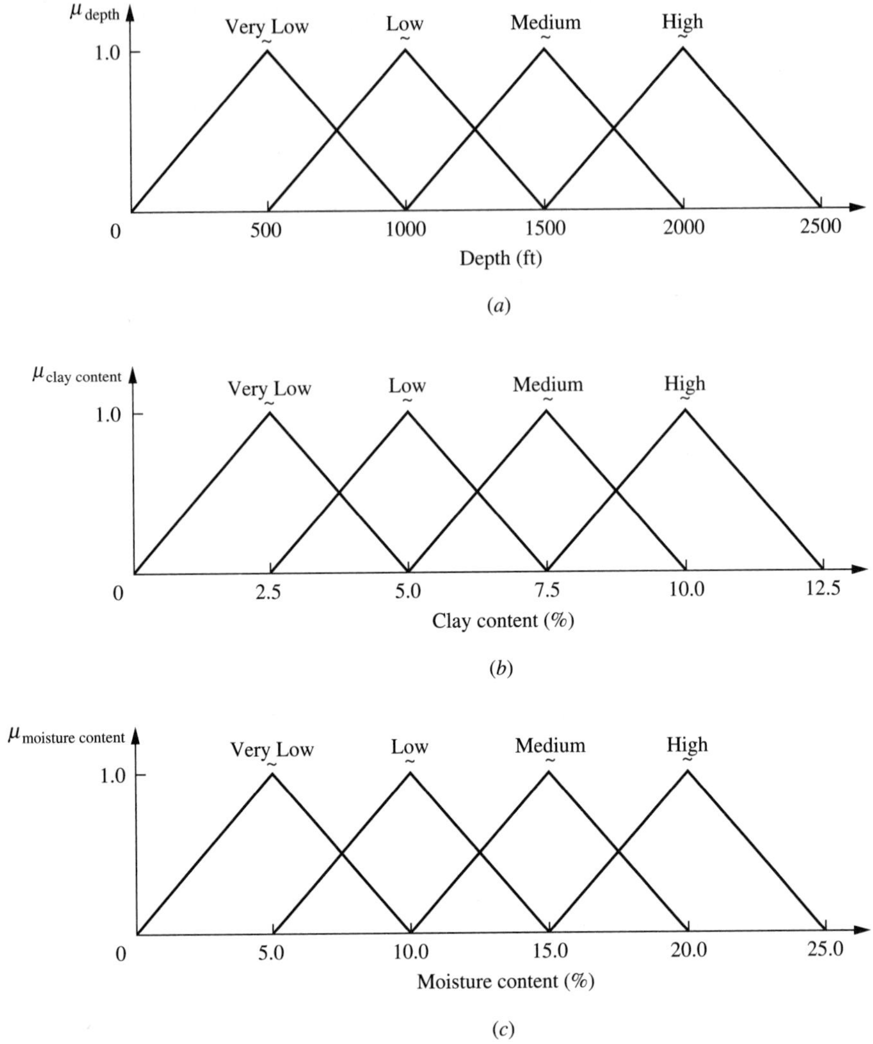

FIGURE P12.7

12.7. Transuranic waste will be stored at a southeastern New Mexico site known as WIPP. The site has underlying strata of rock salt, which is well-known for its healing and creeping properties. Healing is the tendency of a material to close fractures or other openings, and creep is the capacity of the material to deform under constant load. The radioactive wastes are stored in rooms excavated deep underground. Because of the creep of the ceiling, these rooms will eventually collapse, thus permanently sealing the wastes in place. The creep properties of salt depend on the depth, moisture content, and clay content of the salt at the location being considered. Rock salt from specified depths was studied through numerous tests conducted at various labs nationwide. These data comprise the known patterns. Hence, each pattern has three features. Now, the possibility of locating a room at a certain depth is being investigated.

We wish to determine the creep properties at that depth of salt with a certain clay and moisture content. Membership functions for each of the patterns are shown in Fig. P12.7. Find which known pattern the unknown pattern matches the best. The features for the unknown pattern are given by the crisp singletons

$$B = \{\text{depth} = 1750 \text{ ft}, \ \text{clay content} = 6.13\%, \ \text{moisture content} = 12.5\%\}$$

The weights given to the features are W = {0.5, 0.3, 0.2}.

12.8. Using the same known pattern as in Problem 12.7, and using fuzzy features for the new pattern, find which known pattern matches the new pattern most closely. Features for the new pattern are as follows:

$$\underset{\sim}{B}_{\text{depth}} = \left\{\frac{0}{1700} + \frac{0.5}{1725} + \frac{1}{1750} + \frac{0.5}{1775} + \frac{0}{1800}\right\}$$

$$\underset{\sim}{B}_{\text{clay content}} = \left\{\frac{0}{5.5} + \frac{0.5}{5.813} + \frac{1.0}{6.13} + \frac{0.5}{6.44} + \frac{0}{6.75}\right\}$$

$$\underset{\sim}{B}_{\text{moisture content}} = \left\{\frac{0}{11.0} + \frac{0.5}{11.75} + \frac{1}{12.5} + \frac{0.5}{13.25} + \frac{0}{14.0}\right\}$$

12.9. In the evaluation of high-definition film, different lots of film are classified using three features: peel moment, imaging quality (spot size), and environmental sturdiness

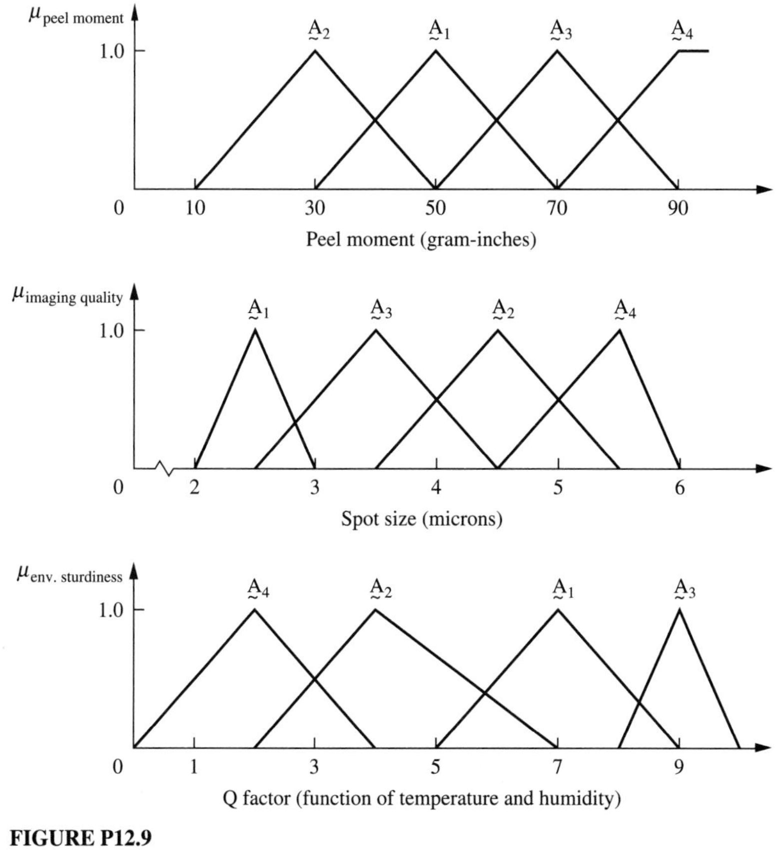

FIGURE P12.9

(Q factor). Each feature is classified into four classes, as shown in Fig. P12.9. The weights assigned to each feature are 0.2, 0.5, and 0.3, respectively. Suppose a new lot is defined by a crisp singleton B = {peel moment = 55, imaging quality = 3.75, environmental sturdiness = 5}. Determine the class to which this new lot belongs.

12.10. A film is clamped on a drum, as shown in Fig. P12.10*a*. The drum rotates at high speed, and a light beam exposes the film as it translates along the direction of the axis of the drum. An automatic focusing scheme keeps the focal length within a few microns. The discontinuity at the clamping areas limits the system performance and the maximum speed of rotation. The drum clamping performance (maximum film slope with calibration film), the edge stiffness of the film (measured with calibration drum), and the servo bandwidth determine the overall performance of the system.

The four standard focusing systems, $\underset{\sim}{A}_1, \ldots, \underset{\sim}{A}_4$, are shown in Figs. P12.10*b*–*d*. The weighting factors are these:

$$\begin{aligned} w_1 \text{ (clamping performance)} &= 0.3 \\ w_2 \text{ (film stiffness)} &= 0.3 \\ w_3 \text{ (servo bandwidth)} &= 0.4 \end{aligned}$$

Classify the process of an auto-focusing system given by the membership functions in Fig. P12.10*e*–*g*.

12.11. Suppose a new computer is introduced into the market and we are interested in determining which current computer most resembles the new one. We will use three features to make the assessment: MIPS, MFLOPS, and storage capacity. The new computer has the rating of 70, 80, 500 for the respective features. The four current computers have the ratings given by triangular fuzzy numbers, $\underset{\sim}{A}_i$, $i = 1, \ldots, 4$,

$$\begin{aligned} \underset{\sim}{A}_1 &= \{50, 80, 200\} \\ \underset{\sim}{A}_2 &= \{120, 160, 600\} \\ \underset{\sim}{A}_3 &= \{100, 70, 340\} \\ \underset{\sim}{A}_4 &= \{90, 120, 700\} \end{aligned}$$

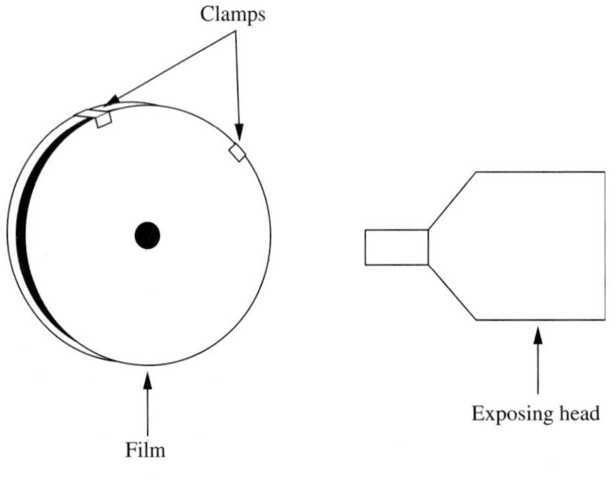

(*a*)

FIGURE P12.10*a*

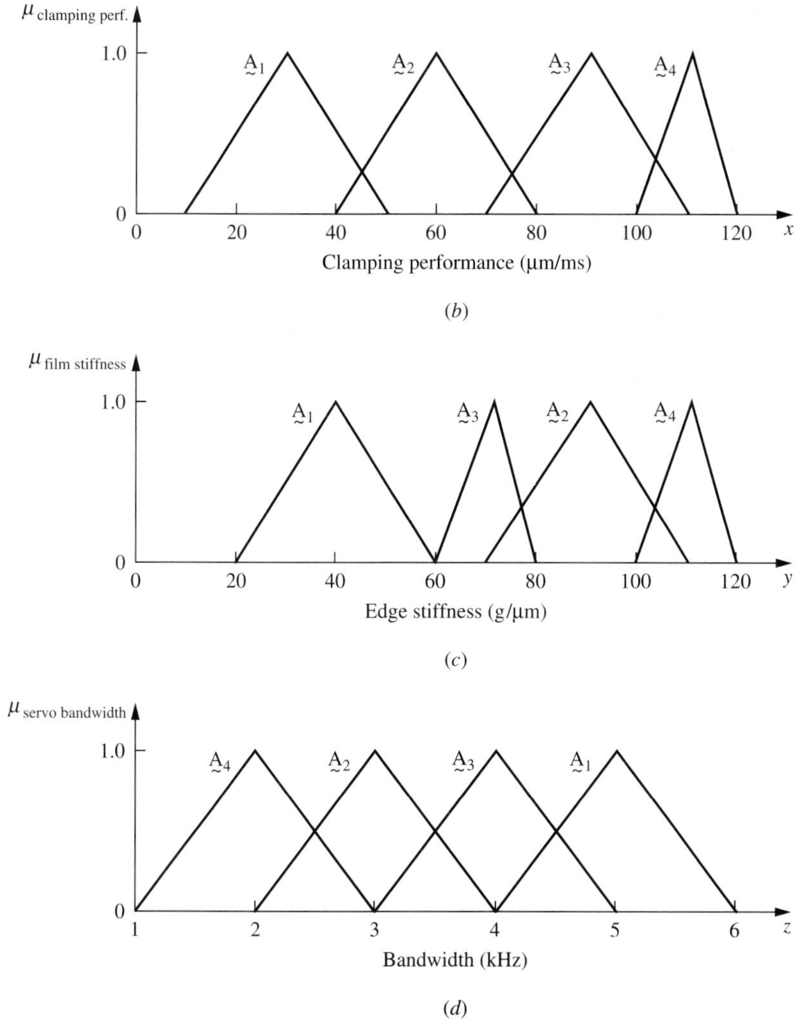

FIGURE P12.10*b–d*

The weights assigned to each of the features are {0.3, 0.4, 0.3}, respectively. Determine which of the old computers the new computer resembles the most.

12.12. A member of the police bomb squad has to be able to assess the type of bomb used in a terrorist activity in order to gain some knowledge that might lead to the capture of the culprit. The most commonly used explosive device is the pipe bomb. Pipe bombs can be made from a variety of explosives ranging from large-grain black powder and gunpowder to more sophisticated compounds, such as PETN or RDX. Identification of the explosive material used in the pipe bomb (after detonation) will tell a bomb squad investigator where the materials might have been purchased, the level of sophistication of the terrorist, and other important identifiers about the criminal.

Four basic types of energetic materials are used in making pipe bombs, each with its own distinctive pattern of post-mortem damage.

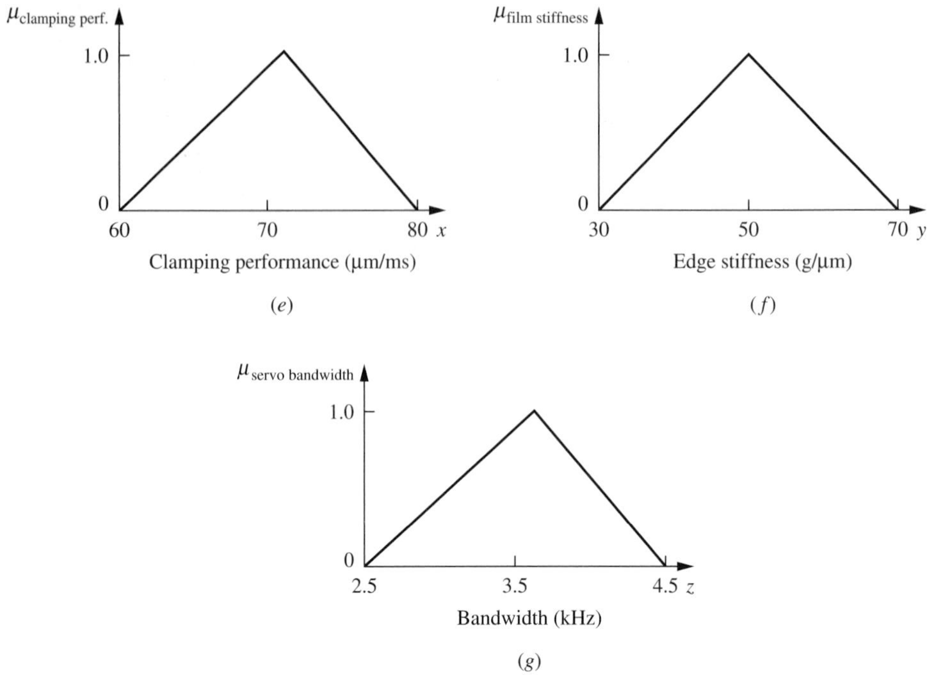

FIGURE P12.10*e–g*

1. *Explosives.* Those that detonate at a velocity equal to the compressional sound speed of the explosive material itself. These materials are by far the most energetic (also the most difficult to acquire) and are characterized (after explosion) by very small pipe fragments, highly discolored (usually bluish in tint) fragments, and extreme collateral damage (especially close to the detonation).
2. *Propellants.* Usually formed from some compound based on nitrocellulose. These materials do not detonate, but burn very rapidly. Usually propellants are formed in special geometric shapes that allow their surface area to remain constant or increase as they burn, thus causing the burning rate to increase until the compound has been completely exhausted. The destructive force of propellant-based pipe bombs is somewhat less than that of true explosives and is characterized by medium fragment size, little discoloration, and moderate collateral damage.
3. *Large-grain black powder.* Has been around since about 600 B.C., when the Chinese discovered the carbon-sulfur-potassium nitrate mixture. The size of black powder grains can vary tremendously, but the geometry is such that the powder always burns down (the burn rate always decreases once the entire surface of the mixture is burning). Although still very deadly, the damage from these types of pipe bombs is less than that of the other two. The residual fragment size is larger, and the discoloration of the fragments is slight.
4. *Gunpowder.* A subclass of black powder, usually considered to be homemade. It is characterized by very large fragments (one or two in number), almost no discoloration, and little collateral damage. Black powder is still very common among terrorists.

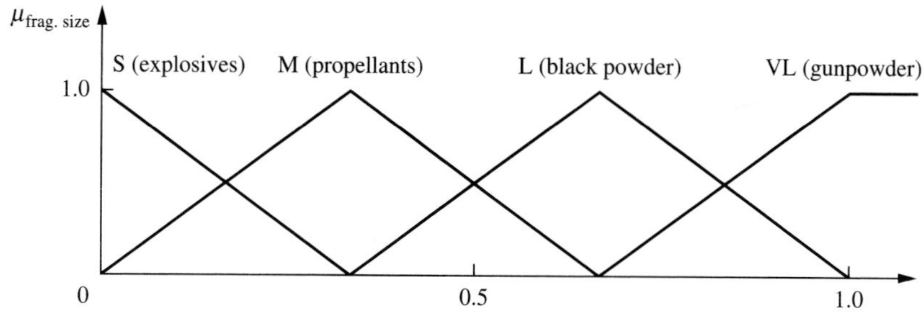

FIGURE P12.12

We can form a table of patterns for each feature:

	Features		
	Fragment size	**Fragment discoloration**	**Damage**
Explosives	Small (S)	High (H)	Extreme (E)
Propellants	Medium (M)	Medium (M)	Large (L)
Black powder	Large (L)	Low (L)	Medium (M)
Gunpowder	Very large (VL)	None (N)	Small (S)

Assume that the membership space for each feature can be partitioned similarly into four sections on a normalized abscissa, as shown in Fig. P12.12. The other two graphs (for discoloration and collateral damage) would look identical to the one in Fig. P12.12 (with different labels). The weights assigned to each feature are 0.5, 0.3, and 0.2, respectively. Now, say a new bombing has taken place and the aftereffects measured over three features are denoted as singletons, given as

$$\text{B} = \left\{ \text{fragment size} = \frac{1}{0.7}, \text{fragment color} = \frac{1}{0.6}, \text{damage} = \frac{1}{0.5} \right\}$$

Determine the composition of the bomb used in the bombing.

12.13. Using the information in Problem 12.12, but selecting a fuzzy input, perform a pattern recognition. The fuzzy input patterns in the form of triangular fuzzy numbers based on the three features are as follows:

$$\underset{\sim}{B}_{\text{frag. size}} = \left\{ \frac{0}{0.43} + \frac{1}{0.5} + \frac{0}{0.57} \right\}$$

$$\underset{\sim}{B}_{\text{frag. color}} = \left\{ \frac{0}{0.53} + \frac{1}{0.6} + \frac{0}{0.67} \right\}$$

$$\underset{\sim}{B}_{\text{damage}} = \left\{ \frac{0}{0.63} + \frac{1}{0.7} + \frac{0}{0.77} \right\}$$

12.14. We intend to recognize preliminary data coming off a satellite. Each of the five data packets has a unique packet header identifier, as follows:

$$\underset{\sim}{A}_1 = \text{satellite performance metrics}$$

$$\underset{\sim}{A}_2 = \text{ground positioning system}$$
$$\underset{\sim}{A}_3 = \text{IR sensor}$$
$$\underset{\sim}{A}_4 = \text{visible camera}$$
$$\underset{\sim}{A}_5 = \text{star mapper}$$

The three header values each set will look for are (1) signal type, (2) terminal number, and (3) data identifier. The weights assigned to each of the headers are 0.3, 0.3, and 0.4, respectively. Let us define the fuzzy pattern as

$$\underset{\sim}{A}_1 = \left\{\frac{0.2}{x_1} + \frac{0.2}{x_2} + \frac{0.6}{x_3}\right\}$$

$$\underset{\sim}{A}_2 = \left\{\frac{0.3}{x_1} + \frac{0.4}{x_2} + \frac{0.7}{x_3}\right\}$$

$$\underset{\sim}{A}_3 = \left\{\frac{0.4}{x_1} + \frac{0.6}{x_2} + \frac{0.8}{x_3}\right\}$$

$$\underset{\sim}{A}_4 = \left\{\frac{0.5}{x_1} + \frac{0.8}{x_2} + \frac{0.9}{x_3}\right\}$$

$$\underset{\sim}{A}_5 = \left\{\frac{0.6}{x_1} + \frac{1.0}{x_2} + \frac{1.0}{x_3}\right\}$$

A data stream given by the crisp singleton,

$$\text{B} = \left\{\frac{1.0}{x_1} + \frac{1.0}{x_2} + \frac{1.0}{x_3}\right\}$$

is received. Determine which of the five different packets we are receiving at the present time.

12.15. Signals are investigated from the following four digital signal-processing plants: $\underset{\sim}{A}_1$ = least mean squares, $\underset{\sim}{A}_2$ = root-mean square, $\underset{\sim}{A}_3$ = Newton's method, and $\underset{\sim}{A}_4$ = steepest descent method. The three ($m = 3$) important parameters that will be considered in each c-space are convergence rate, tracking, and stability. The weights assigned to each of the features are 0.4, 0.4, and 0.2, respectively. The data patterns corresponding to the features are

$$\underset{\sim}{A}_1 = \{0.2, 0.3, 0.8\}$$
$$\underset{\sim}{A}_2 = \{0.4, 0.4, 0.6\}$$
$$\underset{\sim}{A}_3 = \{0.6, 0.2, 0.4\}$$
$$\underset{\sim}{A}_4 = \{0.8, 0.5, 0.2\}$$

The sample data set has the following vector pattern:

$$\underset{\sim}{B} = \{0.5, 0.5, 0.5\}$$

Determine the pattern most closely represented by the sample data set.

12.16. Lube oils are classified by three features: color, viscosity, and flash point. Depending on the values of these features, the lube oil is classified as 100 neutral (100N), 150 neutral (150N), heavy solvent neutral (HSN), and 500 neutral (500N). Among the features, color is the most important, followed by viscosity, then flash point. The reason for this ordering is that it is easier to blend lube oils to obtain correct viscosity and flash point than it is to blend to obtain proper color. Any material not falling into

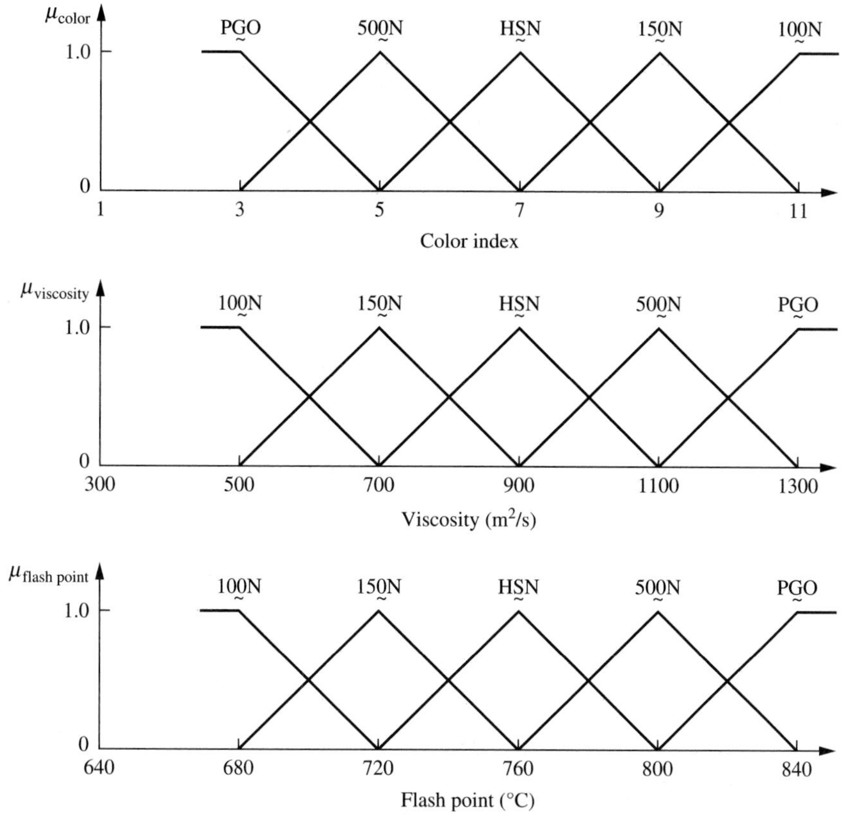

FIGURE P12.16

one of these lube oil categories is downgraded to catalyst cracker feed (PGO), where it is converted to gasoline.

Fuzzy patterns for each of these features is shown in Fig. P12.16. The weights for these features are 0.5 for color, 0.3 for viscosity, and 0.2 for flash point. You received a lab analysis for a sample described by the crisp singleton

$$B = \{\text{color} = 6.5, \text{viscosity} = 825\ \text{m}^2/\text{s}, \text{ flash point} = 750°\text{C}\}$$

Under what category do you classify this sample?

12.17. We wish to determine whether a printed circuit board is built with the correct components. It is very difficult to probe the part and actually measure its value. So, instead we will optically scan the board for various parts to measure the x and y dimensions and to determine the color of the part. We determine the following set of patterns:

	x	y	**Color**
Small resistor	30	50	White (1)
Medium resistor	50	80	White (1)
Capacitor	50	80	Green (0.6)
IC	100	400	Black (0)

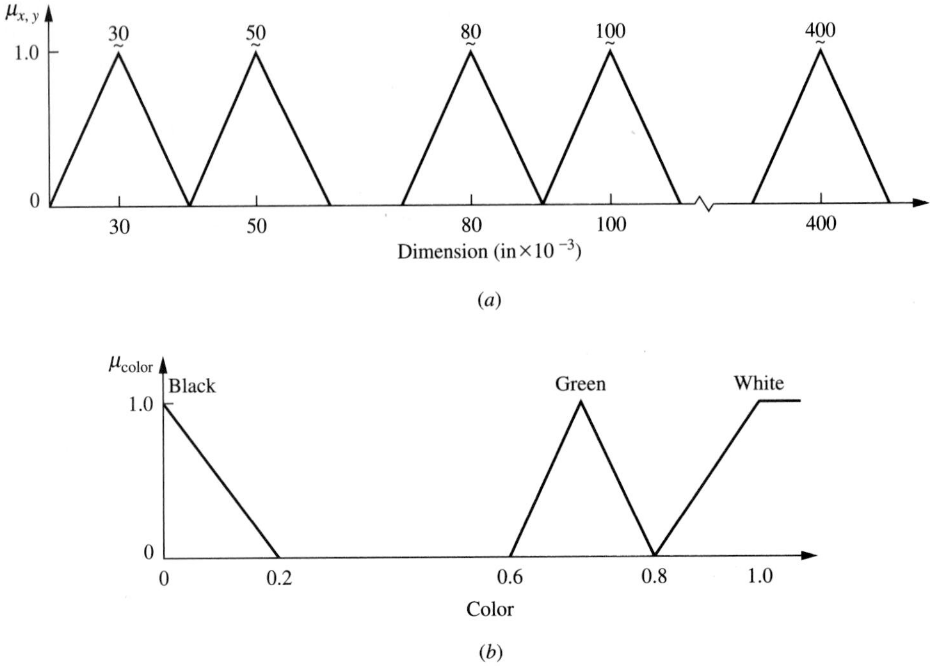

FIGURE P12.17

The weights to be used are $x = 0.3$, $y = 0.3$, color $= 0.4$. The fuzzy set patterns for the x and y features are the same and, along with the color patterns, are shown in Fig. P12.17.

Now, a PCB is scanned to identify what type of component is placed in each position (for the purpose of rejecting the board if a wrong part is placed). The following information for a component is obtained: $x = 32$, $y = 48$, and color $= 0.9$ (gray due to light shadow from a nearby part). Determine which part in the table this unknown component resembles the most.

12.18. Using the same patterns as in Problem 12.17, perform a pattern recognition for fuzzy input pattern as shown in Fig. P12.18.

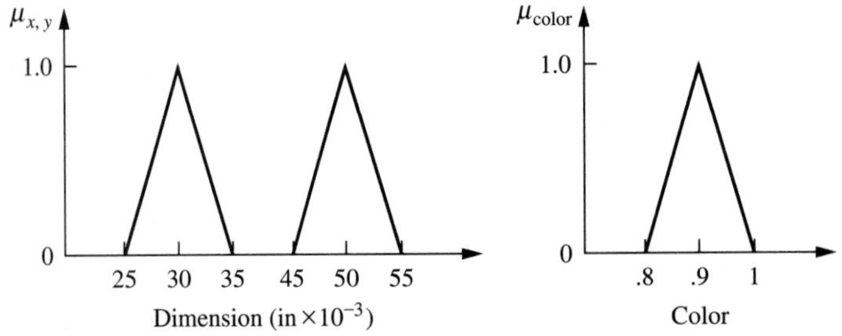

FIGURE P12.18

12.19. A satellite tracking database has the following information for each of the four satellite systems:

	Features		
Satellite	**Signal/noise ratio**	**Wavelength (dm)**	**Range (km)**
LANDSAT	20	10	800
SEASAT	40	13	400
COSMOS	50	15	500
ASTEX	10	10	1000

The fuzzy patterns over the three features obtained by normalizing the values in the table are

$$\begin{aligned} \underset{\sim}{A}_1 &= \{0.17, 0.21, 0.3\} \\ \underset{\sim}{A}_2 &= \{0.33, 0.27, 0.15\} \\ \underset{\sim}{A}_3 &= \{0.42, 0.56, 0.19\} \\ \underset{\sim}{A}_4 &= \{0.08, 0.21, 0.37\} \end{aligned}$$

with weights $w_1 = 0.4$, $w_2 = 0.4$, $w_3 = 0.2$. A new data set, $B = \{1, 0, 0\}$, is provided. To which satellite does this data set correspond?

12.20. Over several years a satellite tracking facility has classified several objects on the universe of signal to noise ratio (SNR), total signal (TS), and radius (R). The fuzzy sets are shown here for four satellites:

$$\text{ASTEX} \quad \underset{\sim}{A}_1 = \left\{ \frac{0.1}{\text{SNR}} + \frac{0.15}{\text{TS}} + \frac{0.2}{\text{R}} \right\}$$

$$\text{DMSP} \quad \underset{\sim}{A}_2 = \left\{ \frac{0.2}{\text{SNR}} + \frac{0.2}{\text{TS}} + \frac{0.3}{\text{R}} \right\}$$

$$\text{SEASAT} \quad \underset{\sim}{A}_3 = \left\{ \frac{0.5}{\text{SNR}} + \frac{0.7}{\text{TS}} + \frac{0.5}{\text{R}} \right\}$$

$$\text{MIR} \quad \underset{\sim}{A}_4 = \left\{ \frac{0.9}{\text{SNR}} + \frac{0.9}{\text{TS}} + \frac{0.9}{\text{R}} \right\}$$

with weights identical to those in Problem 12.19. One night an unknown object is tracked, and the following observation is made:

$$\underset{\sim}{B} = \left\{ \frac{0.3}{\text{SNR}} + \frac{0.3}{\text{TS}} + \frac{0.3}{\text{R}} \right\}$$

Which satellite does the object most closely resemble?

12.21. In evaluating a copier market one needs to divide the machines into classes. Four classes will be used: low volume, medium volume, high volume, and very high volume. The features used to classify the machines are copy rate (copies/min), reliability (copies/service), and finishing features (subjective evaluation). Only one feature of the fuzzy set is given in Fig. P12.21, as the other two are identical to this set.

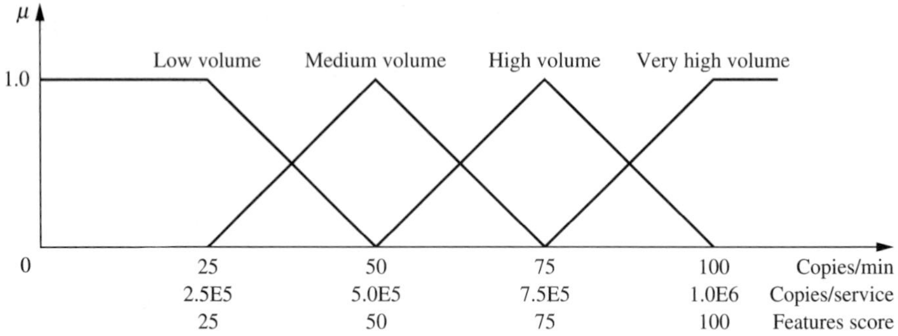

FIGURE P12.21

A new machine introduced into the market has the following rating: copy rate = 88 copies/min, reliability = 5×10^5 copies/service, and finishing features score = 75. Using weights $w_1 = 0.5$, $w_2 = 0.3$, and $w_3 = 0.2$, classify this machine based on the foregoing rating.

12.22. Use the same patterns as in Problem 12.21. But instead of classifying the machine based on crisp singletons, classify using fuzzy ratings as shown in Fig. P12.22.

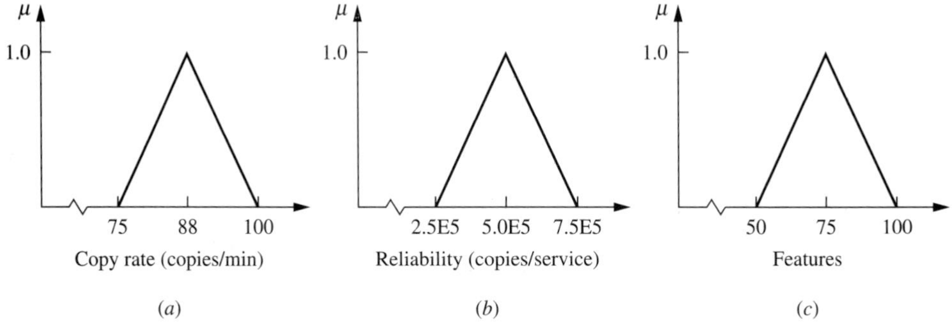

FIGURE P12.22

12.23. A set of patterns indicating the performance of an aluminum smelting cell is developed. The important features are bath temperature T (°C), cell voltage V, and noise N (standard deviation of the cell resistance). The cell conditions (patterns) are described as follows:

- $\underset{\sim}{A}_1$ Cell has a very small anode-cathode distance. Characterized by low temperature, low voltage, and high noise.
- $\underset{\sim}{A}_2$ Cell is in good condition. Characterized by moderately low temperature, moderately low voltage, and low noise.
- $\underset{\sim}{A}_3$ Cell has a very large anode-cathode distance. Characterized by high temperature, high voltage, and low noise.
- $\underset{\sim}{A}_4$ Cell has deposits on bottom cathode. Characterized by moderately high temperature, high voltage, and high noise.

The fuzzy sets are represented by Gaussian membership functions:

$$\underset{\sim}{A}_1 = \left\{\exp\left[-\frac{(T-945)^2}{4^2}\right], \exp\left[-\frac{(V-4.2)^2}{(0.1)^2}\right], \exp\left[-\frac{(N-26)^2}{5^2}\right]\right\}$$

$$\underset{\sim}{A}_2 = \left\{\exp\left[-\frac{(T-950)^2}{4^2}\right], \exp\left[-\frac{(V-4.4)^2}{(0.1)^2}\right], \exp\left[-\frac{(N-6)^2}{2^2}\right]\right\}$$

$$\underset{\sim}{A}_3 = \left\{\exp\left[-\frac{(T-970)^2}{8^2}\right], \exp\left[-\frac{(V-4.8)^2}{(0.1)^2}\right], \exp\left[-\frac{(N-6)^2}{2^2}\right]\right\}$$

$$\underset{\sim}{A}_4 = \left\{\exp\left[-\frac{(T-965)^2}{6^2}\right], \exp\left[-\frac{(V-4.7)^2}{(0.1)^2}\right], \exp\left[-\frac{(N-20)^2}{5^2}\right]\right\}$$

To reflect the relative importance of the features, select $w_1 = 0.3$, $w_2 = 0.5$, and $w_3 = 0.2$. Now a new data sample (measurements from a smelting cell) yields temperature = 953°C, voltage = 4.5 V, and noise = 12 (a data singleton). Classify the operating conditions of the cell.

12.24. Use the same patterns as in Problem 12.23. But, now use a sample comprising fuzzy sets. This is appropriate because measurements such as temperature are subject to substantial error, and electrical signals fluctuate over time as disturbances affect the system. The new sample is represented as the following:

$$\underset{\sim}{B} = \left\{\exp\left[-\frac{(T-957)^2}{3^2}\right], \exp\left[-\frac{(V-4.6)^2}{(0.2)^2}\right], \exp\left[-\frac{(N-16)^2}{3^2}\right]\right\}$$

Classify the operating conditions of the cell based on this information.

12.25. In producing blanks from which gears are made, a machining process first slices a piece off the rod stock, and then drills a hole in the center. The part is then measured for quality control. The critical measurements (features) are d_i = inside diameter, d_0 = outside diameter, and t = thickness (see Fig. P12.25*a*). The "desired" measurements are termed $d_{i\mathrm{D}}$, $d_{o\mathrm{D}}$, and t_{D}. The part will be compared against four standard sets of measurements given in Table P12.25. A desired part has these measurements:

$$d_i = 0.5 \pm 0.001 \text{ inches}$$
$$d_0 = 2.0 \pm 0.003 \text{ inches}$$
$$t = 0.25 \pm 0.004 \text{ inches}$$

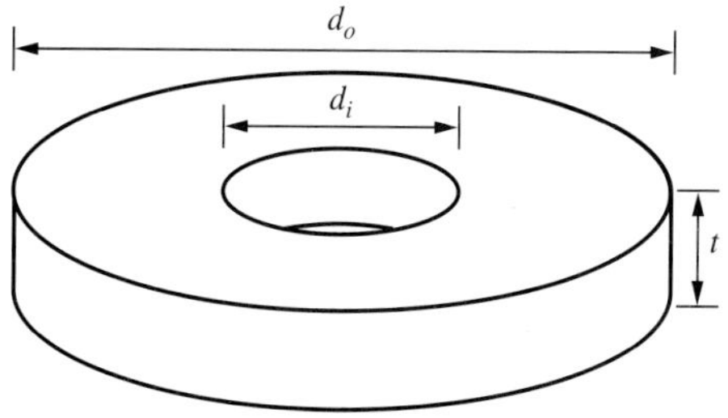

(*a*)

FIGURE P12.25*a*

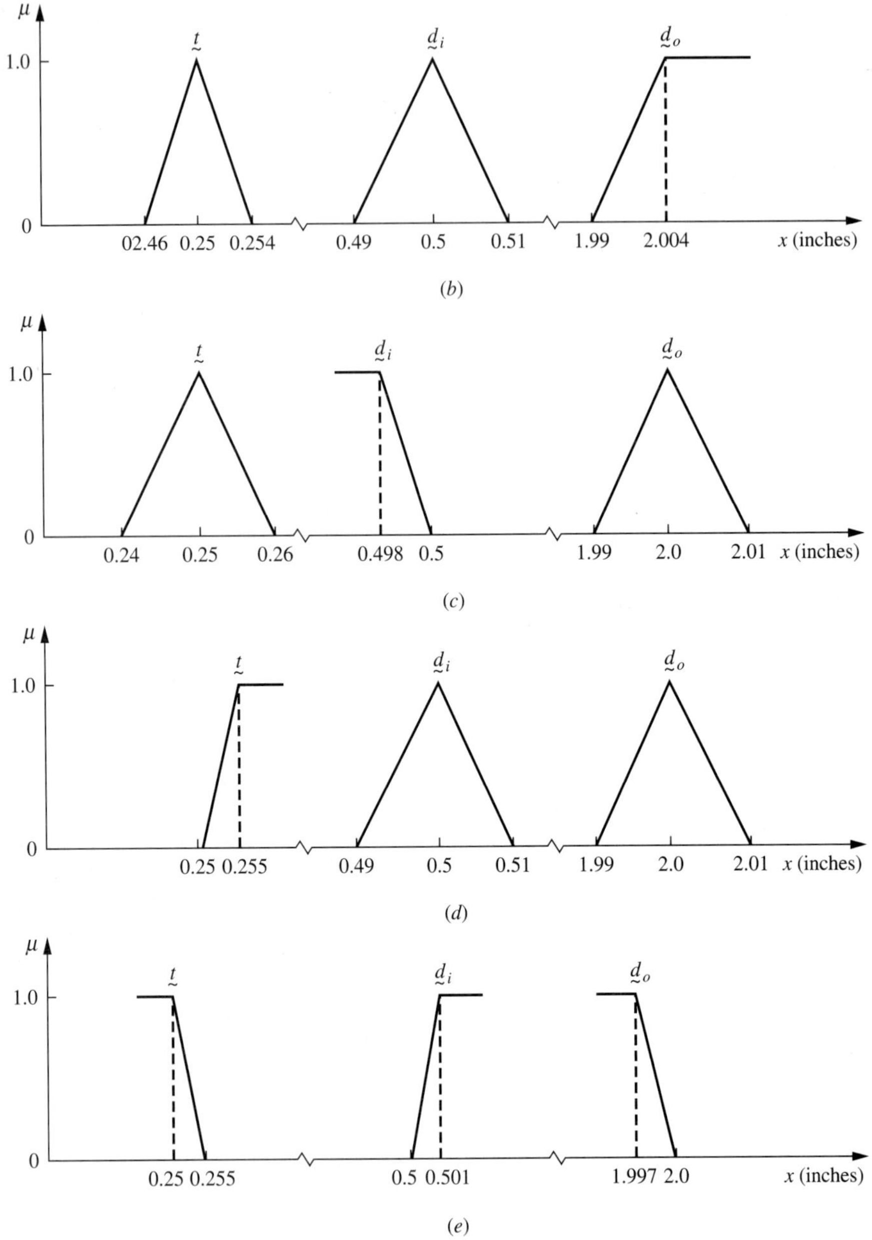

FIGURE P12.25*b–e*

The weighting factors assigned to each of the features are $w_1(d_i) = 0.38$, $w_2(d_o) = 0.34$, and $w_3(t) = 0.28$. A part that was just made has these measurements: $d_i = 0.507$ in., $d_o = 1.998$ in., and $t = 0.251$ in. To which set does this part belong?

12.26. Metal components of airplanes develop cracks, sometimes early in their design life, but usually later. Inspections are planned to look for cracks in the locations where

TABLE P12.25

Set description	Set measurements
Set 1: "d_o too big" (Fig. 12.25*b*)	$d_o > d_{oD}$ $t \approx t_D$ $d_i \approx d_{iD}$ This part can be milled, so its d_o will equal d_{oD}. Action: Send this part to the milling station.
Set 2: "d_i too small"(Fig. 12.25*c*)	$d_i \leq d_{iD}$ $d_o \approx d_{oD}$ $t \approx t_D$ This part can be redrilled to the correct size. Action: Send this part to the drilling station.
Set 3: "t too big"(Fig. 12.25*d*)	$t < t_D$ $d_o \approx d_{oD}$ $d_i \approx d_{iD}$ This part can be ground thinner. Action: Send this part to the grinding station.
Set 4: "Reject"(Fig. 12.25*e*)	$t < t_D$ *or* $d_o < d_{oD}$ *or* $d_i > d_{iD}$ This part cannot be reworked. Action: Send this part to the scrap bin.

they are known to develop. But cracks can develop where they were not anticipated. They can grow and result in the failure of the component or the whole aircraft. We cannot measure the rate at which the crack is growing with a single inspection. We can only record its length, location, and orientation with respect to the principal stresses. The three measures considered are length, location, and orientation. The length is the crack length in millimeters. The location is a subjective measure of criticality to flight safety on a scale from 1 to 10. Orientation is in degrees; 0° is perpendicular to the principal stresses, 90° is in line with it. Gaussian membership functions are assumed for all three measures. The Gaussian membership function is given as

$$\mu_{\underset{\sim}{A}_{ij}}(x) = \exp\left[-\left(\frac{x_j - a_{ij}}{\sigma^2_{a_{ij}}}\right)^2\right]$$

Four fuzzy patterns based on the criticality of the crack are defined as follows:

Level 1: This crack is benign. No concern to flight safety, check it again at one design life—if life extension is needed.

Level 2: Keep an eye on it. Inspect again at the next scheduled depot visit.

Level 3: Caution! Repair crack and develop a repair kit to repair all other similar aircraft in due time.

Level 4: Danger! Replace the cracked component immediately and down all similar aircraft in the fleet until they can be inspected and repaired.

The Gaussian parameters for each feature of the preceding fuzzy patterns are listed in the following table.

	Feature					
	Length, mm		Location		Orientation, deg.	
	a_i	σ	a_i	σ	a_i	σ
Level 1	1	0.3	2	1	45	10
Level 2	10	2	5	1	20	10
Level 3	15	3	7	1	10	5
Level 4	25	4	10	1	0	5

Weights assigned to the features length, location, and orientation are 0.5, 0.3, and 0.2, respectively. An airplane wing root was inspected and a 12 mm crack was found at right angles to the principal far-field stress. The singleton was recorded as B = {length = 12 mm, location = 10 (maximum effect on the safety of flight), orientation = 0°}. On the basis of this information determine what level of action has to be taken.

12.27. After the crack in the wing root of the aircraft in Problem 12.26 was discovered, a thorough inspection of the aircraft was conducted. This inspection involved several locations on the aircraft. The following statistics summarize the findings for a sample, $\underset{\sim}{B}$.

	Feature					
	Length, mm		Location		Orientation, deg.	
	a_i	σ	a_i	σ	a_i	σ
$\underset{\sim}{B}$	8	4	7	3	0	20

According to the fuzzy information given in the table determine the level of action that has to be taken.

12.28. Skis are classified on the basis of three features: weight, performance stiffness, and response times in turns. There are four different types of skis: freestyle, giant slalom (GS), slalom, and all-around. These skis have Gaussian distributions on each of the features and the parameters for the distribution are given in following table. A Gaussian distribution has the form

$$\mu_{\underset{\sim}{A}_{ij}}(x) = \exp\left[-\left(\frac{x_j - a_{ij}}{\sigma^2_{a_{ij}}}\right)^2\right]$$

The fuzzy patterns are defined on a normalized scale as follows:

	Feature					
	Weight		Stiffness		Response time	
	a_i	σ	a_i	σ	a_i	σ
All-around	50	10	40	12	60	7
Slalom	40	3	90	5	75	10
Giant slalom	30	15	80	10	60	10
Freestyle	40	10	20	6	70	3

The weights given to the features are

$$w_{\text{cost}} = 0.3$$
$$w_{\text{stiffness}} = 0.4$$
$$w_{\text{response}} = 0.3$$

A new ski whose features are given on a normalized scale by a crisp singleton,

$$\text{B} = \{\text{weight} = 45, \text{stiffness} = 60, \text{response time} = 65\}$$

is introduced into the market. Determine what type of ski the new ski should be labeled.

12.29. In Problem 12.28 the new ski introduced into the market was given by a crisp singleton. However, given the uncertainty in measurements, it is more appropriate to define a ski by fuzzy parameters. For the same problem, and with the same weights assigned to each of the features, classify the new ski if it is given by a fuzzy set whose membership functions are given by a Gaussian distribution whose parameters (mean and standard deviation) are given in the following table:

	Feature					
	Weight		Stiffness		Response time	
	a_i	σ	a_i	σ	a_i	σ
$\underset{\sim}{\text{B}}$	45	10	60	12	65	20

12.30. A shipping company that runs fast container ships wishes to equip its fleet of freighters with modern computers. These computers will monitor engineering equipment, compile statistics, and control critical equipment. These computers must perform sophisticated engineering computations at a high speed. They should also be able to withstand the maritime environment without breakdowns and must be designed from the ground up for industrial use. Finally, the company desires that the computers be able to perform parallel processing. The reason is that for critical tasks the computer will perform the analysis five times and then check to ensure that all answers are the same (fault-tolerant computing). In order to support a high degree of parallelism the computer must have a great amount of memory (RAM), a large hard drive, and multiple processors. Furthermore, the company wants to be sure that C++ software for parallel processing is available for the computer.

The engineering department decided to develop four pattern templates with three features. The patterns are (1) outstanding, (2) excellent, (3) good, and (4) fair. Each pattern is defined by three features: (1) degree of parallelism, (2) durability, and (3) speed in FLOPS (floating operations per second). The chief engineer established a normalized scale from 0 to 6 for all these features. The weights assigned to the features are 0.4, 0.3, and 0.3, respectively. The membership functions for the patterns are shown in Fig. P12.30 for only one of the features. The membership functions for the other two features are identical to the first feature. Let

$$\underset{\sim}{B} = \{\underset{\sim}{B}_P, \underset{\sim}{B}_D, \underset{\sim}{B}_S\},$$

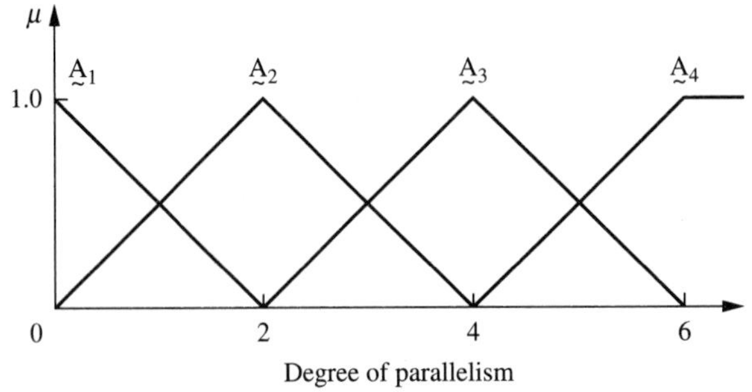

FIGURE P12.30

where $\underset{\sim}{B}_P$ = degree of parallelism
$\underset{\sim}{B}_D$ = durability
$\underset{\sim}{B}_S$ = speed

Determine the status of a new computer whose features are given by the crisp singleton

$$B = \{6.0, 3.4, 2.0\}$$

12.31. Using the same patterns as given in Problem 12.30, classify a new computer that is described by a fuzzy set as follows:

$$\underset{\sim}{B}_P = \left\{ \frac{0}{4.5} + \frac{0.6}{4.7} + \frac{0.8}{5.0} + \frac{1.0}{6.0} \right\}$$

$$\underset{\sim}{B}_D = \left\{ \frac{0}{1.8} + \frac{0.5}{2.2} + \frac{0.8}{2.8} + \frac{1.0}{3.6} + \frac{0.85}{4.0} + \frac{0.8}{4.2} + \frac{0.5}{4.5} + \frac{0}{5.2} \right\}$$

$$\underset{\sim}{B}_S = \left\{ \frac{0}{1.0} + \frac{0.5}{1.4} + \frac{1.0}{2.0} + \frac{0.6}{3.6} + \frac{0.5}{4.0} + \frac{0}{6.0} \right\}$$

12.32. Consider a population that consists of four groups. They will be described as Group I, Group II, Group III, and Group IV. The four groups are representative of four set patterns. The three features associated with each member of each group are (1) height (feet), (2) age (years), and (3) income (dollars). The membership functions for each pattern i are defined by a Gaussian distribution over each feature j. The Gaussian distribution is given by

$$\mu_{\underset{\sim}{A}_{ij}}(x) = \exp\left[-\left(\frac{x_j - a_{ij}}{\sigma^2_{a_{ij}}} \right)^2 \right]$$

where a_{ij} = mean for the ith pattern and jth feature, and σ is the standard deviation for the same. For the pattern recognition consider the following realizations of the parameter sets:

	Feature					
	Height, ft		Age, yr		Income, $K	
	a_i	σ	a_i	σ	a_i	σ
Group I	6.1	0.5	25	3	17	2
Group II	5.9	0.6	31	11	34	6
Group III	5.7	0.65	55	5	39	7
Group IV	5.8	0.8	37	8	41	7

Let the weighting factors be $w_i = \frac{1}{3}$ for $i = 1, 2, 3$. Let the new data sample be an individual given by a set of crisp singletons, B = {height = 6.2 ft, age = 34 yr, income = $45K}. Determine to which group this individual belongs.

12.33. For Problem 12.32 classify an individual whose features are given by a collection of fuzzy sets, $\underset{\sim}{B} = \{\underset{\sim}{B}_1, \underset{\sim}{B}_2, \underset{\sim}{B}_3\}$, where we let $\underset{\sim}{B}_{ij}$ have the feature membership function

$$\mu_{\underset{\sim}{B}_{ij}}(x) = \exp\left[-\left(\frac{x_j - b_{ij}}{\sigma^2_{b_{ij}}}\right)^2\right]$$

The function $\mu_{\underset{\sim}{B}_{ij}}(x)$ is characterized as follows:

	Feature					
	Height, ft		Age, yr		Income, $K	
	b_j	σ	b_j	σ	b_j	σ
$\underset{\sim}{B}$	6.2	0.5	34	5	45	8

Exercises for Syntactic Pattern Recognition

12.34. Generate a fuzzy grammar for the syntactic pattern recognition of an isosceles trapezoid.

12.35. Generate a fuzzy grammar for the syntactic pattern recognition of an equilateral triangle.

12.36. Continue Example 12.13 by developing fuzzy grammars for the pattern recognition of the two electric sources, symbols ⊖ and ⏐⎮.

Exercises for Image Processing

12.37. The accompanying table shows the intensity values (for an 8-bit image) associated with an array of 25 pixels. Use the image enhancement algorithm on these intensity values to enhance the image. Do you recognize the pattern in the image?

111	105	140	107	110
110	132	111	120	105
140	105	105	115	154
137	135	145	150	145
140	118	115	109	148

12.38. The following table shows the intensity values (for an 8-bit image) associated with an array of 25 pixels. Use the image softening algorithm on these intensity values to remove the "salt and pepper" noise (shown as shaded pixels) from the image of the alphabetic character M.

220	30	10	15	250
205	230	**0**	239	230
225	20	225	20	220
217	**256**	30	10	215
220	25	15	**256**	235

CHAPTER 13

FUZZY CONTROL SYSTEMS

The difference between science and the fuzzy subjects is that science requires reasoning, while those other subjects merely require scholarship.

Robert Heinlein
Time Enough for Love, *1973*

A control system is an arrangement of physical components designed to alter, to regulate, or to command, through a *control action,* another physical system so that it exhibits certain desired characteristics or behavior. Control systems are typically of two types: open-loop control systems, in which the control action is independent of the physical system output, and closed-loop control systems (also known as *feedback control systems*), in which the control action depends on the physical system output. Examples of open-loop control systems are a toaster, in which the amount of heat is set by a human, and an automatic washing machine, in which the controls for water temperature, spin-cycle time, and so on are preset by the human. In both these cases the control actions are not a function of the output of the toaster or the washing machine. Examples of feedback control are a room temperature thermostat, which senses room temperature and activates a heating or cooling unit when a certain threshold temperature is reached, and an autopilot mechanism, which makes automatic course corrections to an airplane when heading or altitude deviations from certain preset values are sensed by the instruments in the plane's cockpit.

In order to control any physical variable, we must first measure it. The system for measurement of the *controlled signal* is called a *sensor.* The physical system under control is called a *plant.* In a closed-loop control system, certain forcing signals of the system (called *inputs*) are determined by the responses of the system (called *outputs*). To obtain satisfactory responses and characteristics for the closed-loop control system, it is necessary to connect an additional system, known as a *compensator*, or a *controller*, into the loop. The general form of a closed-loop control system is illustrated in Fig. 13.1 [Phillips and Harbor, 1988].

Control systems are sometimes divided into two classes. If the object of the control system is to maintain a physical variable at some constant value in the presence of disturbances, the system is called a *regulatory* type of control, or a regulator. The room temperature control and autopilot are examples of regulatory controllers. The second class of control systems are *tracking* controllers. In this scheme of control, a physical variable is required to follow or track some desired time function. An example of this type of system is an automatic aircraft landing system, in which the aircraft follows a "ramp" to the desired touchdown point.

The control problem is stated as follows [Phillips and Harbor, 1988]. The output, or response, of the physical system under control (i.e., the plant) is adjusted as required by the *error signal*. The error signal is the difference between the actual response of the plant, as measured by the sensor system, and the desired response, as specified by a *reference input*. In the following section we derive different forms of common mathematical models describing a closed-loop control system.

REVIEW OF CONTROL SYSTEMS THEORY

A mathematical model that describes a wide variety of physical systems is an nth-order ordinary differential equation of the type [Vidyasagar, 1978]

$$\frac{d^n y^{(t)}}{dt^n} = w\left[t, y(t), \dot{y}(t), \ldots, \frac{d^{n-1} y^{(t)}}{dt^{n-1}}, u(t)\right] \tag{13.1}$$

where t is the time parameter, $u(\ )$ is the input function, $w[\ ]$ is a general nonlinear function, and $y(\ )$ is the system output or response function. If we define the auxiliary functions

$$\begin{aligned} x_1(t) &= y(t) \\ x_2(t) &= \dot{y}(t) \\ &\vdots \\ x_n(t) &= \frac{d^{n-1} y(t)}{dt^{n-1}} \end{aligned} \tag{13.2}$$

then the single nth-order equation (13.1) can be equivalently expressed as a system of n first-order equations:

$$\begin{aligned} \dot{x}_1(t) &= x_2(t) \\ \dot{x}_2(t) &= x_3(t) \\ &\vdots \\ \dot{x}_{n-1}(t) &= x_n(t) \\ \dot{x}_n(t) &= w[t, x, (t), x_2(t), \ldots, x_n(t), u(t)] \end{aligned} \tag{13.3}$$

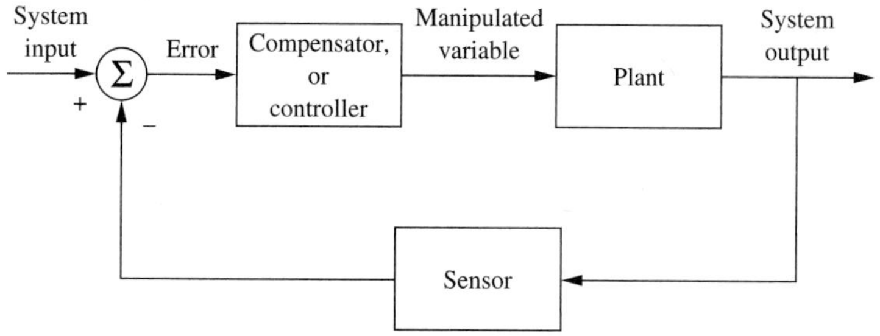

FIGURE 13.1
A closed-loop control system.

Finally, if we define n-vector-valued functions $\mathbf{x}(\,)$ and $\mathbf{f}(\,)$ by

$$\mathbf{x}(t) = [x_1(t), x_2(t), \ldots, x_n(t)]^{\mathrm{T}} \tag{13.4}$$

$$\mathbf{f}(t, \mathbf{x}, u) = [x_2, x_3, \ldots, x_n, w(t, x_1, \ldots, x_n, u)]^{\mathrm{T}} \tag{13.5}$$

where $\mathbf{x}(t)$ is the system state vector at time t (T is standard vector transpose), then the n first-order equations (13.3) can be combined into a first-order vector differential equation:

$$\dot{\mathbf{x}}(t) = \mathbf{f}[t, \mathbf{x}(t), u(t)] \tag{13.6}$$

and the output $y(t)$ is given from Eq. (13.2) as

$$y(t) = [1, 0, \ldots, 0]\mathbf{x}(t) \tag{13.7}$$

Similarly, a system with p inputs, m outputs, and n states will be described, in general, using the vector-valued functions $\mathbf{f}[\,]$ and $\mathbf{g}[\,]$ as

$$\dot{\mathbf{x}}(t) = \mathbf{f}[t, \mathbf{x}(t), \mathbf{u}(t)] \tag{13.8}$$

$$\mathbf{y}(t) = \mathbf{g}[t, \mathbf{x}(t), \mathbf{u}(t)] \tag{13.9}$$

where $\mathbf{u}(t)$ and $\mathbf{y}(t)$ vectors are defined as

$$\mathbf{u}(t) = [u_1(t), u_2(t), \ldots, u_p(t)]^{\mathrm{T}} \tag{13.10}$$

$$\mathbf{y}(t) = [y_1(t), y_2(t), \ldots, y_m(t)]^{\mathrm{T}} \tag{13.11}$$

and are an input vector and an output vector, respectively. The state variables vector $\mathbf{x}(t)$ is defined by Eq. (13.4). Physical systems descriptions based on Eqs. (13.8) and (13.9) are known as state-space representations and $x_1, x_2, \ldots, x_n$ are known as state variables of the system. In the case of nonlinear time-invariant continuous-time systems, Eqs. (13.8) and (13.9) become

$$\dot{\mathbf{x}}(t) = \mathbf{f}[\mathbf{x}(t), \mathbf{u}(t)] \tag{13.12}$$

$$\mathbf{y}(t) = \mathbf{g}[\mathbf{x}(t), \mathbf{u}(t)] \tag{13.13}$$

and for a linear time-invariant system will reduce to the following form:

$$\dot{\mathbf{x}}(t) = \mathrm{A} \cdot \mathbf{x}(t) + \mathrm{B} \cdot \mathbf{u}(t) \tag{13.14}$$

$$\mathbf{y}(t) = \mathrm{C} \cdot \mathbf{x}(t) + \mathrm{D} \cdot \mathbf{u}(t) \tag{13.15}$$

where constants A, B, C, and D are known as system matrices.

A first-order single-input–single-output nonlinear system is described using a discrete-time equation as

$$x_{k+1} = f(x_k, u_k) \tag{13.16}$$

where x_{k+1}, x_k are the values of state at the kth and $(k+1)$th time moments, and u_k is the input at the kth moment. An nth-order single-input–single-output system can be put in the following form:

$$y_{k+n} = f(y_k, y_{k+1}, \ldots, y_{k+n-1}, u_k) \tag{13.17}$$

and for an nth-order, multiple-input–single-output discrete system:

$$y(k+n) = f[y(k), y(k+1), \ldots, y(k+n-1), u_1(k), u_2(k), \ldots, u_p(k)] \tag{13.18}$$

Equations (13.14), (13.15), and (13.17) are illustrated in the inverted pendulum problem, which will be discussed in Example 13.2.

System Identification Problem

The general problem of identifying a physical system based on the measurements of the input, output, and state variables is defined as obtaining functions **f** and **g** in the case of a nonlinear system, and system matrices A, B, C, and D in the case of a linear system. There exist algorithms that adaptively converge to these system parameters based on numerical data taken from input and output variables [Ljung and Soderstrom, 1983]. Fuzzy systems and artificial neural network paradigms are two evolving disciplines for nonlinear system identification problems.

Control System Design Problem

The general problem of feedback control system design is defined as obtaining a generally nonlinear vector-valued function **h**( ), defined as follows [Vadiee, 1993]:

$$\mathbf{u}(t) = \mathbf{h}[t, \mathbf{x}(t), \mathbf{r}(t)] \tag{13.19}$$

where $\mathbf{u}(t)$ is the input to the plant or process, $\mathbf{r}(t)$ is the reference input, and $\mathbf{x}(t)$ is the state vector. The feedback control law **h** is supposed to stabilize the feedback control system and result in a satisfactory performance.

In the case of a time-invariant system with a regulatory type of controller, where the reference input is a constant setpoint, the vast majority of controllers are based on one of the general models given in Eqs. (13.20) and (13.21), that is, either full state feedback or output feedback, as shown in the following:

$$\mathbf{u}(t) = \mathbf{h}[\mathbf{x}(t)] \tag{13.20}$$

$$\mathbf{u}(t) = \mathbf{h}[y(t), \dot{y}, \int y\,dt] \tag{13.21}$$

In the case of a simple single-input–single-output system and a regulatory type of controller, the function **h** takes one of the following forms:

$$\mathbf{u}(t) = K_P \cdot e(t) \tag{13.22}$$

for a proportional, or **P**, controller;

$$\mathbf{u}(t) = K_P \cdot e(t) + K_I \cdot \int e(t)\,dt \tag{13.23}$$

for a proportional-plus-integral, or PI, controller;

$$\mathbf{u}(t) = K_P \cdot e(t) + K_D \cdot \dot{e}(t) \tag{13.24}$$

for a proportional-plus-derivative, or PD, controller (see Example 13.1 for a PD controller);

$$\mathbf{u}(t) = K_P \cdot e(t) + K_I \cdot \int e(t)\,dt + K_D \cdot \dot{e}(t) \tag{13.25}$$

for a proportional-plus-derivative-plus-integral, or PID, controller, where $e(t)$, $\dot{e}(t)$, and $\int e(t)\,dt$ are the output error, error derivative, and error integral, respectively; and

$$\mathbf{u}(t) = -[k_1 \cdot x_1(t) + k_2 \cdot x_2(t) + \cdots + k_n \cdot x_n(t)] \tag{13.26}$$

for a full state-feedback controller.

The problem of control system design is defined as obtaining the generally nonlinear function **h**( ) in the case of nonlinear systems; coefficients K_P, K_I, and K_D in the case of output-feedback systems; and coefficients $k_1, k_2, \ldots k_n$ in the case of a full state-feedback control policy for linear systems. The function **h**( ) in Eqs. (13.20) and (13.21) describes a general nonlinear surface that is known as a control, or decision, surface, discussed in the next section.

Control (Decision) Surface

The concept of a control surface, or decision surface, is central in fuzzy control systems methodology [Vadiee, 1993]. In this section we define this very important concept. The function **h** as defined in Eqs. (13.19), (13.20), and (13.21) is, in general, defining P nonlinear hypersurfaces in an n-dimensional space. For the case of linear systems with output feedback or state feedback it generally is a hyperplane in an n-dimensional space. This surface is known as the control, or decision, surface. The control surface describes the dynamics of the controller and is generally a time-varying nonlinear surface. Owing to unmodeled dynamics present in the design of any controller, techniques should exist for adaptively tuning and modifying the control surface shape.

Fuzzy logic rule-based expert systems use a collection of fuzzy conditional statements derived from an expert knowledge base to approximate and construct the control surface [Mamdani and Gaines, 1981; Kiszka et al., 1985; Sugeno, 1985]. This paradigm of control system design is based on interpolative and approximate reasoning. Fuzzy logic rule-based controllers or system identifiers, are generally model-free paradigms. Fuzzy logic rule-based expert systems are universal nonlinear function approximators, and any nonlinear function (e.g., control surface) of n independent variables and one dependent variable can be approximated to any desired precision.

Alternatively, artificial neural networks are based on analogical learning and try to learn the nonlinear decision surface through adaptive and converging techniques, based on numerical data available from input-output measurements of the system variables and some performance criteria.

Control System Design Stages

In order to obtain the control surface for a nonlinear time-varying real-world complex dynamic system, there are a number of simplifying steps used in modeling a controller for the system. The seven basic steps in designing a controller for a complex physical system are as follows:

1. Large-scale systems are decentralized and decomposed into a collection of decoupled subsystems.
2. The temporal variations of plant dynamics are assumed to be "slowly varying."
3. The nonlinear plant dynamics are locally linearized about a set of operating points (see rotation assumptions in Example 13.2).
4. A set of state variables, control variables, or output features is made available.
5. A simple P, PD, PID (output-feedback), or state-feedback controller is designed for each decoupled system. The controllers are of regulatory type and are fast enough to perform satisfactorily under tracking control situations. Optimal controllers might also prove useful.
6. In addition to uncertainties introduced in the first five steps, there are uncertainties due to external environment. The controller design should be made as close as possible to the optimal one based on the control engineer's knowledge, in the form of input-output numerical observations data and analytic, linguistic, intuitive, and other kinds of information regarding the plant dynamics and the external environment.
7. A supervisory control system, either automatic or a human expert operator, forms an additional feedback control loop to tune and adjust the controller's parameters, in order to compensate for the effects of variations caused by unmodeled dynamics.

Assumptions in a Fuzzy Control System Design

A number of assumptions are implicit in a fuzzy control system design. Six basic assumptions are commonly made whenever a fuzzy logic–based control policy is selected.

1. The plant is observable and controllable: State, input, and output variables are usually available for observation and measurement or computation.
2. There exists a body of knowledge comprised of a set of expert production linguistic rules, engineering common sense, intuition, a set of input/output measurements data, or an analytic model that can be fuzzified and from which rules can be extracted.
3. A solution exists.
4. The control engineer is looking for a "good enough" solution, not necessarily the optimum one.
5. We will design a controller to the best of our available knowledge and within an acceptable range of precision.

6. The problems of stability and optimality are still open problems in fuzzy controller design.

The following sections discuss the procedure for obtaining the control surface, **h**( ), from approximations based on a collection of fuzzy IF-THEN rules that describe the dynamics of the controller. Fuzzy rule–based expert models can also be used to obtain acceptable approximations for the functions **f**( ) and **g**( ) in the case of a system identification problem.

A fuzzy production rule system consists of four structures [Weiss and Donnel, 1979]:

1. A set of rules that represents the policies and heuristic strategies of the expert decision maker.
2. A set of input data assessed immediately prior to the actual decision.
3. A method for evaluating any proposed action in terms of its conformity to the expressed rules, given the available data.
4. A method for generating promising actions and for determining when to stop searching for better ones.

The input data, rules, and output action, or consequence, are generally fuzzy sets expressed as membership functions defined on a proper space. The method used for the evaluation of rules is known as *approximate reasoning,* or *interpolative reasoning,* and is commonly represented by composition of fuzzy relations applied to a fuzzy relational equation.

The control surface, which relates the control action $u(\,)$ to the measured state or output variables, is obtained using these four structures. It is then sampled at a finite number of points, depending on the required resolutions, and a look-up table is constructed. The look-up table thus formed could be downloaded onto a read-only memory chip and would constitute a fixed controller for the plant.

SIMPLE FUZZY LOGIC CONTROLLERS

First-generation (nonadaptive; i.e., the four structures above are fixed) simple fuzzy logic controllers can generally be depicted by a block diagram such as that shown in Fig. 13.2.

The knowledge-base module in Fig. 13.2 contains knowledge about all the input and output fuzzy partitions. It will include the term set and the corresponding membership functions defining the input variables to the fuzzy rule-base system and the output variables, or control actions, to the plant under control.

The steps in designing a simple fuzzy logic control system are as follows:

1. Identify the variables (inputs, states, and outputs) of the plant.
2. Partition the universe of discourse or the interval spanned by each variable into a number of fuzzy subsets, assigning each a linguistic label (subsets include all the elements in the universe).

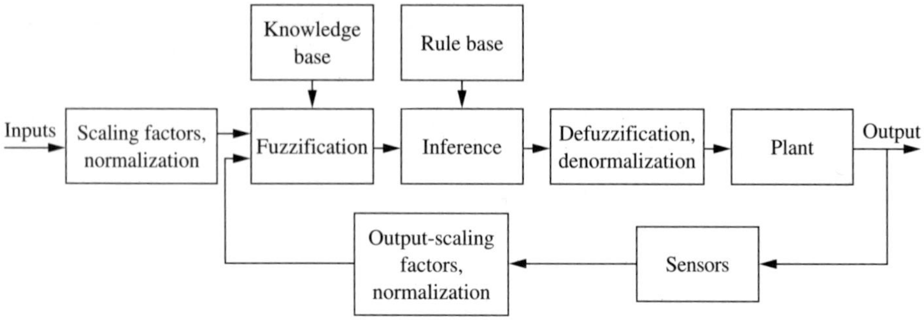

FIGURE 13.2
A simple fuzzy logic control system block diagram.

3. Assign or determine a membership function for each fuzzy subset.
4. Assign the fuzzy relationships between the inputs' or states' fuzzy subsets on the one hand and the outputs' fuzzy subsets on the other hand, thus forming the rule-base.
5. Choose appropriate scaling factors for the input and output variables in order to normalize the variables to the [0, 1] or the [−1, 1] interval.
6. Fuzzify the inputs to the controller.
7. Use fuzzy approximate reasoning to infer the output contributed from each rule.
8. Aggregate the fuzzy outputs recommended by each rule.
9. Apply defuzzification to form a crisp output.

In a nonadaptive simple fuzzy logic controller, the methodology used and the results of the nine steps mentioned above are fixed, whereas in an adaptive fuzzy logic controller, they are adaptively modified based on some adaptation law in order to optimize the controller.

A simple fuzzy logic control system has the following features:

1. Fixed and uniform input- and output-scaling factors.
2. Flat, single-partition rule-base with fixed and noninteractive rules. All the rules have the same degree of certainty and confidence, equal to unity.
3. Fixed membership functions.
4. Limited number of rules, which increases exponentially with the number of input variables.
5. Fixed metaknowledge including the methodology for approximate reasoning, rules aggregation, and output defuzzification.
6. Low-level control and no hierarchical rule structure.

GENERAL FUZZY LOGIC CONTROLLERS

The principal design elements in a general fuzzy logic control system (i.e., nonsimple) are as follows [Lee, 1990]:

1. Fuzzification strategies and the interpretation of a fuzzification operator, or fuzzifier.
2. Knowledge base:
   a. Discretization/normalization of the universe of discourse
   b. Fuzzy partitions of the input and output spaces
   c. Completeness of the partitions
   d. Choice of the membership functions of a primary fuzzy set
3. Rule-base:
   a. Choice of process state (input) variables and control (output) variables
   b. Source of derivation of fuzzy control rules
   c. Types of fuzzy control rules
   d. Consistency, interactivity, and completeness of fuzzy control rules
4. Decision-making logic:
   a. Definition of a fuzzy implication
   b. Interpretation of the sentence connective *and*
   c. Interpretation of the sentence connective *or*
   d. Inference mechanism
5. Defuzzification strategies and the interpretation of a defuzzification operator (defuzzifier)

Adaptation or change in any of the five design parameters above creates an adaptive fuzzy logic control system. If all five are fixed, the fuzzy logic control system is simple and nonadaptive.

SPECIAL FORMS OF FUZZY LOGIC CONTROL SYSTEM MODELS

Most fuzzy logic control system models can be expressed in two different forms: fuzzy rule–based structures and fuzzy relational equations. The most common fuzzy rule–based structures in fuzzy control systems are the five types of fuzzy rule–based system models described in Chapter 9 (Eqs. 9.12–9.21).

Equations (13.27–13.32) demonstrate fuzzy relational equations describing a number of commonly used fuzzy control system models:

1. Consider a discrete first-order system with input u, described in state-space form. The basic fuzzy model of such a system has the following form:

$$x_{k+1} = x_k \circ u_k \circ \underset{\sim}{\mathrm{R}} \quad \text{for } k = 1, 2, \ldots \tag{13.27}$$

where R is the fuzzy system transfer relation, and where $\circ$ is standard composition.

2. A discrete pth-order system with single input u, described in state-space form, is given by the fuzzy system equation as follows:

$$x_{k+p} = x_k \circ x_{k+1} \circ \cdots \circ x_{k+p-1} \circ u_{k+p-1} \circ \underset{\sim}{\mathrm{R}} \quad \text{for } k = 1, 2, \ldots \tag{13.28}$$

$$y_{k+p} = x_{k+p} \tag{13.29}$$

where y_{k+p} is the single output of the system.

3. A second-order system with full state feedback (see Example 13.1) is described as

$$u_k = x_k \circ x_{k-1} \circ \underset{\sim}{R} \quad \text{for } k = 1, 2, \ldots \tag{13.30}$$

$$y_k = x_k \tag{13.31}$$

4. A discrete pth-order single-input–single-output system with full state feedback is represented by the following fuzzy relational equation (termed Sugeno dynamic):

$$u_{k+p} = y_k \circ y_{k+1} \circ \cdots \circ y_{k+p-1} \circ \underset{\sim}{R} \quad \text{for } k = 1, 2, \ldots \tag{13.32}$$

EXAMPLES OF FUZZY CONTROL SYSTEM DESIGN

Most control situations are more complex than we can deal with mathematically. In this situation fuzzy control can be developed, provided a body of knowledge about the control process exists, and formed into a number of fuzzy rules. For example, suppose an industrial process output is given in terms of the pressure. We can calculate the difference between the desired pressure and the output pressure, called the pressure error (e), and we can calculate the difference between the desired rate of change of the pressure, dp/dt, and the actual pressure rate, called the pressure error rate, ($\dot{e}$). Also, assume that knowledge can be expressed in the form of IF-THEN rules such as

IF pressure error (e) is "positive big (PB)" or "positive medium (PM)" and
IF pressure error rate ($\dot{e}$) is "negative small (NS),"
THEN heat input change is "negative medium (NM)."

The linguistic variables defining the pressure error, "PB" and "PM," and the pressure error rate, "NS" and "NM," are fuzzy; but the measurements of both the pressure and pressure rate as well as the control value for the heat (the control variable) ultimately applied to the system are precise (crisp). The schematic in Fig. 13.3 shows this idea. An input to the industrial process (physical system) comes from the controller. The physical system responds with an output, which is sampled and

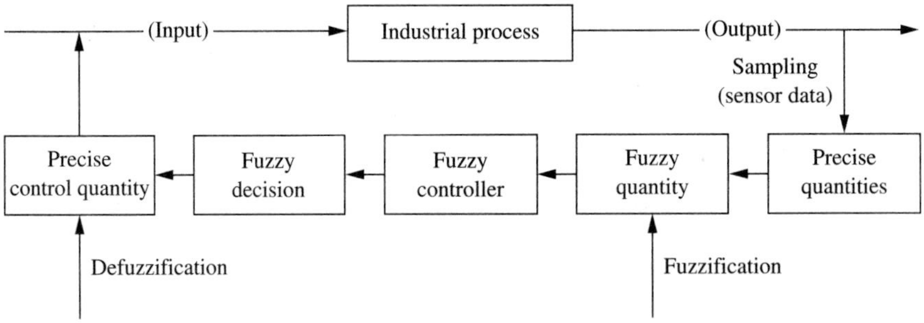

FIGURE 13.3
Typical closed-loop fuzzy control situation.

measured by some device. If the measured output is a crisp quantity it can be fuzzified into a fuzzy set (see Chapter 4). This fuzzy output is then considered as the fuzzy input into a fuzzy controller, which consists of linguistic rules. The output of the fuzzy controller is then another series of fuzzy sets. Since most physical systems cannot interpret fuzzy commands (fuzzy sets), the fuzzy controller output must be converted into crisp quantities using defuzzification methods (also see Chapter 5). These crisp (defuzzified) control-output values then become the input values to the physical system and the entire closed-loop cycle is repeated.

Example 13.1. For some industrial plants a human operator is sometimes more efficient than an automatic controller. These intuitive control strategies, which provide a possible method to handle qualitative information, may be modeled by a fuzzy controller. This example looks at a pressure process controlled by a fuzzy controller. The controller is formed by a number of fuzzy rules, such as: If pressure error is "positive big" or "positive medium," and if the rate of change in the pressure error is "negative small," then heat input change is "negative medium." This example is illustrated in four steps.

Step 1. Value assignment for the fuzzy input and output variables. We will let the error (e) be defined by eight linguistic variables, labeled $A_1, A_2, \ldots, A_8$, partitioned on the error space of $[-e_m, +e_m]$, and the error rate ($\dot{e}$, or de/dt) be defined by seven variables, labeled $B_1, B_2, \ldots, B_7$, partitioned on the error rate space of $[-\dot{e}_m, \dot{e}_m]$. We will normalize these ranges to the same interval $[-a, +a]$ by

$$e_1 = \left(\frac{a}{e_m}\right) \cdot e \tag{13.33}$$

$$\dot{e}_1 = \left(\frac{a}{\dot{e}_m}\right) \cdot \dot{e} \tag{13.34}$$

For the error, the eight fuzzy variables, $\underset{\sim}{A}_i$ $(i = 1, 2, \ldots, 8)$, will conform to the linguistic variables PB, PM, PS, P0, N0, NS, NM, NB. For the error rate, $\dot{e}$, the seven fuzzy variables, $\underset{\sim}{B}_j$ $(j = 1, 2, \ldots, 7)$ will conform to the linguistic variables PB, PM, PS, 0, NS, NM, NB. The membership functions for these quantities will be on the range $[-a, a]$, where $a = 6 (a = 6)$, and are shown in Tables 13.1 and 13.2 (in the tables $x = e$ and $y = \dot{e}$).

The fuzzy output variable, the control quantity (z), will use seven fuzzy variables on the normalized universe, $z = \{-7, -6, -5, \ldots, +7\}$. The control variable will be described by fuzzy linguistic control quantities, $\underset{\sim}{C}_k$, $(k = 1, 2, \ldots, 7)$, which are partitioned on the control universe. Table 13.3 is the normalized control quantity, z, which is defined by seven linguistic variables.

Step 2. Summary of control rules. According to human operator experience, control rules are of the form

If e is $\underset{\sim}{A}_1$ and $\dot{e}$ is $\underset{\sim}{B}_1$, then z is $\underset{\sim}{C}_{11}$.
If e is $\underset{\sim}{A}_1$ and $\dot{e}$ is $\underset{\sim}{B}_2$, then z is $\underset{\sim}{C}_{12}$.
If e is $\underset{\sim}{A}_i$ and $\dot{e}$ is $\underset{\sim}{B}_j$, then z is $\underset{\sim}{C}_{ij}$.

Each rule can be translated into a fuzzy relation, $\underset{\sim}{R}$. Using such an approach will result in linguistic variables, $\underset{\sim}{C}_{ij}$, shown as control entries in Table 13.4.

TABLE 13.1
Membership functions for error (e)*

	x	−6	−5	−4	−3	−2	−1	0−	0+	1	2	3	4	5	6
A_i															
A_1	PB	0	0	0	0	0	0	0	0	0	0	0.1	0.4	0.8	1
A_2	PM	0	0	0	0	0	0	0	0	0	0.2	0.7	1	0.7	0.2
A_3	PS	0	0	0	0	0	0	0	0.3	0.8	1	0.5	0.1	0	0
A_4	P0	0	0	0	0	0	0	0	1	0.6	0.1	0	0	0	0
A_5	N0	0	0	0	0	0.1	0.6	1	0	0	0	0	0	0	0
A_6	NS	0	0	0.1	0.5	1	0.8	0.3	0	0	0	0	0	0	0
A_7	NM	0.2	0.7	1	0.7	0.2	0	0	0	0	0	0	0	0	0
A_8	NB	1	0.8	0.4	0.1	0	0	0	0	0	0	0	0	0	0

*In the case of crisp control the membership values in the shaded boxes become unity and all other values become zero.

Step 3. Conversion between fuzzy variables and precise quantities. From the output of the system we can use an instrument to measure the error (e) and calculate the error rate ($\dot{e}$), both of which are precise numbers. A standard defuzzification procedure to develop membership functions, such as the maximum membership principle (see Chapter 5), can be used to get the corresponding fuzzy quantities ($\underset{\sim}{A}_i$, $\underset{\sim}{B}_i$). Sending the $\underset{\sim}{A}$ and $\underset{\sim}{B}$ obtained from the output of the system to the fuzzy controller will yield a fuzzy

TABLE 13.2*
Membership functions for error rate (de/dt)

	y	−6	−5	−4	−3	−2	−1	0	1	2	3	4	5	6
B_j														
B_1	PB	0	0	0	0	0	0	0	0	0	0.1	0.4	0.8	1
B_2	PM	0	0	0	0	0	0	0	0	0.2	0.7	1	0.7	0.2
B_3	PS	0	0	0	0	0	0	0	0.9	1	0.7	0.2	0	0
B_4	0	0	0	0	0	0	0.5	1	0.5	0	0	0	0	0
B_5	NS	0	0	0.2	0.7	1	0.9	0	0	0	0	0	0	0
B_6	NM	0.2	0.7	1	0.7	0.2	0	0	0	0	0	0	0	0
B_7	NB	1	0.8	0.4	0.1	0	0	0	0	0	0	0	0	0

*In the case of crisp control the membership values in the shaded boxes become unity and all other values become zero.

TABLE 13.3*
Membership functions for the control quantity (z)

	z	−7	−6	−5	−4	−3	−2	−1	0	1	2	3	4	5	6	7
C_k																
C_1	PB	0	0	0	0	0	0	0	0	0	0	0	0.1	0.4	0.8	1
C_2	PM	0	0	0	0	0	0	0	0	0	0.2	0.7	1	0.7	0.2	0
C_3	PS	0	0	0	0	0	0	0	0.4	1	0.8	0.4	0.1	0	0	0
C_4	0	0	0	0	0	0	0	0.5	1	0.5	0	0	0	0	0	0
C_5	NS	0	0	0	0.1	0.4	0.8	1	0.4	0	0	0	0	0	0	0
C_6	NM	0	0.2	0.7	1	0.7	0.2	0	0	0	0	0	0	0	0	0
C_7	NB	1	0.8	0.4	0.1	0	0	0	0	0	0	0	0	0	0	0

*In the case of crisp control the membership values in the shaded boxes become unity and all other values become zero.

action variable $\underset{\sim}{C}$ (control rules) as discussed in step 2. But before implementing the control, we have to enter the precise control quantity z into the system. We need another conversion from $\underset{\sim}{C}$ to z. This can be done by a maximum membership principle, or by a weighted average method (see Chapter 5).

Step 4. Development of control table. When the procedures in step 3 are used for all e and all $\dot{e}$, we obtain a control table as shown in Table 13.5. This table now contains precise numerical quantities for use by the industrial system hardware. If the values in Table 13.5 are plotted, they represent a control surface. Figure 13.4 is the control surface for this example, and Fig. 13.5 would be the control surface for this example if it had been conducted using only crisp sets and operations (for the crisp case, the values in Table 13.5 will be different). The volume under a control surface is proportional to the amount of energy expended by the controller. It can be shown

TABLE 13.4
Control rules (FAM table)

	A_i							
	NB	**NM**	**NS**	**N0**	**P0**	**PS**	**PM**	**PB**
B_j								
PB	PB	PM	NB	NB	NB	NB		
PM	PB	PM	NM	NM	NS	NM		
PS	PB	PM	NS	NS	NS	NS	NM	NB
0	PB	PM	PS	0	0	NS	NM	NB
NS	PB	PM	PS	PS	PS	PS	NM	NB
NM			PS	PS	PM	PM	NM	NB
NB			PB	PB	PB	PB	NM	NB

TABLE 13.5
Control actions

							y						
	−6	**−5**	**−4**	**−3**	**−2**	**−1**	**0**	**1**	**2**	**3**	**4**	**5**	**6**
x													
−6	7	6	7	6	7	7	7	4	4	2	0	0	0
−5	6	6	6	6	6	6	6	4	4	2	0	0	0
−4	7	6	7	6	7	7	7	4	4	2	0	0	0
−3	6	6	6	6	6	6	6	3	2	0	−1	−1	−1
−2	4	4	4	5	4	4	4	1	0	0	−1	−1	−1
−1	4	4	4	5	4	4	1	0	0	0	−3	−2	−1
0-	4	4	4	5	1	1	0	−1	−1	−1	−4	−4	−4
0+	4	4	4	5	1	1	0	−1	−1	−1	−4	−4	−4
1	2	2	2	2	0	0	−1	−4	−4	−3	−4	−4	−4
2	1	1	1	−2	0	−3	−4	−4	−4	−3	−4	−4	−4
3	0	0	0	0	−3	−3	−6	−6	−6	−6	−6	−6	−6
4	0	0	0	−2	−4	−4	−7	−7	−7	−6	−7	−6	−7
5	0	0	0	−2	−4	−4	−6	−6	−6	−6	−6	−6	−6
6	0	0	0	−2	−4	−4	−7	−7	−7	−6	−7	−6	−7

that the fuzzy control surface (Fig. 13.4) will actually fit underneath the crisp control surface (Fig. 13.5), indicating that the fuzzy control expends less energy than the crisp control. Fuzzy control methods, such as this one, have been used for some industrial systems and have achieved significant efficiency [Mamdani, 1974; Pappas and Mamdani, 1976].

In the foregoing example, we did not conduct a simulation of a control process because we do not have a model for the controller. The development of the control surface is derived simply from the control rules and associated membership functions. After the control surface is developed, a simulation can be conducted if a mathematical or linguistic (rule-based) model of the control process is available.

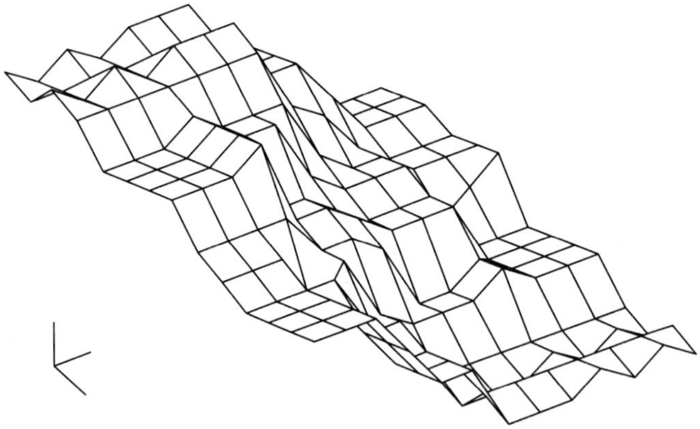

FIGURE 13.4
Control surface for fuzzy process control in Example 13.1.

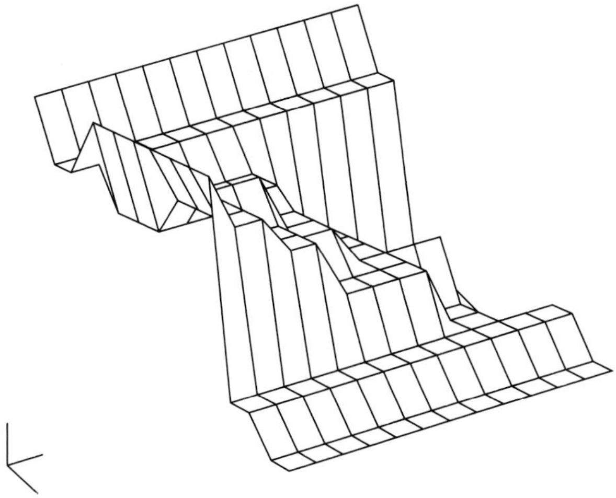

FIGURE 13.5
Control surface for crisp process control in Example 13.1.

CLASSICAL FUZZY CONTROL PROBLEM: INVERTED PENDULUM

Figure 13.6 shows the classic inverted pendulum system, which has been an interesting case in control theory for many years.

Example 13.2. We want to design and analyze a fuzzy controller for the simplified version of the inverted pendulum system shown in Fig. 13.6. The differential equation describing the system is given below [Kailath, 1980; Craig, 1986]:

$$-ml^2\, d^2\theta/dt^2 + (mlg)\sin(\theta) = \tau = u(t) \tag{13.35}$$

where m is the mass of the pole located at the tip point of the pendulum, l is the length of the pendulum, θ is the deviation angle from vertical in the clockwise direction, $\tau = u(t)$ is the torque applied to the pole in the counterclockwise direction ($u(t)$ is the control action), t is time, and g is the gravitational acceleration constant.

Assuming $x_1 = \theta$ and $x_2 = d\theta/dt$ to be the state variables, the state-space representation for the nonlinear system defined by Eq. (13.35) is given by

$$\begin{aligned} dx_1/dt &= x_2 \\ dx_2/dt &= (g/l)\sin(x_1) - (1/ml^2)u(t) \end{aligned}$$

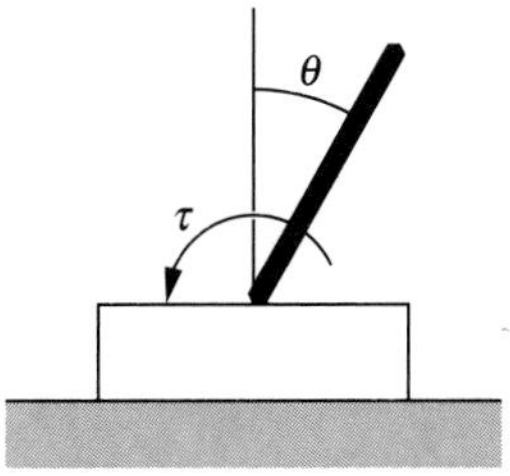

FIGURE 13.6
Inverted pendulum control problem.

It is known that for very small rotations, or θ, we have $\sin(\theta) = \theta$, where θ is measured in radians. This relation is used to linearize the nonlinear state-space equations and we get

$$dx_1/dt = x_2$$
$$dx_2/dt = (g/l)x_1 - (1/ml^2)u(t)$$

If x_1 is measured in degrees and x_2 is measured in degrees per second, by choosing $l = g$ and $m = 180/(\pi \cdot g^2)$, the linearized and discrete-time state-space equations can be represented as matrix difference equations,

$$x_1(k+1) = x_1(k) + x_2(k)$$
$$x_2(k+1) = x_1(k) + x_2(k) - u(k)$$

For this problem we assume the universe of discourse for the two variables to be $-2° \le x_1 \le 2°$ and -5 dps $\le x_2 \le 5$ dps (dps = degrees per second).

Step 1. First we want to construct three membership functions for x_1 on its universe, that is, for the values positive (P), zero (Z), and negative (N), as shown in Fig. 13.7.

We then construct three membership functions for x_2 on its universe, that is, for the values positive (P), zero (Z), and negative (N), as shown in Fig. 13.8.

Step 2. To partition the control space (output), we will construct five membership functions for $u(k)$ on its universe, which is $-24 \le u(k) \le 24$, as shown in Fig. 13.9. (Note that the figure has seven partitions, but we will use only five for this problem.)

Step 3. We then construct nine rules (even though we may not need this many) in a 3×3 FAM table, Table 13.6, for this system, which would involve θ and $\dot{\theta}$ in order to stabilize the inverted pendulum system. The entries in this table are the control actions, $u(k)$.

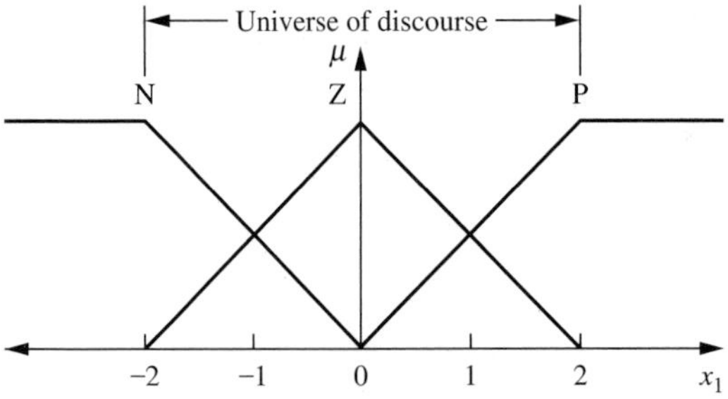

FIGURE 13.7
Input x_1 partitioned.

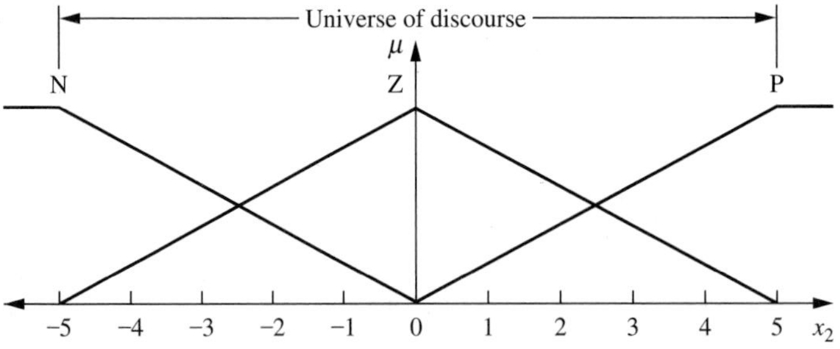

FIGURE 13.8
Input x_2 partitioned.

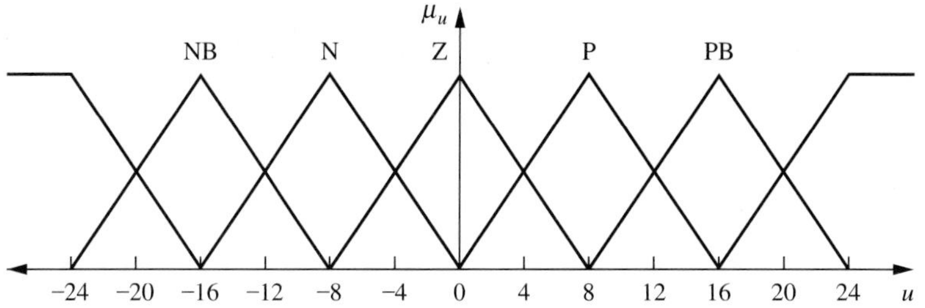

FIGURE 13.9
Output, $u(k)$, partitioned into seven partitions (only five used).

Step 4. Using the rules expressed in Table 13.6, we will now conduct a simulation of this control problem. In conducting the simulation we will use the graphical method presented in Chapter 8 to conduct the fuzzy operations. To start the simulation we will use the following crisp initial conditions:

$$x_1(0) = 1^\circ \quad \text{and} \quad x_2(0) = -4 \text{ dps}$$

Then we will conduct four cycles of simulation using the matrix difference equations, above, for the discrete steps $0 \leq k \leq 3$. Each simulation cycle will result in membership functions for the two input variables. The FAM table will produce a membership function for the control action, $u(k)$. We will defuzzify the membership function for the control action using the centroid method and then use the recursive difference equations to solve for new values of x_1 and x_2. Each simulation cycle after $k = 0$ will begin with the previous values of x_1 and x_2 as the input conditions to the next cycle of the recursive difference equations.

Figures 13.10 and 13.11 show the initial conditions for x_1 and x_2, respectively. From the FAM table (Table 13.6):

If (x_1 = P) and (x_2 = Z), then (u = P) min(0.5, 0.2) = 0.2 (P)
If (x_1 = P) and (x_2 = N), then (u = Z) min(0.5, 0.8) = 0.5 (Z)
If (x_1 = Z) and (x_2 = Z), then (u = Z) min(0.5, 0.2) = 0.2 (Z)
If (x_1 = Z) and (x_2 = N), then (u = N) min(0.5, 0.8) = 0.5 (N)

Figure 13.12 shows the union of the truncated fuzzy consequents for the control variable, u. The final form, with the defuzzified control value, is shown in Fig. 13.13.

TABLE 13.6
FAM table

x_2 / x_1	**P**	**Z**	**N**
P	PB	P	Z
Z	P	Z	N
N	Z	N	NB

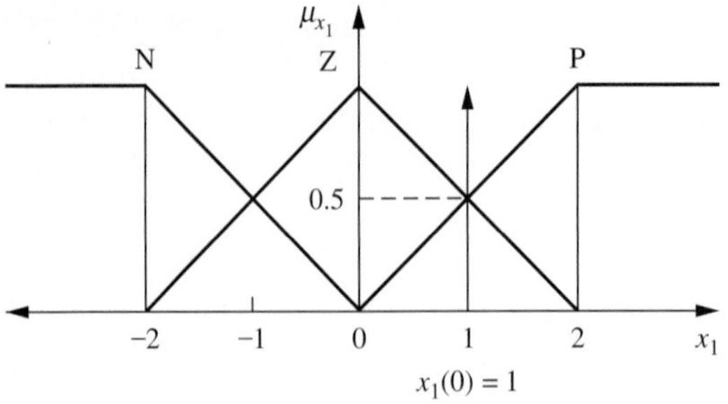

FIGURE 13.10
Initial condition for x_1.

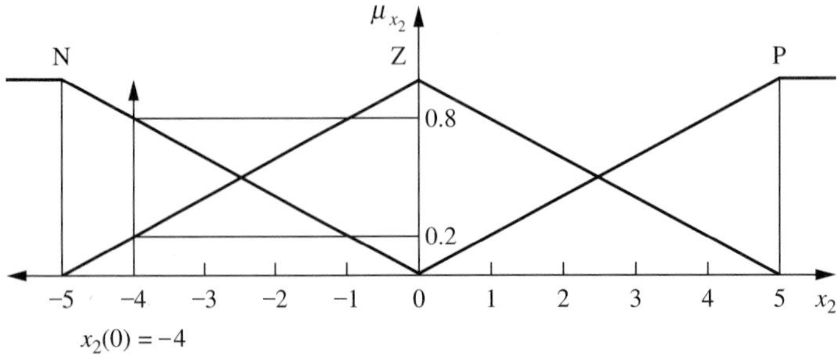

FIGURE 13.11
Initial condition for x_2.

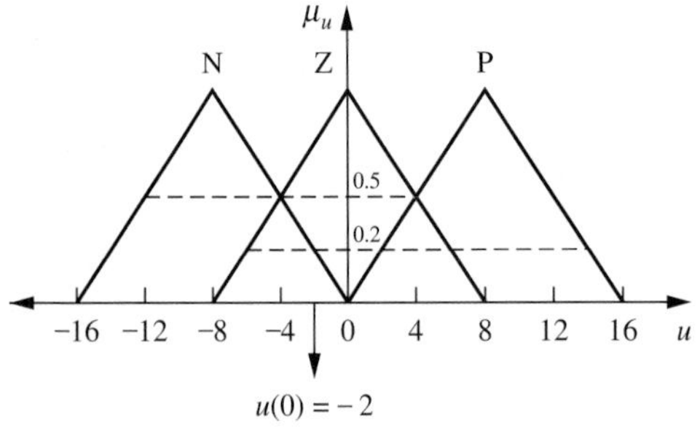

FIGURE 13.12
Union of fuzzy consequents fired by rules.

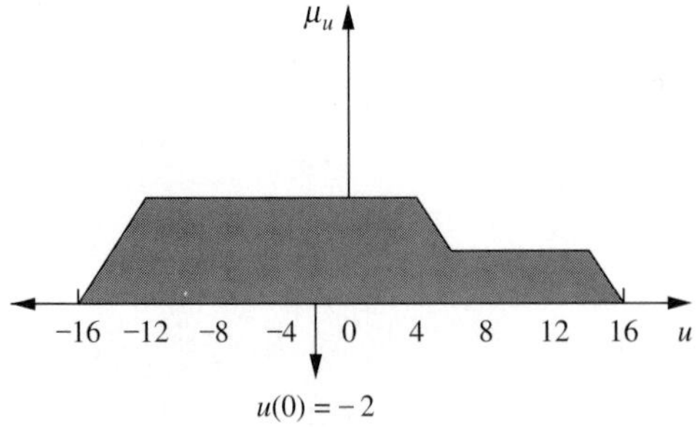

FIGURE 13.13
Union of fuzzy consequents and defuzzified value.

We have completed the first cycle of the simulation. Now we take the value of the defuzzified control variable (i.e., $u = -2$) and, using the system equations, find the initial conditions for next iteration.

$$x_1(1) = x_1(0) + x_2(0) = 1 - 4 = -3$$
$$x_2(1) = x_1(0) + x_2(0) - u(0) = 1 - 4 - (-2) = -1$$

From this we get the initial conditions for the second cycle as $x_1(1) = -3$ and $x_2(1) = -1$, which are shown graphically in Figs. 13.14 and 13.15.
From the FAM table (Table 13.6):

$$\text{If } (x_1 = \text{N}) \text{ and } (x_2 = \text{N}), \text{ then } (u = \text{NB}) \qquad \min(1, 0.2) = 0.2 \text{ (NB)}$$
$$\text{If } (x_1 = \text{N}) \text{ and } (x_2 = \text{Z}), \text{ then } (u = \text{N}) \qquad \min(1, 0.8) = 0.8 \text{ (N)}$$

The union of the fuzzy consequents and the resulting defuzzified output are shown in Fig. 13.16. The defuzzified value is $u = -9.6$.

We now use $u = -9.6$ to find the initial conditions for third cycle iteration.

$$x_1(2) = x_1(1) + x_2(1) = -3 - 1 = -4$$
$$x_2(2) = x_1(1) + x_2(1) - u(1) = -3 - 1 - (-9.6) = +5.6$$

Thus, we get initial conditions $x_1(2) = -4$ and $x_2(2) = 5.6$, which are shown graphically in Figs. 13.17 and 13.18, respectively. From the FAM table (Table 13.6) :

$$\text{If } (x_1 = \text{N}) \text{ and } (x_2 = \text{P}), \text{ then } (u = \text{Z}) \qquad \min(1, 1) = 1 \text{ (Z)}$$

The resulting consequents and defuzzified control variable, $u(z) = 0.0$, are shown in Fig. 13.19.

For the next iteration,

$$x_1(3) = x_1(2) + x_2(2) = -4 + 5.6 = 1.6$$
$$x_2(3) = x_1(2) + x_2(2) - u(2) = -4 + 5.6 - (0.0) = 1.6$$

The initial conditions $x_1(3) = 1.6$ and $x_2(3) = 1.6$ are shown graphically in Figs. 13.20 and 13.21, respectively. From Table 13.6 we get

$$\text{If } (x_1 = \text{Z}) \text{ and } (x_2 = \text{P}), \text{ then } (u = \text{P}) \qquad \min(0.25, 0.3) = 0.3 \text{ (P)}$$
$$\text{If } (x_1 = \text{Z}) \text{ and } (x_2 = \text{Z}), \text{ then } (u = \text{Z}) \qquad \min(0.25, 0.7) = 0.25 \text{ (Z)}$$
$$\text{If } (x_1 = \text{P}) \text{ and } (x_2 = \text{P}), \text{ then } (u = \text{PB}) \qquad \min(0.75, 0.3) = 0.3 \text{ (PB)}$$
$$\text{If } (x_1 = \text{P}) \text{ and } (x_2 = \text{Z}), \text{ then } (u = \text{P}) \qquad \min(0.75, 0.7) = 0.7 \text{ (P)}$$

These conditions are shown graphically in Fig. 13.22, and the defuzzified value is $u = 5.28$.

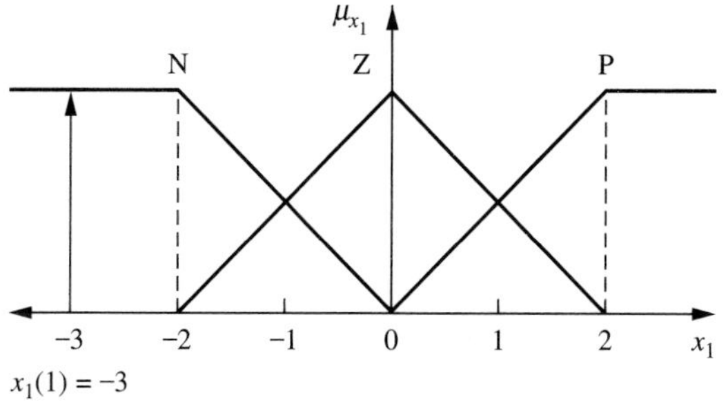

FIGURE 13.14
Initial condition for second cycle for x_1.

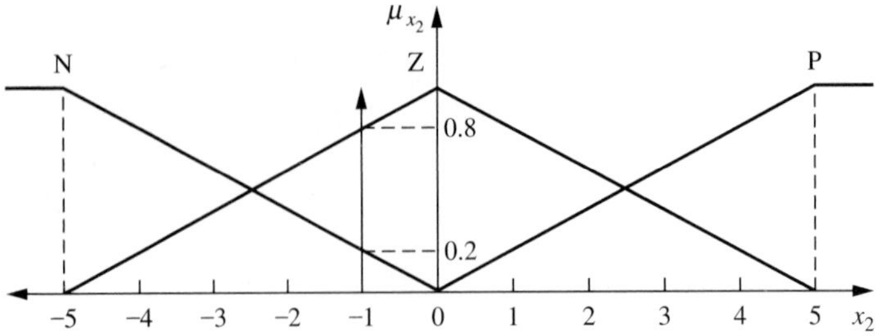

FIGURE 13.15
Initial condition for second cycle for x_2.

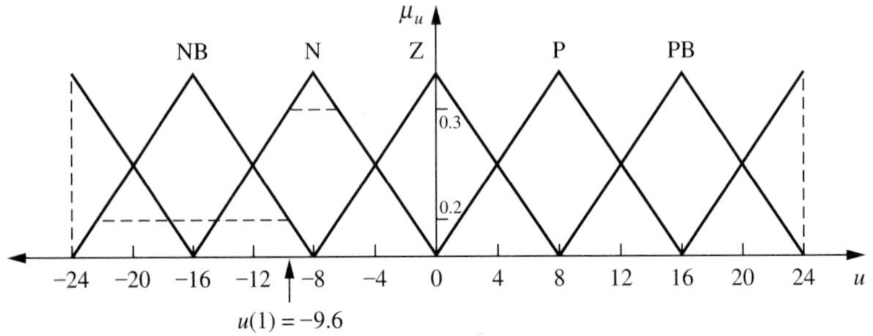

FIGURE 13.16
Truncated consequents and defuzzified output for second cycle.

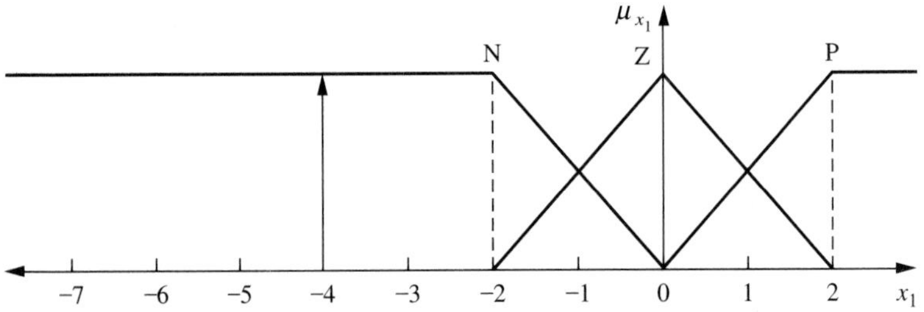

FIGURE 13.17
Initial condition for third cycle for x_1.

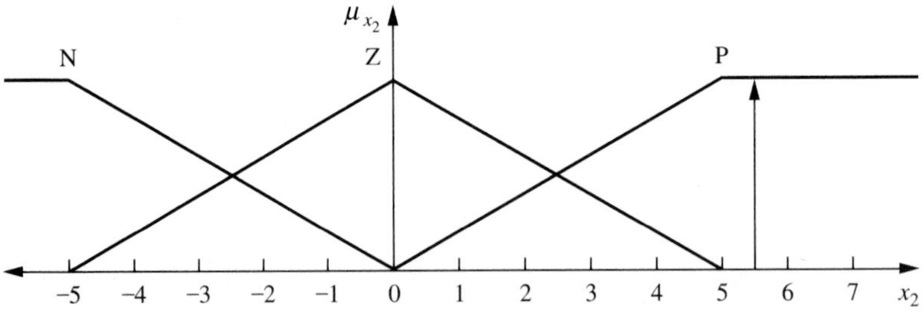

FIGURE 13.18
Initial condition for third cycle for x_2.

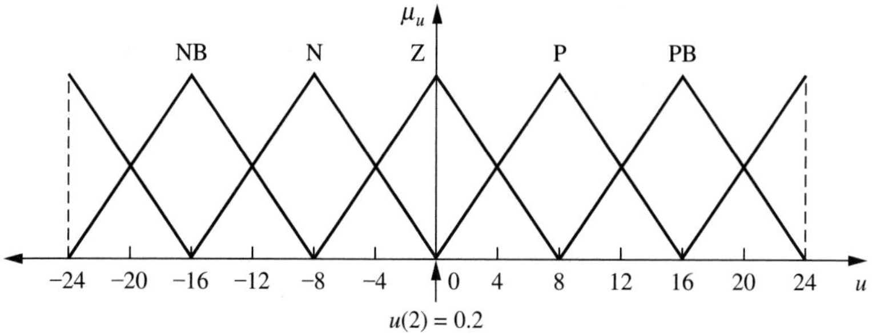

FIGURE 13.19
Defuzzified output for third cycle of simulation.

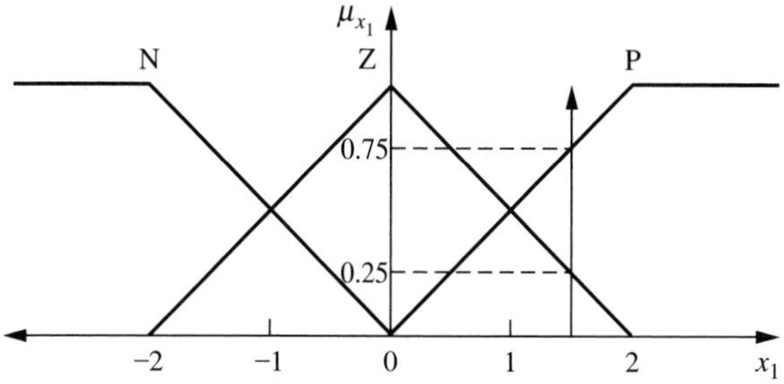

FIGURE 13.20
Initial condition for next iteration for x_1.

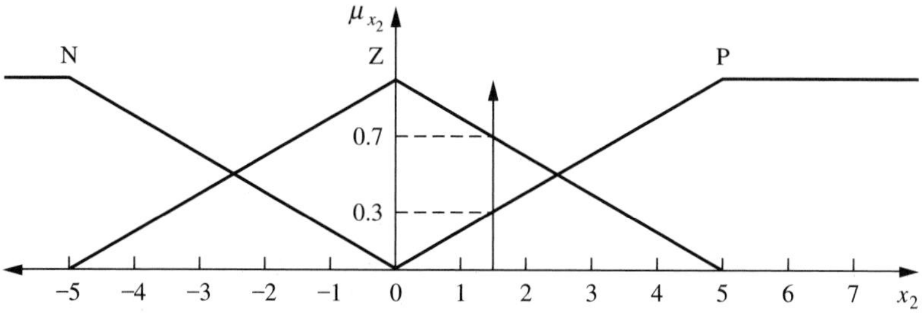

FIGURE 13.21
Initial condition for next iteration for x_2.

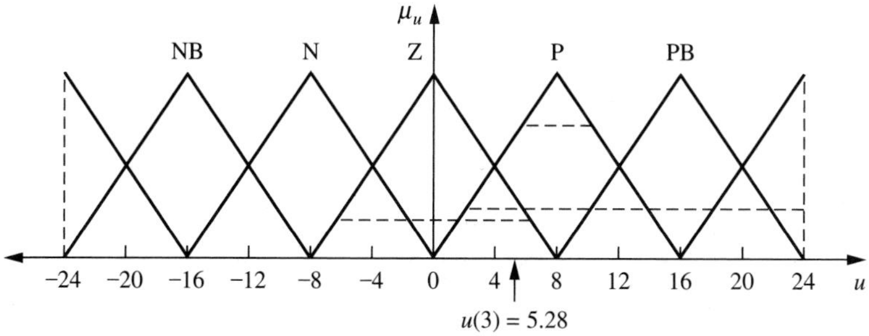

FIGURE 13.22
Defuzzified output for next iteration.

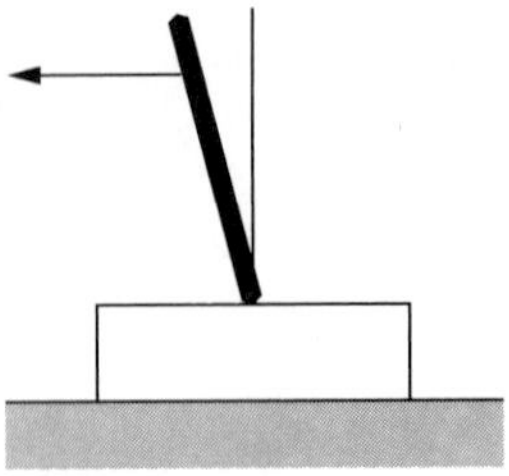

FIGURE 13.23
Initial conditions $x_1(0) = 1°$, $x_2(0) = -4$ dps.

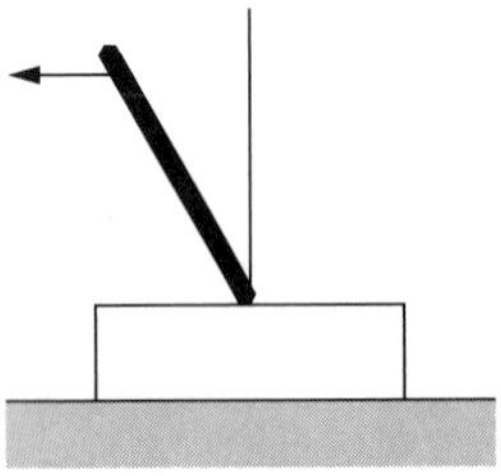

FIGURE 13.24
Iteration 1: $x_1(1) = -3$, $x_2(1) = -1$.

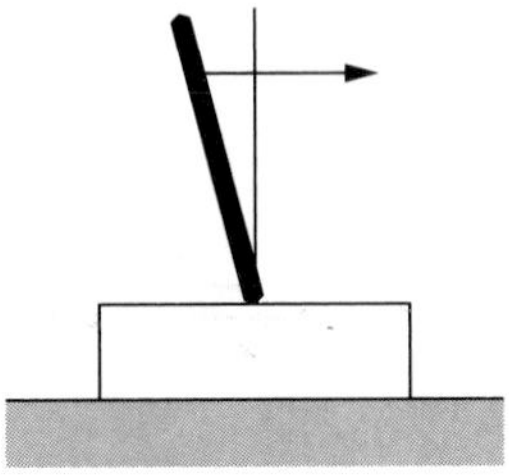

FIGURE 13.25
Iteration 2: $x_1(2) = -4$, $x_2(2) = 5.6$.

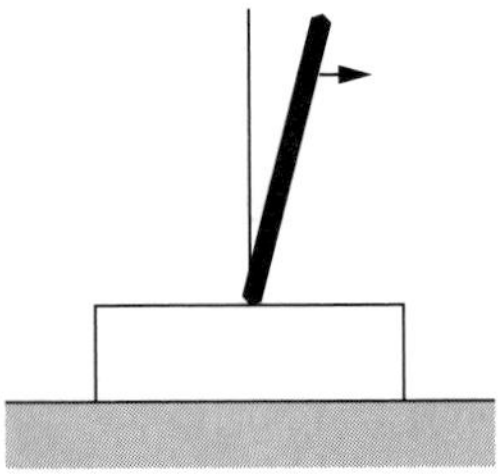

FIGURE 13.26
Iteration 3: $x_1(3) = 1.6$, $x_2(3) = 1.6$.

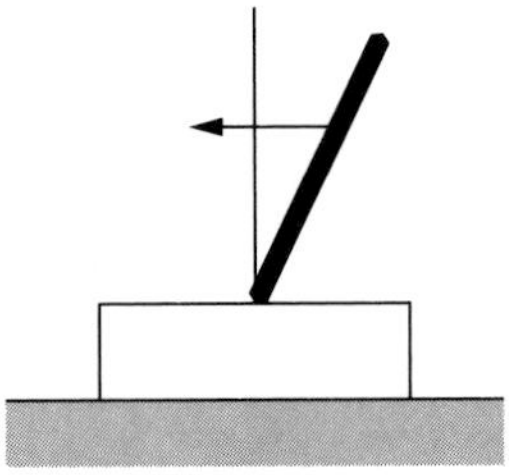

FIGURE 13.27
Iteration 4: $x_1(4) = 3.2$, $x_2(4) = -2.08$.

As before, we will use $u = 5.28$ to find initial conditions for the next iteration,

$$x_1(4) = x_1(3) + x_2(3) = 1.6 + 1.6 = 3.2$$
$$x_2(4) = x_1(3) + x_2(3) - u(3) = 1.6 + 1.6 - (5.28) = -2.08$$

Thus, we get initial conditions $x_1(4) = 3.2$ and $x_2(4) = -2.08$, and $u(4) = 1.12$ (no figure). We conclude the simulation at this point.

Step 5. We now plot the four simulation cycle results for x_1, x_2, and $u(k)$. In Figs. 13.23 to 13.27, the length and direction of the arrow are proportional to the angular velocity and show direction of motion of the pendulum, respectively.

AIRCRAFT LANDING CONTROL PROBLEM

The following example shows the flexibility and reasonable accuracy for another application in fuzzy control.

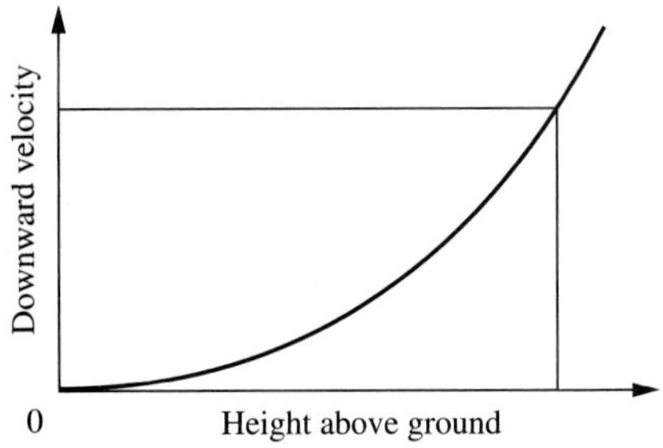

FIGURE 13.28
The desired profile of downward velocity vs. altitude.

Example 13.3. We will conduct a simulation of the final descent and landing approach of an aircraft. The desired profile is shown in Fig. 13.28. The desired downward velocity is proportional to the square of the height. Thus, at higher altitudes, a large downward velocity is desired. As the height (altitude) diminishes, the desired downward velocity gets smaller and smaller. In the limit, as the height becomes vanishingly small, the downward velocity also goes to zero. In this way, the aircraft will descend from altitude promptly but will touch down very gently to avoid damage.

The two state variables for this simulation will be the height above ground, h, and the vertical velocity of the aircraft, v (Fig. 13.29). The control output will be a force that, when applied to the aircraft, will alter its height, h, and velocity, v. The differential control equations are loosely derived as follows. See Fig. 13.30. Mass m moving with velocity v has momentum $p = mv$. If no external forces are applied, the mass will continue in the same direction at the same velocity, v. If a force f is applied over a time

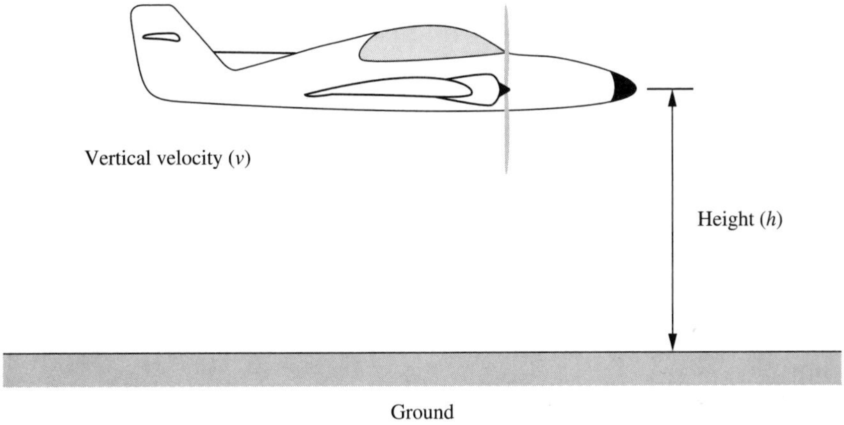

FIGURE 13.29
Aircraft landing control problem.

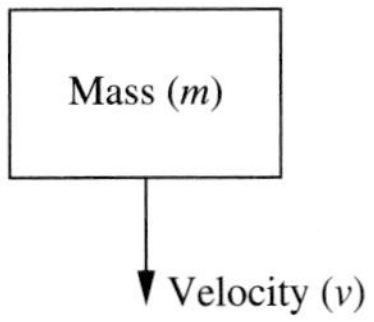

FIGURE 13.30

interval Δt, a change in velocity of $\Delta v = f\Delta t/m$ will result. If we let $\Delta t = 1.0$ (sec) and $m = 1.0$ (lb-sec^2/ft), we obtain $\Delta v = f$ (lb), or the change in velocity is proportional to the applied force.

In difference notation we get

$$v_{i+1} = v_i + f_i$$
$$h_{i+1} = h_i + v_i(1)$$

where v_{i+1} is the new velocity, v_i is the old velocity, h_{i+1} is the new height, and h_i is the old height. These two "control equations" define the new value of the state variables v and h in response to control input and the previous state variable values. Now, following the same procedure as in Example 13.2, step 1, we construct membership functions for the height, h, the vertical velocity, v, and the control force, f:

Step 1. Define membership functions for state variables as shown in Tables 13.7 and 13.8 and Figs. 13.31 and 13.32.

Step 2. Define a membership function for the control output, as shown in Table 13.9 and Fig. 13.33.

Step 3. Define the rules and summarize them in an FAM table (Table 13.10). The values in the FAM table, of course, are the control outputs.

Step 4. Define the initial conditions, and conduct a simulation for four cycles. Since the task at hand is to control the aircraft's vertical descent during approach and landing, we will start with the aircraft at an altitude of 1000 feet, with a downward velocity of -20 ft/s. We will use the following equations to update the state variables for each cycle:

$$v_{i+1} = v_i + f_i$$
$$h_{i+1} = h_i + v_i$$

TABLE 13.7
Membership values for height

	Height (ft)										
	0	**100**	**200**	**300**	**400**	**500**	**600**	**700**	**800**	**900**	**1000**
Large (L)	0	0	0	0	0	0	0.2	0.4	0.6	0.8	1
Medium (M)	0	0	0	0	0.2	0.4	0.6	0.8	1	0.8	0.6
Small (S)	0.4	0.6	0.8	1	0.8	0.6	0.4	0.2	0	0	0
Near zero (NZ)	1	0.8	0.6	0.4	0.2	0	0	0	0	0	0

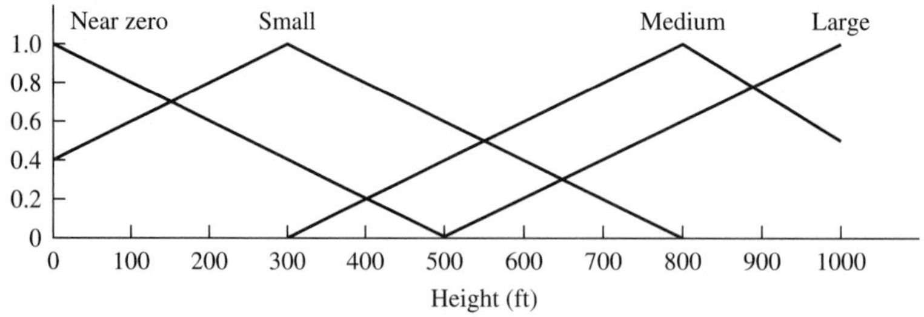

FIGURE 13.31
Height, h, partitioned.

TABLE 13.8
Membership values for velocity

	Vertical velocity (ft/s)												
	−30	**−25**	**−20**	**−15**	**−10**	**−5**	**0**	**5**	**10**	**15**	**20**	**25**	**30**
Up large (UL)	0	0	0	0	0	0	0	0	0	0.5	1	1	1
Up small (US)	0	0	0	0	0	0	0	0.5	1	0.5	0	0	0
Zero (Z)	0	0	0	0	0	0.5	1	0.5	0	0	0	0	0
Down small (DS)	0	0	0	0.5	1	0.5	0	0	0	0	0	0	0
Down large (DL)	1	1	1	0.5	0	0	0	0	0	0	0	0	0

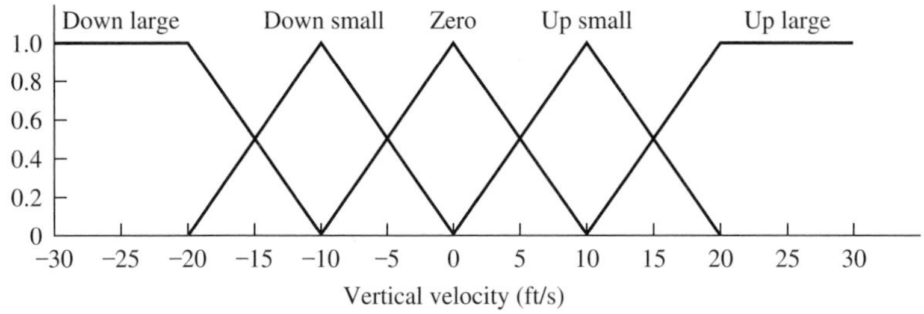

FIGURE 13.32
Velocity, *v*, partitioned.

TABLE 13.9
Membership values for control force

	Output force (lbs)												
	−30	**−25**	**−20**	**−15**	**−10**	**−5**	**0**	**5**	**10**	**15**	**20**	**25**	**30**
Up large (UL)	0	0	0	0	0	0	0	0	0	0.5	1	1	1
Up small (US)	0	0	0	0	0	0	0	0.5	1	0.5	0	0	0
Zero (Z)	0	0	0	0	0	0.5	1	0.5	0	0	0	0	0
Down small (DS)	0	0	0	0.5	1	0.5	0	0	0	0	0	0	0
Down large (DL)	1	1	1	0.5	0	0	0	0	0	0	0	0	0

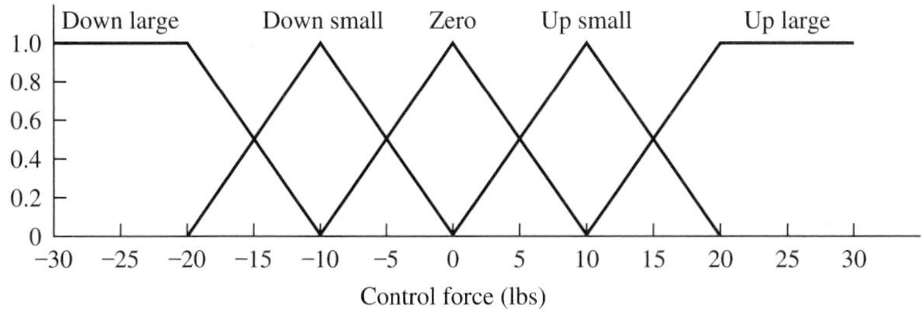

FIGURE 13.33
Control force, *f*, partitioned.

TABLE 13.10
FAM table

	Velocity				
Height	**DL**	**DS**	**Zero**	**US**	**UL**
L	Z	DS	DL	DL	DL
M	US	Z	DS	DL	DL
S	UL	US	Z	DS	DL
NZ	UL	UL	Z	DS	DS

Initial height, h_0: 1000 ft
Initial velocity, v_0: -20 ft/s
Control f_0: to be computed

Height h fires L at 1.0 and M at 0.6.
Velocity v fires only DL at 1.0.

Height		Velocity		Output
L (1.0)	AND	DL (1.0)	$\Rightarrow$	Z (1.0)
M (0.6)	AND	DL (1.0)	$\Rightarrow$	US (0.6)

We defuzzify using the centroid method and get $f_0 = 5.8$ lbs. This is the output force computed from the initial conditions. The results for cycle 1 appear in Fig. 13.34.

Now, we compute new values of the state variables and the output for the next cycle,

$$h_1 = h_0 + v_0 = 1000 + (-20) = 980 \text{ ft}$$
$$v_1 = v_0 + f_0 = -20 + 5.8 = -14.2 \text{ ft/s}$$

Height $h_1 = 980$ ft fires L at 0.96 and M at 0.64
Velocity $v_1 = -14.2$ ft/s fires DS at 0.58 and DL at 0.42

Height		Velocity		Output
L (0.96)	AND	DS (0.58)	$\Rightarrow$	DS (0.58)
L (0.96)	AND	DL (0.42)	$\Rightarrow$	Z (0.42)
M (0.64)	AND	DS (0.58)	$\Rightarrow$	Z (0.58)
M (0.64)	AND	DL (0.42)	$\Rightarrow$	US (0.42)

We find the centroid to be $f_1 = -0.5$ lbs. Results are shown in Fig. 13.35.

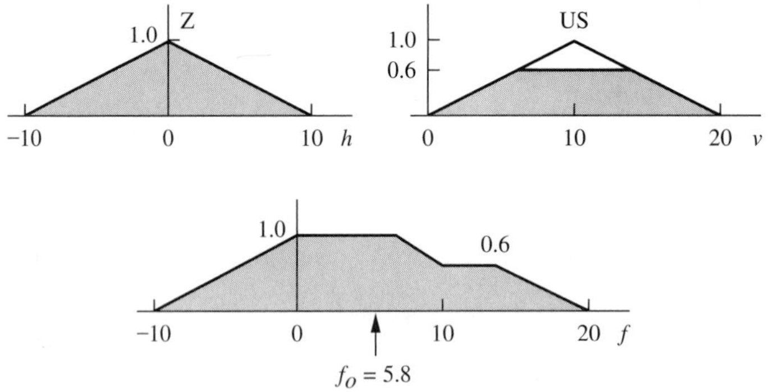

FIGURE 13.34
Truncated consequents and union of fuzzy consequent for cycle 1.

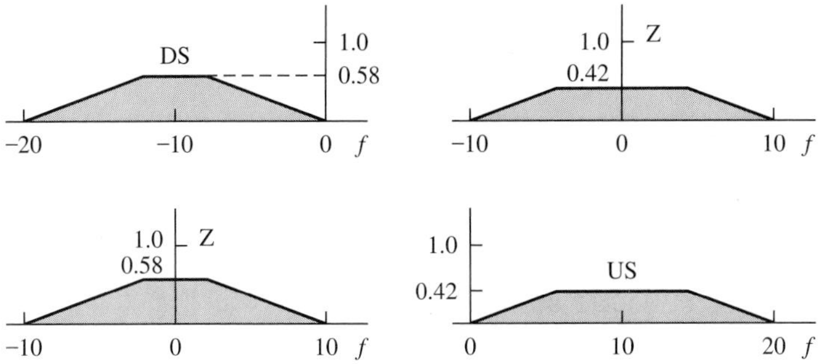

FIGURE 13.35
Truncated consequents for cycle 2.

We compute new values of the state variables and the output for the next cycle.

$$h_2 = h_1 + v_1 = 980 + (-14.2) = 965.8 \text{ ft}$$
$$v_2 = v_1 + f_1 = -14.2 + (-0.5) = -14.7 \text{ ft/s}$$

$$h_2 = 965.8 \text{ ft fires L at 0.93 and M at 0.67}$$
$$v_2 = -14.7 \text{ ft/s fires DL at 0.43 and DS at 0.57}$$

Height		Velocity		Output
L (0.93)	AND	DL (0.43)	⇒	Z (0.43)
L (0.93)	AND	DS (0.57)	⇒	DS (0.57)
M (0.67)	AND	DL (0.43)	⇒	US (0.43)
M (0.67)	AND	DS (0.57)	⇒	Z (0.57)

We find the centroid for this cycle to be $f_2 = -0.4$ lbs. Results appear in Fig. 13.36.

Again, we compute new values of state variables and output:

$$h_3 = h_2 + v_2 = 965.8 + (-14.7) = 951.1 \text{ ft}$$
$$v_3 = v_2 + f_2 = -14.7 + (-0.4) = -15.1 \text{ ft/s}$$

and for one more cycle we get

$$h_3 = 951.1 \text{ ft fires L at 0.9 and M at 0.7}$$
$$v_3 = -15.1 \text{ ft/s fires DS at 0.49 and DL at 0.51}$$

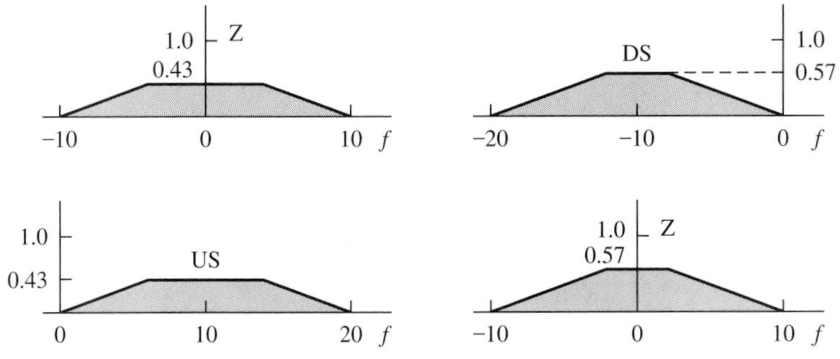

FIGURE 13.36
Truncated consequents for cycle 3.

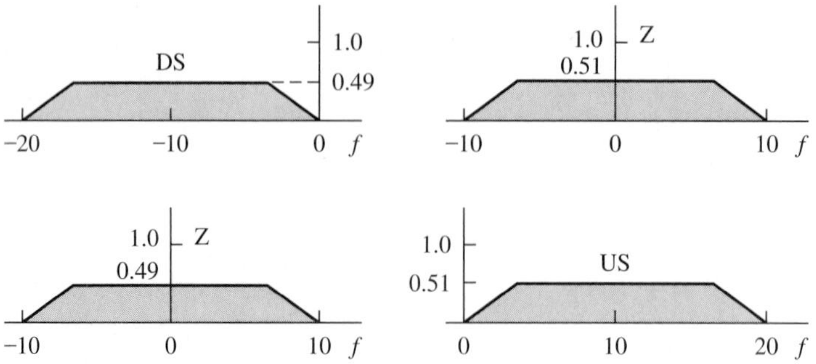

FIGURE 13.37
Truncated consequents for cycle 4.

Height		Velocity		Output
L (0.9)	AND	DS (0.49)	⇒	DS (0.49)
L (0.9)	AND	DL (0.51)	⇒	Z (0.51)
M (0.7)	AND	DS (0.49)	⇒	Z (0.49)
M (0.7)	AND	DL (0.51)	⇒	US (0.51)

The results are shown in Fig. 13.37, with a defuzzified centroid value of $f_3 = 0.3$ lbs.

Now, we compute the final values for the state variables to finish the simulation,

$$h_4 = h_3 + v_3 = 951.1 + (-15.1) = 936.0 \text{ ft}$$
$$v_4 = v_3 + f_3 = -15.1 + 0.3 = -14.8 \text{ ft/s}$$

The summary of the four-cycle simulation results is presented in Table 13.11. If we plot downward velocity vs. altitude (height), we get the descent profile shown in Fig. 13.38, which appears to be a reasonable start at the desired parabolic curve shown at the beginning of the problem.

INDUSTRIAL APPLICATIONS

Two recent papers have provided an excellent review of the wealth of industrial products and consumer appliances that are bringing fuzzy logic applications to the marketplace. One paper describes fuzzy logic applications in a dozen household appliances [Quail and Adnan, 1992] and another deals with a large suite of electronics components in the general area of image processing equipment [Takagi, 1992]. In fact, a recent conference dealt entirely with industrial applications of fuzzy control [Yen et al., 1992].

TABLE 13.11
Summary of four-cycle simulation results

	Cycle 0	Cycle 1	Cycle 2	Cycle 3	Cycle 4
Height, ft	1000.0	980.0	965.8	951.1	936.0
Velocity, ft/s	−20	−14.2	−14.7	−15.1	−14.8
Control force	5.8	0.5	−0.4	0.3	

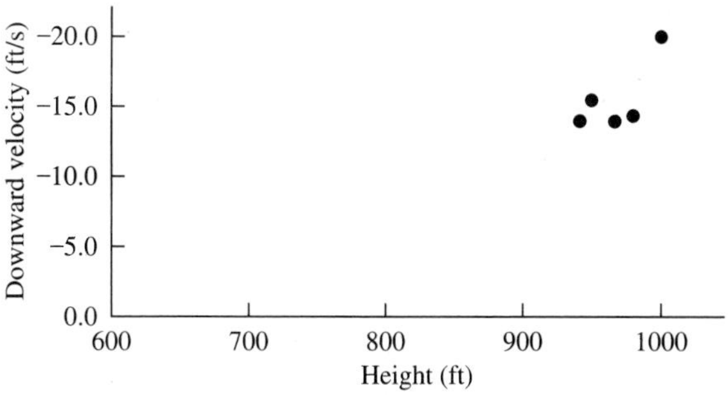

FIGURE 13.38
The profile of downward velocity, v, vs. height, h, using fuzzy logic control.

Few of us could have foreseen the revolution that fuzzy set theory has already produced. Dr. Zadeh himself predicts that fuzzy logic will be part of every appliance when he says that we will "see appliances rated not on horsepower but on IQ" [Rogers and Hoshai, 1990]. In Japan, the revolution has been so strong that "fuzzy logic" has become a common advertising slogan [Reid, 1990]. Whereas the Eastern world equates the word *fuzzy* with a form of computer intelligence, the Western world still largely associates the word derisively within the context of "imprecise or approximate science."

The consumer generally purchases new appliances based on their ability to streamline housework and to use their available time more effectively. Fuzzy logic is being incorporated worldwide in appliances to accomplish these goals, primarily in the control mechanisms designed to make them work. Appliances with fuzzy logic controllers provide the consumer with optimum settings that more closely approximate human perceptions and reactions than those associated with standard control systems. Products with fuzzy logic monitor user-defined settings, then automatically set the equipment to function at the user's preferred level for a given task. For example, fuzzy logic is well-suited to making adjustments in temperature, speed, and other control conditions found in a wide variety of consumer products [Loe, 1991]. A few examples of useful fuzzy logic applications are provided to enlighten the reader to the tremendous potential of this relatively new technology.

Fuzzy Logic Control of Blood Pressure during Anesthesia

In this application [Meier et al., 1992] a fuzzy logic controller was used to control mean arterial pressure (MAP), which was taken as a measure of the depth of anesthesia. The main reason for automating the control of depth of anesthesia is to release the anesthetist so that he or she can devote attention to other tasks as well—such as controlling fluid balance, ventilation, and drug application—that cannot yet be adequately automated and thus to increase the patient's safety. Because a biological process like anesthesia has a nonlinear, time-varying structure and time-varying parameters, modeling it suggests the use of rule-based controllers like fuzzy controllers. The design process here was iterative, and the reference points of the membership functions as well as the linguistic rules were determined by trial and error.

The control rules made use of the error between the desired and the actual values of MAP as well as the integral of the error.

The control loop has the structure shown in Fig. 13.39. There are two different kinds of disturbances: (1) surgical disturbances, (2) measurement noise and artifacts.

Depth of anesthesia is controlled by using a mixture of drugs that are injected intravenously or inhaled as gases. Most of these agents decrease MAP. Among the inhaled gases, isofluorane is widely used, most often in a mixture of 0 to 2 percent by volume of isofluorane in oxygen and/or nitrous oxide. For simulation and controller design purposes the relationship between the inflow concentration of isofluorane, $u(t)$, and the resulting blood pressure, $y(t)$, is modeled as the sum of two first-order terms, each with a pure time delay (the model in Fig. 13.39 includes the patient and also the semiclosed circuit). The step response, $h(t)$, corresponding to a unit step input can be written as follows, where $K_1 = -3$, $K_2 = -7.3$, $t_1 = 23$ s, $t_2 = 101$ s, $\alpha_1 = 0.01$, $\alpha_2 = 0.006$:

$$h(t) = f_1(t) + f_2(t) \tag{13.36a}$$

$$f_1(t) = K_1\alpha_1 \exp[-\alpha_1(t - t_1)] \qquad \text{for } t > t_1 \tag{13.36b}$$

$$f_2(t) = K_2\alpha_2 \exp[-\alpha_2(t - t_2)] \qquad \text{for } t > t_2 \tag{13.36c}$$

It is assumed that $t_1 = c_1 T$ and $t_2 = c_2 T$. The z-transform of the transfer function of the step response in Eq. (13.36a), can be expressed as shown in Eq. (13.37), where $Y(z)$ and $U(z)$ are z-transforms of the output blood pressure and input isofluorane concentration, respectively:

$$\begin{aligned} Y(z) &= \sum_{i=1}^{2} K_i\alpha_i[1 - z^{-1}\exp(-\alpha_i T)]^{-1} z^{-c_i} U(z) \\ &= \frac{b_1 z^{-c_1} + b_2 z^{-1-c_1} + b_3 z^{-c_2} + b_4 z^{-1-c_2}}{1 + a_1 z^{-1} + a_2 z^{-2}} U(z) \end{aligned} \tag{13.37}$$

where
$$\begin{aligned} b_1 &= K_1\alpha_1 \\ b_2 &= -K_1\alpha_1 \exp(-\alpha_2 T) \\ b_3 &= K_2\alpha_2 \\ b_4 &= -K_2\alpha_2 \exp(-\alpha_1 T) \\ a_1 &= -\exp(-\alpha_1 T) - \exp(-\alpha_2 T) \\ a_2 &= \exp(-\alpha_1 T)\exp(-\alpha_2 T) \end{aligned}$$

When Eq. (13.37) is rewritten in recursive form with $T = 10$ s, $a_1 = -1.221$, $a_2 = 0.335$, $b_1 = -0.030$, $b_2 = 0.048$, $b_3 = -0.017$, and $b_4 = 0.041$, we get

$$\begin{aligned} y(kT) = &-a_1 y[(k-1)T] - a_2 y[(k-2)T] + b_1 u[(k-c_1)T] \\ &+ b_2 u[(k-c_1-1)T] + b_1 u[(k-c_2)T] \\ &+ b_4 u[(k-c_2-1)T] \end{aligned} \tag{13.38}$$

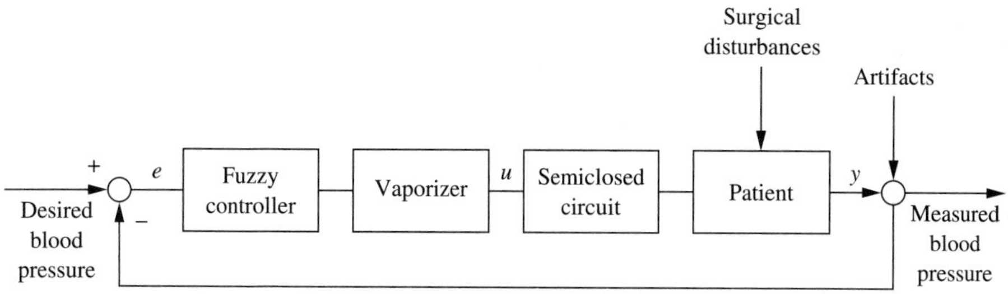

FIGURE 13.39
Block diagram of the control loop for the control of depth of anesthesia (Meier et al., 1992).

For the fuzzy controller, the error $e(t)$ and the integral of error $\int e(t)$ are used for the computation of the control variable $u(t)$. The error $e(t)$ is defined as

$$e(t) = d(t) - y(t) \tag{13.39}$$

where $d(t)$ is the desired blood pressure.

The linguistic rules that describe the anesthetist's actions are given in Table 13.12. The FAM table for these rules is given in Table 13.13. The abbreviations NB, NS, ZE, PS, PB, and PV refer to the linguistic variables "negative big," "negative small," "zero," "positive small," "positive big," and "positive very big," respectively. In Table 13.13, the slash symbol is an exclusive-or. The membership functions used in this case are bell-shaped functions that can be described by the following exponential equation, where ξ is the input value and λ the shifting of the function in relation to zero:

$$\eta = \exp[-\kappa(\xi - \lambda)^2] \tag{13.40}$$

The factor κ determines the width of the bell (Gaussian). The reference points for the membership functions are in Table 13.14. The resulting membership functions are shown in Fig. 13.40, where a value of $\kappa = 5$ is used for all the membership functions for e and for u.

The results for a simulation run using the linguistic rules and reference points provided in Tables 13.12–13.14 are shown in Fig. 13.41. A max-min composition

TABLE 13.12
Linguistic rules

Rule number	Input e	Input $\int e$	Output u
1	NS	—	PB
2	PS	—	PS
3	NB	—	PV
4	PB	—	ZE
5	ZE	ZE	PM
6	ZE	PS	PS
7	ZE	NS	PB
8	—	NB	PV
9	—	PB	ZE

TABLE 13.13
FAM for the nine rules of Table 13.12

$e \backslash \int e$	NB	NS	ZE	PS	PB
NB	PV	PV	PV	PV	PV\ZE
NS	PB\PV	PB	PB	PB	PB\ZE
ZE	PV	PB	PM	PS	ZE
PS	PS\PV	PS	PS	PS	PS\ZE
PB	ZE\PV	ZE	ZE	ZE	ZE

TABLE 13.14
The reference points of the membership functions

	Input e, (mm Hg)	Input $\int e$, (mm Hg · s)	Output	Output u, (%)
NB	−10	−160	ZE	0
NS	−5	−90	PS	1
ZE	0	0	PM	2
PS	5	90	PB	3
PB	10	160	PV	4

and a center of gravity defuzzification method were used here to evaluate the linguistic rules. In Fig. 13.41 at 200 s the set value is raised from 80 mm Hg to 88 mm Hg. At 800 and 1400 s two disturbances are applied. The controller is fast, and only a small overshoot is observed. At 2100 s the set value is reduced to 84 mm Hg.

Fuzzy Logic Applications to Image-Processing Equipment

This section discusses three kinds of equipment in the image-processing area [Takagi, 1992]. The first is an autofocus mechanism for a 35 mm camera. The second is an image stabilization controller for a camcorder. The third is a television.

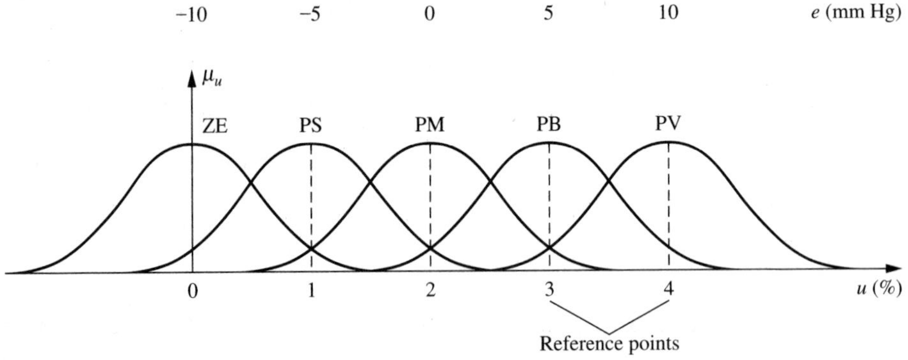

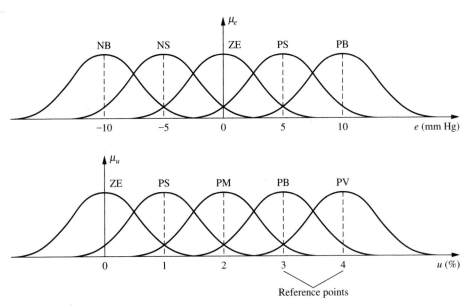

FIGURE 13.40
Membership functions for e and u.

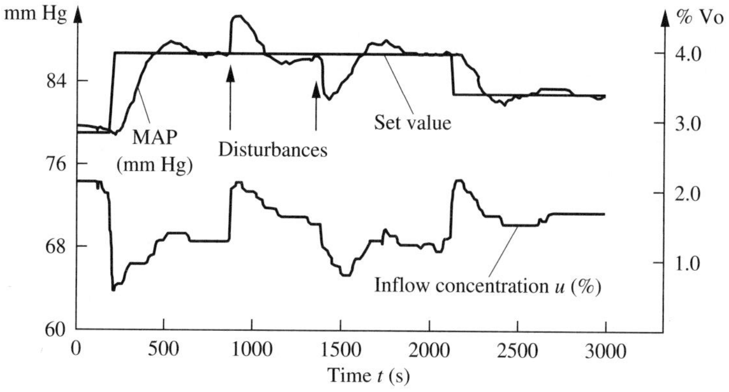

FIGURE 13.41
Simulation run (Meier et al., 1992).

AUTOFOCUS FOR A 35-mm CAMERA. A large Asian camera manufacturer has applied fuzzy logic to decide upon which object in the field of view the camera should be focused, and then to control the autofocus mechanism to focus on that object [Shingu and Nishimori, 1989].

First, distances to three points in the field of view (Fig. 13.42) are measured. Using these data and the relationships between them, fuzzy logic decides where the desired focus lies and then focuses on that point. Table 13.15 shows the rules used in the experimental simulations. The final product may have more rules. In Table 13.15 the symbols L (left), C (center), and R (right) describe the three measured points shown in Fig. 13.42. Also, the symbols Pl, Pc, and Pr denote the plausibility of finding the object of focus at points L, C, or R, respectively.

Comparing Fig. 13.43*a* and 13.43*b*, we see that Fig. 13.43*a* has the main subject on the left, whereas Fig. 13.43*b* has the main subject at the center. However, the relationship is $L < C < R$ in both cases, and both satisfy Rules 2 and 5, listed in Table 13.15. In this case, the decision depends on the values of L, C, and R; and this comparison is done with the help of membership functions. It is more difficult for binary logic rules to model this situation because a large number of rules are required. A few fuzzy rules, however, can easily deal with this problem.

IMAGE STABILIZATION FOR CAMCORDERS. A prominent Asian electronics manufacturer has applied fuzzy logic to determine whether movement of the image in a camcorder image window is due to trembling of the hand holding the camcorder

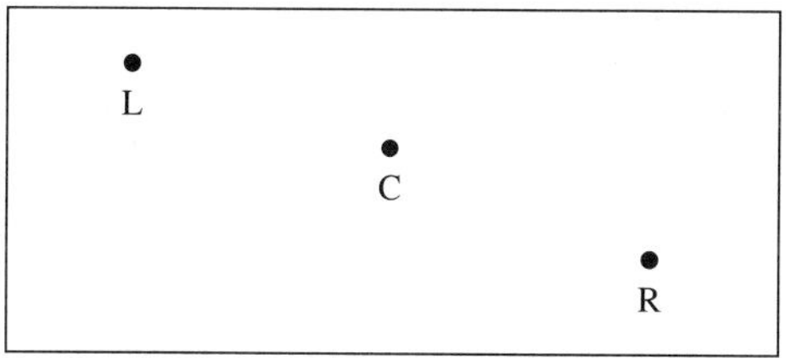

FIGURE 13.42
Three points to which distance is measured.

TABLE 13.15
Fuzzy rules for camera autofocus system

1. If C is near, then Pc is high.
2. If L is near, then Pl is high.
3. If R is near, then Pr is high.
4. If L is far and C is medium and R is near, then Pc is very high.
5. If R is far and C is medium and L is near, then Pc is very high.

or to the inherent motion of the object being photographed [Egusa et al., 1992]. This camera was put on the market in 1990.

As consumer camcorders become lighter and more portable, the problem of dealing with trembling has worsened and needs resolution. To tackle this, a customized LSI (laser sensor image) chip detects the motion vector and a fuzzy system decides if the motion is due to trembling. Digital signal processing is used to compensate if this is indeed the case.

The inputs to the fuzzy system are four motion vectors, each coming from one of the four regions into which the image has been divided, and their rates of change. Each of these four regions is further divided into 30 smaller areas. Two successive frames are compared (see Fig. 13.44) to compute the spatial difference values for each area. These differences are summed over the 30 areas in a region to produce a net difference R_i for region i. For each region, some one-shift vector results in the smallest value of R_i. This minimizing shift is the motion vector $\mathbf{v}_i$ for that region.

Fuzzy reasoning uses the values $\mathbf{v}_i$ as inputs to detect trembling of the hand. When this is the case, and there is no moving object, then the minimum R_i is almost zero. If there is a small moving object in the image and the hand is steady, then the area corresponding to the moving object has a spatial difference value that is different from the surrounding areas. However, the net R_i values should be small.

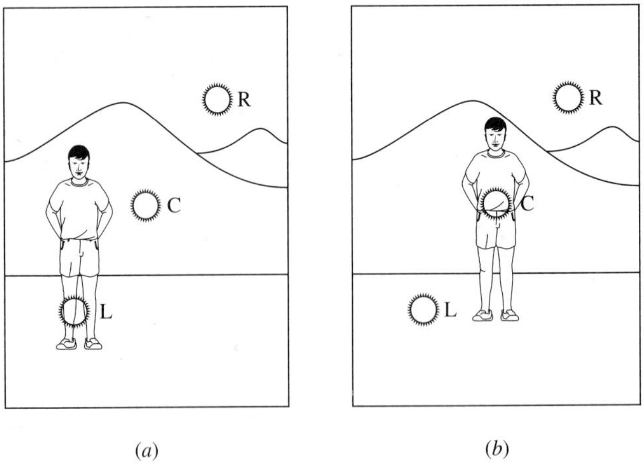

FIGURE 13.43
Example of scenes.

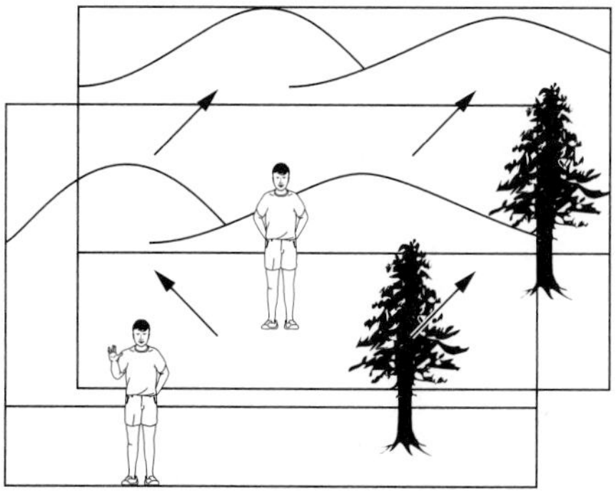

FIGURE 13.44
Example of motion vector.

When there is both hand motion and object motion, then the minimum R_i should be bigger than in the no-moving-object case.

The time derivatives of the R_i values are also used by the fuzzy system. This incorporates information about the size and motion of the moving object. A rule using these may be: *IF the four motion vectors are almost parallel, and their time differential is SMALL, THEN trembling of the hand is occurring and the direction of the trembling is the direction of the moving vectors.*

Once the direction of trembling is known, the frame in the buffer is shifted in the opposite direction to this motion, so that stabilization is achieved. This image processing becomes possible only because all signals used are digitized.

TELEVISION SETS. A well-known Asian manufacturer of consumer electronics used fuzzy inference for controlling the image quality of the reception of a TV set that appeared on the market in autumn 1990 [Hayashi and Yoshida, 1991]. The fuzzy system controls the contrast, brightness, velocity modulation, and sharpness. The input parameters are the ambient brightness in the room and the distance of the viewer from the set. One basic principle used to construct the fuzzy rules can be expressed as "When the room is brightly lit, and the viewer is far away, then the region boundaries on the picture should be sharper and clearer," whereas "If the viewer is close and in a darkened room, then the sharpness should be less, because the high-frequency components are accompanied by noise." Table 13.16 provides some of the other principles used.

The actual rules in the system are more detailed for finer control. There are four membership functions for brightness, three for distance, and seven for the output

TABLE 13.16
Fuzzy inference rules for image quality control of TV set

	Room light		Distance from TV to viewer		
	bright	**dark**	**far**	**mid**	**near**
Contrast	big	small	big	mid	small
Brightness	big	small		same	
Velocity modulation	big	small	big	mid	small
Sharpness	big	small		same	

variables. The inference is by the max-min-gravity method and is implemented by a lookup table in the final product.

Room brightness is computed from a light sensor to give one of eight values. Viewer distance is computed by locating the remote control and has three possible values. The microcontroller uses these to consult the lookup table, and four output values are received. This process is repeated every 50 milliseconds.

Customer Adaptive Fuzzy Control of Home Heating System

A European company introduced fuzzy logic control to a new generation of furnace controllers in a private home heating system [Altrock et al., 1993]. In the system, both the fuzzy logic controller and the conventional control system were implemented on a standard eight-bit microcontroller. The fuzzy logic controller ensures optimum adaptation to changing customer heating demands while using one sensor less than the system of the previous generation.

One type of traditional centralized heating system burns a diesel-type fuel in its furnace to heat the water supply (boiler). From the boiler, the hot water is distributed by a pipe system to individual radiators in the house. To meet the different needs of the customer, the furnace must constantly be adjusting the temperature of the boiler water in relation to the outdoor temperature (heat characteristic). To measure the outdoor temperature, a sensor is installed at the outside of the house. Figure 13.45 depicts the basic scheme of such a system.

The basic structure of a controller for this system is shown in Fig. 13.46. The controller itself realizes an on-off characteristic. If the water temperature in the

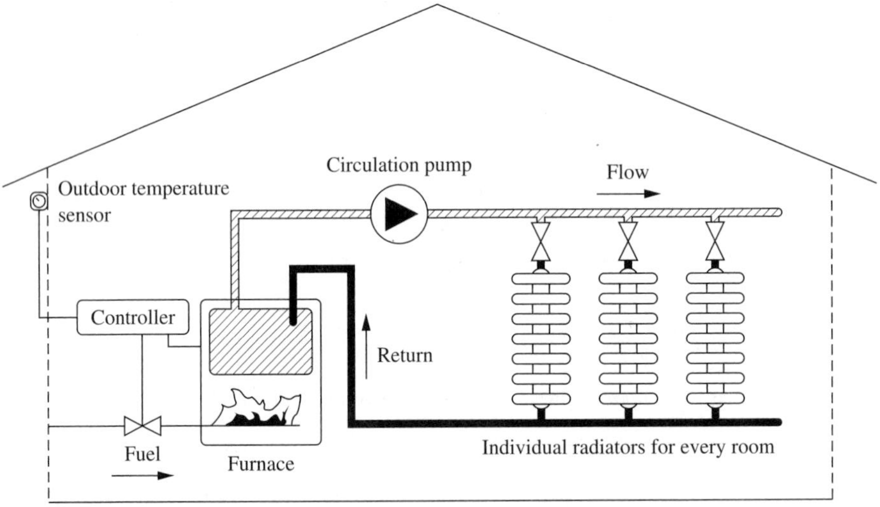

FIGURE 13.45
A centralized heating system (Altrock et al., 1993).

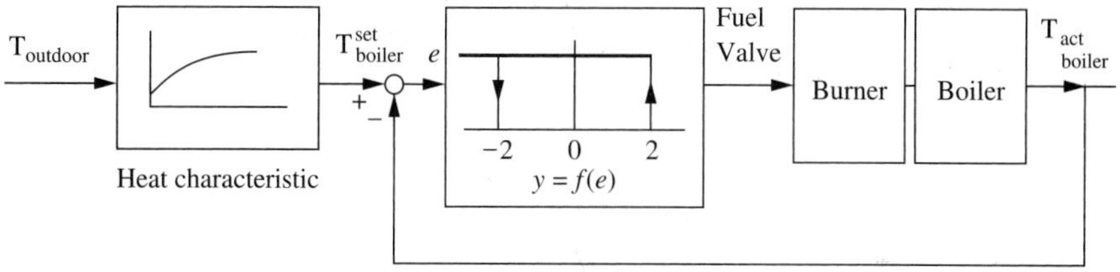

FIGURE 13.46
Conventional furnace controller (Altrock et al., 1993).

furnace drops to 2°C below the set temperature, the fuel valve opens and the ignition system starts the burning process. When the water temperature in the boiler itself rises to 2°C above the set temperature, the fuel valve closes. Traditionally, the boiler set temperature is a parameterized function $T_{\text{boiler}}^{\text{set}} = f(T_{\text{outdoor}})$ defined by European standards in the 1950s. Making the function depend largely on the outdoor temperature was appropriate then, when most houses had only poor thermal insulation. This model is now obsolete owing to the improvement of house insulation. Other factors such as ventilation, door/window openings, and personal lifestyle have to be considered as well.

The most important criteria describing individual customer heat demand patterns come from the actual energy consumption curve of the house, which is measured by the on-off ratio of the burner. An example of such a curve is given in Fig. 13.47. From the curve, four descriptive parameters are derived:

1. Current energy consumption, indicating current load
2. Medium-term tendency (I), indicating heating-up and heating-down phases
3. Short-term tendency (II), indicating disturbances like door/window openings
4. Yesterday's average energy consumption, indicating the house heating level

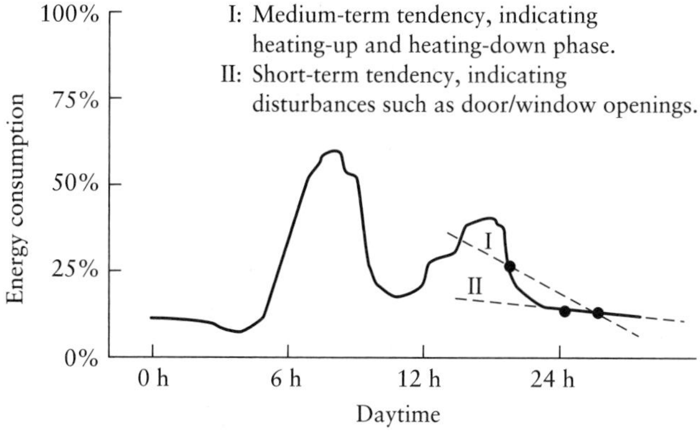

FIGURE 13.47
Actual energy consumption for the house (draft) (Altrock et al., 1993).

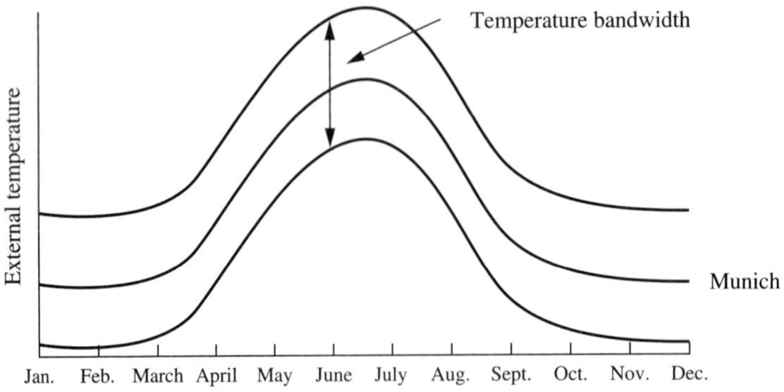

FIGURE 13.48
Average outside temperature in Munich.

These four descriptive parameters were used to form rules heuristically for the determination of the appropriate boiler set temperature. To allow for the formulation of plausibility rules (such as "temperatures below 0°C are rare in August") the appropriate average outdoor temperature for that season is also a system input parameter. These curves are plotted in Fig. 13.48. Since the average temperature curves are given, no outdoor temperature needs to be measured. Hence, the outdoor temperature sensor can be eliminated.

The structure of the new furnace controller is shown in Fig. 13.49. The fuzzy controller uses a total of five inputs, four of which are derived from the energy consumption curve using conventional digital filtering techniques; the fifth is the average outdoor temperature. This input comes from a lookup table within the system clock. The output of the fuzzy system represents the estimated heat requirement of the house and corresponds to the T_{outdoor} value in the conventional controller (Fig. 13.46).

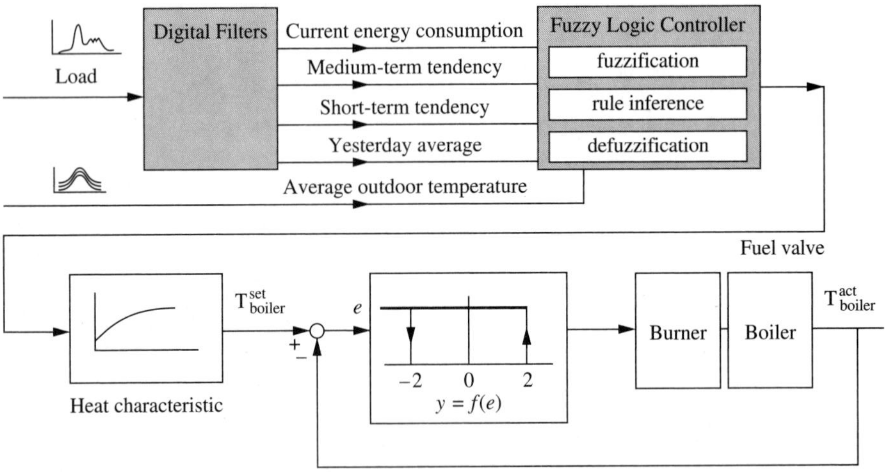

FIGURE 13.49
Schematics of the new furnace controller (Altrock et al., 1993).

The objective of the fuzzy controller is to estimate the actual heat requirement of the house. For this, IF-THEN rules were defined to express the engineering heuristics of the parameter estimation:

IF current energy consumption IS low
  AND medium-term tendency IS increasing
  AND short-term tendency IS decreasing
  AND yesterday's average IS medium
  AND average outside temperature IS very low
THEN estimated heat requirement IS medium-high

In total, 405 rules were defined for the parameter estimation. The inference strategy was extended to allow rules to be associated with a "degree of support," a number between 0 and 1 that expresses the individual importance of each rule with respect to all other rules. This allows for the expression of rules like

IF medium term tendency IS stable
  AND yesterday's average IS very high
THEN estimated heat required IS between high and very high, rather than more high

The inference method used to represent individual degrees of support is based on approximate reasoning and Fuzzy Associative Map (FAM) techniques: after fuzzification, all rule premises are calculated using the minimum operator for the representation of the linguistic AND and the maximum operator for the representation of the linguistic OR. Next, the degree of validity of the premise is weighted with the individual degree of support of the rule, resulting in the degree of truth for the conclusion. In the third step, all conclusions are combined using the maximum operator. The result of these steps is a fuzzy set. The center-of-maximum defuzzification method is used to arrive at a real value from a fuzzy output.

After completion of the design of the fuzzy controller and the definition of linguistic variables, membership function, and rules, the system was compiled to the target hardware, i.e., to 8051 assembly language. With this technology, the fuzzy controller only uses 2.1 bytes of the internal ROM area. By means of matrix rule representation and online development technology, the optimization of a complex fuzzy logic system containing 405 rules was done efficiently. The system performance was evaluated by connecting both the conventional controller and the fuzzy controller to a test house. One such example is shown in Fig. 13.50.

The result of the comparative performance tests showed that the fuzzy controller was highly responsive to the actual heat requirement of the house. It was very reactive to sudden changes in heat demand, like the return of house inhabitants from vacation. Additionally, the elimination of the outdoor temperature sensor saved about \$30 in production costs and even more in installation cost. By setting the boiler set temperature beneath the level typically used by a conventional controller in low-load periods, the fuzzy controller actually saved energy (Altrock et al., 1993).

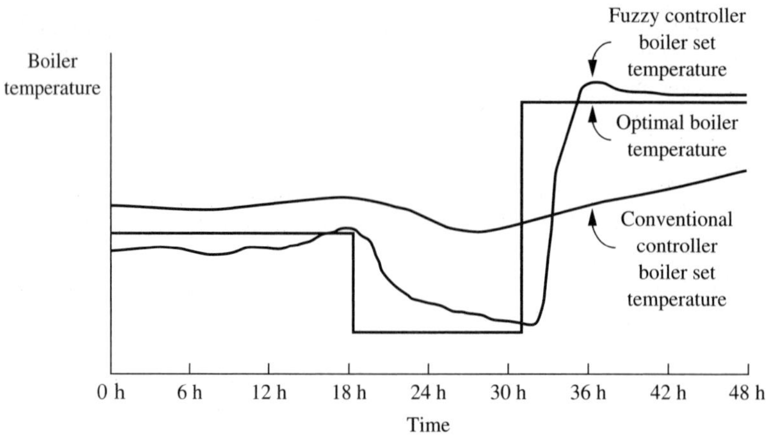

FIGURE 13.50
Conventional performance test (Altrock et al., 1993).

Adaptive Fuzzy Systems

A conventional fuzzy system (Fig. 13.51) usually takes the form of an iteratively adjusting model. In such a system, input values are normalized and converted to fuzzy representations, the model's rule base is executed (in a parallel fashion) to produce a consequent fuzzy region for each solution variable, and the consequent regions are defuzzified (converted to crisp values) to find the expected value of each solution variable. The solution variable is the control. A change is made to the plant, which the sensors pick up and feed back into the fuzzy controller as the next set of input values.

The rules executed by a conventional fuzzy control system are often time-series or process/state-based systems. For example, in control of the action of a steam turbine, the typical rules (Cox, 1993) are

- If temperature($t-1$) is *hot* and temperature(t) is *warm,* then throttle speed(t) is *small open.*
- If [temperature(t) − temperature($t-1$)] is a *small decline,* then throttle speed(t) is *small closed.*

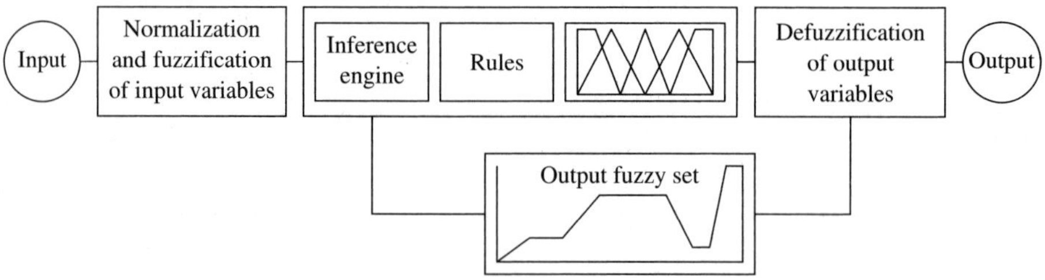

FIGURE 13.51
A conventional fuzzy system.

- If throttle speed ($t - 1$) is above throttle speed (t), then injection pressure is *large increase.*

This conventional system takes sensor readings of the turbine's temperature and pressure in the current time frame (t) and compares them with sensor values taken at time $t - 1$. It then makes changes in the current control surface (adjusts the throttle speed) on the basis of the differences. But it processes its inputs against a fixed set of instructions; it does not adapt.

On the other hand, an adaptive control system adjusts its control surface in accord with varying signal parameters. To see how an adaptive system works, we can look at how to make a simple nonadaptive automobile speed control system (see Fig. 13.52, left) adaptive. In this system, the initial required speed setting causes the throttle controller to open or close the engine throttle at a certain rate (Cox, 1993). A change in load torque caused by a change in the road grade is read as a system disturbance and can modify the throttle specification. The throttle action causes an increase or decrease in the speed of the engine, which is read by the tachometer. If the speed exceeds the setting, then the current speed requirement is reduced. This reduced speed is passed to the throttle controller, and the cycle repeats itself.

The system just described can be made adaptive by adding a facility for changing the step size of the control on a run length metric (Fig. 13.52, right). It solves this problem by modifying the height of the speed signal step by an adaptation rule. This rule intensifies or dilutes the speed signal depending on whether it has changed since it was last read. A typical rule might be to double the step size if there has been no change, and to halve it otherwise.

Although it is an improvement over a nonadaptive system, the system shown in Fig. 13.52 still lacks a means through which the controller can reorganize itself in a permanent way so that it is better able to deal with new classes of stimuli. A combination of adaptive control system with fuzzy logic can realize this system self-organization.

An adaptive fuzzy logic system not only adjusts to time- or process-phased conditions but also changes the supporting system controls. That is, an adaptive system modifies the characteristics of the rules, the topology of the fuzzy sets, and the

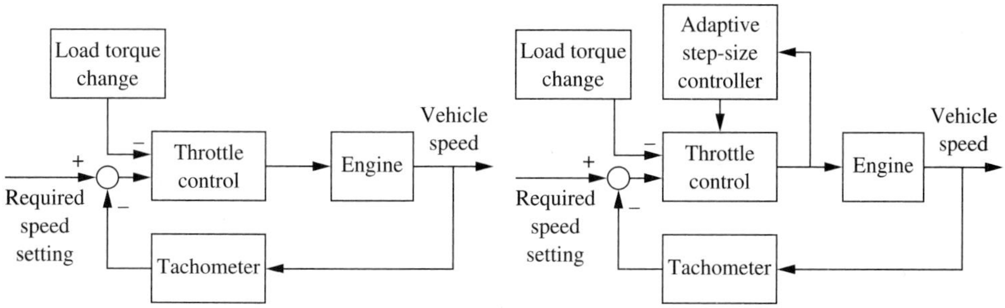

FIGURE 13.52
A nonadaptive conventional car speed control system (left) and an adaptive system (right) (Cox, 1993).

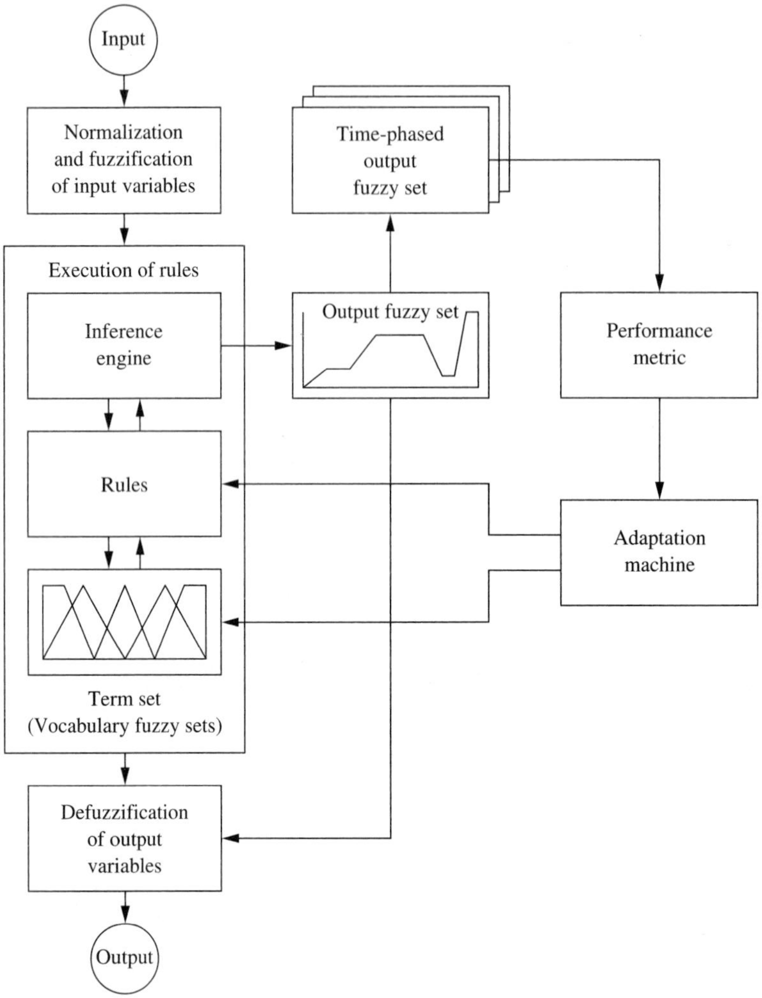

FIGURE 13.53
The two most important adaptation techniques of a fuzzy system (shown by the two outputs of the adaptation machine) are changes in the weighting of the rules and dynamic adjustment of the term set (Cox, 1993).

method of defuzzification based on predictive convergence metrics (or, more simply, on how quickly it is approaching or leaving a goal state).

As Fig. 13.53 shows, an adaptive fuzzy system performs a conventional fuzzy inference, attempting to develop a value for the solution variable from the rules in its knowledge repository. The characteristics of the fuzzy region created for the solution variable are stored in a time- or process-phased buffer where they can be accessed. To produce a feedback signal to the fuzzy model, a performance metric, often an expert system or simply an algorithm that measures the change between sensor measurements, interrogates the current and the stored array of past solutions

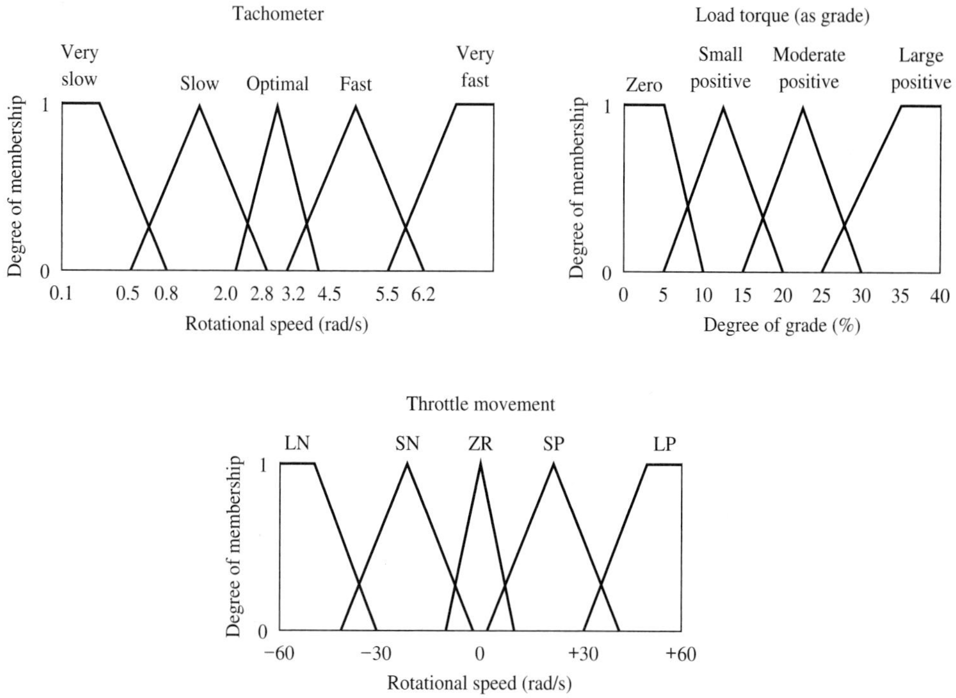

FIGURE 13.54
The inputs to the fuzzy car speed controller are engine speed (tachometer reading) and load torque (caused by the road grade). The output—throttle movement—has five regions: large negative(LN), small negative (SN), zero (ZR), small positive (SP), and large positive (LP) (Cox, 1993).

(as well as, perhaps, a set of statistics associated with the ongoing model). This signal is fed back into the adaptation machine, which decides what changes to make in the underlying fuzzy model.

In a simple adaptive fuzzy controller—for example, the vehicle speed controller mentioned earlier—there are two inputs and one output. The processor reads the current vehicle speed and the load torque produced by the road grade (the two inputs) and adjusts the throttle to increase or decrease the car's speed (the output) (Fig. 13.54).

Throttle control movement is the solution (output) variable from the fuzzy controller. The answer is developed by applying all the rules and then taking the expected value of the resulting fuzzy region. Examples of some rules for the speed controller are

- If tachometer is *very slow* and load torque is *zero,* then throttle is *large positive.*
- If tachometer is *slow* and load torque is *small positive,* then throttle is *small positive.*

The FAM table for these rules is a 4×5 matrix shown in Table 13.17.

TABLE 13.17
The FAM table of the car speed control system

	Tachometer				
Load torque	**Very slow**	**Slow**	**Optimum**	**Fast**	**Very fast**
Zero	LP	SP	ZR	SN	LN
Small positive	LP	SP	ZR	ZR	SN
Moderate positive	LP	SP	SP	SP	ZR
Large positive	LP	LP	LP	SP	SP

In attempting to maintain a consistent control surface response, an autoadaptive fuzzy controller reorganizes the control surface itself to meet the objectives of plant equilibrium and system stability. The controller performs the adaptation by measuring the distance from the equilibrium region in the control surface to regions of increased surface stress—that is, to areas on the surface that appear to be new areas of equilibrium. There are several interconnected means of allowing a fuzzy system to adapt. These include the management of weights attached to the rules, the dynamic hedging of the fuzzy regions, the structural modification of the fuzzy sets, the redefinition of truth in the fuzzy model, and the selection of alternative methods of defuzzification. The first two are the most important.

Rule weights in an adaptive system must be added to the FAM as another dimension. They hold the contribution weights connected with each rule and are factors by which the various primes truths are multiplied as a system adapts to its environment. They may range from 0 up to, but not including, 2. When the weight is less than 1, then the truth of the premise is reduced; when the weight is greater than 1, then the truth of the premise is increased. This weighting determines how much the rule affects the final outcome of the model.

In general, the weight modification algorithm is fairly simple: It consists of multiplying the various weights by an error factor. In supervised training, the error factor is the ratio of the actual system output to the correct output. In an autoadaptive mode, it is the mean square distance from the center of the optimum control region to the center of the system response.

TABLE 13.18
Contribution weight matrix for an adaptive fuzzy model (Cox, 1993)

	Tachometer				
Load torque	**Very slow**	**Slow**	**Optimum**	**Fast**	**Very fast**
Zero	0.70	0.74	1.00	1.11	1.25
Small positive	0.65	0.97	1.08	1.20	1.30
Moderate positive	0.82	0.94	1.11	1.28	1.31
Large positive	0.91	1.03	1.18	1.39	1.44

When an adaptive fuzzy model is initialized, all the weights are set to unity so that no modification of the premise truth is performed. After the adaptive system has run for a while, the weights will display a bias toward the areas where most of the data are located. Table 13.18 shows a contribution weight matrix after a few epochs of running. Note the general bias toward the lower right region of the FAM, indicating that most data are falling within this area. The amplification of the rule weights means that the system is adapting to a change in its central region of control.

Changing the regions is another method for fuzzy adaptive systems. Fuzzy adaptive systems must reorganize their term sets in response to changes in their operating environment. The term, or fuzzy, sets are broadened or narrowed according to the feedback from the previous model output cycle. If the system response in the preceding model execution was above the expected or desired value, then for each fuzzy set in the term set that was accessed during the execution, the domain is slightly narrowed. That is, the left edge of the domain is moved slightly to the right and the right edge is moved slightly to the left. Similarly, if the system response was below expectation, the involved domains are slightly to the left and their right edges are moved slightly to the right (see Fig. 13.55).

The modification of the fuzzy region is constrained by a restriction on the amount of overlap permitted among adjoining fuzzy sets. According to that restriction, for a vertical line drawn through a region of overlap, the truth membership of the points where the line intersects with the fuzzy sets must always total less than or be equal to 1.

Although obviously useful in control systems, adaptive fuzzy systems are particularly appropriate in econometric, marketing, project management, transportation, and field terrain analysis (Cox, 1993). The near future will see a gradual fusing of self-organizing technologies such as fuzzy logic, neural networks, and genetic algorithms. Adaptive fuzzy systems seem a highly suitable foundation for the easy integration and the ergonomic deployment of these techniques. As businesses use fuzzy logic more in addressing highly complicated and computationally deep problems, not only will fuzzy logic gain a wider acceptance by process designers but so too will adaptive fuzzy systems.

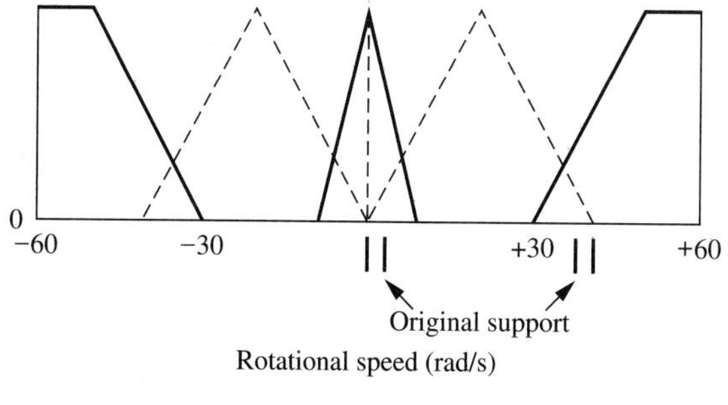

FIGURE 13.55
Adjusting the term set means widening (as shown for the SP region) or narrowing individual fuzzy regions, depending on whether the previous response was below or above the expected value. To avoid overlap problems, the SP region should stay within the dashed lines (Cox, 1993).

SUMMARY

New generations of fuzzy logic controllers are based on the integration of conventional and fuzzy controllers. Fuzzy clustering techniques have also been used to extract the linguistic IF-THEN rules from the numerical data. In general, the trend is toward the compilation and fusion of different forms of knowledge representation for the best possible identification and control of ill-defined complex systems. The two new paradigms—artificial neural networks and fuzzy systems—try to understand a real-world system starting from the very fundamental sources of knowledge, i.e., patient and careful observations, measurements, experience, and intuitive reasoning and judgments, rather than starting from a preconceived theory or mathematical model. Advanced fuzzy controllers use adaptation capabilities to tune the vertices or supports of the membership functions or to add or delete rules to optimize the performance and compensate for the effects of any internal or external perturbations. Learning fuzzy systems try to learn the membership functions or the rules. In addition, principles of genetic algorithms, for example, have been used to find the best string representing an optimum class of input or output symmetrical triangular membership functions (see Chapter 4).

REFERENCES

Altrock, C., H. Arend, B. Krause, C. Steffens, and Behrens-Rommler (1993). *Customer-adaptive fuzzy control of home heating system,* IEEE Press, Piscataway, NJ, pp. 115–119.

Cox, E. (1993). "Adaptive fuzzy systems," *IEEE Spectrum,* February, pp. 27–31.

Craig, J. J. (1986). *Introduction to robotics—mechanics and control,* Addison-Wesley, pp. 173–176.

Egusa, Y., H. Akahori, A. Morimura, and N. Wakami (1992). "An electronic video camera image stabilizer operated on fuzzy theory," *IEEE international conference on fuzzy systems,* IEEE Press, San Diego, pp. 851–858.

Hayashi, H., and K. Yoshida (1991). "... Fuzzy-AI vision C-20ZS101," *Television Technics & Electronics,* No. 39, January, pp. 91–96 (in Japanese).

Kailath, T. (1980). *Linear systems,* pp. 209–211, Prentice Hall.

Kiszka, J. B., M. M. Gupta, and P. N. Nikfrouk (1985). "Some properties of expert control systems," in M. M. Gupta, A. Kendal, W. Bandler, and J. B. Kiszka (eds.), *Approximate reasoning in expert systems,* Publishers B.V. (North-Holland), pp. 283–306, Elsevier Science.

Lee, C. C. (1990). "Fuzzy logic in control systems: fuzzy controllers—part I, part II," *IEEE Trans. on Sys. Man and Cyber.,* vol. 2092, pp. 404–435.

Ljung, L., and T. Soderstrom (1983). *Theory and practice of recursive identification,* MIT Press, Cambridge, MA.

Loe, S. (1991). "SGS-Thomson launches fuzzy-logic research push," *Electronic World News,* August 12, p. 1.

Mamdani, E. (1974). "Application of fuzzy algorithms for control of simple dynamic plant," *Proc. IEEE,* pp. 1585–1588.

Mamdani, E. H., and R. R. Gaines (eds.) (1981). *Fuzzy reasoning and its applications,* Academic Press, London.

Meier, R., J. Nieuwland, A. Zbinden, and S. Hacisalihzade (1992). "Fuzzy logic control of blood pressure during anesthesia," *IEEE Control Syst.,* December, pp. 12–17.

Pappas, C., and E. Mamdani (1976). "A fuzzy logic controller for a traffic junction," Research report, Queen Mary College, London.

Phillips, C. L., and R. D. Harbor (1988). *Feedback control systems,* Prentice Hall, pp. 2–4.

Quail, S., and S. Adnan (1992). "State of the art in household appliances using fuzzy logic," in J. Yen, R. Langari, and L. Zadeh (eds.), *Proceedings of the second international workshop—industrial fuzzy control and intelligent systems,* IEEE Press, College Station, TX, pp. 204–213.

Reid, T. (1990). "The future of electronics looks 'fuzzy'; Japanese firms selling computer logic products," *Washington Post,* Financial Section, p. H-1.
Rogers, M., and Y. Hoshai (1990). "The future looks 'fuzzy,'" *Newsweek,* vol. 115, No. 22.
Shingu, T., and E. Nishimori (1989). "Fuzzy-based automatic focusing system for compact camera," *Proceedings of the third international fuzzy systems association congress,* Seattle, WA, pp. 436–439.
Sugeno, M. (ed.) (1985). *Industrial application of fuzzy control,* North-Holland, New York.
Takagi, H. (1992). "Survey: fuzzy logic applications to image processing equipment," in J. Yen, R. Langari, and L. Zadeh (eds.), *Proceedings of the second international workshop—industrial fuzzy control and intelligent systems,* IEEE Press, College Station, TX, pp. 1–9.
Vadiee, N. (1993). "Fuzzy rule-based expert systems—I," in M. Jamshidi, N. Vadiee, and T. Ross (eds.), *Fuzzy logic and control: software and hardware applications,* Prentice Hall, Englewood Cliffs, NJ, pp. 51–85.
Vidyasagar, M. (1978). *Nonlinear system analysis,* Prentice Hall, Englewood Cliffs, NJ.
Wiess, J. J., and M. L. Donnel (1979). "A general policy capturing device using fuzzy production rules," Gupta, M.M., R.K. Ragade, and R.R. Yager, eds., Advances in fuzzy set theory and applications, North Holland, New York, pp. 589–604.
Yen, J., R. Langari, and L. Zadeh (eds.) (1992). *Proceedings of the second international workshop on industrial fuzzy control and intelligent systems,* IEEE Press, College Station, TX.

PROBLEMS

13.1. The interior temperature of an electrically heated oven is to be controlled by varying the heat input, u, to the jacket. The oven is shown in Fig. P13.1*a*. Let the heat capacities of the oven interior and of the jacket be c_1 and c_2, respectively. Let the interior and the exterior jacket surface areas be a_1 and a_2, respectively. Let the radiation coefficients of the interior and exterior jacket surfaces be r_1 and r_2, respectively. Assume that there is uniform and instantaneous distribution of temperature throughout, and the rate of loss of heat is proportional to area and the excess of temperature over that of the surroundings. If the external temperature is T_0, the jacket temperature is T_1, and the oven interior temperature is T_2, then we have

$$\begin{aligned} c_1\dot{T}_1 &= -a_2r_2(T_1 - T_0) - a_1r_1(T_1 - T_2) + u \\ c_2\dot{T}_2 &= a_1r_1(T_1 - T_2) \end{aligned}$$

Let the state variables be the excess of temperature over the exterior, i.e., $x_1 = T_1 - T_0$ and $x_2 = T_2 - T_0$. With these substituted into the preceding equations we find that

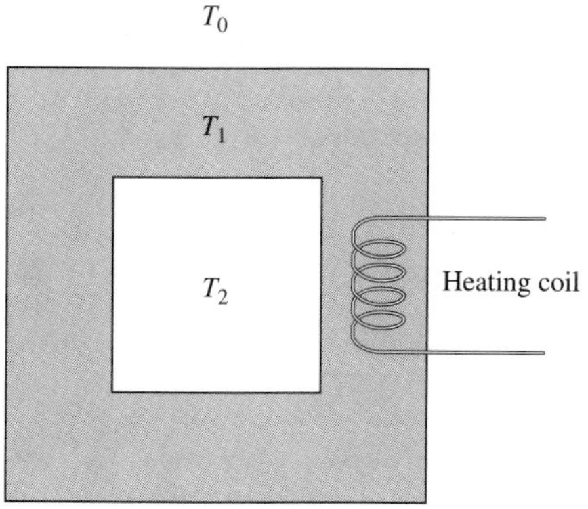

FIGURE P13.1*a*

they can be written as

$$\dot{x}_1 = -\frac{(a_2 r_2 + a_1 r_1)}{c_1} \cdot x_1 + \frac{a_1 r_1}{c_1} \cdot x_2 + \frac{1}{c_1} \cdot u$$

and

$$\dot{x}_2 = \frac{a_1 r_1}{c_2} \cdot x_1 - \frac{a_1 r_1}{c_2} \cdot x_2$$

Assuming that

$$\frac{a_2 r_2}{c_1} = \frac{a_1 r_1}{c_1} = \frac{a_1 r_1}{c_2} = \frac{1}{c_1} = 1$$

we have

$$x_1(t+1) = -2x_1(t) + x_2(t) + u(T)$$

and

$$x_2(t+1) = x_1(t) - x_2(t)$$

The membership functions for each of x_1, x_2, and u, each given on the same universe, are shown in Fig. P13.1*b*. For each of the variables, membership functions are taken to be low (L), medium (M), and high (H) temperatures.

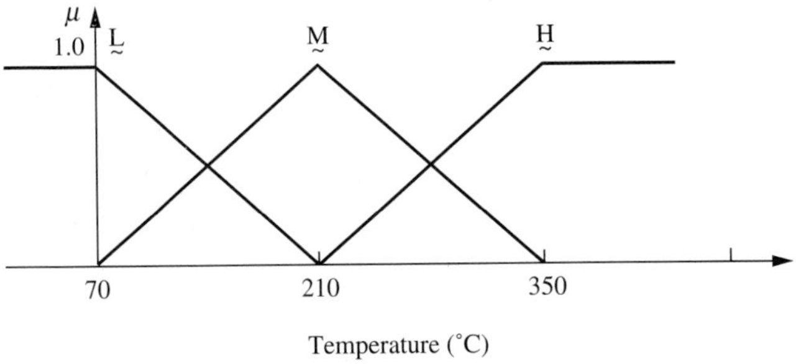

FIGURE P13.1*b*

Using the accompanying FAM table, conduct a graphical simulation of this control problem. The entries in the table are the control actions (u). Conduct at least four simulation cycles similar to Example 13.2. Use initial conditions of $x_1(0) = 80°$, and $x_2(0) = 85°$.

Jacket temperature excess, x_1	Interior temperature excess, x_2		
	L	**M**	**H**
L	H	M	L
M	H	—	L
H	H	M	L

13.2. Conduct a simulation of an automobile cruise control system. The input variables are speed, angle of inclination of the road, and throttle position. Let speed = 0 to

100 (mph), incline = −10 to +10 degrees, and throttle position = 0 to 10. The dynamics of the system are given by the following:

$$T = k_1 v + \theta k_2 + m\dot{v}$$
$$\dot{v} = v(n+1) - v(n)$$
$$T(n) = k_1 v(n) + \theta(n) k_2 + m(v_{n+1} - v_n)$$
$$v_{n+1} = \left(1 - \frac{k_1}{m}\right) v(n) + T(n) - \frac{k_2}{m}\theta(n)$$
$$v_{n+1} = k_a v(n) + [1 \quad -k_b]\begin{bmatrix} T(n) \\ \theta(n) \end{bmatrix}$$

where T = torque
k_1 = viscous friction
v = speed
θ = angle of incline
$k_2 = mg\sin\theta$
$\dot{v}$ = acceleration
m = mass
$k_a = 1 - \frac{k_1}{m}$ and $k_b = \frac{k_2}{m}$

The membership function for speed is determined by the cruise control setting, which we will assume to be 50 mph. The membership functions are shown in Fig. P13.2.

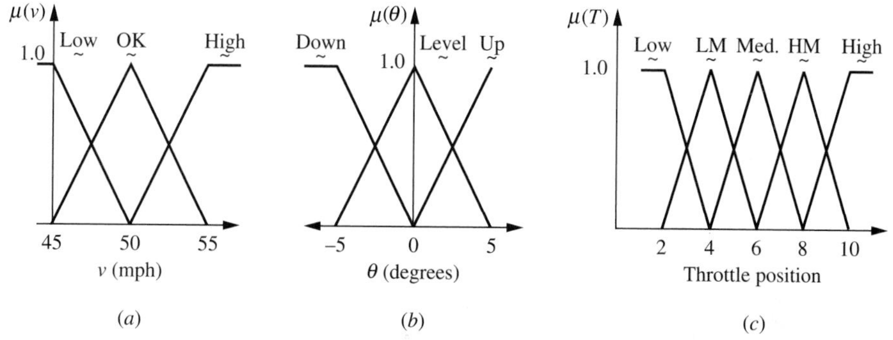

FIGURE P13.2

The FAM table is shown next:

	Inclination of the road		
Speed	**Up**	**Level**	**Down**
High	LM	LM	Low
OK	HM	Medium	LM
Low	High	HM	HM

Use initial conditions of speed = 52.5 mph and angle of incline = −5°. Conduct at least four simulation cycles.

13.3. A printer drum is driven by a brushless DC motor. The moment of inertia of the drum is $J = 0.00185\ \text{kg} \cdot \text{m}^2$. The motor resistance is $R = 1.12\ \Omega$. The torque constant for the motor is $K_T = 0.0363\ \text{N} \cdot \text{m/A}$. The back EMF constant is $k = 0.0363$ V/(rad/s). The equation of the system is

$$J\ddot{\theta} = \frac{K_T(V - \dot{\theta}k)}{R}$$

where
$$\frac{(V - \dot{\theta}k)}{R} = I = \text{motor current}$$
$$\theta = \text{rotational angle}$$
$$V = \text{motor control voltage}$$

The state variables are $x_1 = \theta$ and $x_2 = \dot{\theta}$. Also

$$\ddot{\theta} = \frac{K_T}{JR}V - \frac{K_T k}{JR}\dot{\theta}$$

Now

$$x_2 = \dot{x}_1$$

Therefore,

$$\dot{x}_2 = \frac{K_T}{JR}V - \frac{K_T k}{JR}x_2$$

Substituting in the values of the constants, we find

$$\dot{x}_2 + 0.64x_2 = 17.5\ \text{V}$$

The resulting difference equations will be

$$x_1(k+1) = x_2(k) + x_1(k)$$
$$x_2(k+1) = 17.5V(k) + 0.36x_2(k)$$

The motor can be controlled to run at constant speed or in the position mode. The membership functions for x_1, x_2, and V are shown in Fig. P13.3. The rule-based system is summarized in the following FAM table:

	x_2		
x_1	**Negative**	**Zero**	**Positive**
N	PB	P	N
Z	P	Z	N
P	Z	Z	NB

Using the initial conditions of $x_1 = 7.5°$ and $x_2 = -150$ rad/s and the difference equations, conduct at least four graphical simulation cycles.

13.4. The basic mechanical system behind clocks that are enclosed in glass domes is the torsional pendulum. The general equation that describes the torsional pendulum is

$$J\frac{d^2\theta(t)}{dt^2} = \tau(t) - B\frac{d\theta(t)}{dt} - k\theta(t)$$

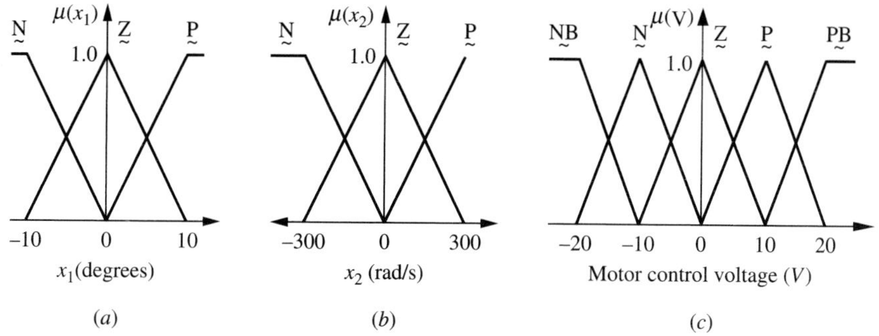

FIGURE P13.3

The moment of inertia of the pendulum bob is represented by J, the elasticity of the brass suspension strip is represented by k, and the friction between the bob and the air is represented by B. The controlling torque $\tau(t)$ is applied at the bob. When this device is used in clocks the actual torque is not applied at the bob but is applied through a complex mechanism at the main spring. The foregoing differential equation is the sum of the torques of the pendulum bob. The numerical values are $J = 1\ \text{kg} \cdot \text{m}^2$, $k = 5\ \text{N} \cdot \text{m/rad}$, and $B = 2\ \text{N} \cdot \text{m} \cdot \text{s/rad}$.

The final differential equation with the foregoing constants incorporated is given by

$$\frac{d^2\theta(t)}{dt^2} + 2\frac{d\theta(t)}{dt} + 5\theta(t) = \tau(t)$$

The state variables are

$$x_1 = \theta(t) \quad \text{and} \quad x_2(t) = \dot{\theta}(t)$$
$$\dot{x}_1 = \dot{\theta}(t) \quad \text{and} \quad \dot{x}_2(t) = \ddot{\theta}(t)$$

Rewriting the differential equation using state variables, we have

$$\dot{x}_2(t) = \tau(t) - 2x_2(t) - 5x_1(t) \tag{P13.4.1}$$
$$\dot{x}_1(t) = x_2(t) \tag{P13.4.2}$$

Using these equations,

$$\dot{x}_1(t) = x_1(k+1) - x_1(k)$$
$$\dot{x}_2(t) = x_2(k+1) - x_2(k)$$

in Eqs. (P13.4.1) and (P13.4.2), and rewriting the equations in terms of θ and $\dot{\theta}$ in matrix form, we have,

$$\begin{bmatrix} \theta(k+1) \\ \dot{\theta}(k+1) \end{bmatrix} = \begin{bmatrix} \theta(k) + \dot{\theta}(k) \\ -5\theta(k) - \dot{\theta}(k) \end{bmatrix} + \begin{bmatrix} 0 \\ \tau(k) \end{bmatrix}$$

The membership values for θ, $\dot{\theta}$, and τ are shown in Fig. P13.4. The rules for the control system are summarized in the accompanying FAM table.

	$\dot{\theta}$		
θ	**Positive**	**Zero**	**Negative**
P	NB	N	Z
Z	Z	Z	Z
N	Z	P	PB

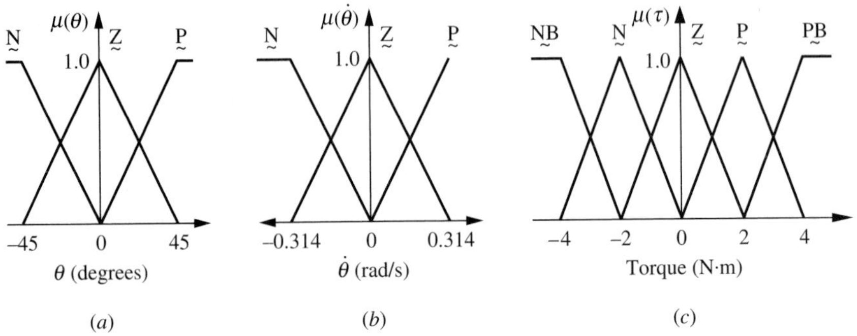

FIGURE P13.4

The initial conditions are given as

$$\theta(0) = 0.7^\circ$$
$$\dot{\theta}(0) = -0.2 \text{ rad/s}$$

Conduct a graphical simulation for the control system.

13.5. On the electrical circuit shown in Fig. P13.5*a* it is desired to control the output current at inductor L_2 by using a variable voltage source, V. By using Kirchoff's voltage law, the differential equation for loop 1 is given in terms of the state variables as

$$\frac{dL_1(t)}{dt} = -2L_1(t) + 2L_2(t) + 2V(t) \qquad \text{(P13.5.1)}$$

and that for loop 2 is

$$\frac{dL_2(t)}{dt} = 0.5L_1(t) - 2L_2(t) \qquad \text{(P13.5.2)}$$

Converting the system of differential equations into a system of difference equations, we get

$$L_1(k+1) = -L_1(k) + 2L_2(k) + 2V(k)$$
$$L_2(k+1) = 0.5L_1(k) - L_2(k)$$

Rewriting the equations in matrix form, we get

$$\begin{bmatrix} L_1(k+1) \\ L_2(k+1) \end{bmatrix} = \begin{bmatrix} -L_1(k) + 2L_2(k) \\ 0.5L_1(k) - L_2(k) \end{bmatrix} + \begin{bmatrix} 2V(k) \\ 0 \end{bmatrix}$$

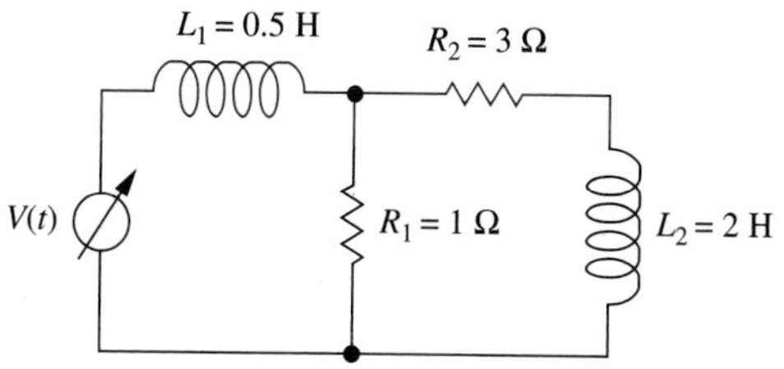

FIGURE P13.5*a*

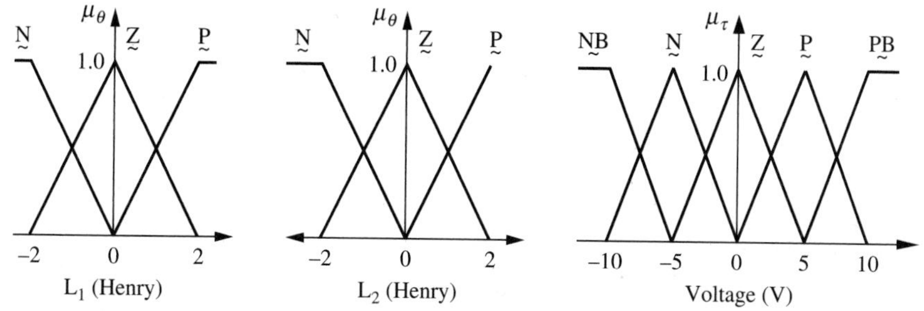

FIGURE P13.5*b*

The membership functions are given in Fig. P13.5*b* and the rules are presented in this table:

	L_2		
L_1	**Negative**	**Zero**	**Positive**
N	PB	P	Z
Z	Z	Z	Z
P	Z	N	NB

The initial conditions are $L_1(0) = 1\ H$ and $L_2(0) = -1\ H$. Conduct a simulation of the system.

13.6. In control system design, a system is studied in the *s* plane by using its Laplace transfer function. Its pole positions in the *s* plane determine whether the system will be stable or unstable. If all the poles are in the left hand side of the $j\omega$ axis, where $j\omega$ is the complex axis of the *s* plane, the system will be stable. If any pole is in the right hand side of the $j\omega$ axis, the system will be unstable. If a pole is on the $j\omega$ axis, the system will have a stable error. Furthermore, the pole position in the left hand side of the $j\omega$ axis will determine the speed with which a system reaches its stability. The farther the pole from the $j\omega$ axis, the faster the system will move toward a position of stability. Assume a system has one pole a, which is in the region of $(-6, 6)$ along the real axis (see Fig. P13.6); the system then has a stability equation as $\exp(at)$. Combining this stability information and fuzzy logic principles, design fuzzy rules and fuzzy membership functions for the system stability regions shown in the figure. In the figure k is a small value near 0. Make fuzzy regions around $-k$, $+1$, and -6 and around $+k$, $+1$, and $+6$.

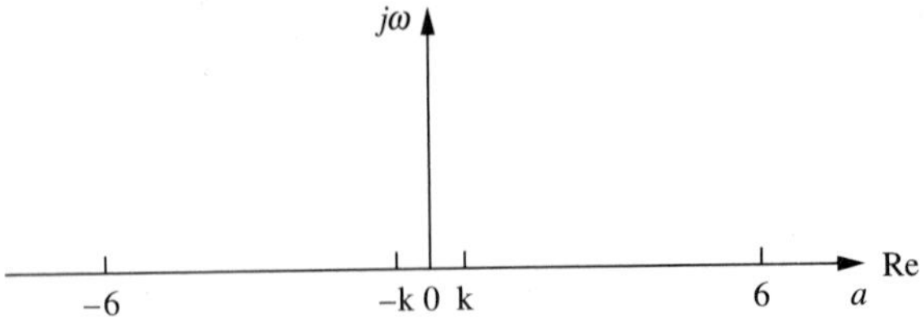

FIGURE P13.6

13.7. A dynamical system is represented by

$$y' + (1/\tau)y = Ax(t)$$

(*a*) Discretize the system's mathematical model into the form of a difference equation, i.e.,

$$y_{n+1} = f(x_n, y_n)$$

(*b*) For $A = 10.0$, $\tau = 0.1$, and a sampling interval $t = 0.01$, design a fuzzy rule-based system that gives values of y_{n+1} and x_n for a given value of y_n. Assume that x_n is varying over an interval between 0 and 1.

***13.8.** A robotic arm is shown in Fig. P13.8. Joints 1 and 2 are passive joints and free to rotate. A force **f** in a direction ϕ is applied at the end of the effector, as shown in the figure. The regions for the angles are defined as $-90° < \theta_1 < 90°$, $-45° < \theta_2 < 45°$, and $-60° < \phi < 60°$. It is desired to park the arm at $\theta_1 = \theta_2 = 0.0$ by a series of movements caused by the force applied at the end-effector.

Design the FAM table and determine the governing system equations for a fuzzy rule-based control system that will take the robotic arm from any orientation on the plane to the parking state.

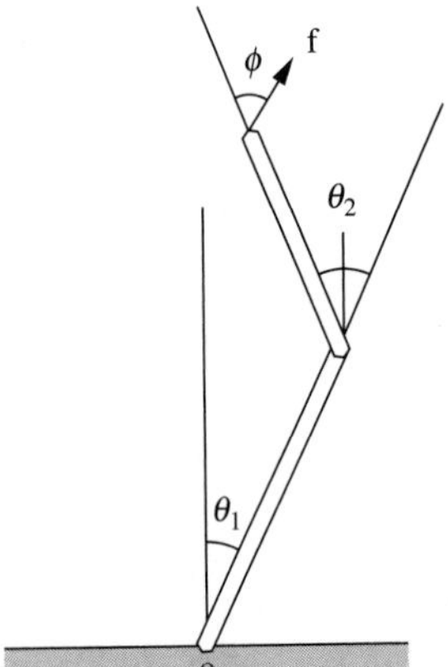

FIGURE P13.8

13.9. To simulate a temperature control system for a room, one must consider the following three controls:

*The derivation of the system equations is a difficult problem in mechanics.

- Opening or closing the window
- Adjusting the heating unit
- Turning the air conditioning on or off

The information that might potentially affect the control system could refer to the current temperature of the room, how heavily it is raining, and the present settings of the three controls (see Fig. P13.9). Each action shall be characterized by a vector containing three quantities:

- The desired window opening in inches
- The heat setting, which may range continuously from 0 (off) to 5 (high)
- The air conditioning state desired, which may be 0 (off) or 1 (on).

Five inputs are going to be used and measured:

- The room temperature in degrees Fahrenheit
- The rate of rainfall in millimeters per hour
- The current window opening in inches
- The current heat setting (between 0 and 5)
- The current air conditioning state (0 or 1).

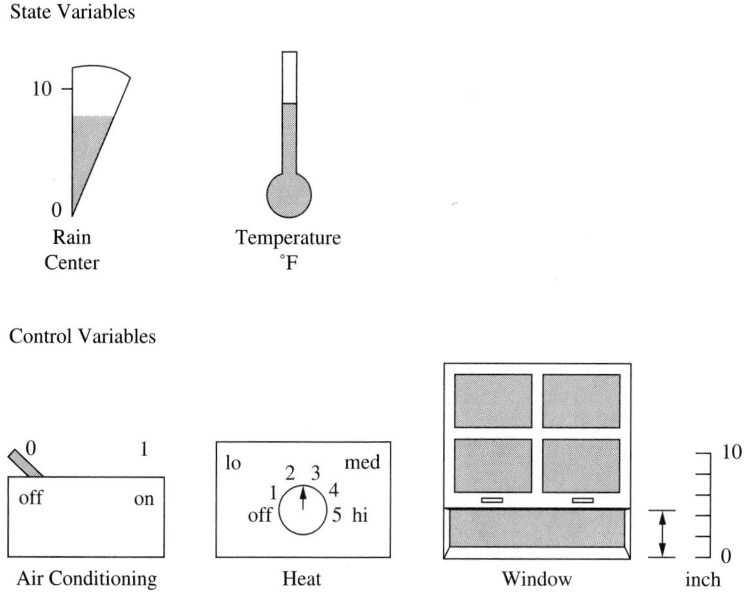

FIGURE P13.9

Design a fuzzy rule-based system to keep the room at a comfortable temperature.

CHAPTER 14

MISCELLANEOUS TOPICS

Fuzzy Wuzzy was a bear, Fuzzy Wuzzy had no hair; then
Fuzzy Wuzzy wasn't fuzzy, was he?

Traditional British poem
Unattributed, nineteenth century

This chapter exposes the reader to a few of the additional applications areas that have been extended with fuzzy logic. These few areas cannot cover the wealth of other applications, but they give to the reader an appreciation of the potential influence of fuzzy logic in almost any technology area. Addressed in this chapter are just three additional applications areas: optimization, inverse fuzzy equations, and linear regression.

FUZZY OPTIMIZATION

Most technical fields, including all those in engineering, involve some form of optimization that is required in the process of design. Since design is an open-ended problem with many solutions, the quest is to find the "best" solution according to some criterion. In fact, almost any optimization process involves trade-offs between costs and benefits because finding optimum solutions is analogous to creating designs—there can be many solutions, but only a few might be optimum, or useful, particularly where there is a generally nonlinear relationship between performance and cost. Optimization, in its most general form, involves finding the most optimum solution from a family of reasonable solutions according to an optimization criterion. For all but a few trivial problems, finding the global optimum (the best optimum solution) can never be guaranteed. Hence, optimization in the last three decades has focused on methods to achieve the best solution per unit computational cost.

In cases where resources are unlimited and the problem can be described analytically and there are no constraints, solutions found by exhaustive search [Akai, 1994] can guarantee global optimality. In effect, this global optimum is found by setting all the derivatives of the criterion function to zero, and the coordinates of the stationary point that satisfy the resulting simultaneous equations represent the solution. Unfortunately, even if a problem can be described analytically there are seldom situations with unlimited search resources. If the optimization problem also requires the simultaneous satisfaction of several constraints and the solution is known to exist on a boundary, then constraint boundary search methods such as Lagrangian multipliers are useful [deNeufville, 1990]. In situations where the optimum is not known to be located on a boundary, methods such as the steepest gradient, Newton-Raphson, and penalty function have been used [Akai, 1994], and some very promising methods have used genetic algorithms [Goldberg, 1989].

For functions with a single variable, search methods such as Golden section and Fibonacci are quite fast and accurate. For multivariate situations, search strategies such as parallel tangents and steepest gradients have been useful in some situations. But most of these classical methods of optimization [Vanderplaats, 1984] suffer from one or more disadvantages: the problem of finding higher-order derivatives of a process, the issue of describing the problem as an analytic function, the problem of combinatorial explosion when dealing with many variables, the problem of slow convergence for small spatial or temporal step sizes, and the problem of overshoot for step sizes too large. In many situations, the precision of the optimization approach is greater than the original data describing the problem, so there is an *impedance mismatch* in terms of resolution between the required precision and the inherent precision of the problem itself.

In the typical scenario of an optimization problem, fast methods with poorer convergence behavior are used first to get the process near a solution point, such as a Newton method, then slower but more accurate methods, such as gradient schemes, are used to converge to a solution. Some current successful optimization approaches are now based on this hybrid idea: fast, approximate methods first, slower and more precise methods second. Fuzzy optimization methods have been proposed as the first steps in hybrid optimization schemes. One of these methods will be introduced here. More methods can be found in Sakawa [1993].

One-Dimensional Optimization

Classical optimization for a one-dimensional (one independent variable) relationship can be formulated as follows. Suppose we wish to find the optimum solution, x^*, which maximizes the objective function $y = f(x)$, subject to the constraints

$$g_i(x) \leq 0, \qquad i = 1, m \tag{14.1}$$

Each of the constraint functions $g_i(x)$ can be aggregated as the intersection of all the constraints. If we let $\mathrm{C}_i = \{x \mid g_i(x) \leq 0\}$, then

$$\mathrm{C} = \mathrm{C}_1 \cap \mathrm{C}_2 \cap \cdots \cap \mathrm{C}_m = \{x \mid g_1(x) \leq 0,\ g_2(x) \leq 0, \ldots, g_m(x) \leq 0\} \tag{14.2}$$

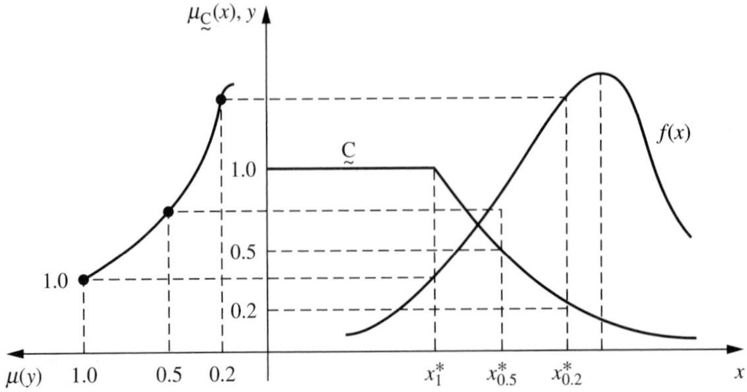

FIGURE 14.1
Function to be optimized, $f(x)$, and fuzzy constraint, C.

which is the feasible domain described by the constraints C_i. Thus, the solution is

$$f(x^*) = \max_{x \in C}\{f(x)\} \tag{14.3}$$

In a real environment, the constraints might not be so crisp, and we could have fuzzy feasible domains (see Fig. 14.1) such as "x could exceed x_0 a little bit." If we use λ-cuts on the fuzzy constraints $\underset{\sim}{C}$, fuzzy optimization is reduced to the classical case. Obviously, the optimum solution x^* is a function of the threshold level λ, as given in Eq. (14.4):

$$f(x^*_\lambda) = \max_{x \in C_\lambda}\{f(x)\} \tag{14.4}$$

Sometimes, the goal and the constraint are more or less contradictory, and some trade-off between them is appropriate. This can be done by converting the objective function $y = f(x)$ into a pseudogoal $\underset{\sim}{G}$ [Zadeh, 1972] with membership function

$$\mu_{\underset{\sim}{G}}(x) = \frac{f(x) - m}{M - m} \tag{14.5}$$

where $m = \inf_{x \in X} f(x)$

$M = \sup_{x \in X} f(x)$

Then the fuzzy solution set $\underset{\sim}{D}$ is defined by the intersection

$$\underset{\sim}{D} = \underset{\sim}{C} \cap \underset{\sim}{G} \tag{14.6}$$

membership is described by

$$\mu_{\underset{\sim}{D}}(x) = \min\{\mu_{\underset{\sim}{C}}(x), \mu_{\underset{\sim}{G}}(x)\} \tag{14.7}$$

and the optimum solution will be x^* with the condition

$$\mu_{\underset{\sim}{D}}(x^*) \geq \mu_{\underset{\sim}{D}}(x) \qquad \text{for all } x \in X \tag{14.8}$$

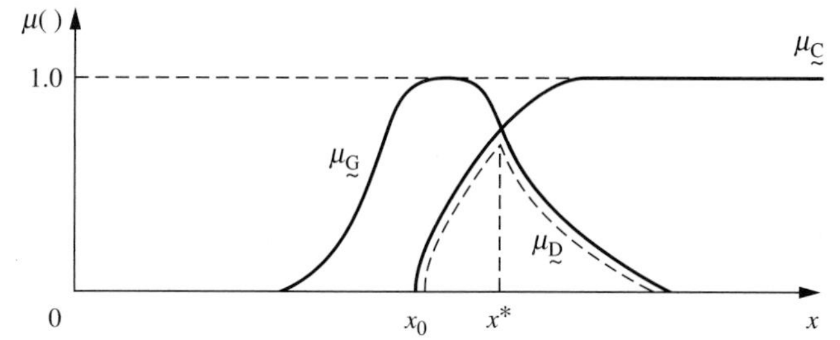

FIGURE 14.2
Membership functions for goal and constraint (Zadeh, 1972).

where $\mu_{\underset{\sim}{C}}(x)$, $\underset{\sim}{C}$, x should be substantially greater than x_0. Figure 14.2 shows this situation.

Example 14.1. Suppose we have a deterministic function given by

$$f(x) = xe^{(1-x/5)}$$

for the region $0 \leq x \leq 5$, and a fuzzy constraint given by

$$\mu_{\underset{\sim}{C}}(x) = \begin{cases} 1 & 0 \leq x \leq 1 \\ \dfrac{1}{1+(x-1)^2} & x > 1 \end{cases}$$

Both of these functions are illustrated in Fig. 14.3. We want to determine the solution set $\underset{\sim}{D}$ and the optimum solution x^*, i.e., find $f(x^*) = y^*$. In this case we have $M = \sup[f(x)] = 5$ and $m = \inf[f(x)] = 0$, hence Eq. (14.5) becomes

$$\mu_{\underset{\sim}{G}}(x) = \frac{f(x)-0}{5-0} = \frac{x}{5}e^{(1-x/5)}$$

which is also shown in Fig. 14.3. The solution set membership function, using Eq. (14.7), then becomes

$$\mu_{\underset{\sim}{D}}(x) = \begin{cases} \dfrac{x}{5}e^{(1-x/5)} & 0 \leq x \leq x^* \\ \dfrac{1}{1+(x-1)^2} & x > x^* \end{cases}$$

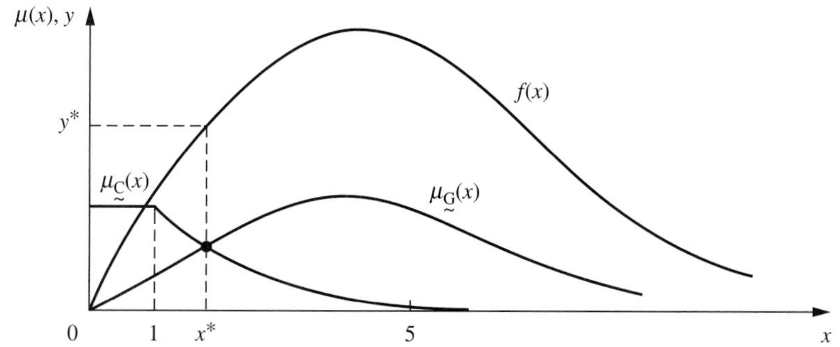

FIGURE 14.3
Problem domain for Example 14.1.

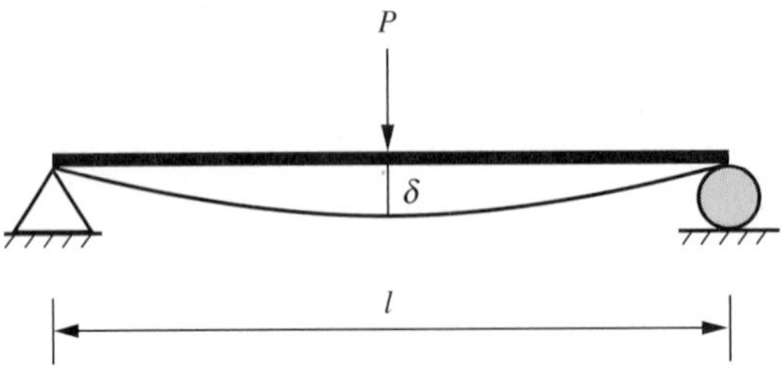

FIGURE 14.4
Simply supported beam with a transverse concentrated load.

and the optimum solution x^*, using Eq. (14.8), is obtained by finding the intersection

$$\frac{x^*}{5}e^{(1-x^*/5)} = \frac{1}{1+(x-1)^2}$$

and is shown in Fig. 14.3.

When the goal and the constraint have unequal importance the solution set $\underset{\sim}{D}$ can be obtained by the convex combination, i.e.,

$$\mu_{\underset{\sim}{D}}(x) = \alpha\mu_{\underset{\sim}{C}}(x) + (1-\alpha)\mu_{\underset{\sim}{G}}(x) \tag{14.9}$$

The single-goal formulation expressed in Eq. (14.9) can be extended to the multiple-goal case as follows. Suppose we want to consider n possible goals and m possible constraints. Then the solution set, $\underset{\sim}{D}$, is obtained by the aggregate intersection, i.e., by

$$\underset{\sim}{D} = \left(\bigcap_{i=1,m} \underset{\sim}{C}_i\right) \cap \left(\bigcap_{j=1,n} \underset{\sim}{G}_j\right) \tag{14.10}$$

Example 14.2. A beam structure is supported at one end by a hinge and at the other end by a roller. A transverse concentrated load P is applied at the middle of the beam, as in Fig. 14.4. The maximum bending stress caused by P can be expressed by the equation $\sigma_b = Pl/w_z$, where w_z is a coefficient decided by the shape and size of a beam and l is the beam's length. The deflection at the centerline of the beam is $\delta = Pl^3/(48EI)$ where E and I are the beam's modulus of elasticity and cross-sectional moment of inertia, respectively. If $0 \le \delta \le 2$ mm, and $0 \le \sigma_b \le 60$ MPa, the constraint conditions are these: the span length of the beam,

$$l = \begin{cases} l_1 & 0 \le l_1 \le 100 \text{ m} \\ 200 - l_1 & 100 < l_1 \le 200 \text{ m} \end{cases}$$

and the deflection,

$$\delta = \begin{cases} 2 - \delta_1 & 0 \le \delta_1 \le 2 \text{ mm} \\ 0 & \delta_1 > 2 \text{ mm} \end{cases}$$

To find the minimum P for this two-constraint and two-goal problem (the goals are the stress, σ_b, and the deflection, δ), we first find the membership function for the two goals and two constraints.

1. The μ_{G_1} for bending stress σ_b is given as follows:

$$P(0) = 0, \quad P(60\text{ MPa}) = \frac{w_z 60}{l}, \quad \text{and} \quad P(\sigma_b) = \frac{w_z\sigma_b}{l}; \quad \text{thus,} \quad \mu_{G_1} = \frac{\sigma_b}{60}$$

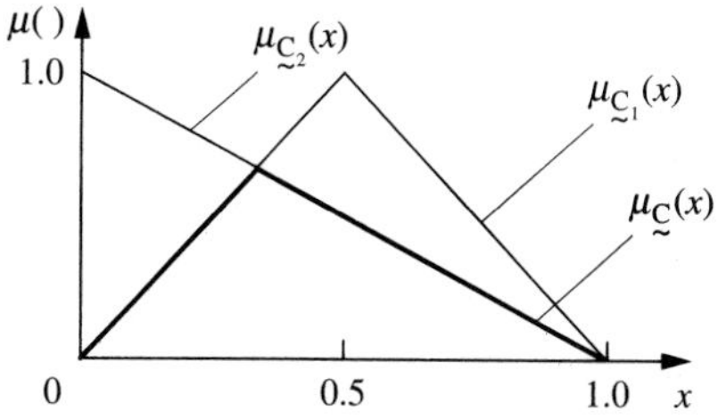

FIGURE 14.5
Minimum of two constraint functions.

To change the argument in μ_{G_1} into a unitless form, let $x = \sigma_b/60$, where $0 \leq x \leq 1$. Therefore, $\mu_{G_1}(x) = x$ when $0 \leq x \leq 1$.

2. The μ_{G_2} for deflection δ is as follows:

$$P(\delta) = \frac{48EI\delta}{l^3}, \quad P(0) = 0, \quad P(2) = \frac{48EI \times 2}{l^3} \quad \text{thus,} \quad \mu_{G_2} = \frac{\delta}{2}$$

Let $x = \delta/2$, so that the argument of μ_{G_2} is unitless. Therefore, $\mu_{G_2} = x, 0 \leq x \leq 1$.

3. Using Eq. (14.10), we combine $\mu_{G_1}(x)$ and $\mu_{G_2}(x)$ to find $\mu_G(x)$:

$$\mu_G(x) = \min(\mu_{G_1}(x), \mu_{G_2}(x)) = x \qquad 0 \leq x \leq 1$$

4. The fuzzy constraint function μ_{C_1} for the span is

$$\mu_{C_1}(x) = \begin{cases} 2x & 0 \leq x \leq 0.5 \\ 2 - 2x & 0.5 < x \leq 1 \end{cases}$$

where $x = l_1/200$. Therefore, the constraint function will vary according to a unitless argument x.

5. The fuzzy constraint function μ_{C_2} for the deflection δ can be obtained in the same way as in point 4:

$$\mu_{C_2}(x) = \begin{cases} 1 - x & 0 \leq x \leq 1 \\ 0 & x > 1 \end{cases}$$

where $x = \delta/2$.

6. The fuzzy constraint function $\mu_C(x)$ for the problem can be found by the combination of $\mu_{C_1}(x)$ and $\mu_{C_2}(x)$, using Eq. (14.10):

$$\mu_C(x) = \min(\mu_{C_1}(x), \mu_{C_2}(x))$$

and $\mu_C(x)$ is shown as the bold line in Fig. 14.5.

Now, the optimum solutions P can be found by using Eq. (14.10):

$$D = (G \cap C)$$

$$\mu_D(x) = \mu_C(x) \wedge \mu_G(x)$$

The optimum value can be determined graphically, as seen in Fig. 14.6, to be $x^* = 0.5$. From this, we can obtain the optimum span length, $l = 100$ m, optimum deflection, $\delta = 1$ mm, and optimum bending stress, $\sigma_b = 30$ MPa. The minimum load P is

$$P = \min\left(\frac{\sigma_b w_z}{l}, \frac{48EI\delta}{l^3}\right)$$

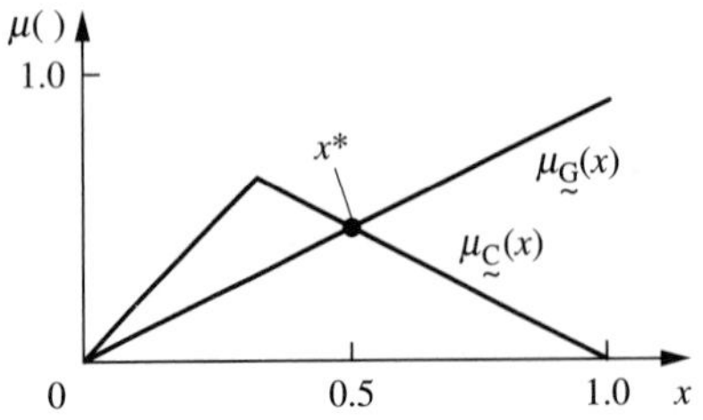

FIGURE 14.6
Graphical solution to minimization problem of Example 14.2.

Suppose that the importance factor for the goal function $\mu_G(x)$ is 0.4. Then the solution for this same optimization problem can be determined using Eq. (14.9) as

$$\mu_D = 0.4\mu_G + 0.6\mu_C$$

where μ_C can be expressed by the function (see Fig. 14.5)

$$\mu_C(x) = \begin{cases} 2x & 0 \le x \le \frac{1}{3} \\ 1 - x & \frac{1}{3} < x \le 1 \end{cases}$$

Therefore,

$$0.6\mu_C(x) = \begin{cases} 1.2x & 0 \le x \le \frac{1}{3} \\ 0.6 - 0.6x & \frac{1}{3} < x \le 1 \end{cases}$$

and $0.4\mu_G(x) = 0.4x$. The membership function for the solution set, from Eq. (14.9), then is

$$\mu_D(x) = \begin{cases} 1.6x & 0 \le x \le \frac{1}{3} \\ 0.6 - 0.2x & \frac{1}{3} < x \le 1 \end{cases}$$

The optimum solution for this is $x^* = 0.33$, which is shown in Fig. 14.7.

Fuzzy Maximum and Minimum

Perhaps the simplest formulation of an optimization problem is to compare two numbers and find the larger or smaller of them [Dubois and Prade, 1980]. In this case, consider numbers that are fuzzy. Such an idea was briefly introduced at the beginning of Chapter 10. For two discrete fuzzy numbers $\underset{\sim}{I}$, defined on universe X, and $\underset{\sim}{J}$, defined on universe Y (see Chapters 4 and 5 for the definition of a fuzzy number),

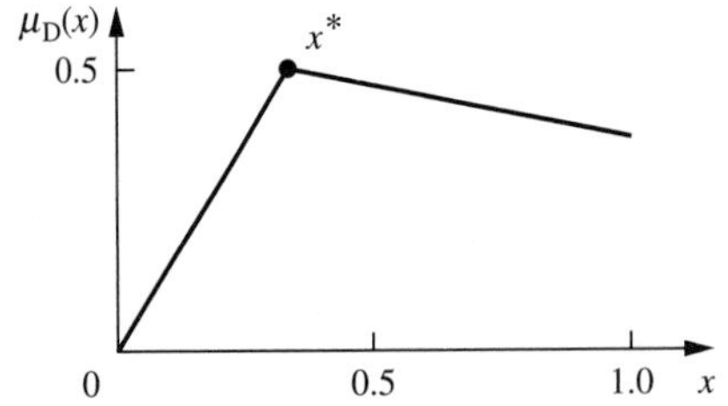

FIGURE 14.7
Solution of Example 14.2 considering an importance factor.

the fuzzy maximum is defined as

$$\mu_{\max(\underset{\sim}{I},\underset{\sim}{J})}(z) = \max_{z=\max(x,y)} \{\min(\mu_{\underset{\sim}{I}}(x), \mu_{\underset{\sim}{J}}(y))\} \tag{14.11}$$

where the max operator is replaced by the supremum if $\underset{\sim}{I}$ is continuous. Similarly, their fuzzy minimum is defined as

$$\mu_{\min(\underset{\sim}{I},\underset{\sim}{J})}(z) = \min_{y=\min(x,y)} \{\max(\mu_{\underset{\sim}{I}}(x), \mu_{\underset{\sim}{J}}(y))\} \tag{14.12}$$

where the min operator is replaced by the infimum if $\underset{\sim}{J}$ is continuous. The regions for the maximum and the minimum are shown in Fig. 14.8. In this figure we see the following relationships by region.

For $z < z_1$	$\mu_{\max(\underset{\sim}{I},\underset{\sim}{J})}(z) = \mu_{\underset{\sim}{I}}(x)$	$\mu_{\min(\underset{\sim}{I},\underset{\sim}{J})}(z) = \mu_{\underset{\sim}{J}}(y)$
For $z_1 < z < z_2$	$\mu_{\max(\underset{\sim}{I},\underset{\sim}{J})}(z) = \mu_{\underset{\sim}{J}}(y)$	$\mu_{\min(\underset{\sim}{I},\underset{\sim}{J})}(z) = \mu_{\underset{\sim}{I}}(x)$
For $z_2 < z < z_3$	$\mu_{\max(\underset{\sim}{I},\underset{\sim}{J})}(z) = \mu_{\underset{\sim}{J}}(y)$	$\mu_{\min(\underset{\sim}{I},\underset{\sim}{J})}(z) = \mu_{\underset{\sim}{I}}(x)$
For $z > z_3$	$\mu_{\max(\underset{\sim}{I},\underset{\sim}{J})}(z) = \mu_{\underset{\sim}{I}}(x)$	$\mu_{\min(\underset{\sim}{I},\underset{\sim}{J})}(z) = \mu_{\underset{\sim}{J}}(y)$

Example 14.3. Suppose we have two discrete-valued fuzzy numbers, $\underset{\sim}{I}$ and $\underset{\sim}{J}$, defined by

$$\underset{\sim}{I} = \left\{\frac{0.2}{0} + \frac{0.5}{1} + \frac{1}{2} + \frac{0.7}{3} + \frac{0.4}{4} + \frac{0.1}{5}\right\}$$

$$\underset{\sim}{J} = \left\{\frac{0}{0} + \frac{0.4}{1} + \frac{0.8}{2} + \frac{1}{3} + \frac{0.5}{4} + \frac{0.3}{5} + \frac{0.1}{6}\right\}$$

For this problem the universes X, Y, and Z are the real line. Because this problem deals with discrete quantities we can use matrices to illustrate the computations. First, we seek the fuzzy maximum. The following matrix shows the pairwise maximum values, $z = \max(x, y)$; the row labels are the x values, and the column labels are the y values.

$$z = \max(x, y) = \begin{array}{c} \\ 0 \\ 1 \\ 2 \\ 3 \\ 4 \\ 5 \end{array} \begin{array}{c} \begin{matrix} 0 & 1 & 2 & 3 & 4 & 5 & 6 \end{matrix} \\ \begin{bmatrix} 0 & 1 & 2 & 3 & 4 & 5 & 6 \\ 1 & 1 & 2 & 3 & 4 & 5 & 6 \\ 2 & 2 & 2 & 3 & 4 & 5 & 6 \\ 3 & 3 & 3 & 3 & 4 & 5 & 6 \\ 4 & 4 & 4 & 4 & 4 & 5 & 6 \\ 5 & 5 & 5 & 5 & 5 & 5 & 6 \end{bmatrix} \end{array}$$

In Eq. (14.11) we see that the internal operation is the pairwise minimum operation between elements in $\underset{\sim}{I}$ and $\underset{\sim}{J}$. The following matrix contains the pairwise minimum membership values:

$$\min(\mu_{\underset{\sim}{I}}(x), \mu_{\underset{\sim}{J}}(y)) = \begin{array}{c} \\ 0 \\ 1 \\ 2 \\ 3 \\ 4 \\ 5 \end{array} \begin{array}{c} \begin{matrix} 0 & 1 & 2 & 3 & 4 & 5 & 6 \end{matrix} \\ \begin{bmatrix} 0 & .2 & .2 & .2 & .2 & .2 & .1 \\ 0 & .4 & .5 & .5 & .5 & .3 & .1 \\ 0 & .4 & .8 & 1 & .5 & .3 & .1 \\ 0 & .4 & .7 & .7 & .5 & .3 & .1 \\ 0 & .4 & .4 & .4 & .4 & .3 & .1 \\ 0 & .1 & .1 & .1 & .1 & .1 & .1 \end{bmatrix} \end{array}$$

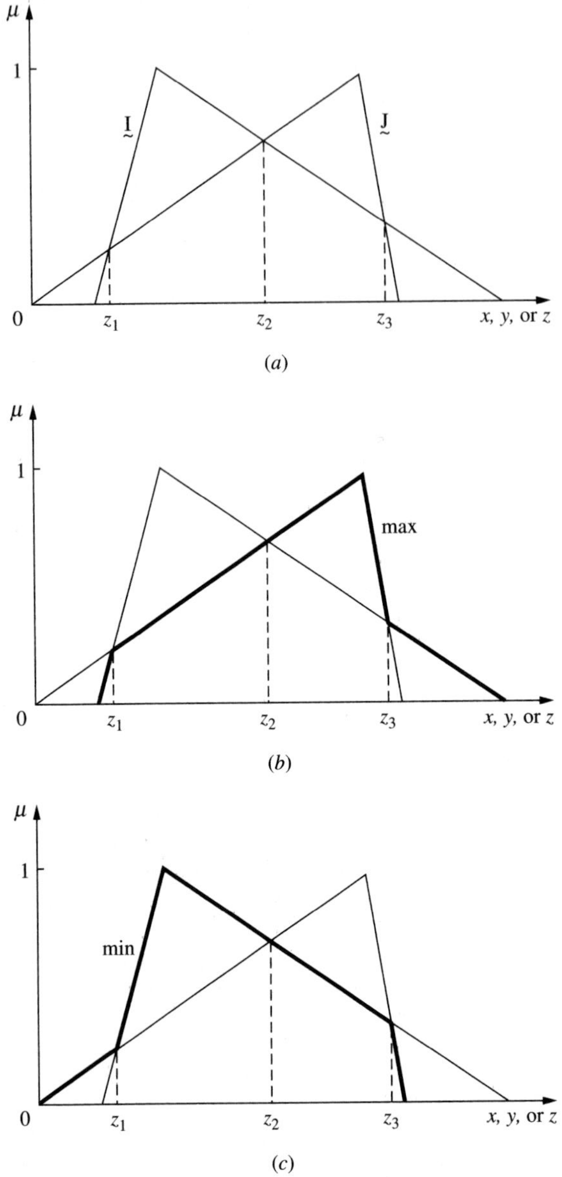

FIGURE 14.8
Fuzzy maximum and minimum: (*a*) fuzzy numbers I and J; (*b*) maximum; (*c*) minimum.

We see in the first matrix that there is only one occurrence of the number $z = 0$, only three occurrences of the number $z = 1$, five occurrences of the number $z = 2$, and so on. To perform the max operation (or supremum) in the outer loop of Eq. (14.11), we take the largest membership value occurring for each value of z. For example, the value $z = 1$ occurs at the three points $(x, y) = (0, 1), (1, 0)$, and $(1, 1)$. We take the maximum of the three associated minimum values, 0, 0.4, and 0.2, as shown in the shaded areas in the two matrices above; the max would be 0.4. For the value $z = 4$ we take the maximum of the nine associated minimum values, 0, 0.4, 0.4, 0.4, 0.4, 0.2, 0.5,

0.5, and 0.5, also given in the shaded areas in the above matrix; the maximum is 0.5. Continuing for the other values z, we get for the fuzzy maximum

$$\mu_{\max(\underset{\sim}{\mathrm{I}},\underset{\sim}{\mathrm{J}})}(z) = \left\{ \frac{0}{0} + \frac{0.4}{1} + \frac{0.8}{2} + \frac{1}{3} + \frac{0.5}{4} + \frac{0.3}{5} + \frac{0.1}{6} \right\}$$

In a similar fashion we can construct the fuzzy minimum. The following matrix shows the pairwise maximum values, $z = \min(x, y)$; the row labels are the x values and the column labels are the y values.

$$z = \min(x, y) = \begin{array}{c} \\ 0 \\ 1 \\ 2 \\ 3 \\ 4 \\ 5 \end{array} \begin{array}{c} \begin{array}{ccccccc} 0 & 1 & 2 & 3 & 4 & 5 & 6 \end{array} \\ \begin{bmatrix} 0 & 0 & 0 & 0 & 0 & 0 & 0 \\ 0 & 1 & 1 & 1 & 1 & 1 & 1 \\ 0 & 1 & 2 & 2 & 2 & 2 & 2 \\ 0 & 1 & 2 & 3 & 3 & 3 & 3 \\ 0 & 1 & 2 & 3 & 4 & 4 & 4 \\ 0 & 1 & 2 & 3 & 4 & 5 & 5 \end{bmatrix} \end{array}$$

In Eq. (14.12) we see that the internal operation is the pairwise maximum operation between elements in $\underset{\sim}{\mathrm{I}}$ and $\underset{\sim}{\mathrm{J}}$. The following matrix contains the pairwise maximum membership values:

$$\max(\mu_{\underset{\sim}{\mathrm{I}}}(x), \mu_{\underset{\sim}{\mathrm{J}}}(y)) = \begin{array}{c} \\ 0 \\ 1 \\ 2 \\ 3 \\ 4 \\ 5 \end{array} \begin{array}{c} \begin{array}{ccccccc} 0 & 1 & 2 & 3 & 4 & 5 & 6 \end{array} \\ \begin{bmatrix} .2 & .4 & .8 & 1 & .5 & .3 & .2 \\ .5 & .5 & .8 & 1 & .5 & .5 & .5 \\ 1 & 1 & 1 & 1 & 1 & 1 & 1 \\ .7 & .7 & .8 & 1 & .7 & .7 & .7 \\ .4 & .4 & .8 & 1 & .5 & .4 & .4 \\ .1 & .4 & .8 & 1 & .5 & .3 & .1 \end{bmatrix} \end{array}$$

We see in the third matrix in this example, that of $z = \min(x, y)$, there are 12 occurrences of the number $z = 0$, 10 occurrences of the number $z = 1$, eight occurrences of the number $z = 2$, and so on. To perform the min operation (or infimum) in the outer loop of Eq. (14.12), we take the smallest membership value occurring for each value of z. For example, there are four occurrences of the value $z = 4$. We take the minimum of the four associated maximum values, 0.5, 0.5, 0.4, and 0.4, as shown in the shaded areas in the above matrix; the min would be 0.4. For the number $z = 2$ we take the minimum of the eight associated maximum values, 0.8, 0.8, 0.8, 1, 1, 1, 1, and 1, also shown in the shaded areas in the above matrix; the min would be 0.8. Continuing for the other values for z, we get for the fuzzy minimum

$$\mu_{\min(\underset{\sim}{\mathrm{I}},\underset{\sim}{\mathrm{J}})}(z) = \left\{ \frac{0.1}{0} + \frac{0.4}{1} + \frac{0.8}{2} + \frac{0.7}{3} + \frac{0.4}{4} + \frac{0.1}{5} + \frac{0}{6} \right\}$$

INVERSE FUZZY RELATIONAL EQUATIONS

Suppose we have a standard fuzzy relational equation of the form, $\underset{\sim}{\mathrm{B}} = \underset{\sim}{\mathrm{A}} \circ \underset{\sim}{\mathrm{R}}$. In the normal situation we have a fuzzy relation $\underset{\sim}{\mathrm{R}}$ from either rules or data, and the information contained in $\underset{\sim}{\mathrm{A}}$ is also known from data or is assumed. The determination of $\underset{\sim}{\mathrm{B}}$ usually is accomplished through some form of composition. Suppose, however, that we know $\underset{\sim}{\mathrm{B}}$ and $\underset{\sim}{\mathrm{R}}$, and we are interested in finding $\underset{\sim}{\mathrm{A}}$. If the relational equation were linear, we would find the inverse of the equation, i.e., $\underset{\sim}{\mathrm{A}} = \underset{\sim}{\mathrm{B}} \circ \underset{\sim}{\mathrm{R}}^{-1}$.

Unfortunately, a fuzzy relational equation is not linear, and the inverse cannot provide a unique solution, in general, or any solution in some situations. In fact, the inverse is difficult to find for most situations [Terano et al., 1992].

A fuzzy relational equation can be expressed in expanded form by

$$\begin{aligned}(a_1 \wedge r_{11}) \vee (a_2 \wedge r_{12}) \vee \cdots (a_n \wedge r_{1n}) &= b_1 \\ (a_1 \wedge r_{21}) \vee (a_2 \wedge r_{22}) \vee \cdots (a_n \wedge r_{2n}) &= b_2 \\ &\cdots \\ (a_1 \wedge r_{m1}) \vee (a_2 \wedge r_{m2}) \vee \cdots (a_n \wedge r_{mn}) &= b_m\end{aligned} \tag{14.13}$$

where a_i, r_{ij}, and b_i are membership values for $\underset{\sim}{A}$, $\underset{\sim}{R}$, and $\underset{\sim}{B}$, respectively.

To solve $\underset{\sim}{A} = \{a_1, a_2, \ldots, a_n\}$ given r_{ij} and b_j ($i = 1, 2, \ldots, n$ and $j = 1, 2, \ldots, m$), we can use a method reported by Tsukamoto and Terano [1977] that produces interval values for the solution. In this approach there are two standard definitions that first must be presented: a fuzzy equality and a fuzzy inequality. An *equality* is expressed as

$$\text{Equality} \qquad a \wedge r = b \tag{14.14}$$

The inverse solution for a in the equality in Eq. (14.14), given r and b are known, is generally an interval number and is denoted by an operator $b \Diamond r$,

$$b \Diamond r = \begin{cases} b & r > b \\ [b, 1] & r = b \\ \varnothing & r < b \end{cases} \tag{14.15}$$

An *inequality* is defined as

$$\text{Inequality} \qquad a \wedge r \leq b \tag{14.16}$$

For the inequality (14.16) the inverse solution for a is also an interval number, denoted by the operator $b \hat{\Diamond} r$, and is given by

$$b \hat{\Diamond} r = \begin{cases} [0, b] & r > b \\ [0, 1] & r \leq b \end{cases} \tag{14.17}$$

A typical row of the standard fuzzy relational equation system [see Eqs. (14.13)], say the first row, can be represented in a simpler form,

$$(a_1 \wedge r_1) \vee (a_2 \wedge r_2) \vee \cdots (a_n \wedge r_n) = b \tag{14.18}$$

The expression in Eq. (14.18) can be subdivided into n *equalities* of the type

$$(a_1 \wedge r_1) = b, \qquad (a_2 \wedge r_2) = b, \qquad \ldots, \qquad (a_n \wedge r_n) = b \tag{14.19}$$

and n *inequalities* of the type

$$(a_1 \wedge r_1) \leq b, \qquad (a_2 \wedge r_2) \leq b, \qquad \ldots, \qquad (a_n \wedge r_n) \leq b \tag{14.20}$$

A solution, represented as an interval vector, to the fuzzy relational Eq. (14.18) exists if and only if there is at least one equality and no more than $(n-1)$ inequalities in the solution. That is, the inverse solution for a_i in Eq. (14.18), denoted W_i, is

$$a_i = \mathrm{W}_1 \quad \text{or} \quad \mathrm{W}_2 \quad \text{or} \quad \ldots \quad \text{or} \quad \mathrm{W}_n \tag{14.21}$$

where

$$\mathrm{W}_i = (b \hat{\Diamond} r_1, \ldots, b \hat{\Diamond} r_{i-1}, b \Diamond r_i, b \hat{\Diamond} r_{i+1}, \ldots, b \hat{\Diamond} r_n) \tag{14.22}$$

Note the ith term in Eq. (14.22) is an equality and the other terms are inequalities.

Example 14.4 [Dong, 1987]. Suppose we want to solve for a_i ($i = 1, 2, 3, 4$) in the single inverse fuzzy equation

$$(a_1 \wedge 0.7) \vee (a_2 \wedge 0.8) \vee (a_3 \wedge 0.6) \vee (a_4 \wedge 0.3) = 0.6$$

Making use of expressions (14.19)–(14.20), we subdivide the single fuzzy equation into $n = 4$ equalities:

$$\begin{aligned} Y_{\text{eq}} &= \{b \Diamond r_1, b \Diamond r_2, \ldots, b \Diamond r_n\} \\ &= \{0.6 \Diamond 0.7, 0.6 \Diamond 0.8, 0.6 \Diamond 0.6, 0.6 \Diamond 0.3\} \\ &= \{0.6, 0.6, [0.6, 1], \varnothing\} \qquad \text{(4 } \textit{equality} \text{ values)} \end{aligned}$$

and into $n = 4$ inequalities,

$$\begin{aligned} Y_{\text{ineq}} &= \{b \hat{\Diamond} r_1, b \hat{\Diamond} r_2, \ldots, b \hat{\Diamond} r_n\} \\ &= \{0.6 \hat{\Diamond} 0.7, 0.6 \hat{\Diamond} 0.8, 0.6 \hat{\Diamond} 0.6, 0.6 \hat{\Diamond} 0.3\} \\ &= \{[0, 0.6], [0, 0.6], [0, 1], [0, 1]\} \qquad \text{(4 } \textit{inequality} \text{ values)} \end{aligned}$$

Then Eq. (14.22) provides for the $n = 4$ potential solutions, W_i, where $i = 1, 2, 3, 4$:

$$\begin{aligned} \mathrm{W}_1 &= \{0.6, [0, 0.6], [0, 1], [0, 1]\} && \text{(position one is the first equality value)} \\ \mathrm{W}_2 &= \{[0, 0.6], 0.6, [0, 1], [0, 1]\} && \text{(position two is the second equality value)} \\ \mathrm{W}_3 &= \{[0, 0.6], [0, 0.6], [0.6, 1], [0, 1]\} && \text{(position three is the third equality value)} \\ \mathrm{W}_4 &= \{[0, 0.6], [0, 0.6], [0, 1], \varnothing\} = \varnothing && \text{(position four is the fourth equality value)} \end{aligned}$$

where all values are the inequalities, except those equalities noted specifically. Equation (14.21) provides for the aggregated solution in interval form; note that W_4 does not contribute to the aggregated solution because it has a value of null. Hence,

$$a_i = \mathrm{W}_1 \text{ or } \mathrm{W}_2 \text{ or } \mathrm{W}_3$$

Its maximum solution is $a_{\max} = \{0.6, 0.6, 1, 1\}$. Its minimum solutions are

$$\begin{aligned} a_{\min} &= \{0.6, 0, 0, 0\} && \text{from } \mathrm{W}_1 \\ &= \{0, 0.6, 0, 0\} && \text{from } \mathrm{W}_2 \\ &= \{0, 0, 0.6, 0\} && \text{from } \mathrm{W}_3 \end{aligned}$$

Now suppose instead of a single equation we want to find the solution for an equation set, that is, a collection of m simultaneous equations of the form given in Eqs. (14.13). Then the solution set will consist of an m-set of n equalities, expressed in an $m \times n$ matrix (denoted Y) and an m-set of n inequalities, also expressed in an $m \times n$ matrix (denoted $\hat{\mathrm{Y}}$).

$$\text{Y} = \begin{bmatrix} b_1 \diamond r_{11} & b_1 \diamond r_{12} & \dots & b_1 \diamond r_{1n} \\ b_2 \diamond r_{21} & b_2 \diamond r_{22} & \dots & b_2 \diamond r_{2n} \\ & \dots & & \\ b_m \diamond r_{m1} & b_m \diamond r_{m2} & \dots & b_m \diamond r_{mn} \end{bmatrix} \quad (14.23)$$

$$\hat{\text{Y}} = \begin{bmatrix} b_1 \hat{\diamond} r_{11} & b_1 \hat{\diamond} r_{12} & \dots & b_1 \hat{\diamond} r_{1n} \\ b_2 \hat{\diamond} r_{21} & b_2 \hat{\diamond} r_{22} & \dots & b_2 \hat{\diamond} r_{2n} \\ & \dots & & \\ b_m \hat{\diamond} r_{m1} & b_m \hat{\diamond} r_{m2} & \dots & b_m \hat{\diamond} r_{mn} \end{bmatrix} \quad (14.24)$$

Taking an element for each row from Y and replacing the corresponding element in $\hat{\text{Y}}$, we get an array solution for each element, ij, in the $m \times n$ equation matrix.

$$\text{W}^*_{(i_1, i_2, \dots, i_m)} = \begin{bmatrix} b_1 \hat{\diamond} r_{11} & \dots & b_1 \diamond r_{1i} & \dots & b_1 \hat{\diamond} r_{1n} \\ b_2 \hat{\diamond} r_{21} & \dots & b_2 \diamond r_{2i} & \dots & b_2 \hat{\diamond} r_{2n} \\ & & \dots & & \\ b_m \hat{\diamond} r_{m1} & \dots & b_m \diamond r_{mi} & \dots & b_m \hat{\diamond} r_{mn} \end{bmatrix} = (w^*_{ij}) \quad (14.25)$$

where indices $i_1 = (1, 2, \dots, n)$, $i_2 = (1, 2, \dots, n)$, and $i_m = (1, 2, \dots, n)$ represent m arrays that are all of length n. Each array is a solution to one row of the original equations [i.e., of Eqs. (14.13)], and we need at most m arrays for the complete solution. Hence, we can have (n^m) solutions, including null solutions, of the type expressed by Eq. (14.25). To establish a sign convention we denote a particular solution, (w^*_{ij}), of the $m \times n$ array by identifying its location (ij) in the array. The complete solution incorporating all $m \times n$ possible solutions will be denoted

$$\text{W}_{(i_1, i_2, \dots, i_m)} = \{w_1, w_2, \dots, w_m\} \quad (14.26)$$

where

$$w_j = \bigcap_i w^*_{ij} \quad (14.27)$$

and where $i = 1, 2, \dots, n$ and $j = 1, 2, \dots, m$.

In the foregoing development, each possible solution $w_1, w_2, \dots, w_m$ in Eq. (14.26) is found by taking the intersection of all the solutions in the jth column, i.e., Eq. (14.27), of the solution arrays described by Eq. (14.25).

Example 14.5 [Dong, 1987]. Suppose we have a system of three simultaneous equations as given here. In this example, $m = 3$ and $n = 3$. Hence, there is a potential for $3^3 = 27$ distinct solutions that need to be explored in order to develop the full solution. Our goal is to find interval values for the three unknown quantities, $\{a_1, a_2, a_3\}$, in the following inverse equations:

$$\{a_1, a_2, a_3\} \circ \begin{bmatrix} 0.3 & 0.5 & 0.2 \\ 0.2 & 0 & 0.4 \\ 0 & 0.6 & 0.1 \end{bmatrix} = [0.2 \quad 0.4 \quad 0.2]$$

To begin the solution process, we need to find the individual equality sets, Y, and inequality sets, $\hat{Y}$, using Eqs. (14.23) and (14.24), respectively. So for the first row of Y we operate b_1 on the first column of the r matrix, i.e., using Eq. (14.15),

$$\{b_1 \diamond r_{11}, b_1 \diamond r_{21}, b_1 \diamond r_{31}\} = \{0.2 \diamond 0.3, 0.2 \diamond 0.2, 0.2 \diamond 0\} = \{0.2, [0.2, 1], \varnothing\}$$

The second row of Y is found, by operating b_2 on the second column of the r matrix, to be

$$\{b_2 \diamond r_{12}, b_2 \diamond r_{22}, b_2 \diamond r_{32}\} = \{0.4 \diamond 0.5, 0.4 \diamond 0, 0.4 \diamond 0.6\} = \{0.4, \varnothing, 0.4\}$$

The third row of Y is found, by operating b_3 on the third column of the r matrix, to be

$$\{b_3 \diamond r_{13}, b_3 \diamond r_{23}, b_3 \diamond r_{33}\} = \{0.2 \diamond 0.2, 0.2 \diamond 0.4, 0.2 \diamond 0.1\} = \{[0.2, 1], 0.2, \varnothing\}$$

Therefore, we have

$$\text{Y} = \begin{bmatrix} 0.2 & [0.2, 1] & \varnothing \\ 0.4 & \varnothing & 0.4 \\ [0.2, 1] & 0.2 & \varnothing \end{bmatrix}$$

To calculate the inequality matrix, $\hat{Y}$, we have the same sequence of operations of elements in the b vector on columns in the r matrix, but we use the operator $\hat{\diamond}$. For the first row of Y we operate b_1 on the first column of the r matrix, i.e., using Eq. (14.17),

$$\{b_1 \hat{\diamond} r_{11}, b_1 \hat{\diamond} r_{21}, b_1 \hat{\diamond} r_{31}\} = \{0.2 \hat{\diamond} 0.3, 0.2 \hat{\diamond} 0.2, 0.2 \hat{\diamond} 0\} = \{[0, 0.2], [0, 1], [0, 1]\}$$

The second row of Y is found, by operating b_2 on the second column of the r matrix, to be

$$\{b_2 \hat{\diamond} r_{12}, b_2 \hat{\diamond} r_{22}, b_2 \hat{\diamond} r_{32}\} = \{0.4 \hat{\diamond} 0.5, 0.4 \hat{\diamond} 0, 0.4 \hat{\diamond} 0.6\} = \{[0, 0.4], [0, 1], [0, 0.4]\}$$

The third row of Y is found, by operating b_3 on the third column of the r matrix, to be

$$\{b_3 \hat{\diamond} r_{13}, b_3 \hat{\diamond} r_{23}, b_3 \hat{\diamond} r_{33}\} = \{0.2 \hat{\diamond} 0.2, 0.2 \hat{\diamond} 0.4, 0.2 \hat{\diamond} 0.1\} = \{[0, 1], [0, 0.2], [0, 1]\}$$

Therefore, we have

$$\hat{\text{Y}} = \begin{bmatrix} [0, 0.2] & [0, 1] & [0, 1] \\ [0, 0.4] & [0, 1] & [0, 0.4] \\ [0, 1] & [0, 0.2] & [0, 1] \end{bmatrix}$$

Now, using Eq. (14.25), we can construct the W^*_{ij} matrices. The first matrix is W^*_{111}, the subscripts denoting that the equality element for the first row of W^*_{111} comes from the *first* position in Y (the first subscript 1) in row 1; the equality element for the second row of W^*_{111} comes from the *first* position in Y (the second subscript 1) in row 2; the equality element for the third row of W^*_{111} comes from the *first* position in Y(the third subscript 1) in row 3. All other elements for W^*_{111} come from the same positions they are in for the matrix $\hat{Y}$. Hence, W^*_{111} looks like

$$\text{W}^*_{111} = \begin{bmatrix} 0.2 & [0, 1] & [0, 1] \\ 0.4 & [0, 1] & [0, 0.4] \\ [0.2, 1] & [0, 0.2] & [0, 1] \end{bmatrix}$$

Finally, using Eqs. (14.26)–(14.27), we take the intersection of the elements in each column of W^*_{111} to get

$$\text{W}_{111} = (\varnothing, [0, 0.2], [0, 0.4]) = \varnothing$$

Since there is a null element in W_{111}, we set the entire value equal to null.

Continuing in a similar fashion, the second matrix is W^*_{112}, with the subscripts denoting that the equality element for the first row of W^*_{112} comes from the *first* position in Y (the first subscript 1) in row 1; the equality element for the second row of W^*_{112} comes from the *first* position in Y (the second subscript 1) in row 2; the equality element for the third row of W^*_{112} comes from the *second* position in Y (the third subscript 2) in row 3. All other elements for W^*_{112} come from the same positions they are in for the matrix $\hat{Y}$. Hence, W^*_{112} looks like

$$W^*_{112} = \begin{bmatrix} 0.2 & [0,1] & [0,1] \\ 0.4 & [0,1] & [0,0.4] \\ [0,1] & 0.2 & [0,1] \end{bmatrix}$$

and using Eqs. (14.26)–(14.27) again, we get

$$W_{112} = (\varnothing, 0.2, [0, 0.4]) = \varnothing$$

because there is at least one null element in W^*_{112}.

Now, moving to other elements in w^*_{ij}, such as W^*_{131}, we get

$$W^*_{131} = \begin{bmatrix} 0.2 & [0,1] & [0,1] \\ [0,0.4] & [0,1] & 0.4 \\ [0.2,1] & [0,0.2] & [0,1] \end{bmatrix}$$

and the resulting solution after performing intersections on the elements in each column is

$$W_{131} = (0.2, [0, 0.2], 0.4)$$

This process continues for the other 24 solutions, e.g., W_{211}, W_{232}, W_{333}, etc., out of the total of 27 (3^3).

Of all 27 possible solutions, only four are nonnull (nonempty). These four are

$$\begin{aligned} W_{131} &= (0.2, [0, 0.2], 0.4) \\ W_{132} &= (0.2, 0.2, 0.4) \\ W_{231} &= (0.2, 0.2, 0.4) \\ W_{232} &= ([0, 0.2], 0.2, 0.4) \end{aligned}$$

By inspection we can see that $W_{131} \supseteq W_{132}$ and $W_{231} \supseteq W_{232}$, and that $W_{231} = W_{132}$; hence, the solution can be expressed by the intervals W_{132} and W_{232}. This solution space in the three dimensions governed by the original coordinates a_1, a_2, and a_3 is shown in Fig. 14.9.

In the figure we see that the minimum solution is given by two points

$$\begin{aligned} a_{\min,1} &= \{a_1, a_2, a_3\} = \{0.2, 0, 0.4\} \\ a_{\min,2} &= \{a_1, a_2, a_3\} = \{0, 0.2, 0.4\} \end{aligned}$$

and that the maximum solution is given by the single point

$$a_{\max} = \{a_1, a_2, a_3\} = \{0.2, 0.2, 0.4\}$$

The entire solution in this three-dimensional example comprises the two edges that are darkened in Fig. 14.9.

It is perhaps clear from Examples 14.4 and 14.5 that when the cardinal numbers n and m are large, the number of analytical solutions becomes exponentially large;

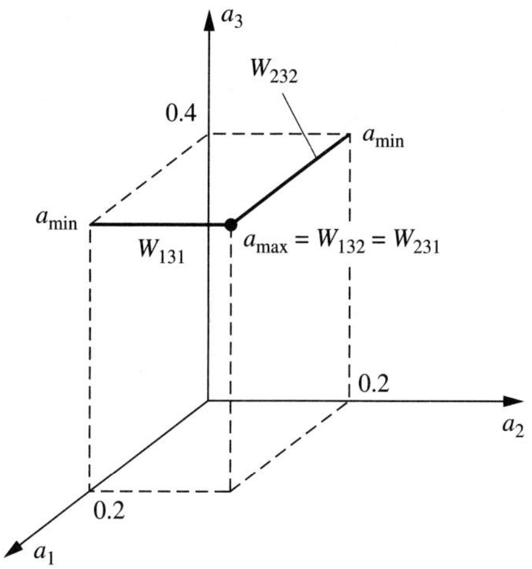

FIGURE 14.9
Solution for Example 14.5.

also, sometimes the final solution can be null. For both these cases other approaches, such as those in pattern recognition, might be more practical. In any case, fuzzy inverses are only approximate; they are not unique in general, even for linear equations.

FUZZY LINEAR REGRESSION

Regression analysis is used to model the relationship between dependent and independent variables. In regression analysis, the dependent variable, y, is a function of the independent variables; and the degree of contribution of each variable to the output is represented by coefficients on these variables. The model is empirically developed from data collected from observations and experiments. A crisp linear regression model is shown in Eq. (14.28),

$$y = f(x, a) = a_0 + a_1 x_1 + a_2 x_2 + \cdots + a_n x_n \tag{14.28}$$

In conventional regression techniques, the difference between the observed values and the values estimated from the model is assumed to be due to observational errors, and the difference is considered a random variable. Upper and lower bounds for the estimated value are established, and the probability that the estimated value will be within these two bounds represents the confidence of the estimate. In other words, conventional regression analysis is probabilistic. But in fuzzy regression, the difference between the observed and the estimated values is assumed to be due to the ambiguity inherently present in the system. The output for a specified input is assumed to be a range of possible values, i.e., the output can take on any of these possible values. Therefore, fuzzy regression is possibilistic in nature. Moreover, fuzzy regression analyses use fuzzy functions to represent the coefficients as opposed to crisp coefficients used in conventional regression analysis [Terano et al., 1992]. Equation (14.29) shows a typical fuzzy linear regression model,

$$\underset{\sim}{Y} = f(x, \underset{\sim}{A}) = \underset{\sim}{A}_1 x_1 + \underset{\sim}{A}_2 x_2 + \cdots + \underset{\sim}{A}_n x_n \tag{14.29}$$

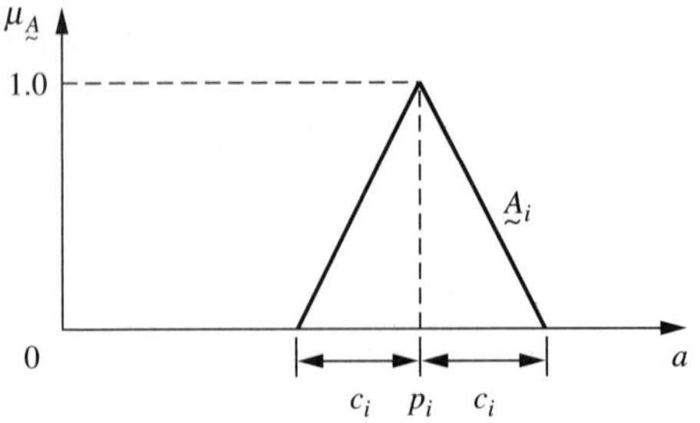

FIGURE 14.10
Membership function for the fuzzy coefficient $\underset{\sim}{A}$.

where $\underset{\sim}{A}_i$ is the ith fuzzy coefficient (usually a fuzzy number).

Fuzzy regression estimates a range of possible values that are represented by a possibility distribution (a more rigorous definition of possibilities is given in Chapter 15), known as the membership function. Membership functions are formed by assigning a specific membership value (degree of belonging) to each of the estimated values (Fig. 14.10). Such membership functions are also defined for the coefficients of the independent variables. Triangular membership functions for the fuzzy coefficients, like those shown in Fig. 14.10, allow for the solution to be found via a linear programming formulation; other membership functions for the coefficients require alternative approaches [Kikuchi and Nanda, 1991].

The membership function $\mu_{\underset{\sim}{A}}$ for each of the coefficients is expressed as

$$\mu_{\underset{\sim}{A}_i}(a_i) = \begin{cases} 1 - \dfrac{|\, p_i - a_i \,|}{c_i} & p_i - c_i \le x_i \le p_i + c_i \\ 0 & \text{otherwise} \end{cases} \tag{14.30}$$

The fuzzy function $\underset{\sim}{A}$ is a function of two parameters, p and c, known as the middle value and the spread, respectively. The spread denotes the fuzziness of the function. The figure shows the membership function for a fuzzy number "approximately p_i." A more detailed explanation of membership functions, fuzzy numbers, and operations on fuzzy numbers is given in Chapters 4 and 5. The fuzzy parameters $\underset{\sim}{A} = \{\underset{\sim}{A}_1, \ldots, \underset{\sim}{A}_n\}$ can be denoted in the vector form of $\underset{\sim}{A} = \{\mathbf{p}, \mathbf{c}\}$, where $\mathbf{p} = (p_1, \ldots, p_n)$ and $\mathbf{c} = (c_1, \ldots, c_n)$. Therefore, the output is a revised version of Eq. (14.29),

$$\underset{\sim}{Y} = (p_1, c_1)x_1 + (p_2, c_2)x_2 + \cdots + (p_n, c_n)x_n$$

The membership function for the output fuzzy parameter, $\underset{\sim}{Y}$, is given by

$$\mu_{\underset{\sim}{Y}}(y) = \begin{cases} \max(\min_i[\mu_{\underset{\sim}{A}_i}(a_i)]) & \{a \mid y = f(x, a)\} \ne \varnothing \\ 0 & \text{otherwise} \end{cases} \tag{14.31}$$

Substituting Eq. (14.30) into Eq. (14.31), we get [see Tanaka et al., 1982]

$$\mu_{\underset{\sim}{Y}}(y) = \begin{cases} 1 - \dfrac{|\, y - \sum_{i=1}^{n} p_i x_i|}{\sum_{i=1}^{n} c_i \,|\, x_i \,|} & x_i \ne 0 \\ 1 & x_i = 0, \quad y = 0 \\ 0 & x_i = 0, \quad y \ne 0 \end{cases} \tag{14.32}$$

The foregoing equations are applied to m data sets that can be obtained from sampling. The output and the input data can be either fuzzy or nonfuzzy. Table 14.1 shows an example of the data sets for the nonfuzzy data. In the table, y_j is the output for the jth sample, and x_{ij} is the ith input variable for the jth sample.

The Case of Nonfuzzy Data

Tanaka et al. [1982] have determined the solution to the regression model by converting it to a linear programming problem (this is not the only approach, as is discussed shortly). For nonfuzzy data the objective of the regression model is to determine the optimum parameters $\underset{\sim}{A}^*$ such that the fuzzy output set, which contains y_i, is associated with a membership value greater than h, that is,

$$\mu_{\underset{\sim}{Y}_j}(y_j) \geq h, \qquad j = 1, \ldots, m \tag{14.33}$$

The degree h is specified by the user; as h increases, the fuzziness of the output increases [Kikuchi and Nanda, 1991]. Figure 14.11 shows the membership function for the fuzzy output. Equation (14.33) states that the fuzzy output should lie between A and B of Fig. 14.11. In the figure the middle value ($\sum_{i=1}^{n} p_i x_i$) and the spread ($\sum_{i=1}^{n} c_i|x_i|$) are obtained by considering Eq. (14.32), where h is specified by the user.

In regression we seek to find the fuzzy coefficients that minimize the spread of fuzzy output for all the data sets. Equation (14.34) shows the objective function that has to be minimized.

$$O = \min\left\{\sum_{j=1}^{m}\sum_{i=1}^{n} c_i x_{ij}\right\} \tag{14.34}$$

The objective function given in Eq. (14.34) is minimized, subject to two constraints. The constraints are obtained by substituting Eq. (14.32) into Eq. (14.33); they become

$$y_j \geq \sum_{i=1}^{n} p_i x_{ij} - (1-h)\sum_{i=1}^{n} c_i x_{ij} \tag{14.35}$$

TABLE 14.1
An example of the data sets for nonfuzzy data

Sample number, j	Output, y_j	n inputs, x_{ij}
1	y_1	$x_{11}, x_{21}, \ldots, x_{n1}$
⋮	⋮	⋮
m	y_m	$x_{1m}, x_{2m}, \ldots, x_{nm}$

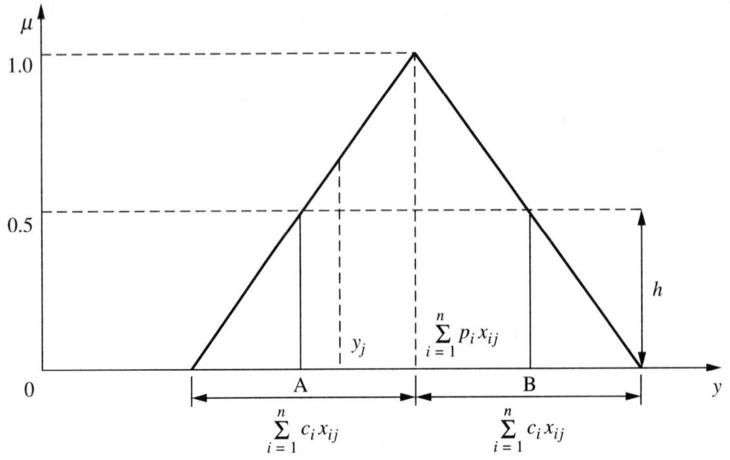

FIGURE 14.11
Fuzzy output function.

and

$$y_j \leq \sum_{i=1}^{n} p_i x_{ij} + (1 - h) \sum_{i=1}^{n} c_i x_{ij} \tag{14.36}$$

Since each data set produces two constraints, there is a total of $2m$ constraints for each data set.

The Case of Fuzzy Data

When human judgment or imprecise measurements are involved in determining the output, the output is seldom a crisp number. The output in such situations is best represented by a fuzzy number as $\underset{\sim}{Y}_j = (y_j, e_j)$, where y_j is the middle value and e_j represents the ambiguity in the output, as seen in Fig. 14.12.

The membership function for the observed fuzzy output is given as

$$\mu_{\underset{\sim}{Y}_j}(y) = 1 - \frac{|y_j - y|}{e_j} \tag{14.37}$$

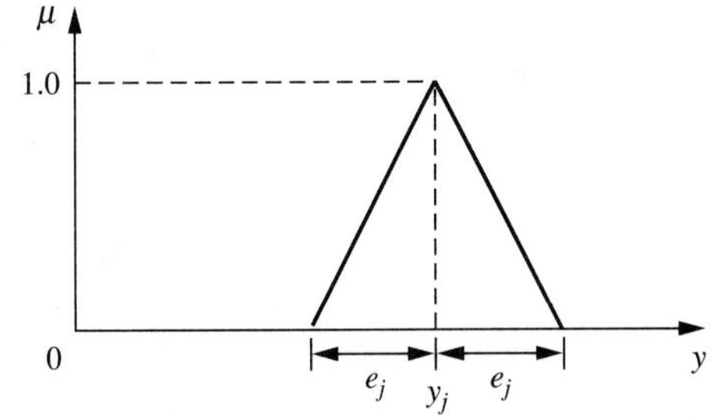

FIGURE 14.12
An example of fuzzy output.

An estimate of this fuzzy output can be obtained from Eq. (14.32) as

$$\mu_{\underset{\sim}{Y}_j^*}(y) = 1 - \frac{\left|y_j - \sum_{i=1}^{n} p_i x_{ij}\right|}{\sum_{i=1}^{n} c_i |x_{ij}|} \qquad \text{for } j = 1, m \tag{14.38}$$

The degree of fitting of the estimated fuzzy output $\underset{\sim}{Y}_j^*$ to the given data $\underset{\sim}{Y}_j$ is determined by h_j, which maximizes h subject to $\underset{\sim}{Y}_j^h \subset Y_j^{h^*}$ where

$$\begin{aligned} \underset{\sim}{Y}_j^h &= \{y \mid \mu_{\underset{\sim}{Y}_j}(y) \geq h\} \\ \underset{\sim}{Y}_j^{h^*} &= \left\{y \mid \mu_{\underset{\sim}{Y}_j^*}(y) \geq h\right\} \end{aligned} \tag{14.39}$$

Figure 14.13 illustrates these concepts. The objective of the fuzzy linear regression model is to determine fuzzy parameters $\underset{\sim}{A}^*$ that minimize the spread subject to the constraint that $h_j \geq H$ for all j, where H is chosen by the user as the degree of fitting of the fuzzy linear model. The jth fitting parameter, h_j, is computed from Fig. 14.13 as

$$h_j = 1 - \frac{\left|y_j - \sum_{i=1}^{n} p_i x_{ij}\right|}{\sum_{i=1}^{n} c_i |x_{ij}| - e_j} \tag{14.40}$$

In summary the objective function to be minimized is

$$O_f = \min\left\{\sum_{j=1}^{m}\sum_{i=1}^{n} c_i x_{ij}\right\} \tag{14.41}$$

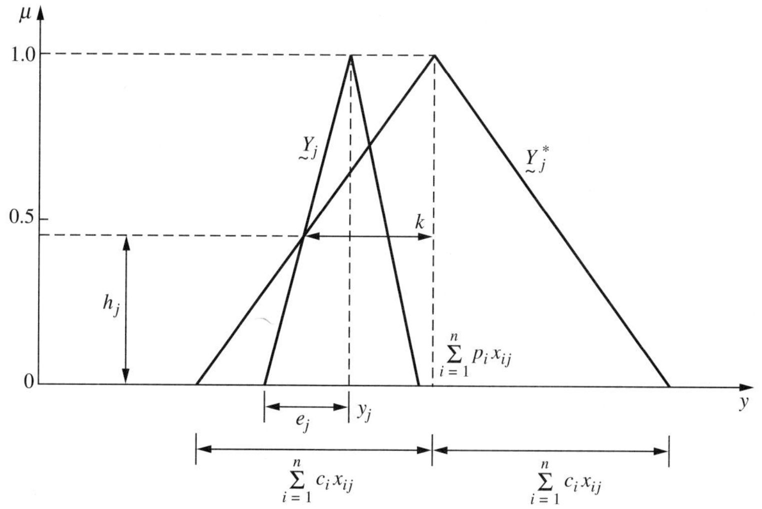

FIGURE 14.13
Degree of fitting of estimated fuzzy output to the given fuzzy output.

subject to the constraints

$$y_j \geq \sum_{i=1}^{n} p_i x_{ij} - (1-H)\sum_{i=1}^{n} c_i x_{ij} + (1-H)e_j \tag{14.42}$$

and

$$y_j \leq \sum_{i=1}^{n} p_i x_{ij} + (1-H)\sum_{i=1}^{n} c_i x_{ij} - (1-H)e_j \tag{14.43}$$

for each data set where $j = 1, \ldots, m$. In Eqs. (14.42) and (14.43) we note that $e_j = y_j - y$. The important equations used in the fuzzy linear regression model for both the nonfuzzy output (y_j) and the fuzzy output ($\underset{\sim}{Y}_j$) cases, along with the equation numbers, are summarized in Table 14.2.

Example 14.6. The concepts of fuzzy regression are illustrated through a trivial one-dimensional problem. Here, we will only use two data points to illustrate the linear regression approach. The data set is shown in Table 14.3. For these two points the equation of the line that runs through them both is $y = 2.9762 + 0.5524x$. We should get these values in our fuzzy analysis since, for two points, we do not have uncertainty in the regression analysis.

Let this data set be represented by a linear regression model, $\underset{\sim}{Y} = \underset{\sim}{A}_0 + \underset{\sim}{A}_1 x$, where the coefficients $\underset{\sim}{A}_0$ and $\underset{\sim}{A}_1$ are fuzzy numbers. For the data set given in the Table 14.3, $i = 1$ and $j = 2$. From Eq. (14.34) the objective function to be minimized is given by

$$O = \left\{\sum_{j=1}^{m}\sum_{i=1}^{n} c_i x_{ij}\right\} = c_0 + (x_{11} + x_{12})c_1 = c_0 + (0.52 + 1.36)c_1$$
$$= c_0 + 1.88c_1$$

Since there are two data sets, the objective function has to be minimized subject to four constraints as shown here:

$$y_1 \geq p_0 + 0.52p_1 - (1-h)(c_0 + 0.52c_1)$$
$$y_1 \leq p_0 + 0.52p_1 + (1-h)(c_0 + 0.52c_1)$$
$$y_2 \geq p_0 + 1.36p_1 - (1-h)(c_0 + 1.36c_1)$$
$$y_2 \leq p_0 + 1.36p_1 + (1-h)(c_0 + 1.36c_1)$$

Substituting the values of y_j $(i = 1, 2)$, and setting $h = 0.5$, we get

$$2.1 \geq p_0 + 0.52p_1 - 0.5c_0 - 0.26c_1$$
$$2.1 \leq p_0 + 0.52p_1 + 0.5c_0 + 0.26c_1$$
$$4.6 \geq p_0 + 1.36p_1 - 0.5c_0 - 0.68c_1$$
$$4.6 \leq p_0 + 1.36p_1 + 0.5c_0 + 0.68c_1$$

The linear programming problem is now solved using the simplex method; a good explanation of this method is available in Hillier and Lieberman [1980]. Since the constraint equations are expressed by inequality relationships, basic variables (D_i) are introduced to convert the inequalities to equations (these variables are equated to the

TABLE 14.2
Summary of equations used in fuzzy linear regression model [Eq. (14.29)], $\underset{\sim}{Y} = f(x, \underset{\sim}{A}) = \underset{\sim}{A}_0 + \underset{\sim}{A}_1 x_1 + \underset{\sim}{A}_2 x_2 + \cdots + \underset{\sim}{A}_n x_n$

Membership functions

$\underset{\sim}{A}$

$$\mu_{\underset{\sim}{A}_i}(a_i) = \begin{cases} 1 - \dfrac{|p_i - a_i|}{c_i} & p_i - c_i \le x_i \le p_i + c_i \\ 0 & \text{otherwise} \end{cases} \qquad \text{Eq. (14.30)}$$

$\underset{\sim}{Y}$

$$\mu_{\underset{\sim}{Y}}(y) = \begin{cases} 1 - \dfrac{\left|y - \sum_{i=1}^{n} p_i x_i\right|}{\sum_{i=1}^{n} c_i |x_i|} & x_i \ne 0 \\ 1 & x_i = 0, y = 0 \\ 0 & x_i = 0, y \ne 0 \end{cases} \qquad \text{Eq. (14.32)}$$

Solution

Nonfuzzy data

Objective function

$$O = \min\left\{\sum_{j=1}^{m} \sum_{i=1}^{n} c_i x_{ij}\right\} \qquad \text{Eq. (14.34)}$$

Constraints

1

$$y_j \ge \sum_{i=1}^{n} p_i x_{ij} - (1 - h) \sum_{i=1}^{n} c_i x_{ij} \qquad \text{Eq. (14.35)}$$

2

$$y_j \le \sum_{i=1}^{n} p_i x_{ij} + (1 - h) \sum_{i=1}^{n} c_i x_{ij} \qquad \text{Eq. (14.36)}$$

Fuzzy data

Objective function

$$O_f = \min\left\{\sum_{j=1}^{m} \sum_{i=1}^{n} c_i x_{ij}\right\} \qquad \text{Eq. (14.41)}$$

Constraints

1

$$y_j \ge \sum_{i=1}^{n} p_i x_{ij} - (1 - H) \sum_{i=1}^{n} c_i x_{ij} + (1 - H) e_j \qquad \text{Eq. (14.42)}$$

2

$$y_j \le \sum_{i=1}^{n} p_i x_{ij} + (1 - H) \sum_{i=1}^{n} c_i x_{ij} - (1 - H) e_j \qquad \text{Eq. (14.43)}$$

TABLE 14.3
Two data sets describing one-dimensional problem

y_i	x_{ij}
2.1	0.52
4.6	1.36

slack in the inequality). Also, since basic variables cannot be negative (this requirement arises owing to the "less than" inequality), artificial variables are introduced to account for the negative sign. The artificial variables are denoted by a bar on top of the letter, i.e., $\overline{D}_i$.

$$2.1 = p_0 + 0.52p_1 - 0.5c_0 - 0.26c_1 + D_1 \quad (1)$$
$$2.1 = p_0 + 0.52p_1 + 0.5c_0 + 0.26c_1 - D_2 + \overline{D}_3 \quad (2)$$
$$4.6 = p_0 + 1.36p_1 - 0.5c_0 - 0.68c_1 + D_4 \quad (3)$$
$$4.6 = p_0 + 1.36p_1 + 0.5c_0 + 0.68c_1 - D_5 + \overline{D}_6 \quad (4)$$

In the simplex method, an additional variable, denoted M, is used to weight the artificial variables (there are two, $\overline{D}_3$ and $\overline{D}_6$) in the objective function. The updated objective function is then

$$O = c_0 + 1.88c_1 + M\overline{D}_3 + M\overline{D}_6$$

where M is a very large number compared with the magnitudes of the numbers in the original data set; in this example M is set at 100. Therefore, the final objective function is

$$-O + c_0 + 1.88c_1 + 100\overline{D}_3 + 100\overline{D}_6 = 0 \quad (0)$$

The basic variables are D_1, $\overline{D}_3$, D_4, and $\overline{D}_6$. Since the stopping rule of the simplex method requires the basic variables to have a coefficient of 0, Eq. (0) is subtracted from M times the equations containing the artificial variables, Eqs. (2) and (4). This procedure works very much like a Gaussian elimination method for solving simultaneous equations. The calculations are shown in Table 14.4.

The values of c_0, c_1, p_0, and p_1 are determined so that $-O$ has the maximum value (or O has the minimum value). The calculations performed to determine c_0, c_1, p_0, and p_1 by the simplex method are shown in Tables 14.4 and 14.5. The simplex method is an iterative process in which the basic variable is replaced by the nonbasic variable that lies in the column of the highest negative coefficient in row 0 (the nonbasic variable in this column becomes an *entering* basic variable). This is the nonbasic variable that would increase the objective function at the fastest rate. For example, in the first block of Table 14.5 (blocks are separated by double horizontal lines), the value of -200 is the largest negative number in equation (0); therefore, p_0 is selected as the entering basic variable.

The basic variable that is replaced by the entering basic variable is called the *leaving* basic variable. The leaving basic variable is determined by dividing y (right-hand side) by the positive coefficients in the column containing the entering basic variable; the row that yields the lowest value contains the leaving basic variable. This is the basic variable that reaches zero first as the entering basic variable is increased. In this example, for block 1 the lowest value (2.1/1, 2.1/1, 4.6/1, 4.6/1) lies in rows 1 and 2; either one of the rows can be chosen for selecting the leaving basic variable.

The column that contains the entering basic variable is called the pivot column and the row containing the leaving basic variable is the pivot row. As the basic variables must have a coefficient of $+1$, the entire pivot row is divided by the pivot number (placed at the intersection of the pivot row and pivot column). The new basic variable is now eliminated from all the other equations by the following formula:

$$\text{New row} = \text{old row} - (\text{pivot column coefficient}) \times \text{new pivot row}$$

TABLE 14.4
First step in the simplex method

Equation	Coefficient of										
	c_0	c_1	p_0	p_1	D_1	D_2	$\overline{D_3}$	D_4	D_5	$\overline{D_6}$	y
Eq. (0)	1	1.88	0	0	0	0	100	0	0	100	0
Eq. (2)	50	26	100	52	0	−100	100	0	0	0	210
Eq. (4)	50	68	100	136	0	0	0	0	−100	100	460
Eq. (0) − Eq. (2) − Eq. (4)	−99	−92.1	−200	−188	0	100	0	0	100	0	−670

Therefore, row (0) of the second block is

$$\begin{array}{rlcccccccccc} & [-99 & -92.1 & -200 & 188 & 0 & 100 & 0 & 0 & 100 & 0 & -670\ ] \\ -(-200)\times & [\ \ 0.5 & 0.26 & 1 & 0.52 & 0 & -1 & 1 & 0 & 0 & 0 & 2.1] \\ \text{New row} = & [\ \ 1 & -40.1 & 0 & -84 & 0 & -100 & 200 & 0 & 100 & 0 & -250\ ] \end{array}$$

Similar calculations are conducted on all the other equations of the first block.

Row (1):

$$\begin{array}{rlcccccccccc} & [-0.5 & -0.26 & 1 & 0.52 & 1 & 0 & 0 & 0 & 0 & 0 & 2.1] \\ -(1)\times & [\ \ 0.5 & 0.26 & 1 & 0.52 & 0 & -1 & 1 & 0 & 0 & 0 & 2.1] \\ \text{New row} = & [-1 & -0.52 & 0 & 0 & 1 & 1 & -1 & 0 & 0 & 0 & 0\ ] \end{array}$$

Row (3):

$$\begin{array}{rlcccccccccc} & [-0.5 & -0.68 & 1 & 1.36 & 0 & 0 & 0 & 1 & 0 & 0 & 4.6] \\ -(1)\times & [\ \ 0.5 & 0.26 & 1 & 0.52 & 0 & -1 & 1 & 0 & 0 & 0 & 2.1] \\ \text{New row} = & [-1 & -0.94 & 0 & 0.84 & 0 & 1 & -1 & 1 & 0 & 0 & 2.5] \end{array}$$

Row (4):

$$\begin{array}{rlcccccccccc} & [0.5 & 0.68 & 1 & 1.36 & 0 & 0 & 0 & 0 & -1 & 1 & 4.6] \\ -(1)\times & [0.5 & 0.26 & 1 & 0.52 & 0 & -1 & 1 & 0 & 0 & 0 & 2.1] \\ \text{New row} = & [0 & 0.42 & 0 & 0.84 & 0 & 1 & -1 & 0 & -1 & 1 & 2.5] \end{array}$$

In Table 14.5 the leaving basic variable in each block of calculations is identified with an asterisk.

From the last column in the final block of Table 14.6, we see that $\underset{\sim}{A}_0 = (p_0, c_0) = (0.55, 0)$ and $\underset{\sim}{A}_1 = (p_1, c_1) = (2.97, 0)$. Substituting these values into $\underset{\sim}{Y} = \underset{\sim}{A}_0 + \underset{\sim}{A}_1 x$ yields

$$\underset{\sim}{Y} = (0.55, 0) + (2.97, 0) \times 0.52 = 2.09$$

which is essentially (to within computational error) the same as the actual y value (i.e., 2.1).

In Example 14.5 the fuzziness in the coefficients is 0; we should get the exact solution because we only had two points in the data set. For nontrivial data sets this outcome is not the case, and the coefficients turn out to be fuzzy sets as represented by a triangular membership function of the form expressed in Fig. 14.10. The following example illustrates this idea.

TABLE 14.5
Calculations of the simplex method

Basic Variable	Equation	Coefficient of O	c_0	c_1	p_0	p_1	D_1	D_2	$\overline{D}_3$	D_4	D_5	$\overline{D}_6$	y
Block 1													
O	(0)	−1	−99	−92.1	−200	188	0	100	0	0	100	0	−670
D_1	(1)	0	−0.5	−0.26	1	0.52	1	0	0	0	0	0	2.1
$^*\overline{D}_3$	(2)	0	0.5	0.26	1	0.52	0	−1	1	0	0	0	2.1
D_4	(3)	0	−0.5	−0.68	1	1.36	0	0	0	1	0	0	4.6
$\overline{D}_6$	(4)	0	0.5	0.68	1	1.36	0	0	0	0	−1	1	4.6
Block 2													
O	(0)	−1	1	−40.1	0	−84	0	−100	200	0	100	0	−250
$^*D_1$	(1)	0	−1	−0.52	0	0	1	1	−1	0	0	0	0
p_0	(2)	0	0.5	0.26	1	0.52	0	−1	1	0	0	0	2.1
D_4	(3)	0	−1	−0.94	0	0.84	0	1	−1	1	0	0	2.5
$\overline{D}_6$	(4)	0	0	0.42	0	0.84	0	1	−1	0	−1	1	2.5
Block 3													
O	(0)	−1	−99	−90.1	0	−84	100	0	100	0	100	0	−250
D_2	(1)	0	−1	−0.52	1	0	1	1	−1	0	0	0	0
p_0	(2)	0	−0.5	−0.26	0	0.52	1	0	0	0	0	0	2.1
D_4	(3)	0	0	−0.42	0	0.84	−1	0	0	1	0	0	2.5
$^*\overline{D}_6$	(4)	0	1	0.94	0	0.84	−1	0	0	0	−1	1	2.5
Block 4													
O	(0)	−1	0	2.94	0	−0.84	1	0	100	0	1	99	−2.5
$^*D_2$	(1)	0	0	0.42	0	0.84	0	1	−1	0	−1	1	2.5
p_0	(2)	0	0	0.21	1	0.94	0.5	0	0	0	−0.5	0.5	3.35
D_4	(3)	0	0	−0.42	0	0.84	−1	0	0	1	0	0	2.5
c_0	(4)	0	1	0.94	0	0.84	−1	0	0	0	−1	1	2.5

* Leaving variable

TABLE 14.6
Final block of simplex calculations

		Coefficient of											
Basic Variable	**Equation**	O	c_0	c_1	p_0	p_1	D_1	D_2	$\overline{D}_3$	D_4	D_5	$\overline{D}_6$	y
O	0	−1	0	3.29	0	0	1	1	99	0	0	2	0
p_1	1	0	0	0.5	0	1	0	1.2	−1.2	0	−1.2	1.2	2.97
p_0	2	0	0	−0.26	1	0	0.5	−1.1	1.1	0	0.62	−0.44	0.55
D_4	3	0	0	−0.84	0	0	−1	−1	1	1	1	−1	0
c_0	4	0	1	0.52	0	0	−1	−1	1	0	0	0	0

Example 14.7 [Kikuchi and Nanda, 1991]. Consider the data set given in Table 14.7. In this case there are five data points; hence there will be $2 \times 5 = 10$ constraints.

The fuzzy linear regression equation, $\underset{\sim}{Y} = \underset{\sim}{A}_0 + \underset{\sim}{A}_1 x_1 + \underset{\sim}{A}_2 x_2$ is used to fit the data set. The objective function to be minimized is

$$O = c_0 + \sum_j x_{1j} c_1 + \sum_j x_{2j} c_2 + M\overline{D}_3 + M\overline{D}_6 + M\overline{D}_9 + M\overline{D}_{12} + M\overline{D}_{15}$$

$$O = c_0 + 3.68c_1 + 2.05c_2 + M\overline{D}_3 + M\overline{D}_6 + M\overline{D}_9 + M\overline{D}_{12} + M\overline{D}_{15}$$

Using an h value of 0.5 and Eqs. (14.35)–(14.36) for each of the m data points, we get the following constraint equations:

$$3.54 = p_0 + 0.84p_1 + 0.46p_2 - 0.5[c_0 + 0.84c_1 + 0.46c_2] + D_1 \quad (1)$$
$$3.54 = p_0 + 0.84p_1 + 0.46p_2 + 0.5[c_0 + 0.84c_1 + 0.46c_2] - D_2 + \overline{D}_3 \quad (2)$$
$$4.05 = p_0 + 0.65p_1 + 0.52p_2 - 0.5[c_0 + 0.65c_1 + 0.52c_2] + D_4 \quad (3)$$
$$4.05 = p_0 + 0.65p_1 + 0.52p_2 + 0.5[c_0 + 0.65c_1 + 0.52c_2] - D_5 + \overline{D}_6 \quad (4)$$
$$4.51 = p_0 + 0.76p_1 + 0.57p_2 - 0.5[c_0 + 0.76c_1 + 0.57c_2] + D_7 \quad (5)$$
$$4.51 = p_0 + 0.76p_1 + 0.57p_2 + 0.5[c_0 + 0.76c_1 + 0.57c_2] - D_8 + \overline{D}_9 \quad (6)$$
$$2.63 = p_0 + 0.7p_1 + 0.3p_2 - 0.5[c_0 + 0.7c_1 + 0.3c_2] + D_{10} \quad (7)$$
$$2.63 = p_0 + 0.7p_1 + 0.3p_2 + 0.5[c_0 + 0.7c_1 + 0.3c_2] - D_{11} + \overline{D}_{12} \quad (8)$$
$$1.9 = p_0 + 0.73p_1 + 0.2p_2 - 0.5[c_0 + 0.73c_1 + 0.2c_2] + D_{13} \quad (9)$$
$$1.9 = p_0 + 0.73p_1 + 0.2p_2 + 0.5[c_0 + 0.73c_1 + 0.2c_2] - D_{14} + \overline{D}_{15} \quad (10)$$

TABLE 14.7
Five data samples

y	x_1	x_2
3.54	0.84	0.46
4.05	0.65	0.52
4.51	0.76	0.57
2.63	0.7	0.3
1.9	0.73	0.2

As explained in Example 14.6, the final objective function equation is obtained by considering the equations containing the artificial variables:

$$\begin{aligned}
&-1(O) - 1.5M(c_0) - 1.84M(c_1) - 1.025M(c_2) - 5M(p_0) - 3.68M(p_1) \\
&- 2.05M(p_2) + 0(D_1) + (1)M(D_2) + 0(\overline{D}_3) + 0(D_4) + (1)M(D_5) + 0(\overline{D}_6) \\
&+ 0(D_7) + (1)M(D_8) + 0(\overline{D}_9) + 0(D_{10}) + (1)M(D_{11}) + 0(\overline{D}_{12}) + 0(D_{13}) \\
&+ (1)M(D_{14}) + 0(\overline{D}_{15}) - 16.63M(y) = 0
\end{aligned} \tag{0}$$

where the zeros in expression (0) are coefficients on the many slack variables that do not appear in the objective equation. The same solution procedure as in Example 14.6 is used to determine the parameters of the coefficients of the fuzzy linear regression model. The fuzzy coefficients computed [Kikuchi and Nanda, 1991] are

$$\underset{\sim}{A}_0 = (1.242, 0); \qquad \underset{\sim}{A}_1 = (0, 1.4); \qquad \underset{\sim}{A}_2 = (5.843, 0)$$

We see that coefficient $\underset{\sim}{A}_1$ is fuzzy because it has a nonzero spread. It can be determined that as h increases, the fuzziness of the output increases (see Problem 14.8 at the end of the chapter).

We conclude this section with a few comments on fuzzy regression. First, the fact that the estimated value for the output variable, y, is given as a fuzzy number represents a drawback when outlier points exist in the data set. The presence of an outlier point makes the spread of the estimate very large since the estimate must cover that point at least to the level of confidence, h. However, Kikuchi and Nanda [1991] have shown that, although the spread of the fuzzy numbers may become large in fuzzy regression, the prototypical values (modal values) of the estimates remain relatively stable. Second, if a high value of h is given, the spread of the estimate of the output, y, increases in order to satisfy the increased measure of goodness of fit. Third, each data point requires two constraint equations; see Eqs. (14.35)–(14.36) or (14.42)–(14.43). The first of each of these equation pairs represents the case when y_j lies in the interval to the left of the prototypical value and the latter represents the case when y_j lies in the interval to the right of the prototypical value. For m data sets this means $2m$ constraints; for large data sets the computational load associated with these equations can become a deterrent to this regression method.

There are, of course, alternatives to the solution of fuzzy regression equations using a linear programming approach. For example, Tanaka et al. [1982] solve the *dual problem* expressed by Eqs. (14.42)–(14.43). Here the number of equations is related to the number of independent variables, n, as opposed to the number of constraints, $2N$. This relation makes the computational burden significantly lower than with conventional linear programming methods for situations where $N \geq n$. In addition to the problem of $N \geq n$, when n changes the entire set of constraints has to be reformulated; this characteristic also limits the utility of the fuzzy regression method. Any linear programming formulation requires that all the unknown variables must be positive, as demonstrated in the examples here. Hence, a prior knowledge of the effect of each variable on the outcome is useful. If ambiguity in the sign of a variable is a feature of the problem, the unknown value must be presented as a linear combination of two positive numbers—yet another growth in the number of equations. To overcome these difficulties, some investigators are using other methods to solve fuzzy regression problems, such as artificial neural networks [Kikuchi and Nanda, 1991].

SUMMARY

This chapter summarizes fuzzy logic applications in the areas of optimization, inverse equations, and regression. This only begins to scratch the surface of the plethora of applications being developed in the rapidly expanding field of fuzzy logic. The reader is referred to the literature for many other applications projects, which are summarized in such recent works or collected bibliographies as Schmucker [1984], Klir and Folger [1988], Jamshidi et al. [1993], McNeill and Freiberger [1993], Kosko [1993], and Cox [1994], or discussed in some of the active international research journals focusing on fuzzy applications such as *Intelligent and Fuzzy Systems, Fuzzy Sets and Systems,* and *IEEE Transactions on Fuzzy Systems.*

REFERENCES

Akai, T. (1994). *Applied numerical methods for engineers,* John Wiley & Sons, New York, chapter 10.

Cox, E. (1994). *The fuzzy systems handbook,* Academic Press Professional, Cambridge, MA.

deNeufville, R. (1990). *Applied systems analysis: engineering planning and technology management,* McGraw-Hill, New York.

Dong, W. (1987). Personal notes.

Dubois, D., and H. Prade (1980). *Fuzzy sets and systems: Theory and applications,* Academic, New York.

Fuzzy Sets and Systems (1994). C. Negoita, L. Zadeh, and H. Zimmerman (eds.), Elsevier Science, Amsterdam, the Netherlands.

Goldberg, D. (1989). *Genetic algorithms in search, optimization and machine learning,* Addison-Wesley, Reading, MA.

Hillier, F., and G. Lieberman (1980). *Introduction to operations research,* Holden-Day, San Francisco.

IEEE Transactions on Fuzzy Systems. IEEE Press, New York.

Intelligent and Fuzzy Systems (1994). M. Jamshidi and T. Ross (eds.), John Wiley & Sons, New York.

Jamshidi, M., N. Vadiee, and T. Ross (eds.) (1993). *Fuzzy logic and control: software and hardware applications,* Prentice Hall, Englewood Cliffs, NJ.

Kikuchi, S., and R. Nanda (1991). "Fuzzy regression analysis using a neural network: application to trip generation model," unpublished manuscript.

Klir, G., and T. Folger (1988). *Fuzzy sets, uncertainty, and information,* Prentice Hall, Englewood Cliffs, NJ.

Kosko, B. (1993). *Fuzzy thinking,* Hyperion Press, New York.

McNeill, D., and P. Freiberger (1993). *Fuzzy logic,* Simon & Schuster, New York.

Sakawa, M. (1993). *Fuzzy sets and interactive multiobjective optimization,* Plenum Press, New York.

Schmucker, K. (1984). *Fuzzy sets, natural language computations, and risk analysis,* Computer Science Press, Rockville, MD.

Tanaka, H., S. Uejima, and K. Asai (1982). "Linear regression analysis with fuzzy model," *IEEE Trans. Syst. Man Cybern.,* vol. SMC-12, No. 6.

Terano, T., K. Asai, and M. Sugeno (1992). *Fuzzy system theory and its applications,* Academic, San Diego.

Tsukamoto, Y., and T. Terano (1977). "Failure diagnosis by using fuzzy logic," *Proc. IEEE Conf. Decision Control,* New Orleans, vol. 2, pp. 1390–1395.

Vanderplaats, G. (1984). *Numerical optimization techniques for engineering design with applications,* McGraw-Hill, New York.

Zadeh, L. (1972). "On fuzzy algorithms," Memo UCB/ERL M-325, University of California, Berkeley.

PROBLEMS

Fuzzy Optimization

14.1. The feedforward transfer function for a unit-feedback control system is $1/(s-1)$, as shown in Fig. P14.1. A unit step signal is input to the system. Determine the minimum error of the system response by using a fuzzy optimization method for the time period, $0 < t < 10$ s and a fuzzy constraint given by

$$u_c(t) = \begin{cases} 1 & 0 \le t \le 1 \\ e^{1-t} & t > 1 \end{cases}$$

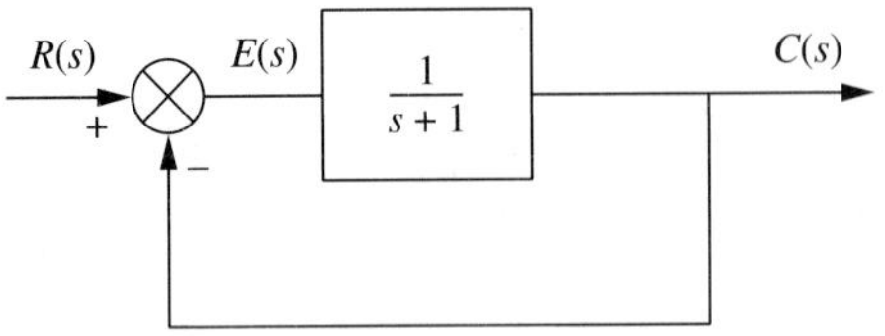

FIGURE P14.1

14.2. A beam structure is forced by an axial load P (Fig. P14.2). When P is increased to its *critical* value, the beam will buckle. Prove that the critical force P to cause buckling can be expressed by a function

$$P = \frac{n^2\pi^2 EI}{L}$$

where EI = stiffness of the beam
L = span length of the beam
n = number of sine waves the beam shape takes when it buckles (assume it to be continuous)

If $0 \le n \le 2$, assume that n is constrained by the fuzzy member function

$$u_c(n) = \begin{cases} 1 - n & 1 \le n \le 2 \\ 0 & n < 1 \end{cases}$$

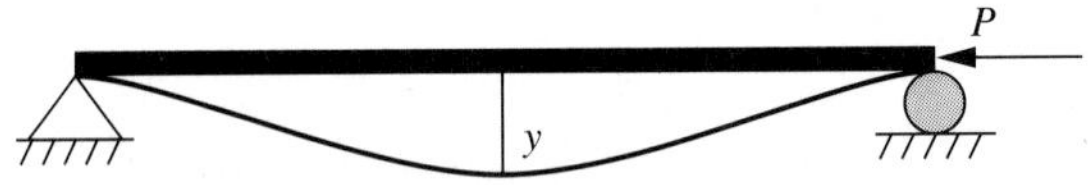

FIGURE P14.2

14.3. Suppose that the beam structure in Problem 14.2 also has a transverse load P applied at the middle of the beam. Then the maximum bending stress can be calculated by the equation $\sigma_b = Pl/4w_z$, where w_z, with units m^3, is a coefficient based on the shape and size of the cross section of the beam, and l is in meters. If $0 \le \sigma_b \le 60$ MPa, and the fuzzy constraint function for σ_b is

$$\mu_c(\sigma_b) = \begin{cases} \dfrac{1}{(x+1)^2} & 0 \le x \le 1 \\ 0 & x > 1 \end{cases}$$

where $x = \dfrac{\sigma}{60 \text{ MPa}}$

combine the conditions given in Problem 14.2 to find the optimum load P, in Newtons. (*Hint:* This problem involves multiple constraints.)

14.4. In the metallurgical industry, the working principle for a cold rolling mill is to extrude a steel strip through two rows of working rollers, as shown in Fig. P14.4. The size of the roller is very important. The stress between the roller and the strip can be expressed by the following function:

$$\sigma_H = 0.564\sqrt{\frac{PE}{LR}}$$

where E = Young's modulus (kN/cm^2)
P = loading force (N)
L = contact length between roll and strip (cm)
R = radius of a roller (cm)

If $\sigma_H = 2.5$ kN/cm^2 and $10 < R < 20$, find the minimum R in which σ_H has a maximum value. The radius R has a fuzzy constraint of

$$u_c(R) = \begin{cases} 1 & 10 \le R \le 15 \\ \dfrac{20 - R}{5} & 15 < R \le 20 \end{cases}$$

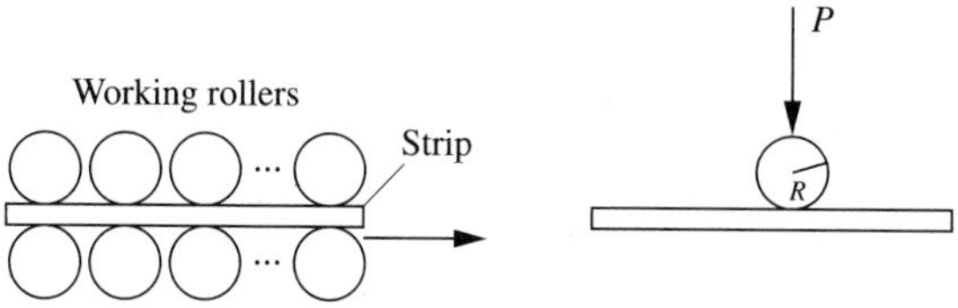

FIGURE P14.4

Inverse fuzzy relational equations

14.5. In a fuzzy relation equation, $\underset{\sim}{A} \circ \underset{\sim}{R} = \underset{\sim}{B}$, r_i is known as (0.5, 0.7, 0.9) and b_i is 0.6. Use the Tsukamoto method for an inverse fuzzy equation to find a_i $(i = 1, 2, 3)$.

14.6. A fuzzy relation has an expression given as

$$\{a_1, a_2\} \circ \begin{bmatrix} 0.4 & 0.6 \\ 0.8 & 0.1 \end{bmatrix} = [0.3 \quad 0.1]$$

Find a_i $(i = 1, 2)$ by using the Tsukamoto method for inverse fuzzy relations.

14.7. A system having a single degree of freedom has the following ordinary differential governing equation:

$$a_1\ddot{x} + a_2\dot{x} + a_3 x = b$$

If b is the input signal, in which $b_1 = 0.5$ and $b_2 = 0.6$, and two sets of response data of x are (0.4, 0.6, 0.8) and (0.5, 0.7, 0.9), respectively, find the system coefficients a_i $(i = 1, 2, 3)$.

Regression

14.8. Show that as the parameter h increases, the fuzziness of the output increases in the five-point regression problem (Example 14.7) [Kikuchi and Nanda, 1991]. Use the simplex method for values of $h = 0.2$ and 0.8.

14.9. Risk assessment is fast becoming the basis of many EPA guidelines that determine whether a site contaminated with hazardous substances needs to be remediated or not. Risk is defined as the likelihood of an adverse health impact to the public due to exposure to environmental hazards. Risk assessment consists of four parts: (*i*) hazard identification; (*ii*) dose-response assessment—assessing the health response to a certain dose (concentration) of the chemical; (*iii*) exposure assessment—assessing the duration and concentration of exposure; and (*iv*) risk characterization—quantification and presentation of risks. Part (*ii*) of the risk assessment process consists of exposing a controlled population of animals to various doses of the chemical and fitting a dose-response curve to the experimental data. The following table comprises the data derived from the tests:

Dose, mg/L	Response
0	0
2	0.02
5	1.0
10	12.3

Assuming that the dose-response relationship can be expressed by a fuzzy linear regression model, $\underset{\sim}{Y} = \underset{\sim}{A}_0 + \underset{\sim}{A}_1 x$, where x represents the dose in mg/L, determine the fuzzy coefficients $\underset{\sim}{A}_0$ and $\underset{\sim}{A}_1$. Use an h value of 0.5.

14.10. In a survey on costs for the construction of new houses, the number of rooms (including bedrooms, kitchen, bathroom, and living rooms) in a house was compared with the material costs of the house. The following table gives the results of the survey:

Cost, $	Number of rooms
10,000	2
25,000	5
100,000	8

Assuming an h of 0.5, determine the coefficients of the fuzzy one-dimensional linear regression model, $\underset{\sim}{Y} = \underset{\sim}{A}_0 + \underset{\sim}{A}_1 x$.

14.11. In fuzzy regression, the output y is a triangular fuzzy number with the spread e_j representing the error in measurement (Fig. P14.11).

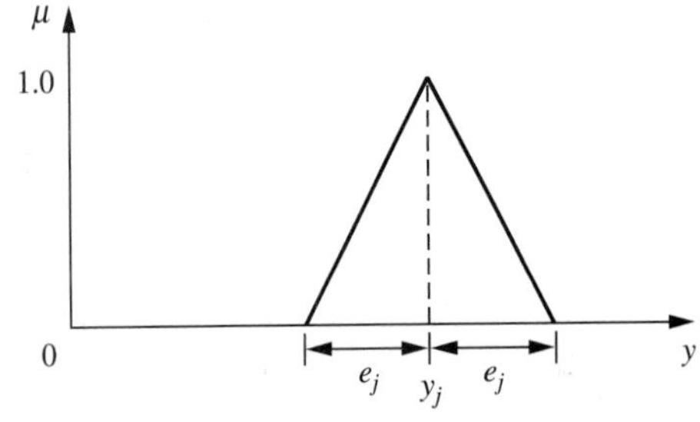

FIGURE P14.11

The accompanying table shows the fuzzy output and the corresponding crisp input:

y, e	x
2.1, 0.2	0.52
4.6, 0.35	1.36

Using Eqs. (14.41) through (14.43) for fuzzy output, and $h = 0.4$, determine the fuzzy coefficients for a simple fuzzy linear regression model, $\underset{\sim}{Y} = \underset{\sim}{A}_0 + \underset{\sim}{A}_1 x$.

CHAPTER 15

FUZZY MEASURES: BELIEF, PLAUSIBILITY, PROBABILITY, AND POSSIBILITY

All traditional logic habitually assumes that precise symbols are being employed. It is therefore not applicable to this terrestrial life but only to an imagined celestial existence.

Bertrand Russell, 1923
British philosopher and Nobel Laureate

Most of this text has dealt with the quantification of various forms of nonnumeric uncertainty. Two prevalent forms of uncertainty are those arising from vagueness and from imprecision. How do vagueness and imprecision differ as forms of uncertainty? Often vagueness and imprecision are used synonymously, but they can differ in the following sense. Vagueness can be used to describe certain kinds of uncertainty associated with linguistic information or intuitive information. Examples of vague information are that the image quality is "good," or that the transparency of an optical element is "acceptable." Imprecision can be associated with quantitative or countable data as well as noncountable data. As an example of the latter, one might say the length of a bridge span is "long." An example of countable imprecision would be to report the length to be 300 meters. If we take a measuring device and measure the length of the bridge 100 times we likely will come up with 100 different values; the differences in the numbers will no doubt be on the order of the precision of the measuring device. Measurements using a 10-meter chain will be less precise than those developed from a laser theodolite. If we plot the bridge lengths on some sort of probit paper and develop a Gaussian distribution to describe the length of this bridge, we could state the imprecision in probabilistic terms. In this case the length of the bridge is uncertain to some degree of precision that is quantified in the language of statistics. Since we are not able to make this measurement an infinite number of times, there is also uncertainty in the statistics describing the bridge length. Hence, imprecision can be used to quantify random variability in quantitative uncertainty and it can also be used to describe a lack of knowledge for descriptive entities (e.g., acceptable transparency, good image quality). Vagueness is usually related to nonmeasurable issues.

This chapter develops the relationships among fuzzy set theory, fuzzy measures theory, probability theory, and evidence theory. All of these theories are related, and all have been used to characterize and model various forms of uncertainty. That they are all related mathematically is an especially crucial advantage in their use in quantifying the uncertainty spectrum because, as more information about a problem becomes available, the mathematical description of uncertainty can easily transform from one theory to the next in the characterization of the uncertainty. This chapter begins by describing the difference between a fuzzy set and a fuzzy measure. It then discusses various forms of fuzzy measures such as belief, plausibility, possibility, and probability. Then there is a discussion on the relationship between a possibility distribution and a fuzzy set. The chapter concludes with a discussion about the similarities in the formal axioms of a probability theory and a fuzzy set theory. Examples are provided to illustrate the various theories.

FUZZY MEASURES

A fuzzy measure describes the vagueness or imprecision in the assignment of an element *a* to two or more crisp sets. Figure 15.1 shows this idea. In the figure the universe of discourse comprises a collection of sets and subsets, or the power set. In a fuzzy measure what we are trying to describe is the vagueness or imprecision in assigning this point to any of the crisp sets on the power set. This notion is not random; the crisp sets have no uncertainty about them. The uncertainty is about the assignment. This uncertainty is usually associated with evidence to establish an assignment. The evidence can be completely lacking—the case of total ignorance—or the evidence can be complete—the case of a probability assignment. Hence, the difference between a fuzzy measure and a fuzzy set on a universe of elements is that, in the former the imprecision is in the assignment of an element to one of two or more crisp sets, and in the latter the imprecision is in the prescription of the boundaries of a set.

BELIEF AND PLAUSIBILITY

There are special forms of fuzzy measures. A form associated with preconceived notions is called a belief measure. A form associated with information that is possible, or plausible, is called a plausibility measure. Specific forms of belief measures

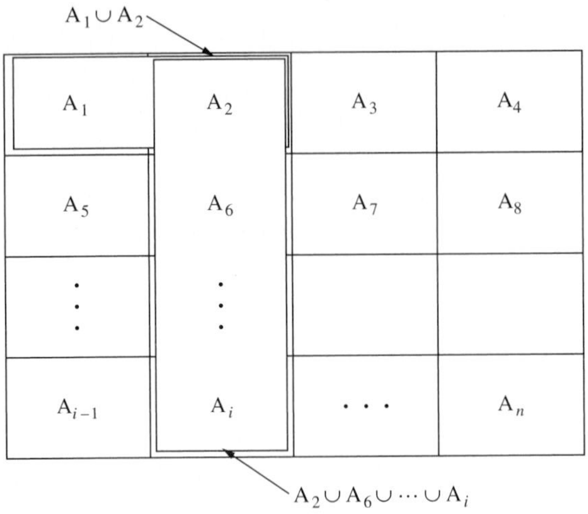

FIGURE 15.1
A fuzzy measure.

and plausibility measures are known as certainty and possibility measures, respectively. The intersection of belief measures and plausibility measures (i.e., where belief equals plausibility) will be shown to be a probability. Fuzzy measures are defined by weaker axioms than probability theory, thus subsuming probability measures as specific forms of fuzzy measures.

Basically, a belief measure is a quantity, denoted bel(A), that expresses the degree of support, or evidence, for a collection of elements defined by one or more of the crisp sets existing on the power set of a universe. The plausibility measure of this collection A is defined as the "complement of the Belief of the complement of A," or

$$\text{pl}(A) = 1 - \text{bel}(\overline{A}) \tag{15.1}$$

Since belief measures are quantities that measure the degree of support for a collection of elements or crisp sets in a universe, it is entirely possible that the belief measure of some set A plus the belief measure of $\overline{A}$ will not be equal to unity (the total belief, or evidence, for all elements or sets on a universe is equal to one, by convention). When this sum equals 1, we have the condition where the belief measure is a probability; that is, the evidence supporting set A can be described probabilistically. The difference between the sum of these two quantities [bel(A) + bel($\overline{A}$)] and 1 is called the ignorance, i.e., ignorance $= 1 - [\text{bel}(A) + \text{bel}(\overline{A})]$. When the ignorance equals 0 we have the case where the evidence can be described by probability measures.

Say we have evidence about a certain prospect in our universe of discourse, evidence of some set occurring or some set being realized, and we have no evidence (zero evidence) of the complement of that event. In probability theory we must assume, because of the excluded middle laws, that if we know the probability of A then the probability of $\overline{A}$ is also known, because we have in all cases involving probability measures, prob(A) + prob($\overline{A}$) $= 1$. This constraint of the excluded middle laws is not a requirement in evidence theory. The probability of $\overline{A}$ also has to be supported with some sort of evidence. If there is no evidence (zero degree of support) for $\overline{\mathrm{A}}$ then the degree of ignorance is large. This distinction between evidence theory and probability theory is important. It will also be shown that this is an important distinction between fuzzy set theory and probability theory.

Fuzzy measures are very useful in quantifying uncertainty that is difficult to measure or that is linguistic in nature. For example, in assessing structural damage in buildings and bridges after an earthquake or hurricane, evidence theory has proven quite successful because what we have are nonquantitative estimates from experts; the information concerning damage is not about how many inches of displacement or microinches per inch of strain the structure might have undergone, but rather is about expert judgment concerning the suitability of the structure for habitation or its intended function. These kinds of judgments are not quantitative; they are qualitative.

The mathematical development for fuzzy measures follows [see, for example, Klir and Folger, 1988]. We begin by assigning a value of membership to each crisp set existing in the power set of a universe, signifying the degree of evidence or belief that a particular element from the universe, say x, belongs in any of the crisp sets on the power set. We will label this membership $g(\mathrm{A})$, where it is a mapping between the power set and the unit interval,

$$g : \mathrm{P}(x) \rightarrow [0, 1] \tag{15.2}$$

and where $\mathrm{P}(x)$ is the power set of crisp sets (see Chapter 2). So, the membership value $g(\mathrm{A})$ represents the degree of available evidence of the belief that a given element x belongs to a crisp subset A.

The collection of these degrees of belief represents the *fuzziness* associated with several *crisp* alternatives. This type of uncertainty, which we call a fuzzy measure, is *different* from the uncertainty associated with the boundaries of a single set, which we call a fuzzy set. Fuzzy measures are defined for a *finite* universal set by at least three axioms, two of which are given here (a third axiom is required for an infinite universal set):

$$\begin{aligned} &\textbf{1. } g(\varnothing) = 0, \qquad g(\mathrm{X}) = 1 \\ &\textbf{2. } g(\mathrm{A}) \le g(\mathrm{B}) \text{ for A,} \qquad \mathrm{B} \in \mathrm{P}(x), \mathrm{A} \subseteq \mathrm{B} \end{aligned} \tag{15.3}$$

The first axiom represents the boundary conditions for the fuzzy measure, $g(\mathrm{A})$. It says that there is no evidence for the null set and there is complete (i.e., unity) membership for the universe. The second axiom represents monotonicity by simply stating that if one set A is completely contained in another set B then the evidence supporting B is at least as great as the evidence supporting the subset A.

A belief measure also represents a mapping from the crisp power set of a universe to the unit interval representing evidence, denoted

$$\mathrm{bel} : \mathrm{P}(\mathrm{X}) \rightarrow [0, 1] \tag{15.4}$$

Belief measures can be defined by adding a third axiom to those represented in Eqs. (15.3), given by

$$\begin{aligned} \mathrm{bel}(\mathrm{A}_1 \cup \mathrm{A}_2 \cup \cdots \cup \mathrm{A}_n) \ge \sum_i \mathrm{bel}(\mathrm{A}_i) - \sum_{i<j} \mathrm{bel}(\mathrm{A}_i \cap \mathrm{A}_j) + \cdots \\ + (-1)^{n+1}\mathrm{bel}(\mathrm{A}_1 \cap \mathrm{A}_2 \cap \cdots \cap \mathrm{A}_n) \end{aligned} \tag{15.5}$$

where there are n crisp subsets on the universe X. For each crisp set $A \in P(X)$, bel(A) is the degree of belief in set A based on available evidence. When the sets A_i in Eq. (15.5) are pairwise disjoint, i.e., where $A_i \cap A_j = \varnothing$, then Eq. (15.5) becomes

$$\text{bel}(A_1 \cup A_2 \cup \cdots \cup A_n) \geq \text{bel}(A_1) + \text{bel}(A_2) + \cdots + \text{bel}(A_n) \quad (15.6)$$

For the special case where $n = 2$, we have two disjoint sets A and $\overline{A}$, and Eq. (15.6) becomes

$$\text{bel}(A) + \text{bel}(\overline{A}) \leq 1 \quad (15.7)$$

A plausibility measure is also a mapping on the unit interval characterizing the total evidence, i.e.,

$$\text{pl} : P(X) \rightarrow [0, 1] \quad (15.8)$$

Plausibility measures satisfy the basic axioms of fuzzy measures, Eqs. (15.3), and one additional axiom [different from Eq. (15.5) for beliefs],

$$\begin{aligned} \text{pl}(A_1 \cap A_2 \cap \cdots \cap A_n) \leq & \sum_i \text{pl}(A_i) - \sum_{i<j} \text{pl}(A_i \cup A_j) + \cdots \\ & + (-1)^{n+1}\text{pl}(A_1 \cup A_2 \cup \cdots \cup A_n) \end{aligned} \quad (15.9)$$

From Eq. (15.1) we have a mutually dual system between plausibility and belief,

$$\begin{aligned} \text{pl}(A) &= 1 - \text{bel}(\overline{A}) \\ \text{bel}(A) &= 1 - \text{pl}(\overline{A}) \end{aligned} \quad (15.10)$$

For the specific case of $n = 2$, i.e., for two disjoint sets A and $\overline{A}$, Eq. (15.10) produces

$$\text{pl}(A) + \text{pl}(\overline{A}) \geq 1 \quad (15.11)$$

By combining Eq. (15.7) and Eq. (15.10) it can be shown that

$$\text{pl}(A) \geq \text{bel}(A) \quad (15.12)$$

Equation (15.12) simply states that for whatever evidence supports set A, its plausibility measure is always at least as great as its belief measure.

We now define another function on the crisp sets ($A \in P(X)$) of a universe, denoted $m(A)$, which can be used to express and determine both belief and plausibility measures. This measure is also a mapping from the power set to the unit interval,

$$m : P(X) \rightarrow [0, 1] \quad (15.13)$$

This measure, called a basic probability assignment (bpa) in the literature, has boundary conditions,

$$m(\varnothing) = 0 \quad (15.14)$$

$$\sum_{A \in P(X)} m(A) = 1 \quad (15.15)$$

The measure $m(A)$ is the degree of belief that a specific element, x, of the universe X belongs to the set A, *but not to any specific subset of* A. In this way $m(A)$ differs from both beliefs and plausibility. It important to remark here, to avoid confusion with standard terms in probability theory, that there is a distinct difference between a basic probability assignment (bpa) and a probability density function (pdf). The former are defined on sets of the power set of a universe (i.e., on $A \in P(X)$), whereas the latter are defined on the singletons of the universe (i.e., on $x \in P(X)$). This difference will be reinforced through some examples in this chapter. To add to the jargon of the literature, the first boundary condition, Eq. (15.14), provides for a *normal* bpa.

The bpa is used to determine a belief measure by

$$\text{bel}(A) = \sum_{B \subseteq A} m(B) \tag{15.16}$$

In Eq. (15.16) note that $m(A)$ is the degree of evidence in set A *alone,* whereas bel(A) is the total evidence in set A *and* all subsets (B) of A. The measure $m(A)$ is used to determine a plausibility measure by

$$\text{pl}(A) = \sum_{B \cap A \neq \emptyset} m(B) \tag{15.17}$$

Equation (15.17) shows that the plausibility of an event A is the total evidence in set A plus the evidence in all sets of the universe that intersect with A (including those sets that are also subsets of A). Hence, the plausibility measure in set A contains all the evidence contained in a belief measure (bel(A)) plus the evidence in sets that intersect with set A. Hence, Eq. (15.12) is verified.

Example 15.1. A certain class of short-range jet aircraft has had, for the shorter fuselage versions, a history of an oscillatory behavior described as *vertical bounce*. This is due to the in-flight flexing of the fuselage about two body-bending modes. Vertical bounce is most noticeable at the most forward and aft locations in the airplane. An acceptable acceleration threshold of ± 0.1g has been set as the point at which aft lower-body vortex generators should be used to correct this behavior. In order to avoid the cost of instrumented flight tests, expert engineers often decide whether vertical bounce is present in the airplane. Suppose an expert engineer is asked to assess the evidence in a particular plane for the following two conditions:

1. Are oscillations caused by other phenomena? (O)
2. Are oscillations characteristic of the vertical bounce? (B)

This universe is a simple one, consisting of the singleton elements O and B. The nonnull [Eq. (15.14) reminds us that the null set contains no evidence, i.e., $m(\emptyset) = 0$] power set then consists simply of the two singletons and the union of these two, $O \cup B$; including the null set there are $2^2 = 4$ elements in the power set. All the elements in the power set are called *focal elements*. Suppose the expert provides the measures of evidence shown in Table 15.1 for each of the focal elements [i.e., she gives $m(A_i)$, for $i = 1, \ldots, 4$]. Note that the sum of the evidences in the $m(A)$ column equals unity, as required by Eq. (15.15). We now want to calculate the degrees of belief and plausibility for this evidence set. Using Eq. (15.16), we find

$$\text{bel}(O) = m(O) = 0.4 \quad \text{and} \quad \text{bel}(B) = m(B) = 0.2$$

as seen in Table 15.1. The singletons O and B have no other subsets in them. Using Eq. (15.16) we find

$$\text{bel}(O \cup B) = m(O) + m(B) + m(O \cup B) = 0.4 + 0.2 + 0.4 = 1$$

TABLE 15.1
Measures of evidence for airplane bounce

Focal element,	Expert		
A_i	$m(A_i)$	$bel(A_i)$	$pl(A_i)$
$\varnothing$	0	0	0
O	0.4	0.4	0.8
B	0.2	0.2	0.6
$O \cup B$	0.4	1	1

as seen in Table 15.1. Using Eq. (15.17), we find

$$\text{pl}(O) = m(O \cap O) + m(O \cap (O \cup B)) = 0.4 + 0.4 = 0.8$$

and

$$\text{pl}(B) = m(B \cap B) + m(B \cap (O \cup B)) = 0.2 + 0.4 = 0.6$$

since sets O and B both intersect with the set $O \cup B$; and finally,

$$\text{pl}(O \cup B) = m((O \cup B) \cap O) + m((O \cup B) \cap B) + m((O \cup B) \cap (O \cup B)) = 0.4 + 0.2 + 0.4 = 1$$

since all sets in the power set intersect with $(O \cup B)$. These quantities are included in the fourth column of Table 15.1. Thus, the engineer *believes* the evidence supporting set O (other oscillations) is at least 0.4 and *possibly* as high as 0.8 (plausibility), and she *believes* the evidence supporting set B (vertical bounce) is at least 0.2 and *possibly* as high as 0.6 (plausibility). Finally, the evidence supporting either of these sets $(O \cup B)$ is full, or complete (i.e., bel = pl = 1).

EVIDENCE THEORY

The material presented in the preceding section now sets the stage for a more complete assessment of evidence, called evidence theory [Shafer, 1976]. Suppose the evidence for certain fuzzy measures comes from more than one source, say two experts. Evidences obtained in the same context (e.g., for sets A_i on a universe X) from two independent sources (e.g., two experts) and expressed by two bpa's (e.g., m_1 and m_2) on some power set P(X) can be combined to obtain a joint bpa, denoted m_{12}, using Dempster's rule of combined evidence [Dempster, 1967]. The procedure to combine evidence is given here in Eqs. (15.18) and (15.19):

$$m_{12}(A) = \frac{\sum_{B \cap C = A} m_1(B) \cdot m_2(C)}{1 - K} \text{ for } A \neq \varnothing \tag{15.18}$$

where the denominator is a normalizing factor such that

$$K = \sum_{B \cap C = \varnothing} m_1(B) \cdot m_2(C) \tag{15.19}$$

Dempster's rule of combination combines evidence in a manner analogous to the way in which joint probability density functions (pdf) in probability theory are calculated from two independent marginal pdf's. We can define a *body of evidence,* then, as a pair (A, m) where A are sets with available evidence m(A).

Example 15.2. If a generator is to run untended, the external characteristic of the shunt machine may be very unsatisfactory, and that of a series even more so, since a source of constant potential difference supplying a varying load current is usually required. The situation is even less satisfactory if the load is supplied via a feeder with appreciable resistance, since this will introduce an additional drop in potential at the load-end of the feeder. What is required is a generator with rising external characteristics, since this would counteract the effect of feeder resistance. Such a characteristic may be obtained from a compound generator. In a compound generator we can get variable induced electromotive force (emf) with increase of load current by arranging the field magnetomotive force (mmf). So, shunt and field windings are used in the generator as shown in Fig. 15.2. In this figure, R_a, R_c, R_f, and R_s are the armature, compound, field, and series resistance, respectively; I_f and I_l are the field and load current, respectively; E_a and V_t are the induced armature and terminal voltages, respectively; and N_f and N_s are the number of turns in the series and field windings, respectively. By arranging the field winding in different combinations (i.e., varying the difference between N_s and N_f), we can get different combinations of compound generators; in particular we can get (1) overcompounded (OC), (2) flatcompounded (FC), and (3) undercompounded (UC) generators. Each type of compounded generator has its own external and internal characteristics.

We can say that these three types comprise a universal set of generators, X. Let us consider two experts, E_1 an electrical engineer and E_2 a marketing manager, called to evaluate the efficiency and performance of a compound generator. We can come to some conclusion that, say for a particular outdoor lighting situation, the machine chosen by the two experts may be different for various reasons. On the one hand, the electrical engineer may think about performance issues like minimizing the error, maintaining

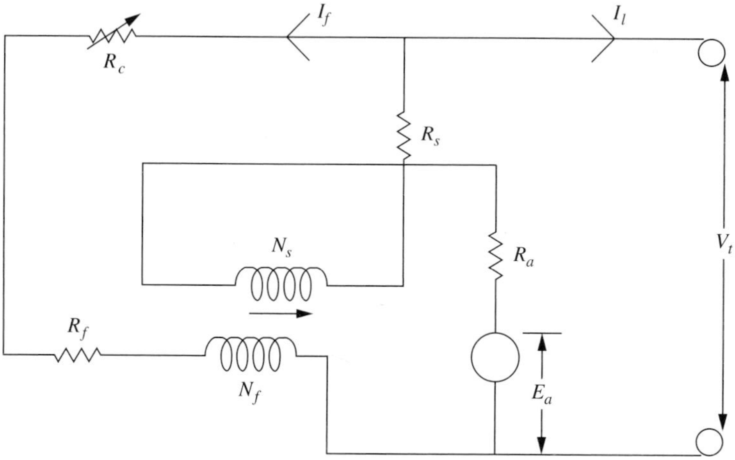

FIGURE 15.2
Electrical diagram of a compound generator.

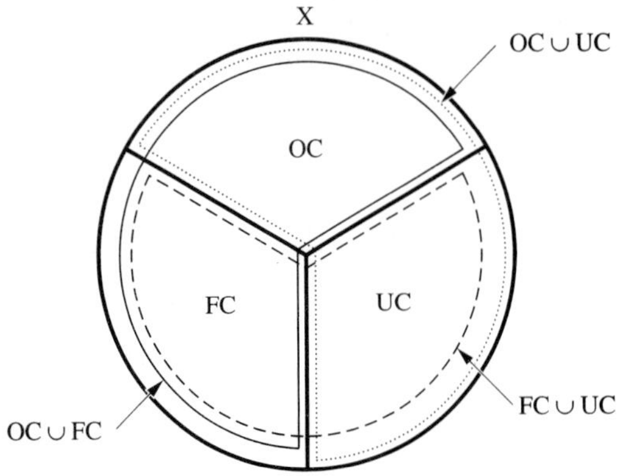

FIGURE 15.3
Universe X of compound generators.

constant voltage, and other electrical problems. On the other hand, the marketing manager may be concerned only with issues like minimizing the cost of running or minimizing maintenance and depreciation costs. In reality both experts may have valid reasons for the selection of a specific machine required for final installation.

Suppose that a company hires these two individuals to help them decide on a specific kind of generator to buy. Each expert is allowed to conduct tests or surveys to collect information (evidence) about the value of each of the three generators. The universe showing the individual sets of the power set is illustrated in Fig. 15.3. The focal elements of the universe in this figure are OC, FC, UC, OC ∪ FC, OC ∪ UC, FC ∪ UC, and OC ∪ FC ∪ UC [hereafter we ignore the null set in determining evidence since this set contains no evidence by definition, Eq. (15.14)], as listed in the first column of Table 15.2. Note that there are $2^3 - 1 = 7$ nonnull elements. Suppose that the two experts, E_1 and E_2, give their information (evidence measures) about each focal element A_i, where $i = 1, 2, \ldots, 7$; i.e., they provide $m_1(A_i)$ and $m_2(A_i)$, respectively (the second and fourth columns in Table 15.2). Note that the sum of the entries in the second and fourth columns equals unity, again guaranteeing Eq. (15.15).

Using this information, we can calculate the belief measures for each expert. For example,

$$\text{bel}_1(\text{OC} \cup \text{FC}) = \sum_{\text{B} \subseteq (\text{OC} \cup \text{FC})} m_1(\text{B}) = m_1(\text{OC} \cup \text{FC}) + m_1(\text{OC}) + m_1(\text{FC})$$

$$= 0.15 + 0 + 0.05 = 0.20$$

TABLE 15.2
Focal elements and evidence for compound generators

Focal elements, A_i	Expert 1 $m_1(A_i)$	Expert 1 $\text{bel}_1(A_i)$	Expert 2 $m_2(A_i)$	Expert 2 $\text{bel}_2(A_i)$	Combined evidence $m_{12}(A_i)$	Combined evidence $\text{bel}_{12}(A_i)$
OC	0	0	0	0	.01	.01
FC	.05	.05	.15	.15	.21	.21
UC	.05	.05	.05	.05	.09	.09
OC ∪ FC	.15	.20	.05	.20	.12	.34
OC ∪ UC	.05	.10	.05	.20	.06	.16
FC ∪ UC	.10	.20	.20	.40	.20	.50
OC ∪ FC ∪ UC	.60	1	.50	1	.30	1

$$\text{bel}_2(\text{FC} \cup \text{UC}) = \sum_{\text{B} \subseteq (\text{FC} \cup \text{UC})} m_2(\text{B}) = m_2(\text{FC} \cup \text{UC}) + m_2(\text{FC}) + m_2(\text{UC})$$
$$= 0.20 + 0.15 + 0.05 = 0.40$$

The remaining calculated belief values for the two experts are shown in the third and fifth columns of Table 15.2.

Using the Dempster rule of combination [Eqs. (15.18)–(15.19)], we can calculate the combined evidence (m_{12}) measures (calculations will be to two significant figures). First we must calculate the normalizing factor, K, using Eq. (15.19). In calculating this expression we need to sum the multiplicative measures of all those focal elements whose intersection is the null set; i.e., all those focal elements that are disjoint.

$$\begin{aligned} K = {} & m_1(\text{FC})m_2(\text{OC}) + m_1(\text{FC})m_2(\text{UC}) + m_1(\text{FC})m_2(\text{OC} \cup \text{UC}) + m_1(\text{OC})m_2(\text{FC}) \\ & + m_1(\text{OC})m_2(\text{UC}) + m_1(\text{OC})m_2(\text{FC} \cup \text{UC}) + m_1(\text{UC})m_2(\text{FC}) \\ & + m_1(\text{UC})m_2(\text{OC}) + m_1(\text{UC})m_2(\text{FC} \cup \text{OC}) + m_1(\text{FC} \cup \text{OC})m_2(\text{UC}) \\ & + m_1(\text{FC} \cup \text{UC})m_2(\text{OC}) + m_1(\text{OC} \cup \text{UC})m_2(\text{FC}) = 0.03 \end{aligned}$$

Hence $1 - K = 0.97$. So, for example, combined evidence on the set FC can be calculated using Eq. (15.18):

$$\begin{aligned} m_{12}(\text{FC}) = {} & \{m_1(\text{FC})m_2(\text{FC}) + m_1(\text{FC})m_2(\text{OC} \cup \text{FC}) + m_1(\text{FC})m_2(\text{FC} \cup \text{UC}) \\ & + m_1(\text{FC})m_2(\text{FC} \cup \text{UC} \cup \text{OC}) + m_1(\text{OC} \cup \text{FC})m_2(\text{FC}) \\ & + m_1(\text{OC} \cup \text{FC})m_2(\text{FC} \cup \text{UC}) + m_1(\text{FC} \cup \text{UC})m_2(\text{FC}) \\ & + m_1(\text{FC} \cup \text{UC})m_2(\text{OC} \cup \text{FC}) + m_1(\text{FC} \cup \text{UC} \cup \text{OC})m_2(\text{FC})\}/0.97 \\ = {} & \{.05 \times .15 + .05 \times .05 + .05 \times .2 + .05 \times .5 + .15 \times .15 + .15 \times .2 \\ & + .1 \times .15 + .1 \times .05 + .6 \times .15\}/.97 = 0.21 \end{aligned}$$

Similarly, for the combined event FC ∪ UC, we get

$$\begin{aligned} m_{12}(\text{FC} \cup \text{UC}) = {} & \{m_1(\text{FC} \cup \text{UC})m_2(\text{FC} \cup \text{UC}) + m_1(\text{FC} \cup \text{UC})m_2(\text{FC} \cup \text{OC} \cup \text{UC}) \\ & + m_1(\text{FC} \cup \text{OC} \cup \text{UC})m_2(\text{FC} \cup \text{UC})\}/0.97 \\ = {} & \{.1 \times .2 + .1 \times .5 + .6 \times .2\}/0.97 = 0.20 \end{aligned}$$

Finally, using the combined evidence measures, m_{12}, we can calculate the combined belief measures (bel_{12}). For example, for OC ∪ FC we have

$$\text{bel}_{12}(\text{OC} \cup \text{FC}) = m_{12}(\text{OC} \cup \text{FC}) + m_{12}(\text{OC}) + m_{12}(\text{FC}) = 0.12 + 0.01 + 0.21 = 0.34$$

The remaining calculated values are shown in Table 15.2.

PROBABILITY MEASURES

When the additional belief axiom [Eq. (15.5)] is replaced with a stronger axiom (illustrated for only two sets, A and B),

$$\text{bel}(\text{A} \cup \text{B}) = \text{bel}(\text{A}) + \text{bel}(\text{B}), \ \text{A} \cap \text{B} = \varnothing \qquad (15.20)$$

we get a *probability measure*. Let us now introduce a formal definition for a probability measure in the context of an evidence theory.

If we have a basic probability assignment (bpa) for a singleton, x, denoted $m(x) = \text{bel}(x)$ and we have $m(\text{A}) = 0$ for all subsets A of the power set, P(X), that are *not* singletons, then $m(x)$ is a probability measure. A probability measure is also a mapping of some function, say $p(x)$, to the unit interval, i.e.,

$$p : x \rightarrow [0, 1] \tag{15.21}$$

To conform to the literature we will let $m(x) = p(x)$ to denote $p(x)$ as a probability measure. The mapping $p(x)$ then maps evidence only on singletons to the unit interval. The key distinction between a probability measure and either a belief or plausibility measure, as can be seen from Eq. (15.21), is that a probability measure arises when all the evidence is on singletons only, i.e., only on elements x; whereas, when we have some evidence on subsets that are not singletons, we cannot have a probability measure and will have only belief and plausibility measures [both, because they are duals—see Eqs. (15.10)]. If we have a probability measure, we will then have

$$\text{bel}(\text{A}) = \text{pl}(\text{A}) = p(\text{A}) = \sum_{x \in \text{A}} p(x) \qquad \text{for all A} \in \text{P(X)} \tag{15.22}$$

where set A is simply a collection of singletons; this would define the probability of set A. Equation (15.22) reveals that the belief, plausibility, and probability of a set A are all equal for a situation involving probability measures. Moreover, Eqs. (15.7) and (15.11) become a manifestation of the excluded middle laws (see Chapter 2) for a probability measure:

$$\text{pl}(\text{A}) = p(\text{A}) = \text{bel}(\text{A}) \rightarrow p(\text{A}) + p(\overline{\text{A}}) = 1 \tag{15.23}$$

Example 15.3. Two quality control experts from PrintLaser Inc. are trying to determine the source of scratches on the media that exits the sheet feeder of a new laser printer already in production. One possible source is the upper arm and the other source is media sliding on top of other media (e.g., paper on paper). We shall denote the following focal elements:

W denotes scratches from wiper arm.

M denotes scratches from other media.

The experts provide their assessments of evidence supporting each of the focal elements as follows:

Focal elements	Expert 1, m_1	Expert 2, m_2
W	.6	.3
M	.4	.7
W ∪ M	0	0

We want to determine the beliefs, plausibilities, and probabilities for each nonnull focal element. We can see that evidence is only available on the singletons, W and M. We find the following relationships for the first expert:

$$\text{bel}_1(\text{W}) = m_1(\text{W}) = 0.6$$
$$\text{bel}_1(\text{M}) = m_1(\text{M}) = 0.4$$

$$\begin{aligned}
\text{bel}_1(\text{W} \cup \text{M}) &= m_1(\text{W}) + m_1(\text{M}) + m_1(\text{W} \cup \text{M}) = 0.6 + 0.4 + 0 = 1 \\
\text{pl}_1(\text{W}) &= m_1(\text{W}) + m_1(\text{W} \cup \text{M}) = 0.6 + 0 = 0.6 \\
\text{pl}_1(\text{M}) &= m_1(\text{M}) + m_1(\text{W} \cup \text{M}) = 0.4 + 0 = 0.4 \\
\text{pl}_1(\text{W} \cup \text{M}) &= m_1(\text{W}) + m_1(\text{M}) + m_1(\text{W} \cup \text{M}) = 0.6 + 0.4 + 0 = 1
\end{aligned}$$

We note that $\text{bel}_1(\text{W}) = \text{pl}_1(\text{W})$, $\text{bel}_1(\text{M}) = \text{pl}_1(\text{M})$, and $\text{bel}_1(\text{W} \cup \text{M}) = \text{pl}_1(\text{W} \cup \text{M})$. From Eq. (15.23), these are all probabilities. Hence, $p_1(\text{W}) = 0.6$, $p_1(\text{M}) = 0.4$, and $p_1(\text{W} \cup \text{M}) = p(\text{W}) + p(\text{M}) = 0.6 + 0.4 = 1$ (this also follows from the fact that the probability of the union of disjoint events is the sum of their respective probabilities). In a similar fashion for the second expert we find

$$p_2(\text{W}) = 0.3, \qquad p_2(\text{M}) = 0.7 \qquad \text{and} \qquad p_2(\text{W} \cup \text{M}) = 0.3 + 0.7 = 1$$

POSSIBILITY AND NECESSITY MEASURES

Suppose we have a collection of some or all of the subsets on the power set of a universe that have the property: $\text{A}_1 \subset \text{A}_2 \subset \text{A}_3 \subset \cdots \subset \text{A}_n$. With this property these sets are said to be *nested* [Shafer, 1976]. When the elements of a set, or universe, having evidence are nested, then we say that the belief measures, $\text{bel}(\text{A}_i)$, and the plausibility measures, $\text{pl}(\text{A}_i)$, represent a *consonant* body of evidence. By consonant we mean that the evidence allocated to the various elements of the set (subsets on the universe) does *not* conflict; i.e., the evidence is free of dissonance.

For a consonant body of evidence, we have the following relationships [Klir and Folger, 1988] for two different sets on the power set of a universe, i.e., for $\text{A}, \text{B} \in \text{P}(\text{X})$:

$$\text{bel}(\text{A} \cap \text{B}) = \min[\text{bel}(\text{A}), \text{bel}(\text{B})] \tag{15.24}$$

$$\text{pl}(\text{A} \cup \text{B}) = \max[\text{pl}(\text{A}), \text{pl}(\text{B})] \tag{15.25}$$

The expressions in Eqs. (15.24)–(15.25) indicate that the belief measure of the intersection of two sets is the smaller of the belief measures of the two sets and the plausibility measure of the union of these two sets is the larger of the plausibility measures of the two sets.

In the literature consonant belief and plausibility measures are referred to as necessity (denoted η) and possibility (denoted π) measures, respectively. Equations (15.24) and (15.25) become, respectively for all $\text{A}, \text{B} \in \text{P}(\text{X})$,

$$\eta(\text{A} \cap \text{B}) = \min[\eta(\text{A}), \eta(\text{B})] \tag{15.26}$$

$$\pi(\text{A} \cup \text{B}) = \max[\pi(\text{A}), \pi(\text{B})] \tag{15.27}$$

For a consonant body of evidence, the dual relationships expressed in Eqs. (15.10) then take the forms,

$$\begin{aligned}
\pi(\text{A}) &= 1 - \eta(\overline{\text{A}}) \\
\eta(\text{A}) &= 1 - \pi(\overline{\text{A}})
\end{aligned} \tag{15.28}$$

Since the necessity and possibility measures are dual relationships, the discussion to follow focuses only on one of these, possibility. If necessity measures are desired, they can always be derived with the expressions in Eq. (15.28).

We now define a possibility distribution function as a mapping of the singleton elements, x, in the universe, X, to the unit interval, i.e.,

$$r : \mathrm{X} \rightarrow [0, 1] \tag{15.29}$$

This mapping will be related to the possibility measure, $\pi(\mathrm{A})$, through the relationship

$$\pi(\mathrm{A}) = \max_{x \in \mathrm{A}} r(x) \tag{15.30}$$

for each $\mathrm{A} \in \mathrm{P(X)}$ [see Klir and Folger, 1988, for a proof]. Now, a possibility distribution can be defined as an ordered sequence of values,

$$\mathbf{r} = (\rho_1, \rho_2, \rho_3, \ldots, \rho_n) \tag{15.31}$$

where $\rho_i = r(x_i)$ and where $\rho_i \geq \rho_j$ for $i < j$. The *length* of the ordered possibility distribution given in Eq. (15.31) is the number n. Every possibility measure also can be characterized by the n-tuple, denoted as a basic distribution [Klir and Folger, 1988],

$$\mathbf{m} = (\mu_1, \mu_2, \mu_3, \ldots, \mu_n) \tag{15.32a}$$

$$\sum_{i=1}^{n} \mu_i = 1 \tag{15.32b}$$

where $\mu_i \in [0, 1]$ and $\mu_i = m(\mathrm{A}_i)$. Of course, the sets A_i are nested as is required of all consonant bodies of evidence. From Eq. (15.17) and the relationship

$$\rho_i = r(x_i) = \pi(x_i) = \mathrm{pl}(x_i) \tag{15.33}$$

it can be shown [Klir and Folger, 1988] that

$$\rho_i = \sum_{k=i}^{n} \mu_k = \sum_{k=i}^{n} m(\mathrm{A}_k) \tag{15.34}$$

or, in a recursive form,

$$\mu_i = \rho_i - \rho_{i+1} \tag{15.35}$$

where $\rho_{n+1} = 0$ by convention. Equation (15.35) produces a set of equations of the form

$$\begin{aligned} \rho_1 &= \mu_1 + \mu_2 + \mu_3 + \cdots + \mu_n \\ \rho_2 &= \qquad \mu_2 + \mu_3 + \cdots + \mu_n \\ \rho_3 &= \qquad\qquad \mu_3 + \cdots + \mu_n \\ &\qquad\qquad\qquad \cdots \\ \rho_n &= \qquad\qquad\qquad\qquad \mu_n \end{aligned} \tag{15.36}$$

Example 15.4. Suppose we had a computer vision application where a robot was to "look" at a scene and determine whether the scene contained a pentagon, an octagon, or a circle. Moreover, suppose that two different algorithms were being used for classification of the geometric shapes. We will define the sets and the experts for this situation as follows:

A = {circles}   Expert 1 = Algorithm 1
B = {octagons}   Expert 2 = Algorithm 2
C = {pentagons}

Hence, the robot vision system uses an algorithm to assign a basic probability assignment (bpa) to the focal element in question. We construct a table showing the bpa's on the power set P(X), for experts 1 and 2.

Focal Elements	**A**	**B**	**C**	**Expert 1, m₁**	**Expert 2, m₂**	
∅	0	0	0	0	0	
C	0	0	1	0	.1	x_1
B	0	1	0	.2	.1	x_2
B ∪ C	0	1	1	.3	0	x_3
A	1	0	0	0	.1	x_4
A ∪ C	1	0	1	0	0	x_5
A ∪ B	1	1	0	0	.3	x_6
A ∪ B ∪ C	1	1	1	.5	.4	x_7

For the situation just described, Expert 1 makes the basic assignment from the available evidence that

$\mathbf{m}_1(B) = 0.2$ Scene contains octagons

$\mathbf{m}_1(B \cup C) = 0.3$ Scene contains octagons or pentagons

$\mathbf{m}_1(A \cup B \cup C) = 0.5$ Scene contains octagons or pentagons or circles

By inspection we can see that the focal elements on m_2 do not represent a nested series of sets, since both sets C and B have evidence, but $C \not\subset B$. However, inspection of the focal elements on m_1 reveals a nested situation since all the focal elements that have evidence (B, B ∪ C, and A ∪ B ∪ C) are nested, i.e., $B \subset (B \cup C) \subset (A \cup B \cup C)$. In this case the basic distribution (for nonnull focal elements) for Expert 1 (m_1) is

$$\mathbf{m}_1 = (A_1, A_2, A_3, A_4, A_5, A_6, A_7) = (\mu_1, \mu_2, \mu_3, \mu_4, \mu_5, \mu_6, \mu_7) = (0, .2, .3, 0, 0, 0, .5)$$

Using Eq. (15.34) produces a possibility distribution of the form

$$\mathbf{r}_1 = (1, 1, .8, .5, .5, .5, .5)$$

The focal elements for Expert 1 could be arranged in a nesting diagram as shown in Fig. 15.4. This nesting diagram provides both the basic distribution (nonzero entities) and the possibility distribution for the various focal elements.

For Expert 2 the situation represents a case where the subsets having evidence are not nested and the plausibility measure leads to a position of conflict. This arises when $\mathbf{m}_2(A) \neq 0$, $\mathbf{m}_2(B) \neq 0$, and $A \cap B = \varnothing$.

Nesting of focal elements can be an important physical attribute of a body of evidence. Consider the following two examples where physical nesting is an important feature of the engineering system.

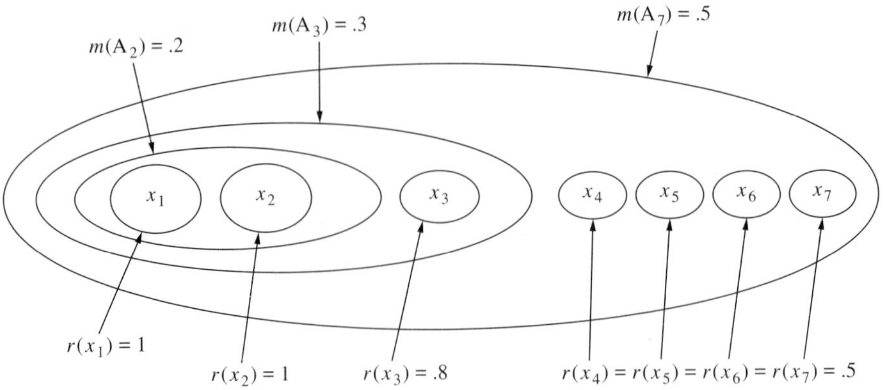

FIGURE 15.4
Nesting diagram, Example 15.4.

Example 15.5. In an airplane flight test program, hundreds of hours of data representing thousands of individual flight conditions and data sets are gathered. The timely identification and retrieval of specific sets, subsets, and individual data conditions are extremely important. These data can be selected and grouped using possibility theory for an enhanced database search function.

Suppose we have five focal elements, A, B, C, D, and E, where

$A \cup B \cup C \cup D \cup E$: represents all of the flight test data
E: represents data conditions required for the certification agencies of various governments
$A \cup B \cup C \cup D$: represents data collected to satisfy internal company requirements
D: represents data gathered for updating the flight crew training simulator
C: represents data gathered to verify the functionality of various airplane subsystems
$A \cup B$: represents data that relate overall airplane performance: fuel flow, fuel quantity, thrust level, field length requirements
B: represents performance data used to verify compliance with contractual performance guarantees such as fuel burn, range, and payload capability
A: represents performance data used to update the airplane performance database used for supporting future contract guarantees and answer in-service performance-related customer inquiries

The nesting of the various focal elements could be represented as shown in Fig. 15.5. We might have evidence that suggests the following basic distribution:

$$\mathbf{m} = (0, .2, 0, .5, .3)$$

where
$$A_2 = A \cup B$$
$$A_4 = A \cup B \cup C \cup D \text{ (or } A_4 = A_2 \cup C \cup D)$$
$$A_5 = A \cup B \cup C \cup D \cup E \text{ (or } A_5 = A_4 \cup E)$$

Here we have nesting of the form $A_2 \subset A_4 \subset A_5$. Then Eq. (15.34) can be used to determine the appropriate possibility distribution for this consonant body of evidence,

$$\mathbf{r} = (1, 1, .8, .8, .3)$$

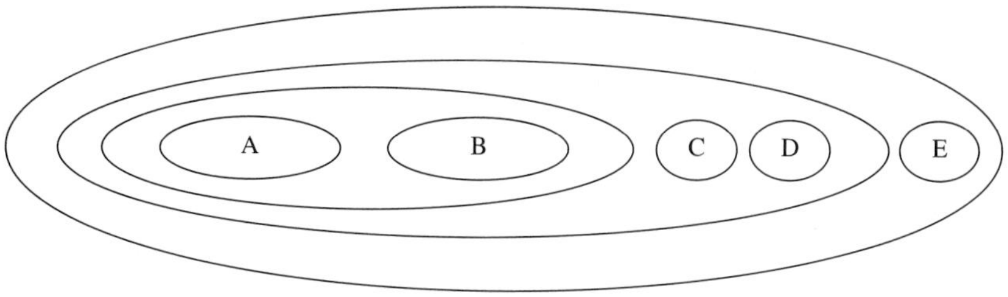

FIGURE 15.5
Nesting diagram, Example 15.5.

Example 15.6. Suppose there are seven nodes in a communication network X, labeled x_1 through x_7 and represented by boxes. Of these seven nodes, one is causing a problem. The company network expert is asked for an opinion on which node is causing the communications problem. The network expert aggregates these nodes into sets, as given in the accompanying table. The third column in the table represents the expert's basic distribution (basic probability assignments), and the last column is the possibility distribution found from Eq. (15.34).

Set A	Aggregation of focal elements	$\mu_n = \mu(A_n)$	ρ_i
A_1	x_1	.4	1
A_2	$x_1 \cup x_2$	.2	.6
A_3	$x_1 \cup x_2 \cup x_3$	0	.4
A_4	$x_1 \cup x_2 \cup x_3 \cup x_4$	.1	.4
A_5	$x_1 \cup x_2 \cup x_3 \cup x_4 \cup x_5$	0	.3
A_6	$x_1 \cup x_2 \cup x_3 \cup x_4 \cup x_5 \cup x_6$	.2	.3
A_7	$x_1 \cup x_2 \cup x_3 \cup x_4 \cup x_5 \cup x_6 \cup x_7$	.1	.1

The physical significance of this nesting (shown in Fig. 15.6) can be described as follows. In the network expert's belief, node x_1 is causing the problem. This node has new hardware and is an experimental CPU. For these reasons, the network expert places the highest belief on this set (set A_1). The next set with nonzero belief (supporting evidence), A_2, is comprised of the union, x_1 or x_2. The network expert has less belief that node x_1 or x_2 is causing the problem. Node x_2 has new hardware as well, but has a trusted CPU. The next set with nonzero belief, A_4, is nodes x_1 or x_2 or x_3 or x_4. The network expert has even less belief that the problem is caused by this set. The expert reasons that x_3 or x_4 has trusted hardware and CPUs. The next set with nonzero belief, A_6, is nodes x_1 or x_2 or x_3 or x_4 or x_5 or x_6. The network expert has slightly more belief that this set is the problem than set A_4, but much less than the initial set A_1, the reasons being that there are two new programmers using these nodes for testing communications software. The final set with evidence is the union of all seven nodes. The expert has little belief that this set is the problem because node x_7 is usually turned off.

Note that the first element of any ordered possibility distribution, ρ_1, is always equal to unity, i.e., $\rho_1 = 1$. This fact is guaranteed by Eq. (15.32*b*). The smallest possibility distribution of length n has the form $\mathbf{r} = (1, 0, 0, 0, \ldots, 0)$, where there

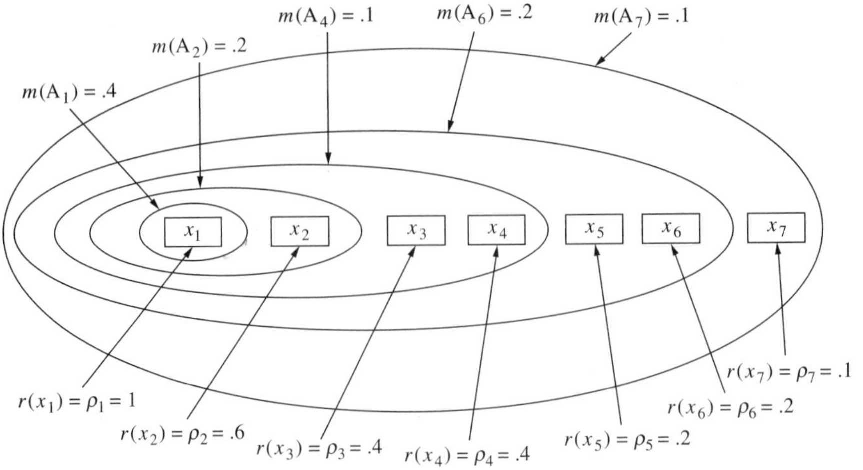

FIGURE 15.6
Nesting diagram, Example 15.6.

are $(n-1)$ zeros after a value of unity in the distribution. The associated basic distribution would have the form $\mathbf{m} = (1, 0, 0, 0, \ldots, 0)$. In this case there would be only one focal element with evidence, and it would have all the evidence. This situation represents *perfect evidence;* there is no uncertainty involved in this case.

Alternatively, the largest possibility distribution of length n has the form $\mathbf{r} = (1, 1, 1, 1, \ldots, 1)$, where all values are unity in the distribution. The associated basic distribution would have the form $\mathbf{m} = (0, 0, 0, 0, \ldots, 1)$. In this case all the evidence is on the focal element comprising the entire universe, i.e., $\mathrm{A}_n = x_1 \cup x_2 \cup \cdots \cup x_n$; hence, we know nothing about any specific focal element in the universe except the universal set. This situation is called *total ignorance.* In general, the larger the possibility distribution, the less specific the evidence and the more ignorant we are of making any conclusions.

Since possibility measures are special cases of plausibility measures and necessity measures are special cases of belief measures, we can relate possibility measures and necessity measures to probability measures. Equation (15.23) shows that, when all the evidence in a universe resides solely on the singletons of the universe, the belief and plausibility measures become probability measures. In a like fashion it can be shown that the plausibility measure approaches the probability measure from an upper bound and that the belief measure approaches the probability measure from a lower bound; the result is a range around the probability measure [Yager and Filev, 1994],

$$\mathrm{bel}(\mathrm{A}) \leq p(\mathrm{A}) \leq \mathrm{pl}(\mathrm{A}) \tag{15.37}$$

Example 15.7. Probabilities can be determined by finding point-valued quantities and then determining the relative frequency of occurrence of these quantities. In determining the salvage value of older computers, the age of the computer is a key variable. This variable is also important in assessing depreciation costs for the equipment. Sometimes there is uncertainty in determining the age of computers if their purchase records are lost or if the equipment was acquired through secondary acquisitions or trade. Suppose the age of five computers is known, and we have no uncertainty; here the ages are point-valued quantities.

Computer	Age, months
1	26
2	21
3	33
4	24
5	30

With the information provided in the table we could answer the following question: What percentage of the computers have an age in the range of 20 to 25 months, i.e., what percentage of the ages fall in the interval [20, 25]? This is a countable answer of $\frac{2}{5}$, or 40 percent.

Now suppose that the age of the computers is not known precisely, but rather each age is assessed as an interval. Now the ages are set-valued quantities, as follows:

Computer	Age, months
1	[22, 26]
2	[20, 22]
3	[30, 35]
4	[20, 24]
5	[28, 30]

With this information we can only assess possible solutions to the question just posed: What percentage of the computers possibly fall in the age range of [20, 25] months? Because the ages of the computers are expressed in terms of ranges (or sets on the input space), the solution space of percentages will also have to be expressed in terms of ranges (or sets on the solution space).

To approach the solution we denote the query range as Q, i.e., Q = [20, 25] months. We denote the age range of the *i*th computer as D_i. Now, we can determine the certainty and possibility ranges using the following rules:

1. Age (i) is certain if $D_i \subset Q$.
2. Age (i) is possible if $D_i \cap Q \neq \varnothing$.
3. Age (i) is not possible if $D_i \cap Q = \varnothing$.

The first rule simply states that the age is certainly in the query range if the age range of the *i*th computer is completely contained within the query range. The second rule states that the age is possibly in the query range if the age range of the *i*th computer and the query range have a nonnull intersection, i.e., if they intersect at any age. The third rule states that the age is not possible if the age range of the *i*th computer and the query range have no age in common, i.e., their intersection is null. We should note here that a solution that is certain is necessarily possible (certainty implies possibility), but the converse is not always true (things that are possible are not always certain). Hence, the set of certain quantities is a subset of the set of possible quantities. In looking at the five computers and using the three rules already given, we determine the following relationships:

Computer	η or π
1	Possible
2	Certainty
3	Not possible
4	Certainty
5	Not possible

In the table we see that, of the five computers, two have age ranges that are certainly (denoted $\eta(\mathrm{Q})$) in the query interval and three have age ranges that are possibly (denoted $\pi(\mathrm{Q})$) in the query interval (one possible and two certain). We will denote the solution as the response to the query, or resp(Q). This will be an interval-valued quantity as indicated earlier, or

$$\mathrm{resp(Q)} = [\eta(\mathrm{Q}), \pi(\mathrm{Q})] = \left[\tfrac{2}{5}, \tfrac{3}{5}\right]$$

Hence, we can say that the answer to the query is "Certainly 40 percent and possibly as high as 60 percent." We can also see that the range represented by resp(Q) represents a lower bound and an upper bound to the actual point-valued probability (which was determined earlier to be 40 percent) as indicated by Eq. (15.37).

In Example 15.7 all computer ranges were used in determining the percentages for possibilities and certainties (i.e., all five). In evidence theory null values are *not* counted in the determination of the percentages as seen in the normalization constant, K, expressed in Eq. (15.19). This characteristic can lead to fallacious responses, as illustrated in the following example.

Example 15.8. Suppose again we wish to determine the age range of computers, this time expressed in units of years. In this case we ask people to tell us the age of their own computer (PC). These responses are provided in the accompanying table. In the table the null symbol, $\varnothing$, indicates that the person queried has no computer.

Person	Age of PC, yrs
1	[3, 4]
2	$\varnothing$
3	[2, 3]
4	$\varnothing$
5	$\varnothing$

Let us now ask the question: What percentage of the computers have an age in the range Q = [2, 4] years? Using the rules given in Example 15.7, we see that we have two certainties (hence, we have two possibilities) and three null values. If we include the null values in our count, the solution is

$$\text{resp}(Q) = [\eta(Q), \pi(Q)] = \left[\tfrac{2}{5}, \tfrac{2}{5}\right] = \tfrac{2}{5}$$

In this case, we have not used a normalization process because we have counted the null values. If we decide to neglect the null values (hence, we normalize as the Dempster rule of combination suggests), then the solution is

$$\text{resp}(Q) = [\eta(Q), \pi(Q)] = \left[\tfrac{2}{2}, \tfrac{2}{2}\right] = 1$$

We see a decidedly different result when normalization is used.

A graphical interpretation of the evidence theory developed by Dempster and Shafer is provided in what is called the ball-box analogy [Zadeh, 1986].

Example 15.9 [Zadeh, 1984]. Suppose the king of country X believes a submarine, S, is in the territorial waters of X. The king summons n experts, $E_1, E_2, \ldots, E_n$, to give him advice on the location of the submarine, S. The n experts each provide his or her assessment of the location of S; call these possible locations $L_1, L_2, \ldots, L_m, \ldots, L_n$, where $m \leq n$. Here, L_i are subsets of the territorial waters, X. To be more specific, experts $E_1, E_2, \ldots, E_m$ say that S is in $L_1, L_2, \ldots, L_m$ and experts $E_{m+1}, \ldots, E_{n-1}, E_n$ say that S is *not* located in the territorial waters of X, i.e., $L_{m+1} = L_{m+2} = \cdots = L_n = \varnothing$. So there are $(m - n)$ experts who say that S is not in the territorial waters. Now the king asks, "Is S in a *subset* A of our territorial waters?" Figure 15.7 shows possible location regions, L_i, and the query region of interest, region A.

If we denote E_i as the location proffered by the ith expert, we have the following two rules:

1. $E_i \subset A$ implies that it is certain that $S \in A$.
2. $E_i \cap A \neq \varnothing$ implies that it is possible that $S \in A$.

We note again that a certainty is contained in the set of possibilities, i.e., certainty implies possibility. We further assume that the king aggregates the opinions of his

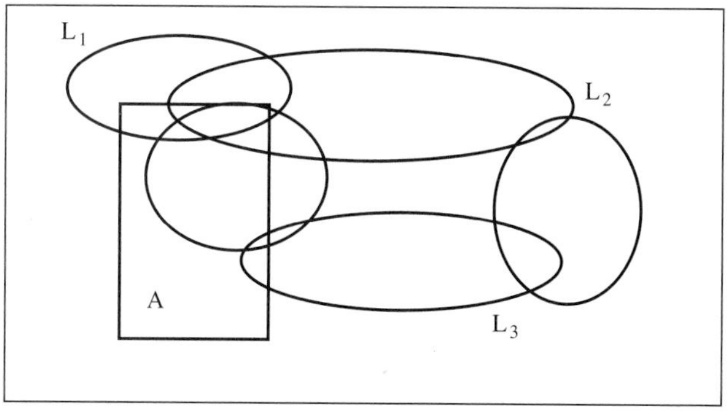

FIGURE 15.7
Ball-box analogy of the Dempster-Shafer evidence theory.

experts by averaging. Thus, if k out of n experts vote for rule number 1, then the average certainty $= k/n$; if l out of n (where $l \geq k$) experts vote for rule number 2, then the average possibility $= l/n$. Finally, if the judgment of those experts who think there is no submarine anywhere in the territorial waters is ignored, the average certainty and possibility will be k/m and l/m, respectively (where $m \leq n$). Ignoring the opinion of those experts whose L_i is the null set corresponds to the normalization [Eq. (15.19)] in the Dempster-Shafer evidence theory.

The Dempster rule of combination may lead to counterintuitive results because of the normalization issue. The reason for this [Zadeh, 1984] is that normalization throws out evidence that asserts the object under consideration does not exist, i.e., is null or empty (∅). The following example from the medical sciences illustrates this idea very effectively.

Example 15.10 [Zadeh, 1984]. A patient complaining of a severe headache is examined by two doctors (doctor 1 and doctor 2). The diagnosis of doctor 1 is that the patient has either meningitis (M) with probability 0.99 or a brain tumor (BT) with probability 0.01. Doctor 2 agrees with doctor 1 that the probability of a brain tumor (BT) is 0.01, but she disagrees with doctor 1 on the meningitis; instead doctor 2 feels that there is a probability of 0.99 that the patient just has a concussion (C). The following table shows the evidence for each of the focal elements in this universe for each of the doctors (m_1 and m_2) as well as the calculated values for the combined evidence measures (m_{12}). In the table there is evidence only on the singletons, M, BT, and C; hence, all the measures are probability measures (the doctors provided their opinions in terms of probability).

Focal element	$\mathbf{m_1}$	$\mathbf{m_2}$	$\mathbf{m_{12}}$
M	.99	0	0
BT	.01	.01	1
C	0	.99	0
M ∪ BT	0	0	
M ∪ C	0	0	
BT ∪ C	0	0	
M ∪ BT ∪ C	0	0	

The combined evidence measures are calculated as follows. First, by using Eq. (15.19) the normalization constant, K, is calculated. Then use of Eq. (15.18) produces values for m_{12}. For example, m_{12} for a brain tumor is found as

$$m_{12}(\text{BT}) = \frac{.01(.01)}{1 - \{.99[.01 + .99] + .01[0 + .99] + 0[0 + .01]\}} = \frac{.0001}{1 - .9999} = 1$$

One can readily see that the combined measures $m_{12}(\text{C})$ and $m_{12}(\text{M})$ will be zero from the fact that

$$m_1(\text{C}) = m_2(\text{M}) = 0$$

The table reveals that using the Dempster-Shafer rule of combination [in Shafer (1976), null values are not allowed in the definition of belief functions but do enter in the rule of combination of evidence] results in a combined probability of 1 that the patient has a brain tumor when, in fact, both doctors agreed individually that it was only one chance in a hundred! What is even more confusing is that the same conclusion (i.e., $m_{12}(\mathrm{BT}) = 1$) results regardless of the probabilities associated with the other possible diagnoses.

In Example 15.10 it appears that the normalization process *suppressed* expert opinion; but is this omission mathematically allowable? This question leads to the conjecture that the rule of combination cannot be used until it is ascertained that the bodies of evidence are conflict-free; that is, at least one parent relation exists that is absent of conflict. In particular, under this assertion, it is not permissible to combine distinctly different bodies of evidence. In the medical example, the opinions of both doctors reveal some missing information about alternative diagnoses. A possibility theory might suggest an alternative approach to this problem in which the incompleteness of information in the knowledge base propagates to the conclusion and results in an interval-valued, possibilistic answer. This approach addresses rather than finesses (like excluding null values) the problem of incomplete information.

POSSIBILITY DISTRIBUTIONS AS FUZZY SETS

Belief structures that are nested are called consonant. A fundamental property of consonant belief structures is that their plausibility measures are possibility measures. As suggested by DuBois and Prade [1988], possibility measures can be seen to be formally equivalent to fuzzy sets. In this equivalence, the membership grade of an element x corresponds to the plausibility of the singleton consisting of that x, that is, a consonant belief structure is equivalent to a fuzzy set F of X where $F(x) = pl(\{x\})$.

A problem in equating consonant belief structures with fuzzy sets is that the combination of two consonant belief functions using Dempster's rule of combination in general does not necessarily lead to a consonant result [Yager, 1993]. Hence, since Dempster's rule is essentially a conjunction operation, the intersection of two fuzzy sets interpreted as consonant belief structures may not result in a valid fuzzy set (i.e., a consonant structure).

Example 15.11 [Yager, 1993]. Suppose we have a universe comprised of five singletons, i.e.,

$$X = \{x_1, x_2, x_3, x_4, x_5\}$$

and we have evidence provided by two experts. The accompanying table provides the experts' degrees of belief about specific subsets of the universe, X.

Focal elements	Expert 1, $m_1(A_i)$	Expert 2, $m_2(B_i)$
$A_1 = \{x_1, x_2, x_3\}$	0.7	
$A_2 = X$	0.3	
$B_1 = \{x_3, x_4, x_5\}$		0.8
$B_2 = X$		0.2

Because $A_1 \subset A_2$ and $B_1 \subset B_2$ we have two consonant (nested) belief structures represented by A and B. Using Dempster's rule of combination and applying Eqs. (15.18)–(15.19), we have for any set D on the universe X

$$m(D) = \frac{1}{1-k} \sum_{A_i \cap B_j = D} m_1(A_i) \cdot m_2(B_j)$$

and

$$k = \sum_{A_i \cap B_j = \varnothing} m_1(A_i) \cdot m_2(B_j)$$

Since there are two focal elements in each experts' belief structures, we will have $2^2 = 4$ belief structures in the combined evidence case, which we will denote as m. We note for these data that we get a value of $k = 0$, because there are no intersections between the focal elements of A and B that result in the null set. For example, the intersection between A_1 and B_1 is the singleton, x_3. Then, we get

$$D_1 = A_1 \cap B_1 = \{x_3\} \qquad m(D_1) = 0.56 \qquad (\text{i.e., } 0.7 \times 0.8)$$

$$D_2 = A_1 \cap B_2 = \{x_1, x_2, x_3\} \qquad m(D_2) = 0.14 \qquad (\text{i.e., } 0.7 \times 0.2)$$

$$D_3 = A_2 \cap B_1 = \{x_3, x_4, x_5\} \qquad m(D_3) = 0.24 \qquad (\text{i.e., } 0.3 \times 0.8)$$

$$D_4 = A_2 \cap B_2 = X \qquad m(D_4) = 0.06 \qquad (\text{i.e., } 0.3 \times 0.2)$$

For the focal elements D_i we note that $D_1 \subset D_2 \subset D_4$ and $D_1 \subset D_3 \subset D_4$, but we do not have $D_2 \subset D_3$ or $D_3 \subset D_2$. Hence, the combined case is not consonant (i.e., not completely nested).

Yager [1993] has developed a procedure to prevent the situation illustrated by Example 15.11 from occurring; that is, a method is available to combine consonant possibility measures where the result is also a consonant possibility measure. However, this procedure is very lengthy to describe and is beyond the scope of this text; the reader is referred to the literature [Yager, 1993] to learn this method.

Another interpretation of a possibility distribution as a fuzzy set was proposed by Zadeh [1978]. He defined a possibility distribution as a fuzzy restriction that acts as an elastic constraint on the values that may be assigned to a variable. In this case the possibility distribution represents the degrees of membership for some linguistic variable, but the membership values are strictly monotonic as they are for an ordered possibility distribution. For example, let $\underset{\sim}{A}$ be a fuzzy set on a universe X, and let the membership value, μ, be a variable on X that assigns a "possibility" that an element of x is in $\underset{\sim}{A}$. So we get

$$\pi(x) = \mu_{\underset{\sim}{A}}(x) \tag{15.38}$$

Zadeh points out that the possibility distribution is nonprobabilistic and is used primarily in natural language applications. There is a loose relationship, however, between the two through a possibility/probability consistency principle [Zadeh, 1978]. In sum, what is *possible* may not be *probable*, but what is *impossible* is inevitably *improbable*.

Example 15.12. Let $\underset{\sim}{A}$ be a fuzzy set defined on the universe of columns needed to support a building. Suppose there are 10 columns altogether, and we start taking columns away until the building collapses; we record the number of columns at the time the building collapses. Let $\underset{\sim}{A}$ be the fuzzy set defined by the number of columns "possibly" needed, out of 10 total, just before the structure fails. The structure most certainly needs at least three columns to stand (imagine a stool). After that the number of columns required for the building to stand is a fuzzy issue (because of the geometric layout of the columns, the weight distribution, etc.), but there is a possibility it may need more than three. The following fuzzy set may represent this possibility:

$$\underset{\sim}{A} = \left\{ \frac{1}{1} + \frac{1}{2} + \frac{1}{3} + \frac{.9}{4} + \frac{.6}{5} + \frac{.3}{6} + \frac{.1}{7} + \frac{0}{8} + \frac{0}{9} + \frac{0}{10} \right\}$$

A probability distribution on the same universe may look something like the following table, where u is the number of columns prior to collapse, and $p(u)$ is the probability that u is the number of columns at collapse:

u	1	2	3	4	5	6	7	8	9	10
$p(u)$	0	0	.1	.5	.3	.1	0	0	0	0

As seen, although it is *possible* that one column will sustain the building, it is not *probable*. Hence, a high degree of possibility does not imply a high probability, nor does a low degree of probability imply a low degree of possibility.

Belief measures and plausibility measures overlap when they both become probability measures. However, possibility, necessity, and probability measures do not overlap with one another except for one special measure: the measure of one focal element that is a singleton. These three measures become equal when one element of the universal set is assigned a value of unity, and all other elements in the universe are assigned a value of zero. This measure represents *perfect evidence* [Klir and Folger, 1988].

EPILOGUE

In the preface of this book is a discussion of the relationships and historical confusion between probability theory and fuzzy set theory. It seems fitting that this book should conclude by coming full circle to that same discussion. A paper by Gaines [1978] does an eloquent job of addressing this issue. Historically, probability and fuzzy sets have been presented as distinct theoretical foundations for reasoning and decision making in situations involving uncertainty. Yet when one examines the underlying axioms of both probability and fuzzy set theories, the two theories differ by only one axiom in a total of sixteen axioms needed for a complete foundation! The material that follows is a brief summary of Gaines's paper, which established a common basis for both forms of logic of uncertainty in which a basic uncertainty logic is defined in terms of valuation on a lattice of propositions. Addition of the axiom of the excluded middle to the basic logic gives a standard probability logic. Alternatively, addition of a requirement for strong truth-functionality gives a fuzzy logic.

In this discussion fuzzy logic is taken to be a multivalued extension of Boolean logic based on fuzzy set theory in which truth values are extended from the endpoints of the interval [0, 1] to range through the entire interval. The normal logical operations are defined in terms of arithmetic operations on these values; the values are regarded as degrees of membership to truth. The logic operations and associated arithmetic operations are those of conjunction (involving the minimum operator), disjunction (involving the maximum operator), and negation (involving a subtraction from unity). Gaines [1978] points out that the use of the max and min operations in fuzzy logic is not sufficient to distinguish it from that of probability theory—both operators arise naturally in the calculation of the conjunction and disjunction of probabilistic events. Our association of addition and multiplication as natural operations upon probabilities comes from our frequent interest in statistically independent events, not from the logic of probability itself.

In developing a basic uncertainty logic, we begin first by defining a lattice consisting of a universe of discourse, X, a maximal element T, a minimal element F, a conjunction, $\bigwedge$, and a disjunction, $\bigvee$. This lattice will be denoted L(X, T, F, $\bigwedge$, $\bigvee$). For the axioms (or postulates) to follow, lowercase letters, x, y, and z denote specific elements of the universe X within the lattice. The following fifteen axioms completely specify a basic uncertainty logic.

The basic uncertainty logic begins with the lattice L satisfying idempotency,

1. For all $x \in \mathrm{L}$ $\quad x \bigvee x = x \bigwedge x = x$

commutativity,

2. For all $x, y \in \mathrm{L}$ $\quad x \bigvee y = y \bigvee x \quad$ and $\quad x \bigwedge y = y \bigwedge x$

associativity,

3. For all $x, y, z \in \mathrm{L}$ $\quad x \bigvee (y \bigvee z) = (x \bigvee y) \bigvee z$

absorption,

4. For all $x, y \in \mathrm{L}$ $\quad x \bigvee (x \bigwedge y) = x, x \bigwedge (x \bigvee y) = x$

and the definition of the maximal and minimal elements,

5. For all $x \in \mathrm{L}$ $\quad x \bigvee \mathrm{T} = \mathrm{T} \quad$ and $\quad x \bigwedge \mathrm{T} = x$

The usual order relation may also be defined,

6. For all $x, y \in \mathrm{L}$ $\quad x \leq y$ if there exists a $z \in \mathrm{L}$ such that $y = x \bigvee z$

Now suppose that every element of the lattice L is assigned a truth value (for various applications this truth value would be called a probability, degree of belief, etc.) in the interval [0, 1] by a continuous, order-preserving function, $p : \mathrm{L} \rightarrow [0, 1]$, with constraints,

7. $p(\mathrm{F}) = 0; p(\mathrm{T}) = 1$
8. For all $x, y \in \mathrm{L}$ $\quad x \leq y$ then $p(x) \leq p(y)$

and an additivity axiom,

9. For all $x, y \in \mathrm{L}$ $\quad p(x \wedge y) + p(x \vee y) = p(x) + p(y)$

We note that for p to exist we have to have

$$p(x \wedge y) \leq \min[p(x), p(y)] \leq \max[p(x), p(y)] \leq p(x \vee y)$$

Now a logical equivalence (or congruence) is defined by

10. For all $x, y \in \mathrm{L}$ $\quad x \leftrightarrow y$ if $p(x \wedge y) = p(x \vee y)$

The general structure provided by the first 10 axioms is common to virtually all logics. To finalize Gaines's basic uncertainty logic we now need to define implication and negation, for it is largely the definition of these two operations that distinguish among various multivalued logics [Gaines, 1978]. We also note that postulate 9 still holds when the outer inequalities become equalities—a further illustration that the additivity of probability-like valuations is completely compatible with, and closely related to, the minimum and maximum operations of fuzzy logic.

To define implication and negation we make use of a metric on the lattice L that measures *distance* between the truth values of two different propositions. This is based on the notion that logically equivalent propositions should have a zero distance between them. So, we define a distance measure,

11. For all $x, y \in \mathrm{L}$ $\quad d(x, y) = p(x \vee y) - p(x \wedge y)$

where $d(x, x) = 0, 0 \leq d(x, y) \leq 1$, and $d(x, y) + d(y, z) = d(x, z)$. This axiom is shown schematically in Fig. 15.8.

Therefore, a measure of equivalence between two elements can be 1 minus the distance between them, or

12. For all $x, y \in \mathrm{L}$ $\quad p(x \leftrightarrow y) = 1 - d(x, y) = 1 - p(x \vee y) + p(x \wedge y)$

Hence, if $d = 0$, the two elements x and y are equivalent, as seen in Fig. 15.9.

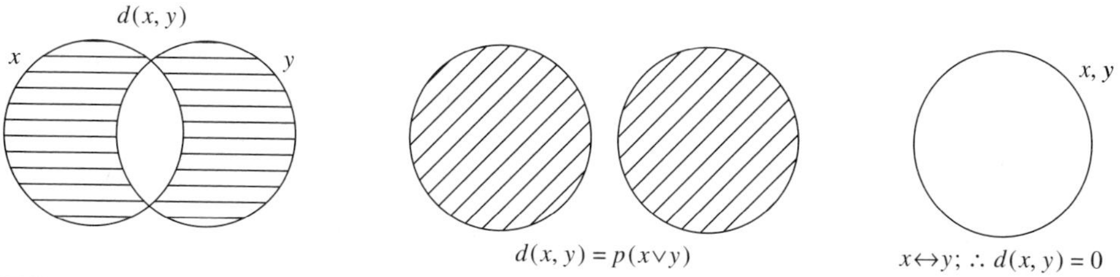

FIGURE 15.8
Venn diagrams on distance measure, $d(x, y)$.

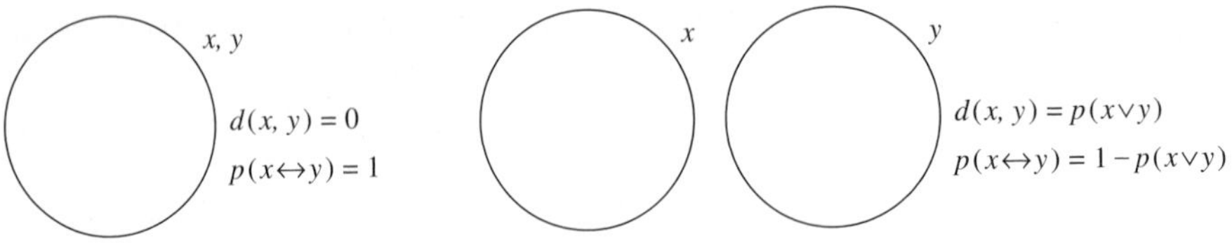

FIGURE 15.9
Venn diagram on equivalent and maximally nonequivalent elements.

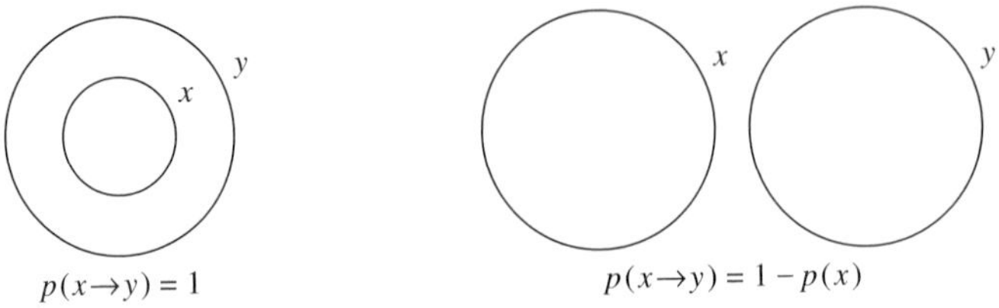

FIGURE 15.10
Venn diagram for implication.

To measure the strength of an implication, we measure a distance between x and $x \wedge y$, as seen in Fig. 15.10 (other forms of the implication operator exist; see Chapter 2 for a classical form).

13. For all $x, y \in \mathrm{L}$

$$\begin{aligned} p(x \rightarrow y) &= p(x \leftrightarrow x \wedge y) = 1 - d(x, x \wedge y) \\ &= 1 - p(x) + p(x \wedge y) \\ &= 1 + p(y) - p(x \vee y) = 1 - d(y, x \vee y) \end{aligned}$$

Negation can now be defined in terms of equivalence and implication as

14. For all $x \in \mathrm{L}$ $\quad p(\overline{x}) = p(x \leftrightarrow \mathrm{F}) = 1 - p(x) = 1 - d(x, \mathrm{F})$

We note by combining axioms 9 and 14 that if element y is replaced by element $\overline{x}$, we get

$$p(x \vee \overline{x}) + p(x \wedge \overline{x}) = p(x) + p(\overline{x}) = 1$$

Finally, we add a postulate of distributivity,

15. For all $x, y, z \in \mathrm{L}$ $\quad x \wedge (y \vee z) = (x \wedge y) \vee (x \wedge z)$

which will prove useful for the two specializations of this basic uncertainty logic to be described in the remaining paragraphs.

Axioms 1 through 15 provide for a basic distributive uncertainty logic. The addition of a special sixteenth (denoted 16.1) axiom, known in the literature as the law of the excluded middle (see Chapter 2),

16.1. For all $x \in \mathrm{L}$ $\quad p(x \vee \overline{x}) = 1$

leads to Rescher's standard probability logic [Rescher, 1969].

Alternatively, if we add another special sixteenth (denoted 16.2) axiom to the basic axioms 1 through 15, we get a special form of fuzzy logic,

16.2. For all $x, y \in \mathrm{L}$ $\quad \{p(x \rightarrow y) = 1 \vee p(y \rightarrow x) = 1\}$

known in the literature as the Lukasiewicz infinite-valued logic [Rescher, 1969]. Axiom 16.2 is called the *strong truth-functionality*, or *strict implication*, in the literature.

It should be pointed out that a *weaker* form of axiom 16.2, or

$$\text{For all } x, y \in \mathrm{L} \qquad p((x \rightarrow y) \vee (y \rightarrow x)) = 1$$

is embraced by both a probability and a fuzzy logic [Gaines, 1978].

In summary, the preceding material has established a formal relationship between probability and fuzzy logics and has illuminated axiomatically that their common features are more substantial than their differences. It should be noted that, whereas the additivity axiom 9 is common to both a probability and a fuzzy logic, it is rejected in the Dempster-Shafer theory of evidence [Gaines, 1978], and it often presents difficulties to humans in their reasoning. For a probability logic the law of the excluded middle (or its dual, the law of contradiction, i.e., $p(x \wedge \overline{x}) = 0$) *must* apply; for a fuzzy logic it *may* or *may not* apply.

There are at least two reasons why the law of the excluded middle might be inappropriate for some problems. First, people may have a high degree of belief about a number of possibilities in a problem. A proposition should not be "crowded out" just because it has a large number of competing possibilities. The difficulties people have in expressing beliefs consistent with the axioms of a probability logic are sometimes manifested in the rigidity of the law of the excluded middle [Wallsten and Budescu, 1983]. Second, the law of the excluded middle results in an inverse relationship between the information content of a proposition and its probability. For example, in a universe of n singletons, as more and more evidence becomes available on each of the singletons, the relative amount of evidence on any one diminishes [Blockley, 1983]. This characteristic makes axiom 16.1 inappropriate as a measure for modeling uncertainty in many situations.

Finally, rather than debate what is the correct set of axioms to use (i.e., which logic structure) for a given problem, one should look closely at the problem, determine which propositions are vague or imprecise and which ones are statistically independent or mutually exclusive, and use these considerations to apply a proper uncertainty logic, with or without the law of the excluded middle. By examining a problem so closely as to determine these relationships, one finds out more about the structure of the problem in the first place. Then the assumption of a strong truth-functionality (for a fuzzy logic) could be viewed as a computational device that simplifies calculations, and the resulting solutions would be presented as ranges of values that most certainly form bounds around the true answer if the assumption is not reasonable. A choice of whether a fuzzy logic is appropriate is, after all, a question of balancing the model with the nature of the uncertainty contained within it. Problems without an underlying physical model, problems involving a complicated weave of technical, social, political, and economic factors, and problems with incomplete, ill-defined, and inconsistent information where conditional probabilities cannot be supplied or rationally formulated perhaps are candidates for fuzzy logic applications. Perhaps, then, with additional algorithms like fuzzy logic, those in the technical and engineering professions will realize that such difficult issues can now be modeled in their designs and analyses.

REFERENCES

Blockley, D. (1983). "Comments on 'Model uncertainty in structural reliability,' by Ove Ditlevsen," *J. Struct. Safety,* vol. 1, pp. 233–235.

Dempster, A. (1967). "Upper and lower probabilities induced by a multivalued mapping," *Ann. Math. Stat.,* vol. 38, pp. 325–339.

DuBois, D., and H. Prade (1988). *Possibility theory,* Plenum, New York.

Gaines, B. (1978). "Fuzzy and probability uncertainty logics," *Inf. Control,* vol. 38, pp. 154–169.

Klir, G., and T. Folger (1988). *Fuzzy sets, uncertainty, and information,* Prentice Hall, Englewood Cliffs, NJ.

Rescher, N. (1969). *Many-valued logic,* McGraw-Hill, New York.

Shafer, G. (1976). *A mathematical theory of evidence,* Princeton University Press, Princeton, NJ.

Wallsen, T., and D. Budescu (1983). *Manage. Sci.,* vol. 29, no. 2, pp. 167.

Yager, R. (1993). "Aggregating fuzzy sets represented by belief structures," *J. Intelligent Fuzzy Syst.,* vol. 1, no. 3, pp. 215–224.

Yager, R., and D. Filev (1994). "Template-based fuzzy systems modeling," *J. Intelligent Fuzzy Syst.,* vol. 2, no. 1, pp. 39–54.

Zadeh, L. (1978). "Fuzzy sets as a basis for a theory of possibility," *Fuzzy Sets Syst.,* vol. 1, pp. 3–28.

Zadeh, L. (1984). "Review of the book *A mathematical theory of evidence,* by Glenn Shafer," *AI Mag.,* Fall, pp. 81–83.

Zadeh, L. (1986). "Simple view of the Dempster-Shafer theory of evidence and its implication for the rule of combination," *AI Mag.,* Summer, pp. 85–90.

PROBLEMS

15.1. In structural dynamics a particular structure that has been subjected to a shock environment may be in either of the fuzzy sets "damaged" or "undamaged," with a certain degree of membership over the magnitude of the shock input. If there are two crisp sets, functional (F) and nonfunctional (NF), then a fuzzy measure would be the evidence that a particular system that has been subjected to shock loading is a member of functional systems or nonfunctional systems. Given the evidence from two experts shown here for a particular structure, find the beliefs and plausibilities for the focal elements.

Focal elements	m_1	m_2	bel_1	bel_2	pl_1	pl_2
F	.3	.2				
NF	.6	.6				
F ∪ NF	.1	.2				

15.2. Each year several camera engineers go to the International Security Conference in New York to examine the cameras of their competitors. To make it easier to classify a competitor's CCD camera in terms of cost and performance, firms generally try to figure out whose CCD (charge-coupled device) imager is used in the camera. Let us assume that the universe of CCD manufacturers consists of companies R, C, and D. Expert 1 and Expert 2 come across competitor C's camera at the show. Competitor C's camera is enclosed so that the experts cannot see which CCD imager is being used in the camera. Expert 1 and Expert 2 each examine the video output of the camera on a video monitor under several lighting conditions. Each expert assigns a degree of belief for each of the focal elements in the universe, i.e., all the sets in the power set of CCD manufacturers. Let R, C, and D denote subsets of our universe X (the set of all CCD manufacturers) that contain the set of CCDs manufactured by companies R, C, and D, respectively. In general Expert 1 thinks it is more likely to be a company R CCD, and Expert 2 thinks it is more likely to be a company C CCD. The complete bpa's for each expert are given in the accompanying table.

(*a*) Calculate belief measures for focal elements R, C, R ∪ D, and R ∪ C.

(*b*) Calculate belief measures for focal elements D, C, D ∪ C, and R ∪ D ∪ C.

(*c*) Calculate plausibility measures for focal elements R, C, R ∪ D, and R ∪ C.

(*d*) Calculate plausibility measures for focal elements D, C, D ∪ C, and R ∪ D ∪ C.

(*e*) Calculate combined evidence measures, m_{12}, for focal elements R, D, and C.

(*f*) Calculate combined evidence measures, m_{12}, for focal elements R ∪ D and R ∪ C.

Focal elements	Expert 1, m_1	Expert 2, m_2
∅	0	0
R	.07	.07
D	.05	.1
C	.03	.08
R ∪ D	.2	.2
R ∪ C	.1	.05
D ∪ C	.05	.1
R ∪ D ∪ C	.5	.4

15.3. Suppose you have found an old radio (vacuum tube type) in your grandparents' attic and you're interested in determining its age. The make and model of the radio are unknown to you; without this information you cannot find in a collector's guide the year in which the radio was produced. Here, the year of manufacture is assumed to be within a particular decade. You've asked two antique radio collectors for their opinion on the age. The evidence provided by the collectors is fuzzy. Assume the following questions:

1. Was the radio produced in the 1920s?
2. Was the radio produced in the 1930s?
3. Was the radio produced in the 1940s?

Let R, D, and W denote subsets of our universe set P—the set of radio-producing years called the 1920s (Roaring 20s), the set of radio-producing years called the 1930s (Depression years), and the set of radio-producing years called the 1940s (War years), respectively. The radio collectors provide bpa's as given in the accompanying table.

Focal elements	Collector 1 m_1	bel_1	pl_1	Collector 2 m_2	bel_2	pl_2	Combined evidence m_{12}	bel_{12}	pl_{12}
R	.05	.05	.8	.15	.15	.85	.1969		
D	.1	.1		.1	.1				
W	0	0		0	0				
R ∪ D	.2	.35	1	.25	.5	1	.2677		
R ∪ W	.05			.05					
D ∪ W	.1			.05					
R ∪ D ∪ W	.5			.4					

(*a*) Calculate the missing belief values for the two collectors.
(*b*) Calculate the missing plausibility values for the two collectors.
(*c*) Calculate the missing combined evidence values.
(*d*) Calculate the missing combined belief and plausibility values.

15.4. The quality control for welded seams in the hulls of ships is a major problem. Ultrasonic defectoscopy is frequently used to monitor welds, as is x-ray photography. Ultrasonic defectoscopy is faster but less reliable than x-ray photography. Perfect identification of flaws in welds is dependent on the experience of the person reading the signals. An abnormal signal occurs for three possible types of situations. Two of these are flaws in welds: a cavity (C) and a cinder inclusion (I); the former is the more dangerous. Another situation is due to a loose contact of the sensor probe (L), which is not a defect in the welding seams but an error in measuring. Suppose we have two experts, each using a different weld monitoring method, who are asked to identify the defects in an important welded seam. Their responses in terms of bpa's are given in the table. Calculate the missing portions of the table.

Focal elements	Expert 1 m_1	bel_1	pl_1	Expert 2 m_2	bel_2	pl_2	Combined evidence m_{12}	bel_{12}	pl_{12}
C	.3	.3	.85	.2	.2		.4	.4	
I	.05	.05		.1	.1		.15	.15	
L	.05	.05		.05	.05				
C ∪ I	.2	.55		.15	.45		.16	.71	
C ∪ L	.05	.4	.95	.05	.3				
I ∪ L	.05	.5		.15	.3				
C ∪ I ∪ L	.3	1		.3	1				

15.5. You are an aerospace engineer who wishes to design a bang-bang control system for a particular spacecraft using thruster jets. You know that it is difficult to get a good feel for the amount of thrust that these jets will yield in space. Gains of the control system depend on the amount of the force the thrusters yield. Thus, you pose a region of three crisp sets that are defined with respect to specific gains. Each set will correspond to a different gain of the control system. See Fig. P15.5.

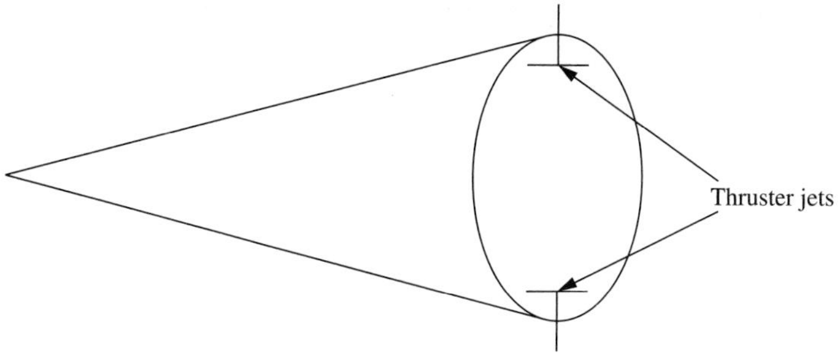

FIGURE P15.5

You can use an initial estimate of the force you get from the thrusters, but you can refine it in real time utilizing different gains for the control system. You can get a force estimate and a belief measure for that estimate for a specific set. Suppose you define the following regions for the thrust values, where thrust is in pounds:

A_1 applies to a region $.8 \leq$ thrust value $\leq .9$.
A_2 applies to a region $.9 \leq$ thrust value ≤ 1.0.
A_3 applies to a region $1.0 \leq$ thrust value ≤ 1.1.

Two expert aerospace engineers have been asked to provide evidence measures reflecting their degree of belief for the various force estimates. These bpa's along with calculated belief measures are given here. Calculate the combined belief measure for each focal element in the table.

Focal elements	Expert 1 m_1	Expert 1 bel_1	Expert 2 m_2	Expert 2 bel_2
A_1	.1	.1	0	0
A_2	.05	.05	.05	.05
A_3	.05	.05	.1	.1
$A_1 \cup A_2$	.05	.2	.05	.1
$A_1 \cup A_3$	.05	.2	.15	.25
$A_2 \cup A_3$	.1	.2	.05	.2
$A_1 \cup A_2 \cup A_3$	.6	1	.6	1

15.6. Consider an information retrieval problem. You have a set of engineering articles that you are interested in classifying by type of magazine or journal in which they appeared. Your crisp sets are things like articles that appear in medical journals, articles that appear in computer magazines, etc. What is in each set is very well defined, i.e., their boundaries are crisp. Let the universe be the set of all documents contained in your databases; there are documents/articles from medical journals (M), news magazines (N), and computer magazines (C). *Note:* We will assume each document is in only one of these sets or none of them. Assume you have two information retrieval systems (your experts) that can both examine a document from your database and assign basic probabilities m_1 and m_2 that indicate the degree of evidence that support the notion that the document belongs to one of the sets, C, M, N, C∪M, C∪N, M∪N, and C∪M∪N (i.e., the focal elements). Calculate the missing values in the following table.

Focal elements	Expert 1			Expert 2			Combined evidence		
	m_1	bel$_1$	pl$_1$	m_2	bel$_2$	pl$_2$	m_{12}	bel$_{12}$	pl$_{12}$
C	.07	.07	.44	.15	.15	.1	.18	.18	
M	.01	.01	.2	0	0	.4			
N	.03	.03		.05	.05				
C∪M	.12	.2		.05	.2				
C∪N	.1	.2		.2	.4				
M∪N	.4	.44		.05	.1				
C∪M∪N	.27	1		.5	1				

15.7. Consider the DC series generator shown in Fig. P15.7. Let R_a = armature resistance, R_s = field resistance, and R = load resistance. The voltage generated across the terminal is given by $V_l = E_g - (I_s R_s + I_a R_a)$. *Note:* If there is no assignment for R, i.e., if the value of R is infinity, then the generator will not build up because of an open circuit. Also R_a can have a range of values from a low value to a high value. To generate different load voltages required, we can assign values for R, R_a, and R_s in different ways to get the voltage. They are very much interrelated and the generated voltage need not have a unique combination of R, R_a, and R_s. Hence, nesting of focal elements for these resistances does have some physical significance.

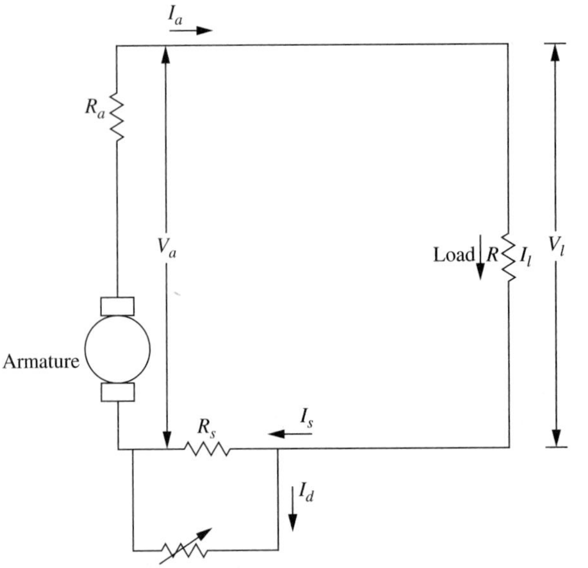

FIGURE P15.7

Let the basic probability assignment for the elements of universe X $\{R, R_a, R_s\}$ be as shown in the accompanying table:

X	$\mathbf{m}_1$	$\mathbf{m}_2$
$\varnothing$	0	0
R_a	.1	0.1
R	.1	0
R_s	.1	.5
$R_a \cup R_s$	.3	.3
$R \cup R_a$	.1	0
$R_s \cup R$	.2	0
$R_a \cup R \cup R_s$	.1	.1

(*a*) Does $\mathbf{m}_1$ or $\mathbf{m}_2$ represent a possibility measure?

(*b*) If either of the evidence measures (or both) is nested, find the possibility distributions.

15.8. A general problem in biophysics is to segment volumetric MRI (magnetic resonance imaging) data of the head given a new set of MRI data. We use a "model head" that has already had the structures in the head (mainly brain and brain substructures) labeled. We can use the model head to help in segmenting the data from the new head by assigning bpa's to each voxel (a voxel is a three-dimensional pixel) in the new MRI data set, based on what structures contain, or are near, the corresponding voxel in the model head. In this example a nested subset corresponds to the physical containment of a head structure within another structure. We will select a voxel for which the bpa's form a consonant body of evidence.

X = {structures in the MRI data}
H = head
B = brain
N = neocortex
L = occipital lobe
C = calcarine fissure

Thus, we have $C \subset L \subset N \subset B \subset H$. For voxel V we have a basic distribution of

$$m = (\mu_C, \mu_L, \mu_N, \mu_B, \mu_H) = (.1, .1, .2, .3, .3)$$

Find the corresponding possibility distribution and draw the nesting diagram.

15.9. A test and diagnostics capability is being developed for a motion control subsystem that consists of the following hardware: a motion control IC (integrated circuit), an interconnect between motion control IC, an H-switch current driver, an interconnect between H-switch current driver, a motor, and an optical encoder.

Elements of motion control subsystem

X_1 = motion control IC
X_2 = interconnect-1

X_3 = H-switch current driver
X_4 = interconnect-2
X_5 = motor
X_6 = optical encoder

If a motion control subsystem failure exists, a self-test could describe the failure in the following bpa: $m = (0.2, 0, 0.3, 0, 0, 0.5)$. This nested structure is based on the level of hardware isolation of the diagnostic software. This isolation is hierarchical in nature. You first identify a motion control subsystem failure $m(A_6)$ that includes a possibility of any component failure $(x_1, x_2, x_3, x_4, x_5, x_6)$. The test then continues and, owing to isolation limitations, a determination can be made of the failure possibility consisting of $m(A_3)$, subset (x_1, x_2, x_3), followed by the ability to isolate to an x_1 failure if x_1 is at fault. Basic probability assignments are constructed from empirical data and experience.

Find the associated possibility distribution and draw the nesting diagram.

15.10. Design of a geometric traffic route can be described by four roadway features: a corner, a curve, a U-turn, and a circle. The traffic engineer can use four different evaluation criteria (expert guidance) to use in the design process:

$\mathbf{m}_1$ = criteria: fairly fast, short distance, arterial road, low slope points
$\mathbf{m}_2$ = criteria: slow, short distance, local road, low slope points
$\mathbf{m}_3$ = criteria: fast, long distance, ramp-type road, medium slope points
$\mathbf{m}_4$ = criteria: very fast, medium distance, highway, medium slope points

Corner	Curve	U-turn	Circle	m_1	m_2	m_3	m_4
0	0	0	1	0	0	.2	.1
0	0	1	0	.1	.1	0	0
0	0	1	1	0	0	0	0
0	1	0	0	.3	.2	.3	.4
0	1	0	1	0	0	.5	.5
0	1	1	0	.1	.1	0	0
0	1	1	1	0	0	0	0
1	0	0	0	.2	.3	0	0
1	0	0	1	0	0	0	0
1	0	1	0	.1	.1	0	0
1	0	1	1	0	0	0	0
1	1	0	0	.1	.1	0	0
1	1	0	1	0	0	0	0
1	1	1	0	.1	.1	0	0
1	1	1	1	0	0	0	0

Using the fifteen $(2^4 - 1)$ focal elements shown in the accompanying table, determine which, if any, of the four evidence measures $(m_1–m_4)$ results in an ordered possibility distribution.

15.11. This situation involves an error being present in the output voltage of an analog amplifier. Since there are several components involved in this amplifier circuit, we can conclude which ones *might* contribute to the overall error. The circuit in Fig. P15.11

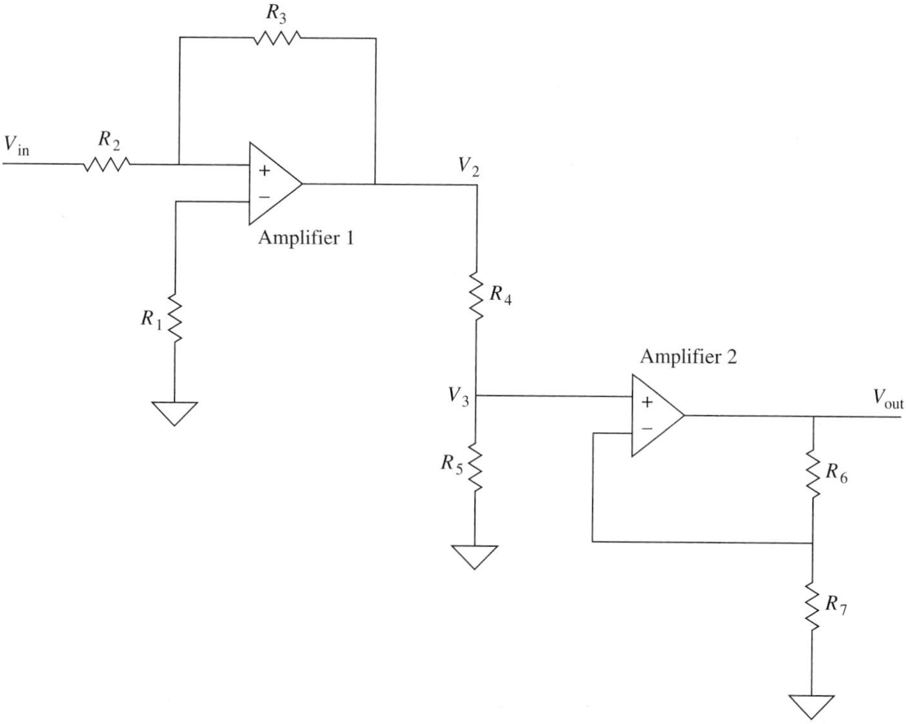

FIGURE P15.11

will be used to illustrate this problem. The individual elements of this problem and their associated bpa's are as follows:

$$
\begin{aligned}
A_1 &= V_2 \text{ error due to } R_2 = .15 \\
A_2 &= V_2 \text{ error due to } R_3 = .15 \\
A_3 &= V_2 \text{ error due to Amplifier 1} = .05 \\
A_4 &= V_3 \text{ error due to } R_4 = .15 \\
A_5 &= V_3 \text{ error due to } R_5 = .15 \\
A_6 &= V_{out} \text{ error due to Amplifier 2} = .05 \\
A_7 &= V_{out} \text{ error due to } R_6 = .15 \\
A_8 &= V_{out} \text{ error due to } R_7 = .15
\end{aligned}
$$

(*a*) From the information provided determine the possibility distribution.
(*b*) For this situation discuss the physical reasons for nesting of the elements.

15.12. Given a communication link with a sender, receiver, and interconnecting link, an error in a message could occur at the sender, receiver, or on the interconnecting link. Combinations such as an error on the link that is not corrected by the receiver are also possible. Let S, R, and L represent sources of error in the sender, receiver, and link, respectively. If E is the universe of error sources, then

$$P(E) = (\varnothing, \{S\}, \{R\}, \{L\}, \{S, R\}, \{S, L\}, \{R, L\}, \{S, R, L\})$$

Now assume each source has its own expert and each of these provides their basic assignment of the actual source of an error as follows:

S	R	L	m_S	m_R	m_L
0	0	0	0	0	0
0	0	1	.4	.3	0
0	1	0	.2	0	0
0	1	1	.2	0	0
1	0	0	0	.5	.5
1	0	1	.1	.2	0
1	1	0	0	0	.4
1	1	1	.1	0	.1

Indicate which experts, if any, have evidence that is consonant. For each of these, do the following:

(*a*) Determine the possibility distribution.

(*b*) Draw the nesting diagram.

(*c*) Give the physical significance of the nesting.

15.13. Show that the weak truth-functionality, i.e.,

$$\text{For all } x, y \in \text{L} \qquad p((x \to y) \vee (y \to x)) = 1$$

applies to both a probability and a

(*a*) Lukasiewicz fuzzy logic (use strict implication, defined in this chapter, Axiom 16.2).

(*b*) Zadeh fuzzy logic (use classical implication, defined in Chapter 2).

INDEX